Abel

Crystal Structures

CRYSTAL STRUCTURES

Second Edition

Ralph W. G. Wyckoff, *University of Arizona, Tucson, Arizona*

VOLUME 5
The Structures of Aliphatic Compounds

INTERSCIENCE PUBLISHERS

a division of John Wiley & Sons, New York • London • Sydney

Preface

The large number of organic crystal structures determined over the last years has made increasingly unsatisfactory the classification of aliphatic compounds used in the loose-leaf edition. In the search for a better unity drastic rearrangements have been made in this volume.

The choice of structures to be analyzed has not been, and one may hope never will be, undertaken according to the kind of generally comprehensive scheme essential to a classification of the innumerable compounds of organic chemistry. Instead they are selected either to find an answer to a specific question of structural chemistry or because of real or fancied relationships within a group of compounds. There is at this time no simple, rational scheme for arranging these analyzed structures. While proceeding in a general sort of way from the simple to the more complex, we shall continue considering together those substances which derive from a single type, such for instance as the substituted ammonium compounds, the derivatives of urea, the amino acids and numerous metal chelates.

This volume initiates one important change in the method of description. Throughout the first edition of *Crystal Structures*, as in the earlier *Structures of Crystals*, illustrations were prepared using left-hand axes. Right-hand axes have now been so universally adopted that there seems little justification for continuing the older convention. All new drawings for this and succeeding volumes therefore will use right-hand axes. Since it is not practical to remake all the older drawings, this necessarily leads to some confusion; but it has been minimized by stating in the legends for each illustration the axial sequence.

A start has also been made towards stating the probable accuracy of a structure. Such an indication of accuracy is necessarily based on the author's evaluation of his determination; the compiler has only intervened, when more than one analysis has been made, to the extent of choosing that set of data which seems to him as probably the more accurate. A complete description of the structure of an organic crystal would now state not only atomic parameters but the accuracy of each and the estimates of the vibrational excursions of each kind of atom. Such information is of course supplied only by the latest determinations. Values of standard deviations, σ, can be added to the parameter tables without appreciably expanding them and this has been done with the most recently prepared material for this volume. No attempt has, however, been made to supplement the earlier descriptions even when these estimates exist, and there is no intention at

v

this time to add tables that will describe the thermal and vibrational motions in a structure. The bulk of the present volume and the accelerating pace of structure determinations for organic compounds forbids any extension in the scope of these descriptions, however desirable it might be.

Contents of Volume 5

The Structures of Aliphatic Compounds

Contents

Volume 1

The Elements and Compounds RX and RX$_2$

Volume 2

Inorganic Compounds RX$_n$, R$_n$MX$_2$, R$_n$MX$_3$

Volume 3

Inorganic Compounds R$_x$(MX$_4$)$_y$, R$_x$(M$_n$X$_p$)$_y$, Hydrates and Ammoniates

Volume 4

Miscellaneous Inorganic Compounds, Silicates and Intermetallic Compounds (*in preparation*)

Chapter XIV

THE STRUCTURES OF ALIPHATIC COMPOUNDS

A. METHANE DERIVATIVES

1. Simple Derivatives

XIV,a1. *Methane*, CH_4, was several times examined in earlier years. It is solid below ca. 89°K., forming cubic crystals with a tetramolecular unit for which

$$a_0 = 5.84 \text{ A. } (13.9°\text{K.})$$

The value of a_0 appears to diminish with rising temperature between this point and 17°K. and to rise again to the above value at 18°K.

The unit cube contains four molecules which in all probability are in the face-centered distribution 000; F.C. The data are not, however, precise enough to show if the hydrogen atoms are in positions determined by symmetry, i.e., if the molecules are "rotating."

XIV,a2. Crystals of *iodoform*, CHI_3, are hexagonal with a unit cell of the dimensions:

$$a_0 = 6.818 \text{ A.,} \qquad c_0 = 7.524 \text{ A.}$$

Its two molecules have been given an atomic arrangement based on C_6^6 ($P6_3$) with iodine atoms in the general positions:

(6c) $xyz;$ $\quad y-x,\bar{x},z;$ $\quad \bar{y},x-y,z;$
$\bar{x},\bar{y},z+{}^1/_2;$ $x-y,x,z+{}^1/_2;$ $y,y-x,z+{}^1/_2$
with $x = $ ca. 0.35, $y = $ ca. 0.04, $z = 0$

The carbon atoms are in

(2b) ${}^1/_3 \, {}^2/_3 \, u;$ ${}^2/_3,{}^1/_3,u+{}^1/_2$

with a parameter that has been chosen as $u = $ ca. 0.596.

1

In this structure (Fig. XIVA,1) the molecules are tetrahedral with C–I = 2.18 A. Between molecules the shortest I–I = 3.98 A.

This is the same arrangement possessed by the boron halides BCl_3, BBr_3 and BI_3 and by PI_3 (**V,b16**).

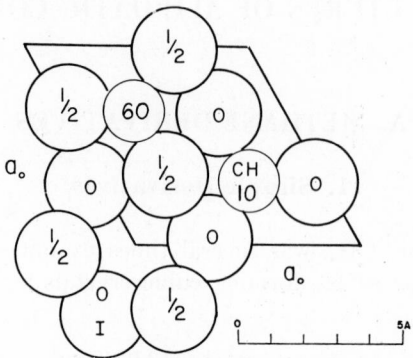

Fig. XIVA,1. The hexagonal structure of iodoform projected along its c_0 axis. Right-hand axes.

XIV,a3. *Iodoform* combines with *sulfur* to yield an addition compound with the formula $CHI_3 \cdot 3S_8$. Its rhombohedral crystals have a unimolecular cell of the dimensions:

$$a_0 = 14.12 \text{ A.}, \qquad \alpha = 118°55'$$

For the trimolecular hexagonal unit:

$$a_0' = 24.32 \text{ A.}, \qquad c_0' = 4.44 \text{ A.} \pm 0.5\%$$

The space group has been chosen as C_{3v}^5 (*R3m*) with atoms in the hexagonal positions:

C: (3a) $00u$; rh   with $u = 0.0963$
I: (9b) $u\bar{u}v$; $u\,2u\,v$; $2\bar{u}\,\bar{u}\,v$; rh
           with $u = -0.0487$, $v = 0.0000$
S(1): (9b)   with $u = -0.1291$, $v = -0.1973$
S(2): (9b)   with $u = -0.2407$, $v = -0.2572$

The other sulfur atoms are in

(18c) xyz; $\bar{y},x-y,z$; $y-x,\bar{x},z$; $\bar{y}\bar{x}z$; $x,x-y,z$; $y-x,y,z$; rh

with, for S(3): $x = -0.0810$, $y = 0.2158$, $z = 0.0394$
    S(4): $x = -0.0889$, $y = 0.2810$, $z = -0.2191$
    S(5): $x = -0.1587$, $y = 0.2928$, $z = -0.0410$

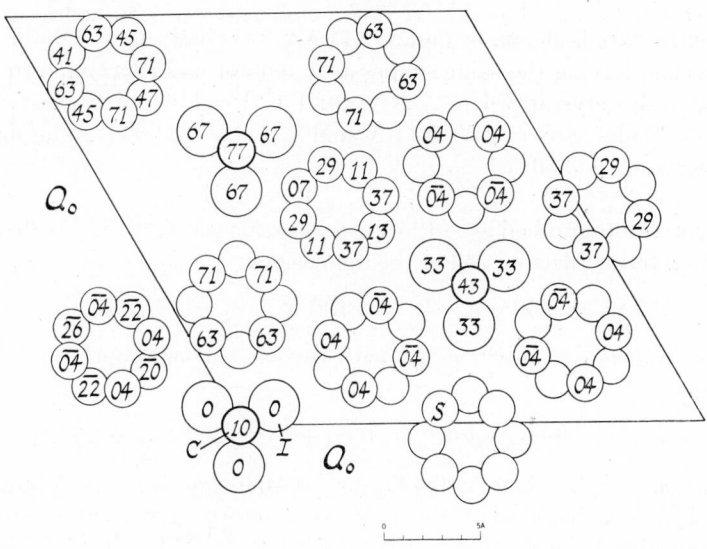

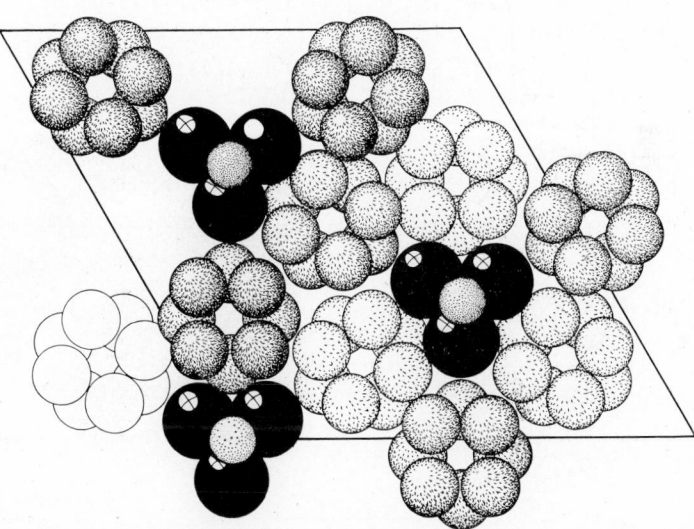

Fig. XIVA,2a (top). The hexagonal structure of $CHI_3 \cdot 3S_8$ projected along its c_0 axis. Right-hand axes.

Fig. XIVA,2b (bottom). A packing drawing of the hexagonal $CHI_3 \cdot 3S_8$ arrangement viewed along its c_0 axis. The iodine atoms are black, the sulfur atoms fine-line shaded.

The structure is shown in Figure XIVA,2. It is built up of iodoform and S_8 molecules having the same shapes as in iodoform (**XIV,a2**) and in sulfur (**II,n1**). In this crystal C–I = 2.10 A. and I–C–I = 116°; in its chair-shaped sulfur molecules, S–S = 2.00–2.11 A. and S–S–S = 102–111°. The shortest S–I separation is 3.50 A.

XIV,a4. Photographed at −125°C., *methyl chloride*, CH_3Cl, is orthorhombic with a tetramolecular cell of the dimensions:

$$a_0 = 6.495 \text{ A.;} \quad b_0 = 5.139 \text{ A.;} \quad c_0 = 7.523 \text{ A.}$$

The space group is $C_{2v}^{12}(Cmc2_1)$ with carbon and chlorine atoms in the positions:

$$(4a) \quad 0uv; \; {}^1/_2, u+{}^1/_2, v; \; 0, \bar{u}, v+{}^1/_2; \; {}^1/_2, {}^1/_2-u, v+{}^1/_2$$

For carbon, $u = 0.380$, $v = 0.422$; for chlorine they are $u = 0.1351$, $v = 0.250$. It is concluded that the most satisfactory agreement is obtained if one set of hydrogens is placed in $(4a)$ with $u = 0.573$, $v = 0.362$, and the remaining eight in general positions:

$$(8b) \quad \begin{array}{ll} xyz; & \bar{x}yz; \\ \bar{x}, \bar{y}, z+{}^1/_2; & x, \bar{y}, z+{}^1/_2; \\ x+{}^1/_2, y+{}^1/_2, z; & {}^1/_2-x, y+{}^1/_2, z; \\ {}^1/_2-x, {}^1/_2-y, z+{}^1/_2; & x+{}^1/_2, {}^1/_2-y, z+{}^1/_2 \end{array}$$

with $x = 0.137$, $y = 0.357$, $z = 0.504$

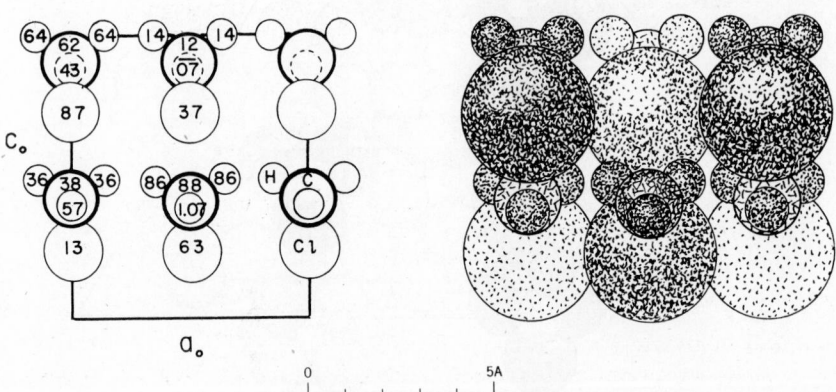

Fig. XIVA,3a (left). The orthorhombic structure of CH_3Cl projected along its b_0 axis. Left-hand axes.

Fig. XIVA,3b (right). A packing drawing of the orthorhombic CH_3Cl structure viewed along its b_0 axis. The chlorine atoms are the large, the hydrogen atoms the very small dotted circles; the carbon atoms are line shaded.

The resulting arrangement is illustrated in Figure XIVA,3. It leads to a C–Cl separation in the molecule of 1.80 A. Between molecules the contacts are H–Cl = 3.00 A. and more.

XIV,a5. *Methyl alcohol*, CH_3OH, has a λ-point transition at 157.8°K. and its structure has been examined above and below this temperature. Between 158 and 175°K., where it melts, its crystals have a tetramolecular, orthorhombic unit with the edge lengths:

$$a_0 = 6.43 \text{ A.}; \quad b_0 = 7.24 \text{ A.}; \quad c_0 = 4.67 \text{ A. } (-110°\text{C.})$$

The space group is V_h^{17} (*Cmcm*) with both carbon and oxygen atoms in special positions:

$$(4c) \quad \pm(0 \ u \ ^1/_4; \ ^1/_2, u + ^1/_2, ^1/_4)$$

For the carbon atoms, u has been determined to be 0.214; for the oxygen, it is 0.4105. The structure is indicated in Figure XIVA,4.

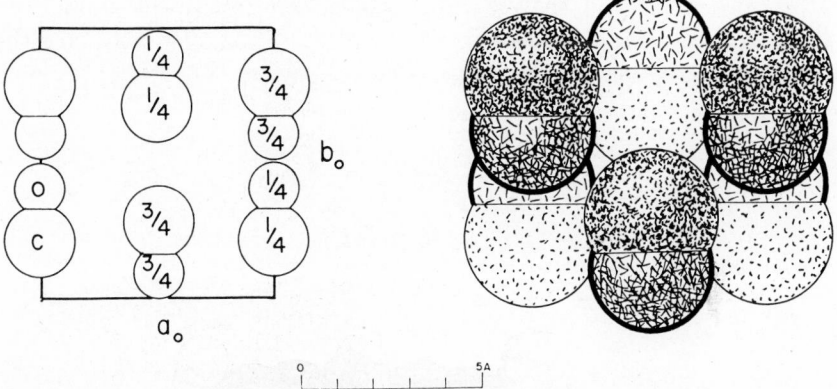

Fig. XIVA,4a (left). The orthorhombic structure of the higher-temperature form of CH_3OH projected along its c_0 axis.   Left-hand axes.

Fig. XIVA,4b (right). A packing drawing of the orthorhombic structure of high-temperature CH_3OH viewed along its c_0 axis.   The methyl carbon is dotted; the hydroxyl group is heavily outlined and line shaded.

Cooling crystals of this high-temperature form to $-160°$C. invariably resulted in twinned individuals whose structure could not be established with complete certainty. It was considered, however, that the following structure accounts for all but a few faint reflections. The bimolecular unit is monoclinic with the dimensions:

$$a_0 = 4.59 \text{ A.}; \quad b_0 = 4.68 \text{ A.}; \quad c_0 = 4.92 \text{ A.}; \quad \beta = 97°30'(-180°\text{C.})$$

Both carbon and oxygen atoms were placed in $(2e)$ $\pm(u \ ^1/_4 \ v)$ of $C_{2h}{}^2$ $(P2_1/m)$ with $u(C) = 0.25$, $v(C) = 0.16$ and $u(O) = 0.37$, $v(O) = 0.43$.

XIV,a6. There is an addition compound of *methyl alcohol* and *bromine*, $2CH_3OH \cdot Br_2$, which crystallizes with monoclinic symmetry. Its eight molecules are in a cell of the dimensions:

$$a_0 = 16.07 \ A.; \quad b_0 = 11.16 \ A.; \quad c_0 = 8.16 \ A.; \quad \beta = 92°18'$$

The space group has been chosen as $C_{2h}{}^5(P2_1/c)$ with all atoms in the positions:

$$(4e) \ \pm(xyz; \ x, ^1/_2 - y, z + ^1/_2)$$

The determined parameters are listed in Table XIVA,1.

In this arrangement (Fig. XIVA,5), the light element positions must be considered as approximate. Both bromine and alcohol molecules are

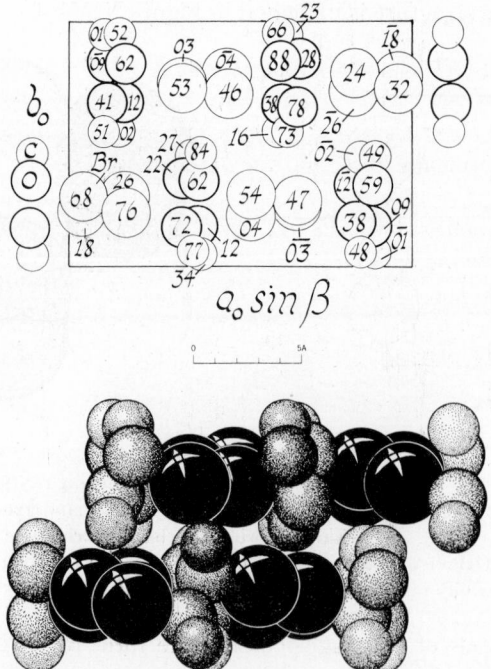

Fig. XIVA,5a (top). The monoclinic structure of $2CH_3OH \cdot Br_2$ projected along its c_0 axis.   Right-hand axes.

Fig. XIVA,5b (bottom). A packing drawing of the monoclinic structure of $2CH_3OH \cdot Br_2$ seen along its c_0 axis.   The bromine atoms are large and black; the oxygen atoms are somewhat larger than the dotted carbon atoms.

TABLE XIVA,1
Parameters of the Atoms in $2CH_3OH \cdot Br_2$

Atom	x	y	z
Br(1)	0.0362	0.2291	0.1845
Br(2)	0.1669	0.2846	0.2603
Br(3)	0.5333	0.2107	0.0399
Br(4)	0.6688	0.2542	-0.0297
O(1)	-0.117	0.159	0.086
O(2)	0.335	0.352	0.319
O(3)	0.375	0.159	0.116
O(4)	0.834	0.330	-0.119
C(1)	-0.109	0.043	-0.010
C(2)	0.359	0.449	0.270
C(3)	0.383	0.053	0.343
C(4)	0.846	0.450	-0.019

present. In CH_3OH, C–O = 1.27 and 1.61 A.; in Br_2 the atomic separation is 2.29 A. They are linked together by O–H–O bridges of length 2.62 A. and there is a short O–Br = 2.78 A. The distribution of the four atoms close to oxygen is tetrahedral and the line Br–Br–O is nearly straight (177°).

XIV,a7. *Pentaerythritol*, $C(CH_2OH)_4$, is one of the simpler substitution products of methane for which structures have been determined. After much early uncertainty the following arrangement was established. Its tetragonal crystals have a bimolecular unit which according to a recent redetermination has the dimensions:

$$a_0 = 6.083 \pm 0.002 \text{ A.}, \qquad c_0 = 8.726 \pm 0.002 \text{ A.}$$

Atoms are in the following positions of the low symmetry space group S_4^2 ($I\bar{4}$):

C(1): (2a)   000; $^1/_2$ $^1/_2$ $^1/_2$
C(2): (8g) xyz; $\bar{x}\bar{y}z$; $y\bar{x}\bar{z}$; $\bar{y}x\bar{z}$; B.C.
          with $x = 0.1628$, $y = 0.1264$, $z = 0.1059$
O: (8g)   with $x = 0.3166$, $y = 0.2475$, $z = 0.0188$

When examined by neutron diffraction the hydrogen atoms have been given the parameters:

Atom	x	y	z
H(OH)	0.229	0.108	0.003
H(1)	0.240	0.013	0.164
H(2)	0.065	0.254	0.177

This is a body-centered grouping of $C(CH_2OH)_4$ molecules (Fig. XIVA,6) in which each molecule consists of a tetrahedral distribution of (CH_2OH) radicals about a central carbon atom. Within a molecule, $C(1)–CH_2 = 1.548 \pm 0.011$ A. and $CH_2–OH = 1.425 \pm 0.014$ A. Between adjacent molecules, contact is made by $CH_2–OH = $ ca. 3.59 A. and the large $OH–OH = $ ca. 3.47 A.

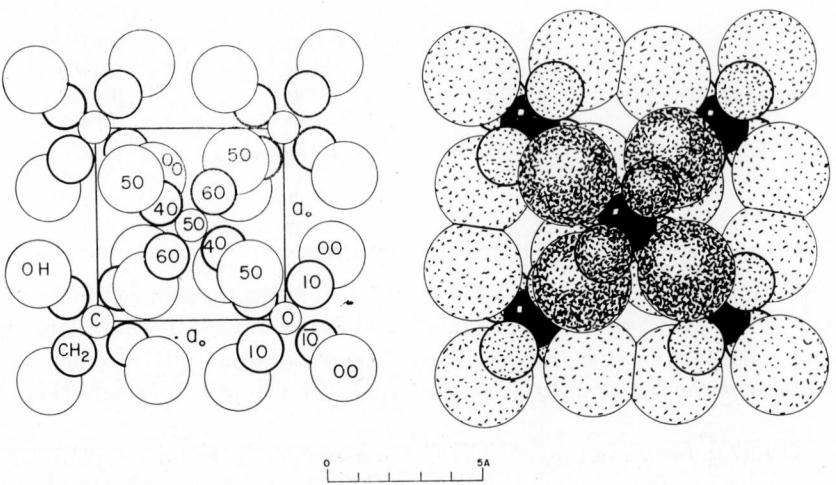

Fig. XIVA,6a (left). The tetragonal structure of pentaerythritol projected along its c_0 axis.   Left-hand axes.

Fig. XIVA,6b (right). A packing drawing of the tetragonal structure of $C(CH_2OH)_4$ seen along its c_0 axis.   The large circles are hydroxyls, the small ones carbon.   For the central carbon of the molecules, these are black.

It has been suggested that the differences among accurate cell data obtained by different investigators may be attributed to impurities in the crystals measured.

At temperatures between 179.5 and 260.5°C. pentaerythritol is cubic with a tetramolecular unit having

$$a_0 = 8.963 \text{ A. } (230°\text{C.})$$

The molecules are in a face-centered array and it is proposed that at this elevated temperature they have a statistical orientation which makes them effectively spherical.

XIV,a8. *Pentaerythritol tetranitrate*, $C(CH_2ONO_2)_4$, is, like pentaerythritol itself (**XIV,a7**), tetragonal with a bimolecular unit which in this case has the dimensions:

$$a_0 = 9.33 \text{ A.}, \qquad c_0 = 6.66 \text{ A.}$$

Its space group is V_d^4 ($P\bar{4}2_1c$) with carbon atoms at the molecular centers in

$$(2a) \quad 000; \, {}^1/_2 \, {}^1/_2 \, {}^1/_2$$

All the other atoms are in the general positions:

$$(8e) \quad xyz; \; \bar{x}\bar{y}z; \; {}^1/_2-x, y+{}^1/_2, {}^1/_2-z; \; x+{}^1/_2, {}^1/_2-y, {}^1/_2-z;$$
$$\bar{y}x\bar{z}; \; y\bar{x}\bar{z}; \; y+{}^1/_2, x+{}^1/_2, z+{}^1/_2; \; {}^1/_2-y, {}^1/_2-x, z+{}^1/_2$$

with the following recently redetermined parameters:

Atom	x	y	z
CH_2	0.1207	0.0636	0.1270
O(1)	0.1608	−0.0483	0.2681
N	0.2870	−0.0165	0.3716
O(2)	0.3158	−0.1050	0.4951
O(3)	0.3500	0.0927	0.3240

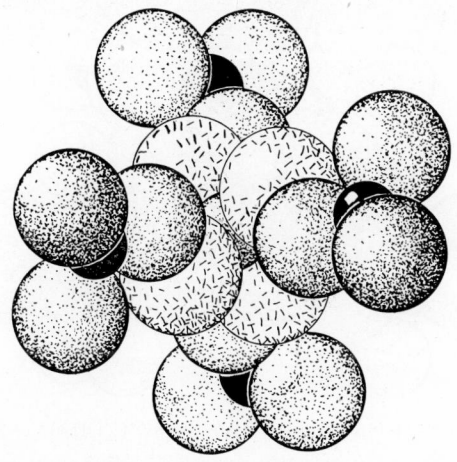

Fig. XIVA,7. A packing drawing of the molecule of $C(CH_2ONO_2)_4$. The CH_2 groups are line shaded; both the central carbon atom and the larger oxygen atoms are dotted. The nitrogen atoms are black.

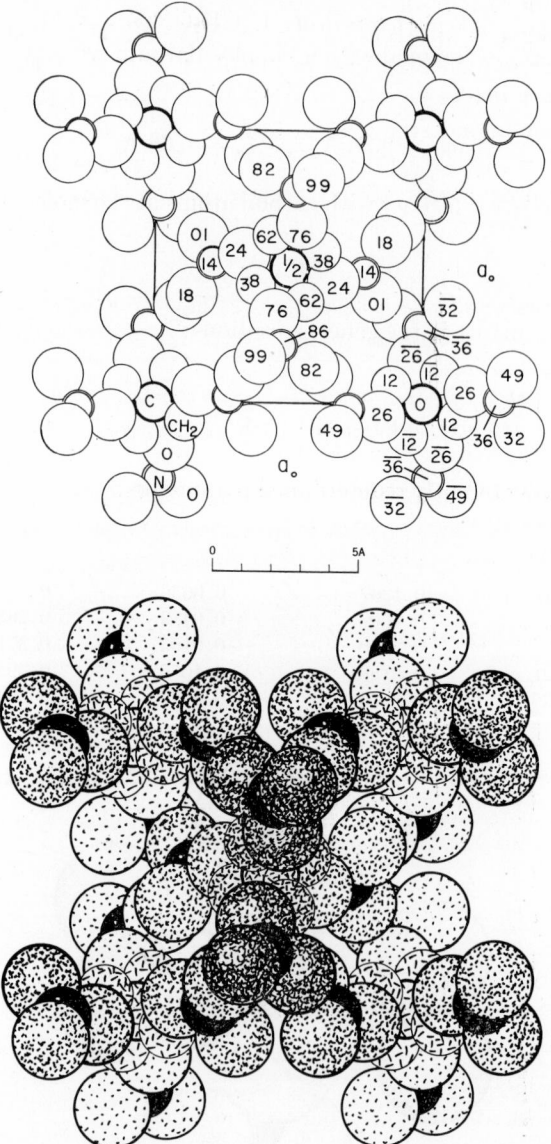

Fig. XIVA,8a (top). The tetragonal structure of $C(CH_2ONO_2)_4$ projected along its c_0 axis.   Left-hand axes.

Fig. XIVA,8b (bottom). A packing drawing of the tetragonal $C(CH_2ONO_2)_4$ arrangement seen along its c_0 axis.   The nitrogen atoms are black and the carbons line shaded. The oxygen atoms are larger and dotted.

The molecule that results is shown in Figure XIVA,7. Its pseudo-body-centered packing is apparent in Figure XIVA,8. Within a molecule the significant interatomic distances are C–CH$_2$ = 1.537 A., CH$_2$–O(1) = 1.462 A., O(1)–N = 1.404 A., N–O(2) = 1.203 A., N–O(3) = 1.225 A.; important angles between atoms are CH$_2$–C–CH$_2$ = 113°, C–CH$_2$–O(1) = 106°, CH$_2$–O(1)–N = 113° = O(1)–N–O(2), O(1)–N–O(3) = 117°.

This structure is very similar to that of modification I of C(SCH$_3$)$_4$ (**XIV,a10**).

XIV,a9. *Pentaerythritol tetraacetate*, C(CH$_2$O·COCH$_3$)$_4$, is, like the crystals of the two foregoing paragraphs, tetragonal. Its cell is also bimolecular body-centered but with

$$a_0 = 12.00 \text{ A.,} \qquad c_0 = 5.506 \text{ A.}$$

it has a different shape and its space group, C$_{4h}^4$ (*P*4$_2$/*n*), has higher symmetry. The central carbon atoms of the molecules are in the special positions:

$$(2a) \quad 000; \ {}^1/_2 \, {}^1/_2 \, {}^1/_2$$

of this group; all other atoms are in general positions:

$$(8g) \quad xyz; \ \bar{x}\bar{y}z; \ x+{}^1/_2, y+{}^1/_2, {}^1/_2-z; \ {}^1/_2-x, {}^1/_2-y, {}^1/_2-z;$$
$$\bar{y}x\bar{z}; \ y\bar{x}\bar{z}; \ {}^1/_2-y, x+{}^1/_2, z+{}^1/_2; \ y+{}^1/_2, {}^1/_2-x, z+{}^1/_2$$

with the following parameters:

Atom	x	y	z
CH$_2$(2)	0.095	0.040	0.161
C(3)	0.270	0.120	0.125
CH$_3$(4)	0.366	0.133	0.950
O(1)	0.181	0.070	0.000
O(2)	0.280	0.120	0.367

The significant atomic separations within the resulting molecules, which have the shape shown in Figure XIVA,9, are C–CH$_2$(2) = C(3)–CH$_3$(4) = 1.52 A., CH$_2$–O(1) = C(3)–O(1) = 1.41 A., C(3)–O(2) = 1.33 A. It will be noticed that this arrangement makes O(1) and O(2) different and results in contact between the acetate radicals and the central molecular complex through O(1) only.

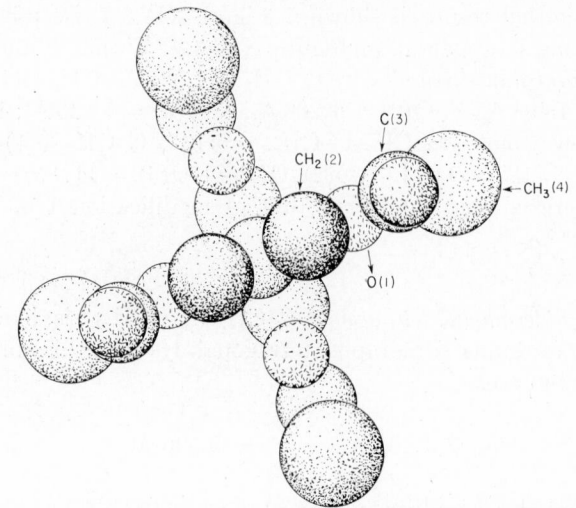

Fig. XIVA,9. A packing drawing of a molecule of C(CH₂OCOCH₃)₄. The oxygen atoms are line shaded; all are designated as in the text.

XIV,a10. *Tetramethyl orthothiocarbonate*, $C(SCH_3)_4$, occurs in three modifications each of which has been investigated with x-rays.

The *modification I*, stable below 23.2°C., is tetragonal with a bimolecular unit of the dimensions:

$$a_0 = 8.536 \text{ A.,} \qquad c_0 = 6.949 \text{ A.}$$

Atoms have been assigned the following positions of V_d^4 ($P\bar{4}2_1c$):

$$C: (2a) \quad 000; \ ^1/_2 \, ^1/_2 \, ^1/_2$$

All other atoms are in the general positions:

$$(8e) \quad xyz; \ \bar{x}\bar{y}z; \ ^1/_2-x,y+^1/_2, \ ^1/_2-z; \ x+^1/_2, \ ^1/_2-y,^1/_2-z;$$
$$\bar{y}x\bar{z}; \ y\bar{x}\bar{z}; \ y+^1/_2,x+^1/_2, z+^1/_2; \ ^1/_2-y, \ ^1/_2-x,z+^1/_2$$

For sulfur these parameters were found to be $x = 0.149$, $y = 0.091$, $z = 0.15$, and for CH_3, $x = 0.25$, $y = -0.04$, $z = 0.32$. The significant interatomic distances within a molecule of this structure (Fig. XIVA,10) are C–S = 1.81 A., S–CH₃ = 1.8 A., and S–S = 2.95 A. The shortest distance between molecules is CH₃–CH₃ = ca. 4 A.

This is substantially the same structure as that just described (**XIV,a8**) for $C(CH_2ONO_2)_4$.

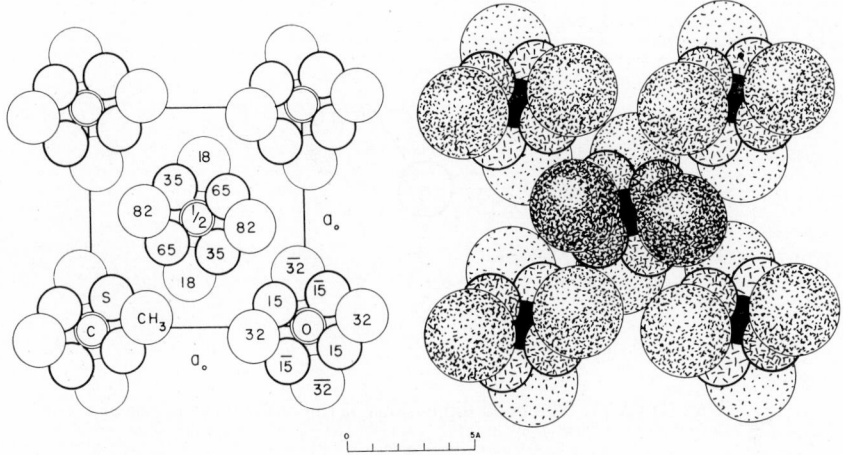

Fig. XIVA,10a (left). The tetragonal structure of form I of $C(SCH_3)_4$ projected along its c_0 axis.   Left-hand axes.

Fig. XIVA,10b (right). A packing drawing of the tetragonal structure of modification I of $C(SCH_3)_4$ seen along its c_0 axis.   The central carbon atoms are black; the methyl groups are larger and dotted.   The atoms of sulfur are line shaded.

Modification II, stable between 23.2 and 45.5°C., also is tetragonal with a bimolecular cell that is somewhat shorter in the a_0 direction and has a longer c_0 axis. This cell has the dimensions:

$$a_0 = 8.17 \text{ A.,} \qquad c_0 = 7.96 \text{ A.}$$

As in the first, lower-temperature, form, the molecules have their centers at 000 and $^1/_2$ $^1/_2$ $^1/_2$. Piezoelectricity is lost in passing from modification I to modification II and there is therefore an increase in crystal class. A structure has been proposed based on the existence of two kinds of molecule which agree in the positions of their CH_3 radicals but have the sulfur atoms turned 90° with respect to one another. A more detailed study is evidently needed.

Modification III, stable from above 45°C. to the melting point at 66°C., is cubic with a bimolecular unit of the edge length:

$$a_0 = 8.15 \text{ A.}$$

It is proposed that the degree of disorder corresponding in modification II to two orientations of the CS_4 tetrahedra here is extended further.

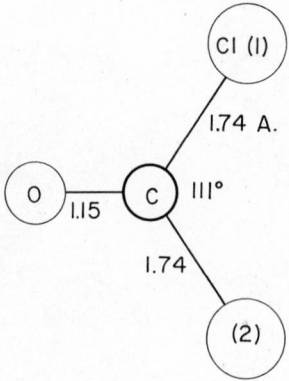

Fig. XIVA,11. The bond dimensions of the molecule of phosgene.

XIV,a11. *Phosgene*, $COCl_2$, at $-160°C$. is tetragonal with a cell containing 16 molecules and having the edge lengths:

$$a_0 = 15.82 \text{ A.}, \qquad c_0 = 5.72 \text{ A.}$$

Atoms are in general positions of C_{4h}^6 ($I4_1/a$):

(16*f*) $\pm(xyz;$ $x,y+^1/_2,\bar{z};$
                $y+^1/_4,^3/_4-x,^3/_4-z;$ $^3/_4-y,x+^3/_4,^3/_4-z);$ B.C.

with the parameters:

Atom	x	y	z
Cl(1)	0.0394	-0.1417	0.150
Cl(2)	0.1038	-0.0528	-0.250
O	0.1982	-0.1306	0.039
C	0.1295	-0.1126	-0.004

The resulting molecule has bond dimensions (Fig. XIVA,11) that are in acceptable agreement with those found from electron diffraction in the vapor. The general molecular packing in the crystal is shown in Figure XIVA,12. Between molecules the closest atomic approaches are Cl–Cl = 3.64 A. and Cl–O = 3.12 A.

XIV,a12. Crystals of *trans-bis-nitrosomethane*, $(CH_3NO)_2$, are orthorhombic with a tetramolecular unit of the edge lengths:

$$a_0 = 7.25 \text{ A.}; \quad b_0 = 9.38 \text{ A.}; \quad c_0 = 6.27 \text{ A.}, \qquad \text{all} \pm 0.01 \text{ A.}$$

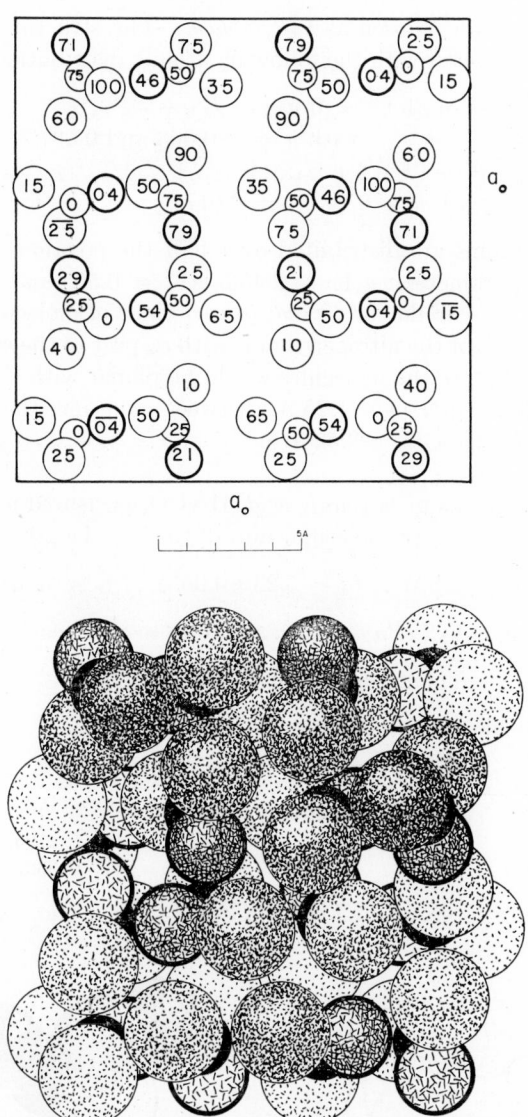

Fig. XIVA,12a (top). The tetragonal structure of COCl₂ projected along its c_0 axis. The larger, light circles are chlorine; the small ones are carbon. Heavily ringed circles are oxygen. Left-hand axes.

Fig. XIVA,12b (bottom). A packing drawing of the tetragonal COCl₂ arrangement seen along its c_0 axis. The carbon atoms are black, the chlorine atoms large and dotted. Oxygen atoms are heavily outlined and line shaded.

The space group was chosen as V_h^{17} (*Cmcm*). The structure is thought to be partially disordered with the following atomic distribution:

O in two sets of $(4c)$ $\pm (0\ u\ ^1/_4;\ ^1/_2, u + ^1/_2, ^1/_4)$
with $u = -0.063$ and 0.297

C in $(8g)$ $\pm (u\ v\ ^1/_4;\ \bar{u}\ v\ ^1/_4;\ u + ^1/_2, v + ^1/_2, ^1/_4;\ ^1/_2 - u, v + ^1/_2, ^1/_4)$
with $u = 0.243$, $v = 0.617$

The nitrogen atoms are distributed over half the positions of two sets of $(8g)$ with the parameters: $u(1) = 0.450$, $v(1) = 0.565$ and $u(2) = 0.550$, $v(2) = 0.669$. This would correspond to two kinds of molecule which differ in the distribution of the nitrogen atoms with respect to the other atoms.

In such a structure the molecules would be planar with C–C = 1.57 A., N–N = 1.22 A., and N–O = 1.25 A.; between molecules the shortest separation would be CH_3–CH_3 = 3.52 A.

XIV,a13. Crystals of *isocyanic acid*, HNCO, measured at $-130°$C. are orthorhombic with a tetramolecular unit of the edge lengths:

$$a_0 = 10.82 \pm 0.08\ \text{A.};\quad b_0 = 5.23 \pm 0.05\ \text{A.};\quad c_0 = 3.57 \pm 0.03\ \text{A.}$$

The space group is V_h^{16} (*Pnma*) with all atoms in the special positions:

$$(4c)\quad \pm (u\ ^1/_4\ v;\ u + ^1/_2, ^1/_4, ^1/_2 - v)$$

Their determined parameters are:

Atom	u	v
N	0.0648	0.1108
C	0.1670	0.2195
O	0.2705	0.3367

The structure is shown in Figure XIVA,13. In it the NCO groups are

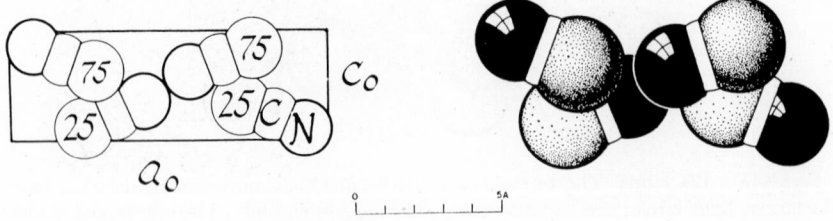

Fig. XIVA,13a (left). The orthorhombic structure of isocyanic acid, HNCO, projected along its b_0 axis. Right-hand axes.

Fig. XIVA,13b (right). A packing drawing of the orthorhombic HNCO structure viewed along its b_0 axis. The nitrogen atoms are black, the oxygen atoms larger and dotted. The carbon atoms appear as unshaded bands between the two.

straight with N–C = 1.183 A. and C–O = 1.184 A. The shortest inter-molecular distance in the same reflecting plane is N–O = 3.19 A.; between planes it is N–N = 3.07 A., N–O = 3.17 A., and C–O = 3.03 A. Hydrogen positions were not established, but it was thought that these atoms are probably between the nitrogen atoms since the other separations would involve improbable bond angles. If, however, the bond is N–H$\cdots$N, there is hydrogen disorder expressed by placing half atoms in general positions of V_h^{16} (or the true space group is the lower subgroup C_{2v}^9).

XIV,a14. Crystals of *cyanamide*, $NH_2C{\equiv}N$, are orthorhombic with a unit containing eight molecules and having the edge lengths:

$$a_0 = 7.06 \text{ A.}; \quad b_0 = 6.82 \text{ A.}; \quad c_0 = 9.03 \text{ A.}$$

The space group is V_h^{15} (*Pbca*) with atoms in the positions:

$$(8c) \quad \pm (xyz; \; {}^1\!/_2{-}x,y{+}{}^1\!/_2,z; \; x,{}^1\!/_2{-}y,z{+}{}^1\!/_2; \; x{+}{}^1\!/_2,y,{}^1\!/_2{-}z)$$

The parameters, established for all but the H(2) atoms, are:

Atom	x	y	z
C	0.136	0.088	0.275
N(1)	0.139	0.175	0.404
N(2)	0.133	0.011	0.162
H(1)	0.135	0.292	0.345

The resulting structure is shown in Figure XIVA,14. In its molecules, N(1)–C = 1.31 A. and C–N(2) = 1.15 A.; these three atoms are in an almost perfectly straight line. Between molecules the shortest C–C = 3.56 A.; the shortest N(1)–N(2) between molecules, presumably involving a hydrogen bond, is 3.04 A., and others are 3.16 and 3.22 A.

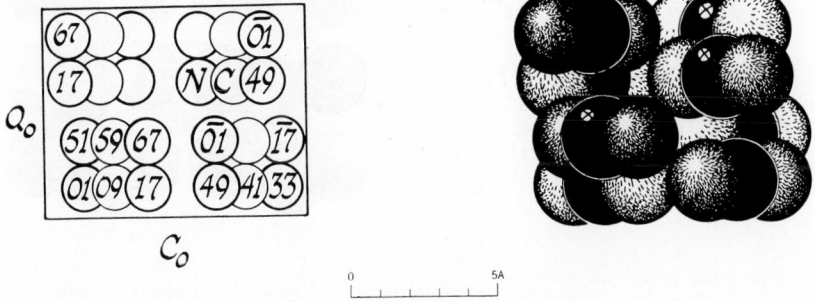

Fig. XIVA,14a (left). The orthorhombic structure of cyanamide, NH_2CN, projected along its b_0 axis. Right-hand axes.

Fig. XIVA,14b (right). A packing drawing of the orthorhombic cyanamide arrangement seen along its b_0 axis. The carbon atoms are black, the nitrogen atoms line shaded.

2. Amino- and Nitro-Compounds

XIV,a15. At $-150°C$. crystals of *methyl amine*, CH_3NH_2, are ortho-rhombic with a cell containing eight molecules and having the edges:

$$a_0 = 5.75 \text{ A.}; \quad b_0 = 6.18 \text{ A.}; \quad c_0 = 13.61 \text{ A.}$$

The space group is V_h^{15} (*Pcab*) with all atoms in the general positions:

$$(8c) \quad \pm(xyz; \; x+\tfrac{1}{2},\bar{y},z+\tfrac{1}{2}; \; x,y+\tfrac{1}{2},\tfrac{1}{2}-z; \; x+\tfrac{1}{2},\tfrac{1}{2}-y,z)$$

The established carbon and nitrogen parameters are listed in Table XIVA,2, together with those assigned to the hydrogen atoms.

The resulting structure is shown in Figure XIVA,15. Amino hydrogen positions have been based on the assumption that the N–N separations of

TABLE XIVA,2
Parameters of the Atoms in Methyl Amine at $-150°C$.

Atom	x	y	z
C	0.216	0.004	0.106
N	0.257	0.065	0.210
Amino hydrogens			
H(1)	0.257	0.220	0.235
H(2)	0.416	0.024	0.235
Methyl hydrogens (calculated)			
H(3)	0.36	0.07	0.06
H(4)	0.21	−0.17	0.10
H(5)	0.05	0.07	0.08

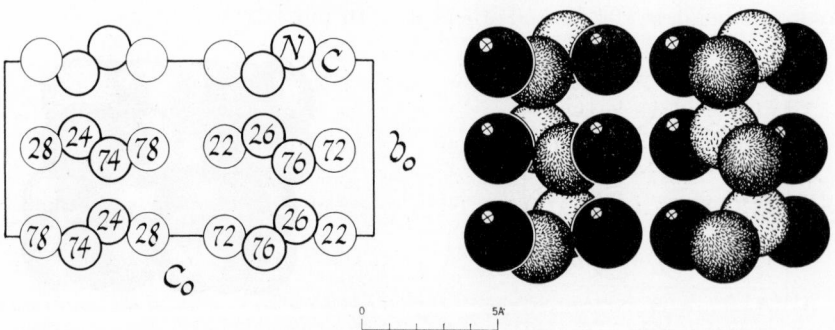

Fig. XIVA,15a (left). The orthorhombic structure of CH_3NH_2 projected along its a_0 axis. Right-hand axes.

Fig. XIVA,15b (right). A packing drawing of the orthorhombic CH_3NH_2 structure viewed along its a_0 axis. The carbon atoms are black, the nitrogen atoms line shaded.

3.18 and 3.27 A. involve hydrogen bonds, on Fourier projections including a difference projection on a_0c_0, and on the assumption that N–H = 1.01 A. For the methyl hydrogens, the specific positions chosen assume a C–H separation of 1.09 A and a van der Waals radius of 1.20 A. These parameters lead to H–H separations of 2.4–2.5 A. between adjacent methyl groups and support the probability that in this crystal the methyl groups are not rotating.

The λ-point transition observed in methyl amine does not result in an atomic rearrangement. At $-185°C.$,

$$a_0 = 5.73 \text{ A.}; \quad b_0 = 6.11 \text{ A.}; \quad c_0 = 13.51 \text{ A.}$$

XIV,a16. Crystals of the complex *methylamine–boron trifluoride*, $CH_3NH_2 \cdot BF_3$, are monoclinic with a bimolecular unit of the dimensions:

$$a_0 = 5.06 \text{ A.}; \quad b_0 = 7.28 \text{ A.}; \quad c_0 = 5.81 \text{ A.}; \quad \beta = 101°31'$$

The space group has been chosen as C_{2h}^2 ($P2_1/m$) with atoms in the positions:

$$(2e) \quad \pm(u\ ^1/_4\ v)$$
$$(4f) \quad \pm(xyz;\ x,^1/_2-y,z)$$

The selected positions for the atoms and their parameters are as follows:

Atom	Position	x	y	z
C	(2e)	−0.252	$^1/_4$	0.415
N	(2e)	−0.180	$^1/_4$	0.152
B	(2e)	0.134	$^1/_4$	0.050
F(1)	(2e)	0.172	$^1/_4$	−0.190
F(2)	(4f)	0.244	0.095	0.129

All atoms of the molecule in this structure (Fig. XIVA,16) are coplanar except for one set of fluorines. Within the BF_3 part, B–F = 1.37 and 1.38 A., and F–B–F = 110°6′ and 110°54′. In the CH_3NH_2 part, C–N = 1.50 A. Between the two parts, N–B = 1.58 A., C–N–B = 113°54′, and N–B–F = 108°6′ and 108°30′.

XIV,a17. The corresponding *trimethylamine–boron trifluoride* complex, $(CH_3)_3N \cdot BF_3$, is of higher symmetry. Its hexagonal unit, and that of the isomorphous borine compound, have the dimensions:

$$(CH_3)_3N \cdot BF_3: a_0 = 9.34 \text{ A.}, \qquad c_0 = 6.10 \text{ A.}$$
$$(CH_3)_3N \cdot BH_3: a_0 = 9.33 \text{ A.}, \qquad c_0 = 5.90 \text{ A.}$$

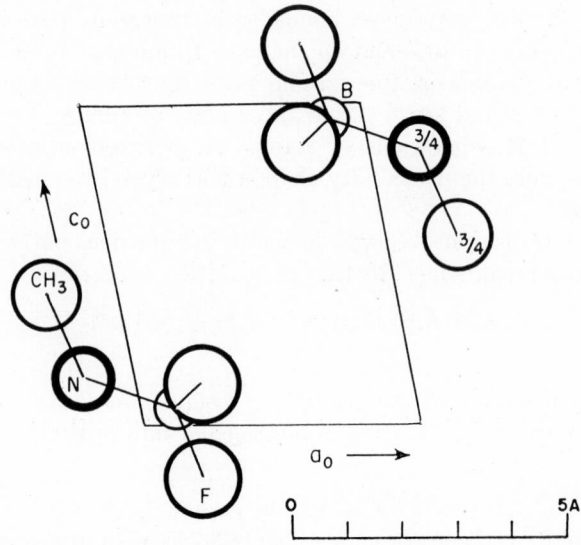

Fig. XIVA,16. The monoclinic structure of $CH_3NH_2 \cdot BF_3$ projected along its b_0 axis. Except for one set of fluorine atoms, [F(2)], all atoms of a molecule are in a plane normal to b_0. Left-hand axes.

The space lattice is rhombohedral, and the rhombohedral cells corresponding to the foregoing are:

$$(CH_3)_3N \cdot BF_3: a_0' = 5.76 \text{ A.}, \qquad \alpha = 108°16'$$
$$(CH_3)_3N \cdot BH_3: a_0' = 5.74 \text{ A.}, \qquad \alpha = 109°$$

These rhombohedra contain one, the corresponding hexagonal cells three molecules.

The space group is C_{3v}^5 ($R3m$). A determination of atomic positions has been made for the fluoride; in terms of the hexagonal cell, atoms are in

$(3a)$ $00u;$ $^1/_3, ^2/_3, u + ^2/_3;$ $^2/_3, ^1/_3, u + ^1/_3$
$(9b)$ $u\bar{u}v;$ $u\,2u\,v;$ $2\bar{u}\,\bar{u}\,v;$ rh

The determined positions and parameters are the following:

Atom	Position	u	v
CH_3	$(9b)$	0.090	0.334
N	$(3a)$	0.270	—
B	$(3a)$	0.010	—
F	$(9b)$	−0.080	−0.073

In the resulting structure (Fig. XIVA,17), CH_3–N = 1.50 A. and CH_3–

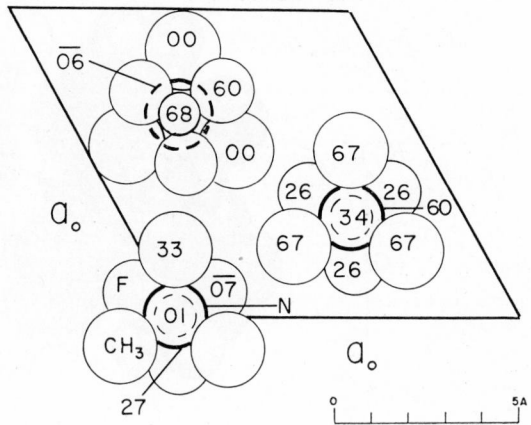

Fig. XIVA,17. The hexagonal structure of $(CH_3)_3N \cdot BF_3$ projected along its c_0 axis. The smallest circles are the atoms of boron.   Left-hand axes.

N–CH$_3$ = 114°. In the BF$_3$ part of the molecule, B–F = 1.39 A. and F–B–F = 107°. Between the molecular parts, N–B = 1.585 A., N–B–F = 112°, and CH$_3$–N–B = 105°.

XIV,a18.   The addition compound *trimethyl amine–iodine monochloride*, $(CH_3)_3N \cdot ICl$, forms orthorhombic crystals whose unit, containing eight molecules, has the cell edges:

$$a_0 = 11.52 \text{ A.}; \quad b_0 = 11.08 \text{ A.}; \quad c_0 = 10.77 \text{ A.} \ (-20°C.)$$

The space group is V_h^{15} (*Pbca*) with all atoms in the positions:

$$(8c) \quad \pm (xyz; \ 1/2-x, y+1/2, z; \ x, 1/2-y, z+1/2; \ x+1/2, y, 1/2-z)$$

The determined parameters are those of Table XIVA,3.

TABLE XIVA,3
Parameters of the Atoms in $(CH_3)_3N \cdot ICl$

Atom	x	y	z
I	0.0982	0.1345	0.1005
Cl	−0.028	0.246	−0.053
N	0.205	0.033	0.248
C(1)	0.270	0.126	0.330
C(2)	0.126	−0.033	0.323
C(3)	0.290	−0.033	0.176

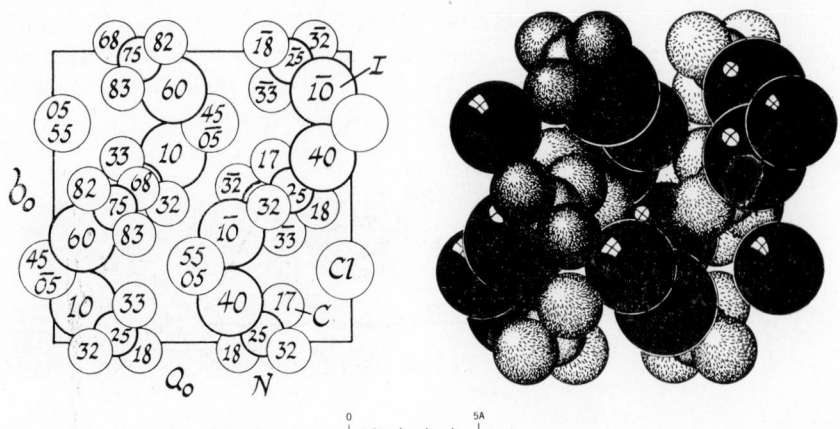

Fig. XIVA,18a (left). The orthorhombic structure of $(CH_3)_3N \cdot ICl$ projected along its c_0 axis. Right-hand axes.

Fig. XIVA,18b (right). A packing drawing of the orthorhombic structure of $(CH_3)_3N \cdot ICl$ projected along its c_0 axis. The largest black circles are iodine, the somewhat smaller ones chlorine. The carbon atoms are line shaded, the nitrogen atoms dotted and more heavily outlined.

As Figure XIVA,18 indicates, the molecule can be thought of as a direct addition of its two components, with N–I–Cl essentially a straight line. Within this molecule, I–Cl = 2.52 A., N–I = 2.30 A., N–C = 1.42–1.55 A. Around the nitrogen atom the distribution is tetrahedral. For this determination $R = 0.09$.

XIV,a19. Crystals of the addition compound *trimethyl amine–iodine*, $(CH_3)_3N \cdot I_2$, are orthorhombic with a tetramolecular unit of the edge lengths:

$$a_0 = 11.37 \text{ A.}; \quad b_0 = 8.30 \text{ A.}; \quad c_0 = 8.34 \text{ A.}$$

The space group is V_h^{16} (*Pnma*) with atoms in the positions:

(4c) $\pm (u \; 1/4 \; v; \; u+1/2, 1/4, 1/2-v)$
(8d) $\pm (xyz; \; 1/2-x, y+1/2, z+1/2; \; x, 1/2-y, z; \; x+1/2, y, 1/2-z)$

The chosen positions and parameters are as follows:

Atom	Position	x	y	z
I(1)	(4c)	0.613	1/4	0.498
I(2)	(4c)	0.448	1/4	0.752
N	(4c)	0.741	1/4	0.289
C(1)	(8d)	0.812	0.101	0.302
C(3)	(4c)	0.686	1/4	0.133

The resulting structure is shown in Figure XIVA,19. The I–I distance is 2.83 A. and N–I = 2.27 A.; these three atoms are collinear. Within the amine, C–N = 1.44 or 1.48 A. The intermolecular distances are said to agree with the sums of the usual van der Waals atomic radii.

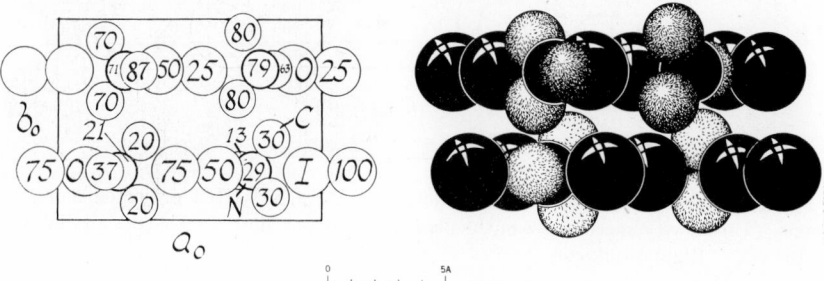

Fig. XIVA,19a (left). The orthorhombic structure of $(CH_3)_3N \cdot I_2$ projected along its c_0 axis. Right-hand axes.

Fig. XIVA,19b (right). A packing drawing of the orthorhombic structure of $(CH_3)_3N \cdot I_2$ seen along its c_0 axis. The iodine atoms are the large, the nitrogen atoms the somewhat smaller black circles. Atoms of carbon are line shaded.

XIV,a20. *O-Methyl hydroxylamine hydrochloride*, $CH_3ONH_2 \cdot HCl$, is orthorhombic with a unit containing eight molecules and having the edge lengths:

$$a_0 = 7.36 \text{ A.}; \quad b_0 = 17.95 \text{ A.}; \quad c_0 = 6.17 \text{ A.}$$

The space group is V_h^{18} (*Bbcm*) with all atoms in the special positions:

$$(8f) \quad \pm(uv0; \, u,^1/_2-v,^1/_2; \, u+^1/_2,v,^1/_2; \, u+^1/_2,^1/_2-v,0)$$

Established parameters are the following:

Atom	u	v
Cl	0.7972	0.1775
O	0.3876	0.0751
N	0.3805	0.1543
C	0.2024	0.0463

The resulting structure is shown in Figure XIVA,20. In the small planar molecules, C–O = 1.46 A., O–N = 1.42 A., and C–O–N = 109°. Each nitrogen has around it three chlorine atoms, that in the plane with N–Cl = 3.10 A., and two others above and below with N–Cl = 3.17 A. It is presumed that these involve hydrogen bonds. For this determination R = 0.276.

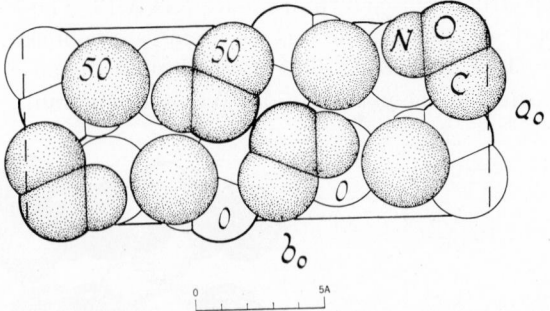

Fig. XIVA,20. The orthorhombic structure of $CH_3ONH_2 \cdot HCl$ projected along its c_0 axis. Unshaded molecules are in the plane of the paper, dotted molecules are halfway up the cell. Right-hand axes.

XIV,a21. Crystals of *trimethylamine oxide*, $(CH_3)_3NO$, are monoclinic with a tetramolecular unit of the dimensions:

$$a_0 = 10.154 \text{ A.}; \quad b_0 = 8.793 \text{ A.}; \quad c_0 = 5.006 \text{ A.}; \quad \beta = 91°2'$$

The space group is C_{2h}^3 ($C2/m$) and atoms are in

$$\begin{aligned}
(4i) &\quad \pm (u0v; \ u+{}^1/_2,{}^1/_2,v), \text{ and} \\
(8j) &\quad \pm (xyz; \ x\bar{y}z; \ x+{}^1/_2,y+{}^1/_2,z; \ x+{}^1/_2,{}^1/_2-y,z)
\end{aligned}$$

with the following parameters:

Atom	Position	x	y	z
O	$(4i)$	0.2041	0	0.3985
N	$(4i)$	0.2094	0	0.6758
C(1)	$(4i)$	0.3469	0	0.7728
C(2)	$(8j)$	0.1418	0.1373	0.7779

Hydrogen positions were obtained by assuming that $C-H = 1.0$ A. and $H-C-H = 110°$, and that one set of hydrogen atoms is in $(4i)$.

Atom	Position	x	y	z
H(1,1)	$(4i)$	0.348	0	0.963
H(1,2)	$(8j)$	0.391	0.083	0.691
H(2,1)	$(8j)$	0.146	0.133	0.981
H(2,2)	$(8j)$	0.044	0.131	0.699
H(2,3)	$(8j)$	0.184	0.230	0.683

The resulting arrangement is shown in Figure XIVA,21. In the molecule, $N-O = 1.388$ A., $N-C(1) = 1.470$ A., and $N-C(2) = 1.484$ A.; the final $C-H$ values lie between 0.95 and 1.06 A. The angles between C, N, and O

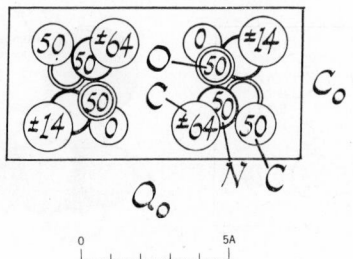

Fig. XIVA,21. The monoclinic structure of $(CH_3)_3NO$ projected along its b_0 axis. Right-hand axes.

are either 109° or 110°. Between molecules, hydrogen atoms are as close as 2.17 A. to one another; almost as short a separation is an O–H = 2.46 A.

XIV,a22. *Trimethylamine oxide hydrochloride*, $(CH_3)_3NO \cdot HCl$, is orthorhombic with a tetramolecular unit that has the edge lengths:

$$a_0 = 14.27 \text{ A.}; \quad b_0 = 5.40 \text{ A.}; \quad c_0 = 7.61 \text{ A.,} \quad \text{all} \pm 0.05 \text{ A.}$$

The space group is V_h^{16} (*Pnam*) with atoms in the positions:

$(4c)$ $\pm(u\ v\ ^1/_4;\ u+^1/_2,^1/_2-v,^1/_4)$

$(8d)$ $\pm(xyz;\ ^1/_2-x,y+^1/_2,z+^1/_2;\ x,y,^1/_2-z;\ x+^1/_2,^1/_2-y,z)$

The parameters are those of Table XIVA,4.

TABLE XIVA,4
Parameters of the Atoms in $(CH_3)_3NO \cdot HCl$

Atom	x	$\sigma(x)$	y	$\sigma(y)$	z	$\sigma(z)$
Cl	0.1034	0.0002	0.1388	0.0006	$^1/_4$	—
O	−0.2232	0.0008	−0.3422	0.0009	$^1/_4$	—
N	−0.1437	0.0004	0.4983	0.0012	$^1/_4$	—
C(1)	−0.0631	0.0006	−0.3303	0.0014	$^1/_4$	—
C(2)	0.1449	0.0008	0.6544	0.0027	0.9115	0.0010
H(0)	0.266	0.019	0.453	0.042	$^1/_4$	—
H(1,1)	0.005	0.017	0.419	0.039	$^1/_4$	—
H(1,2)	0.072	0.006	0.200	0.054	0.863	0.030
H(2,1)	0.207	0.009	0.742	0.014	0.909	0.016
H(2,2)	0.091	0.008	0.786	0.018	0.888	0.007
H(2,3)	0.148	0.016	0.528	0.011	0.007	0.010

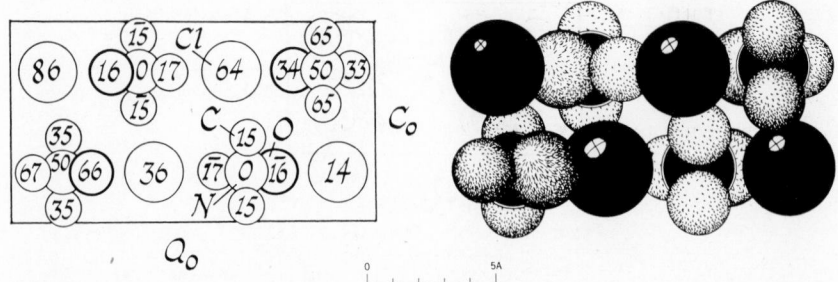

Fig. XIVA,22a (left). The orthorhombic structure of $(CH_3)_3NO \cdot HCl$ projected along its b_0 axis. Right-hand axes.

Fig. XIVA,22b (right). A packing drawing of the orthorhombic structure of $(CH_3)_3NO \cdot HCl$ seen along its b_0 axis. The large black circles are the chlorine atoms; the somewhat smaller ones are nitrogen. The dashed oxygen atoms are slightly larger than the hooked carbon.

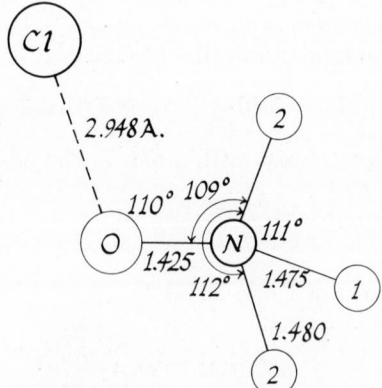

Fig. XIVA,23. Bond dimensions of the $[(CH_3)_3NOH]^+$ ion and its associated Cl^- ion in $(CH_3)_3NO \cdot HCl$.

The structure is shown in Figure XIVA,22. Bond dimensions of the $[(CH_3)_3NOH]^+$ ion are given in Figure XIVA,23. Both types of ion are in layers which, repeated along c_0, have their contacts through the C(2) methyl groups. Between ions, C(2)–2Cl = 3.70 and 3.84 A., C(2)–C(1) = 3.90 A., and C(2)–O = 3.34 A.; the oxygen and chlorine atoms in a plane are connected by a hydrogen bond with Cl–H–O = 2.948 A.

XIV,a23. Crystals of *cis-dinitrosomethane*, $(CH_3ON)_2$, are orthorhombic, with a tetramolecular unit of the edge lengths:

$$a_0 = 6.03 \text{ A.;} \quad b_0 = 12.81 \text{ A.;} \quad c_0 = 5.41 \text{ A.}$$

The space group is V^4 ($P2_12_12_1$) with all atoms in the positions:

(4a) $xyz;$ $^1/_2-x,\bar{y},z+^1/_2;$ $x+^1/_2,^1/_2-y,\bar{z};$ $\bar{x},y+^1/_2,^1/_2-z$

The determined parameters are those of Table XIVA,5.

TABLE XIVA,5
Parameters of the Atoms in $(CH_3ON)_2$

Atom	x	y	z
C(1)	0.6934	0.3128	0.2074
C(2)	0.0797	0.4321	0.3102
N(1)	0.7366	0.3485	0.4563
N(2)	0.9104	0.4074	0.5005
O(1)	0.5975	0.3243	0.6366
O(2)	0.9444	0.4367	0.7280

The structure is shown in Figure XIVA,24. It is made up of planar mole-cules which have the bond dimensions of Figure XIVA,25. Between mole-cules the shortest atomic separations are O–O = 3.12 A., N–O = 3.14 A., and CH_3–O = 3.31 A. For this structure R = 0.156 and the standard deviations of the atomic positions are ca. 0.02 A.

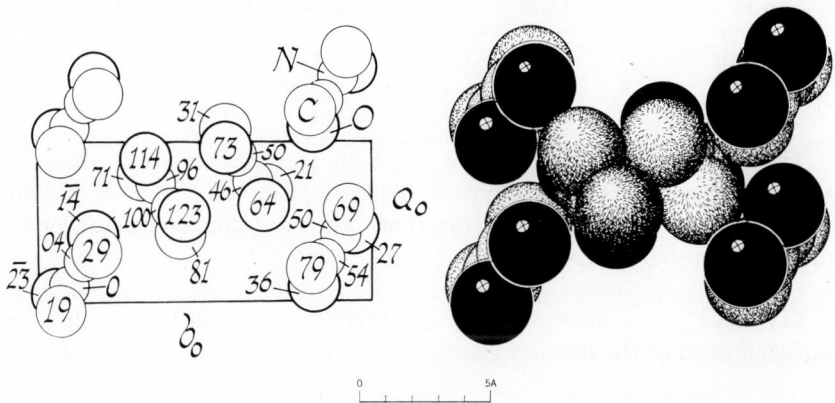

Fig. XIVA,24a (left). The orthorhombic structure of *cis* $(CH_3ON—)_2$ projected along its c_0 axis. Right-hand axes.

Fig. XIVA,24b (right). A packing drawing of the orthorhombic structure of *cis* $(CH_3ON—)_2$ viewed along its c_0 axis. The carbon atoms are black, the oxygens fine-line shaded. Very little of the dotted nitrogen atoms shows.

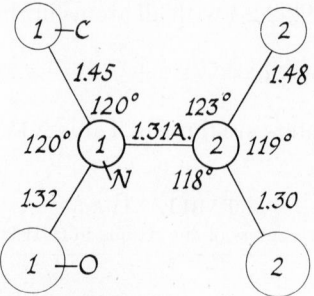

Fig. XIVA,25. The bond dimensions of the molecule of *cis* (CH₃ON—)₂.

XIV,a24. *Azochloramide*, [ClNC(NH₂)N=]₂, is monoclinic with a bi-molecular unit of the dimensions:

$$a_0 = 3.70 \pm 0.02 \text{ A.}; \quad b_0 = 10.43 \pm 0.05 \text{ A.}; \quad c_0 = 9.01 \pm 0.05 \text{ A.}$$
$$\beta = 97°50' \pm 20'$$

The space group is C_{2h}^5 $(P2_1/c)$ with all atoms in the positions:

$$(4e) \quad \pm (xyz; \, x, {}^1/_2 - y, z + {}^1/_2)$$

Parameters have been established as follows:

Atom	x	σ_x, A.	y	σ_y, A.	z	σ_z, A.
Cl	0.429	0.010	0.1358	0.0038	0.1376	0.0055
N(1)	0.069	0.028	0.4517	0.011	−0.0270	0.015
N(2)	0.289	0.029	0.2590	0.013	0.0150	0.016
N(3)	0.139	0.028	0.3738	0.012	0.2229	0.017
C	0.172	0.034	0.3549	0.013	0.0822	0.017

The structure is shown in Figure XIVA,26. Its planar molecules have the bond dimensions of Figure XIVA,27. There appears to be an N–H–N bond of length 2.96 A. tying them together. The next longer intermolecular distance is an N(3)–Cl = 3.32 A.

XIV,a25. *Azodicarbonamide*, [NH₂C(O)N]₂, is monoclinic with a bi-molecular cell of the dimensions:

$$a_0 = 3.57 \pm 0.01 \text{ A.}; \quad b_0 = 9.06 \pm 0.02 \text{ A.}; \quad c_0 = 7.00 \pm 0.02 \text{ A.}$$
$$\beta = 94°50' \pm 15'$$

The space group is C_{2h}^5 $(P2_1/n)$ with all atoms in the positions:

$$(4e) \quad \pm (xyz; \, x + {}^1/_2, {}^1/_2 - y, z + {}^1/_2)$$

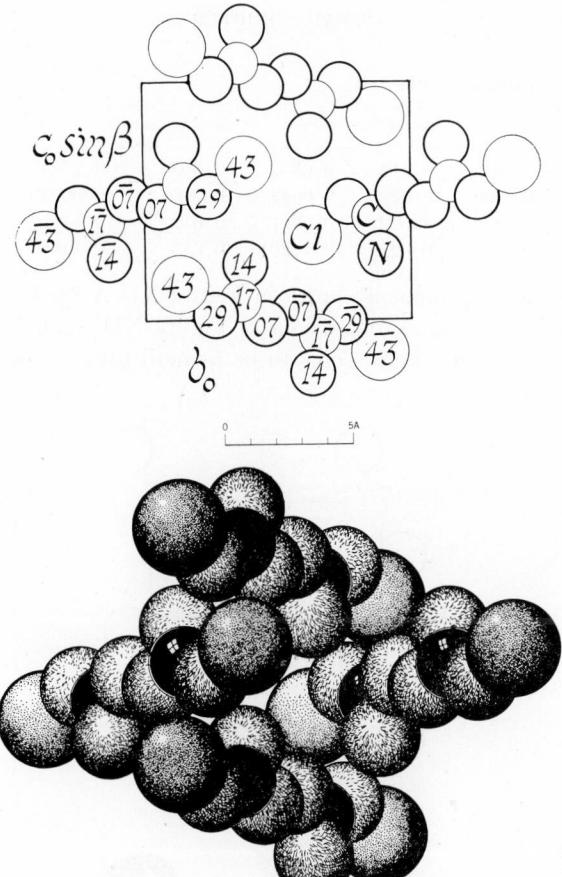

Fig. XIVA,26a (top). The monoclinic structure of azochloramide projected along its a_0 axis. Right-hand axes.

Fig. XIVA,26b (bottom). A packing drawing of the monoclinic azochloramide structure viewed along its a_0 axis. The chlorine atoms are the largest dotted circles; the carbon atoms are black. The nitrogen atoms, of intermediate size, are line shaded.

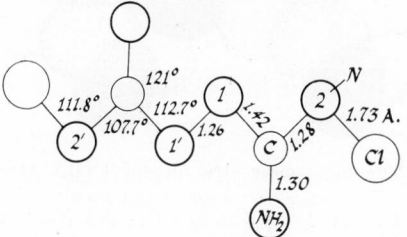

Fig. XIVA,27. Bond dimensions of the azochloramide molecule.

The determined parameters are:

Atom	x	σ_x, A.	y	σ_y, A.	z	σ_z, A.
O	−0.282	0.019	0.0932	0.009	0.2814	0.012
N(1)	0.054	0.022	0.0648	0.010	−0.0022	0.014
N(2)	−0.082	0.022	0.2927	0.012	0.1190	0.014
C	−0.122	0.027	0.1520	0.013	0.1460	0.015

This atomic arrangement is shown in Figure XIVA,28. It is built up of molecules having the bond dimensions of Figure XIVA,29. Each half of a molecule is planar, but there seems to be a small rotation of these halves

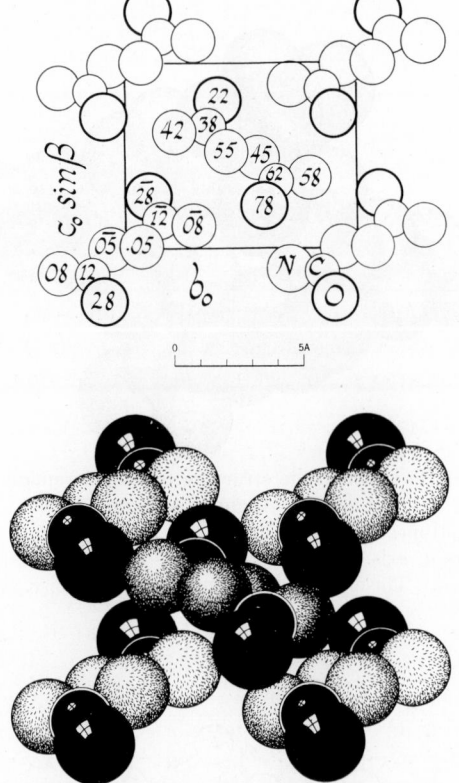

Fig. XIVA,28a (top). The monoclinic structure of $[NH_2C(O)N=]_2$ projected along its a_0 axis. Right-hand axes.

Fig. XIVA,28b (bottom). A packing drawing of the monoclinic azodicarbonamide arrangement seen along its a_0 axis. The oxygens are the larger, the carbons the smaller black circles. Atoms of nitrogen are line shaded.

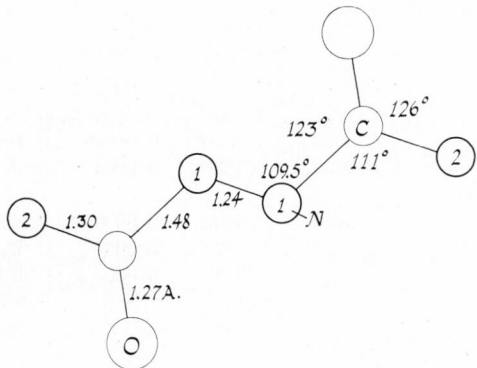

Fig. XIVA,29. The bond dimensions of the azodicarbonamide molecule.

about the C–N(1) bonds. It is thought that the molecules are connected together by hydrogen bonds to form sheets nearly parallel to (101), the sheets being linked by van der Waals forces. The hydrogen bonds are N(2)–H–N(1) = 3.31 A. and N(2)–H–O = 2.87 A. Between molecules the shortest distances not considered to involve hydrogen bonds are N(1)–O = 2.97 A., C–O = 3.11 A., N(1)–C = 3.14 A., and N(1)–N(1) = 3.18 A.

XIV,a26. *Biuret,* $NH_2C(O)NHC(O)NH_2$, crystallizes as a kind of clathrate with a variable amount of water. For this hydrate the symmetry is monoclinic with a tetramolecular cell having the dimensions:

$$a_0 = 3.630 \pm 0.003 \text{ A.}; \quad b_0 = 17.78 \pm 0.01 \text{ A.}; \quad c_0 = 9.18 \pm 0.01 \text{ A.}$$
$$\beta = 119°30' \pm 10'$$

The space group is C_{2h}^5 ($P2_1/c$) with atoms in the general positions:

$$(4e) \quad \pm (xyz; \; x, {}^1\!/_2 - y, z + {}^1\!/_2)$$

The determined parameters are those of Table XIVA,6.

The structure is shown in Figure XIVA,30. As it suggests, water is located in a sort of tunnel that runs through the crystal parallel to the a_0 axis; crystallized at room temperature, about 80% of the water positions [involving O(3)] are filled. Bond dimensions of the biuret molecule are given in Figure XIVA,31. Each half of the molecule involving NH_2CONH— atoms is planar and the two halves make an angle of 5°33' with one another. The molecules are in layers parallel to (100) and held together by N–H–O bonds 2.76–3.01 A. in length; there are only van der Waals forces holding the sheets together. The water molecules in the tunnels have O–N distances of 3.17, 3.30 A., and more; the shortest O(3)–O separation is 2.86 A.

Parameters of the Atoms in Biuret Hydrate

Atom	x	$\sigma(x)$	y	$\sigma(y)$	z	$\sigma(z)$
C(1)	0.5760	0.0039	0.2638	0.0005	0.3566	0.0010
C(2)	0.5827	0.0040	0.3852	0.0005	0.4965	0.0010
N(1)	0.5719	0.0031	0.1900	0.0004	0.3807	0.0008
N(2)	0.6107	0.0032	0.3070	0.0004	0.4891	0.0008
N(3)	0.5541	0.0029	0.4260	0.0004	0.3659	0.0008
O(1)	0.5781	0.0026	0.2892	0.0003	0.2294	0.0007
O(2)	0.6058	0.0027	0.4123	0.0003	0.6259	0.0007
O(3)	0.2016	0.0041	0.0440	0.0005	0.4616	0.0011
H(1)	0.573	—	0.172	—	0.485	—
H(2)	0.583	—	0.156	—	0.293	—
H(3)	0.600	—	0.274	—	0.573	—
H(4)	0.549	—	0.399	—	0.267	—
H(5)	0.500	—	0.481	—	0.369	—
H(6)	0.492	—	0.060	—	0.344	—
H(7)	Undetermined	—	—	—	—	—

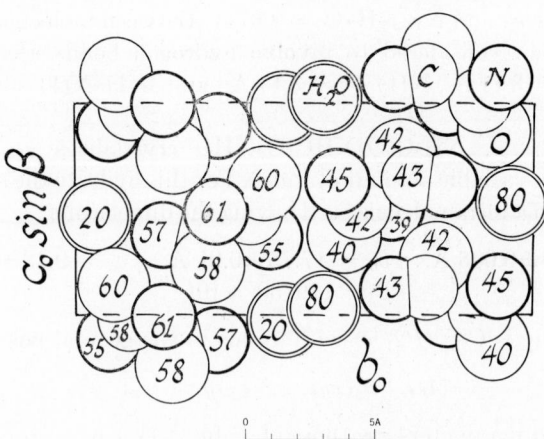

Fig. XIVA,30. The monoclinic structure of biuret, $NH_2C(O)NHC(O)NH_2$, projected along its a_0 axis. Right-hand axes.

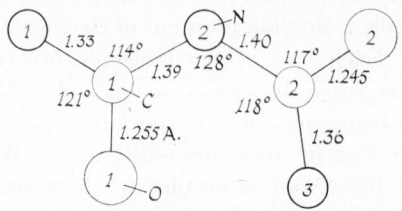

Fig. XIVA,31. Bond dimensions of the biuret molecule.

XIV,a27. *Dimethyl amino boron dichloride*, $N(CH_3)_2BCl_2$, forms mono-clinic crystals which have a tetramolecular unit of the dimensions:

$$a_0 = 11.001 \text{ A.}; \quad b_0 = 8.415 \text{ A.}; \quad c_0 = 6.724 \text{ A.}; \quad \beta = 119°30'$$

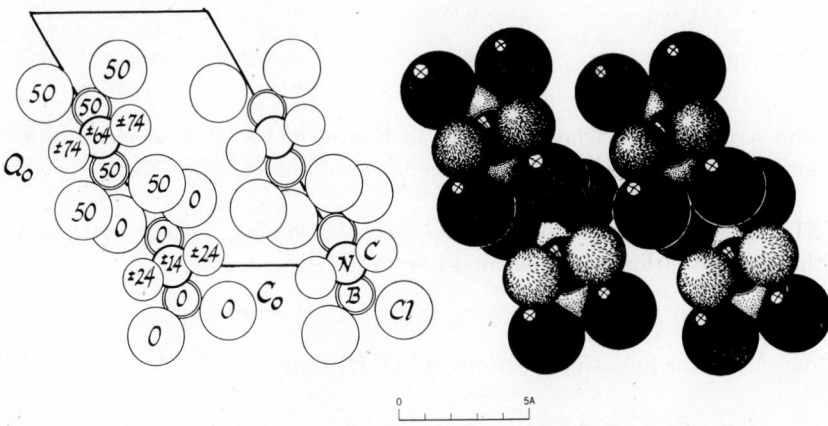

Fig. XIVA,32a (left). The monoclinic structure of $N(CH_3)_2BCl_2$ projected along its b_0 axis. Right-hand axes.

Fig. XIVA,32b (right). A packing drawing of the monoclinic structure of $N(CH_3)_2BCl_2$ seen along its b_0 axis. The chlorine atoms are the large, the nitrogen atoms the smaller black circles. The carbon atoms are line shaded. Boron atoms are pie-shaped, dotted segments between pairs of chlorine atoms.

The space group is $C_{2h}{}^3$ ($C2/m$) with atoms in the positions:

(4g) $\quad \pm (0u0; \ ^1/_2, u+^1/_2, 0)$

(4i) $\quad \pm (u0v; \ u+^1/_2, ^1/_2, v)$

(8j) $\quad \pm (xyz; \ x\bar{y}z; \ x+^1/_2, y+^1/_2, z; \ x+^1/_2, ^1/_2-y, z)$

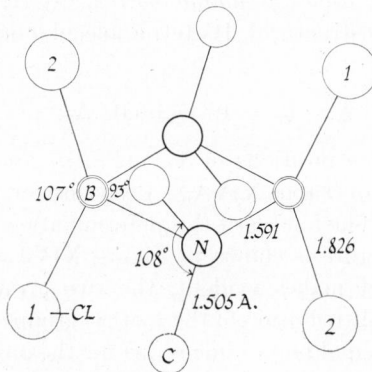

Fig. XIVA,33. Bond dimensions in the molecule of $N(CH_3)_2BCl_2$.

The determined parameters are the following:

Atom	Position	x	y	z
Cl(1)	(4i)	0.1639	0	0.8182
Cl(2)	(4i)	0.2658	0	0.3139
N	(4g)	0	0.1373	0
B	(4i)	0.1080	0	0.0342
C	(8j)	0.0415	0.2418	0.2051

The structure as a whole is shown in Figure XIVA,32. It is made up of dimeric molecules which have the dimensions of Figure XIVA,33.

XIV,a28. The trimer of *dimethyl phosphinoborine*, $[(CH_3)_2PBH_2]_3$, is orthorhombic with a tetramolecular unit of the edge lengths:

$$a_0 = 11.16 \text{ A.}; \quad b_0 = 13.16 \text{ A.}; \quad c_0 = 10.53 \text{ A.}$$

Atoms are in the following positions of V_h^{16} ($Pnma$):

(4c) $\pm (u \ ^1/_4 \ v; \ u+^1/_2, ^1/_4, ^1/_2-v)$
(8d) $\pm (xyz; \ x+^1/_2, ^1/_2-y, ^1/_2-z; \ x, ^1/_2-y, z; \ x+^1/_2, y, ^1/_2-z)$

with the parameters of Table XIVA,7. The hydrogen positions could be identified but only approximately determined from the experimental data.

The resulting structure is shown in Figure XIVA,34. It is built up of molecules which have as core a six-membered P_3B_3 chair-type molecule resembling that of cyclohexane. An indication of the shape of this molecule can be gained from Figure XIVA,35 which also records its principal bond dimensions. Molecules are in contact through CH_3–CH_3 and CH_3–BH with distances that range upwards from 4.0 A.

The corresponding *dimethyl amino borine*, $[(CH_3)_2NBH_2]_3$, is nearly though not exactly isostructural. Its tetramolecular orthorhombic unit has the dimensions:

$$a_0 = 11.20 \pm 0.03 \text{ A.}; \quad b_0 = 13.17 \pm 0.04 \text{ A.}; \quad c_0 = 8.07 \pm 0.02 \text{ A.}$$

Atoms are in the same positions of V_h^{16} ($Pnma$) used for the phosphine with the parameters of Table XIVA,8. Positions for the hydrogen atoms were not established. The final R of this determination is 0.19.

The resulting structure is shown in Figure XIVA,36. As a comparison with Figure XIVA,34 makes evident, the two arrangements differ significantly only in the distribution of the methyl groups about the N(1) and P(5) atoms. The principal bond dimensions for the amino borine are given in Figure XIVA,37.

TABLE XIVA,7
Positions and Parameters of the Atoms in $[(CH_3)_2PBH_2]_3$

Atom	Position	x	y	z
P(1)	(8d)	0.2345	0.1281	0.0822
P(5)	(4c)	0.0712	$1/4$	0.8818
B(2)	(8d)	0.1513	0.1244	0.9200
B(6)	(4c)	0.3236	$1/4$	0.1236
C(3)	(8d)	0.1257	0.0975	0.2102
C(4)	(8d)	0.3340	0.0199	0.0864
C(7)	(4c)	0.0226	$1/4$	0.7155
C(8)	(4c)	0.9219	$1/4$	0.9584

Hydrogen Positions

Atom	Bonding atom	x	y	z
H(1)	B(2)	0.217	0.117	0.853
H(2)	B(2)	0.075	0.067	0.933
H(3)	B(6)	0.417	$1/4$	0.050
H(4)	B(6)	0.340	$1/4$	0.233
H(5)	C(7)	0.083	$1/4$	0.677
H(6)	C(7)	0.992	0.192	0.733
H(7)	C(8)	0.900	0.300	0.933
H(8)	C(8)	0.925	$1/4$	0.033
H(9)	C(3)	0.126	0.476	0.200
H(10)	C(3)	0.083	0.358	0.233
H(11)	C(3)	0.167	0.417	0.283
H(12)	C(4)	0.367	0.458	0.167
H(13)	C(4)	0.400	0.467	0.033
H(14)	C(4)	0.333	0.542	0.083

TABLE XIVA,8
Positions and Parameters of the Atoms in $[(CH_3)_2NBH_2]_3$

Atom	Position	x	y	z
B(1)	(8d)	0.147	0.152	0.591
B(2)	(4c)	0.256	$1/4$	0.342
N(1)	(4c)	0.079	$1/4$	0.645
N(2)	(8d)	0.186	0.149	0.404
C(1)	(8d)	0.285	0.067	0.382
C(2)	(8d)	0.093	0.124	0.283
C(3)	(4c)	-0.049	$1/4$	0.598
C(4)	(4c)	0.068	$1/4$	0.822

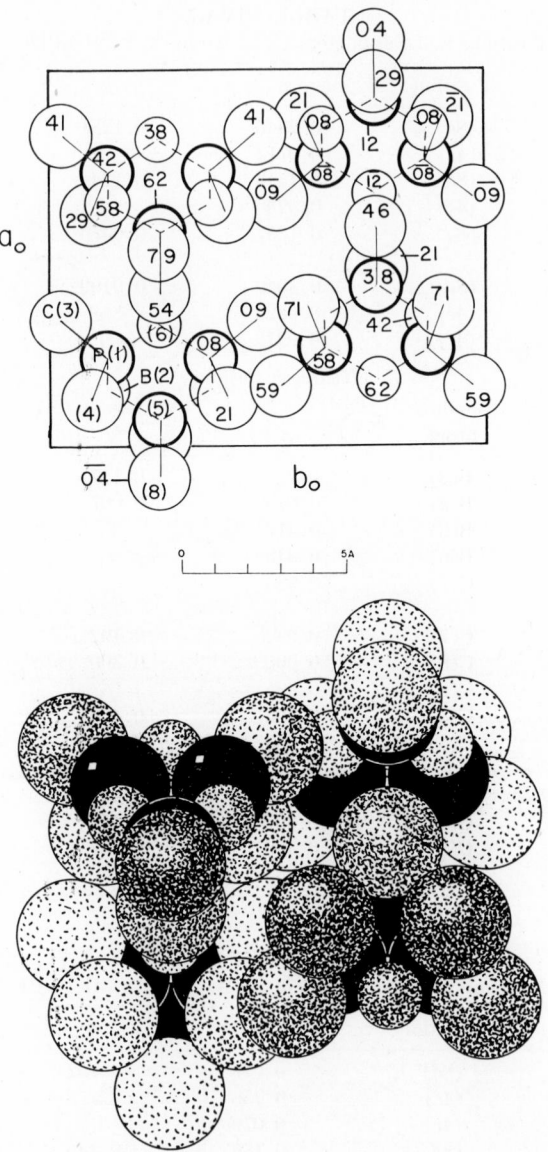

Fig. XIVA,34a (top). The orthorhombic structure of $[(CH_3)_2PBH_2]_3$ projected along its c_0 axis. Boron atoms are the small, lightly outlined circles. Left-hand axes.

Fig. XIVA,34b (bottom). A packing drawing of the orthorhombic structure of $[(CH_3)_2PBH_2]_3$ viewed along its c_0 axis. The phosphorus atoms are black. Carbon atoms are the large, boron atoms the small dotted circles.

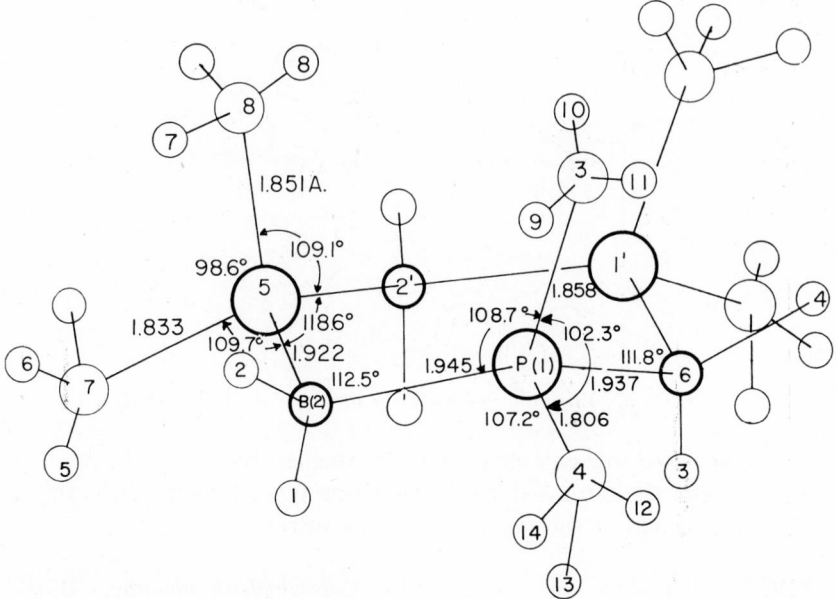

Fig. XIVA,35. Bond dimensions in the trimeric molecule of $[(CH_3)_2PBH_2]_3$.

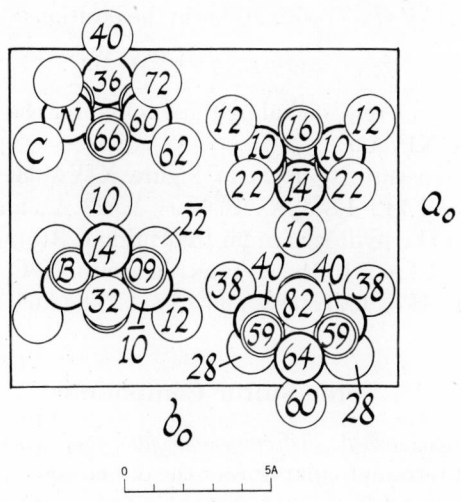

Fig. XIVA,36. The orthorhombic structure of $[(CH_3)_2NBH_2]_3$ projected along its c_0 axis.   Right-hand axes.

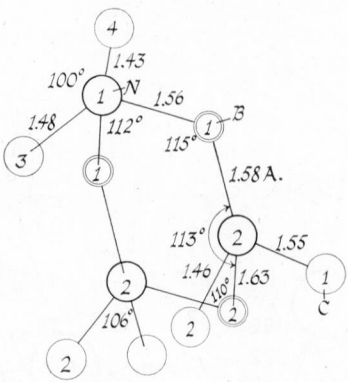

Fig. XIVA,37. Bond dimensions in the molecule of $[(CH_3)_2NBH_2]_3$.

It is to be noted that this chair-like B_3N_3 ring can be imagined as an inorganic analog of the saturated cyclohexane ring just as the flat B_3N_3 ring in $B_3(NH)_3Cl_3$ (**IX,c51**) is the analog of the benzene ring.

XIV,a29. Crystals of *diacetonitrile dodecahydrodecaborane*, $B_{10}H_{12}$-$(CH_3CN)_2$, are monoclinic with a tetramolecular unit of the dimensions:

$$a_0 = 7.81 \pm 0.01 \text{ A.}; \quad b_0 = 11.31 \pm 0.02 \text{ A.}; \quad c_0 = 14.18 \pm 0.03 \text{ A.}$$
$$\beta = 96°52' \pm 15'$$

The space group is C_{2h}^6 (*I2/c*) with atoms in the positions:

$$(8f) \quad \pm (xyz; \ x,\bar{y},z+^1/_2); \text{ B.C.}$$

The determined parameters, including those found for the hydrogen atoms, are given in Table XIVA,9.

The resulting structure is shown in Figure XIVA,38. In the cage-like molecules, B–B = 1.742–1.881 A., B–N = 1.523 A., and B–H = 1.06–1.18 A. (except for the hydrogen atom that bridges B(5) and B(10), where B–H = 1.22 and 1.23 A.). In the acetonitrile group, N–C = 1.137 A., C–C = 1.446 A., and C–H = 0.77, 0.98, and 1.08 A. The final R for this structure is 0.15.

3. Some Sulfur Compounds

XIV,a30. *Dithiodimethyl dodecahydrodecaborane*, $B_{10}H_{12}[S(CH_3)_2]_2$, is monoclinic with a tetramolecular unit of the dimensions:

$$a_0 = 11.83 \text{ A.}; \quad b_0 = 10.78 \text{ A.}; \quad c_0 = 12.74 \text{ A.}, \quad \text{all } \pm 0.02 \text{ A.}$$
$$\beta = 95°10'$$

TABLE XIVA,9
Parameters of the Atoms in $B_{10}H_{12}(CH_3CN)_2$

Atom	x	y	z
B(5)	0.066	−0.037	0.372
B(2)	−0.115	0.037	0.317
B(6)	−0.139	−0.115	0.330
B(7)	0.222	−0.037	0.281
B(1)	0.092	0.088	0.297
N	−0.266	−0.136	0.402
C(1)	−0.360	−0.151	0.456
C(2)	−0.478	−0.171	0.526
H(5)	0.100	−0.028	0.450
H(2)	−0.202	0.078	0.364
H(6)	−0.090	−0.192	0.300
H(7)	0.357	−0.031	0.299
H(1)	0.150	0.179	0.325
H(5–10)	0.158	−0.105	0.333
H(Me1)[a]	−0.444	−0.119	0.592
H(Me2)	−0.454	−0.251	0.548
H(Me3)	−0.543	−0.119	0.517

[a] Me applies to hydrogen atoms belonging to the methyl groups.

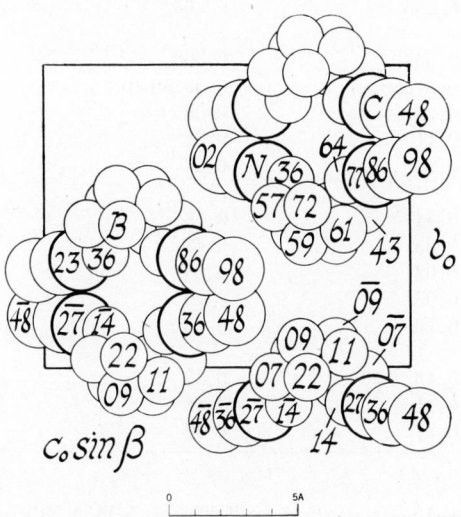

Fig. XIVA,38. The monoclinic structure of $B_{10}H_{12}(CH_3CN)_2$ projected along its a_0 axis.   Right-hand axes.

All atoms are in the general positions of the space group C_{2h}^5 $(P2_1/c)$:

$$(4e) \quad \pm (xyz; \; x,{}^1/_2-y,z+{}^1/_2)$$

with the parameters of Table XIVA,10.

TABLE XIVA,10
Parameters of the Atoms in $B_{10}H_{12}[S(CH_3)_2]_2$

Atom	x	y	z
S(1)	0.2424 ± 0.0003	0.6965 ± 0.0002	0.9760 ± 0.0002
S(2)	0.2648	0.0874 ± 0.0003	0.7904
C(1a)	0.355 ± 0.001	0.673 ± 0.001	0.080 ± 0.001
C(1b)	0.111	0.686	0.043
C(2a)	0.394	0.047	0.872
C(2b)	0.157	0.029	0.872
B(1)	0.303	0.484	0.687
B(2)	0.227	0.577	0.767
B(3)	0.150	0.471	0.687
B(4)	0.236	0.337	0.689
B(5)	0.358	0.526	0.814
B(6)	0.243	0.541	0.901
B(7)	0.109	0.503	0.816
B(8)	0.119	0.339	0.765
B(9)	0.255	0.263	0.812
B(10)	0.368	0.360	0.761
H(1)	0.31 ± 0.01	0.53 ± 0.01	0.60 ± 0.01
H(2)	0.22	0.68	0.76
H(3)	0.11	0.51	0.60
H(4)	0.23	0.27	0.63
H(5)	0.44	0.58	0.83
H(6)	0.25	0.47	0.96
H(7)	0.02	0.54	0.83
H(8)	0.03	0.29	0.75
H(9)	0.26	0.28	0.90
H(10)	0.45	0.33	0.75
H(7,8)	0.11	0.40	0.84
H(5,10)	0.37	0.42	0.84

As the structural representation of Figure XIVA, 39 indicates, the molecules are stacked in rows along the b_0 axis. Individual molecules have the bond lengths of Figure XIVA,40. The decaborane part of the molecule has

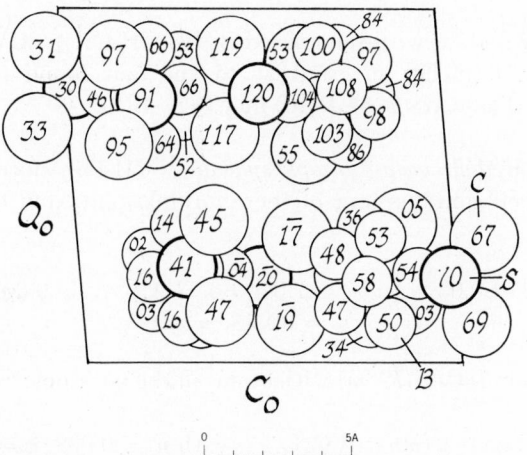

Fig. XIVA,39. The monoclinic structure of $B_{10}H_{12}[S(CH_3)_2]_2$ projected along its b_0 axis. Right-hand axes.

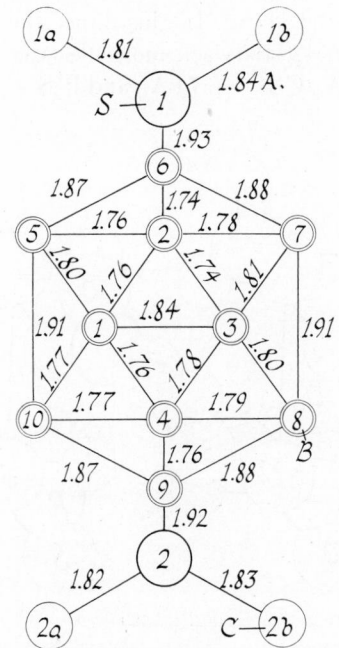

Fig. XIVA,40. Bond lengths in the molecule of $B_{10}H_{12}[S(CH_3)_2]_2$.

C_2^v symmetry and, as with the related $B_{10}H_{12}(CH_3CN)_2$ (**XIV,a29**), has but two hydrogen branches in contrast to the four found for decaborane itself (**V,f30**). These are B(5)–B(10) and B(7)–B(8).

XIV,a31. *Methyl meta-dithiophosphonate*, CH_3PS_2, forms monoclinic crystals. Its cell contains four of these formula units and has the dimensions:

$$a_0 = 6.793 \pm 0.03 \text{ A.}; \quad b_0 = 7.046 \pm 0.03 \text{ A.}; \quad c_0 = 9.207 \pm 0.04 \text{ A.}$$
$$\beta = 92°18' \pm 30'$$

The space group is C_{2h}^3 ($I2/m$) with atoms in the positions:

P: (4i) $\pm(u0v)$; B.C.   with $u = 0.098$, $v = 0.142$
C: (4i)   with $u = -0.062$, $v = 0.297$
S(2): (4i)   with $u = 0.375$, $v = 0.202$
S(1): (4g) $\pm(0u0)$; B.C.   with $u = 0.225$

This structure (Fig. XIVA,41) contains molecules that consist of dimers of the formula as stated above. In this dimer the bonds have the dimensions of Figure XIVA,42. Between molecules the shortest atomic separations are S–S = 3.43 A., C–S = 3.74 A., and P–S = 3.81 A.

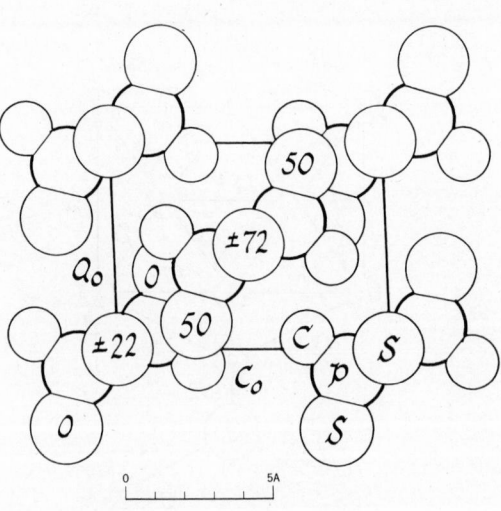

Fig. XIVA,41. The monoclinic structure of CH_3PS_2 projected along its b_0 axis.   Right-hand axes.

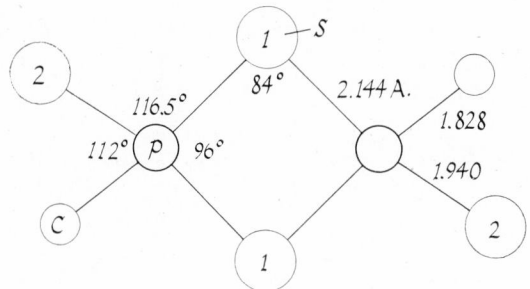

Fig. XIVA,42. Bond dimensions of the $(CH_3PS_2)_2$ dimer.

XIV,a32. Crystals of *hexachloro dimethyl trisulfide*, $(CCl_3)_2S_3$, are monoclinic with a tetramolecular unit of the dimensions:

$$a_0 = 9.338 \text{ A.}; \quad b_0 = 5.890 \text{ A.}; \quad c_0 = 20.069 \text{ A., all } \pm 0.005 \text{ A.}$$
$$\beta = 91°59'$$

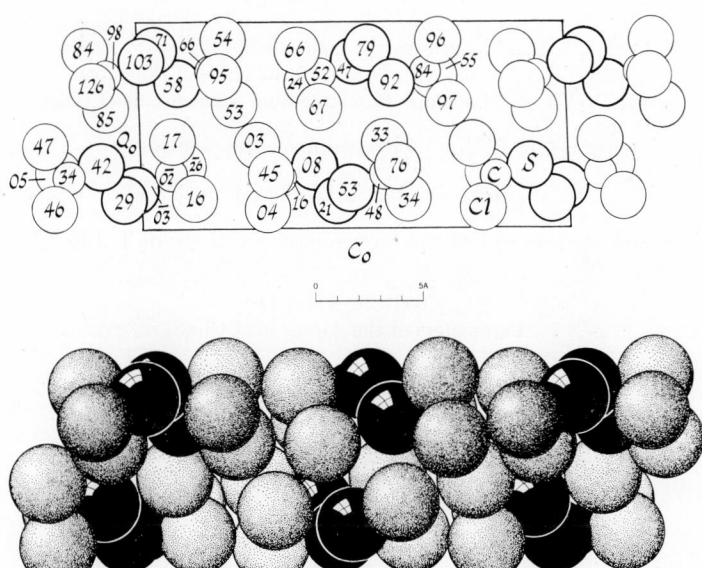

Fig. XIVA,43a (top). The monoclinic structure of $(CCl_3)_2S_3$ projected along its b_0 axis. Right-hand axes.

Fig. XIVA,43b (bottom). A packing drawing of the monoclinic structure of $(CCl_3)_2S_3$ seen along its b_0 axis. The sulfur atoms are black, the chlorine atoms dotted. The carbon atoms do not show.

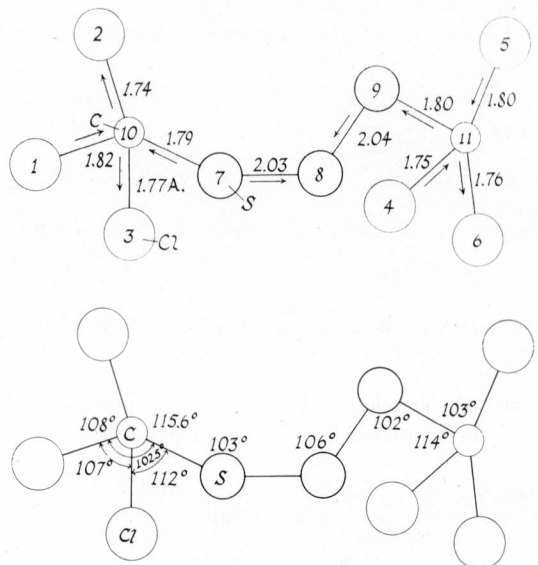

Fig. XIVA,44a (top). Bond lengths in the molecule of $(CCl_3)_2S_3$. Arrows indicate the direction of bonds with respect to the plane of the paper.
Fig. XIVA,44b (bottom). Bond angles in the molecule of $(CCl_3)_2S_3$.

The space group is C_{2h}^5 $(P2_1/c)$ with atoms in the positions:

$$(4e) \quad \pm(xyz;\ x,\tfrac{1}{2}-y,z+\tfrac{1}{2})$$

The averaged final values of the parameters are those of Table XIVA,11.

TABLE XIVA,11
Parameters of the Atoms in $(CCl_3)_2S_3$

Atom	x	σ_x, A.	y	σ_y, A.	z	σ_z, A.
Cl(1)	0.3937	0.0069	0.0288	0.0104	0.2768	0.0069
Cl(2)	0.2540	—	0.4507	—	0.3073	—
Cl(3)	0.0886	—	0.0413	—	0.2956	—
Cl(4)	0.4225	—	0.3272	—	0.5748	—
Cl(5)	0.2899	—	0.7582	—	0.6032	—
Cl(6)	0.1348	—	0.3377	—	0.6206	—
S(7)	0.3006	0.0060	0.0821	0.0095	0.4085	0.0060
S(8)	0.1324	—	0.2127	—	0.4573	—
S(9)	0.1936	—	0.5298	—	0.4871	—
C(10)	0.2525	0.0249	0.1615	0.0353	0.3240	0.0248
C(11)	0.2628	—	0.4786	—	0.5705	—

The resulting structure is shown in Figure XIVA,43. It is made up of molecules which have the bond dimensions of Figure XIVA,44.

XIV,a33. *Dicyanotrisulfide*, NCS_3CN, is orthorhombic with a tetra-molecular unit of the dimensions:

$$a_0 = 10.14 \text{ A.}; \quad b_0 = 12.82 \text{ A.}; \quad c_0 = 4.35 \text{ A.}$$

The space group is V_h^{16} (*Pnma*) with one set of sulfur atoms in

$$S(1):(4c) \quad \pm(u \; {}^1/_4 \; v; \; u+{}^1/_2, {}^1/_4, {}^1/_2-v)$$
$$\text{with } u = 0.523, \; v = 0.479$$

The other atoms are in the general positions:

$$(8d) \quad \pm(xyz; \; {}^1/_2-x,y+{}^1/_2,z+{}^1/_2; \; x,{}^1/_2-y,z; \; x+{}^1/_2,y,{}^1/_2-z)$$

with the parameters:

Atom	x	y	z
S(2)	0.435	0.122	0.250
C	0.297	0.109	0.465
N	0.198	0.104	0.619

The resulting structure is shown in Figure XIVA,45. Its molecules have

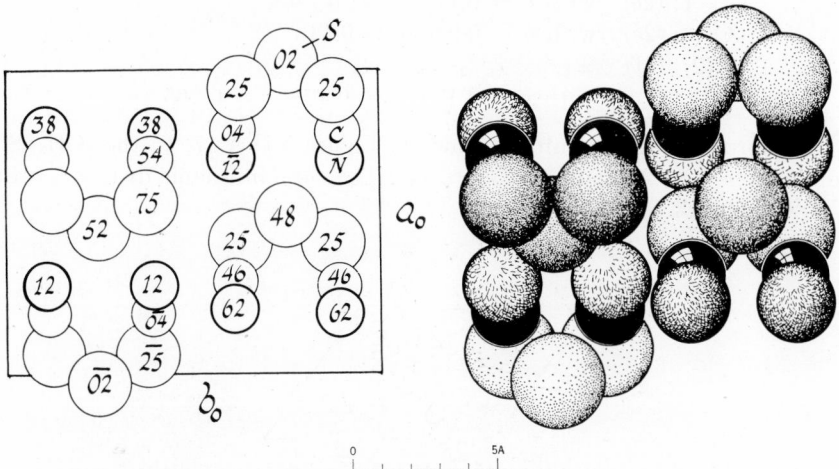

Fig. XIVA,45a (left). The orthorhombic structure of NCS_3CN projected along its c_0 axis.   Right-hand axes.

Fig. XIVA,45b (right). A packing drawing of the orthorhombic structure of NCS_3CN seen along its c_0 axis.   The large dotted circles are sulfur; the black circles are carbon. The nitrogen atoms are line shaded and more heavily ringed.

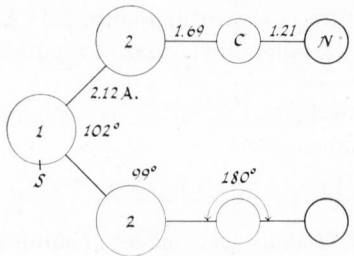

Fig. XIVA,46. Bond dimensions in the molecule of NCS_3CN.

the bond dimensions of Figure XIVA,46. Between molecules the smallest atomic separation is the rather short $S(2)-N = 3.12$ A.

XIV,a34. Crystals of *trimethyl sulfonium iodide*, $(CH_3)_3SI$, are monoclinic with a bimolecular unit of the dimensions:

$$a_0 = 5.944 \text{ A.}; \quad b_0 = 8.003 \text{ A.}; \quad c_0 = 8.922 \text{ A.}, \quad \text{all } \pm0.010 \text{ A.}$$
$$\beta = 126°32' \pm 20'$$

The space group is probably C_{2h}^2 ($P2_1/m$) with atoms in the positions:

$$S: (2e) \quad \pm(u\,{}^1/_4\,v) \qquad \text{with } u = 0.305, v = 0.332$$
$$I: (2e) \quad \text{with } u = 0.1144, v = 0.7409$$
$$C(1): (2e) \quad \text{with } u = 0.926, v = 0.143$$
$$C(2): (4f) \quad \pm(xyz;\ x,{}^1/_2-y,z)$$
$$\text{with } x = 0.430, y = 0.429, z = 0.265$$

The structure that results is shown in Figure XIVA,47. In the $(CH_3)_3S$ ion, $C-S = 1.83$ A. and $C-S-C = 103°$. The close interionic distances are $I-C = 3.80$–4.32 A.

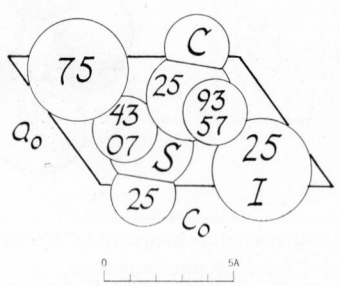

Fig. XIVA,47. The monoclinic structure of $(CH_3)_3SI$ projected along its b_0 axis. Right-hand axes.

XIV,a35. *Dimethyl sulfone*, $(CH_3)_2SO_2$, forms orthorhombic crystals. There are four molecules in its unit cell which has the edge lengths:

$$a_0 = 7.360 \text{ A.}; \quad b_0 = 8.035 \text{ A.}; \quad c_0 = 7.340 \text{ A.}$$

The space group is V_h^{17} (*Cmcm*) with atoms in the positions:

S: (4c) $\pm(0\ u\ ^1/_4;\ ^1/_2,u+^1/_2,^1/_4)$ with $u = 0.1490$

C: (8f) $\pm(0uv;\ 0,u,^1/_2-v;\ ^1/_2,u+^1/_2,v;\ ^1/_2,u+^1/_2,^1/_2-v)$
       with $u = 0.287,\ v = 0.440$.

O: (8g) $\pm(u\ v\ ^1/_4;\ \bar{u}\ v\ ^1/_4;\ u+^1/_2,v+^1/_2,^1/_4;\ ^1/_2-u,v+^1/_2,^1/_4)$
       with $u = 0.168,\ v = 0.056$

The atomic arrangement is shown in Figure XIVA,48. In the individual molecules, the sulfur atoms are tetrahedrally surrounded by two methyl groups and two oxygen atoms with S–C = 1.778 A. and S–O = 1.445 A. Between molecules the shortest atomic separations are a C–C and a C–O = 3.54 A.

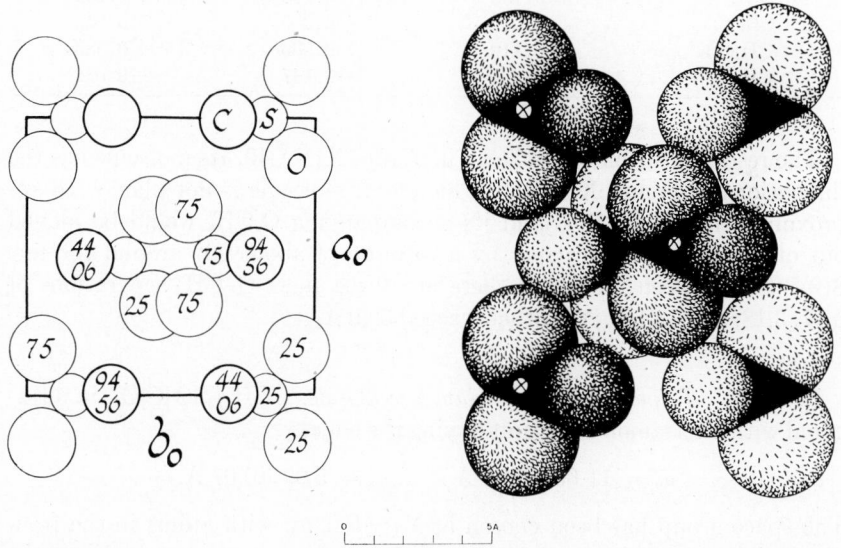

Fig. XIVA,48a (left). The orthorhombic structure of $(CH_3)_2SO_2$ projected along its c_0 axis. Right-hand axes.

Fig. XIVA,48b (right). A packing drawing of the orthorhombic $(CH_3)_2SO_2$ arrangement seen along its c_0 axis. The sulfur atoms are pie-shaped and black; the oxygens are the large line shaded circles. The smaller, hook shaded circles are carbon.

XIV,a36. *Dimethyl sulfonyl disulfide*, $(CH_3SO_2S)_2$, is monoclinic with a tetramolecular cell of the dimensions:

$$a_0 = 5.52 \text{ A.}; \quad b_0 = 15.78 \text{ A.}; \quad c_0 = 10.05 \text{ A.}; \quad \beta = 97°36'$$

Atoms have been placed in general positions of C_{2h}^5 $(P2_1/c)$:

$$(4e) \quad \pm(xyz; \ x,\tfrac{1}{2}-y,z+\tfrac{1}{2})$$

with the parameters of Table XIVA,12.

TABLE XIVA,12
Parameters of the Atoms in $(CH_3SO_2S)_2$

Atom	x	y	z
S(1)	0.353	0.020	0.286
S(2)	0.212	−0.097	0.340
S(3)	−0.062	−0.120	0.183
S(4)	0.087	−0.205	0.056
O(1)	0.605	0.008	0.367
O(2)	0.322	0.030	0.146
O(3)	−0.139	−0.236	−0.034
O(4)	0.230	−0.255	0.155
CH₃(1)	0.119	0.086	0.333
CH₃(2)	0.302	−0.147	−0.022

The resulting structure is shown in Figure XIVA,49. Its molecule has the dimensions of Figure XIVA,50. The actual molecule is not planar; to approximate its shape, S(4) and its accompanying O_2CH_3 would be moved out of the plane of the paper by a rotation of about 90° around the line S(2)S(3). Between molecules there are three short $O–CH_3$ separations of 3.17, 3.18, and 3.21 A.; the others exceed 3.40 A.

XIV,a37. *Trimethyl oxosulfonium perchlorate*, $[(CH_3)_3SO]ClO_4$, is tetragonal with a tetramolecular unit having the edges:

$$a_0 = 11.66 \pm 0.01 \text{ A.}, \qquad c_0 = 5.99 \pm 0.01 \text{ A.}$$

The space group has been chosen as V_d^3 $(P\bar{4}2_1m)$ with atoms in the positions:

Cl(1): (2a) 000; $\tfrac{1}{2}\ \tfrac{1}{2}\ 0$
Cl(2): (2c) $0\ \tfrac{1}{2}\ u;\ \tfrac{1}{2}\ 0\ \bar{u}$ with $u = 0.9141$
  S: (4e) $u,u+\tfrac{1}{2},v;\ \bar{u},\tfrac{1}{2}-u,v;\ u+\tfrac{1}{2},\bar{u},\bar{v};\ \tfrac{1}{2}-u,u,\bar{v}$
      with $u = 0.2208, v = -0.6133$

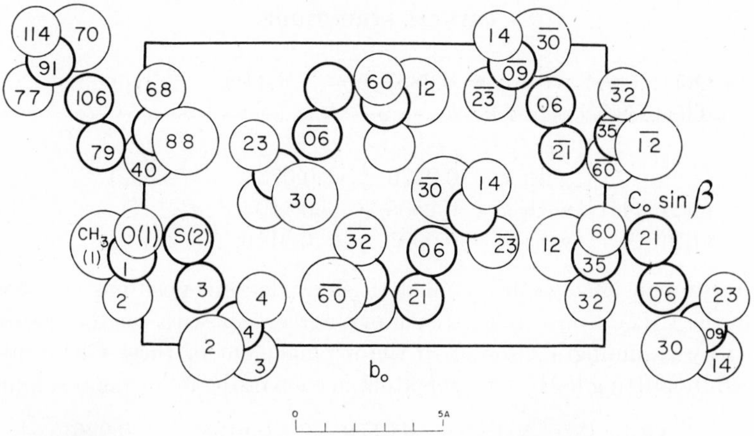

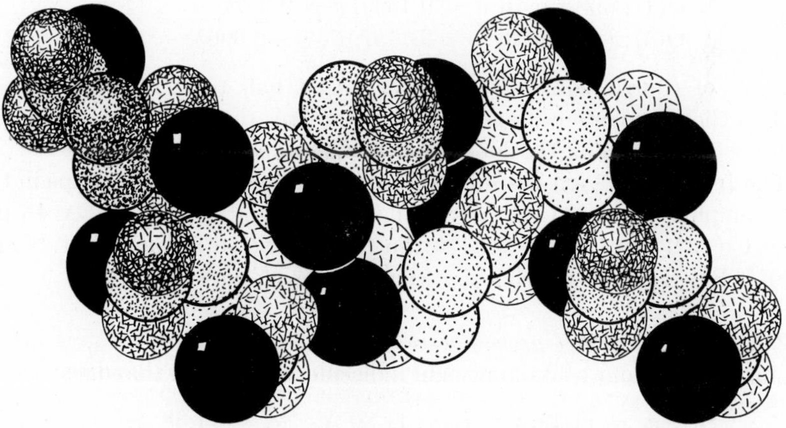

Fig. XIVA,49a (top). The monoclinic structure of $(CH_3SO_2S)_2$ projected along its a_0 axis. Left-hand axes.

Fig. XIVA,49b (bottom). A packing drawing of the monoclinic $(CH_3SO_2S)_2$ arrangement viewed along its a_0 axis. The methyl groups are black, the oxygens line shaded. The atoms of sulfur are heavily ringed and dotted.

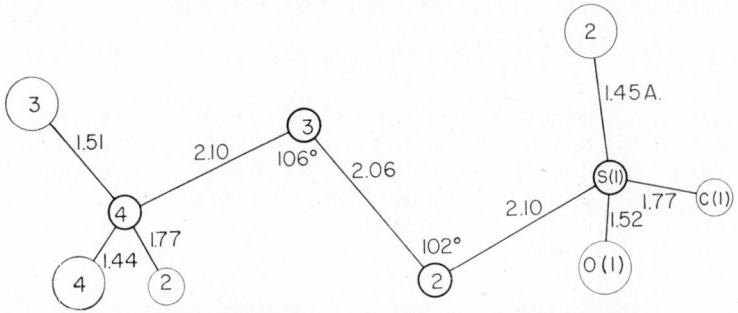

Fig. XIVA,50. Bond dimensions in the molecule of $(CH_3SO_2S)_2$.

O(1): (4e)   with $u = 0.2857$, $v = -0.4566$
C(1): (8f) xyz; $\bar{x}\bar{y}z$; $^1/_2-x,y+^1/_2,\bar{z}$; $x+^1/_2,^1/_2-y,\bar{z}$;
             $\bar{y}x\bar{z}$; $yx\bar{z}$; $y+^1/_2,x+^1/_2,z$; $^1/_2-y,^1/_2-x,z$
             with $x = 0.2576$, $y = 0.0735$, $z = 0.5821$
C(2): (4e)   with $u = 0.2505$, $v = 0.1103$
O(2): (8f)   with $x = 0.0997$, $y = 0.0116$, $z = 0.1313$

A three-dimensional Fourier failed to show definite positions for oxygen about the Cl(2) atoms, but satisfactory agreement with the data was obtained by assuming a disorder in the arrangement of these ClO_4 ions expressed by putting half an oxygen atom in each of the following positions:

$^1/_2O(3)$: (8f)   with $x = 0.0178$, $y = 0.6198$, $z = 0.9261$
$^1/_2O(4)$: (4e)   with $u = 0.4529$, $v = 0.2977$
$^1/_2O(5)$: (4e)   with $u = 0.4457$, $v = -0.0965$

This is, of course, equivalent to saying that half the $Cl(2)O_4$ ions have each of these two orientations in the crystal. It is the same kind of disorder that was also found for the fluoborate (**XIV,a38**).

The $[(CH_3)_3SO]^+$ ion has substantially the same tetrahedral shape in the two compounds; here C–S $= 1.73$ and 1.75 A., and S–O $= 1.42$ A. In the fixed $Cl(1)O_4$ ion, Cl–O $= 1.41$ A.; in the disordered anion, it is 1.41 and 1.49 A. Between ions the shortest separation is a C–O(4) $= 3.17$ A.

XIV,a38.   *Trimethyl oxosulfonium fluoborate*, $[(CH_3)_3SO]BF_4$, is orthorhombic with a unit containing eight molecules and having the edges:

$$a_0 = 11.49 \text{ A.}; \quad b_0 = 11.80 \text{ A.}; \quad c_0 = 11.58 \text{ A.}$$

The space group was chosen as V_h^{14} (*Pbcn*). Atoms have been put in the general positions:

(8d) $\pm(xyz; x+^1/_2,y+^1/_2,^1/_2-z; ^1/_2-x,y+^1/_2,z; x,\bar{y},z+^1/_2)$

Some of the atoms, as listed below, are in fixed positions:

Atom	x	y	z	Average σ
S	0.2175	0.2028	0.0439	0.0004
O	0.2929	0.2812	−0.0146	0.0011
C(1)	0.0702	0.2302	0.0180	0.0015
C(2)	0.2431	0.0619	0.0014	0.0015
C(3)	0.2375	0.2034	0.1947	0.0015
F(3)	0.0090	0.5560	0.3377	0.0018
F(6)	0.4895	0.5157	0.3396	0.0018

The analysis indicated disorder in the positions of the remaining two fluorine atoms and the boron atom, and it was considered this could be expressed by saying that there was half a fluorine atom in each of the following positions:

Atom	x	y	z
$^1/_2$F(1)	0.070	0.408	0.269
$^1/_2$F(2)	0.111	0.459	0.224
$^1/_2$F(4)	0.438	0.367	0.257
$^1/_2$F(5)	0.386	0.437	0.230

If the boron atoms are equidistant from four fluorines, they are in the half positions:

Atom	x	y	z
$^1/_2$B(1)	0.005	0.488	0.245
$^1/_2$B(2)	0.492	0.453	0.249

In the $[(CH_3)_3SO]^+$ ion, the sulfur atom is tetrahedrally surrounded by three carbon atoms (with S–C = 1.76 A.) and the oxygen atom at a distance of 1.44 A. With the parameters as stated above, the B–F separations lie between 1.18 and 1.40 A.

XIV,a39. The compound $(CH_3)_2NSO_2N(CH_3)_2$ forms orthorhombic crystals. Their unit cell contains eight molecules and has the edges:

$$a_0 = 11.76 \text{ A.}; \quad b_0 = 5.68 \text{ A.}; \quad c_0 = 22.03 \text{ A.}$$

The space group has been found to be V_h^{18} (*Cmca*). Three of the atoms are in

$$(8f) \quad \pm (0uv; \; ^1/_2,\bar{u},v+^1/_2; \; ^1/_2,u+^1/_2,v; \; 0,^1/_2-u,v+^1/_2)$$

with $u = 0.015$, $v = 0.126$ for S, $u = -0.101$, $v = 0.184$ for O(1) and $u = -0.120$, $v = 0.072$ for O(2). The rest of the atoms are in the general positions:

$$(16g) \quad \pm (xyz; \qquad x\bar{y}\bar{z};$$
$$x,y+^1/_2,^1/_2-z; \; x,^1/_2-y,z+^1/_2;$$
$$x+^1/_2,y+^1/_2,z; \; x+^1/_2,^1/_2-y,\bar{z};$$
$$x+^1/_2,y,^1/_2-z; \; x+^1/_2,\bar{y},z+^1/_2)$$

with the following parameters:

Atom	x	y	z
N	0.114	0.172	0.125
C(1)	0.144	0.292	0.068
C(2)	0.147	0.306	0.179

The resulting structure is shown in Figure XIVA,51. The bond dimensions of its molecules are those of Figure XIVA,52.

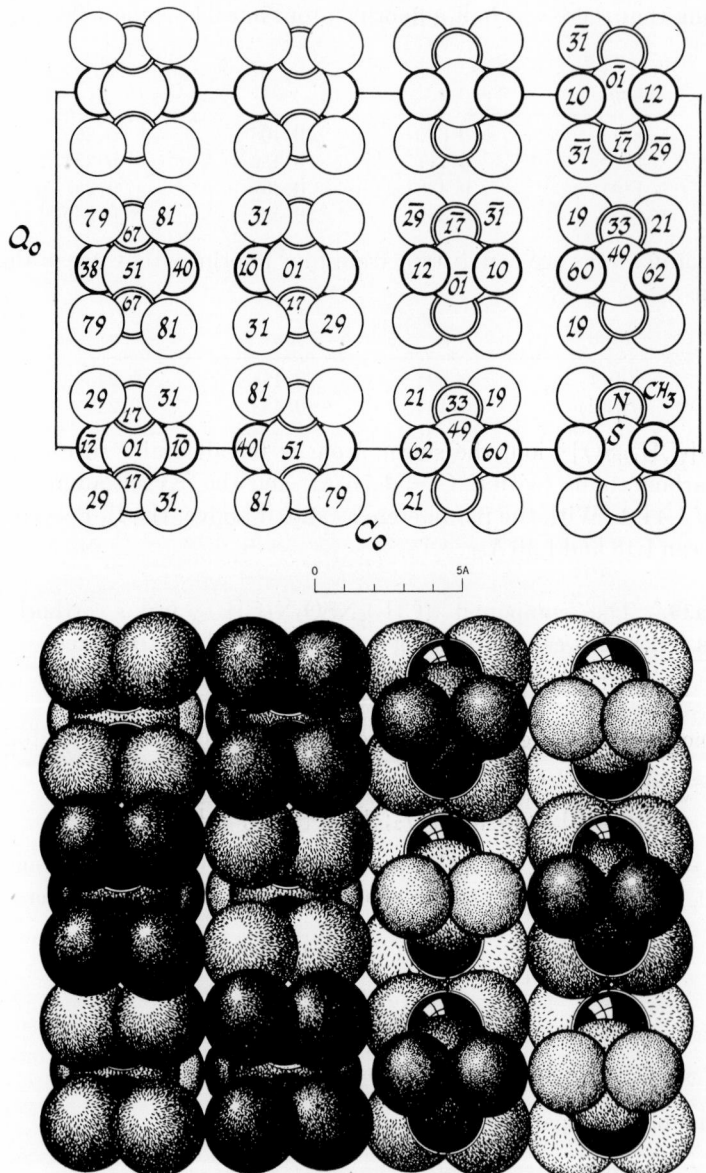

Fig. XIVA,51a (top). The orthorhombic structure of (CH₃)₂NSO₂N(CH₃)₂ projected along its b_0 axis. Right-hand axes.

Fig. XIVA,51b (bottom). A packing drawing of the orthorhombic (CH₃)₂NSO₂N-(CH₃)₂ arrangement seen along its b_0 axis. The nitrogen atoms are black; sulfur atoms are the largest hook-shaded circles. The oxygens are dotted and heavily outlined.

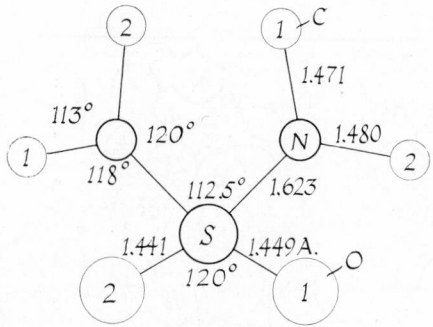

Fig. XIVA,52. Bond dimensions in the molecule of $(CH_3)_2NSO_2N(CH_3)_2$.

For the hydrogen atoms the following parameters have been selected from difference Fourier analysis:

Atom	x	y	z
H(1)	0.130	0.167	0.042
H(2)	0.128	0.220	0.211
H(3)	0.217	0.333	0.067
H(4)	0.117	0.443	0.067
H(5)	0.217	0.317	0.181
H(6)	0.130	0.440	0.187

XIV,a40. Crystals of *tetramethyl thiuram monosulfide*, $[(CH_3)_2NC(S)S-]_2$, are monoclinic with a tetramolecular unit of the dimensions:

$$a_0 = 9.72 \text{ A.;} \quad b_0 = 7.53 \text{ A.;} \quad c_0 = 13.60 \text{ A.;} \quad \beta = 110°8'$$

TABLE XIVA,13
Parameters of the Atoms in $[(CH_3)_2NC(S)S-]_2$

Atom	x	y	z
S(1)	0.212	0.600	0.358
S(2)	0.458	0.610	0.599
S(3)	0.208	0.440	0.595
N(1)	0.127	0.354	0.405
N(1')	0.413	0.775	0.420
C(1)	0.094	0.299	0.297
C(2)	0.007	0.259	0.427
C(3)	0.178	0.485	0.466
C(1')	0.360	0.835	0.312
C(2')	0.556	0.865	0.460
C(3')	0.370	0.655	0.475

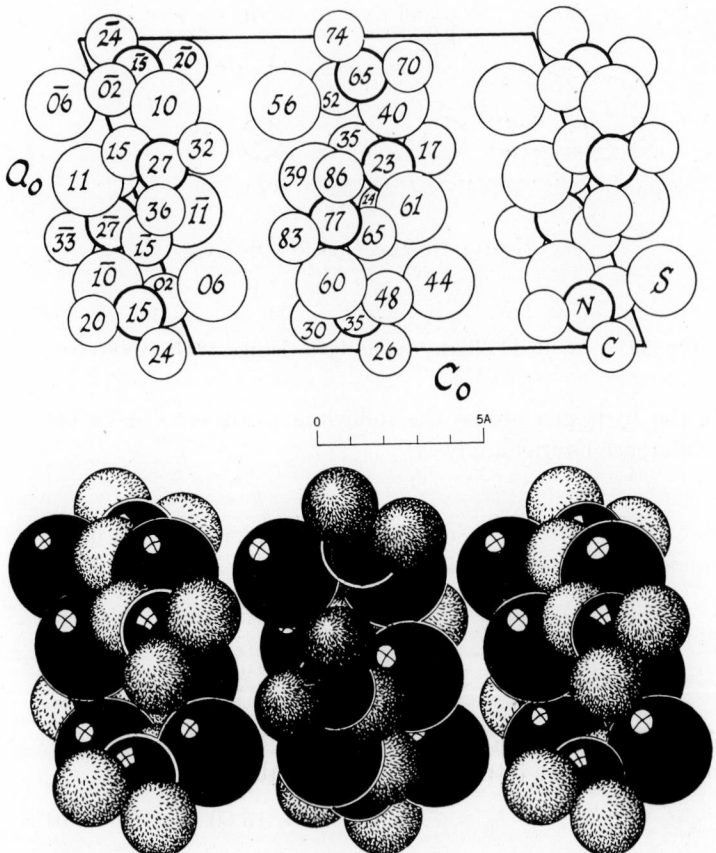

Fig. XIVA,53a (top). The monoclinic structure of tetramethyl thiuram monosulfide projected along its b_0 axis. Right-hand axes.

Fig. XIVA,53b (bottom). A packing drawing of the monoclinic $[(CH_3)_2NC(S)S—]_2$ arrangement viewed along its b_0 axis. The sulfur atoms are the large, the nitrogen atoms the smaller black circles. Atoms of carbon are line shaded.

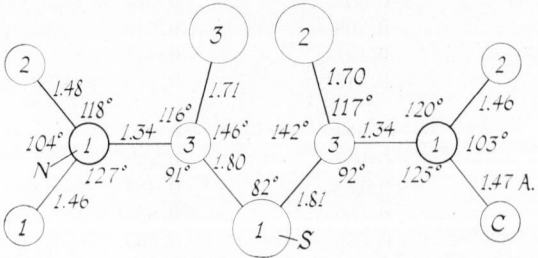

Fig. XIVA,54. Bond dimensions in the molecule of $[(CH_3)_2NC(S)S—]_2$.

The space group is C_{2h}^5 ($P2_1/c$) with atoms in the positions:

$$(4e) \quad \pm (xyz; \ x,{}^1/_2-y,z+{}^1/_2)$$

The stated parameters are listed in Table XIVA,13.

The resulting structure, shown in Figure XIVA,53, is built up of molecules having the bond dimensions of Figure XIVA,54.

4. Some Phosphorus Compounds

XIV,a41. The *tetrameric phosphorus carbon trifluoride*, $(PCF_3)_4$, is tetragonal with a unit containing two of these molecules in a cell of the edge lengths:

$$a_0 = 10.100 \pm 0.020 \text{ A.}, \qquad c_0 = 6.397 \pm 0.015 \text{ A.}$$

The space group is D_{4h}^{15} ($P4_2/nmc$) with atoms in the positions:

$$(8g) \quad 0uv; \ u0\bar{v}; \ {}^1/_2,u+{}^1/_2,{}^1/_2-v; \ u+{}^1/_2,{}^1/_2,v+{}^1/_2;$$
$$0\bar{u}v; \ \bar{u}0\bar{v}; \ {}^1/_2,{}^1/_2-u,{}^1/_2-v; \ {}^1/_2-u,{}^1/_2,v+{}^1/_2$$
$$(16h) \quad xyz; \ \bar{x}\bar{y}z; \ x+{}^1/_2,y+{}^1/_2,{}^1/_2-z; \ {}^1/_2-x,{}^1/_2-y,{}^1/_2-z,$$
$$\bar{x}yz; \ x\bar{y}z; \ {}^1/_2-x,y+{}^1/_2,{}^1/_2-z; \ x+{}^1/_2,{}^1/_2-y,{}^1/_2-z;$$
$$\bar{y}x\bar{z}; \ y\bar{x}\bar{z}; \ {}^1/_2-y,x+{}^1/_2,z+{}^1/_2; \ y+{}^1/_2,{}^1/_2-x,z+{}^1/_2;$$
$$yx\bar{z}; \ \bar{y}\bar{x}\bar{z}; \ y+{}^1/_2,x+{}^1/_2,z+{}^1/_2; \ {}^1/_2-y,{}^1/_2-x,z+{}^1/_2$$

The established parameters are the following:

Atom	Position	x	$\sigma(x)$	y	$\sigma(y)$	z	$\sigma(z)$
P	$(8g)$	0	—	0.1475	0.0003	−0.0529	0.0005
C	$(8g)$	0	—	0.2536	0.0013	0.1859	0.0025
F(1)	$(8g)$	0	—	0.1943	0.0010	0.3706	0.0013
F(2)	$(16h)$	0.3304	0.0008	0.1048	0.0008	−0.1851	0.0013

The resulting structure is shown in Figure XIVA,55. Its molecules have the bond dimensions of Figure XIVA,56. The P_4 ring is not planar, but has a torsion angle of 34°. Between molecules there are F–F separations of 3.23 A. and more. The final R of this determination is 0.08.

XIV,a42. *Pentameric phosphorus carbon trifluoride*, $(PCF_3)_5$, has monoclinic symmetry with a tetramolecular unit of the dimensions:

$$a_0 = 9.87 \pm 0.01 \text{ A.}; \quad b_0 = 9.78 \pm 0.01 \text{ A.}; \quad c_0 = 16.67 \pm 0.06 \text{ A.}$$
$$\beta = 103°0' \quad (-100°C.)$$

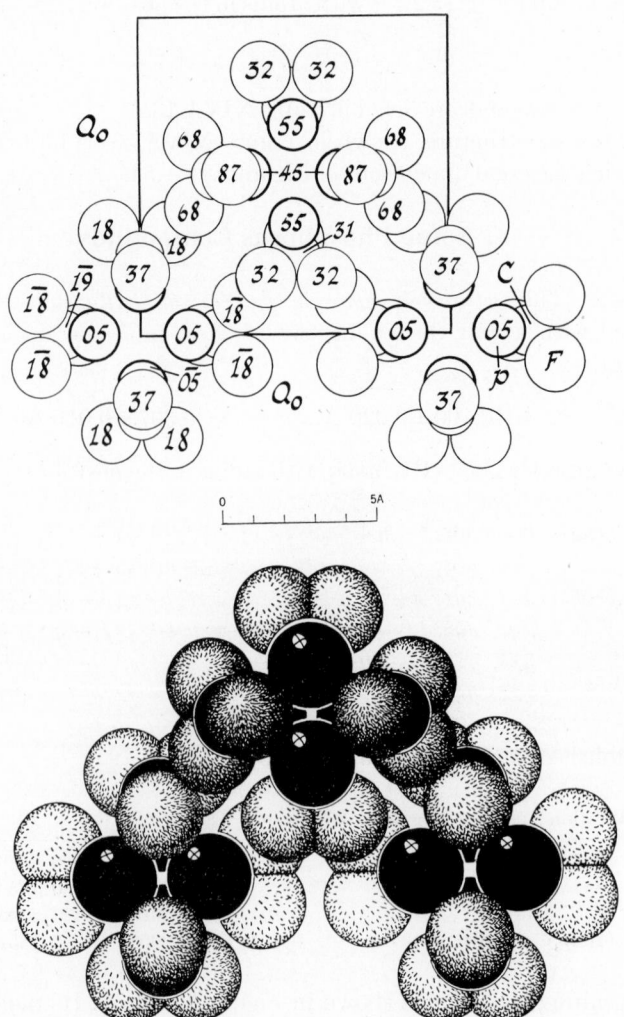

Fig. XIVA,55a (top). The tetragonal structure of $(PCF_3)_4$ projected along its c_0 axis. Right-hand axes.

Fig. XIVA,55b (bottom). A packing drawing of the tetragonal $(PCF_3)_4$ structure seen along its c_0 axis. The atoms of fluorine are line shaded. The phosphorus atoms are black, and the carbon atoms, where they show, are also black.

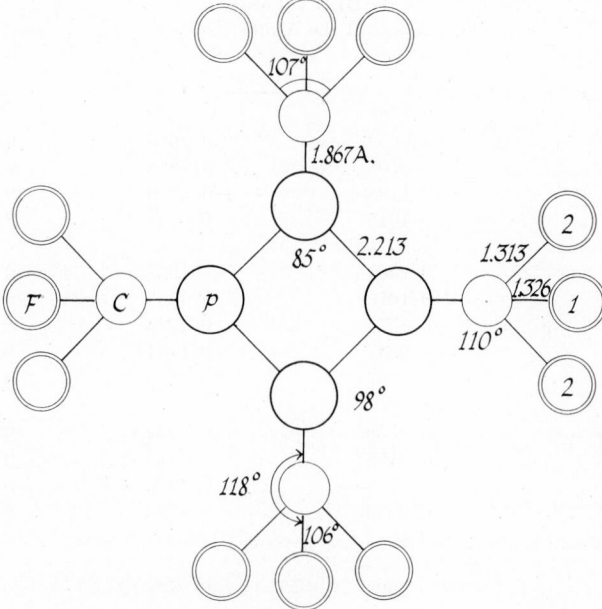

Fig. XIVA,56. Bond dimensions in the molecule of (PCF₃)₄.

Atoms are in the general positions of C_{2h}^5 $(P2_1/n)$:

$$(4e) \quad \pm(xyz; \; x+{}^1/_2,{}^1/_2-y,z+{}^1/_2)$$

with the parameters of Table XIVA,14.

The structure as shown in Figure XIVA,57 is made up of molecules having the dimensions of Figure XIVA,58. Like the tetramer (**XIV,a41**), they consist of a phosphorus ring, each atom of which bears a CF₃ radical. The ring is far from planar and has such a shape that the angles across P(2) and P(5) are 10° smaller than those across the other phosphorus atoms. The average C–F = 1.346±0.032 A. The CF₃ radical attached to P(4) is turned towards the center of the ring to a degree that brings one of its fluorine atoms within 3.04 A. of P(2), a value shorter than the usual van der Waals separation of 3.25 A.

This polymer has a much lower melting point (−33°C.) than the tetramer (66.3°C.).

TABLE XIVA,14
Parameters of the Atoms in $(PCF_3)_5$[a]

Atom	x	y	z
P(1)	0.1203	0.2622	0.2046
P(2)	−0.0659	0.2655	0.0997
P(3)	−0.2108	0.1865	0.1759
P(4)	−0.1138	−0.0099	0.2231
P(5)	0.1015	0.0635	0.2652
C(1)	0.2404	0.1985	0.1404
C(2)	−0.1070	0.4553	0.1131
C(3)	−0.3577	0.1236	0.0905
C(4)	−0.0847	−0.1141	0.1305
C(5)	0.0767	0.1376	0.3674
F(1a)	0.2559	0.2922	0.0841
F(1b)	0.2168	0.0783	0.1024
F(1c)	0.3738	0.1879	0.1915
F(2a)	−0.0956	0.4953	0.1933
F(2b)	−0.0082	0.5301	0.0860
F(2c)	−0.2303	0.4821	0.0721
F(3a)	−0.4420	0.2295	0.0609
F(3b)	−0.4356	0.0327	0.1194
F(3c)	−0.3197	0.0680	0.0223
F(4a)	−0.0457	−0.0397	0.0699
F(4b)	0.0196	−0.2019	0.1576
F(4c)	−0.1969	−0.1824	0.0973
F(5a)	−0.0195	0.2321	0.3619
F(5b)	0.0491	0.0380	0.4149
F(5c)	0.1998	0.1945	0.4078

[a] Average probable errors: ±0.005 A. for P, ±0.022 A. for C, and ±0.014 A. for F.

XIV,a43. *Octamethyl cyclo tetraphosphonitrile*, $[PN(CH_3)_2]_4$, is tetragonal with a tetramolecular unit of the edge lengths:

$$a_0 = 15.705 \pm 0.010 \text{ A.}, \qquad c_0 = 6.425 \pm 0.005 \text{ A.}$$

The space group is C_{4h}^6 ($I4_1/a$) with all atoms in the positions:

$$(16f) \quad \pm (xyz; \qquad\qquad x, y+1/2, \bar{z};$$
$$3/4 - y, x+1/4, z+1/4; \; y+1/4, 1/4 - x, z+1/4); \text{ B.C.}$$

The determined parameters, including those chosen for the hydrogen atoms, are shown in Table XIVA,15.

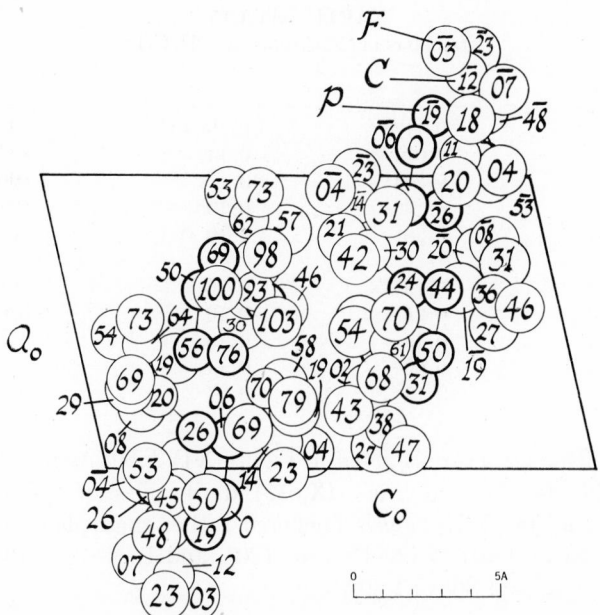

Fig. XIVA,57. The monoclinic structure of (PCF₃)₅ projected along its b_0 axis. Right-hand axes.

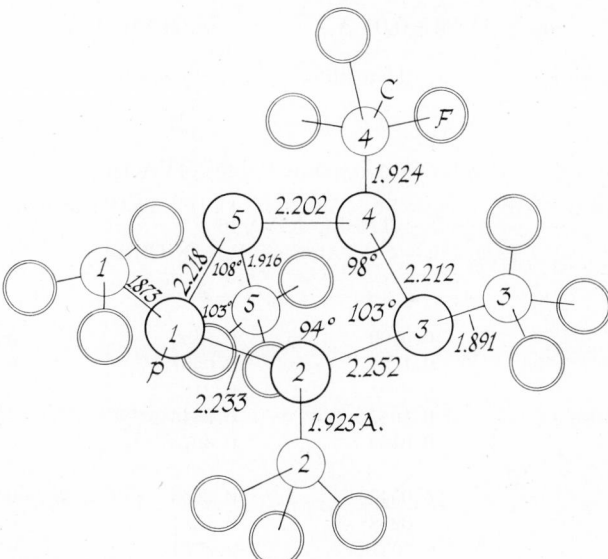

Fig. XIVA,58. Bond dimensions in the molecule of (PCF₃)₅.

TABLE XIVA,15
Parameters of the Atoms in $[PN(CH_3)_2]_4$

Atom	x	y	z
P	0.1123	0.1848	0.1574
N	0.034	0.141	0.041
C(1)	0.204	0.161	−0.002
C(2)	0.136	0.135	0.404
H(1)	0.26	0.19	0.08
H(2)	0.19	0.20	−0.15
H(3)	0.21	0.10	−0.03
H(4)	0.19	0.16	0.47
H(5)	0.09	0.14	0.52
H(6)	0.15	0.07	0.38

The resulting structure is given in Figure XIVA,59. As in the dimethyl-amide (**XIV,a44**) and chlorine (**IX,e13**) derivatives of the tetramer of phosphonitrile, the P_4N_4 ring is puckered rather than plane. In the ring, P–N = 1.591 or 1.601 A., N–P–N = 120°, and P–N–P = 132°. For the side chains, P–C = 1.808 or 1.802 A.

XIV,a44. *Tetrameric phosphonitrilic dimethylamide,* $P_4N_4[N(CH_3)_2]_8$, is tetragonal with a bimolecular cell of the edge lengths:

$$a_0 = 13.00 \pm 0.01 \text{ A.,} \qquad c_0 = 8.59 \pm 0.01 \text{ A.}$$

The space group is $S_4{}^2$ $(I\overline{4})$ with atoms in the positions:

$$(8g) \quad xyz; \; \bar{x}\bar{y}z; \; y\bar{x}\bar{z}; \; \bar{y}x\bar{z}; \; \text{B.C.}$$

The determined parameters are listed in Table XIVA,16.

TABLE XIVA,16
Parameters of the Atoms in $P_4N_4[N(CH_3)_2]_8$[a]

Atom	x	y	z
P	0.0500	0.1480	0.0208
N(1)	0.1254	0.0567	0.0603
N(2)	0.0394	0.2107	0.1919
N(3)	0.1047	0.2343	−0.0972
C(1)	−0.0169	0.3075	0.1845
C(2)	0.0543	0.2865	−0.2203
C(3)	0.0258	0.1511	0.3324
C(4)	0.2051	0.2774	−0.0496

[a] Calculated H parameters are stated in the original as well as values of σ in A.

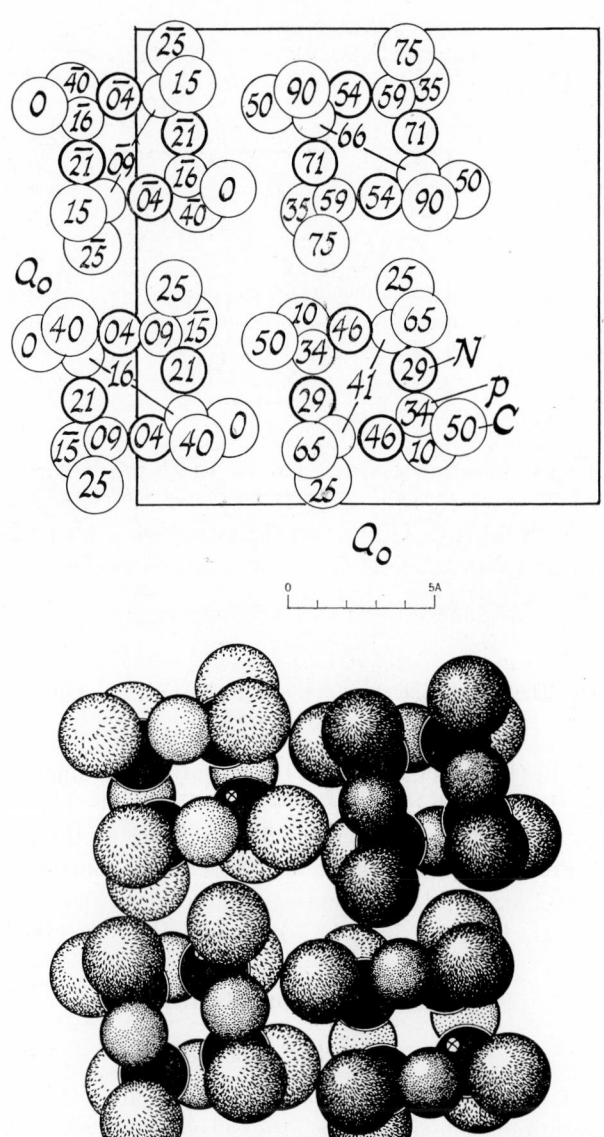

Fig. XIVA,59a (top). The tetragonal structure of $[PN(CH_3)_2]_4$ projected along its c_0
axis. Right-hand axes.

Fig. XIVA,59b (bottom). A packing drawing of the tetragonal $[PN(CH_3)_2]_4$ arrangement seen along its c_0 axis. The phosphorus atoms are black, the nitrogen atoms
heavily outlined and dotted. Atoms of carbon are the larger line shaded circles.

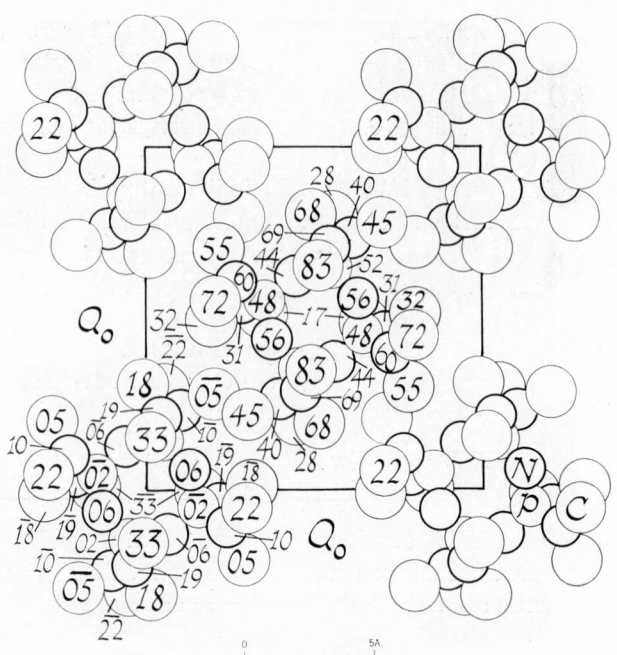

Fig. XIVA,60a. The tetragonal structure of $P_4N_4[N(CH_3)_2]_8$ projected along its c_0 axis. Right-hand axes.

The structure (Fig. XIVA,60) is a body-centered arrangement of molecules having the bond dimensions of Figure XIVA,61. They contain the same kind of puckered eight-membered P_4N_4 ring as the corresponding chloride (**IX,e13**) with bond lengths like those in the more recent study of this substance (1962: H,M&V). Between molecules the shortest interatomic distances are CH_3–CH_3 = 3.73 and 3.76 A., and CH_3–N = 4.00 A.

5. Metallo-Compounds and Salts

a. Metallo-Compounds

XIV,a45. Crystals of *methyl lithium*, CH_3Li, are cubic with a unit containing eight molecules and having the cell edge:

$$a_0 = 7.241 \pm 0.01 \text{ A.}$$

The space group has been chosen as T_d^3 ($I\bar{4}3m$). Both the carbon and the lithium atoms are placed in positions:

(8c) $uuu; u\bar{u}\bar{u}; \bar{u}u\bar{u}; \bar{u}\bar{u}u;$ B.C.
      with $u(Li) = 0.125 \pm 0.006$, $u(C) = 0.320 \pm 0.003$

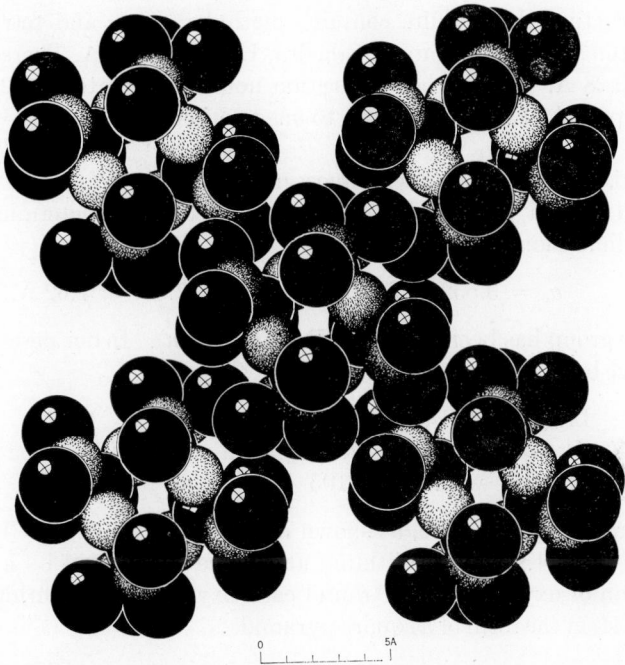

Fig. XIVA,60b. A packing drawing of the tetragonal $P_4N_4[N(CH_3)_2]_8$ arrangement viewed along its c_0 axis.   Both the carbon and the slightly smaller phosphorus atoms are black.   The nitrogen atoms are heavily outlined and dotted.

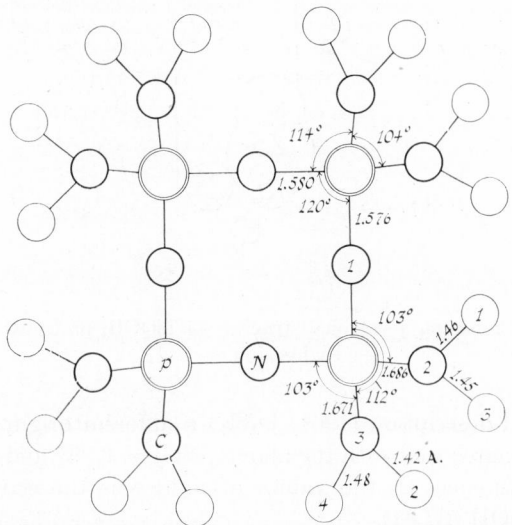

Fig. XIVA,61. Bond dimensions in the molecule of $P_4N_4[N(CH_3)_2]_8$.

The structure that results contains methyl radicals and tetrahedra of lithium atoms. In the lithium tetrahedra, Li–Li = 2.56 A. There is a close Li–C = 2.28 A. and another separation not much greater (2.52 A.). The closest approach of methyl radicals to one another is 3.69 A.

XIV,a46. According to preliminary notes *lithium methoxide*, $LiOCH_3$, is tetragonal with a bimolecular unit stated in the two determinations to have the dimensions:

$$a_0 = 3.552 \text{ or } 3.627 \text{ A.,} \qquad c_0 = 7.687 \text{ or } 7.62 \text{ A.}$$

The space group has been chosen as D_{4h}^7 ($P4/nmm$). In one description the atoms have been given the positions:

Li: (2a) $\pm (1/4 \ 3/4 \ 0)$
C: (2c) $\pm (1/4 \ 1/4 \ u)$ with $u = 0.29$
O: (2c) with $u = 0.105$

In the resulting structure, as shown in Figure XIVA,62, $O–CH_3$ = 1.42 A. and Li–O = 1.95 A. The lithium atoms are surrounded by a distorted tetrahedron of oxygen atoms. Around each oxygen are four lithium atoms and one CH_3 in the form of a square pyramid.

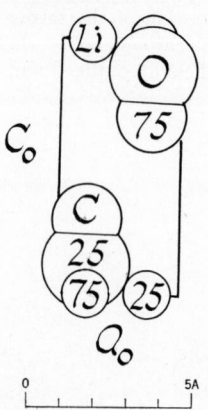

Fig. XIVA, 62. The simple tetragonal structure of $LiOCH_3$ projected along an a_0 axis Right-hand axes.

In the second description (1961: D&K) a different origin was used, but the values of u agree well with the above [u(C) = 0.295 and u(O) = 0.09]. Except for differences in the values of u, this is the structure already described for LiOH (**III,e1**).

XIV,a47. *Potassium methylate*, $KOCH_3$, is tetragonal with a unit that is bimolecular and has the edges:

$$a_0 = 3.949 \pm 0.005 \text{ A.}, \qquad c_0 = 8.768 \pm 0.01 \text{ A.}$$

The space group is D_{4h}^7 $(P4/nmm)$ with atoms in the positions:

K: (2c) $0\ ^1/_2\ u;\ ^1/_2\ 0\ \bar{u}$ with $u = 0.345$
O: (2c)   with $u = 0.655$
C: (2c)   with $u = 0.815$

The structure that results is shown in Figure XIVA,63. Each potassium atom has five oxygen neighbors with K–O = 2.66 or 2.80 A. The separation O–C = 1.40 A., and K–K = 3.68 A. It is thought probable that the methyl groups are rotating.

This arrangement may be compared with that of $K_2O_2C_2$ (**XIV,b46**).

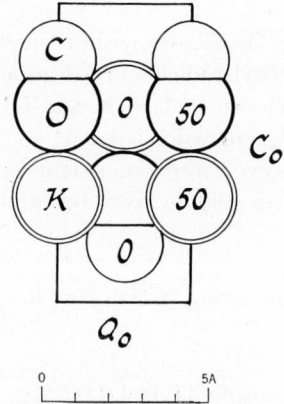

Fig. XIVA,63. The tetragonal structure of $KOCH_3$ projected along an a_0 axis.

XIV,a48. Crystals of *dimethyl beryllium*, $(CH_3)_2Be$, are orthorhombic with a tetramolecular unit of the dimensions:

$$a_0 = 6.14 \text{ A.}; \quad b_0 = 11.53 \text{ A.}; \quad c_0 = 4.18 \text{ A.}$$

The space group has been chosen as V_h^{26} $(Ibam)$ with atoms in the positions:

Be: (4a) $\pm(0\ 0\ ^1/_4)$; B.C.
CH_3: (8j) $\pm(uv0;\ \bar{u}\ v\ ^1/_2)$; B.C.
      with $u = 0.182, v = 0.101$

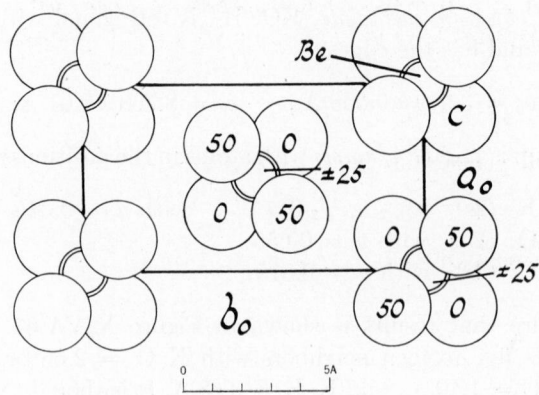

Fig. XIVA,64. The orthorhombic structure of Be(CH$_3$)$_2$ projected along its c_0 axis. Right-hand axes.

This is a structure (Fig. XIVA,64) within which the atoms are grouped in chains. In a chain, methyl radicals are tetrahedrally distributed about beryllium atoms, with CH$_3$–Be = 1.92 A. and CH$_3$–Be–CH$_3$ = 114°. Between chains the CH$_3$–CH$_3$ separation is 4.1 A.

It is to be noted that this is the same structure as that found for SiS$_2$ (**IV,e14**) and BeCl$_2$, the a_0 and b_0 axes being interchanged in the two descriptions.

Recently *dimethyl magnesium*, (CH$_3$)$_2$Mg, has been shown to be isostructural, with:

$$a_0 = 6.00 \pm 0.03 \text{ A.}; \quad b_0 = 11.48 \pm 0.05 \text{ A.}; \quad c_0 = 5.45 \pm 0.03 \text{ A.}$$

For the carbon atoms in (8j), $u = 0.210 \pm 0.003$ and $v = 0.120 \pm 0.002$.

Positions have also been assigned the hydrogen atoms in this compound. There is considered to be disorder in their arrangement which can be expressed by putting half atoms in two sets of (8j) [with $u = 0.345, v = 0.077$ and $u' = 0.145, v' = 0.180$] and in two sets of the general positions

$$(16k) \quad \pm(xyz; \ xy\bar{z}; \ x,\bar{y},{}^1/_2-z; \ x,\bar{y},z+{}^1/_2); \text{ B.C.}$$
with $x = 0.195, y = 0.155, z = 0.135$
and $x' = 0.295, y' = 0.102, z' = 0.135$

The final $R = 0.117$.

XIV,a49. A structure has been proposed for *mono-methyl chlor mercury mercaptan*, CH_3SHgCl. It is monoclinic with a tetramolecular cell of the dimensions:

$$a_0 = 7.45 \text{ A.}; \quad b_0 = 7.37 \text{ A.}; \quad c_0 = 7.82 \text{ A.}; \quad \beta = 86°24'$$

All atoms are in general positions of C_{2h}^5 ($P2_1/c$):

$$(4e) \quad \pm(xyz; \, x, {}^1/_2 - y, z + {}^1/_2)$$

with the following parameters:

Atom	x	y	z
Hg	0.47	0.11	0.20
S	0.26	0.11	0.95
Cl	0.66	0.11	0.43
CH₃	0.07	0.25	0.00

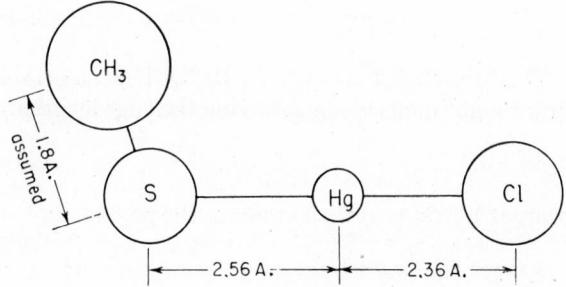

Fig. XIVA,65. Bond lengths in the molecule of CH_3SHgCl.

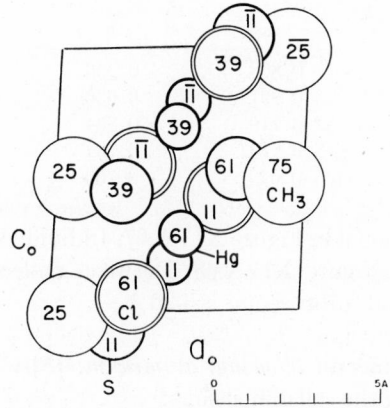

Fig. XIVA,66 The monoclinic structure of CH_3SHgCl projected along its b_0 axis. Left-hand axes.

These lead to molecules shaped as in Figure XIVA,65 and packed as indicated in Figure XIVA,66. The significant separations between adjacent molecules are S–Cl = 3.40 A., Hg–S = 2.78 A., CH_3–Cl = 3.30 A., and CH_3–CH_3 = 3.91 A.

The corresponding *mono-ethyl chlor mercury mercaptan*, C_2H_5SHgCl, is isostructural, with:

$$a_0 = 9.34 \text{ A.}; \quad b_0 = 7.45 \text{ A.}; \quad c_0 = 7.81 \text{ A.}; \quad \beta = 82°30'$$

Positions were not assigned the ethyl radicals, but the other atoms were given the following parameters:

Atom	x	y	z
Hg	0.475	0.125	0.20
S	0.29	0.125	0.00
Cl	0.64	0.125	0.38

XIV,a50. *Mercury methyl mercaptide*, $Hg(SCH_3)_2$, is orthorhombic with a unit containing eight molecules and having the edge lengths:

$$a_0 = 19.80 \pm 0.01 \text{ A.}; \quad b_0 = 7.58 \pm 0.02 \text{ A.}; \quad c_0 = 7.80 \pm 0.02 \text{ A.}$$

The space group is V_h^{15} (*Pbca*) with atoms in the positions:

$$(8c) \quad \pm (xyz; \; 1/2-x, y+1/2, z; \; x, 1/2-y, z+1/2; \; x+1/2, y, 1/2-z)$$

Parameters are as follows:

Atom	x	y	z
Hg	0.808	0.750	0.422
S(1)	0.897	0.636	0.254
S(2)	0.719	0.864	0.590
C(1)	0.973	0.731	0.312
C(2)	0.643	0.769	0.532

The structure is shown in Figure XIVA,67. Individual molecules have the bond dimensions of Figure XIVA,68. Between molecules the closest approach of mercury and sulfur atoms is 3.25 A.

XIV,a51. The dimer of *trimethyl aluminum*, $[Al(CH_3)_3]_2$, forms monoclinic crystals having the cell dimensions:

$$a_0 = 13.0 \text{ A.}; \quad b_0 = 6.96 \text{ A.}; \quad c_0 = 14.7 \text{ A.}; \quad \beta = 125°$$

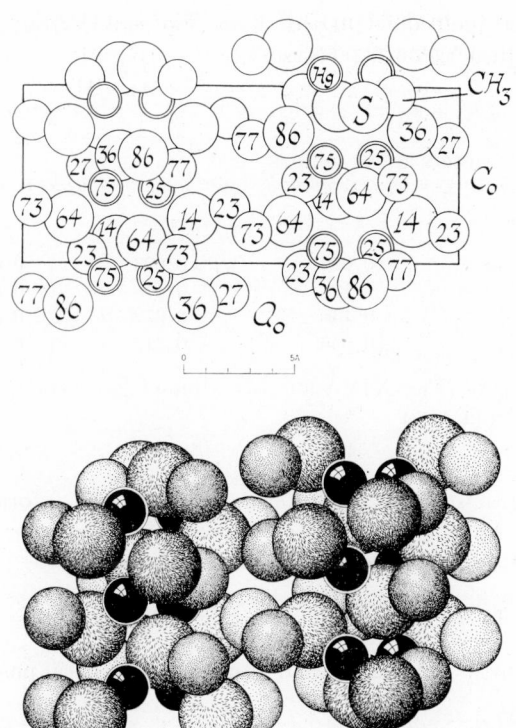

Fig. XIVA,67a (top). The orthorhombic structure of Hg(SCH₃)₂ projected along its b_0 axis. Right-hand axes.

Fig. XIVA,67b (bottom). A packing drawing of the orthorhombic Hg(SCH₃)₂ arrangement seen along its b_0 axis. The mercury atoms are black; the sulfur atoms are the large ringed, the carbon the smaller dotted circles.

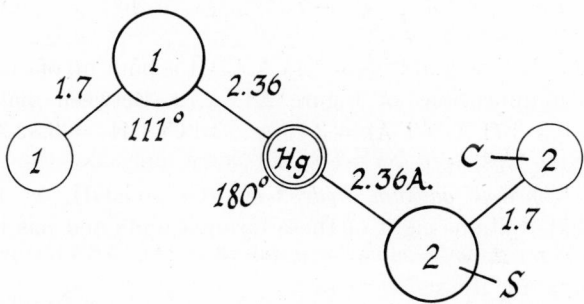

Fig. XIVA,68. Bond dimensions of the molecule of Hg(SCH₃)₂.

Four dimers are contained in this unit. The space group is C_{2h}^6 ($C2/c$) with all atoms in the general positions:

$$(8f) \quad \pm (xyz; \ x,\bar{y},z+{}^1\!/_2; \ x+{}^1\!/_2,y+{}^1\!/_2,z; \ x+{}^1\!/_2,{}^1\!/_2-y,z+{}^1\!/_2)$$

The chosen parameters are as follows:

Atom	x	y	z
Al	0.028	0.073	0.091
C(1)	0.121	0.148	0.006
C(2)	0.148	−0.072	0.230
C(3)	−0.056	0.317	0.084

In this structure (Fig. XIVA,69), the dimers have the dimensions indicated in Figure XIVA,70.

XIV,a52. *Aluminum methyl dichloride*, $AlCH_3Cl_2$, forms monoclinic crystals which have eight of these molecular elements in a unit of the dimensions:

$$a_0 = 11.92 \text{ A.}; \quad b_0 = 6.92 \text{ A.}; \quad c_0 = 12.53 \text{ A.}; \quad \beta = 109°55'$$

The space group is C_{2h}^6 ($C2/c$) with atoms in the general positions:

$$(8f) \quad \pm (xyz; \ x,\bar{y},z+{}^1\!/_2; \ x+{}^1\!/_2,y+{}^1\!/_2,z; \ x+{}^1\!/_2,{}^1\!/_2-y,z+{}^1\!/_2)$$

The parameters were determined to be the following:

Atom	x	y	z	σ, A.
Al	0.081	−0.190	0.044	0.007
Cl(1)	0.115	0.117	0.003	0.008
Cl(2)	−0.094	0.367	0.083	0.008
C	0.157	−0.240	0.204	0.030

The structure is shown in Figure XIVA,71. It is built up of dimers which have the bond dimensions of Figure XIVA,72. Between molecules the shortest Cl–Cl = 3.71 A., Cl–Al = 3.78 A., and Cl–CH₃ = 3.82 A.

XIV,a53. *Dimethyl gallium hydroxide*, $(CH_3)_2GaOH$, is monoclinic with a cell that contains eight of these formula units and has the dimensions:

$$a_0 = 8.62 \pm 0.01 \text{ A.}; \quad b_0 = 12.14 \pm 0.04 \text{ A.}; \quad c_0 = 8.50 \pm 0.02 \text{ A.}$$
$$\beta = 92°3' \pm 6'$$

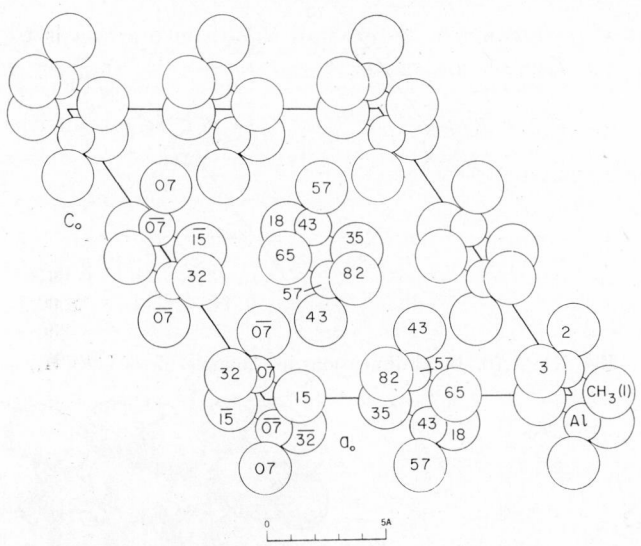

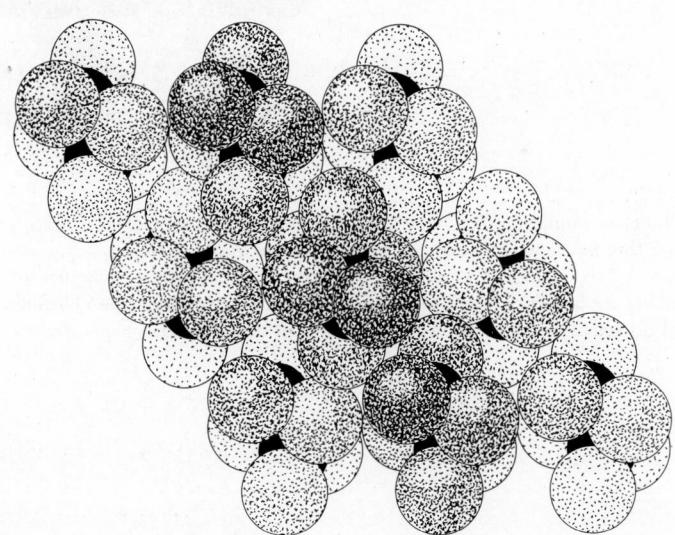

Fig. XIVA,69a (top). The monoclinic structure of [Al(CH₃)₃]₂ projected along its b_0 axis.   Left-hand axes.

Fig. XIVA,69b (bottom). A packing drawing of the monoclinic [Al(CH₃)₃]₂ arrangement seen along its b_0 axis.   The aluminum atoms are black, the methyl groups large and dotted.

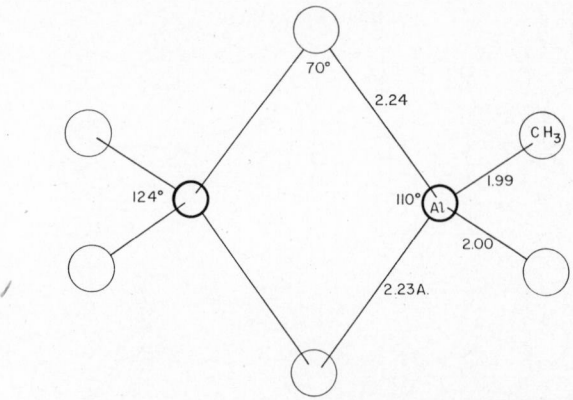

Fig. XIVA,70. Bond dimensions in the molecule of [Al(CH₃)₃]₂.

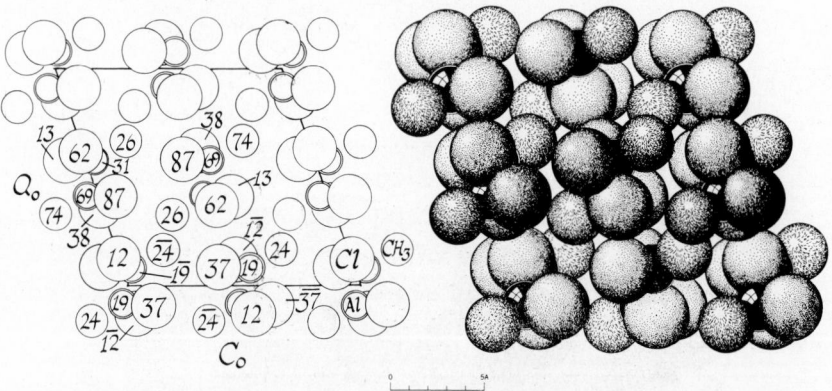

Fig. XIVA,71a (left). The monoclinic structure of AlCH₃Cl₂ projected along its b_0 axis. The close similarity between this arrangement and that of [Al(CH₃)₃]₂ is seen by comparing this figure with Figure XIVA,69. Right-hand axes.

Fig. XIVA,71b (right). A packing drawing of the monoclinic structure of AlCH₃Cl₂ viewed along its b_0 axis. The atoms of aluminum are black; the chlorine atoms are large and dotted; those of carbon are smaller and line shaded.

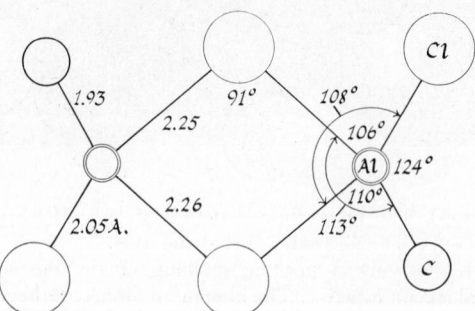

Fig. XIVA,72. Bond dimensions in the dimeric molecule [AlCH₃Cl₂] .

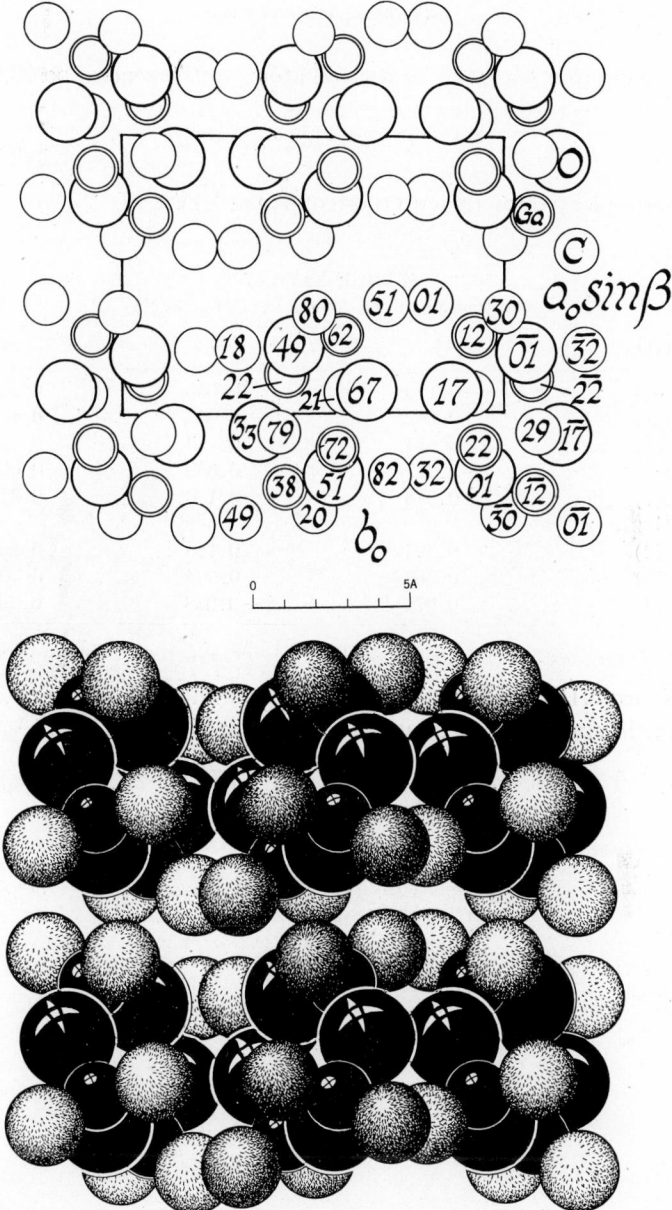

Fig. XIVA,73a (top). The monoclinic structure of $(CH_3)_2GaOH$ projected along its c_0 axis. Right-hand axes.

Fig. XIVA,73b (bottom). A packing drawing of the monoclinic $(CH_3)_2GaOH$ arrangement seen along its c_0 axis. The gallium atoms are the small, the oxygen atoms the large black circles. The carbon atoms are line shaded.

The space group is C_{2h}^5 ($P2_1/c$) with all atoms in the general positions:

$$(4e) \quad \pm(xyz; \; x,{}^1\!/_2-y,z+{}^1\!/_2)$$

The determined parameters are those of Table XIVA,17.

TABLE XIVA,17
Parameters of the Atoms in $(CH_3)_2GaOH$

Atom	x	y	z
Ga(1)	−0.120	0.065	0.221
Ga(2)	0.282	0.073	0.122
O(1)	0.074	0.133	0.167
O(2)	−0.217	0.052	0.006
C(1)	−0.213	0.196	0.318
C(2)	0.394	0.188	0.006
C(3)	0.366	−0.004	0.303
C(4)	−0.078	−0.088	0.295

As Figure XIVA,73 indicates, this is a molecular structure made up of tetramers having the bond dimensions shown in Figure XIVA,74.

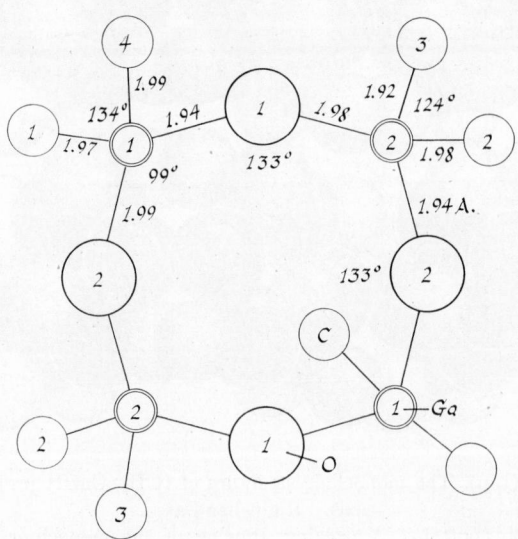

Fig. XIVA,74. Bond dimensions of the tetrameric molecule of $(CH_3)_2GaOH$.

XIV,a54. Crystals of *trimethyl indium*, $(CH_3)_3In$, are tetragonal with a unit containing eight molecules and having the cell edges:

$$a_0 = 13.24 \pm 0.01 \text{ A.}, \qquad c_0 = 6.44 \text{ A.}$$

The space group is C_{4h}^4 $(P4_2/n)$ with all atoms presumably in the general positions:

$(8g) \qquad \pm(xyz; \; {}^1/_2 - x, {}^1/_2 - y, z; \; {}^1/_2 - y, x, {}^1/_2 - z; \; y, {}^1/_2 - x, {}^1/_2 - z)$

The stated parameters are as follows:

Atom	x	y	z
In	0.2140	0.0038	0.4124
C(1)	0.1420	0.1282	0.2684
C(2)	0.1710	0.9620	0.7086
C(3)	0.3422	0.9282	0.2785

According to this structure, the C(2) and C(3) atoms [for instance, C(2) in $^1/_2 - x, ^1/_2 - y, z$ and C(3) in $y + ^1/_2, \bar{x}, z + ^1/_2$] of different monomeric groups are less than 2.5 A. from one another. This improbably short distance, and an apparent misprint in the general coordinates $(8g)$ as stated in the original paper, make further work desirable.

XIV,a55. Structures have been proposed for *dimethyl thallic bromide*, $Tl(CH_3)_2Br$, and the isomorphous *chloride* and *iodide*. They are tetragonal with elongated bimolecular units of the dimensions:

$$Tl(CH_3)_2Cl: a_0 = 4.29 \text{ A.}, \; c_0 = 14.015 \text{ A.}$$
$$Tl(CH_3)_2Br: a_0 = 4.47 \text{ A.}, \; c_0 = 13.78 \text{ A.}$$
$$Tl(CH_3)_2I: a_0 = 4.778 \text{ A.}, \; c_0 = 13.43 \text{ A.}$$

For the bromide and iodide, atoms have been found to be in the following special positions of D_{4h}^{17} $(I4/mmm)$:

$$Tl: (2a) \quad 000; \; ^1/_2 \; ^1/_2 \; ^1/_2$$
$$I(Br): (2b) \quad 0 \; 0 \; ^1/_2; \; ^1/_2 \; ^1/_2 \; 0$$
$$CH_3: (4e) \quad \pm(00u; \; ^1/_2, ^1/_2, u + ^1/_2)$$
$$\text{with } u = \text{ca. } 0.15$$

In the iodide this structure (Fig. XIVA,75) leads to a Tl–CH_3 separation of 2.01 A. and to the rather large CH_3–CH_3 distance of 4.17 A.

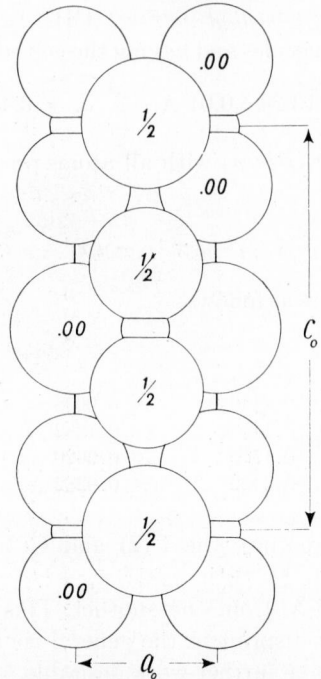

Fig. XIVA,75. The tetragonal structure of $Tl(CH_3)_2I$ projected along an a_0 axis. The atoms of iodine are the largest circles; the $Tl(CH_3)_2$ groups are dumbbell shaped. Left-hand axes.

XIV,a56. The two compounds *trimethyl platinum chloride*, $(CH_3)_3PtCl$, and *tetramethyl platinum*, $Pt(CH_3)_4$, form cubic crystals. Their unit cubes, which contain eight molecules, have the edges:

$$Pt(CH_3)_3Cl: a_0 = 10.55 \text{ A.}$$
$$Pt(CH_3)_4: a_0 = 10.145 \text{ A.}$$

The atomic arrangements in both crystals are considered to be in accord with the space group T_d^3 ($I\bar{4}3m$).

It was established that the platinum atoms in both these crystals and the chlorine atoms in $Pt(CH_3)_3Cl$ are in

$$(8c) \quad uuu; \ \bar{u}\bar{u}u; \ \bar{u}u\bar{u}; \ u\bar{u}\bar{u}; \ \text{B.C.}$$

For $Pt(CH_3)_4$, $u(Pt) = 0.380$; for the chloride, $u(Pt) = 0.375$, and $u(Cl) = 0.11$.

It is considered that the established positions of the platinum and chlorine atoms in $Pt(CH_3)_3Cl$ require that the molecule be a tetramer which then

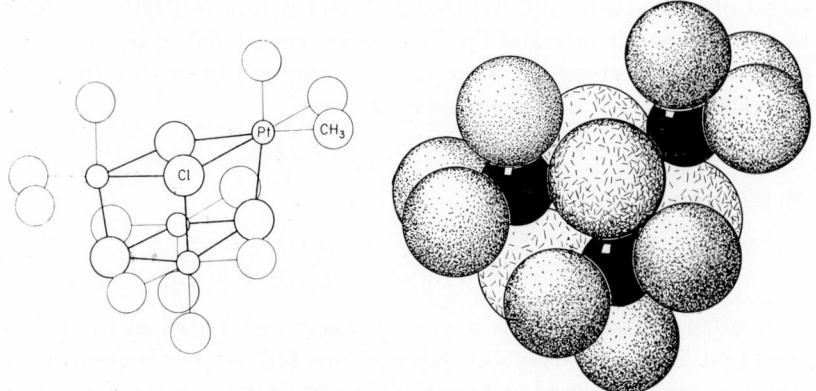

Fig. XIVA,76a (left). A perspective drawing of the tetrameric molecule of $Pt(CH_3)_3Cl$. Molecules of $Pt(CH_3)_4$ have the same shape, with an additional CH_3 replacing Cl.
Fig. XIVA,76b (right). A packing drawing of the tetrameric molecule of $Pt(CH_3)_3Cl$.
The chlorine atoms are line shaded; the platinum atoms are black.

might have the shape given in Figure XIVA,76. The 24 carbon atoms per cell have been assigned the positions:

$$(24g) \quad uuv; \; u\bar{u}\bar{v}; \; \bar{u}u\bar{v}; \; \bar{u}\bar{u}v; \; \text{B.C.}; \text{tr}$$

where, if the Pt–CH_3 separation is chosen as the sum of the covalent radii, $u(C)$ can be ca. 0.375 and v, ca. 0.18.

For $Pt(CH_3)_4$, the analogous structure would put CH_3 groups in place of the Cl atoms with $u = 0.11$, as well as in $(24g)$ with $u = 0.375$, $v = 0.18$.

These structures are body-centered groupings of the tetrameric molecules. Within a chloride tetramer: Pt–Cl = 2.48 A., Pt–Pt = 3.73 A., and Cl–Cl = 3.28 A. Each platinum atom has about it six octahedrally coordinated methyl or chlorine neighbors. In the proposed molecule (Fig. XIVA,76), each chlorine atom is equidistant from three platinum atoms and it should be pointed out that the same thing is true of a quarter of the CH_3 radicals in $Pt(CH_3)_4$.

XIV,a57. *Trimethyl tin fluoride*, $(CH_3)_3SnF$, is orthorhombic with a tetramolecular unit of the edge lengths:

$$a_0 = 4.32 \pm 0.01 \text{ A.}; \quad b_0 = 10.85 \pm 0.02 \text{ A.}; \quad c_0 = 12.84 \pm 0.02 \text{ A.}$$

The space group was chosen as V_h^{16} (*Pmcn*). Tin atoms are in the positions:

$$(4c) \quad \pm (^1/_4 \, u \, v; \; ^1/_4, ^1/_2 - u, v + ^1/_2)$$
$$\text{with } u = 0.0654, \; v = 0.2114$$

One set of carbon atoms [C(1)] is also in (4c) with $u = 0.065$, $v = 0.368$. The other carbon atoms and the fluorine are considered to be disordered. For these carbon atoms the disorder was expressed by saying that there are half atoms in each of the following positions:

$$(8d) \quad \pm (xyz;\ x+\tfrac{1}{2},y+\tfrac{1}{2},\tfrac{1}{2}-z;\ \tfrac{1}{2}-x,y,z;\ x,\tfrac{1}{2}-y,z+\tfrac{1}{2})$$

with the parameters:

$$\tfrac{1}{2}\,C(2): x = 0.20,\ y = 0.232,\ z = 0.131$$
$$\tfrac{1}{2}\,C(3): x = 0.20,\ y = -0.105,\ z = 0.131$$

The fluorine electron density is very diffusely distributed in the Fourier summation. This is thought to indicate that fluorine lies somewhere between tin atoms so placed as to give chains running along the a_0 direction.

XIV,a58. Crystals of *α-dimethyl tellurium dichloride*, $(CH_3)_2TeCl_2$, are monoclinic with a tetramolecular cell of the dimensions:

$$a_0 = 9.552\ \text{A.};\quad b_0 = 6.180\ \text{A.};\quad c_0 = 11.314\ \text{A.},\qquad \text{all } \pm 0.010\ \text{A.}$$
$$\beta = 97°54' \pm 6'$$

The space group is C_{2h}^5 ($P2_1c$) with atoms in the positions:

$$(4e) \quad \pm (xyz;\ x,\tfrac{1}{2}-y,z+\tfrac{1}{2})$$

The parameters, and their standard deviations, are as follows:

Atom	x	$\sigma(x)$	y	$\sigma(y)$	z	$\sigma(z)$
Te	0.1758	0.0001	0.1084	0.0003	0.1645	0.0001
Cl(1)	0.3571	0.0008	0.3934	0.0016	0.2144	0.0007
Cl(2)	0.0176	0.0008	0.7806	0.0016	0.1163	0.0006
C(1)	0.2586	0.0028	0.9545	0.0042	0.3227	0.0019
C(2)	0.3131	0.0027	0.9767	0.0062	0.0533	0.0021

In this atomic arrangement (Fig. XIVA,77), the molecules have the same type of structure as in the several similar benzene derivatives that have been studied (**XIV,b7** and **8** loose-leaf); they have been described as trigonal bipyramids with one equatorial position vacant. Bond dimensions are those of Figure XIVA,78. Between molecules each tellurium atom is closest to three Cl(2) atoms with Te–Cl(2) = 3.46 and 3.52 A. If these are considered to represent a loose bonding, the structure becomes a system of molecular sheets parallel to (100). There are also Cl(2)–C(1) separations of 3.46 A.; all other atomic distances exceed 3.70 A. The final $R = 0.099$.

XIV,a59. Crystals of *bis tetramethyl disilanilene dioxide*, $[(CH_3)_4Si_2O]_2$, are monoclinic with a bimolecular unit of the dimensions:

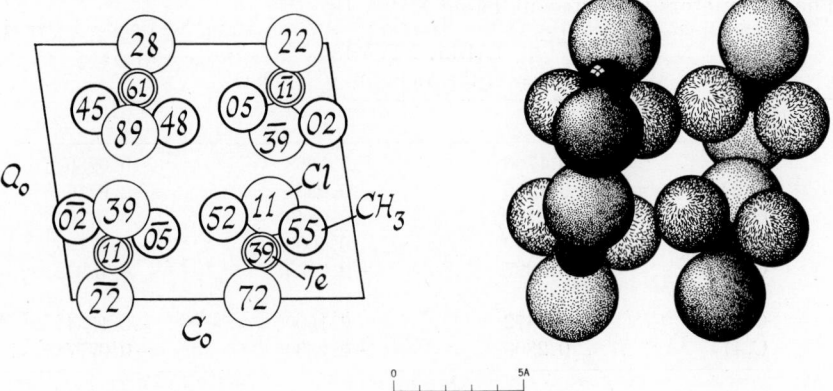

Fig. XIVA,77a (left). The monoclinic structure of $(CH_3)_2TeCl_2$ projected along its b_0 axis. Right-hand axes.

Fig. XIVA,77b (right). A packing drawing of the monoclinic $(CH_3)_2TeCl_2$ arrangement seen along its b_0 axis. The tellurium atoms are black. Chlorine atoms are large and dotted; the carbon atoms are smaller and line shaded.

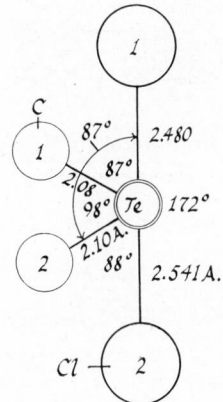

Fig. XIVA,78. Bond dimensions of the $(CH_3)_2TeCl_2$ molecule.

$$a_0 = 7.67 \pm 0.02 \text{ A.}; \quad b_0 = 6.64 \pm 0.02 \text{ A.}; \quad c_0 = 17.39 \pm 0.03 \text{ A.}$$
$$\beta = 111°0' + 12'$$

The space group is C_{2h}^5 ($P2_1/c$) with all atoms in the general positions:

$$(4e) \quad \pm (xyz; \ x, \tfrac{1}{2} - y, z + \tfrac{1}{2})$$

The parameters are listed in Table XIVA,18.

TABLE XIVA,18
Parameters of the Atoms in $[(CH_3)_4Si_2O]_2$

Atom	x	y	z
Si(1)	0.1700	0.1262	0.0986
Si(2)	0.1400	−0.0233	−0.0716
O	0.2170	0.0845	0.0170
C(1)	0.2000	0.4000	0.1179
C(2)	0.3397	−0.0135	0.1894
C(3)	0.1472	0.1500	−0.1534
C(4)	0.2800	−0.2590	−0.0777

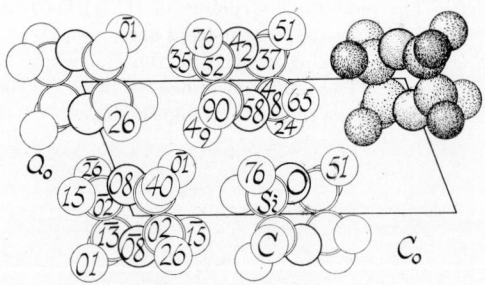

Fig. XIVA,79. The monoclinic structure of $[(CH_3)_4Si_2O]_2$ projected along its b_0 axis.
Right-hand axes.

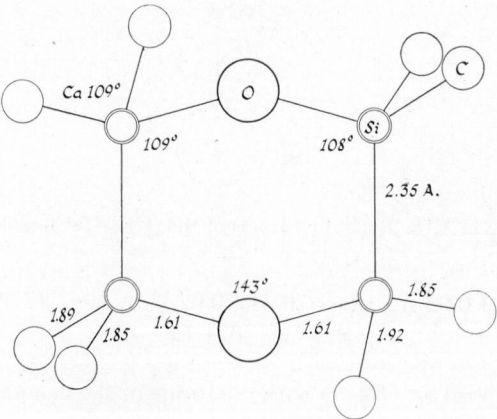

Fig. XIVA,80. Bond dimensions in the molecule of $[(CH_3)_4Si_2O]_2$.

As Figure XIVA,79 indicates, the molecule is a six-sided Si_4O_2 ring with two methyl groups attached to each silicon atom; it has the dimensions of Figure XIVA,80. The four silicon atoms are coplanar with oxygen atoms displaced from this plane. The final R is 0.201 for $(0kl)$ and 0.135 for $(h0l)$ reflections.

XIV,a60. *Octamethyl cyclo tetrasiloxane*, $(CH_3)_8Si_4O_4$, is dimorphous with a low-temperature form that is tetragonal. Its tetramolecular unit has the edges:

$$a_0 = 16.10 \pm 0.02 \text{ A.}, \qquad c_0 = 6.47 \pm 0.01 \text{ A. } (-50°C.)$$

The space group is C_{4h}^4 $(P4_2/n)$ with atoms in the positions:

$$(8g) \quad \pm (xyz;\ x+{}^1/_2,y+{}^1/_2,\bar{z};\ \bar{y},x+{}^1/_2,z+{}^1/_2;\ y+{}^1/_2,\bar{x},z+{}^1/_2)$$

The parameters as determined are listed in Table XIVA,19.

In this structure (Fig. XIVA,81), Si–O = 1.64–1.66 A. and Si–C = 1.90–1.95 A. In the chair-like S_4O_4 ring, Si–O–Si = 141.5 and 143.5°, and O–Si–O = 105 and 113°; C–Si–C = 102 and 110°.

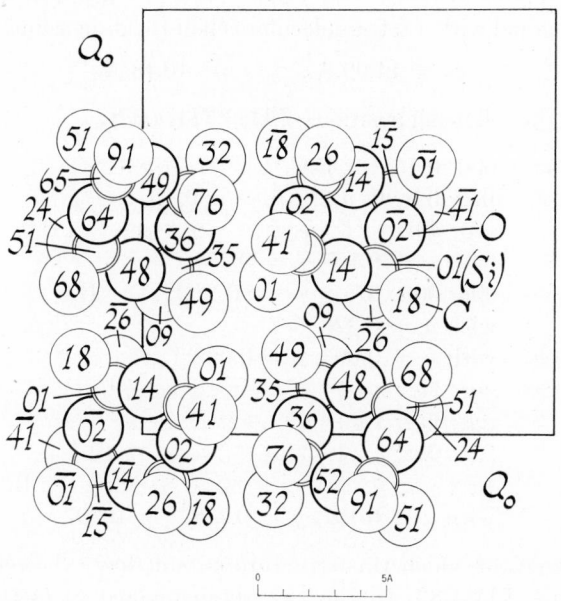

Fig. XIVA,81. The tetragonal structure of low $(CH_3)_8Si_4O_4$ projected along its c_0 axis. Right-hand axes.

TABLE XIVA,19
Parameters of the Atoms in Low $(CH_3)_8Si_4O_4$

Atom	x	$\sigma_{x,y}$, A.	y	z	σ_z, A.
Si(1)	0.1098	0.01	0.0652	0.149	0.02
Si(2)	0.0677	0.01	−0.1148	−0.007	0.02
O(1)	0.0183	0.02	0.1095	0.145	0.04
O(2)	0.1135	0.02	−0.0233	0.018	0.04
C(1)	0.1597	0.04	0.0472	0.413	0.08
C(2)	0.1885	0.04	0.1365	0.014	0.08
C(3)	0.0504	0.04	−0.1643	0.265	0.08
C(4)	0.1423	0.04	−0.1792	−0.168	0.08

At $-16.3°$C. there is a transition to a high-temperature form of the similar cell dimensions:

$$a_0 = 16.10 \text{ A.}, \qquad c_0 = 6.83 \text{ A.}$$

The space group becomes C_{4h}^6 ($I4_1/a$) and there is disorder in the structure which, however, has not been established in detail.

XIV,a61. The substance *octamethyl spiro-[5·5]pentasiloxane*, $(CH_3)_8$-Si_5O_6, is tetragonal with a tetramolecular cell of the dimensions:

$$a_0 = 14.09 \text{ A.}, \qquad c_0 = 10.18 \text{ A.}$$

Atoms are in the following positions of D_{4h}^{19} ($I4/amd$):

Si(1): (4a)   000; 0 $^1/_2$ $^1/_4$; B.C.

Si(2): (16h) $0uv;\ 0\bar{u}v;\ 0,\ u+^1/_2,^1/_4-v;\ 0,^1/_2-u,^1/_4-v,$
              $u0\bar{v};\ \bar{u}0\bar{v};\ u,^1/_2,v+^1/_4;\ \ \ \ \ \ \bar{u},^1/_2,v+^1/_4;$ B.C.
              with $u = 0.107$, $v = 0.255$

O(1): (8e) $00u;\ 00\bar{u};\ 0,^1/_2,u+^1/_4;\ 0,^1/_2,^1/_4-u;$ B.C.
              with $u = 0.326$

O(2): (16h)   with $u = 0.093$, $v = 0.098$

CH₃: (32i) $xyz;\ \bar{x}\bar{y}z;\ x,y+^1/_2,^1/_4-z;\ \bar{x},^1/_2-y,^1/_4-z;$
              $\bar{x}yz;\ x\bar{y}z;\ \bar{x},y+^1/_2,^1/_4-z;\ x,^1/_2-y,^1/_4-z;$
              $yx\bar{z};\ \bar{y}\bar{x}\bar{z};\ y,x+^1/_2,z+^1/_4;\ \bar{y},^1/_2-x,z+^1/_4;$
              $\bar{y}x\bar{z};\ y\bar{x}\bar{z};\ \bar{y},x+^1/_2,z+^1/_4;\ y,^1/_2-x,z+^1/_4;$ B.C.
              with $x = 0.107$, $y = 0.178$, $z = 0.307$

The molecules of which this structure is built have the general appearance of Figure XIVA,82. In the two six-membered Si–O rings, Si–O = 1.61–1.67 A., with the angle O–Si–O = 105 or 108° and Si–O–Si = 129 or 134°. The Si–CH₃ separation is 1.88 A.

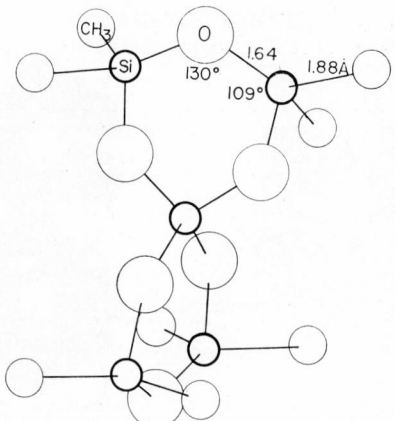

Fig. XIVA,82. Bond dimensions and general shape of the molecule of $(CH_3)_8Si_5O_6$.

It is stated that better intensity agreement involving reflections of high indices is obtained if the methyl groups are considered to be pivoting about the silicon atoms.

XIV,a62. The compound *octa(methyl silsesquioxane)*, $(CH_3SiO_{1.5})_8$, is hexagonal (rhombohedral) with three of these molecules in the hexagonal cell. For this unit:

$$a_0' = 12.498 \pm 0.010 \text{ A.}, \qquad c_0' = 13.087 \text{ A.}$$

The space group is C_{3i}^2 ($R\bar{3}$) with atoms in the positions:

Si(1): (6c) $\pm (00u)$; rh   with $u = 0.2985$
C(1): (6c)   with $u = 0.1559$

All other atoms are in the general positions:

(18f) $\pm (xyz; \bar{y}, x-y, z; y-x, \bar{x}, z)$; rh

with the following parameters:

Atom	x	y	z
Si(2)	0.0168	0.2098	0.4334
O(1)	0.0070	0.1261	0.3371
O(2)	0.1368	0.2472	0.5018
C(2)	0.0209	0.3444	0.3586

In this structure (Fig. XIVA,83a and b) the molecules have the general shape of Figure XIVA,83c. Within them, Si–O = 1.603–1.620 A. and C–Si = 1.866 or 1.925 A.; the angles O–Si–O = 108–112.5° and Si–O–Si = 145°.

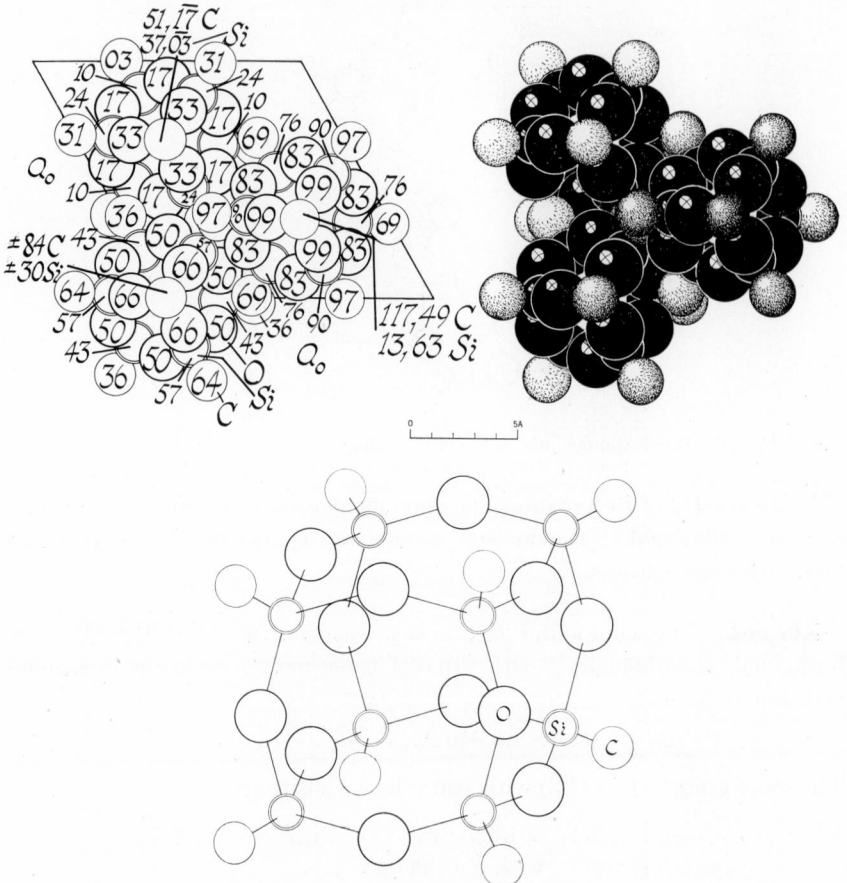

Fig. XIVA,83a (top left). The hexagonal structure of $(CH_3SiO_{1.5})_8$ projected along its c_0 axis.   Right-hand axes.

Fig. XIVA,83b (top right). A packing drawing of the hexagonal $(CH_3SiO_{1.5})_8$ structure viewed along its c_0 axis.   The oxygen atoms are the large, the silicon atoms the smaller black circles.   Carbon atoms are line shaded.

Fig. XIVA,83c (bottom). A drawing to indicate the general shape of the large molecules of $(CH_3SiO_{1.5})_8$.

Between molecules the shortest C–O = 3.37 A. and the shortest C–Si = 3.86 A. The corresponding unimolecular rhombohedron has the dimensions:

$$a_0 = 8.432 \text{ A.}, \qquad \alpha = 95°39'$$

XIV,a63.   The compound *tetramethyl-N,N'-bis-trimethyl silyl cyclodisilazane*, $[(CH_3)_2Si]_2[N_2Si_2(CH_3)_6]$, forms monoclinic crystals whose bimolecular units have the dimensions:

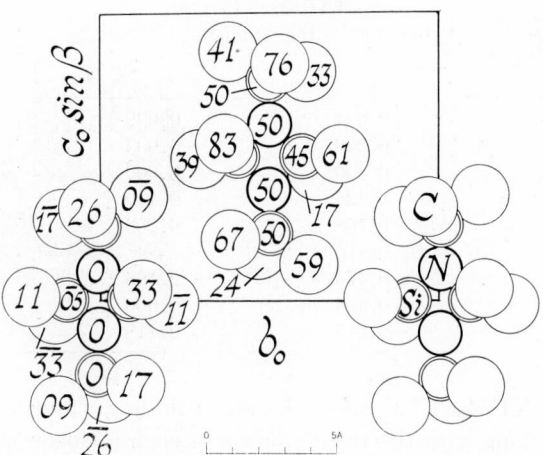

Fig. XIVA,84. The monoclinic structure of $[(CH_3)_2Si]_2[N_2Si_2(CH_3)_6]$ projected along its a_0 axis.   Right-hand axes.

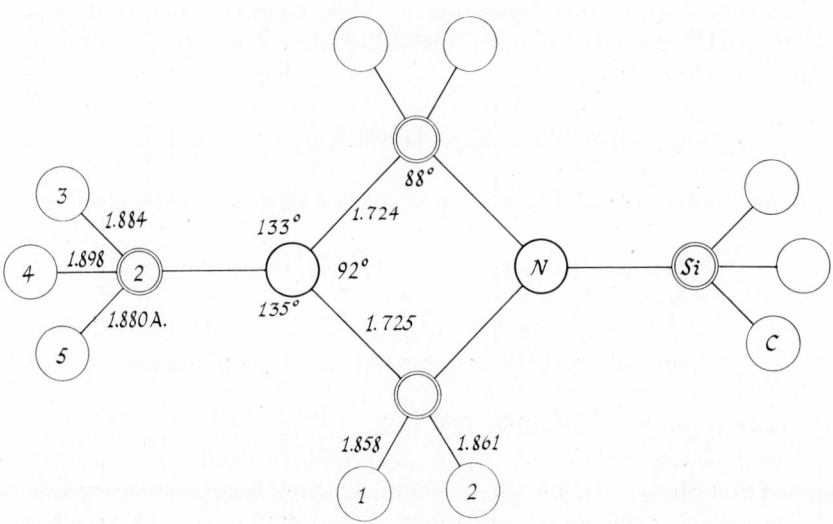

Fig. XIVA,85. Bond dimensions in the molecule of $[(CH_3)_2Si]_2[N_2Si_2(CH_3)_6]$.

$$a_0 = 6.759 \pm 0.02 \text{ A.}; \quad b_0 = 13.181 \pm 0.04 \text{ A.}; \quad c_0 = 11.225 \pm 0.03 \text{ A.}$$
$$\beta = 104°23' \pm 30'$$

The space group is C_{2h}^5 $(P2_1/n)$ with atoms in the general positions:

$$(4e) \quad \pm (xyz; \ x+\tfrac{1}{2}, \tfrac{1}{2}-y, z+\tfrac{1}{2})$$

The parameters have been given the values stated in Table XIVA,20.

TABLE XIVA,20
Parameters of the Atoms in $(CH_3)_{10}Si_4N_2$

Atom	x	y	z
Si(1)	0.054	0.090	0.008
Si(2)	0.000	−0.003	0.259
N	0.000	0.000	0.107
C(1)	0.326	0.130	0.052
C(2)	−0.108	0.206	−0.017
C(3)	−0.173	−0.106	0.291
C(4)	0.263	−0.030	0.362
C(5)	−0.090	0.118	0.317

As Figure XIVA,84 indicates, the molecule has a four-membered N_2Si_2 core which, along with the third valence of each nitrogen atom, is planar. Bond dimensions are shown in Figure XIVA,85. Between molecules, closest contacts are between methyl groups, with CH_3–CH_3 = 4.3 A. or more.

XIV,a64. *Tetramethyl stibonium tetrakis trimethyl siloxy aluminate*, $[(CH_3)_4Sb]Al[OSi(CH_3)_3]_4$, forms orthorhombic crystals having a bimolecular unit of the cell edges:

$$a_0 = 13.412 \text{ A.}; \quad b_0 = 11.884 \text{ A.}; \quad c_0 = 9.899 \text{ A.}$$

The space group was chosen as V_h^{13} (*Pmmn*) with atoms in the positions:

$$(2a) \quad \pm(^1/_4 \ ^1/_4 \ u)$$
$$(4e) \quad \pm(^1/_4 \ u \ v; \ ^1/_4, ^1/_2 - u, v)$$
$$(4f) \quad \pm(u \ ^1/_4 \ v; \ ^1/_2 - u, ^1/_4, v)$$
$$(8g) \quad \pm(xyz; \ ^1/_2 - x, y, z; \ x, ^1/_2 - y, z; \ ^1/_2 - x, ^1/_2 - y, z)$$

Parameters are given in Table XIVA,21.

The structure is shown in Figure XIVA,86. In deducing it, it was assumed that Sb–O = 1.79 A., Si–C = 1.87 A.; Si–O is approximately 1.56 A.

The following compounds are known to be isostructural. Their cells are:

$[(CH_3)_4Sb]Ga[OSi(CH_3)_3]_4$:
$$a_0 = 13.450 \text{ A.}; \quad b_0 = 11.936 \text{ A.}; \quad c_0 = 9.869 \text{ A.}$$

$[(CH_3)_4Sb]Fe[OSi(CH_3)_3]_4$:
$$a_0 = 13.508 \text{ A.}; \quad b_0 = 11.953 \text{ A.}; \quad c_0 = 9.816 \text{ A.}$$

TABLE XIVA,21
Positions and Parameters of the Atoms in $[(CH_3)_4Sb]Al[OSi(CH_3)_3]_4$

Atom	Position	x	y	z
Sb	$(2a)$	$1/4$	$1/4$	0.193
Al	$(2a)$	$1/4$	$1/4$	0.723
Si(1)	$(4e)$	$1/4$	0.504	0.834
Si(2)	$(4f)$	0.475	$1/4$	0.613
O(1)	$(4e)$	$1/4$	0.373	0.827
O(2)	$(4f)$	0.359	$1/4$	0.619
C(1)	$(4e)$	$1/4$	0.569	0.662
C(2)	$(4f)$	0.524	$1/4$	0.436
C(3)	$(8g)$	0.361	0.560	0.927
C(4)	$(8g)$	0.530	0.376	0.698
C(5)	$(4f)$	0.384	$1/4$	0.065
C(6)	$(4e)$	$1/4$	0.401	0.321

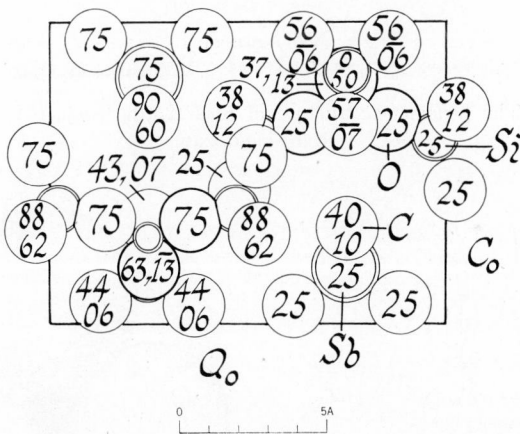

Fig. XIVA,86. The orthorhombic structure of $[(CH_3)_4Sb]Al[OSi(CH_3)_3]_4$ projected along its b_0 axis. The smallest doubly ringed circle is an aluminum atom. Right-hand axes.

XIV,a65. The three *trimethyl antimony halides* are hexagonal. Their bimolecular units have the cell edges:

$$Sb(CH_3)_3Cl_2: a_0 = 7.28 \text{ A.}, \qquad c_0 = 8.44 \text{ A.}$$
$$Sb(CH_3)_3Br_2: a_0 = 7.38 \text{ A.}, \qquad c_0 = 8.90 \text{ A.}$$
$$Sb(CH_3)_3I_2: a_0 = 7.53 \text{ A.}, \qquad c_0 = 9.59 \text{ A.}$$

A structure has been deduced for the *bromide* which places atoms in the following positions of D_{3h}^4 ($P\bar{6}2c$):

Sb: (2c) $\pm (^1/_3 \ ^2/_3 \ ^1/_4)$

Br: (4f) $\pm (^1/_3 \ ^2/_3 \ u; \ ^1/_3, ^2/_3, ^1/_2 - u)$ with $u = -0.046$

CH$_3$: (6h) $u \ v \ ^1/_4; \ \bar{v}, u-v, ^1/_4; \ v-u, \bar{u}, ^1/_4;$
                $v \ u \ ^3/_4; \ \bar{u}, v-u, ^3/_4; \ u-v, \bar{v}, ^3/_4$
                with $u = 0.069, v = 0.710$

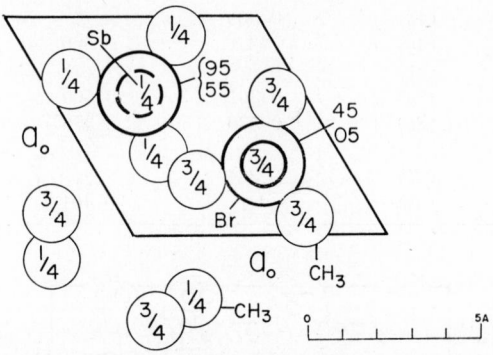

Fig. XIVA,87. The hexagonal structure of Sb(CH$_3$)$_3$Br$_2$ projected along its c_0 axis. Right-hand axes.

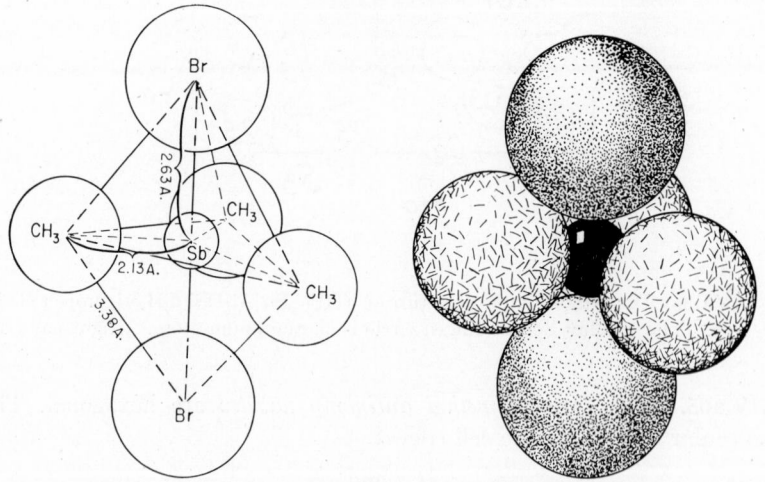

Fig. XIVA,88a (left). A drawing to show the general shape and the bond dimensions of the Sb(CH$_3$)$_3$Br$_2$ molecule.

Fig. XIVA,88b (right). A packing drawing of the Sb(CH$_3$)$_3$Br$_2$ molecule. The antimony atom is black, the bromine atoms large and dotted.

This is a structure (Fig. XIVA,87) composed of $Sb(CH_3)_3Br_2$ molecules which are hexagonally close-packed except for the reduced value of the axial ratio. The molecule itself has the shape of a trigonal bipyramid with an antimony atom at its center, the bromine atoms at its apices, and the three methyl groups at the corners of the base (Fig. XIVA,88). Within such a molecule, $Sb-Br = 2.63$ A., and $Sb-CH_3 = 2.13$ A. Between adjacent molecules, significant atomic separations are $Br-Br = 3.63$ A. and $Br-CH_3 = 3.38$ A.

Atomic parameters have not been determined for the other two halides with this structure.

XIV,a66. The double compound *antimony pentachloride oxytrimethyl phosphorus*, $SbCl_5 \cdot PO(CH_3)_3$, is orthorhombic with a tetramolecular unit of the edge lengths:

$$a_0 = 16.66 \pm 0.02 \text{ A.}; \quad b_0 = 8.205 \pm 0.01 \text{ A.}; \quad c_0 = 8.865 \pm 0.01 \text{ A.}$$

The space group is V_h^{16} (*Pnma*) with atoms in the positions:

(4c) $\pm (u \; ^1/_4 \; v; \; u+^1/_2,^1/_4,^1/_2-v)$

(8d) $\pm (xyz; \; ^1/_2-x,y+^1/_2,z+^1/_2; \; x,^1/_2-y,z; \; x+^1/_2,y,^1/_2-z)$

The positions and determined parameters of the several atoms are those of Table XIVA,22.

TABLE XIVA,22
Positions and Parameters of the Atoms in $SbCl_5 \cdot PO(CH_3)_3$

Atom	Position	x	y	z
Sb	(4c)	0.1454	$^1/_4$	0.0710
Cl(1)	(4c)	0.2615	$^1/_4$	0.9269
Cl(2)	(4c)	0.0258	$^1/_4$	0.2150
Cl(3)	(4c)	0.2240	$^1/_4$	0.2871
Cl(6)	(8d)	0.1424	0.9641	0.0541
P	(4c)	0.0794	$^1/_4$	0.7074
O	(4c)	0.0749	$^1/_4$	0.8890
C(1)	(4c)	0.4800	$^1/_4$	0.8800
C(2)	(8d)	0.1397	0.0707	0.6375

The resulting structure is shown in Figure XIVA,89. It is made up of molecules having the bond dimensions of Figure XIVA,90. Between molecules, the C–Cl and Cl–Cl separations range upwards from 3.62 and 3.66 A., respectively.

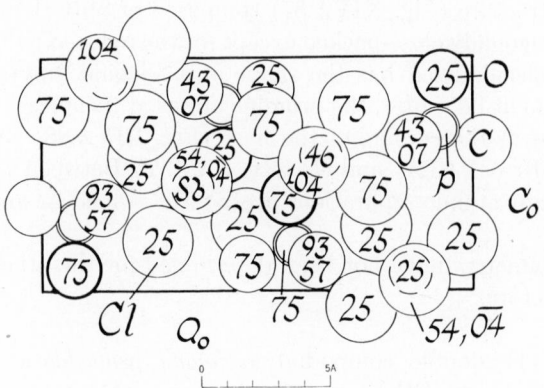

Fig. XIVA,89. The orthorhombic structure of $SbCl_5 \cdot PO(CH_3)_3$ projected along its b_0 axis. Right-hand axes.

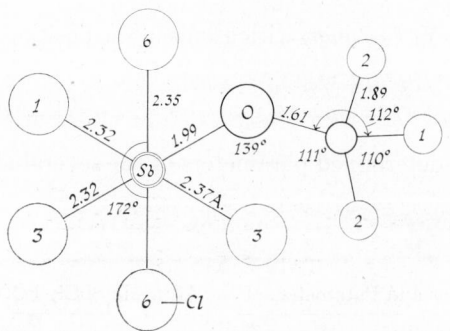

Fig. XIVA,90. Bond dimensions in the molecule of $SbCl_5 \cdot PO(CH_3)_3$.

The compound $SbCl_5 \cdot POCl_3$ is isostructural with a unit that has the dimensions:

$$a_0 = 16.42 \pm 0.01 \text{ A.}; \quad b_0 = 8.06 \pm 0.01 \text{ A.}; \quad c_0 = 8.93 \pm 0.02 \text{ A.}$$

XIV,a67. Crystals of *tellurium dimethane thiosulfonate*, $Te(S_2O_2CH_3)_2$, are monoclinic with a tetramolecular cell of the dimensions:

$$a_0 = 11.43 \text{ A.}; \quad b_0 = 5.29 \text{ A.}; \quad c_0 = 16.32 \text{ A.}; \quad \beta = 91°$$

Atoms have been placed in general positions of C_{2h}^5 $(P2_1/n)$:

$$(4e) \quad \pm(xyz; \, x+\tfrac{1}{2}, \tfrac{1}{2}-y, \, z+\tfrac{1}{2})$$

with the parameters of Table XIVA,23.

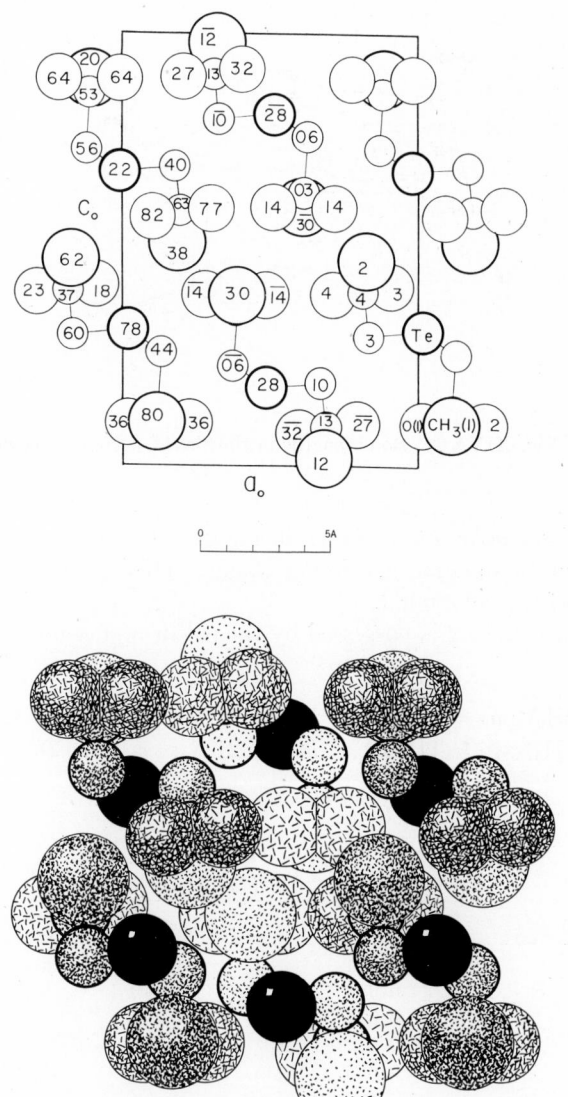

Fig. XIVA,91a (top). The monoclinic structure of $Te(S_2O_2CH_3)_2$ projected along its b^0 axis.   Left-hand axes.

Fig. XIVA,91b (bottom). A packing drawing of the monoclinic $Te(S_2O_2CH_3)_2$ arrangement viewed along its b_0 axis.   The tellurium atoms are black; the methyl groups are large and dotted.   Sulfur atoms are small and heavily ringed; the atoms of oxygen are line shaded.

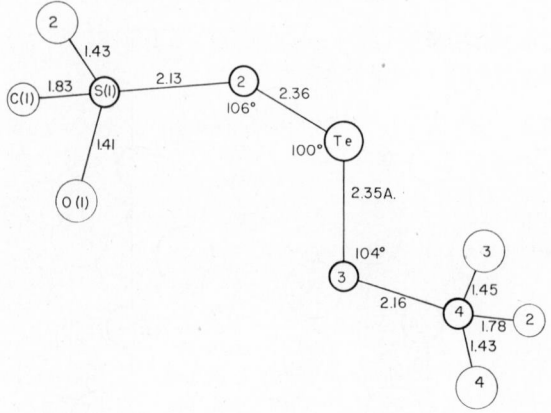

Fig. XIVA,92. Bond lengths in the $Te(S_2O_2CH_3)_2$ molecule.

The resulting structure is shown in Figure XIVA,91. Its molecules (Fig. XIVA,92) are nonplanar and doubly kinked. This molecule is effectively a *trans* form of the compound.

The same structure is possessed by the sulfur and selenium compounds:

$$S(S_2O_2CH_3)_2: a_0 = 11.33 \text{ A.;} \quad b_0 = 5.21 \text{ A.;} \quad c_0 = 16.14 \text{ A.;} \quad \beta = 91°$$
$$Se(S_2O_2CH_3)_2: a_0 = 11.38 \text{ A.;} \quad b_0 = 5.21 \text{ A.;} \quad c_0 = 16.23 \text{ A.;} \quad \beta = 91°$$

TABLE XIVA,23
Parameters of the Atoms in $Te(S_2O_2CH_3)_2$

Atom	x	y	z
Te	0.484	0.279	0.182
S(1)	0.383	−0.031	0.362
S(2)	0.371	−0.058	0.232
S(3)	0.672	0.101	0.195
S(4)	0.686	−0.134	0.087
O(1)	0.487	−0.140	0.393
O(2)	0.279	−0.142	0.393
O(3)	0.592	−0.317	0.078
O(4)	0.792	−0.271	0.099
$CH_3(1)$	0.384	0.301	0.393
$CH_3(2)$	0.680	0.119	0.015

b. Metal Amines, Arsines, etc.

XIV,a68. *Cuprous chloride azomethane,* $Cu_2Cl_2 \cdot N_2(CH_3)_2$, is triclinic with a unimolecular cell of the dimensions:

$$a_0 = 6.867 \pm 0.018 \text{ A.}; \; b_0 = 7.029 \pm 0.018 \text{ A.}; \; c_0 = 3.821 \pm 0.014 \text{ A.}$$
$$\alpha = 97°7' \pm 3'; \qquad \beta = 95°15' \pm 6'; \qquad \gamma = 111°45' \pm 3'$$

The space group is taken as C_i^1 ($P\overline{1}$) with all atoms in the positions $(2i)$ $\pm$ (xyz). The determined parameters are as follows:

Atom	x	σ_x, A.	y	σ_y, A.	z	σ_z, A.
Cu	0.0737	0.003	0.1932	0.003	−0.1655	0.004
Cl	0.2486	0.005	0.1462	0.005	0.3604	0.005
N	−0.0477	0.016	0.4030	0.013	−0.0201	0.016
C	−0.2665	0.019	0.3207	0.020	0.0494	0.024

The structure (Fig. XIVA,93) is made up of double CuCl chains parallel to the c_0 axis, and of azomethane molecules. Each copper atom has two chlorine atoms belonging to its own chain, with Cu–Cl = 2.319 and 2.368 A., and a third chlorine belonging to the other half of the double chain, with Cu–Cl = 2.547 A. In the azomethane molecules which tie these chains together along the b_0 direction, N–N = 1.257 A., N–C = 1.459 A., and N–N–C = 118°. Nitrogen atoms are closest to the CuCl chains with Cu–N = 1.993 A.

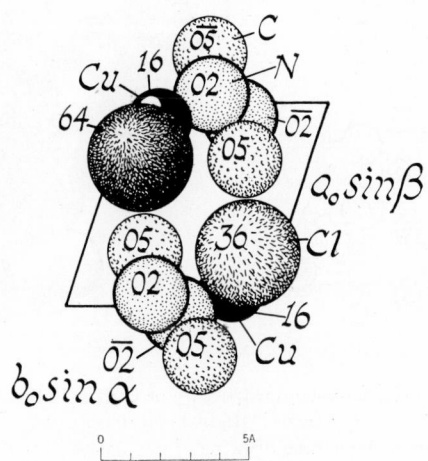

Fig. XIVA,93. The triclinic structure of $Cu_2Cl_2 \cdot N_2(CH_3)_2$ projected along its c_0 axis. Right-hand axes.

XIV,a69.   A structure has been described for *cupric bromide tetra methyl-amine*, $CuBr_2 \cdot 4CH_3NH_2$, which gives the orthorhombic crystals a bimolecular unit of the dimensions:

$$a_0 = 10.83 \text{ A.}; \quad b_0 = 9.18 \text{ A.}; \quad c_0 = 6.74 \text{ A.}; \quad \beta = 118°7'$$

The space group has been chosen as C_{2h}^5 $(P2_1/a)$ with atoms in the positions

$$Cu: (2a) \quad 000; \; {}^1/_2 \, {}^1/_2 \, 0$$

All other atoms have been put in the general positions:

$$(4e) \quad \pm (xyz; \; x+{}^1/_2, {}^1/_2-y, z)$$

with the following parameters:

Atom	x	y	z
Br	0.142	0.147	−0.267
N(1)	0.183	−0.033	0.317
N(2)	−0.042	0.192	0.133
C(1)	0.300	0.033	0.300
C(2)	0.017	0.317	0.067

The structure that results is shown in Figure XIVA,94. The copper atoms have around them a square of nitrogen atoms, with Cu–N = 2.12 or 2.14 A.; the nearest bromide ions are at a distance of 3.17 A.

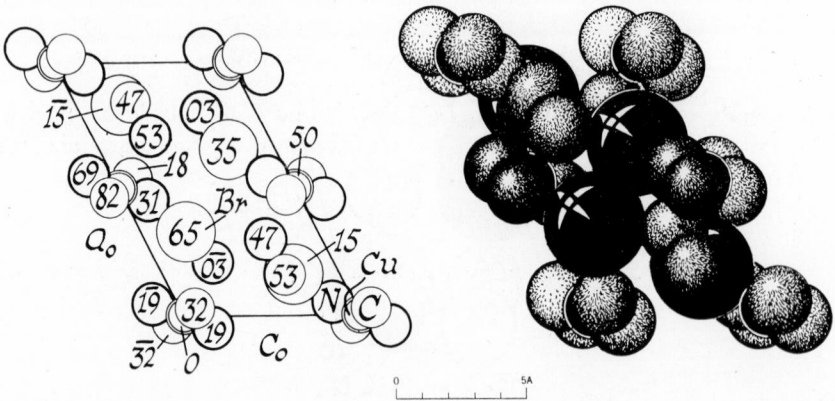

Fig. XIVA,94a (left). The monoclinic structure of $CuBr_2 \cdot 4CH_3NH_2$ projected along its b_0 axis.   Right-hand axes.

Fig. XIVA,94b (right). A packing drawing of monoclinic $CuBr_2 \cdot 4CH_3NH_2$ seen along its b_0 axis.   The bromine atoms are the large, the copper atoms the small black circles. The carbon atoms are line shaded; the nitrogens are more heavily outlined and curl shaded.

XIV,a70.   *Aluminum hydride di trimethylamine,* $AlH_3 \cdot 2N(CH_3)_3$, crystallizes with orthorhombic symmetry. Its tetramolecular unit has the edge lengths:

$$a_0 = 10.10 \text{ A.}; \quad b_0 = 8.84 \text{ A.}; \quad c_0 = 12.94 \text{ A.}$$

The space group has been chosen as V_h^{18} (*Cmca*) with a disordered structure that places the aluminum atoms in

$$(4a) \quad 000; \; 0\,\tfrac{1}{2}\,\tfrac{1}{2}; \; \tfrac{1}{2}\,\tfrac{1}{2}\,0; \; \tfrac{1}{2}\,0\,\tfrac{1}{2}$$

and the nitrogen atoms in

$$(8f) \quad \pm(0uv; \; \tfrac{1}{2},u,\tfrac{1}{2}-v; \; \tfrac{1}{2},u+\tfrac{1}{2},v; \; 0,u+\tfrac{1}{2},\tfrac{1}{2}-v)$$
$$\text{with } u = 0.1874, \; v = 0.1092$$

Half atoms of carbon are in the following positions, corresponding to two orientations of the trimethylamine groups:

$\tfrac{1}{2}\,C(1):$ $(8f)$ with $u = 0.3255, \; v = 0.0466$
$\tfrac{1}{2}\,C(3):$ $(8f)$ with $u = 0.1412, \; v = 0.2172$
$\tfrac{1}{2}\,C(2):$ $(16g)$ $\pm(xyz; \qquad\qquad x\bar{y}\bar{z};$
$\qquad\qquad\qquad\qquad x,y+\tfrac{1}{2},\tfrac{1}{2}-z; \; x,\tfrac{1}{2}-y,z+\tfrac{1}{2};$
$\qquad\qquad\qquad\qquad x+\tfrac{1}{2},y+\tfrac{1}{2},z; \; x+\tfrac{1}{2},\tfrac{1}{2}-y,\bar{z};$
$\qquad\qquad\qquad\qquad x+\tfrac{1}{2},y,\tfrac{1}{2}-z; \; x+\tfrac{1}{2},\bar{y},z+\tfrac{1}{2})$
$\qquad\qquad$ with $x = 0.1186, \; y = 0.1661, \; z = 0.1715$
$\tfrac{1}{2}\,C(4):$ $(16g)$ with $x = 0.1188, \; y = 0.2909, \; z = 0.1016$

In spite of the disorder and the incompleteness in knowledge of structure that results, it is evident that the compound is a monomer and that aluminum and its two attached nitrogen atoms are collinear, with Al–N = 2.18 A. According to the assigned parameters, N–C = 1.46–1.51 A.

XIV,a71.   *Gallium hydride trimethylamine,* $GaH_3 \cdot N(CH_3)_3$, forms hexagonal, rhombohedral, crystals whose unimolecular rhombohedron has the dimensions:

$$a_0 = 5.91 \text{ A.}, \qquad \alpha = 106°25'$$

The space group is C_{3v}^5 (*R3m*) with atoms in the positions:

Ga: $(1a)$ uuu $u = 0$ (arbitrary)
N: $(1a)$ with $u = 0.294$
C: $(3b)$ $uuv; \; uvu; \; vuu$ with $u = 0.462, \; v = 0.215$

For the corresponding trimolecular hexagonal cell:

$$a_0' = 9.465 \text{ A.}, \qquad c_0' = 6.75 \text{ A.}$$

The coordinates of the atoms in this cell are

Ga: (3a)   00u; rh   with $u = 0$
N: (3a)   with $u = 0.294$
C: (9b) $u\bar{u}v; u 2u v; 2\bar{u} \bar{u} v$; rh
           with $u = 0.082, v = 0.379$

In this structure (Fig. XIVA,95), the Ga–N distance is 1.97 ± 0.09 A.; the bond dimensions of the amine group are C–N = 1.47 A. and C–N–C = 105°.

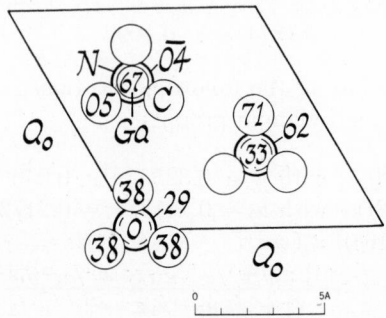

Fig.XIVA,95. The hexagonal structure of $GaH_3 \cdot N(CH_3)_3$ projected along its c_0 axis
Right-hand axes.

XIV,a72. Crystals of the somewhat analogous *auric bromide trimethyl phosphine*, $AuBr_3 \cdot P(CH_3)_3$, are built up of monomeric molecules. The unit orthorhombic cell has the edges:

$$a_0 = 9.06 \text{ A.}; \quad b_0 = 9.85 \text{ A.}; \quad c_0 = 22.05 \text{ A.}$$

The atoms of its eight molecules are all in general positions of V_h^{15} (*Pbca*):

(8c) $\pm (xyz; x+\frac{1}{2},\frac{1}{2}-y,\bar{z}; \bar{x},y+\frac{1}{2},\frac{1}{2}-z; \frac{1}{2}-x,\bar{y},z+\frac{1}{2})$

with the parameters listed in Table XIVA,24.

In this molecular structure (Figs. XIVA,96 and 97), the three bromine atoms and the phosphorus atom of the $P(CH_3)_3$ component have a planar, square distribution about the gold atom, with Au–Br = Au–P = 2.50 A. The $P(CH_3)_3$ groups, which have been located in the structure according to the demands of best packing, are trigonal pyramids with phosphorus atoms at the apices and P—C = 1.87 A. (by assumption). Contact between adjacent molecules is through CH_3–Br separations, the shortest of which is 3.43 A.

TABLE XIVA,24
Parameters of the Atoms in $AuBr_3 \cdot P(CH_3)_3$

Atom	x	y	z
Au	0.250	0.267	0.125
Br(1)	0.250	0.101	0.212
Br(2)	0.250	0.433	0.038
Br(3)	−0.026	0.267	0.125
P	0.526	0.267	0.125
C(1)	0.625	0.431	0.119
C(2)	0.625	0.197	0.193
C(3)	0.625	0.172	0.065

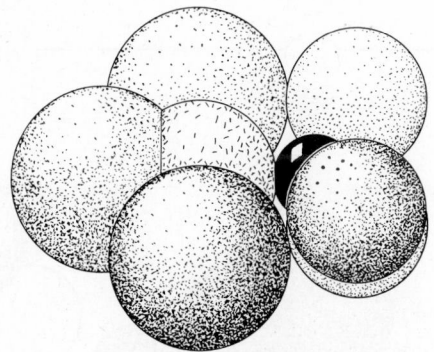

Fig. XIVA,96. A packing drawing of the molecule of $AuBr_3 \cdot P(CH_3)_3$. The larger dotted circles are bromine; the smaller are the methyl radicals. The black circle is phosphorus and the larger line shaded one is gold.

XIV,a73. *Diiodomethyl arsine*, CH_3AsI_2, is monoclinic with a large unit containing eight molecules and having the dimensions:

$a_0 = 14.45$ A.; $\quad b_0 = 4.60$ A.; $\quad c_0 = 19.97$ A.; $\quad \beta = 114°20'$ (5–10°C.)

The space group was chosen as C_{2h}^6 ($C2/c$) with all atoms in the general positions

$$(8f) \quad \pm (xyz;\ x,\bar{y},z+^1/_2;\ x+^1/_2,y+^1/_2,z;\ x+^1/_2,^1/_2-y,z+^1/_2)$$

The determined parameters are

Atom	x	y	z	σ
As	0.2045	0.3804	0.1501	0.010
I(1)	0.0590	0.0436	0.1384	0.006
I(2)	0.3475	0.0349	0.1615	0.006
C	0.156	0.380	0.037	0.07

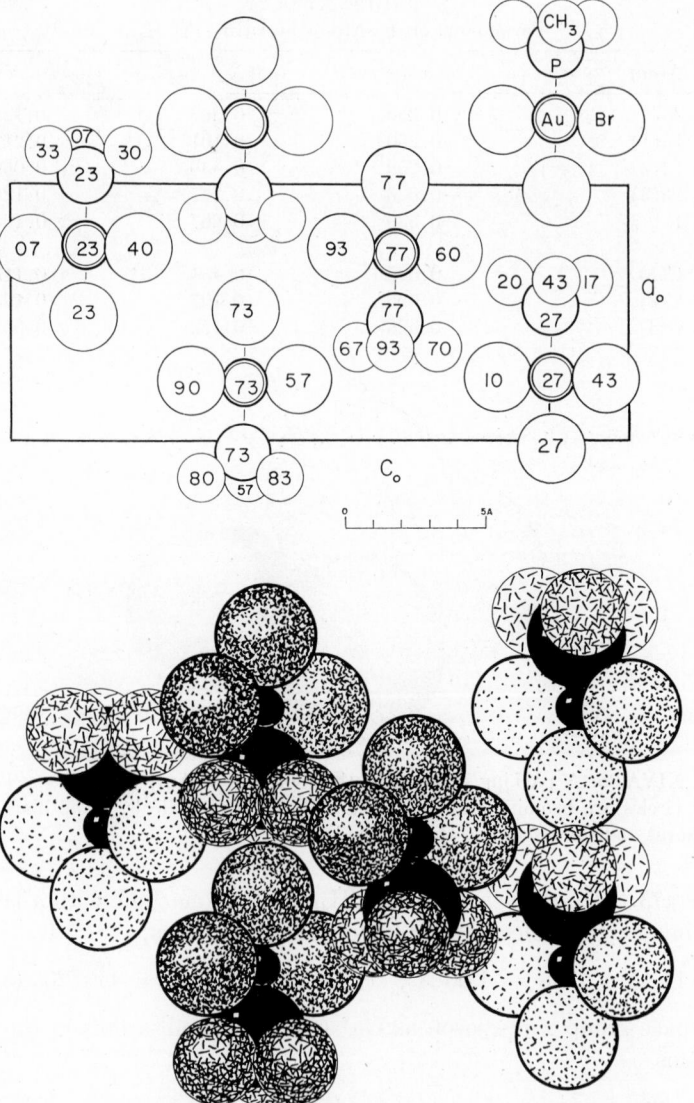

Fig. XIVA,97a (top). The orthorhombic structure of AuBr$_3$·P(CH$_3$)$_3$ projected along its b_0 axis. Left-hand axes.

Fig. XIVA,97b (bottom). A packing drawing of the orthorhombic AuBr$_3$·P(CH$_3$)$_3$ arrangement seen along its b_0 axis. The gold are the small, the phosphorus the larger black circles. The bromine atoms are the large, heavily ringed circles; the methyl groups are line shaded.

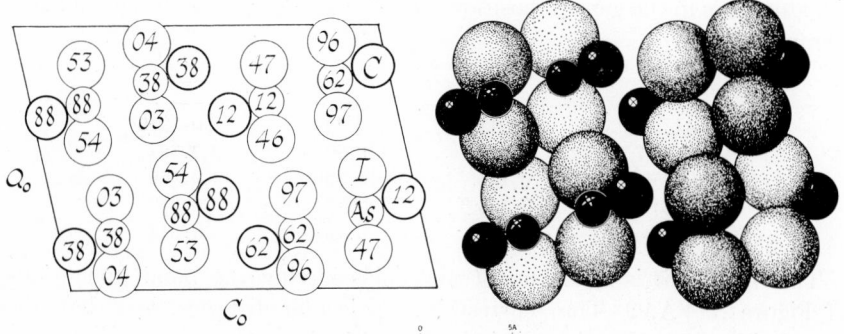

Fig. XIVA,98a (left). The monoclinic structure of CH_3AsI_2 projected along its b_0 axis. Right-hand axes.

Fig. XIVA,98b (right). A packing drawing of the monoclinic structure of CH_3AsI_2 seen along its b_0 axis. The large iodine atoms are dotted. The large black circles are carbon atoms; the smaller are arsenic.

In this structure (Fig. XIVA,98), the arsenic atom has about it two iodine atoms, with As–I = 2.54 A., and one carbon atom (As–C = 2.07 A.). The C–As–I angles are right and I–As–I = 104°. Between molecules the shortest C–C = 3.77 A. and the shortest I–I = 3.98 A.

XIV,a74. *Cyano dimethyl arsine* (cacodyl cyanide), $(CH_3)_2AsCN$, forms triclinic crystals which have a bimolecular cell of the dimensions:

$$a_0 = 6.31 \text{ A.}; \ b_0 = 8.02 \text{ A.}; \ c_0 = 6.27 \text{ A.}$$
$$\alpha = 110°0'; \ \beta = 119°45'; \ \gamma = 81°47'$$

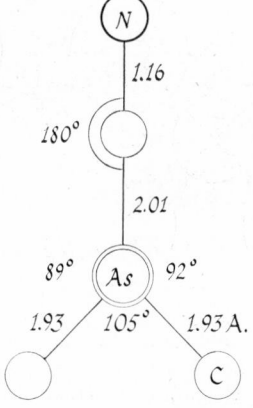

Fig. XIVA,99. Bond dimensions in the molecule of $(CH_3)_2AsCN$.

All atoms are in the general positions of C_i^1 ($P\bar{1}$), $\pm(xyz)$, with the following parameters

Atom	x	y	z
As	0.2115	0.1468	0.3316
C(1)	0.378	0.272	0.698
C(2)	0.134	0.329	0.169
C(3)	−0.106	0.174	0.339
N	−0.290	0.191	0.344

This structure is made up of molecules having the bond dimensions of Figure XIVA,99. The shortest intermolecular distance, less than the sum of the van der Waals radii, is As–N = 3.18 A.

XIV,a75. Crystals of *dibromo bis-(trimethyl arsine)* *μ-dibromodipalladium* are built up of dimers of $As(CH_3)_3 \cdot PdBr_2$. Obtained from dioxane they are tetragonal with units having

$$a_0 = 16.6 \text{ A.}, \qquad c_0 = 7.48 \text{ A.}$$

and containing four molecules of the dimer. Atoms have been placed in the following positions of C_{4h}^5 ($I4/m$):

$$(8h) \quad \pm(uv0; v\bar{u}0); \text{ B.C.}$$
$$(16i) \quad \pm(xyz; \bar{x}\bar{y}z; \bar{y}xz; y\bar{x}z); \text{ B.C.},$$

with the parameters of Table XIVA,25.

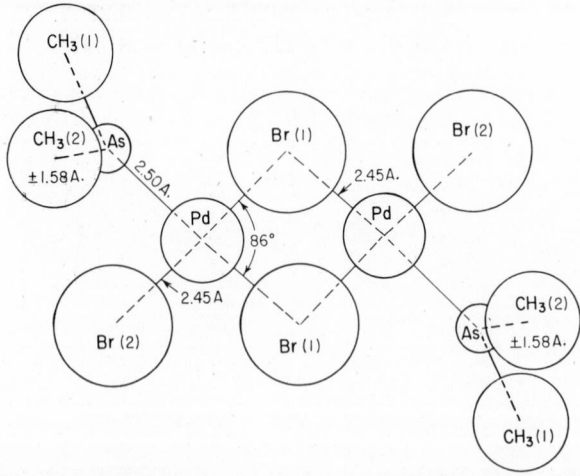

Fig. XIVA,100. Bond dimensions in the dimeric molecule of $As(CH_3)_3 \cdot PdBr_2$. The values within methyl circles indicate their distances above and below the plane of the rest of the molecule.

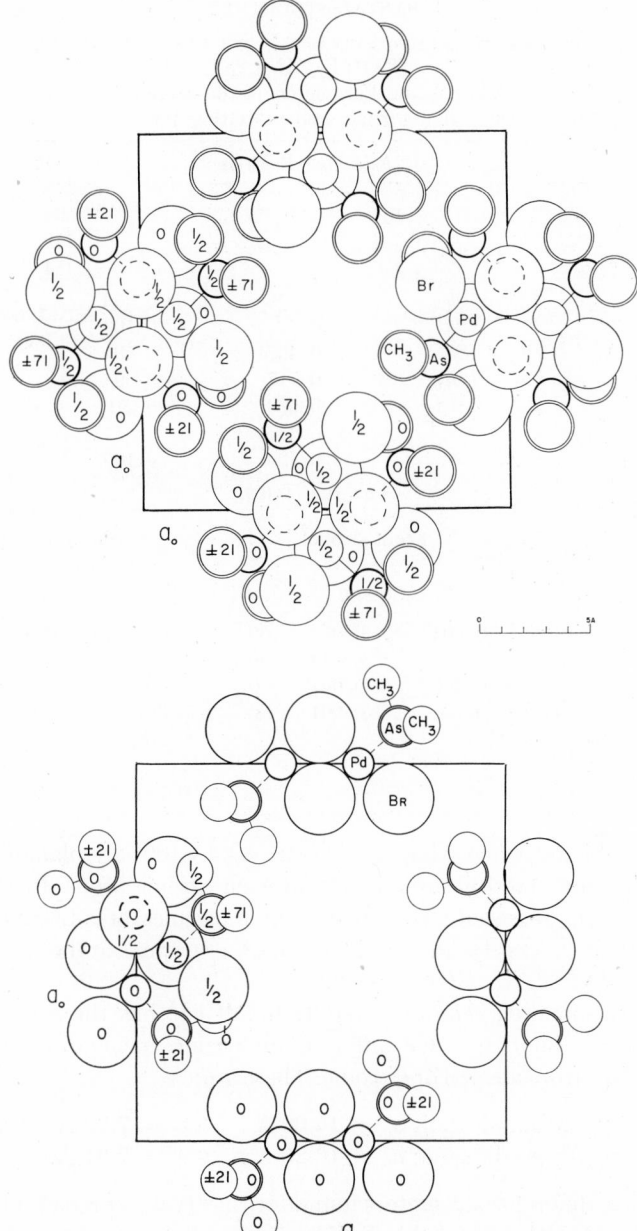

Fig. XIVA,101a (top). The tetragonal structure of As(CH$_3$)$_3 \cdot$PdBr$_2$ projected along its c_0 axis. Left-hand axes.

Fig. XIVA,101b (bottom). Part of the projection of Fig. XIVA,101a showing only the molecules in the paper plane (plus part of one molecule at $c_0 = $ $^1/_2$). This brings out more clearly the dimeric character of the molecules.

TABLE XIVA,25
Positions and Parameters of the Atoms in
$[As(CH_3)_3 \cdot PdBr_2]_2$ and $[As(CH_3)_3 \cdot PdCl_2]_2$

$[As(CH_3)_3 \cdot PdBr_2]_2$

Atom	Position	From dioxane		
		x	y	z
Pd	(8h)	0.608	0.003	0
As	(8h)	0.715	0.109	0
Br(1)	(8h)	0.497	0.101	0
Br(2)	(8h)	0.717	−0.094	0
C(1)	(8h)	0.669	0.219	0
C(2)	(16i)	0.783	0.100	0.212

$[As(CH_3)_3 \cdot PdCl_2]_2$

Atom	Position	From alcohol			From dioxane		
		x	y	z	x	y	z
Pd	(8h)	0.604	0.004	0	0.604	0.000	0
As	(8h)	0.716	0.114	0	0.716	0.106	0
Cl(1)	(8h)	0.496	0.098	0	0.500	0.097	0
Cl(2)	(8h)	0.716	−0.091	0	0.708	−0.096	0
C(1)	(8h)	—	—	—	0.669	0.218	0
C(2)	(16i)	—	—	—	0.785	0.100	0.217

The dimeric molecules that result (Fig. XIVA,100) are planar except for two of the three methyl groups around each arsenic atom. In the crystal as a whole these essentially square, flat molecules are distributed (Fig. XIVA,101) in layers stacked one above another normal to the c_0 axis.

The corresponding *chloride*, $[As(CH_3)_3 \cdot PdCl_2]_2$, has the same structure but has been found to have different dimensions depending on whether it crystallizes from alcohol or dioxane. These are:

From alcohol: $a_0 = 16.00$ A., $c_0 = 7.22$ A.
From dioxane: $a_0 = 16.25$ A., $c_0 = 7.31$ A.

The slightly different parameters found for the crystals prepared from these two solvents are those of Table XIVA,25.

It is proposed that the difference in the dimensions of these crystals is due to their containing some solvent molecules when crystallized from dioxane, but not when crystallized from alcohol. It is further suggested that these dioxane molecules are rotating in the empty channels that run parallel to the c_0 axis through the origin and the base centers of the unit cell.

XIV,a76.   *Tetramethyl arsonium bromide*, $As(CH_3)_4Br$, is hexagonal with a bimolecular unit of the edge lengths:

$$a_0 = 7.04 \text{ A.,} \qquad c_0 = 9.94 \text{ A.}$$

The space group has been found to be C_{6v}^4 ($P6_3mc$) with atoms in the positions:

As: (2b) $^1/_3 \, ^2/_3 \, u; \, ^2/_3, ^1/_3, u + ^1/_2$ with $u = 0.4090$
Br: (2b)   with $u = 0.0058$
C(1): (2b)   with $u = 0.600$
C(2): (6c) $u\bar{u}v; \qquad u \, 2u \, v; \qquad 2\bar{u} \, \bar{u} \, v;$
$\bar{u},u,v+^1/_2; \, \bar{u},2\bar{u},v+^1/_2; \, 2u,u,v+^1/_2$
with $u = 0.188, \, v = 0.354$

The structure, as shown in Figure XIVA,102, is best considered as a ZnO arrangement of its $As(CH_3)_4^+$ and Br^- ions, each being tetrahedrally surrounded by four of those of opposite sign. Three of the four As–C separations are 1.85 A., the fourth is 1.90 A., and C–As–C = 107 or 112°. Between ions, Br–C = 3.89 or 4.03 A.

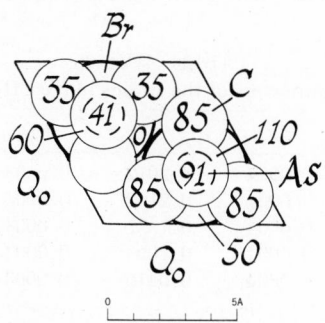

Fig. XIVA,102. The hexagonal structure of $As(CH_3)_4Br$ projected along its c_0 axis. Right-hand axes.

XIV,a77.   The yellow form of *arsenomethane*, $(AsCH_3)_5$, which melts at 12°C., is monoclinic with a tetramolecular cell of the dimensions:

$$a_0 = 8.89 \text{ A.;} \quad b_0 = 12.54 \text{ A.;} \quad c_0 = 11.55 \text{ A.;} \quad \beta = 101°46'$$

The space group is C_{2h}^5 ($P2_1/n$) with all atoms in the positions:

(4e) $\pm(xyz; \, x+^1/_2, ^1/_2-y, z+^1/_2)$

The parameters are those listed in Table XIVA,26.

The resulting structure (Fig. XIVA,103) is molecular with the molecules consisting of five-membered rings of arsenic atoms to each of which a methyl group is attached in a position far from the general plane of the ring. Bond dimensions of the nonplanar molecule are given in Figure XIVA,104.

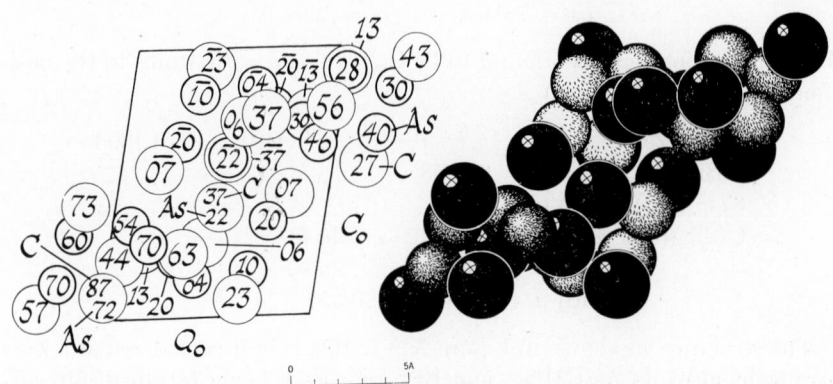

Fig. XIVA,103a (left). The monoclinic structure of $(AsCH_3)_5$ projected along its b_0 axis. Right-hand axes.

Fig. XIVA,103b (right). A packing drawing of the monoclinic $(AsCH_3)_5$ arrangement viewed along its b_0 axis. The methyl groups are black, the arsenic atoms line shaded.

TABLE XIVA,26
Parameters of the Atoms in $(AsCH_3)_5$

Atom	x	$\sigma(x)$	y	$\sigma(y)$	z	$\sigma(z)$
As(1)	0.3263	0.0004	0.2012	0.0003	0.2154	0.0003
As(2)	0.4962	0.0004	0.2162	0.0003	0.4094	0.0003
As(3)	0.7440	0.0004	0.1977	0.0003	0.3534	0.0003
As(4)	0.6928	0.0004	0.1003	0.0003	0.1677	0.0003
As(5)	0.4287	0.0004	0.0419	0.0003	0.1414	0.0003
CH₃(1)	0.152	0.004	0.133	0.003	0.265	0.003
CH₃(2)	0.496	0.004	0.370	0.003	0.409	0.003
CH₃(3)	0.819	0.004	0.073	0.003	0.454	0.003
CH₃(4)	0.652	0.004	0.226	0.003	0.069	0.003
CH₃(5)	0.454	0.004	0.944	0.003	0.278	0.003

XIV,a78. Crystals of the so-called *cacodyl disulfide*, $[(CH_3)_2AsS]_2$, are triclinic with a bimolecular unit of the dimensions:

$$a_0 = 6.34 \pm 0.01 \text{ A.}; \quad b_0 = 7.11 \pm 0.01 \text{ A.}; \quad c_0 = 11.35 \pm 0.02 \text{ A.}$$
$$\alpha = 100°14' \pm 5'; \quad \beta = 95°46' \pm 5'; \quad \gamma = 89°55' \pm 5'$$

The space group is C_i^1 ($P\bar{1}$) with atoms in the positions $(2i) \pm (xyz)$. The determined parameters are those of Table XIVA,27.

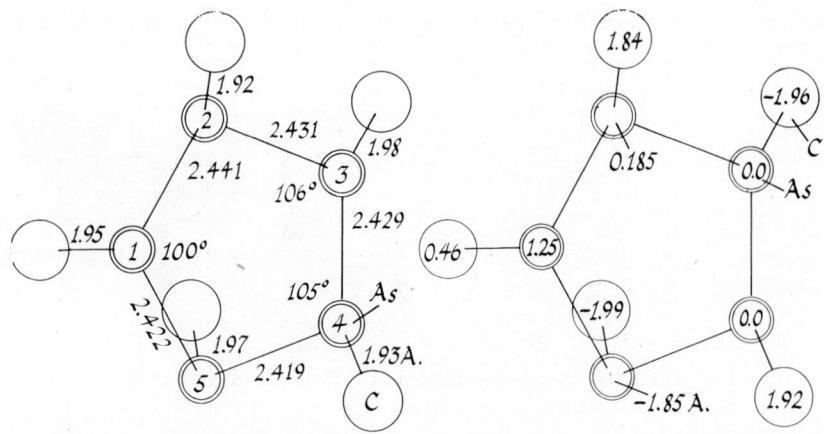

Fig. XIVA, 104a (left). Bond dimensions in the molecule of $(AsCH_3)_5$.
Fig. XIVA, 104b (right). The $(AsCH_3)_5$ molecule of Fig. XIVA, 104a with number that indicate the distances of atoms above and below the plane of the paper.

In this structure (Fig. XIVA,105), the molecules have the bond dimensions of Figure XIVA,106.   The structure as found here shows that the compound is in fact dimethylarsino dimethyldithioarsenate with a tetrahedral distribution of two sulfur and two methyl groups around the pentavalent arsenic, and a trigonal pyramidal arrangement of two methyls and a sulfur about the trivalent arsenic.   Between molecules the shortest atomic separation is an As–As = 3.24A.

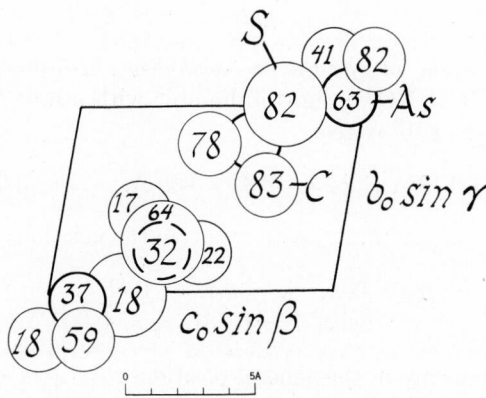

Fig. XIVA,105. The triclinic structure of $[(CH_3)_2AsS]_2$ projected along its a_0 axis. Right-hand axes.

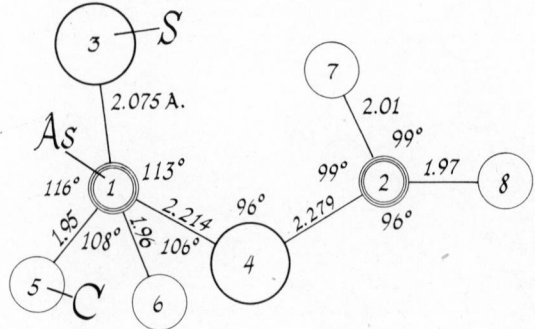

Fig. XIVA,106. Bond dimensions in the molecule of [(CH₃)₂AsS]₂.

TABLE XIVA,27
Parameters of the Atoms in [(CH₃)₂AsS]₂

Atom	x	σ_x, A.	y	σ_y, A.	z	σ_z, A.
As(1)	0.3161	0.0024	0.2240	0.0024	0.3668	0.0025
As(2)	0.3700	0.0025	−0.0619	0.0025	0.1033	0.0026
S(3)	0.6419	0.0074	0.2461	0.0073	0.3645	0.0074
S(4)	0.1789	0.0069	−0.0498	0.0065	0.2641	0.0068
C(5)	0.2180	0.025	0.2229	0.026	0.5238	0.025
C(6)	0.1664	0.026	0.4265	0.025	0.2969	0.023
C(7)	0.5950	0.027	−0.2441	0.027	0.1467	0.027
C(8)	0.1786	0.026	−0.2432	0.026	−0.0100	0.027

c. Salts, etc.

XIV,a79. Crystals of *potassium methylene bis-nitrosohydroxylamine*, $K_2[ONN(O)CH_2N(O)NO]$, are orthorhombic with a unit containing eight molecules and having the edges:

$$a_0 = 26.40 \pm 0.13 \text{ A.}; \quad b_0 = 12.81 \pm 0.06 \text{ A.}; \quad c_0 = 3.95 \pm 0.02 \text{ A.}$$

The space group is C_{2v}^{19} (*Fdd2*) with atoms in the positions:

C: (8a)   00u; ¹/₄,¹/₄,u+¹/₄; F.C.
with $u = 0.751$, $\sigma_u = 0.007$ A.

All the other atoms are in the general positions:

(16b) xyz; $\bar{x}\bar{y}z$; ¹/₄−x,y+¹/₄,z+¹/₄; x+¹/₄,¹/₄−y,z+¹/₄; F.C.

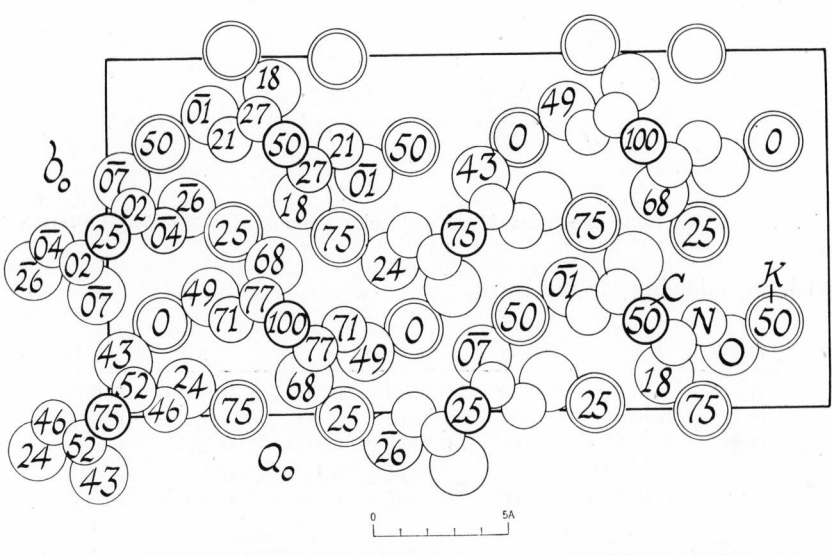

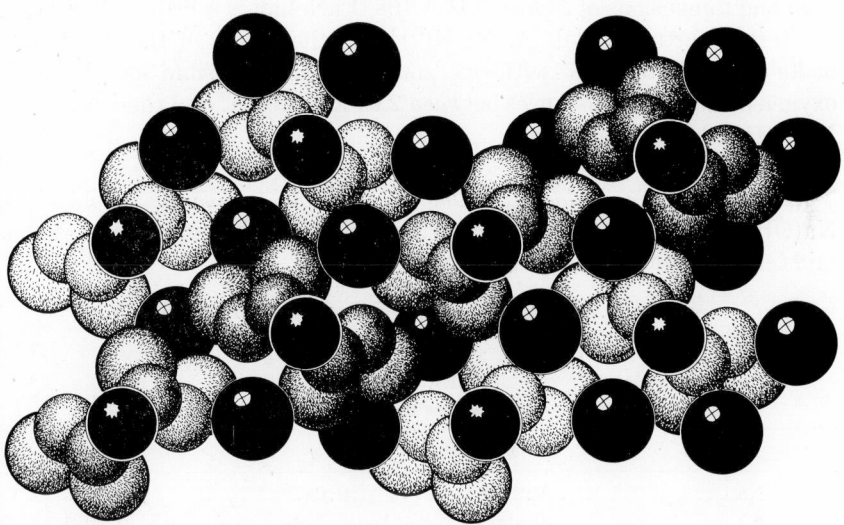

Fig. XIVA,107a (top). The orthorhombic structure of K₂[ONN(O)CH₂N(O)NO] projected along its c_0 axis. Right-hand axes.

Fig. XIVA,107b (bottom). A packing drawing of the orthorhombic structure of K₂[ONN(O)CH₂N(O)NO] seen along its c_0 axis. The potassium atoms are the larger, the carbon atoms the smaller black circles. Of the other atoms the oxygens are larger, line shaded, and more heavily outlined.

Fig. XIVA,108. Bond dimensions in the molecular anion of $K_2[ONN(O)CH_2N(O)NO]$.

with the following parameters:

Atom	x	σ_x, A.	y	σ_y, A.	z	σ_z, A.
K	0.0746	0.0007	0.2608	0.0007	0.0000	0.0013
O(1)	0.0194	0.003	0.1510	0.003	0.427	0.004
O(2)	0.1040	0.003	0.0714	0.003	0.238	0.004
N(1)	0.0343	0.003	0.0613	0.003	0.518	0.004
N(2)	0.0772	0.003	0.0157	0.003	0.456	0.006

The atomic arrangement is shown in Figure XIVA,107. The anions have the bond dimensions of Figure XIVA,108. Each half is sensibly planar [the greatest departure is 0.04 A. for N(2)] and the planes of the two halves make an angle of 75°6′ with one another. The potassium ions have six oxygen neighbors at distances between 2.64 and 3.04 A. and there is also an N(2) atom 3.27 A. distant.

XIV,a80. *Sodium hydroxymethane sulfinate dihydrate* (rongalite), $Na(OHCH_2SO_2) \cdot 2H_2O$, forms orthorhombic crystals having a unit that contains eight molecules. Its dimensions are:

$$a_0 = 6.78 \text{ A.}; \quad b_0 = 10.835 \text{ A.}; \quad c_0 = 15.97 \text{ A.}$$

TABLE XIVA,28
Parameters of the Atoms in $Na(OHCH_2SO_2) \cdot 2H_2O$

Atom	x	y	z
Na	0.2751	0.0923	0.2557
S	0.1063	0.2741	0.1014
C	0.0432	0.1439	0.0339
O(1)	0.0904	0.1681	−0.0498
O(2)	−0.0299	0.3724	0.0681
O(3)	0.0270	0.2287	0.1840
$H_2O(1)$	0.4765	0.4352	0.2081
$H_2O(2)$	−0.0140	0.6019	0.1391

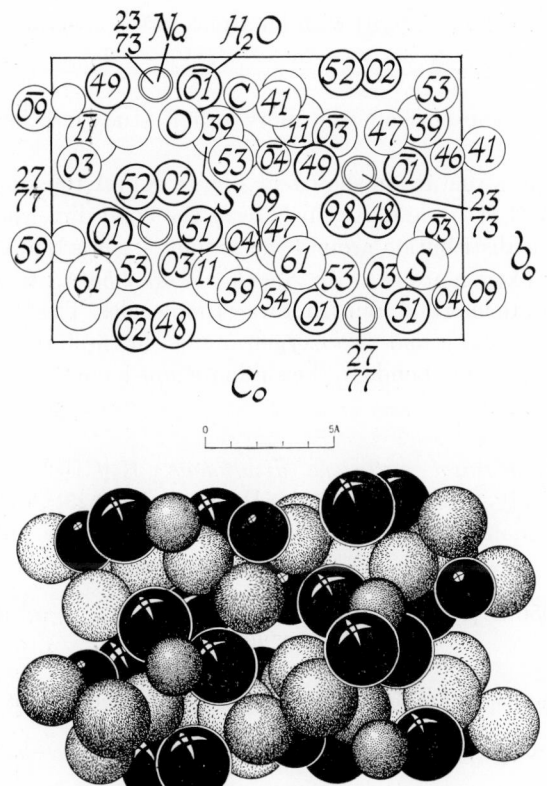

Fig. XIVA,109a (top). The orthorhombic structure of $Na(OHCH_2SO_2) \cdot 2H_2O$ projected along its a_0 axis. Right-hand axes.

Fig. XIVA,109b (bottom). A packing drawing of the orthorhombic $Na(OHCH_2SO_2) \cdot 2H_2O$ arrangement seen along its a_0 axis. Of the black circles the larger are the water molecules, the smaller the atoms of carbon. Atoms of sodium are small, dotted, and more heavily outlined. The sulfur atoms are the largest of the dotted circles; the oxygens are of intermediate size and short-line shaded.

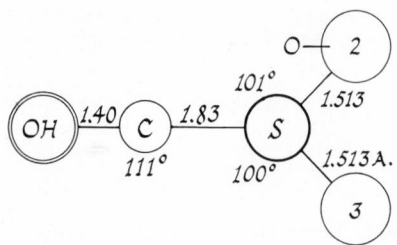

Fig. XIVA,110. Bond dimensions in the anion of $Na(OHCH_2SO_2)$.

The space group is V_h^{15} (*Pbca*) with all atoms in the general positions:

$$(8c) \quad \pm (xyz; \; x+{}^1/_2,{}^1/_2-y,\bar{z}; \; \bar{x},y+{}^1/_2,{}^1/_2-z; \; {}^1/_2-x,\bar{y},z+{}^1/_2)$$

The parameters chosen from the more recent study are listed in Table XIVA,28.

The resulting structure is shown in Figure XIVA,109. It contains pyramidal $(OHCH_2SO_2)^-$ anions which have the bond dimensions of Figure XIVA,110. The distribution around the sodium atom is that of a distorted octahedron of oxygen atoms, with Na–O = 2.45–2.53 A.; two of these are from the sulfinate ions and four from water molecules. The latter have four neighbors, two sodium and two oxygen atoms, in the form of a distorted tetrahedron. Hydrogen bonds to the sulfinate ions have the lengths O–H–O = 2.63 and 2.87 A.

XIV,a81. *Potassium methylene disulfonate,* $K_2[CH_2(SO_3)_2]$, has the structure already described for $K_2NH(SO_3)_2$ (**VII,b23**) and for $K_2S_2O_7$ (**IX,d10**). Its monoclinic crystals have a tetramolecular unit of the dimensions:

$$a_0 = 12.556 \text{ A.}; \quad b_0 = 7.773 \text{ A.}; \quad c_0 = 7.253 \text{ A.}, \quad \text{all} \pm 0.005 \text{ A.}$$
$$\beta = 90°30' \pm 1'$$

The space group is C_{2h}^6 (*C2/c*) with all atoms in the general positions:

$$(8f) \quad \pm (xyz; \; x,\bar{y},z+{}^1/_2; \; x+{}^1/_2,y+{}^1/_2,z; \; x+{}^1/_2,{}^1/_2-y,z+{}^1/_2)$$

The chosen parameters are those of Table XIVA,29.

TABLE XIVA,29
Parameters of the Atoms in $K_2[CH_2(SO_3)_2]$

Atom	x	y	z
K	0.152	0.132	−0.146
S	0.396	−0.194	0.140
O(1)	0.443	−0.295	−0.008
O(2)	0.327	−0.057	0.075
O(3)	0.347	−0.300	0.281
C	${}^1/_2$	−0.080	${}^1/_4$
H	0.536	−0.004	0.150

The resulting anions have the bond dimensions of Figure XIVA,111; the structure is that pictured in Figure VIIB,22.

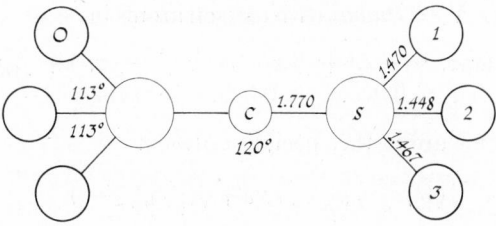

Fig. XIVA,111. Bond dimensions in the anion of $K_2[CH_2(SO_3)_2]$.

XIV,a82. *Sodium methane thiosulfonate monohydrate,* $NaS_2O_2CH_3 \cdot H_2O$, is orthorhombic with a tetramolecular cell of the edge lengths:

$$a_0 = 6.49 \text{ A.}; \quad b_0 = 5.55 \text{ A.}; \quad c_0 = 16.23 \text{ A.}$$

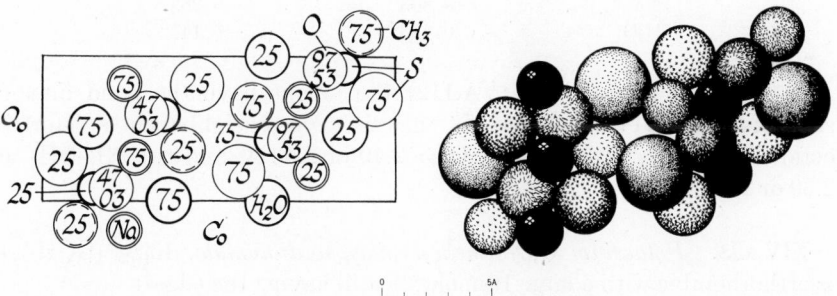

Fig. XIVA,112a (left). The orthorhombic structure of $Na[S_2O_2CH_3] \cdot H_2O$ projected along its b_0 axis. Right-hand axes.

Fig. XIVA,112b (right). A packing drawing of the orthorhombic $Na[S_2O_2CH_3] \cdot H_2O$ structure seen along its b_0 axis. The sodium atoms are black, the methyl groups only slightly larger and hook shaded. The large dotted circles are sulfur; the more heavily outlined, cross-shaded circles of intermediate size are the water molecules.

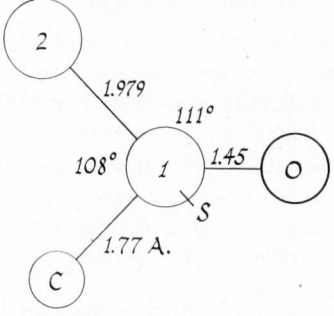

Fig. XIVA,113. Bond dimensions of the anion in $Na[S_2O_2CH_3] \cdot H_2O$.

The space group is V_h^{16} (*Pnma*) with oxygen atoms in

(8d) $\pm (xyz; \; 1/2-x,y+1/2,z+1/2; \; x,1/2-y,z; \; x+1/2,y,1/2-z)$
    with $x = 0.076$, $y = 0.029$, $z = 0.197$

All the other atoms are in the special positions:

(4c) $\pm (u \; 1/4 \; v; \; u+1/2,1/4,1/2-v)$

with the following parameters:

Atom	u	v
S(1)	0.0756	0.1492
S(2)	0.3030	0.0680
C	−0.161	0.095
Na	−0.305	−0.263
H_2O	0.488	−0.141

In this structure (Fig. XIVA,112) the anions have the bond dimensions of Figure XIVA,113. The Na^+ has six oxygen neighbors in a distorted octahedral array, with four Na–O = 2.40 or 2.43 A. and two Na–OH$_2$ = 2.39 or 2.46 A.

XIV,a83. *Potassium O,O-dimethyl phosphordithionate*, $K[S_2P(OCH_3)_2]$, is orthorhombic with a large 16-molecule cell having the edges:

$a_0 = 17.21 \pm 0.04$ A.; $b_0 = 17.50 \pm 0.04$ A.; $c_0 = 11.45 \pm 0.03$ A.

The space group is V_h^{24} (*Fddd*) with atoms in the positions:

K: (16f) $0u0; \; 0\bar{u}0; \; 1/4,1/4-u,1/4; \; 1/4,u+1/4,1/4;$ F.C.
    with $u = 0.1207$, $\sigma = 0.0004$.

P: (16e) $u00; \; \bar{u}00; \; 1/4-u,1/4,1/4; \; u+1/4,1/4,1/4;$ F.C.
    with $u = 0.1816$, $\sigma = 0.0007$.

S: (32h) $xyz; \; \bar{x}y\bar{z}; \; 1/4-x,1/4-y,1/4-z; \; x+1/4,1/4-y,z+1/4;$
    $x\bar{y}\bar{z}; \; \bar{x}\bar{y}z; \; 1/4-x,y+1/4,z+1/4; \; x+1/4,y+1/4,1/4-z;$ F.C.
    with $x = 0.1231$, $y = 0.0077$, $z = 0.8536$
    $\sigma(x) = 0.0005$, $\sigma(y) = 0.0004$, $\sigma(z) = 0.0009$

O: (32h)   with $x = 0.2515$, $y = 0.0612$, $z = 0.0268$
    $\sigma(x) = 0.0035$, $\sigma(y) = 0.0016$, $\sigma(z) = 0.0030$

C: (32h)   with $x = 0.2776$, $y = 0.1039$, $z = 0.9120$
    $\sigma(x) = 0.0029$, $\sigma(y) = 0.0020$, $\sigma(z) = 0.0050$

Part of the structure that results is shown in Figure XIVA,114. The anion has the bond dimensions of Figure XIVA,115. Each potassium atom

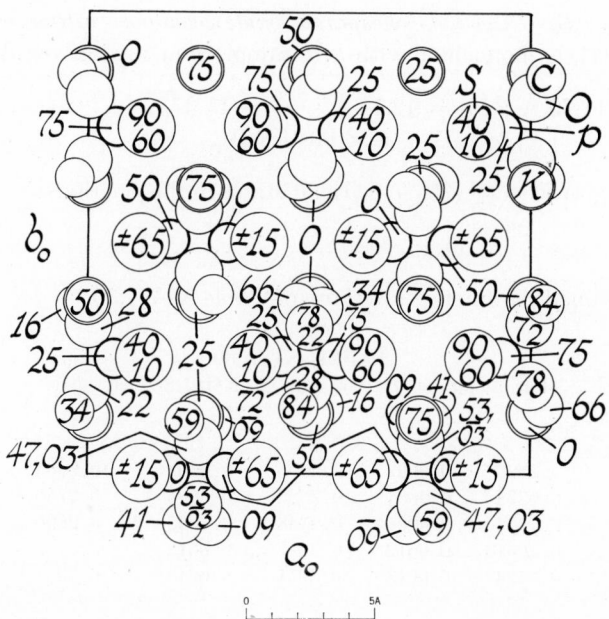

Fig. XIVA,114. Part of the contents of the complicated orthorhombic unit of K[S₂P-(OCH₃)₂] projected along its c_0 axis. Right-hand axes.

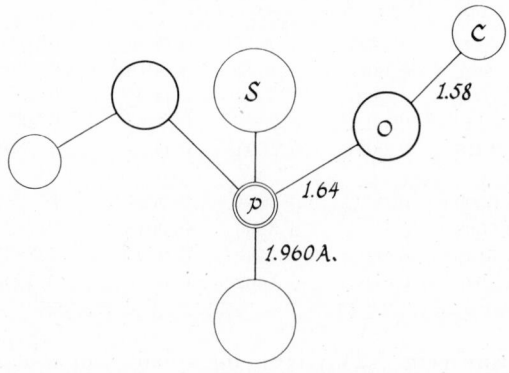

Fig. XIVA,115. Bond lengths in the anion of K[S₂P(OCH₃)₂].

is surrounded by six sulfur and two oxygen atoms having the form of a distorted tetragonal antiprism, with K–O = 2.82 A. and K–S = 3.35, 3.45, and 3.51 A.

XIV,a84. *Bis(trimethyl phosphine oxide)cobaltous nitrate,* $Co(NO_3)_2 \cdot 2[(CH_3)_3PO]$, is monoclinic with a tetramolecular unit of the dimensions:

$$a_0 = 11.70 \text{ A.}; \quad b_0 = 12.11 \text{ A.}; \quad c_0 = 11.37 \text{ A.}, \quad \text{all } \pm 0.02 \text{ A.}$$
$$\beta = 93°0' \pm 10'$$

The space group is C_{2h}^5 $(P2_1/c)$ with all atoms in the positions:

$$(4e) \quad \pm (xyz; \; x, \tfrac{1}{2} - y, z + \tfrac{1}{2})$$

The determined parameters are those of Table XIVA,30.

TABLE XIVA,30
Parameters of the Atoms in $Co(NO_3)_2 \cdot 2[(CH_3)_3PO]$

Atom	x	$\sigma(x)$	y	$\sigma(y)$	z	$\sigma(z)$
Co(1)	0.2396	0.0002	0.4095	0.0002	0.2566	0.0003
P(2)	0.0274	0.0004	0.2385	0.0005	0.2750	0.0006
P(3)	0.4474	0.0004	0.2348	0.0005	0.2690	0.0006
O(4)	0.1031	0.0012	0.3247	0.0012	0.2189	0.0014
O(5)	0.3524	0.0013	0.3051	0.0013	0.3231	0.0016
O(6)	0.3007	0.0011	0.4142	0.0012	0.0825	0.0015
O(7)	0.1794	0.0013	0.5341	0.0014	0.1227	0.0017
O(8)	0.2439	0.0017	0.5288	0.0017	−0.0521	0.0024
O(9)	0.1913	0.0013	0.4715	0.0013	0.4265	0.0016
O(10)	0.3230	0.0016	0.5479	0.0016	0.3428	0.0021
O(11)	0.2829	0.0021	0.6054	0.0020	0.5118	0.0027
N(12)	0.2402	0.0015	0.4922	0.0016	0.0468	0.0022
N(13)	0.2691	0.0020	0.5403	0.0020	0.4314	0.0026
C(14)	0.0449	0.0021	0.2341	0.0021	0.4336	0.0029
C(15)	−0.1215	0.0021	0.2695	0.0021	0.2307	0.0027
C(16)	0.0593	0.0020	0.0984	0.0021	0.2289	0.0028
C(17)	0.5608	0.0017	0.3233	0.0018	0.2183	0.0023
C(18)	0.5145	0.0019	0.1489	0.0020	0.3830	0.0024
C(19)	0.3946	0.0017	0.1502	0.0018	0.1496	0.0022

In this structure (Fig. XIVA,116), the cobalt atoms have an irregular sixfold coordination with each metal atom surrounded in planar fashion by two oxygen atoms belonging to each of the two nitrate groups and by oxygen atoms from the two $P(CH_3)_3O$ radicals. The Co–O distances range between 2.14 and 2.23 A. for the nitrate oxygens, and Co–O = 1.92 and 1.95 A. for the others. In the $(CH_3)_3PO$ radicals, P–C = 1.82 A. and P–O = 1.53 A. to produce an approximately tetrahedral distribution.

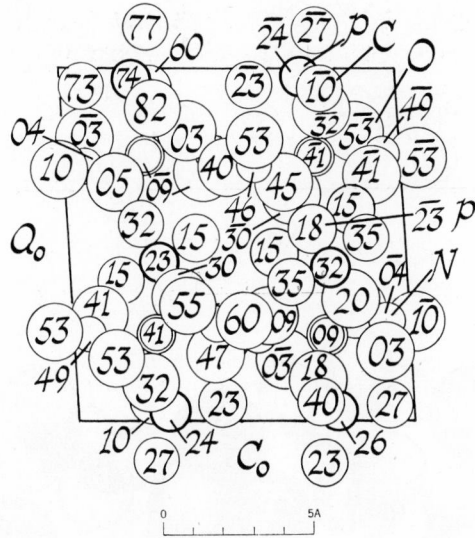

Fig. XIVA,116. A projection along its b_0 axis of most but not all the atoms in the monoclinic structure of $Co(NO_3)_2 \cdot 2[(CH_3)_3PO]$. Cobalt atoms are the small, doubly ringed circles. Right-hand axes.

XIV,a85. Crystals of *cadmium chloride bis-biuret*, $CdCl_2[OC(NH_2)-NHC(NH_2)O]_2$, are monoclinic with a bimolecular unit of the dimensions:

$$a_0 = 3.704 \pm 0.005 \text{ A.}; \quad b_0 = 19.96 \pm 0.03 \text{ A.}; \quad c_0 = 8.20 \pm 0.01 \text{ A.}$$
$$\beta = 111°6'$$

The space group is C_{2h}^5 ($P2_1/c$) with the cadmium atoms in

$$(2a) \quad 000; 0 \; {}^1/_2 \; {}^1/_2$$

and all other atoms in

$$(4e) \quad \pm (xyz; \; x, {}^1/_2 - y, z + {}^1/_2)$$

The determined parameters are those of Table XIVA,31.

The atomic arrangement is shown in Figure XIVA,117. The planar biuret molecule has the *trans* configuration and bond dimensions of Figure XIVA,118; its shape agrees with that found for the hydrate (1961: H,Y&F). Cadmium atoms are octahedrally surrounded by four chloride ions, with Cd–Cl = 2.55 and 2.62 A., and by two O(1) atoms with Cd–O = 2.34 A.; the angles do not depart by more than 2° from 90°. The molecules are held together by hydrogen bonds between nitrogen and chlorine or oxygen, with

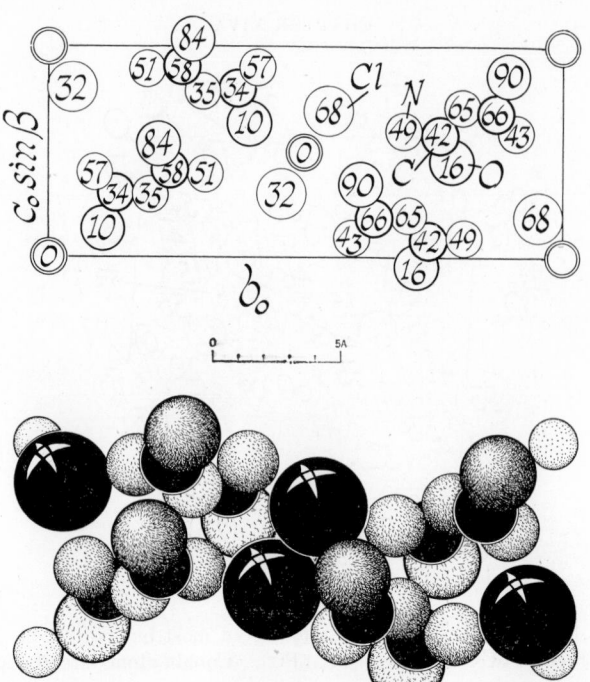

Fig. XIVA,117a (top). The monoclinic structure of $CdCl_2 \cdot [OC(NH_2)NHC (NH_2)O]_2$ projected along its a_0 axis. The doubly ringed circles are cadmium. Right-hand axes.

Fig. XIVA,117b (bottom). A packing drawing of the monoclinic $CdCl_2 \cdot [OC(NH_2)-NHC(NH_2)O]_2$ arrangement seen along its a_0 axis. The chlorine atoms are the large, the carbon atoms the smaller black circles. The cobalt atoms are small and dotted; the nitrogen atoms are slightly larger and hook shaded. Atoms of oxygen are larger, more heavily outlined, and line shaded.

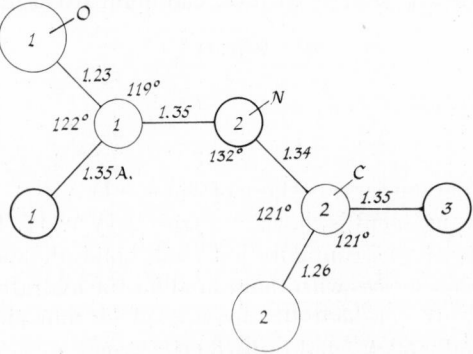

Fig. XIVA,118. Bond dimensions of the biuret molecule in $CdCl_2 \cdot [OC(NH_2)NHC-(NH_2)O]_2$.

TABLE XIVA,31
Parameters of Atoms in CdCl₂ bis-Biuret

Atom	x	σ_x, A.	y	σ_y, A.	z	σ_z, A.
Cl	0.3234	0.005	0.0446	0.005	−0.2053	0.011
O(1)	0.104	0.007	0.105	0.017	0.135	0.029
O(2)	0.842	0.020	0.214	0.025	0.543	0.031
N(1)	0.572	0.024	0.092	0.022	0.400	0.039
N(2)	0.354	0.022	0.197	0.023	0.284	0.029
N(3)	0.510	0.023	0.304	0.033	0.393	0.024
C(1)	0.337	0.031	0.130	0.038	0.268	0.048
C(2)	0.579	0.017	0.238	0.023	0.410	0.045

N(2)–O(2) = 2.82 A., N(3)–O(1) = 2.99 A., N(3)–O(2) = 3.06 A., and N(3)–Cl = 3.14 A.

The mercury salt, $HgCl_2[OC(NH_2)NHC(NH_2)O]_2$ is isostructural. For it:

$$a_0 = 3.768 \pm 0.004 \text{ A.}; \quad b_0 = 20.56 \pm 0.02 \text{ A.}; \quad c_0 = 8.16 \pm 0.01 \text{ A.}$$
$$\beta = 109°6'$$

XIV,a86. The complex *potassium copper(II) dibiuret tetrahydrate,* $K_2Cu[NHC(O)NHC(O)NH_2]_2 \cdot 4H_2O$ is monoclinic with a bimolecular unit of the dimensions:

$$a_0 = 3.843 \pm 0.004 \text{ A.}; \quad b_0 = 13.321 \pm 0.008 \text{ A.}; \quad c_0 = 13.916 \pm 0.008 \text{ A.}$$
$$\beta = 94°5' \pm 15'$$

The space group is C_{2h}^5 $(P2_1/n)$ with atoms in the positions:

$$\text{Cu: } (2a) \quad 000; \, \frac{1}{2}\,\frac{1}{2}\,\frac{1}{2}$$

All the other atoms are in the general positions:

$$(4e) \quad \pm (xyz; \, x+\frac{1}{2}, \frac{1}{2}-y, z+\frac{1}{2})$$

with the parameters of Table XIVA,32.

The atomic arrangement is that of Figure XIVA,119. In it each copper atom is surrounded by an almost perfect square of nitrogen atoms from two biuret ions, with Cu–N = 1.93 A. Each potassium ion has around it six oxygen atoms at the corners of a trigonal prism, with K–O = 2.67–2.89 A. The biuret molecule has the bond dimensions of Figure XIVA,120. It is not strictly planar but the atoms C(1), O(1), N(1), N(2) are coplanar and so are C(2), O(2), N(2), N(3). Their two planes make an angle of 4°56′ with

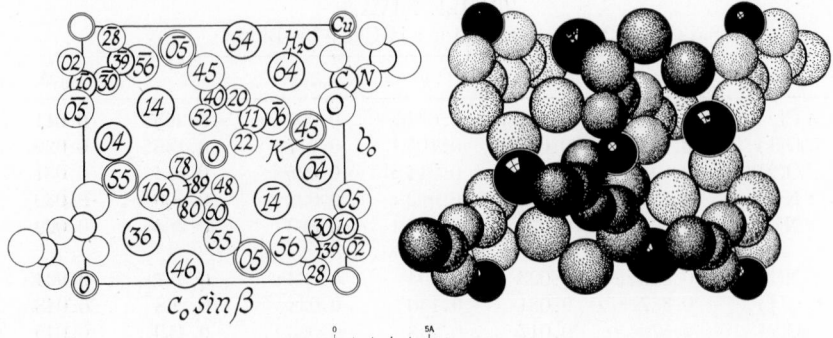

Fig. XIVA,119a (left). The monoclinic structure of $K_2Cu[NHC(O)NHC(O)NH_2]_2 \cdot 4H_2O$
projected along its short a_0 axis.   Right-hand axes.

Fig. XIVA,119b (right). A packing drawing of the monoclinic potassium cupric
dibiuret tetrahydrate structure seen along its a_0 axis.   The potassium atoms are
the larger, the copper atoms the smaller black circles.   Water molecules and oxygen
atoms are of the same [large] size, the water molecules being heavily ringed and hook
shaded and the other oxygens line shaded.   Of the somewhat smaller carbon and nitrogen
atoms, the nitrogens are dotted and the carbons line shaded and more heavily outlined.

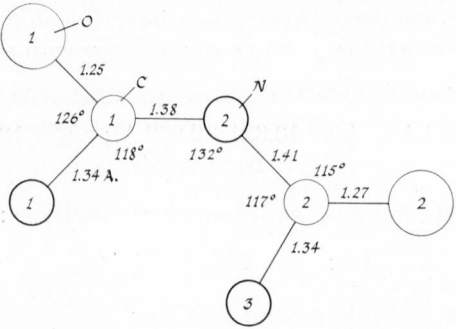

Fig. XIVA,120. Bond dimensions of the biuret anion in the potassium cupric salt.

one another. There are hydrogen bonds between the water molecules and
biuret oxygen atoms (2.73–2.81 A.) and between $N(3)$ and water oxygens
(3.05 A.)

There is a definite difference between the biuret molecule in this crystal
and in the preceding compound with $CdCl_2$ (**XIV,a85**). It is expressed by the
interchange of $O(2)$ and $NH_2(3)$ and is apparent from a comparison of Figures XIVA,118 and 120.

TABLE XIVA,32
Parameters of Atoms in $K_2Cu[NHC(O)NHC(O)NH_2]_2 \cdot 4H_2O$

Atom	x	$\sigma(x)$	y	$\sigma(y)$	z	$\sigma(z)$
K	0.5503	0.0004	0.4030	0.0001	−0.1387	0.0001
C(1)	0.3882	0.0017	0.1301	0.0005	0.1469	0.0006
C(2)	0.0968	0.0016	0.2226	0.0004	0.0021	0.0006
N(1)	0.2800	0.0016	0.0396	0.0004	0.1146	0.0005
N(2)	0.3034	0.0015	0.2123	0.0004	0.0893	0.0005
N(3)	−0.0251	0.0015	0.1386	0.0004	−0.0408	0.0005
O(1)	0.5557	0.0014	0.1477	0.0004	0.2251	0.0004
O(2)	0.0491	0.0014	0.3125	0.0004	−0.0272	0.0004
$H_2O(3)$	0.9379	0.0019	0.4469	0.0004	0.1160	0.0005
$H_2O(4)$	0.8621	0.0017	0.3259	0.0005	0.2786	0.0005
H(1)	0.320	—	−0.010	—	0.170	—
H(2)	0.395	—	0.274	—	0.117	—
H(3)	−0.217	—	0.150	—	−0.094	—
H(4)	0.978	—	0.400	—	0.066	—
H(5)	0.911	—	0.405	—	0.172	—
H(6)	0.752	—	0.263	—	0.259	—
H(7)	0.733	—	0.337	—	0.334	—

XIV,a87. *Nitroso(dimethyl dithiocarbonato)cobalt*, $Co(NO)[S_2CN(CH_3)_2]_2$, forms monoclinic crystals that have a tetramolecular unit of the dimensions:

$$a_0 = 6.49 \pm 0.02 \text{ A.}; \quad b_0 = 13.51 \pm 0.04 \text{ A.}; \quad c_0 = 16.95 \pm 0.05 \text{ A.}$$
$$\beta = 116°30'$$

The space group is C_{2h}^5 ($P2_1/c$) with all atoms in the general positions

$$(4e) \quad \pm (xyz; \ x, \tfrac{1}{2}-y, z+\tfrac{1}{2})$$

The determined parameters are listed in Table XIVA,33.

The resulting structure is shown in Figure XIVA,121. Its molecules have the bond dimensions of Figure XIVA,122. The central cobalt atom at the apex of a flat, nearly square pyramid, is 0.54 A. above the plane of the atoms of sulfur. The NO radical is above the cobalt atom, with Co–N = ca. 1.70 A. and with the axis of the NO bond making an angle of 134–140° with the axis of the pyramid. Atoms of the $S_2CN(CH_3)_2$ molecule lie in the plane which also includes the cobalt atom.

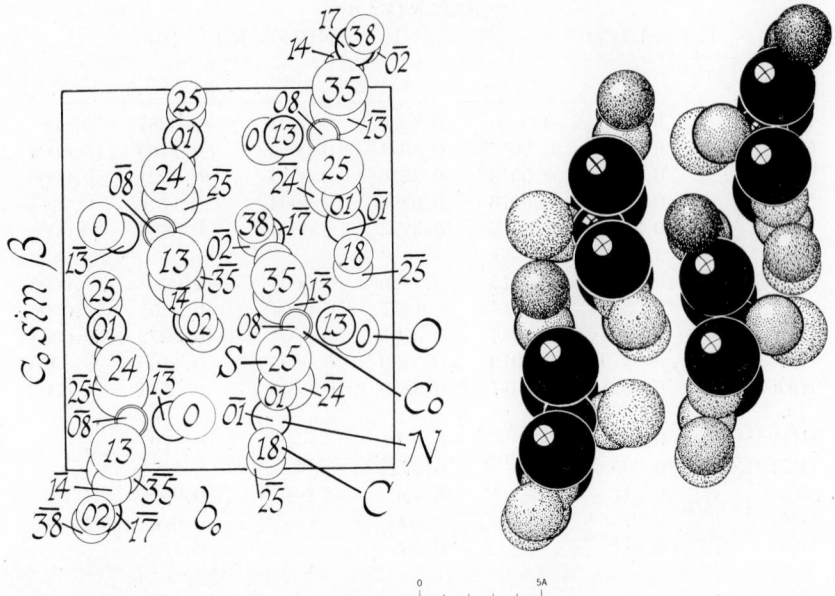

Fig. XIVA,121a (left). The monoclinic structure of Co(NO)[S₂CN(CH₃)₂]₂ projected
along its a_0 axis.   Right-hand axes.

Fig. XIVA,121b (right). A packing drawing of the monoclinic Co(NO)[S₂CN(CH₃)₂]₂
arrangement seen along its a_0 axis.   The cobalt atoms are the small, the sulfur atoms the
large black circles.   Of the dotted circles, the nitrogen atoms are more heavily outlined
than the carbon.   Atoms of oxygen are line shaded.

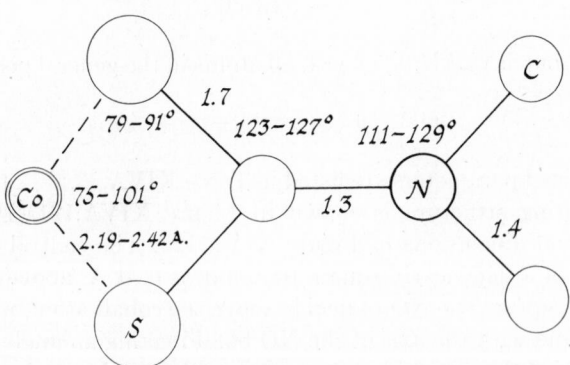

Fig. XIVA,122. Bond dimensions of the [S₂CN(CH₃)]₂ anion in Co(NO)[S₂CN(CH₃)₂]₂.

TABLE XIVA,33

Parameters of the Atoms in Co(NO)[S₂CN(CH₃)₂]₂

Atom	x	y	z
Co	−0.0796	0.2028	0.1289
S(1)	−0.2549	0.1694	0.2102
S(2)	0.2377	0.1715	0.2656
S(3)	0.1350	0.1586	0.0487
S(4)	−0.3457	0.1527	−0.0037
C(1)	−0.014	0.153	0.294
C(4)	−0.140	0.135	−0.039
N(1)	0.013	0.130	0.367
N(2)	−0.171	0.102	−0.113
N(3)	−0.126	0.327	0.129
O	0.002	0.386	0.142
CH₃(2)	0.254	0.116	0.448
CH₃(3)	−0.176	0.113	0.390
CH₃(5)	0.016	0.100	−0.135
CH₃(6)	−0.381	0.076	−0.187

6. Derivatives and Salts of Formic Acid

XIV,a88. At −50°C. the orthorhombic crystals of *formic acid*, HCOOH, have a tetramolecular cell of the dimensions:

$$a_0 = 10.23 \text{ A.}; \quad b_0 = 3.64 \text{ A.}; \quad c_0 = 5.34 \text{ A.}$$

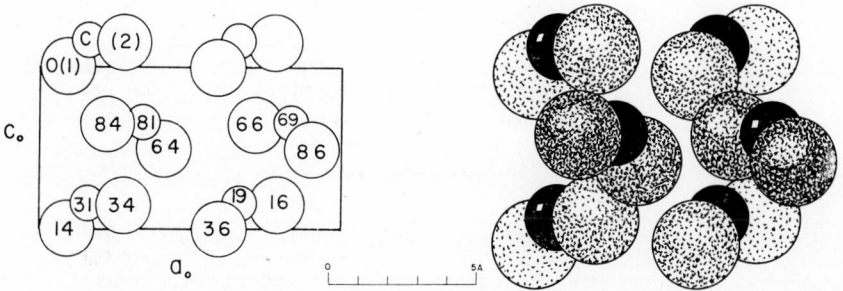

Fig. XIVA,123a (left). The orthorhombic structure of HCOOH projected along its b_0 axis. Left-hand axes.

Fig. XIVA,123b (right). A packing drawing of the orthorhombic HCOOH structure seen along its b_0 axis. The carbon atoms are black, the oxygens dotted.

Atoms are in general positions of C_{2v}^9 (Pna):

$$(4a) \quad xyz; \bar{x},\bar{y},z+^1/_2; \; ^1/_2-x,y+^1/_2,z+^1/_2; \; x+^1/_2,^1/_2-y,z$$

with the parameters:

Atom	x	y	z
C	0.157	0.306	0.161
O(1)	0.089	0.140	0.000
O(2)	0.276	0.337	0.147

In the resulting molecule the C–O separations are nearly equal (1.23 and 1.26 A.) and the angle O–C–O = 123°. The two C–O separations differ by about 0.15 A. from those in the vapor molecule as measured by electron diffraction. The molecular packing in the crystal (Fig. XIVA,123) points to the existence of strings of molecules tied together by hydrogen bonds of length 2.58 A.

XIV,a89. Crystals of *formamide*, $HCONH_2$, measured at $-50°C$., are monoclinic with a tetramolecular unit of the dimensions:

$$a_0 = 3.69 \text{ A.}; \quad b_0 = 9.18 \text{ A.}; \quad c_0 = 6.87 \text{ A.}; \quad \beta = 98°$$

Atoms are in the following general positions of C_{2h}^5 $(P2_1/n)$:

$$(4e) \quad \pm (xyz; \; x+^1/_2,^1/_2-y,z+^1/_2)$$

The established parameters are listed in Table XIVA,34.

TABLE XIVA,34
Parameters of the Atoms in Formamide

Atom	x	y	z
O	0.434	−0.067	0.244
N	0.330	0.159	0.128
C	0.310	0.059	0.261
Assumed hydrogen positions			
H(1)	0.231	0.262	0.176
H(2)	0.420	0.124	−0.010
H(3)	0.184	0.085	0.389

The bond lengths and angles of the resulting molecule are shown in Figure XIVA,124. The distribution of these molecules within the crystal is indicated in Figure XIVA,125. The molecules themselves are tied together to form puckered sheets by hydrogen bonds between nitrogen and oxygen;

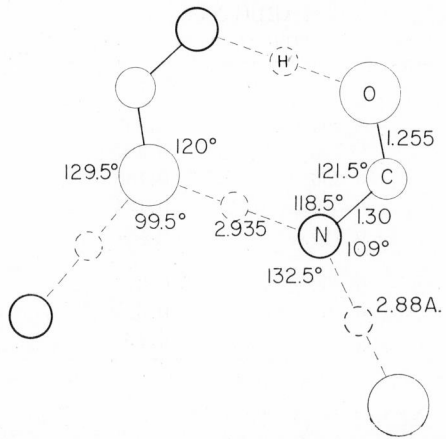

Fig. XIVA,124. Bond dimensions and intermolecular hydrogen-bond dimensions in HCONH₂.

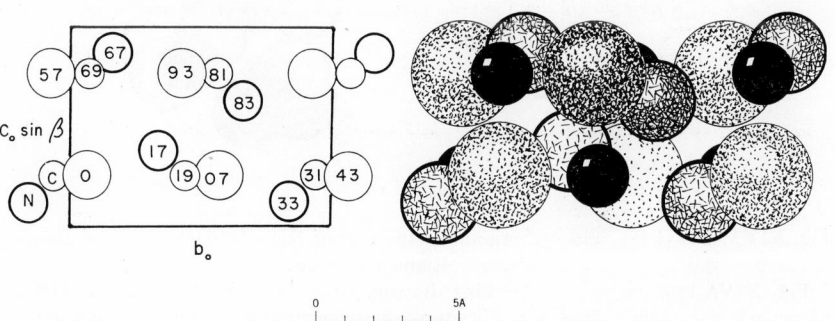

Fig. XIVA,125a (left). The monoclinic structure of HCONH₂ projected along its a_0 axis. Left-hand axes.

Fig. XIVA,125b (right). A packing drawing of the monoclinic structure of HCONH₂ viewed along its a_0 axis. The carbon atoms are black; the nitrogen atoms are heavily outlined and line shaded. The largest, dotted circles are oxygen.

the lengths of these bonds are 2.935 and 2.88 A. Attention should be drawn to the short C–N = 1.30 A. which is considered to indicate a strong double-bond character. Assumed positions for the hydrogen atoms are also listed in Table XIVA,34.

XIV,a90. Crystals of *formamidoxime*, NH_2CHNOH, are orthorhombic with a tetramolecular unit of the edge lengths:

$$a_0 = 8.22 \text{ A.}; \quad b_0 = 7.36 \text{ A.}; \quad c_0 = 4.78 \text{ A.}$$

124   CRYSTAL STRUCTURES

TABLE XIVA,35
Parameters of the Atoms in Formamidoxime

Atom	x	y	z
C	0.294	0.297	0.242
N(1)	0.426	0.401	0.225
N(2)	0.256	0.160	0.083
O	0.379	0.134	−0.120
H(1)	0.518	0.391	0.082
H(2)	0.202	0.317	0.396
H(3)	0.38	0.07	−0.13
H(4)	0.36	0.18	−0.15

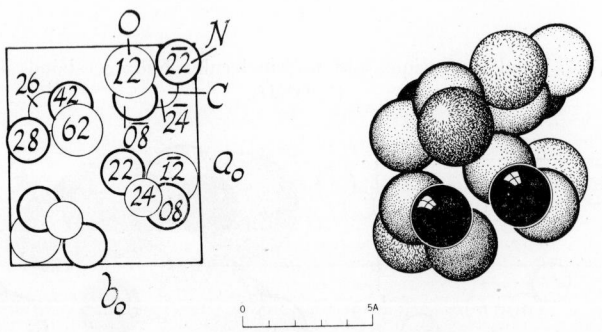

Fig. XIVA,126a (left). The orthorhombic structure of NH_2CHNOH projected along its c_0 axis. Right-hand axes.

Fig. XIVA,126b (right). A packing drawing of the orthorhombic NH_2CHNOH arrangement seen along its c_0 axis. The atoms of carbon are black; the nitrogen atoms are heavily outlined and dotted. The oxygen atoms are slightly larger and fine-line shaded.

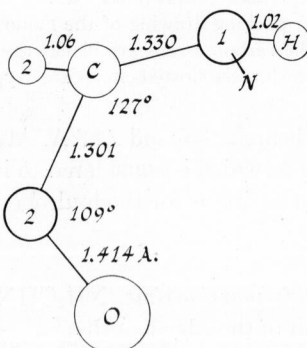

Fig. XIVA,127. Bond dimensions in the molecule of formamidoxime.

The space group is V^4 $(P2_12_12_1)$ with all atoms in the positions:

(4a) $xyz;$ $^1/_2-x,\bar{y},z+^1/_2;$ $x+^1/_2,^1/_2-y,\bar{z};$ $\bar{x},y+^1/_2,^1/_2-z$

The determined parameters are those of Table XIVA,35.

The structure is shown in Figure XIVA,126. Its molecules have the bond dimensions of Figure XIVA,127. The positions of the H(3) and H(4) atoms are uncertain; all others are coplanar. The shortest intermolecular separations are N–N = 3.12 A., and N–O = 2.81 and 3.01 A.

XIV,a91. *Diformyl hydrazine*, OCHNHNHCHO, is monoclinic with a bimolecular cell of the dimensions:

$$a_0 = 8.939\pm0.004 \text{ A.}; \quad b_0 = 6.253\pm0.002 \text{ A.}; \quad c_0 = 3.565\pm0.002 \text{ A.}$$
$$\beta = 112°30'\pm6'$$

The space group is C_{2h}^5 $(P2_1/a)$ with all atoms in the positions:

(4e) $\pm(xyz; x+^1/_2,^1/_2-y,z)$

Their parameters have been established as:

Atom	x	y	z
N	0.0030	0.1046	−0.0633
C	0.1389	0.2148	0.1088
O	0.2606	0.1477	0.3758
H(1)	0.130	0.355	0.010
H(2)	−0.075	0.160	−0.240

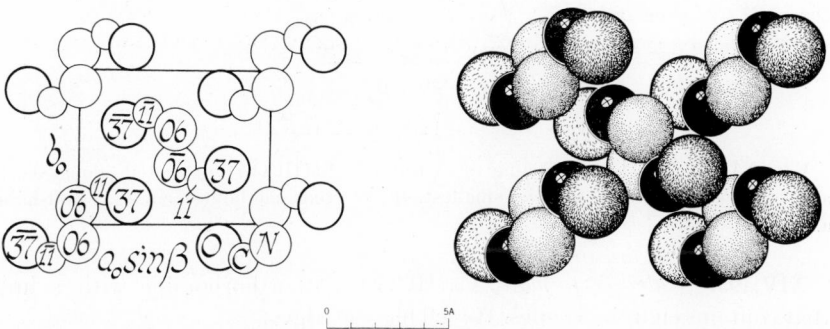

Fig. XIVA,128a (left). The monoclinic structure of diformyl hydrazine, OCHNH·NHCHO, projected along its c_0 axis. Right-hand axes.

Fig. XIVA,128b (right). A packing drawing of the monoclinic OCHNH·NHCHO structure viewed along its c_0 axis. The carbon atoms are black and the nitrogen atoms dotted. The larger oxygen circles are heavily outlined and line shaded.

The structure is shown in Figure XIVA,128. In these molecules, N–N = 1.392 A., N–C = 1.325 A., and C–O = 1.214 A. The hydrogen coordinates correspond to a C–H = 0.94 A. and N–H = 0.82 A. Between molecules there is an N–H–O bond of 2.799 A.

XIV,a92. Crystals of *sodium formate*, Na(HCOO), are monoclinic with a tetramolecular unit of the dimensions:

$$a_0 = 6.19 \text{ A.}; \quad b_0 = 6.72 \text{ A.}; \quad c_0 = 6.49 \text{ A.}; \quad \beta = 121°42'$$

The carbon and sodium atoms have been placed in special positions (4e) of C_{2h}^6 (C2/c):

$$(4e) \quad \pm (0 \; u \; {}^1/_4; \; {}^1/_2, u + {}^1/_2, {}^1/_4)$$
with $u(\text{Na}) = -0.139$ and $u(\text{C}) = 0.283$

The oxygen atoms are in general positions:

$$(8f) \quad \pm (xyz; \; \bar{x}, y, {}^1/_2 - z; \; x + {}^1/_2, y + {}^1/_2, z; \; {}^1/_2 - x, y + {}^1/_2, {}^1/_2 - z)$$
with $x = 0.043$, $y = 0.194$, and $z = 0.185$.

In this structure (Fig. XIVA,129), C–O within a formate ion is 1.27 A. and each sodium ion is surrounded by six oxygen atoms belonging to five HCOO ions at distances between 2.35 and 2.50 A.

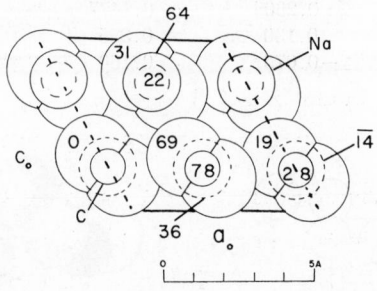

Fig. XIVA,129. The monoclinic structure of Na(HCOO) projected along its b_0 axis. The carbon atoms are the smallest, the oxygens the largest circles. Left-hand axes.

XIV,a93. *Calcium formate*, Ca(HCOO)$_2$, is orthorhombic with a unit that contains eight molecules. Its cell has the edges:

$$a_0 = 10.163 \text{ A.}; \quad b_0 = 13.381 \text{ A.}; \quad c_0 = 6.271 \text{ A. } (18°\text{C.})$$

Atoms are in the following general positions of V_h^{15} (Pcab):

$$(8c) \quad \pm (xyz; \; x + {}^1/_2, y, {}^1/_2 - z; \; {}^1/_2 - x, y + {}^1/_2, z; \; x, {}^1/_2 - y, z + {}^1/_2)$$

with the parameters of Table XIVA,36.

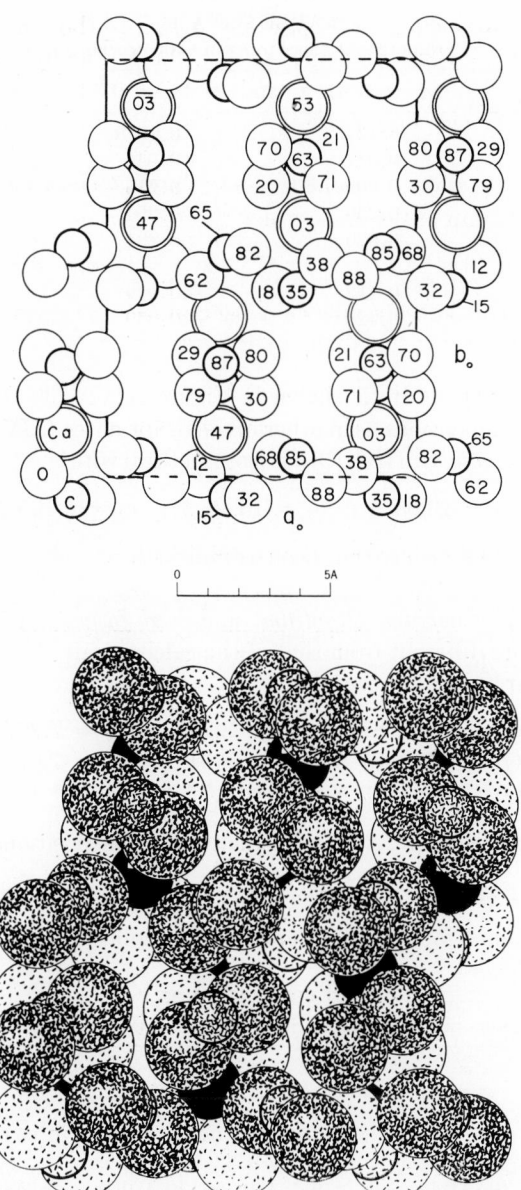

0 |___|___|___|___|___| 5A

Fig. XIVA,130a (top). The orthorhombic structure of Ca(HCOO)₂ projected along its c_0 axis. Left-hand axes.

Fig. XIVA,130b (bottom). A packing drawing of the orthorhombic Ca(HCOO)₂ arrangement viewed along its c_0 axis. The calcium atoms are black and the carbon atoms are the smallest, heavily outlined circles. The atoms of oxygen are large and dotted.

TABLE XIVA,36
Parameters of the Atoms in Calcium Formate

Atom	x	y	z
Ca	0.1345	0.1078	0.0295
O(1)	0.043	−0.049	0.180
O(2)	0.200	0.016	0.379
O(3)	0.200	0.204	−0.287
O(4)	0.020	0.294	−0.297
C(1)	0.112	−0.049	0.346
C(2)	0.129	0.270	−0.367

The resulting structure is shown in Figure XIVA,130. Its formate ions have the same dimensions as in other bivalent formates (**XIV,a94**).

Cadmium formate, $Cd(HCOO)_2$, is isostructural with:

$$a_0 = 10.03 \text{ A.}; \quad b_0 = 13.18 \text{ A.}; \quad c_0 = 6.135 \text{ A.}$$

Its atomic parameters have not been established.

XIV,a94. The *barium, strontium,* and *lead formates* are isomorphous, with a structure different from that of the calcium salt. Their orthorhombic tetramolecular units have the edges:

$Sr(HCOO)_2$: $a_0 = 6.874$ A.; $b_0 = 8.747$ A.; $c_0 = 7.267$ A. (23°C.)
$Ba(HCOO)_2$: $a_0 = 6.81$ A.; $b_0 = 8.91$ A.; $c_0 = 7.67$ A.
$Pb(HCOO)_2$: $a_0 = 6.53$ A.; $b_0 = 8.77$ A.; $c_0 = 7.42$ A.

The space group is V^4 ($P2_12_12_1$) with all atoms in the positions:

$$(4a) \quad xyz; \; x+1/2, 1/2-y, \bar{z}; \; 1/2-x, y+1/2, 1/2-z; \; \bar{x}, \bar{y}, z+1/2$$

Parameters found for the three crystals are collected in Table XIVA,37,

TABLE XIVA,37
Parameters of the Atoms in Strontium, Barium, and Lead Formates

Atom	x	y	z
Sr (Ba) [Pb]	0.000 (0.0750) [0.0709]	0.0915 (0.0834) [0.0875]	0.000 (0.0417) [0.000]
O(1)	0.145 (0.162) [0.152]	0.154 (0.281) [0.170]	0.358 (0.437) [0.360]
O(2)	0.384 (0.242) [0.399]	0.330 (0.410) [0.345]	0.350 (0.196) [0.300]
O(3)	0.268 (0.342) [0.354]	0.235 (0.166) [0.241]	0.828 (0.805) [0.800]
O(4)	0.130 (0.471) [0.179]	0.370 (0.100) [0.378]	0.057 (0.063) [0.000]
C(1)	0.245 (0.243) [0.294]	0.260 (0.385) [0.272]	0.420 (0.356) [0.404]
C(2)	0.132 (0.388) [0.227]	0.303 (0.073) [0.343]	0.906 (0.920) [0.792]

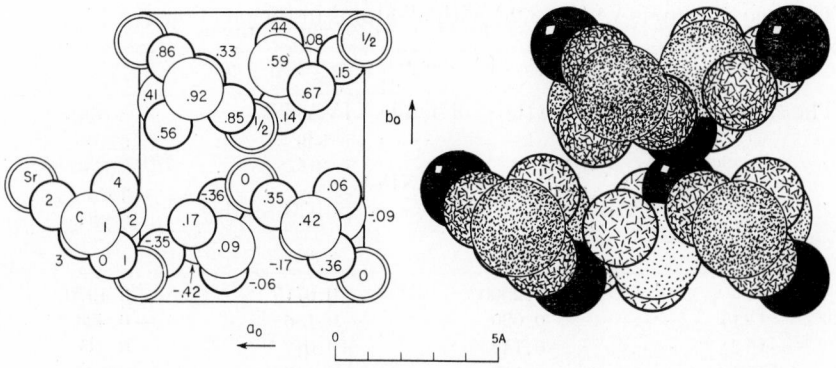

Fig. XIVA,131a (left). The orthorhombic structure of Sr(HCOO)₂ projected along its c_0 axis. Left-hand axes.

Fig. XIVA,131b (right). A packing drawing of the orthorhombic structure of Sr-(HCOO)₂ viewed along its c_0 axis. The strontium atoms are black; the carbon atoms in this case are large and dotted. The atoms of oxygen are heavily outlined and line shaded.

those for the barium salt being given in parentheses and for the lead salt in brackets.

The structure, as illustrated by Sr(HCOO)₂, is given in Figure XIVA,131. In the formate ion, C–O = 1.24–1.26 A. and O–C–O = 127°.

XIV,a95. Crystals of *strontium formate dihydrate*, Sr(HCOO)₂·2H₂O, are orthorhombic with a tetramolecular unit of the dimensions:

$a_0 = 7.332 \pm 0.001$ A.; $b_0 = 12.040 \pm 0.001$ A.; $c_0 = 7.144 \pm 0.002$ A.

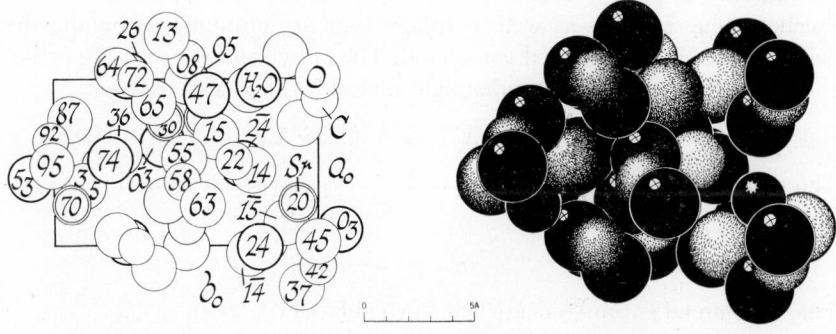

Fig. XIVA,132a (left). The orthorhombic structure of Sr(HCOO)₂·2H₂O projected along its c_0 axis. Right-hand axes.

Fig. XIVA,132b (right). A packing drawing of the orthorhombic structure of Sr-(HCOO)₂·2H₂O seen along its c_0 axis. The oxygen atoms are black and the carbon atoms dotted. The water molecules are more heavily outlined and fine-line shaded.

130 CRYSTAL STRUCTURES

The space group is V^4 ($P2_12_12_1$) with all atoms in the positions:

(4a) $xyz;\ ^1/_2-x,\bar{y},z+^1/_2;\ x+^1/_2,^1/_2-y,\bar{z};\ \bar{x},y+^1/_2,^1/_2-z$

The chosen parameters are those of Table XIVA,38.

TABLE XIVA,38
Parameters of the Atoms in $Sr(HCO_2)_2\cdot2H_2O$

Atom	x	y	z
Sr	0.2500	0.0715	0.1970
C(1)	0.020	0.186	−0.223
C(2)	−0.142	−0.012	0.417
O(1)	0.112	0.116	−0.147
O(2)	−0.061	0.259	−0.144
O(3)	0.022	−0.005	0.450
O(4)	−0.247	0.063	0.372
H₂O(1)	0.089	−0.092	0.031
H₂O(2)	−0.025	0.221	0.241

This structure is shown in Figure XIVA,132. Each strontium atom is surrounded by eight oxygen atoms, three of which are from water molecules; these Sr–O distances lie between 2.40 and 2.66 A.

A preliminary note has since been published (1961: G&F) which states results of another determination. There seems to be an error in the parameters given in this paper since they do not lead to a formate ion of reasonable dimensions.

XIV,a96. *Copper formate*, $Cu(HCOO)_2$, has a modification that forms excellent blue crystals as well as others that are obtained as bundles by dehydration of the hydrated compound. The anhydrous crystals are orthorhombic with a unit that contains eight molecules and has the edges:

$a_0 = 14.195$ A.; $b_0 = 8.955$ A.; $c_0 = 6.218$ A., all ±0.005 A.

The space group has been found to be V_h^{15} (*Pbca*) with all atoms in the positions:

(8c) $\pm(xyz;\ x+^1/_2,^1/_2-y,\bar{z};\ x,^1/_2-y,z+^1/_2;\ ^1/_2-x,\bar{y},z+^1/_2)$

The determined parameters are listed in Table XIVA,39.

The structure as a whole is shown in Figure XIVA,133. As is usual, the cupric ions have about them four close neighbors at the corners of an approximate square. These four Cu–O distances are 1.93–1.99 A.; three of the

TABLE XIVA,39
Parameters of the Atoms in $Cu(HCO_2)_2$

Atom	x	y	z
Cu	0.1152	0.0588	0.0065
O(1)	0.0893	0.2427	0.1621
O(2)	0.2223	0.0262	0.1950
O(3)	0.0067	0.0781	−0.1902
O(4)	0.1493	−0.1207	−0.1459
C(1)	0.0184	0.3226	0.1613
C(2)	0.2778	0.1353	0.2336

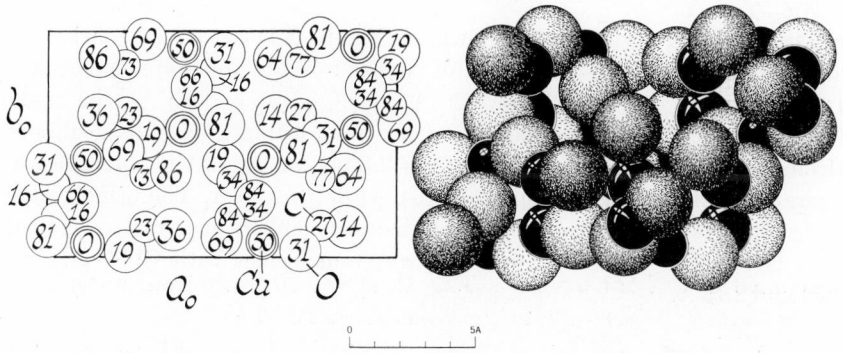

Fig. XIVA,133a (left). The orthorhombic structure of $Cu(HCOO)_2$ projected along its c_0 axis. Right-hand axes.

Fig. XIVA,133b (right). A packing drawing of the orthorhombic $Cu(HCOO)_2$ arrangement viewed along its c_0 axis. The copper atoms are the smaller, the carbon atoms the larger black circles. Oxygen atoms are line shaded.

O–Cu–O angles $= 88°$, and the fourth is $95°$. Above and below the copper atoms are oxygens at distances of 2.40 A. In the formate groups, $C(2)$–O $=$ 1.28 A. and $C(1)$–O $=$ 1.24 or 1.30 A.; the angles O–C–O $= 122°$.

XIV,a97. Monoclinic crystals of *cupric formate tetrahydrate*, $Cu(HCOO)_2 \cdot 4H_2O$, have the bimolecular unit:

$$a_0 = 8.18 \text{ A.}; \quad b_0 = 8.15 \text{ A.}; \quad c_0 = 6.35 \text{ A.}; \quad \beta = 101°5'$$

The space group is C_{2h}^5 $(P2_1/a)$ [and not the previously assigned C_{2h}^3]. Copper atoms are in

$$(2a) \quad 000; \ {}^1/_2 \ {}^1/_2 \ 0$$

All others are in general positions:

$$(4e) \quad \pm (xyz; \; x+{}^1/_2, {}^1/_2-y, z)$$

with the following parameters:

Atom	x	y	z
O(1)	0.206	−0.092	−0.080
O(2)	0.117	0.210	0.086
O(3),H_2O	0.077	−0.101	0.353
O(4),H_2O	0.086	0.349	0.483
C	0.238	0.270	0.018

The resulting structure is shown in Figure XIVA,134. It consists of alternate sheets, along the c_0 axis, of copper formate and water molecules. In the formate ion, C–O(1) = 1.26 A. and C–O(2) = 1.25 A.; the angle O–C–O = 120°. Each copper atom has four formate oxygen neighbors (from four different ions) with Cu–O = 2.00 and 2.01 A. Above and below it are two water oxygen atoms with Cu–O(3) = 2.36 A. The other water molecule does not approach copper but has as neighbors one formate oxygen with O(4)–O(2) = 2.82 A. and two water molecules with O(4)–O(3) = 2.74 and 2.78 A.

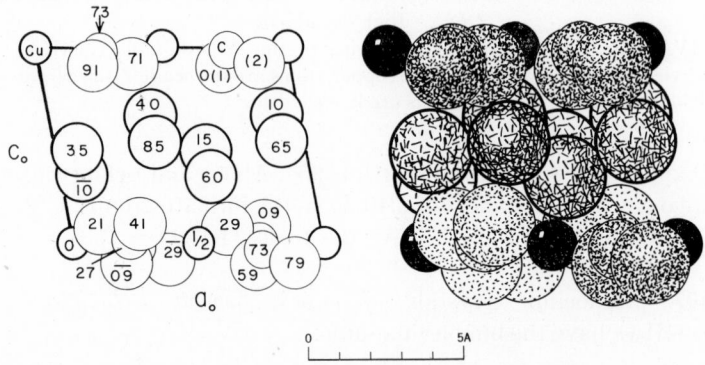

Fig. XIVA,134a (left). The monoclinic structure of Cu(HCOO)$_2$·4H$_2$O projected along its b_0 axis. Water molecules are the large, heavily outlined circles. Left-hand axes.

Fig. XIVA,134b (right). A packing drawing of the monoclinic structure of Cu-(HCOO)$_2$·4H$_2$O seen along its b_0 axis. The copper atoms are black. Of the largest circles, the carboxyl oxygens are dotted; the water molecules are heavily outlined and line shaded. The carbon atoms of intermediate size are dotted.

XIV,a98. *Nickel formate dihydrate*, Ni(HCOO)$_2$·2H$_2$O, forms monoclinic crystals. Their tetramolecular cells have the dimensions:

$$a_0 = 8.60 \text{ A.}; \quad b_0 = 7.06 \text{ A.}; \quad c_0 = 9.21 \text{ A.}; \quad \beta = 96°50'$$

The space group is C$_{2h}^5$ ($P2_1/c$) with the nickel atoms in the positions:

$$\text{Ni(1): } (2a) \quad 000; \; 0 \; ^1/_2 \; ^1/_2$$
$$\text{Ni(2): } (2d) \quad ^1/_2 \; 0 \; ^1/_2; \; ^1/_2 \; ^1/_2 \; 0$$

and all other atoms in the general positions:

$$(4e) \quad \pm (xyz; \; x,^1/_2-y,z+^1/_2)$$

The determined parameters are those of Table XIVA,40, values for hydrogen being only approximate.

TABLE XIVA,40
Parameters of Atoms in Ni(HCOO)$_2$·2H$_2$O

Atom	x	$\sigma(x)$	y	$\sigma(y)$	z	$\sigma(z)$
O(1)	0.4091	0.0006	0.2791	0.0007	0.1021	0.0005
O(2)	0.4036	0.0006	0.1069	0.0007	0.3011	0.0005
O(3)	0.2948	0.0006	0.8391	0.0007	0.4981	0.0005
O(4)	0.0644	0.0006	0.7271	0.0008	0.0717	0.0006
H$_2$O(1)	0.0866	0.0007	0.1049	0.0010	0.2002	0.0006
H$_2$O(2)	0.2163	0.0006	0.4684	0.0007	0.4286	0.0005
C(1)	0.4627	0.0008	0.2257	0.0010	0.2284	0.0007
C(2)	0.1750	0.0009	0.6159	0.0011	0.0604	0.0008
H(1)	0.206	—	0.106	—	0.243	—
H(2)	0.020	—	0.110	—	0.290	—
H(3)	0.236	—	0.613	—	0.457	—
H(4)	0.287	—	0.132	—	0.010	—
H(5)	0.556	—	0.306	—	0.247	—
H(6)	0.194	—	0.470	—	0.094	—

The resulting arrangement is shown in Figure XIVA,135. In the formate ions, C–O = 1.222–1.278 A. The nickel atoms are octahedrally coordinated, with Ni–O = 2.03–2.10 A. Around the Ni(1) atoms these neighbors are all formate oxygens; around Ni(2) they are both formate and water oxygen atoms.

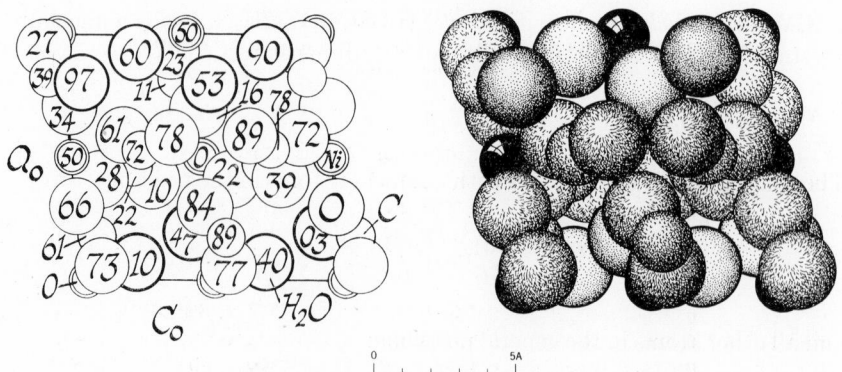

Fig. XIVA,135a (left). The monoclinic structure of Ni(HCOO)$_2 \cdot$2H$_2$O projected along its b_0 axis. Right-hand axes.

Fig. XIVA,135b (right). A packing drawing of the monoclinic Ni(HCOO)$_2 \cdot$2H$_2$O arrangement seen along its b_0 axis. The nickel atoms are black. Of the largest circles, those that are line shaded are the carboxyl oxygens; the water molecules are more heavily outlined and dotted. The carbon atoms are somewhat smaller and hook shaded.

Two divalent metal formates, *magnesium* and *manganous formates dihydrate*, Mg(HCOO)$_2 \cdot$2H$_2$O and Mn(HCOO)$_2 \cdot$2H$_2$O, are isostructural. Their units have the dimensions:

Mg(HCOO)$_2 \cdot$2H$_2$O:

$a_0 = 8.69$ A.; $b_0 = 7.18$ A.; $c_0 = 9.39$ A., all ± 0.02 A.

$\beta = 97°36' \pm 30'$

Mn(HCOO)$_2 \cdot$2H$_2$O:

$a_0 = 8.86$ A.; $b_0 = 7.29$ A.; $c_0 = 9.60$ A., all ± 0.02 A.

$\beta = 97°42' \pm 30'$

An origin different from that employed with the nickel salt was chosen for the description of their atomic arrangements but after a shift of origin their parameters are those of Table XIVA,41. They are for the most part close to those of Table XIVA,40.

For the magnesium compound these lead to the bond lengths: Mg–O = 2.05–2.13 A., C–O = 1.25–1.28 A. As in the nickel salt, one oxygen atom of each formate ion is close to a magnesium atom; the other is hydrogen-bonded to a water oxygen, with O–H–O = 2.80 or 2.84 A.

XIV,a99. *Gadolinium formate*, Gd(HCOO)$_3$, is rhombohedral with a unimolecular cell of the dimensions:

$$a_0 = 6.17 \text{ A.,} \qquad \alpha = 115°30'$$

TABLE XIVA,41
Parameters of Atoms in $Mg(HCOO)_2 \cdot 2H_2O$, and, in parentheses,
$Mn(HCOO)_2 \cdot 2H_2O$

Atom	x	y	z	σ, A.
C(1)	0.465 (0.462)	0.236 (0.220)	0.230 (0.226)	0.022
C(2)	0.177 (0.172)	0.613 (0.621)	0.061 (0.067)	0.022
O(1)	0.416 (0.416)	0.276 (0.266)	0.101 (0.102)	0.016
O(2)	0.404 (0.404)	0.113 (0.102)	0.304 (0.296)	0.016
O(3)	0.293 (0.287)	0.850 (0.846)	0.500 (0.506)	0.016
O(4)	0.070 (0.061)	0.728 (0.723)	0.074 (0.080)	0.016
H₂O(1)	0.220 (0.233)	0.477 (0.485)	0.438 (0.434)	0.016
H₂O(2)	0.083 (0.090)	0.109 (0.107)	0.199 (0.204)	0.016

Many years ago the atoms were assigned the following positions of C_{3v}^5 $(R3m)$:

$$Gd: (1a) \quad 000$$

All other atoms are in:

$$(3b) \quad uuv; \; uvu; \; vuu$$
$$u(C) = 0.85, \; v(C) = 0.43$$
$$u(O,1) = 0.81, \; v(O,1) = 0.19$$
$$u(O,2) = 0.58, \; v(O,2) = 0.33$$

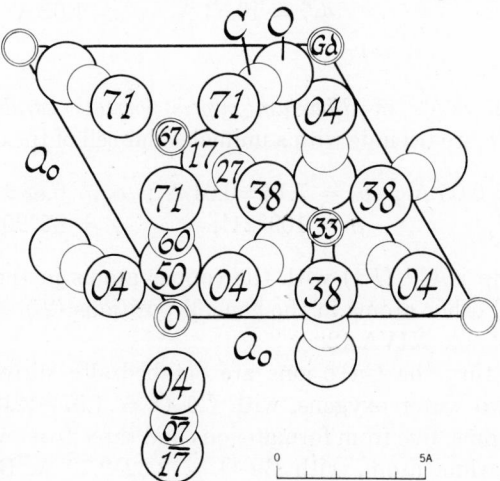

Fig. XIVA,136. The hexagonal structure of $Gd(HCOO)_3$ projected along its c_0 axis. Right-hand axes.

The corresponding trimolecular hexagonal unit has the dimensions:

$$a_0' = 10.44 \text{ A.,} \qquad c_0' = 3.98 \text{ A.}$$

In terms of these axes the atomic positions are:

Gd: ($3a$)   000; rh

C: ($9b$) $u\bar{u}v$; $u\ 2u\ v$; $2\bar{u}\ \bar{u}\ v$; rh
           with $u = 0.14, v = 0.71$

O(1): ($9b$)   with $u = 0.21, v = 0.60$

O(2): ($9b$)   with $u = 0.08, v = 0.50$

As Figure XIVA,136 indicates, this leads to a structure in which gadolinium atoms have nine oxygen neighbors, with Gd–O = 2.30 or 2.50 A. In the formate ions, C–O(1) = 1.27 A. and C–O(2) = 1.33 A.; the angle O–C–O = 121°. Between anions the shortest O–O = 2.61 A.

The following other rare earth formates have been shown to have similar cells, and undoubtedly to be isostructural:

$$\text{Ce(HCOO)}_3: \ a_0 = \ 6.31 \text{ A.,} \quad \alpha = 115°36'$$
$$a_0' = 10.67 \text{ A.,} \ c_0' = 4.08 \text{ A.}$$
$$\text{Nd(HCOO)}_3: \ a_0 = \ 6.27 \text{ A.,} \quad \alpha = 115°36'$$
$$a_0' = 10.61 \text{ A.,} \ c_0' = 4.06 \text{ A.}$$
$$\text{Pr(HCOO)}_3: \ a_0 = \ 6.28 \text{ A.,} \quad \alpha = 115°36'$$
$$a_0' = 10.63 \text{ A.,} \ c_0' = 4.07 \text{ A.}$$
$$\text{Sm(HCOO)}_3: \ a_0 = \ 6.22 \text{ A.,} \quad \alpha = 115°36'$$
$$a_0' = 10.53 \text{ A.,} \ c_0' = 4.02 \text{ A.}$$

XIV,a100. Crystals of *dibarium cupric formate tetrahydrate*, CuBa_2-$\text{(HCOO)}_6 \cdot 4\text{H}_2\text{O}$, are triclinic with a unimolecular cell of the dimensions:

$$a_0 = 8.75 \pm 0.05 \text{ A.;} \quad b_0 = 7.16 \pm 0.04 \text{ A.;} \quad c_0 = 6.88 \pm 0.04 \text{ A.}$$
$$\alpha = 99°3'; \qquad \beta = 109°21'; \qquad \gamma = 82°20', \qquad \text{all} \pm 15'$$

The space group is C_i^1 ($P\bar{1}$) with the copper atoms in the positions ($1g$) $0\ ^1/_2\ ^1/_2$ and all other atoms in the general positions ($2i$) $\pm(xyz)$ with the parameters of Table XIVA,42.

In this structure the Cu^{2+} ions are octahedrally surrounded by four formate and two water oxygens, with Cu–O = 1.97−2.18 A. There are eight oxygen atoms, five from formate ions and three from water molecules, around each barium atom, with Ba–O = 2.62–2.79 A. Between oxygen atoms from the anions and water, there are O–O separations from 2.52 A. upwards. The dimensions of the formate ions agree with those found for other salts.

TABLE XIVA,42
Parameters of the Atoms in $CuBa_2(HCOO)_6 \cdot 4H_2O$

Atom	x	y	z
Ba	0.35	0.233	0.40
C(1)	0.386	0.051	0.090
C(2)	0.90	0.144	0.124
C(3)	0.248	0.716	0.360
O(1)	0.308	0.191	−0.024
O(2)	0.536	0.030	0.172
O(3)	0.014	0.272	0.252
O(4)	0.756	0.181	0.034
O(5)	0.328	0.856	0.45
O(6)	0.168	0.626	0.45
$H_2O(1)$	0.823	0.700	0.378
$H_2O(2)$	0.500	0.508	0.296

XIV,a101. Crystals of *sodium iodide tri dimethylformamide*, $NaI \cdot 3(CH_3)_2NCHO$, are hexagonal with a bimolecular unit of the edge lengths·

$$a_0 = 11.75 \pm 0.03 \text{ A.}, \qquad c_0 = 6.55 \pm 0.02 \text{ A.}$$

TABLE XIVA,43
Crystals with the NH_3CH_3I Structure

Crystal	a_0, A.	c_0, A.	u (halogen)
$NH_3 \cdot nC_4H_9Br$	5.02	15.23	—
$NH_3 \cdot nC_4H_9Cl$	5.02	14.85	—
$NH_3 \cdot nC_4H_9I$	ca. 5.18	15.30	—
$NH_3 \cdot nC_5H_{11}Br$	5.00	16.95	—
$NH_3 \cdot nC_5H_{11}Cl$	5.01	16.69	—
$NH_3 \cdot nC_5H_{11}I$	ca. 5.18	17.42	0.100
$NH_3 \cdot nC_6H_{13}Br$	4.93	19.78	—
$NH_3 \cdot nC_6H_{13}Cl$	4.98	19.55	—
$NH_3 \cdot nC_6H_{13}I$	ca. 5.18	19.50	—
$NH_3 \cdot nC_7H_{15}Cl$	4.96	21.09	—
$NH_3 \cdot nC_7H_{15}I$ (impure)	ca. 5.18	ca. 21.82	—
$NH_3 \cdot nC_8H_{17}I$	ca. 5.18	23.70	0.070
$NH_3 \cdot nC_{10}H_{21}I$	ca. 5.18	28.09	0.066
$NH_3 \cdot nC_{11}H_{23}I$	ca. 5.18	30.46	0.055
$NH_3 \cdot nC_{12}H_{25}I$	ca. 5.18	31.24	0.055

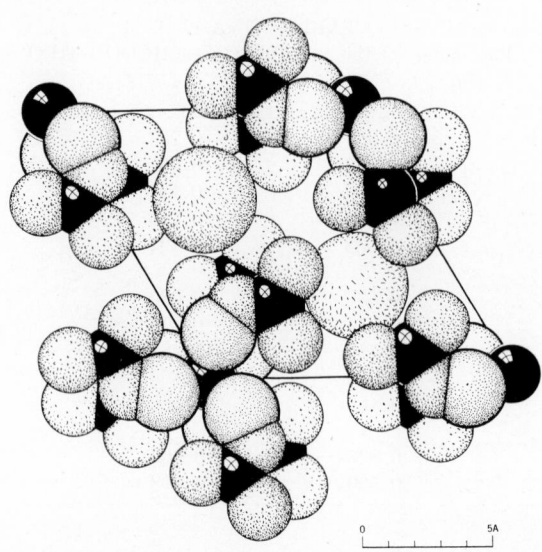

Fig. XIVA,137. The hexagonal structure of NaI·3(CH₃)₂NCHO projected along its
c_0 axis.   The (CH₃)₂NCHO molecules lie in two layers that are either $1/4$ or $3/4$ along
the c_0 axis.   In these molecules the oxygen atoms are the large, heavily outlined, dotted
circles; the carbon atoms are somewhat smaller.   The nitrogen atoms are the black
segments.   The black circles, at the corners of the unit, are the sodium atoms.   The
largest line-shaded circles are I⁻ ions.

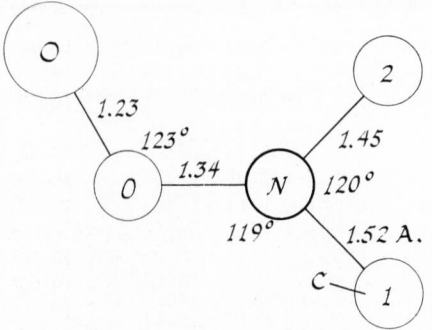

Fig. XIVA,138. Bond dimensions in the (CH₃)₂NCHO molecule as it occurs in its
addition compound with NaI.

The space group is D_{3h}^4 ($P\bar{6}2c$) with atoms in the positions:

Na: (2a)   000; 0 0 $1/2$

I: (2d) $2/3$ $1/3$ $1/4$; $1/3$ $2/3$ $3/4$

The other atoms are in the special positions:

$$(6h) \quad u\,v\,^1/_4;\ \bar{v},u-v,\ ^1/_4;\ v-u,\bar{u},^1/_4;$$
$$v\,u\,^3/_4;\ \bar{u},v-u,\ ^3/_4;\ u-v,\bar{v},^3/_4$$

with the following parameters:

Atom	u	v
C(0)	0.2674	0.2319
C(1)	0.3960	0.4753
C(2)	0.1426	0.3464
N	0.2660	0.3456
O	0.1676	0.1229

The structure is shown in Figure XIVA,137. In the planar dimethyl-formamide molecules the bond dimensions are those of Figure XIVA,138. Each sodium has six oxygen neighbors in a configuration that is neither octahedral nor prismatic, with Na–O $= 2.40$ A. The atoms of iodine are next to methyl radicals.

7. Methyl Ammonium Salts

XIV,a102. Though *monomethyl ammonium chloride*, NH_3CH_3Cl, has a structure different from that of the bromide and iodide, all are tetragonal. The chloride was originally described in terms of a simple unimolecular cell of the dimensions: $a_0 = 4.28$ A., $c_0 = 5.13$ A. Re-examination has, how-ever, pointed to a larger bimolecular unit of the dimensions:

$$a_0 = 6.04\ \text{A.}, \qquad c_0 = 5.05\ \text{A.}$$

The originally chosen structure was a CsCl arrangement with a chlorine atom at the origin and the carbon and nitrogen atoms of the (CH_3NH_3) ion in $^1/_2\,^1/_2\,u$, where $u(C)$ was probably at ca. 0.50 and $u(N)$ at ca. 0.24. The new structure found for this compound differs from the foregoing mainly in that, whereas in the unimolecular structure all (CH_3NH_3) groups were pointed the same way along the c_0 axis, in the new structure half of them are pointed upwards, the other half downwards. In this atomic arrangement, which formally is the same as that of PbO (**III,e1**) and PH_4I (**III,b3**), atoms are distributed according to the following special positions of D_{4h}^7 ($P4/nmm$):

Cl: $(2a)$ $000;\ ^1/_2\,^1/_2\,0$
N: $(2c)$ $0\,^1/_2\,u;\ ^1/_2\,0\,\bar{u}$ with $u = 0.198$
C: $(2c)$ with $u = 0.488$

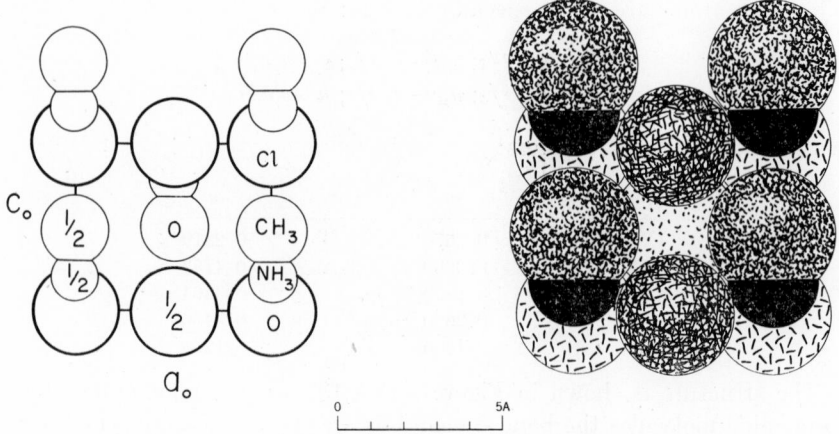

Fig. XIVA,139a (left). The tetragonal structure of CH_3NH_3Cl projected along an a_0 axis.

Fig. XIVA,139b (right). A packing drawing of the tetragonal CH_3NH_3Cl arrangement seen along an a_0 axis. The nitrogen atoms are black; the chlorine atoms are line shaded. Methyl groups are equally large but dotted.

In this grouping (Fig. XIVA,139), N–C = 1.465 A., C–Cl = 3.90 or 3.97 A., and N–Cl = 3.18 A.

The *monopropyl ammonium halides* have tetragonal unit cells resembling that of NH_3CH_3Cl. The atomic arrangement is probably similar though their cations must be "rotating." Bimolecular units have the dimensions:

$NH_3C_3H_7Cl$: $a_0 = 6.220$ A., $c_0 = 7.377$ A.
$NH_3C_3H_7Br$: $a_0 = 6.497$ A., $c_0 = 7.380$ A.
$NH_3C_3H_7I$: $a_0 = 6.931$ A., $c_0 = 7.332$ A.

At temperatures below $-100°C$. the chloride is monoclinic (**XIV,c3**).

XIV,a103. The bimolecular tetragonal unit found for *monomethyl ammonium iodide*, NH_3CH_3I, has a different shape from that of the chloride:

$$a_0 = 5.11 \text{ A.,} \qquad c_0 = 8.97 \text{ A.}$$

Iodine atoms are, however, in the same special positions of D_{4h}^7 ($P4/nmm$) used for the chloride:

I: (2c) $0 \; 1/2 \; u; \; 1/2 \; 0 \; \bar{u}$ with $u = 0.195$

Because of the overwhelming scattering powers of these atoms, positions for nitrogen and carbon could not be established in this early work, but it

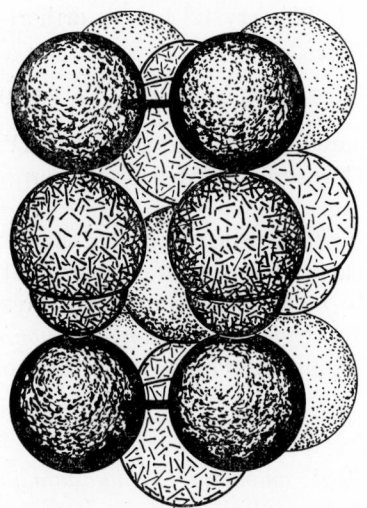

Fig. XIVA,140. A perspective drawing of the tetragonal arrangement in CH_3NH_3I, the corner of the cell shown here being displaced by $1/2\ a_0$ from the origin as taken in the text. The iodine atoms are dotted. The NH_3CH_3 cations are line shaded, the larger sphere being the methyl radical.

was pointed out that satisfactory interatomic distances result if these atoms also are in special positions (2c), with $u(N)$ = ca. 0.79 and $u(C)$ = ca. 0.64. Such a structure, as illustrated in Figure XIVA,140, is a simple distortion of the NaCl (III,a1) arrangement.

The corresponding *bromide*, NH_3CH_3Br, has been re-examined in detail. Its unit has the edge lengths:

$$a_0 = 5.09\ A., \qquad c_0 = 8.76\ A.$$

For it the carbon and nitrogen as well as bromine parameters could be determined. They are $u(Br)$ = 0.183, $u(C)$ = 0.635, and $u(N)$ = 0.804, values close to those given above for the iodide. In the bromide, then, N–C = 1.48 A.; the interionic distances are N–Br = 3.31 and 3.40 A., and C–Br = 3.95 A.

All higher mono alkyl halides except the ethyl and propyl compounds appear to be isomorphous with the iodide and to have the bimolecular cell dimensions of Table XIVA,43. It is noteworthy that the bases of all these units remain constant with increase in chain length. The high symmetry and simple unit cells found for these crystals are understandable only if the carbon and nitrogen atoms of their long zigzag chains "rotate" about the c_0 axis.

A unit cell has been determined for one longer-chained compound which is not isomorphous with the foregoing, *normal monooctadecyl ammonium chloride*, $NH_3 \cdot nC_{18}H_{37}Cl$. This crystal has an orthorhombic pseudotetragonal cell of the dimensions:

$$a_0 = 5.45 \text{ A.}; \quad b_0 = 5.40 \text{ A.}; \quad c_0 = 69.4 \text{ A.}$$

Its doubled cell, compared with those of Table XIVA,43, would contain four molecules; and its low symmetry has been interpreted as evidence that its chains are not rotating.

A preliminary study has also been reported of these long chain substituted ammonium halides at lower temperatures where "rotation" of the chains might be expected to be more or less completely arrested. In the course of this it was found that at $-80°C$. the *amyl ammonium chloride*, $NH_3 \cdot n\text{-}C_5H_{11}Cl$, has a large tetramolecular unit with the dimensions:

$$a_0 = 7.03 \text{ A.}, \quad c_0 = 16.70 \text{ A.}$$

XIV,a104. *Monomethyl ammonium chlorostannate*, $(NH_3CH_3)_2SnCl_6$, and the corresponding chloroplatinate, have structures that are rhombohedral distortions of the K_2PtCl_6 structure (**IX,c10**). Their unimolecular rhombohedral cells have the dimensions:

$[NH_3CH_3]_2SnCl_6:$ $a_0 = 8.42$ A., $\alpha = 50°14'$

$[NH_3CH_3]_2PtCl_6:$ $a_0 = 8.31$ A., $\alpha = 48°46'$

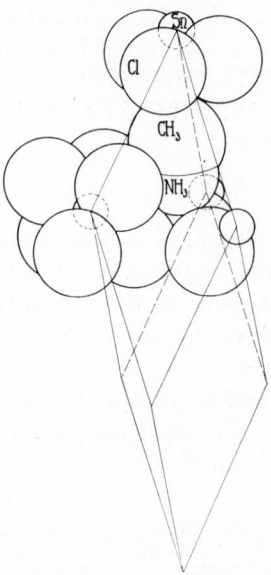

Fig. XIVA,141. A drawing which shows the packing of some of the atoms in the unit rhombohedron of $(NH_3CH_3)_2SnCl_6$.

Metal atoms are in the origin ($1a$) 000 and chlorine atoms in the following positions of D_{3d}^5 ($R\bar{3}m$):

($6h$) $\pm(uuv; uvu; vuu)$ with $u = 0.22$, $v = -0.26$

for the chlorostannate. The positions of the light atoms were not determined experimentally because of the overwhelming effects of the tin, but it was pointed out that both nitrogen and carbon undoubtedly are in ($2c$) $\pm(uuu)$ and that the customary atomic separations result if $u(N) = $ ca. 0.27 and $u(C) = $ ca. 0.21. For the trimolecular hexagonal cells:

$(NH_3CH_3)_2SnCl_6$: $a_0' = 7.148$ A., $c_0' = 22.016$ A.
$(NH_3CH_3)_2PtCl_6$: $a_0' = 6.861$ A., $c_0' = 21.913$ A.

Part of a unit rhombohedron of this structure is shown in Figure XIVA,-141 which brings out the effect of the elongated (NH_3CH_3) ions in separating from one another the packed layers of $SnCl_6$ ions that lie normal to the threefold axis.

XIV,a105. The structural modifications of *monomethyl ammonium aluminum alum*, $(CH_3NH_3)Al(SO_4)_2 \cdot 12H_2O$, have already been discussed in **X,g26.** More recently the β-form has been described as having a partially disordered arrangement. In this study the cell edge is given as

$$a_0 = 12.502 \text{ A.}$$

Except for the $(CH_3NH_2)^+$ ions, all atoms have been put in the same positions as before of T_h^6 ($Pa3$) but with the following somewhat different and presumably more accurate values:

Atom	Position	x	y	z
S	($8c$)	0.3329	0.3329	0.3329
O(1)	($8c$)	0.2639	0.2639	0.2639
O(2)	($24d$)	0.2808	0.3491	0.4372
$H_2O(1)$	($24d$)	0.5029	0.5024	0.1507
$H_2O(2)$	($24d$)	0.2288	0.3331	0.0540

It was concluded that the $(CH_3NH_2)^+$ cations do not lie on threefold axes but have their atoms statistically distributed about them in some of the positions of ($24d$), so as to have six different configurations, with $x = 0.4449$, $y = 0.5264$, $z = 0.5348$.

On cooling to 170°K. this alum becomes orthorhombic with a tetramolecular cell of the edge lengths:

$$a_0 = 12.57 \text{ A.}; \quad b_0 = 12.33 \text{ A.}; \quad c_0 = 12.38 \text{ A.}$$

TABLE XIVA,44

Parameters of the Atoms in Low-Methyl Ammonium Aluminum Alum

Atom	x	y	z
Al	0.006	0.259	0.256
C	0.047	0.770	0.280
N	0.045	0.268	0.711
S(1)	0.163	0.088	0.915
S(2)	0.171	0.581	0.085
O(1)	0.136	0.040	0.811
O(2)	0.240	0.022	0.978
O(3)	0.064	0.096	0.976
O(4)	0.199	0.202	0.904
O(5)	0.169	0.528	0.190
O(6)	0.225	0.510	0.011
O(7)	0.061	0.610	0.046
O(8)	0.231	0.684	0.097
O[H$_2$O,Al(1)]	0.153	0.257	0.236
O[H$_2$O,Al(2)]	0.145	0.748	0.759
O[H$_2$O,Al(3)]	0.012	0.105	0.258
O[H$_2$O,Al(4)]	0.003	0.405	0.247
O[H$_2$O,Al(5)]	0.007	0.242	0.397
O[H$_2$O,Al(6)]	0.000	0.256	0.098
O[H$_2$O,CH$_3$NH$_3$(1)]	0.228	0.302	0.591
O[H$_2$O,CH$_3$NH$_3$(2)]	0.154	0.456	0.801
O[H$_2$O,CH$_3$NH$_3$(3)]	0.235	0.808	0.425
O[H$_2$O,CH$_3$NH$_3$(4)]	0.174	0.988	0.182
O[H$_2$O,CH$_3$NH$_3$(5)]	0.057	0.086	0.521
O[H$_2$O,CH$_3$NH$_3$(6)]	0.053	0.577	0.472
H[H$_2$O,Al(1)]	0.210	0.237	0.305
H[H$_2$O,Al(2)]	0.199	0.277	0.175
H[H$_2$O,Al(3)]	0.196	0.722	0.693
H[H$_2$O,Al(4)]	0.193	0.774	0.825
H[H$_2$O,Al(5)]	0.076	0.057	0.235
H[H$_2$O,Al(6)]	0.048	0.953	0.781
H[H$_2$O,Al(7)]	0.070	0.455	0.220
H[H$_2$O,Al(8)]	0.059	0.542	0.768
H[H$_2$O,Al(9)]	0.026	0.180	0.450
H[H$_2$O,Al(10)]	0.024	0.698	0.960

(*continued*)

TABLE XIVA,44 (*continued*)

Atom	x	y	z
H[H₂O,Al(11)]	0.025	0.192	0.048
H[H₂O,Al(12)]	0.021	0.680	0.547
H[H₂O,CH₃NH₃(1)]	0.006	0.012	0.502
H[H₂O,CH₃NH₃(2)]	0.136	0.060	0.507
H[H₂O,CH₃NH₃(3)]	0.247	0.385	0.557
H[H₂O,CH₃NH₃(4)]	0.243	0.264	0.012
H[H₂O,CH₃NH₃(5)]	0.225	0.488	0.752
H[H₂O,CH₃NH₃(6)]	0.183	0.480	0.885
H[H₂O,CH₃NH₃(7)]	0.004	0.503	0.502
H[H₂O,CH₃NH₃(8)]	0.139	0.552	0.490
H[H₂O,CH₃NH₃(9)]	0.249	0.757	0.490
H[H₂O,CH₃NH₃(10)]	0.244	0.893	0.445
H[H₂O,CH₃NH₃(11)]	0.249	0.010	0.743
H[H₂O,CH₃NH₃(12)]	0.202	0.002	0.108
H[CH₃(1)]	0.120	0.784	0.340
H[CH₃(2)]	0.033	0.840	0.245
H[CH₃(3)]	0.083	0.710	0.225
H[NH₃(1)]	0.119	0.275	0.665
H[NH₃(2)]	0.017	0.330	0.755
H[NH₃(3)]	0.080	0.205	0.775

All atoms are in the general positions of C_{2v}^5 ($Pca2_1$):

(4a) $xyz; \bar{x},\bar{y},z+1/2; 1/2-x,y,z+1/2; x+1/2,\bar{y},z$

The determined parameters are those of Table XIVA,44. As stated here, the positions of the hydrogen atoms belonging to the methyl ammonium ions but not to the water molecules were deduced from the projections.

This arrangement is a distortion of the cubic high-temperature structures expressing the fact that at low temperatures the $(CH_3NH_2)^+$ ion has its atoms in fixed positions. As in the forms with higher symmetry, there are six water molecules octahedrally distributed around each aluminum atom, with Al–H₂O ranging between 1.76 and 1.96 A. The six water molecules around the fixed $(CH_3NH_2)^+$ ion fall into two groups of three each, with average distances from the NH₂ end of the cations of 2.98 and 4.21 A., and from the CH₃ end of 3.24 and 4.19 A.

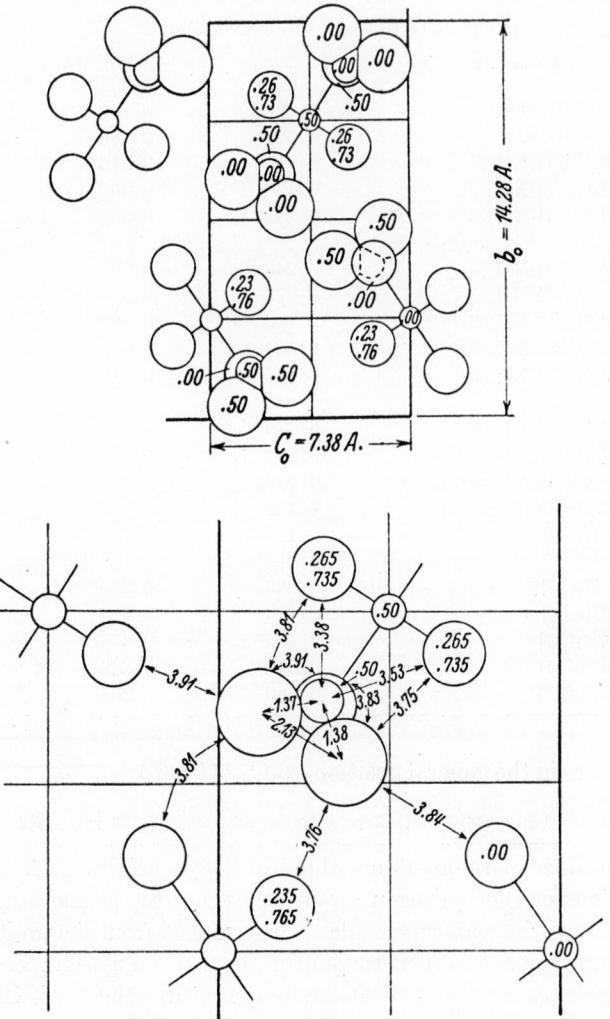

Fig. XIVA,142a (top). A projection of some of the atoms of the orthorhombic [NH$_2$(CH$_3$)$_2$]$_2$SnCl$_6$ structure along its a_0 axis. Methyl radicals, as the largest circles, are connected by smaller wedge-shaped NH$_2$ segments. Four of the six chlorine atoms of the octahedral SnCl$_6^{2-}$ anions are joined to the small tin atoms by light lines. Left-hand axes.

Fig. XIVA.142b (bottom). A few of the atoms in the projection of Fig. XIVA,142a showing the important interionic distances.

XIV,a106.   The orthorhombic *dimethyl ammonium chlorostannate*, [NH$_2$-(CH$_3$)$_2$]$_2$SnCl$_6$, has a structure that is a severe distortion of the K$_2$PtCl$_6$ arrangement (**IX,c10**). Its bimolecular unit has the dimensions:

$$a_0 = 7.26 \text{ A.}; \quad b_0 = 14.28 \text{ A.}; \quad c_0 = 7.38 \text{ A.}$$

Atoms have been placed in the following positions of C$_{2v}^7$ (*Pmn*):

> (2a)   0uv; $^1/_2,\bar{u},v+^1/_2$
> (4b)   xyz; $\bar{x}yz$; $^1/_2-x,\bar{y},z+^1/_2$; $x+^1/_2,\bar{y},z+^1/_2$

with the parameters listed in Table XIVA,45.

The structure that results (Figs. XIVA,142 and 143) is a very tight packing of the ions involved; this entails a serious distortion of the CaF$_2$ grouping, but the observed atomic separations are in accord with those found in similar crystals.

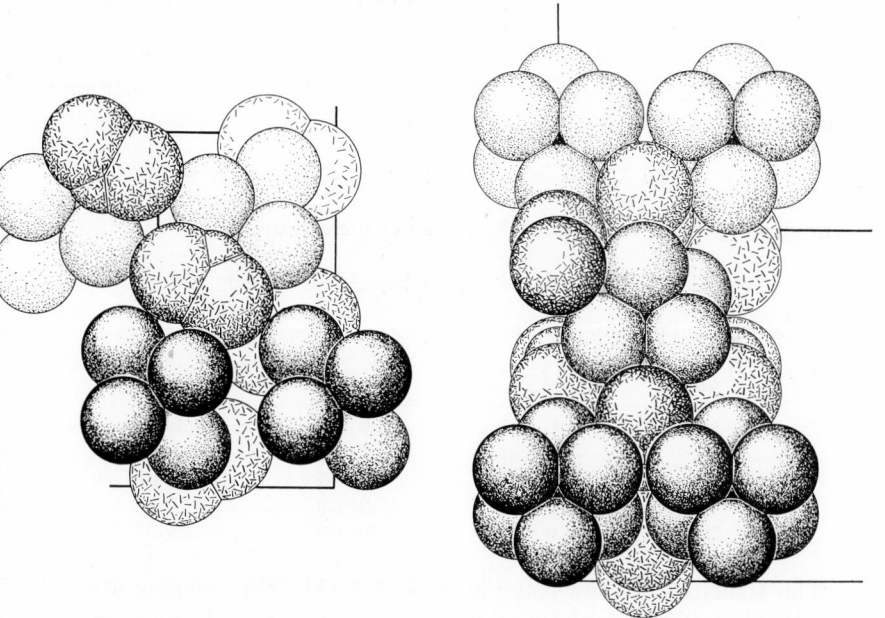

Fig. XIVA,143a (left). A packing drawing of the atoms in the orthorhombic [NH$_2$-(CH$_3$)$_2$]$_2$SnCl$_6$ structure that are shown in Fig. XIVA,142a. The NH$_2$(CH$_3$)$_2^+$ cations are line shaded; the tin atoms are too small to show between the large dotted chlorine atoms.

Fig. XIVA,143b (right). A packing drawing of the orthorhombic [NH$_2$(CH$_3$)$_2$]$_2$SnCl$_6$ structure seen along its c_0 axis. The shading is the same as in Fig. XIVA,143a, but in this case complete SnCl$_6^{2-}$ ions are shown. The tin atoms are visible as tiny black triangles.

TABLE XIVA,45
Positions and Parameters of the Atoms in $[NH_2(CH_3)_2]_2SnCl_6$

Atom	Position	x	y	z
Sn	(2a)	0	0.250	0
Cl(1)	(2a)	0	0.390	0.180
Cl(2)	(2a)	0	0.110	−0.180
Cl(3)	(4b)	0.235	0.185	0.190
Cl(4)	(4b)	0.235	0.315	−0.190
N(1)	(2a)	0	0.620	0.690
N(2)	(2a)	0	0.880	0.310
CH₃(1)	(2a)	0	0.605	0.875
CH₃(2)	(2a)	0	0.895	0.125
CH₃(3)	(2a)	0	0.530	0.625
CH₃(4)	(2a)	0	0.970	0.375

XIV,a107. Crystals of the compound *dimethyl ammonium bromide–bromine*, $[(CH_3)_2NH_2Br]_2 \cdot Br_2$, are monoclinic with a bimolecular unit of the dimensions:

$$a_0 = 20.00 \text{ A.}; \quad b_0 = 6.01 \text{ A.}; \quad c_0 = 5.38 \text{ A.}; \quad \beta = 94°36'$$

The space group is C_2^3 ($C2$) with atoms in the positions:

$$(4c) \quad xyz; \; \bar{x}y\bar{z}; \; x+{}^1/_2, y+{}^1/_2, z; \; {}^1/_2-x, y+{}^1/_2, \bar{z}$$

The determined parameters are as follows:

Atom	x	y	z
Br(1)	0.050	0.000	0.143
Br(2)	0.180	−0.002	0.473
N	0.157	0.467	0.714
C(1)	0.083	0.449	0.630
C(2)	0.183	0.459	0.983

The structure is shown in Figure XIVA,144. The bromine atoms are practically collinear, with Br(1)–Br(1) = 2.42 A. and Br(1)–Br(2) = 3.03 A. Within a methyl ammonium group, C–N = 1.51 A.; its nearest bromine is at the distance Br–N = 3.15 A. Other bromine neighbors are 3.49 A. away.

The *dimethyl ammonium chloride–iodine*, $[(CH_3)_2NH_2Cl]_2 \cdot I_2$, is isostructural, with the unit:

$$a_0 = 20.43 \text{ A.}; \quad b_0 = 5.88 \text{ A.}; \quad c_0 = 5.45 \text{ A.}; \quad \beta = 94°18'$$

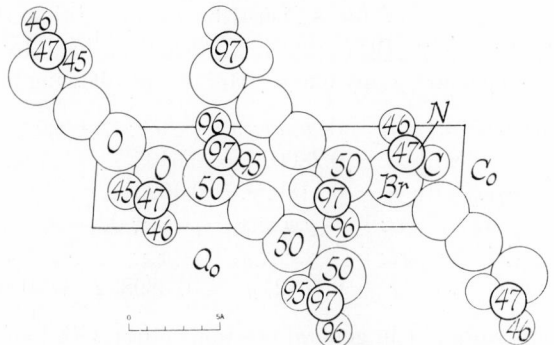

Fig. XIVA,144. The monoclinic structure of $[(CH_3)_2NH_2Br]_2 \cdot Br_2$ projected along its b_0 axis. Right-hand axes.

The atomic parameters are:

Atom	x	y	z
I	0.055	0.000	0.162
Cl	0.186	0.001	0.493
N	0.166	0.480	0.688
C(1)	0.099	0.551	0.672
C(2)	0.188	0.456	0.952

Significant interatomic distances are: I–I = 2.74 A., I–Cl = 3.11 A., Cl–N = 3.05 or 3.28 A., N–C = 1.47 A.

XIV,a108. *Tetramethyl ammonium perchlorate*, $N(CH_3)_4ClO_4$, has the general type of structure described for PH_4I (**III,b3**). According to a recent detailed study, however, there is considerable disorder in the orientation of its ClO_4 anions. This disordered arrangement has been described in terms of a cell which has its origin shifted (with respect to that used in **III,b3**) to contain a center of symmetry. The bimolecular tetragonal unit has the dimensions:

$$a_0 = 8.343 \text{ A.}, \qquad c_0 = 5.982 \text{ A.}$$

Atoms are in the positions of D_{4h}^7 (P/nmm):

N: (2a) $^3/_4\ ^1/_4\ 0;\ ^1/_4\ ^3/_4\ 0$

Cl: (2c) $\pm(^1/_4\ ^1/_4\ u)$ with $u = 0.4180$

C: (8i) $\pm(^1/_4\ u\ v;\ ^1/_4, ^1/_2 - u, v;\ u\ ^1/_4\ v;\ ^1/_2 - u, ^1/_4, v)$
with $u = -0.1062, v = -0.1419$

The experimental data were not satisfied by assuming that the ClO_4 ions had spherical symmetry. Best agreement was obtained by considering that the oxygens instead occupy one-quarter of the following positions:

O(1):   (8j) $\pm (uuv; \frac{1}{2}-u,u,v; u,\frac{1}{2}-u,v; \frac{1}{2}-u,\frac{1}{2}-u,v)$
          with $u = 0.2800, v = 0.6405$
O(2):   (8j)   with $u = 0.1180, v = 0.4023$
O(3): (16k) $\pm (xyz; \frac{1}{2}-x,y,z; x, \frac{1}{2}-y,z; \frac{1}{2}-x, \frac{1}{2}-y,z;$
          $yxz; \frac{1}{2}-y,x,z; y, \frac{1}{2}-x,z; \frac{1}{2}-y, \frac{1}{2}-x,z)$
          with $x = 0.1682, y = 0.3898, z = 0.2954$

Hydrogen atoms were put in general positions (16k) with half atoms given the parameters:

$$\frac{1}{2} H(1): x = 0.650, y = 0.075, z = 0.172$$
$$\frac{1}{2} H(2): x = 0.700, y = 0.125, z = 0.274$$
$$\frac{1}{2} H(3): x = 0.700, y = 0.025, z = 0.070$$

In such a structure, C–N = 1.47 A., Cl–O(1) = 1.39 A., Cl–O(2) = 1.56 A., and Cl–O(3) = 1.54 A. The $N(CH_3)_4$ tetrahedron is almost exactly regular. The shortest interionic distance is C–O = 3.04 A.

This structure resembles that of trimethyl oxosulfonium perchlorate, described in **XIV,a37**.

Tetraethyl ammonium iodide, $N(C_2H_5)_4I$, has a bimolecular tetragonal unit of similar shape with:

$$a_0 = 8.87 \pm 0.02 \text{ A.,} \qquad c_0 = 6.95 \text{ A.}$$

The iodine atoms were early found to be in a body-centered distribution and recently a reexamination has led to the following partial structure. Atoms have been placed in the positions of S_4^2 ($I\bar{4}$):

I: (2a) $000; \frac{1}{2} \frac{1}{2} \frac{1}{2}$
N: (2c) $0 \frac{1}{2} \frac{1}{4}; \frac{1}{2} 0 \frac{3}{4}$
C: (8g) $xyz; \bar{x}\bar{y}z; y\bar{x}\bar{z}; \bar{y}x\bar{z};$ B.C.
          with $x = 0.367, y = 0.048$
N: (8g)   with $x = 0.442, y = 0.190$

The z parameters were not obtained for the carbon and nitrogen.

XIV,a109.   The basic structure of *tetramethyl ammonium dichloroiodide*, $N(CH_3)_4ICl_2$, has already been described in **VI,13**. A recent study has, however, resulted in more accurate parameters and a somewhat different

I–Cl separation. Newly measured dimensions of the bimolecular tetragonal unit are:

$$a_0 = 9.35 \text{ A.}, \qquad c_0 = 5.94 \text{ A.}$$

Atoms are in the following positions of $V_d{}^3$ ($P\bar{4}2_1m$):

N: (2b) $0\ 0\ ^1/_2;\ ^1/_2\ ^1/_2\ ^1/_2$

I: (2c) $0\ ^1/_2\ u;\ ^1/_2\ 0\ \bar{u}$ with $u = 0.1039$

Cl: (4e) $u,u+^1/_2,v;\ \bar{u},^1/_2-u,v;\ u+^1/_2,\bar{u},\bar{v};\ ^1/_2-u,u,\bar{v}$
with $u = 0.1931,\ v = 0.1055$

C: (8f) $xyz;\ \bar{x}\bar{y}z;\ ^1/_2-x,y+^1/_2,\bar{z};\ x+^1/_2,^1/_2-y,\bar{z};$
$\bar{y}x\bar{z};\ y\bar{x}\bar{z};\ y+^1/_2,x+^1/_2,z;\ ^1/_2-y,^1/_2-x,z$
with $x = 0.108,\ y = 0.063,\ z = 0.350$

These values lead to an N–C $= 1.47$ A.; the ICl_2 anion is linear, with I–Cl $= 2.55$ A., a separation that is considerably greater than that previously found.

This atomic arrangement is shown in Figure VI,13. Bearing in mind that the origin is displaced by $^1/_2c_0$ in the two descriptions, it will be seen that this arrangement and that of $N(CH_3)_4I$ (**III,b3**) have much in common.

XIV,a110. Crystals of *tetramethyl ammonium pentaiodide*, $N(CH_3)_4I_5$, are monoclinic with a tetramolecular unit of the dimensions:

$$a_0 = 13.34 \text{ A.}; \quad b_0 = 13.59 \text{ A.}; \quad c_0 = 8.90 \text{ A.}; \quad \beta = 107°50'$$

The space group is $C_{2h}{}^6$ ($C2/c$) with atoms in its following special and general positions:

(4e) $\pm(0\ u\ ^1/_4;\ ^1/_2,u+^1/_2,^1/_4)$

(8f) $\pm(xyz;\ x,\bar{y},z+^1/_2;\ x+^1/_2,y+^1/_2,z;\ x+^1/_2,^1/_2-y,z+^1/_2)$

TABLE XIVA,46
Positions and Parameters of the Atoms in Tetramethyl Ammonium Pentaiodide

Atom	Position	x	y	z
I(1)	(4e)	0	0.1775	$^1/_4$
I(2)	(8f)	0.1612	0.0202	0.1976
I(3)	(8f)	0.1932	0.3733	0.3177
N	(4e)	0	0.645	$^1/_4$
C(1)	(8f)	−0.044	0.581	0.352
C(2)	(8f)	0.086	0.709	0.352

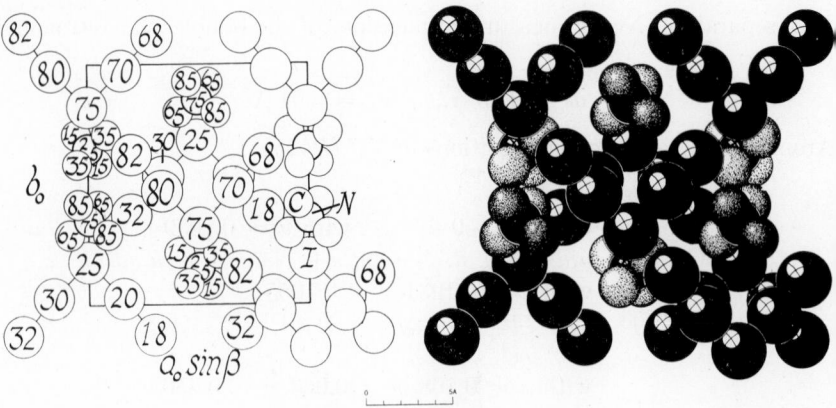

Fig. XIVA,145a (left). The monoclinic structure of $N(CH_3)_4I_5$ projected along its c_0 axis. Right-hand axes.

Fig. XIVA,145b (right). A packing drawing of the monoclinic structure of $N(CH_3)_4I_9$ seen along its c_0 axis. The iodine are the large, the nitrogen the smaller black circles The methyl groups are dotted.

The parameters for the iodine atoms obtained during a recent refinement of this structure (Fig. XIVA,145) are included in Table XIVA,46. The carbon and nitrogen parameters have not been established from the x-ray data but were chosen, in the initial study, to fill the voids that exist between the iodine anions. In doing this, it was assumed that the $N(CH_3)_4$ cations would be tetrahedral in shape and that $C-N = 1.5$ A.

The pentaiodide anions are planar and V-shaped, with $I-I(apex) = 3.17$ A. and the other two $I-I = 2.81$ A. The shortest distance between anions is $I-I = 3.63A$.

TABLE XIVA,47
Parameters of the I Atoms in $N(CH_3)_4I_9$

Atom	x	y	z
I(1)	0.046	0.880	0.586
I(2)	0.191	0.729	0.710
I(3)	0.181	0.084	0.690
I(4)	0.150	0.455	0.665
I(5)	0.041	0.318	0.534
I(6)	0.211	0.058	0.041
I(7)	0.076	0.084	0.190
I(8)	0.249	0.400	0.061
I(9)	0.086	0.416	0.189

XIV,a111. *Tetramethyl ammonium enneaiodide*, $N(CH_3)_4I_9$, is monoclinic with a tetramolecular cell of the dimensions:

$$a_0 = 11.60 \text{ A.}; \quad b_0 = 15.10 \text{ A.}; \quad c_0 = 13.18 \text{ A.}; \quad \beta = 95°25'$$

The space group is $C_{2h}^5(P2_1/n)$ with atoms in the positions:

$$(4e) \quad \pm(xyz; \; x+^1/_2, ^1/_2-y, z+^1/_2)$$

Parameters have been determined for the iodine atoms but not for the cations; they are listed in Table XIVA,47.

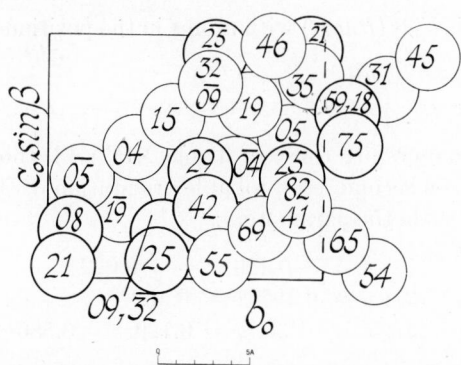

Fig. XIVA,146. The positions of some of the iodine atoms in the monoclinic structure of $N(CH_3)_4I_9$ projected along its a_0 axis.   Right-hand axes.

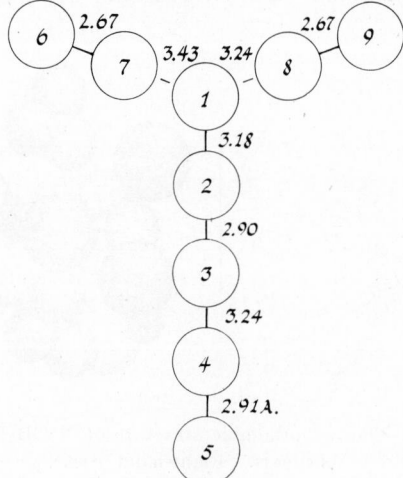

Fig. XIVA,147. The atomic separations that prevail between the iodine atoms in $N(CH_3)_4I_9$.

The distribution of some of the iodine atoms in the unit cell is shown in Figure XIVA,146. It is described as being equivalent to a V-shaped I_5^- ion and two I_2 molecules with the interatomic distances of Figure XIVA,147. The $N(CH_3)_4^+$ ions are considered to lie in the large holes between the sheets of I_5^- anions.

XIV,a112. *Tetramethyl ammonium silver iodide*, $N(CH_3)_4Ag_2I_3$, is orthorhombic with a tetramolecular cell of the edge lengths:

$$a_0 = 17.77 \text{ A.}; \quad b_0 = 10.07 \text{ A.}; \quad c_0 = 7.43 \text{ A.}$$

The space group is V_h^{16} (*Pnam*) with atoms in the positions:

(4c) $\pm (u\ v\ ^1/_4; u+^1/_2, ^1/_2-v, ^1/_4)$

(8d) $\pm (xyz; ^1/_2-x, y+^1/_2, z+^1/_2; x, y, ^1/_2-z; x+^1/_2, ^1/_2-y, z)$

The chosen parameters are those of Table XIVA,48, though it should be noted that there is a second less probable orientation of the methyl groups which would give them the parameters:

C(1): $u = -0.429, v = -0.022$
C(2): $u = 0.197, v = 0.432$
C(3): $x = 0.330, y = 0.129, z = 0.586$

In this structure (Fig. XIVA,148) there is the same sort of Ag_2I_3 chain to be found in $CsAg_2I_3$ (**VII,b29**). Within this chain are four tetrahedrally distributed iodine atoms about each silver atom, with Ag–I = 2.77–3.00 A.

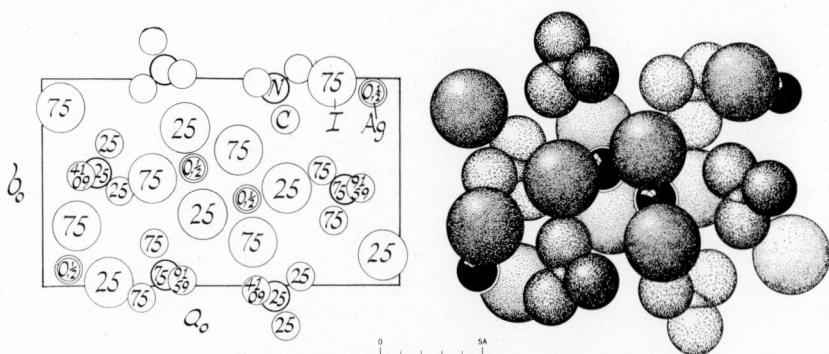

Fig. XIVA,148a (left). The orthorhombic structure of $N(CH_3)_4Ag_2I_3$ projected along its c_0 axis. Right-hand axes.

Fig. XIVA,148b (right). A packing drawing of the orthorhombic $N(CH_3)_4Ag_2I_3$ arrangement seen along its c_0 axis. The silver atoms are black; the iodines are very large and dotted. Atoms of nitrogen are the triangular segments between the line-shaded methyl radicals.

TABLE XIVA,48.   Positions and Parameters of Atoms in $N(CH_3)_4Ag_2I_3$

Atom	Position	x	y	z
Ag	(8d)	0.074	0.075	0.000
I(1)	(4c)	0.185	0.018	1/4
I(2)	(4c)	0.448	0.354	1/4
I(3)	(4c)	−0.096	−0.282	1/4
N	(4c)	−0.348	−0.054	1/4
C(1)	(4c)	−0.319	−0.190	1/4
C(2)	(4c)	0.216	0.461	1/4
C(3)	(8d)	0.393	0.033	0.586

The tetrahedra share three edges and this places silver atoms almost as near to one another, with Ag–Ag = 3.03 A. Every $N(CH_3)_4$ ion has around it two iodine atoms from each of three chains (at ca. 4.7 A.) and six more somewhat more distant (at 5.04 A. and upwards).

XIV,a113. Crystals of *tetramethyl ammonium mercury tribromide*, $N(CH_3)_4HgBr_3$, are monoclinic with a tetramolecular cell of the dimensions:

$$a_0 = 9.05 \text{ A.}; \quad b_0 = 15.90 \text{ A.}; \quad c_0 = 7.94 \text{ A.}; \quad \beta = 93°36'$$

TABLE XIVA,49.   Parameters of the Atoms in $N(CH_3)_4HgBr_3$

Atom	x	y	z
Hg(1)	0.165	0.000	0.025
Hg(2)	0.278	0.046	0.528
Br(1)	0.421	−0.045	−0.023
Br(2)	−0.052	−0.101	−0.007
Br(3)	0.148	0.127	0.223
Br(4)	0.250	−0.106	0.430
Br(5)	0.514	0.126	0.550
Br(6)	0.089	0.100	0.720
N(1)	0.67	−0.13	0.42
N(2)	0.64	0.19	0.07
C(1)	0.57	−0.18	0.35
C(2)	0.59	−0.08	0.51
C(3)	0.74	−0.08	0.30
C(4)	0.77	−0.17	0.50
C(5)	0.56	0.25	−0.01
C(6)	0.56	0.13	0.14
C(7)	0.73	0.23	0.19
C(8)	0.73	0.15	−0.05

156   CRYSTAL STRUCTURES

The space group has been selected as C_2^2 ($P2_1$) with all atoms in the positions:

$$(2a) \quad xyz; \; \bar{x},y+\tfrac{1}{2},\bar{z}$$

The parameters are those of Table XIVA,49.

In the two kinds of HgBr₃ group that exist (Fig. XIVA,149), Hg–Br lies between 2.48 and 2.56 A. A bromine atom of an adjacent group is, however, not very much more distant from mercury (2.92 or 2.94 A.) and

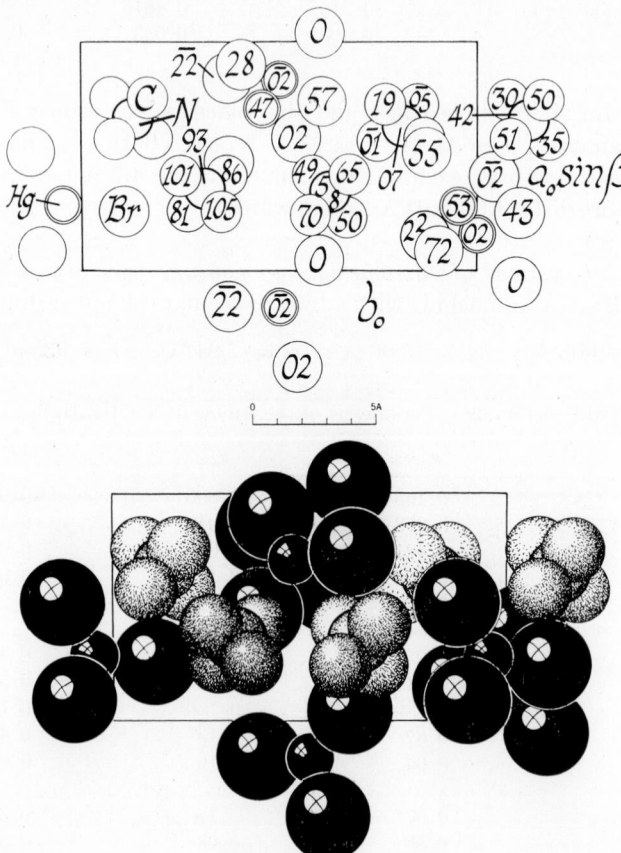

Fig. XIVA,149a (top). The monoclinic structure of N(CH₃)₄HgBr₃ projected along its c_0 axis. Right-hand axes.

Fig. XIVA,149b (bottom). A packing drawing of the monoclinic N(CH₃)₄HgBr₃ arrangement viewed along its c_0 axis. Mercury atoms are the small, bromine atoms the large black circles. The more heavily outlined, and dotted, atoms of nitrogen partially show between the line-shaded methyl groups.

hence the $HgBr_3$ groups can be imagined as lying in strings along the c_0 axis. The N–C separation in the $N(CH_3)_4$ ion could not be accurately determined in the presence of so many heavy atoms, but it is ca. 1.35 A. Between ions the shortest atomic separation is a C–Br = 3.14 A.

The corresponding *chloride*, $N(CH_3)_4HgCl_3$, is isostructural with

$$a_0 = 8.68 \text{ A.}; \quad b_0 = 15.75 \text{ A.}; \quad c_0 = 7.69 \text{ A.}; \quad \beta = 93°0'$$

XIV,a114. *Tetramethyl ammonium tetrachlorozincate*, $[N(CH_3)_4]_2ZnCl_4$ has the K_2SO_4 structure (**VIII,b11**). Its tetramolecular orthorhombic unit has the dimensions:

$$a_0 = 12.268 \pm 0.007 \text{ A.}; \quad b_0 = 8.964 \pm 0.007 \text{ A.}; \quad c_0 = 15.515 \pm 0.012 \text{ A.}$$

Atoms are in (4c) and (8d) of V_h^{16} (*Pnma*):

(4c) $\pm (u \ ^1/_4 \ v; \ u+^1/_2, ^1/_4, ^1/_2-v)$
(8d) $\pm (xyz; \ x+^1/_2, ^1/_2-y, ^1/_2-z; \ x, ^1/_2-y, z; \ x+^1/_2, y, ^1/_2-z)$

They have the parameters of Table XIVA,50.

TABLE XIVA,50
Positions and Parameters of the Atoms in $[N(CH_3)_4]_2ZnCl_4$

Atom	Position	x	y	z
Zn	(4c)	0.2460	$^1/_4$	0.4075
Cl(1)	(4c)	0.0630	$^1/_4$	0.4070
Cl(2)	(8d)	0.3060	0.0445	0.3400
Cl(3)	(4c)	0.3130	$^1/_4$	0.5418
N(1)	(4c)	0.1530	$^1/_4$	0.0975
N(2)	(4c)	0.4945	$^1/_4$	0.8240
C(1)	(4c)	0.2770	$^1/_4$	0.0970
C(2)	(4c)	0.1085	$^1/_4$	0.0020
C(3)	(8d)	0.1115	0.3910	0.1450
C(4)	(4c)	0.4405	$^1/_4$	0.7345
C(5)	(4c)	0.4030	$^1/_4$	0.8935
C(6)	(8d)	0.5660	0.3910	0.8340

The chlorine atoms are tetrahedrally distributed about the zinc atoms, with Zn–Cl = 2.240−2.245 A. In the cations, N–C = 1.521–1.579 A.

The corresponding *cobaltate*, $[N(CH_3)_4]_2CoCl_4$, is isostructural with

$$a_0 = 12.24 \pm 0.03 \text{ A.}; \quad b_0 = 8.92 \pm 0.02 \text{ A.}; \quad c_0 = 15.39 \pm 0.03 \text{ A.}$$

XIV,a115. An approximate structure has been described for *tetramethyl ammonium cupric tetrachloride*, $[N(CH_3)_4]_2CuCl_4$. Its crystals are orthorhombic with a very large unit having the cell edges:

$$a_0 = 36.381 \text{ A.}; \quad b_0 = 9.039 \text{ A.}; \quad c_0 = 15.155 \text{ A.}$$

All but a small number of faint reflections are, however, compatible with a smaller tetramolecular unit having the edges:

$$a_0 = 12.127 \text{ A.}; \quad b_0 = 9.039 \text{ A.}; \quad c_0 = 15.155 \text{ A.}$$

Choosing the space group as V_h^{16} (*Pnma*), atoms have been put in the positions:

(4c) $\pm (u \; ^1/_4 \; v; \; u+^1/_2, ^1/_4, ^1/_2-v)$

(8d) $\pm (xyz; \; ^1/_2-x,y+^1/_2,z+^1/_2; \; x,^1/_2-y,z; \; x+^1/_2,y,^1/_2-z)$

with the parameters listed in Table XIVA,51.

TABLE XIVA,51
Positions and Parameters of the Atoms in $[N(CH_3)_4]_2CuCl_4$

Atom	Position	x	y	z
Cu	(4c)	0.2281	$^1/_4$	0.4028
Cl(1)	(4c)	0.0495	$^1/_4$	0.3700
Cl(2)	(4c)	0.3100	$^1/_4$	0.5320
Cl(3)	(8d)	0.2750	0.029	0.3490
N(1)	(4c)	0.1280	$^1/_4$	0.0970
N(2)	(4c)	0.5050	$^1/_4$	0.8330
C(1)	(4c)	0.2590	$^1/_4$	0.1130
C(2)	(4c)	0.1270	$^1/_4$	-0.0010
C(3)	(8d)	0.0770	0.121	0.1320
C(4)	(4c)	0.4210	$^1/_4$	0.7580
C(5)	(4c)	0.4500	$^1/_4$	0.9150
C(6)	(8d)	0.5710	0.121	0.8280

The available data do not indicate how the precise structure departs from this approximation to it, but it is thought to involve primarily the y coordinates (cf. **XIV,a114**).

XIV,a116. Several tetramethyl ammonium hexachlorides of tetravalent metals, as typified by *tetramethyl ammonium chlorostannate*, $[N(CH_3)_4]_2$-$SnCl_6$, have structures which are essentially the same as that described for

K_2PtCl_6 (**IX,c10**). There are four molecules in the unit cube which, for $[N(CH_3)_4]_2SnCl_6$, has the edge length:

$$a_0 = 12.87 \text{ A.}$$

Atoms are in the following positions of O_h^5 (*Fm3m*):

Sn: (4a) 000; F.C.
Cl: (24e) $\pm(u00; 0u0; 00u)$; F.C. with $u = 0.19$
N: (8c) $\pm(\frac{1}{4}\,\frac{1}{4}\,\frac{1}{4})$; F.C.
C: (32f) $\pm(uuu; u\bar{u}\bar{u}; \bar{u}u\bar{u}; \bar{u}\bar{u}u)$; F.C.
with $u =$ ca. 0.31.

A packing drawing of this undistorted CaF_2-type grouping is shown in Figure XIVA,150.

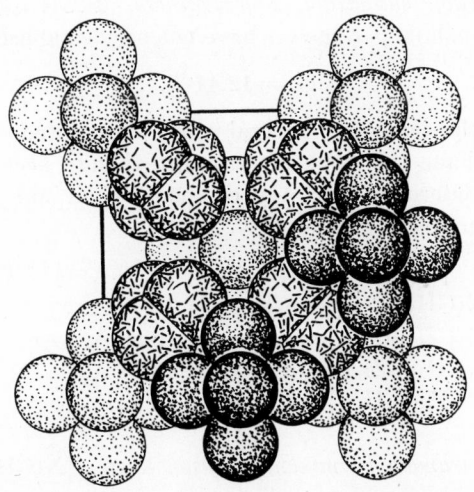

Fig. XIVA,150. A packing drawing of some of the atoms in the unit of $[N(CH_3)_4]_2SnCl_6$ projected along a cube axis. The atoms of chlorine are dotted; the methyl radicals are line shaded. Neither the tin nor the nitrogen atoms show.

Of the other compounds which have been shown to have this arrangement, a particularly accurate study has been made of *tetramethyl ammonium cerium hexachloride*, $[N(CH_3)_4]_2CeCl_6$. Its unit cube has the edge:

$$a_0 = 13.05 \text{ A.}$$

The parameter $u(Cl)$ is 0.1951, and $u(C) = 0.3119$. They lead to the values $Ce–Cl = 2.546$ A. and $N–C = 1.400$ A.

Other compounds having this structure, and their cell edges, are

$$[N(CH_3)_4]_2PtCl_6: a_0 = 12.65 \text{ A.}$$
$$[N(CH_3)_4]_2PuCl_6: a_0 = 12.96 \text{ A.}$$
$$[N(CH_3)_4]_2UCl_6: a_0 = 13.06 \text{ A.}$$

Substantially the same structure has been found for *trimethyl ammonium chlorostannate*, $[NH(CH_3)_3]_2SnCl_6$. For it

$$a_0 = 12.19 \text{ A.}$$

Less than holohedral symmetry prevails for this crystal, however, and the space group is T_h^6 ($Pa3$). For the chlorine atoms in its general positions it has been determined that $x = 0.05$, $y = 0.19$, $z = -0.04$. This corresponds to a slight rotation of the $SnCl_6$ ions around the threefold axes.

Possibly *trimethyl sulfonium chlorostannate*, $[S(CH_3)_3]_2SnCl_6$, has this arrangement though its parameters have not been established. For it

$$a_0 = 12.41 \text{ A.}$$

In addition to the foregoing, several ethyl substituted compounds have been shown to have this type of structure though accurate parameters have not been determined for any of them. Their unit cubes have the following edge lengths:

$$[N(C_2H_5)_4]_2PuCl_6: a_0 = 14.19 \text{ A. } (112°C.)$$
$$[N(CH_3)_3C_2H_5]_2SnCl_6: a_0 = 13.17 \text{ A.}$$
$$[S(CH_3)_2C_2H_5]_2SnCl_6: a_0 = 12.80 \text{ A.}$$
$$[NCH_3(C_2H_5)_3]_2SnCl_6: a_0 = 13.51 \text{ A.}$$
$$[PCH_3(C_2H_5)_3]_2SnCl_6: a_0 = 13.93 \text{ A.}$$

XIV,a117. *Tetramethyl ammonium fluosilicate*, $[N(CH_3)_4]_2SiF_6$, has a structure which is a tetragonal distortion of the arrangements just discussed. Its bimolecular unit has the dimensions:

$$a_0 = 7.88 \text{ A.}, \quad c_0 = 11.19 \text{ A.}$$

The tetramolecular pseudo-cube that is diagonal to this has the cell edges:

$$a_0' = \sqrt{2}\, a_0 = 11.14 \text{ A.}, \quad c_0' = c_0 = 11.19 \text{ A.}$$

Atoms in the bimolecular cell are in the following positions of C_{4h}^5 ($I4/m$):

Si: (1a) 000; B.C.
F(1): (4e) ±(00u); B.C.   with u = 0.155
F(2): (8h) ±(uv0; v\={u}0); B.C.   with u = 0.18, v = 0.12

N: (4*d*) 0 $^1/_2$ $^1/_4$; $^1/_2$ 0 $^1/_4$; B.C.
CH$_3$: (16*i*)   ± (*xyz*; *x̄ȳz*; *ȳxz*; *yx̄z*); B.C.
with *x* = 0.14, *y* = 0.47, *z* = 0.175.

As can be seen from Figure XIVA,151, this is a relatively undistorted CaF$_2$ grouping of SiF$_6$ octahedra and large N(CH$_3$)$_4$ tetrahedra in which the two groups have alternate orientations throughout the structure to provide better packing. The assigned parameters give the following significant atomic separations: Si–F = 1.70–1.73 A., N–C = 1.41 A., CH$_3$–F = 3.39–3.43 A.

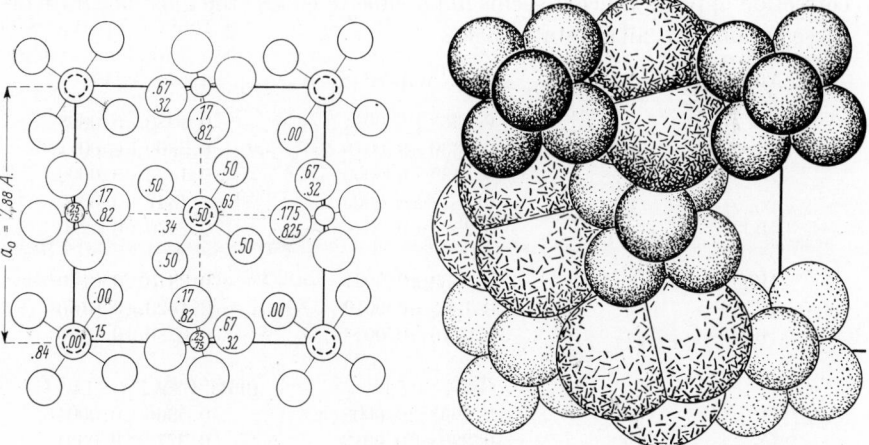

Fig. XIVA,151a (left). The tetragonal [N(CH$_3$)$_4$]$_2$SiF$_6$ structure projected along its c_0 axis. The largest circles are the methyl radicals, the smallest the nitrogen atoms. The fluorine circles are slightly smaller than those of the methyl groups. The atoms of silicon are dashed. Left-hand axes.

Fig. XIVA,151b (right). A packing drawing of the tetragonal structure of [N-(CH$_3$)$_4$]$_2$SiF$_6$ seen along its c_0 axis. The fluorine atoms are dotted; the methyl groups are line shaded. Neither the nitrogen nor the silicon atoms show.

8. Compounds Related to Urea

XIV,a118.   *Urea*, (NH$_2$)$_2$CO, has been repeatedly examined. It has the same structure at room temperature and at −140°C. The symmetry is tetragonal with a bimolecular unit that has the edge lengths:

$$a_0 = 5.662 \pm 0.002 \text{ A.,} \qquad c_0 = 4.716 \pm 0.002 \text{ A.}$$

at room temperature and

$$a_0 = 5.582 \pm 0.002 \text{ A.,} \qquad c_0 = 4.686 \pm 0.002 \text{ A.}$$

at $-140°C$. The space group is $V_d{}^3$ ($P\bar{4}2_1m$) with carbon and oxygen atoms in the special positions:

$$(2c) \quad 0\ {}^1/_2\ u;\ {}^1/_2\ 0\ \bar{u}$$

The atoms of nitrogen and hydrogen are in the positions:

$$(4e) \quad u,{}^1/_2-u,v;\ \bar{u},u+{}^1/_2,v;\ u+{}^1/_2,u,\bar{v};\ {}^1/_2-u,\bar{u},\bar{v}$$

Several accurate determinations have in recent years been made using x-rays and neutrons. The agreement among these results is in general very good, but there are minor differences which depend in part on the type of correction applied. Since it seems impossible to choose the most accurate of these parameters, all are stated below:

Atom	Parameter	1957: W,L&P	1961: S,S&P
O	u	0.5968 ± 0.0011	0.5980 ± 0.0008
C	u	0.3330 ± 0.0010	0.3300 ± 0.0011
N	u	0.1439 ± 0.0009	0.1433 ± 0.0009
N	v	0.1832 ± 0.0005	0.1847 ± 0.0007
H(1)	u	0.2522 ± 0.0026	0.2430 ± 0.0230
H(1)	v	0.2839 ± 0.0021	0.2810 ± 0.0110
H(2)	u	0.1365 ± 0.0019	0.1420 ± 0.0160
H(2)	v	-0.0276 ± 0.0014	0.0280 ± 0.0190
		1964: C&D	1961: S,S&P ($-140°C$.)
O	u	0.5998 ± 0.0008	0.5966 ± 0.0004
C	u	0.3308 ± 0.0012	0.3272 ± 0.0004
N	u	0.1419 ± 0.0007	0.1455 ± 0.0004
N	v	0.1857 ± 0.0010	0.1800 ± 0.0003
H(1)	u	0.2390 ± 0.0100	0.2690 ± 0.0260
H(1)	v	0.2770 ± 0.0120	0.2790 ± 0.0110
H(2)	u	0.1240 ± 0.0070	0.1420 ± 0.0200
H(2)	v	0.0460 ± 0.0150	-0.0280 ± 0.0200

The resulting structure is shown in Figure XIVA,152. Bond lengths found from the different determinations are within 0.01 A. of the following values: C–O = 1.26 A., C–N = 1.34 A.; the angles N–C–N = 118° and N–C–O = 120° to within 1°. Molecules are tied together through hydrogen bonds which have the lengths N–H–O = 2.99 and 3.04 A.

XIV,a119. A partial determination of structure was made long ago of *monomethyl urea*, $NH_2(NHCH_3)CO$. Its tetramolecular orthorhombic unit has the edges:

$$a_0 = 6.89\ A.;\quad b_0 = 6.96\ A.;\quad c_0 = 8.45\ A.$$

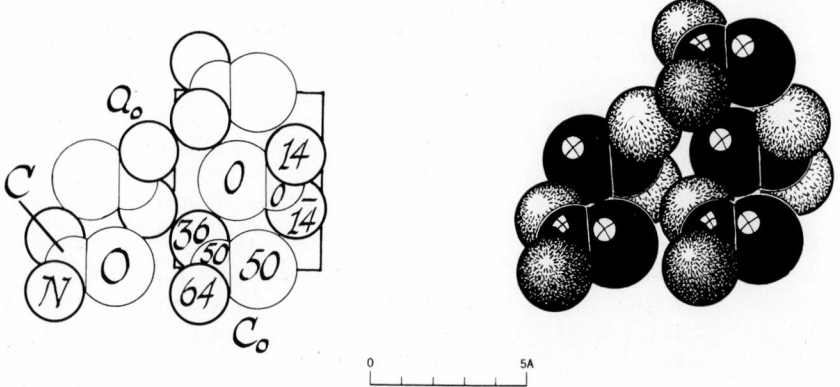

Fig. XIVA,152a (left). The tetragonal structure of urea, CO(NH₂)₂, projected along an a_0
axis. Right-hand axes.

Fig. XIVA,152b (right). A packing drawing of the tetragonal structure of urea seen
along an a_0 axis. The oxygen atoms are the larger, the carbon atoms the smaller black
circles. Nitrogen atoms are line shaded.

All atoms are in general positions of V^4 ($P2_12_12_1$):

$(4a) \quad xyz; x+\frac{1}{2},\bar{y},\frac{1}{2}-z; \bar{x},\frac{1}{2}-y,z+\frac{1}{2}; \frac{1}{2}-x,y+\frac{1}{2},\bar{z}$

The x and y parameters of all atoms are known with some accuracy, but the
z parameters were only estimated. They are

$$C: x = 0.23, y = 0.12, z = \text{ca. } 0.21$$
$$O: x = 0.35, y = 0.04, z = \text{ca. } 0.07$$
$$NH_2: x = 0.20, y = 0.14, z = \text{ca. } 0.37$$
$$NH: x = 0.14, y = 0.18, z = \text{ca. } 0.07$$
$$CH_3: x = 0.17, y = 0.35, z = \text{ca. } 0.02$$

XIV,a120. *Methyl urea nitrate*, $(CH_3NH_2)NH_2CO\cdot NO_3$, is orthorhombic
with a tetramolecular unit of the edge lengths:

$$a_0 = 11.22\pm0.04 \text{ A.}; \quad b_0 = 8.12\pm0.03 \text{ A.}; \quad c_0 = 6.36\pm0.02 \text{ A.}$$

The space group is V_h^{16} (*Pbnm*) with all atoms in the special positions:

$(4c) \quad \pm(u v \,{}^1/_4; {}^1/_2-u,v+{}^1/_2,{}^1/_4)$

According to a preliminary announcement, atoms have the parameters
listed in Table XIVA,52.

The structure is shown in Figure XIVA,153. The planar methyl urea
cation has the dimensions of Figure XIVA,154. The nitrate ions are equi-

TABLE XIVA,52
Parameters of Atoms in Methyl Urea Nitrate

Atom	u	v
O(1)	0.1235	0.5248
O(2)	−0.0200	0.6971
O(3)	0.1587	0.7814
O(4)	−0.0492	0.3129
N(1)	0.0874	0.6668
N(2)	0.1044	0.1350
N(3)	−0.0866	0.0473
C(1)	−0.0085	0.1658
C(2)	−0.2133	0.0749

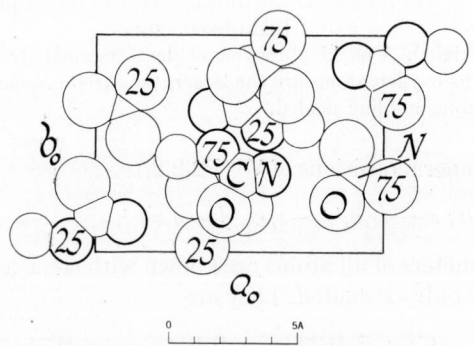

Fig. XIVA,153. The orthorhombic structure of $(CH_3NH_2)NH_2CO \cdot NO_3$ projected along its c_0 axis. Right-hand axes.

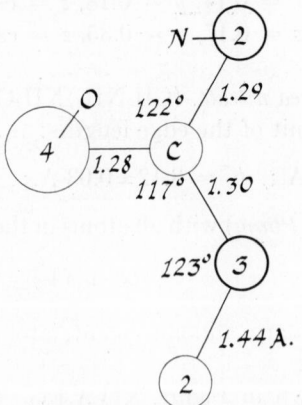

Fig. XIVA,154. Bond dimensions in the cation of $(CH_3NH_2)NH_2CO \cdot NO_3$.

lateral with $N(1)-O = 1.22$ A. and $N(2,3)-O = 1.23$ A.; $O-N-O$ 119–121°. There are considered to be four hydrogen bonds: $O(1)-H-O(4) = 2.59$ A., $N(2)-H-O(3) = 2.91$ and 2.94 A., and $N(3)-H-O(2) = 2.94$ A.

XIV,a121. *Urea phosphate*, $(NH_2)_2CO \cdot H_3PO_4$, is orthorhombic with an eight-molecule cell of the dimensions:

$$a_0 = 17.68 \text{ A.}; \quad b_0 = 7.48 \text{ A.}; \quad c_0 = 9.06 \text{ A.}$$

The space group is V_h^{15} (*Pbca*) with all atoms in the general positions:

$$(8c) \quad \pm (xyz; \ x+^1/_2,^1/_2-y,\bar{z}; \ \bar{x},y+^1/_2,^1/_2-z; \ ^1/_2-x,\bar{y},z+^1/_2)$$

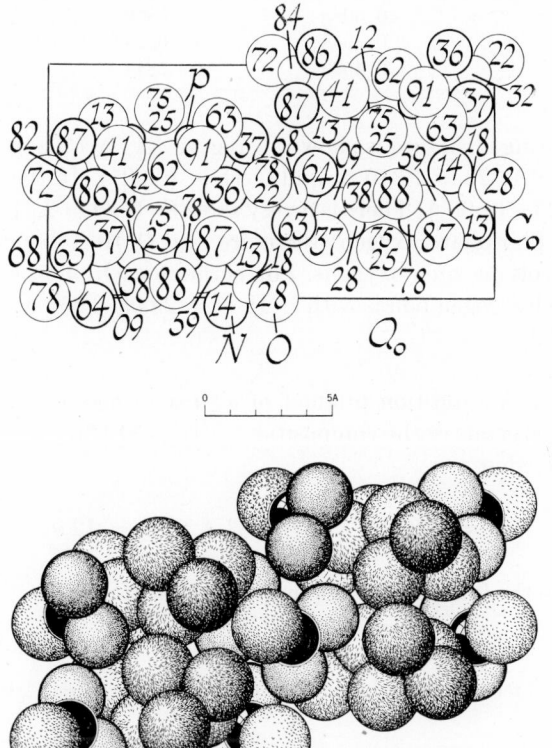

Fig. XIVA,155a (top). The orthorhombic structure of $(NH_2)_2CO \cdot H_3PO_4$ projected along its b_0 axis. Right-hand axes.

Fig. XIVA,155b (bottom). A packing drawing of the orthorhombic $(NH_4)_2CO \cdot H_3PO_4$ viewed along its b_0 axis. The carbon atoms are black, the nitrogen atoms heavily outlined and dotted. The urea oxygens are hooked; those belonging to the phosphate groups are line shaded. Atoms of phosphorus do not show.

The determined parameters are stated in Table XIVA,53.

TABLE XIVA,53
Parameters of the Atoms in $(NH_2)_2CO \cdot H_3PO_4$

Atom	x	y	z
P	0.3111	0.2755	0.3092
O(1)	0.3393	0.0905	0.3618
O(2)	0.2776	0.3847	0.4351
O(3)	0.2464	0.2498	0.1931
O(4)	0.3795	0.3662	0.2392
O(U)	0.4474	0.6327	0.3102
N(1)	0.5085	0.7800	0.4895
N(2)	0.3958	0.6378	0.5431
C	0.4503	0.6814	0.4466

The resulting structure is shown in Figure XIVA,155. In the urea molecule, C–N = 1.323 or 1.340 A. and C–O = 1.290 A. In the phosphate ion, P–O = 1.523 – 1.565 A. There is a very short O–H–O = 2.41 A. between the urea and a phosphate oxygen; there are also O–H–N bonds of 3.103 and 3.167 A. involving urea oxygens. Furthermore each phosphate oxygen has at least two hydrogen bonds with O–H–O = 2.603–2.659 A. and O–H–N = 2.976–3.167 A.

XIV,a122. An addition product of *urea* with *hydrogen peroxide* has orthorhombic crystals of the composition $(NH_2)_2CO \cdot H_2O_2$. Its tetramolecular unit has the edges:

$$a_0 = 6.86 \text{ A.}; \quad b_0 = 4.83 \text{ A.}; \quad c_0 = 12.92 \text{ A.}$$

Atoms are in the following positions of V_h^{14} (*Pnca*):

(4c) $\pm (^1/_4 \ 0 \ u; \ ^1/_4, ^1/_2, u + ^1/_2)$
(8d) $\pm (xyz; \ \bar{x}, y + ^1/_2, ^1/_2 - z; \ x + ^1/_2, ^1/_2 - y, ^1/_2 - z; \ ^1/_2 - x, \bar{y}, z)$

with the parameters:

Atom	Position	x	y	z
O	(4c)	$^1/_4$	0	0.019
C	(4c)	$^1/_4$	0	0.115
N	(8d)	0.131	0.169	0.168
O(H_2O_2)	(8d)	0.159	0.079	0.399

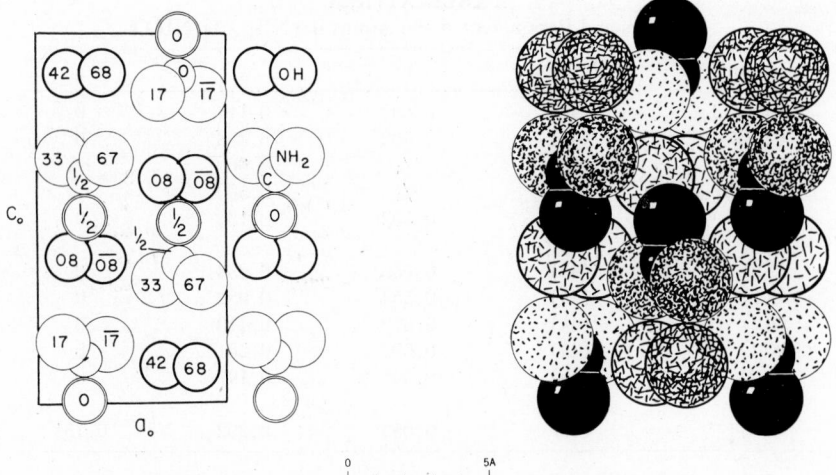

Fig. XIVA,156a (left). The orthorhombic structure of $(NH_4)_2CO \cdot H_2O_2$ viewed along its b_0 axis. Left-hand axes.

Fig. XIVA,156b (right). A packing drawing of the orthorhombic $(NH_4)_2CO \cdot H_2O_2$ arrangement seen along its b_0 axis. The carbon and urea oxygen atoms are black and the NH_2 radicals dotted. Peroxide hydroxyls are line shaded.

As can be seen from Figure XIVA,156, the structure that results has a grouping of urea molecules of the same shape and size as those in crystalline urea, interspersed with molecules of peroxide. In H_2O_2 the O–O distance is 1.46 A. Contact between the two kinds of molecule is through an NH_2–O separation of either 2.94 or 3.04 A. and a separation of 2.63 A. between urea and peroxide oxygens. These intermolecular bonds undoubtedly involve hydrogen.

XIV,a123. Crystals of *urea ammonium chloride*, $(NH_2)_2CO \cdot NH_4Cl$, are orthorhombic with a unit containing eight molecules and having the edges:

$$a_0 = 8.030 \text{ A.}; \quad b_0 = 17.080 \text{ A.}; \quad c_0 = 7.810 \text{ A.}$$

The space group is V_h^{17} ($Pcnm$) with all atoms in the positions:

(4g) $\pm (1/4\ u\ 1/4;\ 1/4\ u\ 3/4)$
(4h) $\pm (u\ v\ 0;\ 1/2-u,v,1/2)$
(8i) $\pm (xyz;\ xy\bar{z};\ x+1/2,\bar{y},1/2-z;\ 1/2-x,y,1/2-z)$

The parameters are stated in Table XIVA,54.

TABLE XIVA,54
Positions and Parameters of the Atoms in $(NH_2)_2CO \cdot NH_4Cl$

Atom	Position	x	y	z
Cl(1)	(4h)	0.032	0.111	0
Cl(2)	(4h)	0.532	0.388	0
NH$_4$(1)	(4g)	$^1/_4$	0.000	$^1/_4$
NH$_4$(2)	(4g)	$^1/_4$	0.500	$^1/_4$
C(1)	(4h)	0.552	0.156	0
C(2)	(4h)	0.056	0.333	0
O(1)	(4h)	0.554	0.082	0
O(2)	(4h)	0.072	0.410	0
N(1)	(4h)	0.692	0.201	0
N(2)	(4h)	0.395	0.197	0
N(3)	(8i)	0.053	0.292	0.151

In this structure (Fig. XIVA,157), the ammonium ions are tetrahedrally surrounded by two oxygens (NH$_4$–O = 2.87 A.), and two chloride ions (NH$_4$–Cl = 3.25 A.). In the urea molecules, O–C = 1.30 A. and C–N = 1.34 or 1.39 A. These amino nitrogens are 3.14 A. or more from the chloride ions.

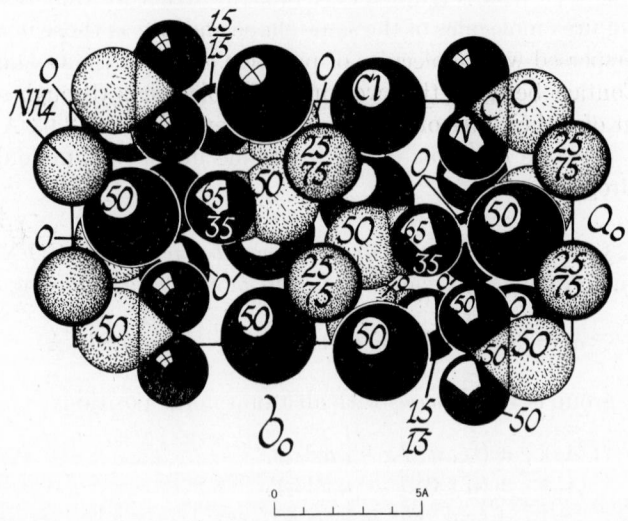

Fig. XIVA,157. The orthorhombic structure of $(NH_2)_2CO \cdot NH_4Cl$ projected along its c_0 axis. Right-hand axes.

XIV,a124. *Urea ammonium bromide*, $(NH_2)_2CO \cdot NH_4Br$, forms monoclinic crystals which have a bimolecular cell of the dimensions:

$$a_0 = 9.03 \text{ A.}; \quad b_0 = 4.79 \text{ A.}; \quad c_0 = 7.10 \text{ A.}; \quad \beta = 101°18'$$

The space group has been chosen as C_2^2 ($P2_1$) with all atoms in the positions:

$$(2a) \quad xyz; \ \bar{x}, y+\tfrac{1}{2}, \bar{z}$$

Determined parameters are the following:

Atom	x	y	z
Br	0.2754	0.2500	0.8201
NH$_4$	0.490	0.250	0.307
NH$_2$(1)	0.098	0.602	0.141
NH$_2$(2)	0.100	0.899	0.406
C	0.174	0.750	0.297
O	0.314	0.750	0.341

The resulting arrangement is shown in Figure XIVA,158. In the urea molecule, NH_2–C = 1.38 and 1.32 A., and C–O = 1.24 A. Ammonium ions are closest to the oxygen atoms with distances of 2.76 and 2.91 A. that would be compatible with hydrogen bonding. The NH_4–Br separation is 3.45 A. and this is substantially the same as the shortest NH_2–Br distance (3.46 A.)

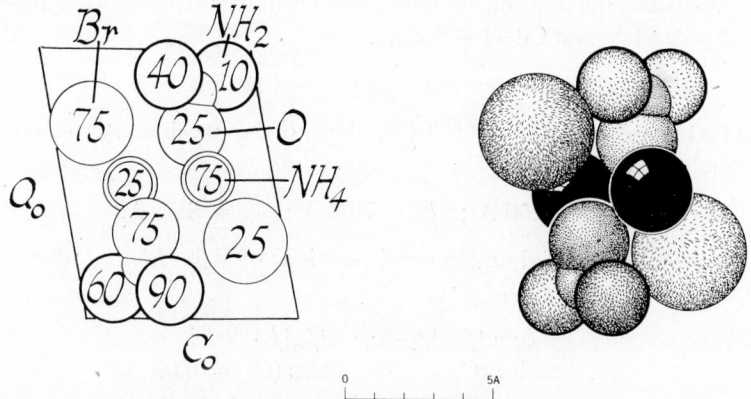

Fig. XIVA,158a (left). The monoclinic structure of $(NH_2)_2CO \cdot NH_4Br$ projected along its b_0 axis. Right-hand axes.

Fig. XIVA,158b (right). A packing drawing of the monoclinic $(NH_2)_2CO \cdot NH_4Br$ arrangement seen along its b_0 axis. The NH_4^+ ions are black; the Br$^-$ are large and fine-line shaded. In the urea molecules the oxygen atoms are dotted; the NH_2 radicals are heavily outlined and hook shaded.

XIV,a125.   *Cadmium chloride diurea*, $CdCl_2 \cdot 2(NH_2)_2CO$, is orthorhombic with a bimolecular unit of the dimensions:

$$a_0 = 9.13 \pm 0.02 \text{ A.}; \quad b_0 = 12.90 \pm 0.02 \text{ A.}; \quad c_0 = 3.72 \pm 0.01 \text{ A.}$$

The space group has been selected as V_h^{19} (*Cmmm*) with atoms in the positions:

Cd: (2a)   000; $^1/_2\, ^1/_2\, 0$

Cl: (8g) $\pm (u\ v\ ^1/_2;\ \bar{u}\ v\ ^1/_2;\ u+^1/_2, v+^1/_2, ^1/_2;\ ^1/_2-u, v+^1/_2, ^1/_2)$
   with $u = 0.205$, $v = 0.012$

It is thought that the atoms of the urea molecules are in some of the positions of

(16r) $\pm (xyz;\ \bar{x}\bar{y}z;\ x+^1/_2, y+^1/_2, z;\ ^1/_2-x, ^1/_2-y, z;$
       $x\bar{y}\bar{z};\ \bar{x}y\bar{z};\ x+^1/_2, ^1/_2-y, \bar{z};\ ^1/_2-x, y+^1/_2, \bar{z})$

with the following parameters:

Atom	x	y	z
O	0.988	0.176	0.993
C	0.071	0.248	0.005
N(1)	0.202	0.234	0.183
N(2)	0.031	0.334	0.779

In such a disordered structure each cadmium atom would be surrounded by a distorted octahedron of four chlorine and two oxygen atoms, with Cd–Cl = 2.64 A. and Cd–O = 2.28 A.

XIV,a126.   *Thiourea*, $(NH_2)_2CS$, has orthorhombic symmetry. Its tetramolecular unit with

$$a_0 = 5.50 \text{ A.}; \quad b_0 = 7.68 \text{ A.}; \quad c_0 = 8.57 \text{ A.}$$

has its atoms in special and general positions of V_h^{16} (*Pbnm*). The carbon and sulfur atoms are in

(4c) $\pm (u\ v\ ^1/_4;\ ^1/_2-u, v+^1/_2, ^1/_4)$
   with $u(C) = -0.1632$, $v(C) = 0.0916$
   and $u(S) = 0.1138$, $v(S) = -0.0073$

The nitrogen atoms are in

(8d) $\pm (xyz;\ x,y,^1/_2-z;\ x+^1/_2, ^1/_2-y, z+^1/_2;\ ^1/_2-x, y+^1/_2, z)$
   with $x = 0.2767$, $y = -0.1322$, $z = -0.1201$

The essentially planar molecules are packed as shown in Figure XIVA,-159. Within a molecule, C–N = 1.329 ± 0.012 A. and C–S = 1.713 ± 0.012 A.

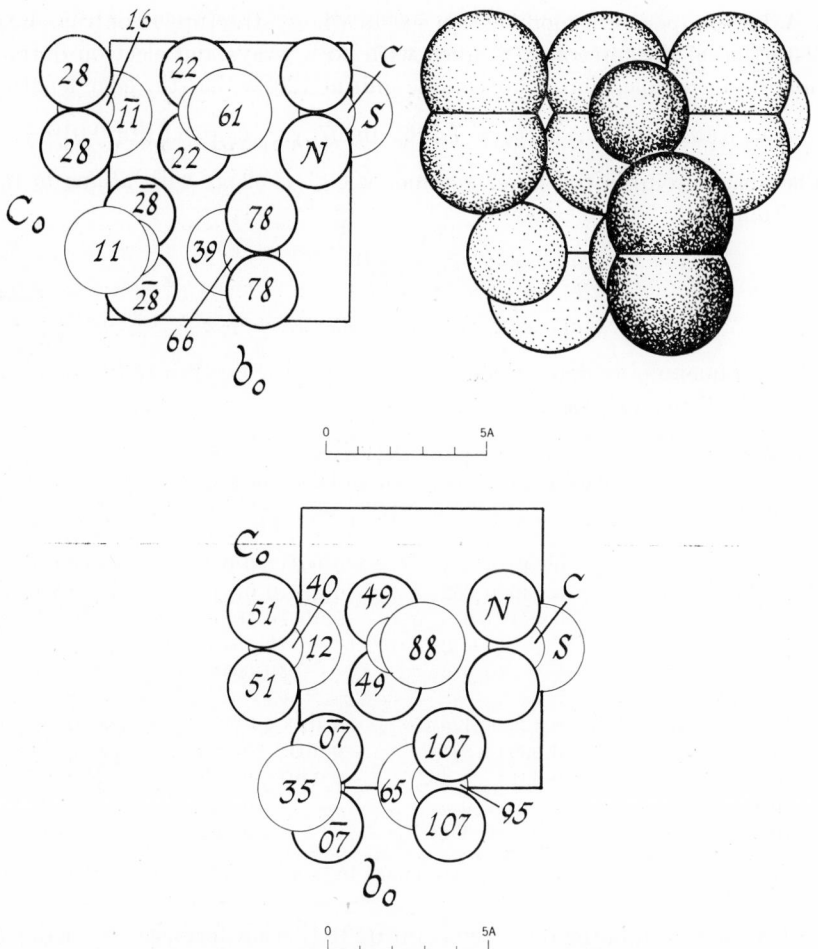

Fig. XIVA,159a (top left). The orthorhombic structure of room-temperature $(NH_2)_2CS$ projected along its a_0 axis. Right-hand axes.

Fig. XIVA,159b (top right). A packing drawing of the orthorhombic thiourea structure seen along its a_0 axis. The large dotted circles are the NH_2 radicals; the small are the atoms of sulfur. Carbon atoms do not show.

Fig. XIVA,159c (bottom). The low-temperature orthorhombic form of $(NH_2)_2CS$ projected along its a_0 axis. Right-hand axes.

The bond angles are S–C–N $= 122.2°$ and N–C–N $= 115.6°$. It has been thought likely that the thermal motion of the hydrogen atoms is so large that their positions do not show in the projections. Contact between molecules is expressed by the separations S–NH_2 = ca. 3.45 A. and NH_2–NH_2 = ca. 3.85 A.

A low-temperature modification exists whose structure is only a small distortion of the foregoing. Studied with both x-rays and electron diffraction, its tetramolecular orthorhombic unit at 120°K. has the edge lengths:

$$a_0 = 5.494 \pm 0.005 \text{ A.}; \quad b_0 = 7.516 \pm 0.007 \text{ A.}; \quad c_0 = 8.519 \pm 0.010 \text{ A.}$$

The space group is the lower symmetry C_{2v}^2 ($Pb2_1m$) with atoms in the positions:

$$(2a) \quad uv0; \quad \bar{u},v+^1/_2,0$$
$$(2b) \quad u \; v \; ^1/_2; \; \bar{u},v+^1/_2,^1/_2$$
$$(4c) \quad xyz; \; xy\bar{z}; \; \bar{x},y+^1/_2,\bar{z}; \; \bar{x},y+^1/_2,z$$

The parameters, as determined by x-ray and by electron diffraction, are shown in Table XIVA,55.

TABLE XIVA,55
Parameters of the Atoms in Low-Thiourea[a]

Atom	Position	x	y	z
S(1)	(2a)	0.349 (0.354)	−0.003 (−0.001)	0.000 (0.000)
C(1)	(2a)	0.049 (0.052)	0.075 (0.071)	0.000 (0.000)
N(1)	(4c)	−0.071 (−0.075)	0.105 (0.105)	0.135 (0.133)
H(1)	(4c)	— (−0.236)	— (0.150)	— (0.133)
H(1′)	(4c)	— (−0.016)	— (0.095)	— (0.237)
S(2)	(2b)	0.125 (0.125)	0.010 (0.017)	0.500 (0.500)
C(2)	(2b)	0.399 (0.384)	−0.103 (−0.108)	0.500 (0.500)
N(2)	(4c)	−0.491 (−0.492)	−0.149 (−0.159)	0.365 (0.367)
H(2)	(4c)	— (−0.327)	— (−0.205)	— (0.367)
H(2′)	(4c)	— (0.458)	— (−0.092)	— (0.263)

[a] Values in parentheses are those determined by electron diffraction.

The change from the room temperature to this low ferroelectric modification involves a small shift and tilt of the molecules marked (2) in the table (Fig. XIVA,159c). In one molecule the S–C separation is 1.748 A., and in the other, 1.729 A.; in both, the C–N distance is 1.34 A.

XIV,a127. Crystals of *tetramethyl thiourea*, [(CH$_3$)$_2$N]$_2$CS, are monoclinic with a tetramolecular unit of the dimensions:

$$a_0 = 5.55 \text{ A.}; \quad b_0 = 12.06 \text{ A.}; \quad c_0 = 11.27 \text{ A.}; \quad \beta = 95°30'$$

The space group is C_{2h}^6 ($C2/c$) with atoms in the positions:

$$\text{S: } (4e) \quad \pm(0 \; u \; ^1/_4; \; ^1/_2, u+^1/_2, ^1/_4) \qquad \text{with } u = 0.103$$
$$\text{C(1): } (4e) \quad \text{with } u = 0.964$$

All the other atoms are in the general positions:

(8*f*) $\pm (xyz; \ x,\bar{y},z+{}^1/_2; \ x+{}^1/_2,y+{}^1/_2,z; \ x+{}^1/_2,{}^1/_2-y,z+{}^1/_2)$

with the following parameters:

Atom	x	y	z
C(2)	0.982	0.800	0.371
C(3)	0.706	0.960	0.403
N	0.883	0.908	0.333

The resulting structure is shown in Figure XIVA,160. In the molecule, S–C = 1.68 A., C(1)–N = 1.37 A., N–CH₃ = 1.47 A. The shortest distance between molecules is S–C = 3.56 A.

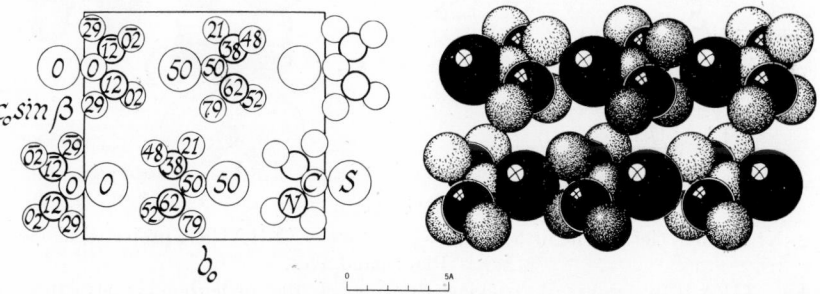

Fig. XIVA,160a (left). The monoclinic structure of tetramethyl thiourea projected along its a_0 axis. Right-hand axes.

Fig. XIVA,160b (right). A packing drawing of the monoclinic structure of tetramethyl thiourea seen along its a_0 axis. The sulfur atoms are the large, the nitrogen atoms the smaller black circles. Atoms of carbon are line shaded.

XIV,a128.   *Thiourea dioxide*, $(NH_2)_2CSO_2$, is orthorhombic with a tetra-molecular unit of the edge lengths:

$$a_0 = 10.133 \ A.; \quad b_0 = 10.655 \ A.; \quad c_0 = 3.924 \ A.$$

The space group is $V_h{}^{16}$ (*Pmnb*) with atoms in the positions:

S: (4*c*) $\pm ({}^1/_4 \ u \ v; \ {}^1/_4,u+{}^1/_2,{}^1/_2-v)$
with $u = 0.46027, \quad v = 0.63527$
$\sigma(u) = 0.00018, \ \sigma(v) = 0.00066$

C: (4*c*)   with $u = 0.1114, \quad v = 0.0976$
$\sigma(u) = 0.0007, \ \sigma(v) = 0.0025$

The other atoms are in

(8*d*) $\pm (xyz; \ x+{}^1/_2,{}^1/_2-y,z+{}^1/_2; \ {}^1/_2-x,y,z; \ x,y+{}^1/_2,{}^1/_2-z)$

with the parameters:

$$O: \quad x = 0.12825, \quad y = 0.40178, \quad z = 0.49324$$
$$\sigma(x) = 0.00049, \quad \sigma(y) = 0.00035, \quad \sigma(z) = 0.00141$$

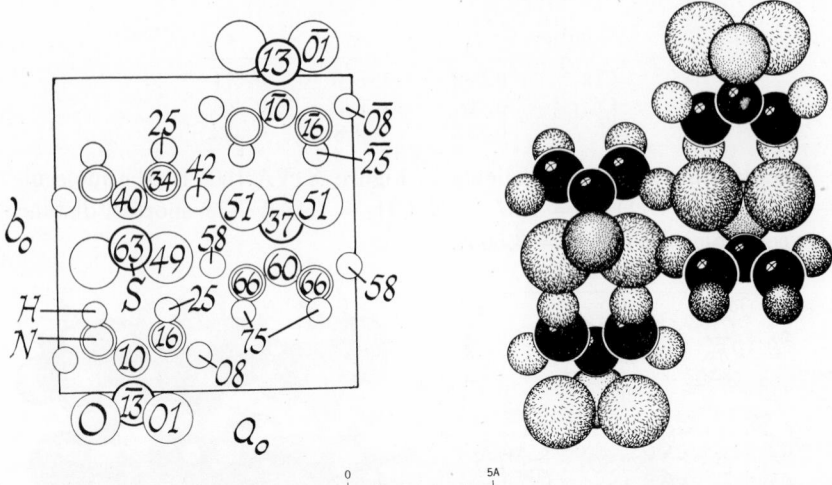

Fig. XIVA,161a (left). The orthorhombic structure of $(NH_2)_2CSO_2$ projected along its c_0 axis. Right-hand axes.

Fig XIVA,161b (right). A packing drawing of the orthorhombic structure of $(NH_2)_2CSO_2$ seen along its c_0 axis. Of the black circles the carbon atoms are slightly smaller than those of nitrogen. The sulfur atoms are heavily outlined and dotted; the oxygens are larger and line shaded. The smallest hook-shaded circles are hydrogen.

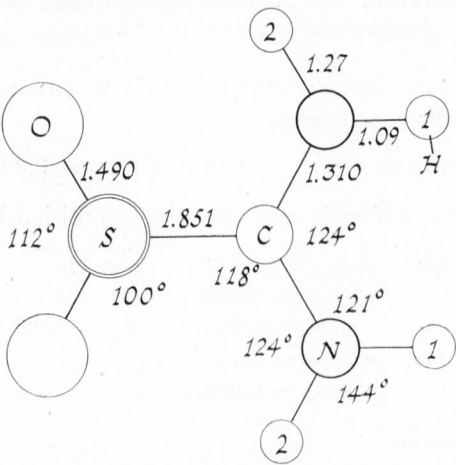

Fig. XIVA,162. Bond dimensions in the molecule of $(NH_2)_2CSO_2$.

$$\text{N:} \quad x = 0.13533, \quad y = 0.16379, \quad z = 0.16020$$
$$\sigma(x) = 0.00053, \; \sigma(y) = 0.00044, \; \sigma(z) = 0.00154$$
$$\text{H(1):} \quad x = 0.1284, \quad y = 0.2603, \quad z = 0.2504$$
$$\sigma(x) = 0.0090, \; \sigma(y) = 0.0062, \; \sigma(z) = 0.0029$$
$$\text{H(2):} \quad x = 0.0253, \quad y = 0.1142, \quad z = 0.0831$$
$$\sigma(x) = 0.0070, \; \sigma(y) = 0.0066, \; \sigma(z) = 0.0025$$

The structure is shown in Figure XIVA,161. Molecules have the bond dimensions of Figure XIVA,162. They are bound together by eight hydrogen bonds per molecule, with N–H–O = 2.84 or 2.85 A.

XIV,a129. *S-Methyl isothiourea sulfate*, $[(NH_2)_2CSCH_3]_2SO_4$, is orthorhombic with a tetramolecular unit of the edge lengths:

$$a_0 = 8.38 \text{ A.}; \; b_0 = 11.32 \text{ A.}; \; c_0 = 12.60 \text{ A.}, \qquad \text{all } \pm 0.01 \text{ A.}$$

The space group is $V_h{}^{14}$ (*Pcan*). The sulfate sulfur atoms are in

$$(4c) \quad \pm (u \; 0 \; {}^1/_4; \; u + {}^1/_2, {}^1/_2, {}^1/_4)$$

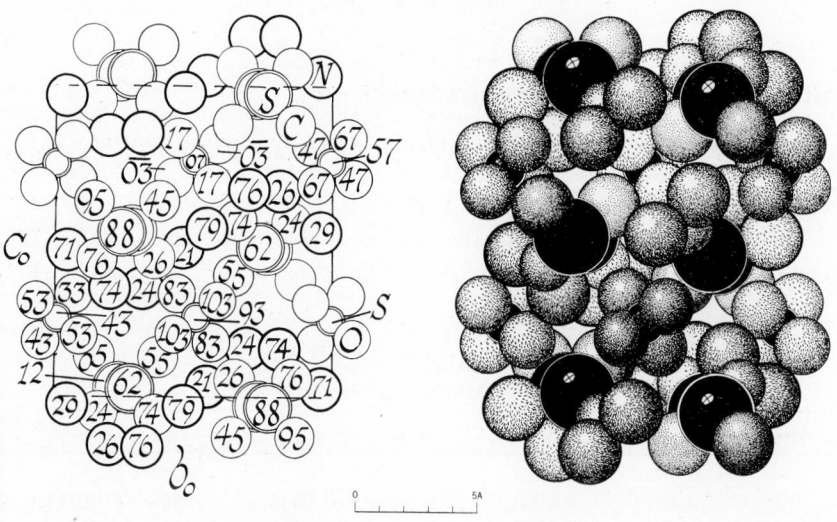

Fig. XIVA,163a (left). The orthorhombic structure of $[(NH_2)_2CSCH_3]_2SO_4$ projected along its a_0 axis. Right-hand axes.

Fig. XIVA,163b (right). A packing drawing of the orthorhombic $[(NH_2)_2CSCH_3]_2SO_4$ arrangement seen along its a_0 axis. In the cation the sulfur atoms are large and black, and the NH_2 radicals are heavily outlined and line shaded, while the CH_3 groups are dotted. The small black sulfur atoms of the $SO_4{}^{2-}$ anions partially show within the tetrahedra of enveloping hook-shaded oxygens.

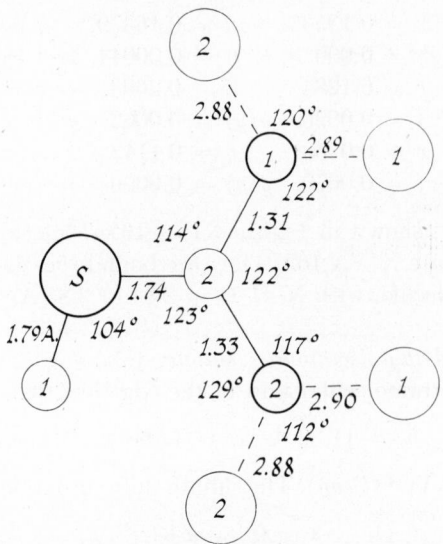

Fig. XIVA,164. Dimensions of the [(NH₂)₂CSCH₃]⁺ ion in the sulfate and the significant
interionic distances.

All the other atoms are in the general positions:

$$(8d) \quad \pm(xyz; \; x+{}^1/_2,y+{}^1/_2,{}^1/_2-z; \; x+{}^1/_2,{}^1/_2-y,z; \; \bar{x},y,z+{}^1/_2)$$

with the parameters of Table XIVA,56.

The atomic arrangement in the crystal is shown in Figure XIVA,163.
The molecular cations in it have the bond dimensions of Figure XIVA,164.

TABLE XIVA,56
Parameters of Atoms in [(NH₂)₂CSCH₃]₂SO₄

Atom	x	$\sigma(x)$	y	$\sigma(y)$	z	$\sigma(z)$
S(1)	0.4293	0.0003	0	—	$^1/_4$	—
S(2)	0.8774	0.0003	0.2354	0.0002	0.5258	0.0002
O(1)	0.5300	0.0006	0.0857	0.0005	0.1934	0.0004
O(2)	0.3284	0.0006	0.0613	0.0005	0.3262	0.0004
N(1)	0.7413	0.0007	0.1940	0.0006	0.3463	0.0005
N(2)	0.7123	0.0007	0.0406	0.0006	0.4663	0.0005
C(1)	0.9466	0.0011	0.1340	0.0008	0.6252	0.0008
C(2)	0.7647	0.0008	0.1479	0.0007	0.4402	0.0005

All but the methyl carbon atoms are practically coplanar and the S(2)–
C(1) bond makes an angle of 18° with this plane. The four hydrogen bonds
between nitrogen and sulfate oxygens are also shown in this figure. The
sulfate anion has its usual regular tetrahedral shape, but with the rather
short S–O = 1.463 ± 0.004 A.

XIV,a130. Crystals of *cuprous chloride tris-thiourea*, $CuCl\cdot3(NH_2)_2CS$,
are tetragonal with a unit containing eight molecules and having the cell
edges:

$$a_0 = 13.41\pm0.01 \text{ A.}, \qquad c_0 = 13.79\pm0.01 \text{ A.}$$

The space group has been chosen as D_4^4 ($P4_12_12$) with all atoms in the general
positions:

(8b) $xyz; \bar{x},\bar{y},z+{}^1/_2; {}^1/_2-y,x+{}^1/_2,z+{}^1/_4; y+{}^1/_2,{}^1/_2-x,z+{}^3/_4;$
  $yx\bar{z}; \bar{y},\bar{x},{}^1/_2-z; {}^1/_2-x,y+{}^1/_2,{}^1/_4-z; x+{}^1/_2,{}^1/_2-y,{}^3/_4-z$

The determined parameters are those of Table XIVA,57.

Part of the resulting structure is shown in Figure XIVA,165. Each copper
atom is tetrahedrally surrounded by the sulfur atoms of four thiourea

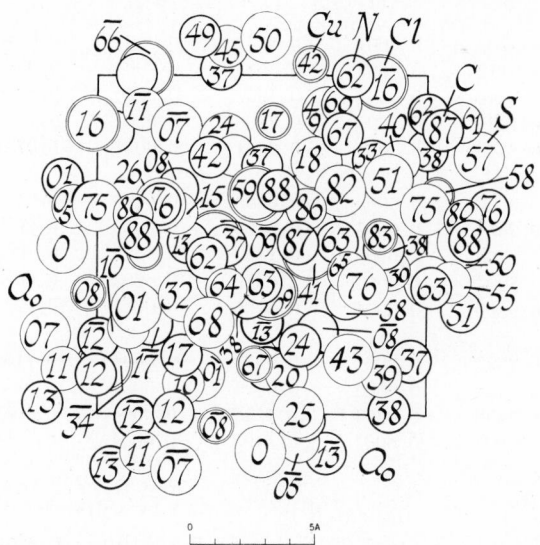

Fig. XIVA,165. Most atoms in the complicated tetragonal structure of $CuCl\cdot3(NH_2)_2CS$
projected along its c_0 axis. Right-hand axes.

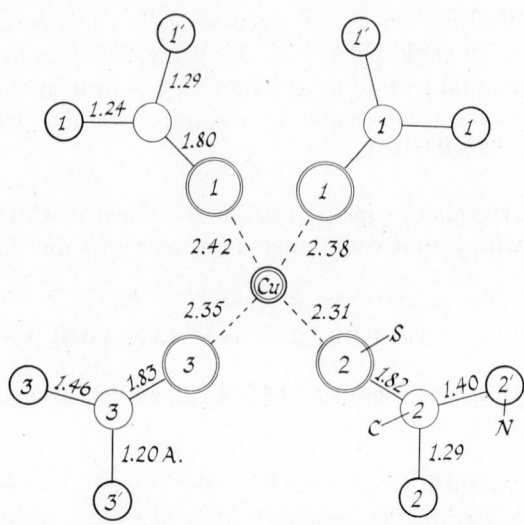

Fig. XIVA,166. Bond dimensions of the thiourea molecules and the copper coordination in CuCl·3(NH₂)₂CS.

TABLE XIVA,57
Parameters of the Atoms in CuCl·3(NH₂)₂CS

Atom	x	$\sigma(x)$	y	$\sigma(y)$	z	$\sigma(z)$
Cu	0.9709	0.0001	0.3571	0.0001	0.0840	0.0001
Cl	0.5112	0.0001	0.3525	0.0001	0.0967	0.0001
S(1)	0.8942	0.0001	0.4962	0.0001	0.0000	0.0001
S(2)	0.1235	0.0001	0.3206	0.0001	0.0143	0.0001
S(3)	0.8449	0.0001	0.2361	0.0001	0.0731	0.0001
C(1)	0.9300	0.0005	0.6153	0.0005	0.0488	0.0004
C(2)	0.2639	0.0005	0.0798	0.0005	0.0978	0.0004
C(3)	0.8903	0.0005	0.1120	0.0005	0.1067	0.0004
N(1)	0.8907	0.0003	0.6882	0.0003	0.0085	0.0002
N(1′)	0.9800	0.0004	0.6105	0.0004	0.1285	0.0003
N(2)	0.2327	0.0004	0.9912	0.0004	0.1169	0.0003
N(2′)	0.2480	0.0004	0.6575	0.0004	0.0857	0.0003
N(3)	0.9990	0.0004	0.1110	0.0004	0.1060	0.0003
N(3′)	0.6604	0.0004	0.5495	0.0004	0.1061	0.0003

groups at the distances indicated in Figure XIVA,166. Each S(1) atom is coordinated to two copper atoms and there thus are formed Cu[SC-(NH₂)₂]₃ chains that spiral along the c_0 axis. Atoms of chlorine are nearest to nitrogen atoms at distances between 3.24 and 3.44 A.

XIV,a131. *Cadmium chloride di thiourea*, $CdCl_2 \cdot 2(NH_2)_2CS$, is ortho-rhombic with a bimolecular unit of the edge lengths:

$$a_0 = 13.07 \text{ A.}; \quad b_0 = 6.48 \text{ A.}; \quad c_0 = 5.80 \text{ A.}$$

The space group has been chosen as C_{2v}^7 ($Pm2_1n$) with atoms in the positions:

(2a) $0uv$; $^1/_2, \bar{u}, v + ^1/_2$
(4b) xyz; $\bar{x}yz$; $^1/_2 - x, \bar{y}, z + ^1/_2$; $x + ^1/_2, \bar{y}, z + ^1/_2$

The positions and parameters are those of Table XIVA,58.

TABLE XIVA,58
Positions and Parameters of the Atoms in $CdCl_2 \cdot 2SC(NH_2)_2$

Atom	Position	x	y	z
Cd	(2a)	0	0.000	0.266
Cl(1)	(2a)	0	0.385	0.243
Cl(2)	(2a)	0	0.932	0.691
S	(4b)	0.169	0.888	0.132
C	(4b)	0.261	0.934	0.318
N(1)	(4b)	0.359	0.931	0.263
N(2)	(4b)	0.242	0.970	0.539

In this arrangement (Fig. XIVA,167), the cadmium atoms are tetra-hedrally surrounded by two chlorine atoms (Cd–Cl = 2.50 or 2.51 A.) and two sulfur atoms (Cd–S = 2.45 A.). In the thiourea molecule, S–C = 1.64 A. and C–N = 1.32 A. The shortest N–Cl = 3.29 A.

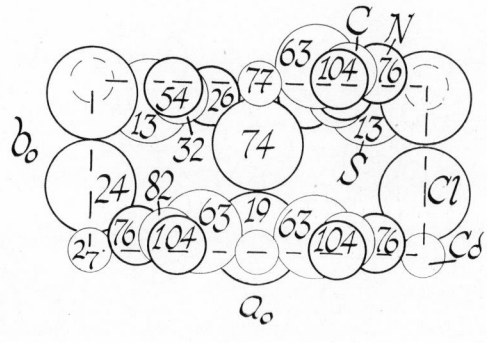

Fig. XIVA,167. The orthorhombic structure of $CdCl_2 \cdot 2(NH_2)_2CS$ projected along its c_0 axis. Right-hand axes.

XIV,a132. *Cadmium formate di thiourea,* $Cd(HCO_2)_2 \cdot 2(NH_2)_2CS$, is orthorhombic with a bimolecular unit of the dimensions:

$$a_0 = 8.000 \pm 0.009 \text{ A.}; \quad b_0 = 17.878 \pm 0.005 \text{ A.}; \quad c_0 = 3.933 \pm 0.007 \text{ A.}$$

According to a brief description, the space group has been chosen as V^3 $(P2_12_12)$ with atoms in the positions:

$$Cd: (2a) \quad 00u; \; {}^1/_2 \, {}^1/_2 \, \bar{u} \qquad \text{with } u = 0.103$$

All other atoms are in the positions:

$$(4c) \quad xyz; \; \bar{x}\bar{y}z; \; x+{}^1/_2, {}^1/_2-y, \bar{z}; \; {}^1/_2-x, y+{}^1/_2, \bar{z}$$

with the parameters listed in Table XIVA,59.

TABLE XIVA,59
Parameters of Atoms in $Cd(HCO_2)_2 \cdot 2SC(NH_2)_2$

Atom	x	y	z
S	0.191	0.062	0.561
O(1)	−0.190	0.100	0.165
O(2)	−0.403	0.163	0.041
N(1)	0.020	0.190	0.569
N(2)	0.270	0.194	0.798
C(1)	0.155	0.163	0.643
C(2)	−0.338	0.107	0.068

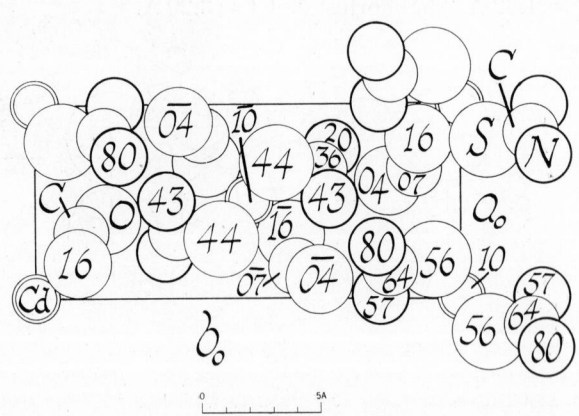

Fig. XIVA,168. The orthorhombic structure of $Cd(HCO_2)_2 \cdot 2(NH_2)_2CS$ projected along its c_0 axis. Right-hand axes.

The resulting structure is shown in Figure XIVA,168. It is described as built up of a series of chains of CdS_4O_2 octahedra which have two sulfur atoms in common. Interatomic distances were not stated, but around the cadmium atoms they are surprisingly great.

XIV,a133. *Zinc chloride di thiourea*, $ZnCl_2 \cdot 2(NH_2)_2CS$, is orthorhombic with a tetramolecular unit of the dimensions:

$$a_0 = 13.065 \text{ A.}; \quad b_0 = 12.722 \text{ A.}; \quad c_0 = 5.890 \text{ A.}, \quad \text{all } \pm 0.005 \text{ A.}$$

The space group is V_h^{16} (*Pnma*) with some atoms in the special positions:

$$(4c) \quad \pm (u \; ^1/_4 \; v; \; u + ^1/_2, ^1/_4, ^1/_2 - v)$$

with for Zn: $u = 0.1668$, $v = 0.3154$
for Cl(1): $u = 0.3409$, $v = 0.2519$
for Cl(2): $u = 0.1458$, $v = -0.2946$

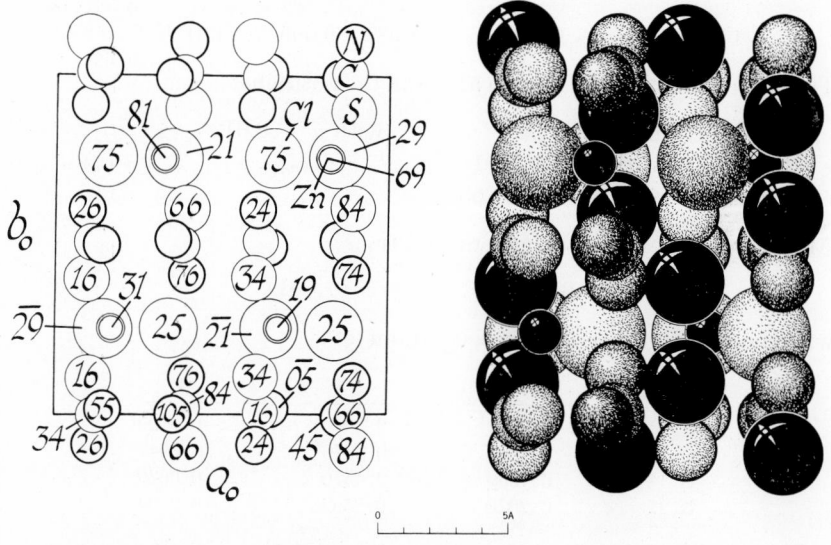

Fig. XIVA,169a (left). The orthorhombic structure of $ZnCl_2 \cdot 2(NH_2)_2CS$ projected along its c_0 axis. Right-hand axes.

Fig. XIVA,169b (right). A packing drawing of the orthorhombic $ZnCl_2 \cdot 2(NH_2)_2CS$ structure seen along its c_0 axis. In the thiourea molecules the sulfur atoms are large and black, the NH_2 radicals heavily outlined and line shaded. The Cl^- ions are large and line shaded; the atoms of zinc are small and black

The other atoms are in the general positions:

(8d) $\pm (xyz;\ ^1/_2 - x, y + ^1/_2, z + ^1/_2;\ x, ^1/_2 - y, z;\ x + ^1/_2, y, ^1/_2 - z)$

with the following parameters:

Atom	x	y	z
S	0.0959	0.0978	0.1560
C	0.1177	−0.0095	0.3447
N(1)	0.1431	0.0062	0.5551
N(2)	0.1043	−0.0995	0.2638

The resulting structure is shown in Figure XIVA,169. The zinc atoms are surrounded by a distorted tetrahedron of two sulfurs (Zn–S = 2.352 A.) and two chlorine atoms (Zn–Cl = 2.314 and 2.328 A.) all ±0.005 A.; the angle S–Zn–S = 111°30′ and S–Zn–Cl = 109 and 110°36′. Within the planar thiourea molecules, S–C = 1.78±0.02 A., C–N = 1.30 and 1.26±0.03 A.

XIV,a134. Crystals of *nickel chloride tetra thiourea*, $NiCl_2 \cdot 4(NH_2)_2CS$, are tetragonal with a bimolecular unit of the dimensions:

$$a_0 = 9.558\ \text{A.,} \qquad c_0 = 8.981 \pm 0.005\ \text{A.} \ (110°\text{K.})$$

The space group is C_4^5 ($I4$) with atoms in the positions:

Ni: (2a) $00u;\ ^1/_2, ^1/_2, u + ^1/_2$ with $u = 0.0120$
Cl(1): (2a)   with $u = 0.2791$
Cl(2): (2a)   with $u = -0.2681$

All other atoms are in the general positions:

(8c) $xyz;\ \bar{x}\bar{y}z;\ \bar{y}xz;\ y\bar{x}z;$ B.C.

with the following recently revised parameters:

Atom	x	y	z
S	0.0284	0.2542	−0.0199
C	0.1545	0.3270	0.0947
N(1)	0.1969	0.4570	0.0620
N(2)	0.2070	0.2592	0.2104

The resulting structure is shown in Figure XIVA,170. It is considered to be molecular in nature with nickel atoms at the approximate centers of distorted octahedra; four of their corners are occupied by sulfur atoms with Ni–S = 2.46 A., the other two by chlorines with Ni–Cl(1) = 2.40 A. and Ni–Cl(2) = 2.52 A. The nickel atoms are not at a center of symmetry and the sulfur square is tilted so that S–Ni–Cl(2) = 83°18′. The thiourea groups themselves are planar, with C–S = 1.73 A., C–N = 1.32 or 1.34 A.,

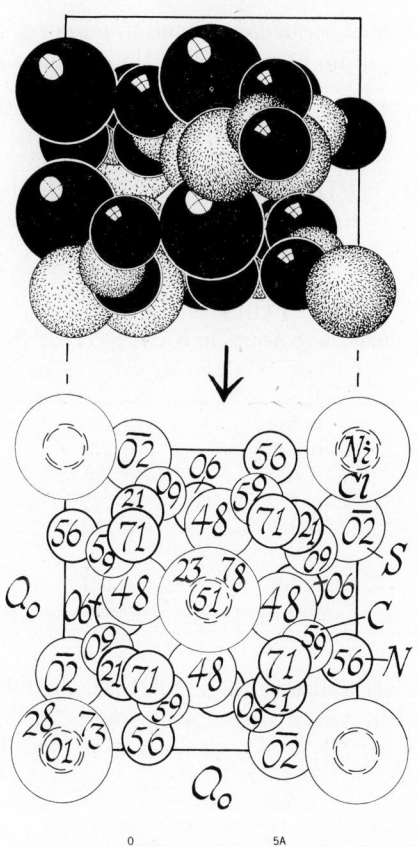

Fig. XIVA,170. Two projections of the tetragonal structure of $NiCl_2 \cdot 4(NH_2)_2CS$. Right-hand axes. In the upper packing drawing, a projection along an a_0 axis, some of the atoms of the unit are shown as viewed from the direction of the arrow. In this drawing the Cl^- ions are large and black and the smaller black circles are the NH_2 radicals. The sulfur atoms are large and line shaded; the carbon atoms are dotted.

and with N–C–N $= 121°$ and N–C–S $= 117$ or $122°$. Each nitrogen atom has a short N–Cl $= 3.23$ or 3.30 A., probably indicative of a hydrogen bonding.

A number of chlorides of other divalent metals undoubtedly have this structure. Their cell dimensions are the following:

Compound	a_0, A.	c_0, A.
$CdCl_2 \cdot 4(NH_2)_2CS$	9.63	9.21
$CoCl_2 \cdot 4(NH_2)_2CS$	9.59	9.07
$FeCl_2 \cdot 4(NH_2)_2CS$	9.64	8.89
$MnCl_2 \cdot 4(NH_2)_2CS$	9.68	8.97

XIV,a135.   Crystals of *nickel thiocyanate di thiourea*, $Ni(NCS)_2 \cdot 2(NH_2)_2$-CS, are triclinic with a unimolecular cell of the dimensions:

$$a_0 = 3.79 \text{ A.}; \quad b_0 = 7.57 \text{ A.}; \quad c_0 = 10.04 \text{ A.}$$
$$\alpha = 92°42'; \quad \beta = 97°48'; \quad \gamma = 104°12'$$

The space group is C_i^1 ($P\bar{1}$) with the nickel atom in the origin ($1a$) 000 and all other atoms in ($2i$) $\pm(xyz)$. The parameters assigned them are listed in Table XIVA,60.

TABLE XIVA,60
Parameters of Atoms in $Ni(NCS)_2 \cdot 2SC(NH_2)_2$

Atom	x	y	z
N(1)	0.083	0.105	0.191
C(1)	0.037	0.179	0.289
S(1)	0.011	0.295	0.423
S(2)	0.392	0.777	0.033
C(2)	0.458	0.704	0.197
N(2)	0.437	0.807	0.303
N(3)	0.577	0.556	0.208

According to this structure (Fig. XIVA,171), the nickel atoms are octahedrally surrounded by two nitrogen atoms with Ni–N = 1.99 A., and by four sulfur atoms with Ni–S = 2.51 or 2.57 A. The thiourea groups are substantially planar.

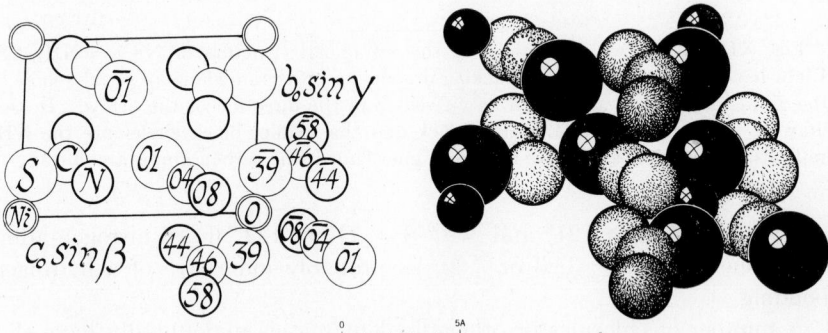

Fig. XIVA,171a (left). The triclinic structure of $Ni(NCS)_2 \cdot 2(NH_2)_2CS$ projected along its a_0 axis. Right-hand axes.

Fig. XIVA,171b (right). A packing drawing of the triclinic structure of $Ni(NCS)_2 \cdot 2$-$(NH_2)_2CS$ seen along its a_0 axis. The sulfur are the large, the nickel the small black circles. Of the dotted circles the nitrogen are more heavily outlined than the carbon.

The following compounds are isostructural:

$Mn(NCS)_2 \cdot 2(NH_2)_2CS$:

$a_0 = 3.95$ A.; $b_0 = 7.72$ A.; $c_0 = 10.04$ A.

$\alpha = 93°54'$; $\beta = 99°6'$; $\gamma = 106°18'$

$Co(NCS)_2 \cdot 2(NH_2)_2CS$:

$a_0 = 3.85$ A.; $b_0 = 7.67$ A.; $c_0 = 10.28$ A.

$\alpha = 92°0'$; $\beta = 99°6'$; $\gamma = 103°30'$

$Cd(NCS)_2 \cdot 2(NH_2)_2CS$:

$a_0 = 4.02$ A.; $b_0 = 7.75$ A.; $c_0 = 10.17$ A.

$\alpha = 90°54'$; $\beta = 99°36'$; $\gamma = 105°6'$

XIV,a136. *Palladous chloride tetra thiourea*, $PdCl_2 \cdot 4(NH_2)_2CS$, is monoclinic with a tetramolecular unit of the dimensions:

$a_0 = 16.89$ A.; $b_0 = 11.18$ A.; $c_0 = 8.89$ A.; $\beta = 91°30'$

The space group is C_{2h}^6 ($C2/c$) with the palladium atoms in the positions:

(4e) $\pm(0 \; u \; 1/4; \; 1/2, u+1/2, 1/4)$ with $u = 0.324$

The other atoms have been assigned the general positions:

(8f) $\pm(xyz; \; x,\bar{y},z+1/2; \; x+1/2,y+1/2,z; \; x+1/2,1/2-y,z+1/2)$

with the parameters of Table XIVA,61.

This structure is made up of $Pd[SC(NH_2)_2]_4^{2+}$ and Cl^- ions. In the cations, the palladium atoms are surrounded by four sulfur atoms at the corners of an approximate square, with Pd–S $= 2.33$ and 2.35 A. and S–Pd–S $= 82$–$97°$. In the thiourea molecules, S–C $= 1.72$ or 1.75 A. and C–N $= 1.32$–1.47 A.

TABLE XIVA,61
Parameters of Atoms in $PdCl_2 \cdot 4(NH_2)_2CS$

Atom	x	y	z
S(1)	0.073	0.482	0.158
S(2)	0.079	0.167	0.155
Cl	0.135	0.182	0.050
N(1)	0.220	0.561	0.242
N(2)	0.189	0.380	0.328
N(3)	0.092	0.078	−0.136
N(4)	0.071	0.288	−0.105
C(1)	0.167	0.476	0.247
C(2)	0.081	0.183	−0.037

There is probably an error in the parameters as given in the original article (and the table) since they result in an improbably short S–Cl = ca. 1.4 A.

XIV,a137. *Plumbous chloride di thiourea*, $PbCl_2 \cdot 2(NH_2)_2CS$, is ortho-rhombic with a tetramolecular unit of the edge lengths:

$$a_0 = 21.20 \pm 0.04 \text{ A.}; \quad b_0 = 4.06 \pm 0.01 \text{ A.}; \quad c_0 = 12.02 \pm 0.02 \text{ A.}$$

The space group is C_{2v}^9 ($Pna2_1$) with all atoms in the positions:

$$(4a) \quad xyz; \; \bar{x},\bar{y},z+^1/_2; \; ^1/_2-x,y+^1/_2,z+^1/_2; \; x+^1/_2,^1/_2-y,z$$

The parameters have been given as those of Table XIVA,62.

TABLE XIVA,62
Parameters of the Atoms in $PbCl_2 \cdot 2(NH_2)_2CS$

Atom	x	$\sigma(x)$	y	$\sigma(y)$	z	$\sigma(z)$
Pb	0.1050	0.0002	0.229	0.001	0.2500	0.0002
Cl(1)	0.232	0.002	0.268	0.004	0.294	0.002
Cl(2)	−0.013	0.002	0.708	0.006	0.238	0.002
S(1)	0.154	0.002	0.725	0.009	0.084	0.002
S(2)	0.095	0.002	0.697	0.007	0.434	0.002
C(1)	0.106	0.004	0.883	0.010	−0.013	0.007
C(2)	0.160	0.009	0.661	0.025	0.527	0.008
N(1)	0.042	0.004	0.963	0.019	−0.007	0.008
N(2)	0.124	0.007	0.894	0.026	−0.121	0.009
N(3)	0.207	0.009	0.872	0.017	0.519	0.006
N(4)	0.150	0.005	0.542	0.030	0.629	0.012

In the resulting atomic arrangement (Fig. XIVA,172) the bond dimensions of the thiourea molecules agree within the limits of this approximate determination with those found in thiourea itself (**XIV,a126**). Each lead atom has a sevenfold coordination: three of its neighbors are chlorine with Pb–Cl = 2.75, 3.17, and 3.28 A., and the other four are sulfur atoms at distances of 2.92, 3.02, 3.04, and 3.10 A. This coordination polyhedron is described as a distorted trigonal prism with the seventh atom beyond one lateral face. Both the Cl(2) and the sulfur atoms thus are associated with two lead atoms and in this way chains are formed running in the b_0 direction. Within a chain, N–Cl = 3.19 and 3.33 A.; between chains, the shortest atomic separations are N–S = 3.29 and 3.37 A., and N–Cl = 3.26, 3.29, and 3.32 A.

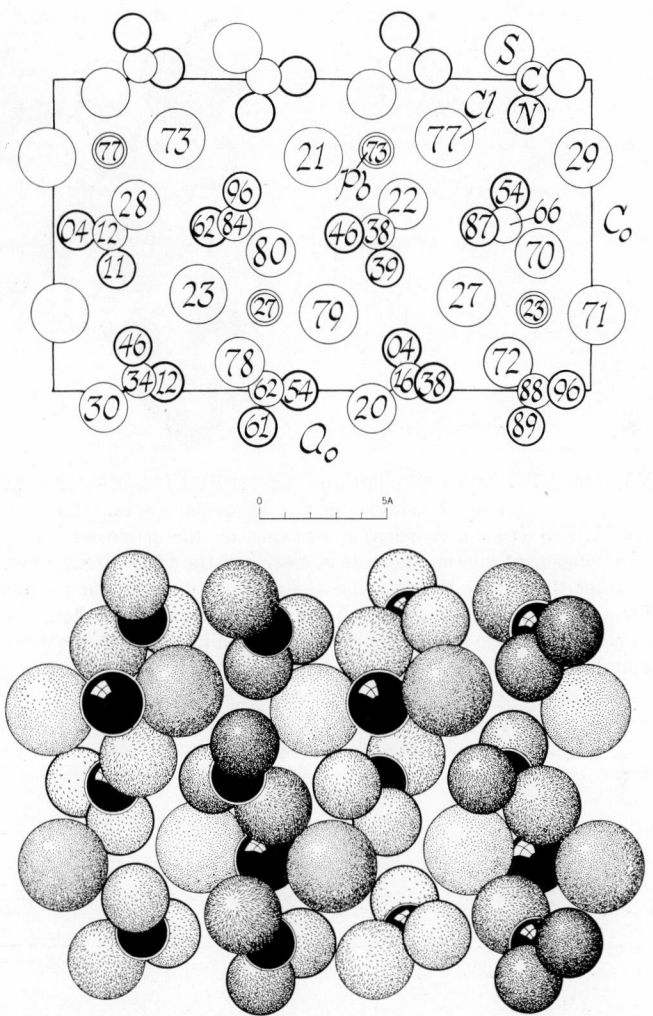

Fig. XIVA,172a (top). The orthorhombic structure of $PbCl_2 \cdot 2(NH_2)_2CS$ projected along its b_0 axis. Right-hand axes.

Fig. XIVA,172b (bottom). A packing drawing of the orthorhombic structure of $PbCl_2 \cdot 2(NH_2)_2CS$ seen along its b_0 axis. The lead atoms are black; the Cl^- ions are large and dotted. In the thiourea molecules, the carbon atoms also are black, the sulfur atoms are line shaded, and the NH_2 radicals are more heavily outlined and hook shaded.

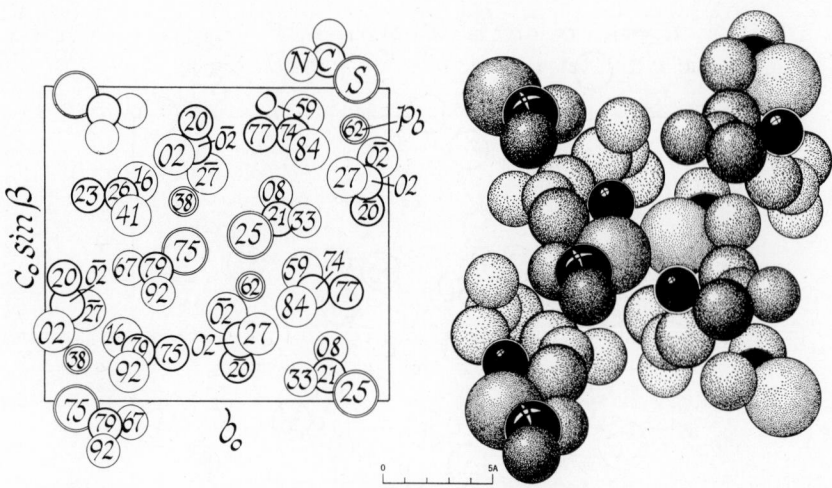

Fig. XIVA,173a (left). The monoclinic structure of Pb(CH₃COO)₂·(NH₂)₂CS projected
along its short a_0 axis. Right-hand axes.

Fig. XIVA,173b (right). A packing drawing of the monoclinic Pb(CH₃COO)₂·-
(NH₂)₂CS arrangement viewed along its a_0 axis. Of the black circles those for lead are
slightly smaller than those representing the carbon atoms of the thiourea molecules.
The sulfur atoms are large and dotted; the NH₂ groups are smaller. In the acetate
ions the oxygen atoms are more heavily outlined and line shaded; the carbon atoms are
smaller and dotted.

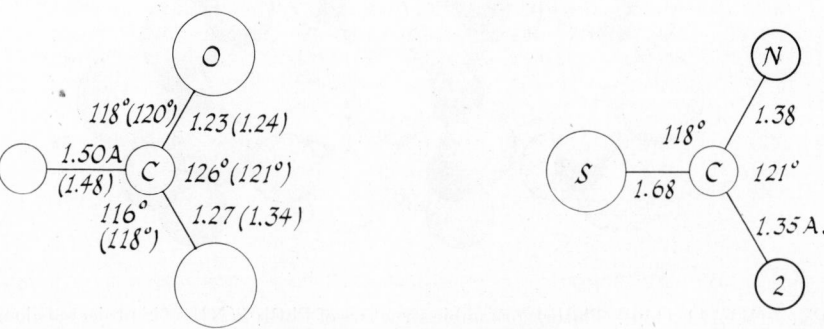

Fig. XIVA,174a (left). Bond dimensions of the two crystallographically different acetate
ions in Pb(CH₃COO)₂·(NH₂)₂CS.
Fig. XIVA,174b (right). Bond dimensions in the thiourea molecule of Pb(CH₃COO)₂·
(NH₂)₂CS.

XIV,a138. Crystals of *lead acetate thiourea*, $Pb(CH_3COO)_2 \cdot (NH_2)_2CS$, are monoclinic with a tetramolecular cell of the dimensions:

$$a_0 = 4.55 \pm 0.01 \text{ A.}; \quad b_0 = 15.81 \pm 0.03 \text{ A.}; \quad c_0 = 14.28 \pm 0.07 \text{ A.}$$
$$\beta = 106°24'$$

The space group is C_{2h}^5 $(P2_1/c)$ with all atoms in the positions:

$$(4e) \quad \pm (xyz; \ x,{}^1/_2-y,z+{}^1/_2)$$

The chosen parameters are listed in Table XIVA, 63.

The structure is shown in Figure XIVA,173. Both the acetate ions and the thiourea molecules are planar with the bond dimensions of Figure XIVA,-174. Lead atoms have eight neighbors with Pb–5 O = 2.37–3.01 A., and Pb–3 S = 3.09–3.34 A.

TABLE XIVA,63
Parameters of Atoms in Lead Acetate Thiourea

Atom	x	σ_x, A.	y	σ_y, A.	z	σ_z, A.
Pb	0.3855	0.0015	0.0972	0.0006	0.1294	0.0007
S	0.7488	0.0073	0.0930	0.0020	−0.0239	0.0024
N(1)	0.674	0.033	0.253	0.018	−0.071	0.017
N(2)	0.924	0.033	0.170	0.018	−0.164	0.017
O(1)	0.163	0.031	0.227	0.018	0.201	0.019
O(2)	0.410	0.031	0.246	0.018	0.088	0.019
O(3)	0.020	0.031	0.031	0.018	0.224	0.019
O(4)	−0.273	0.031	0.116	0.018	0.287	0.019
C(1)	0.791	0.094	0.177	0.055	−0.091	0.044
C(2)	0.233	0.094	0.366	0.055	0.153	0.044
C(3)	0.265	0.094	0.272	0.055	0.147	0.044
C(4)	0.197	0.094	0.058	0.055	0.393	0.044
C(5)	−0.022	0.094	0.069	0.055	0.295	0.044

XIV,a139. *Formamidinium disulfide dibromide monohydrate*, $[(NH_2)_2-CS]_2Br_2 \cdot H_2O$, is monoclinic with a bimolecular unit of the dimensions:

$$a_0 = 8.61 \text{ A.}; \quad b_0 = 5.12 \text{ A.}; \quad c_0 = 12.40 \text{ A.}; \quad \beta = 99°30'$$

The space group is C_{2h}^4 $(P2/c)$ with water oxygens in the positions:

$$(2f) \quad \pm ({}^1/_2 \ u \ {}^1/_4) \quad \text{with } u = 0.711$$

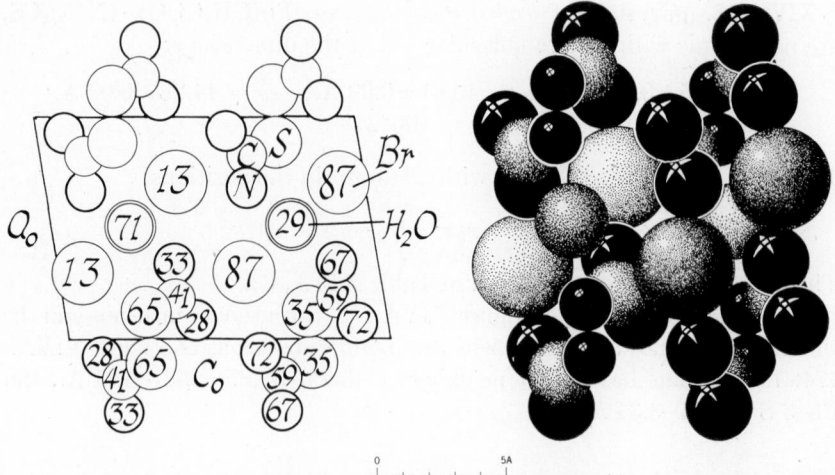

Fig. XIVA,175a (left). The monoclinic structure of [(NH₂)₂CS]₂Br₂·H₂O projected along its b_0 axis. Right-hand axes.

Fig. XIVA,175b (right). A packing drawing of the monoclinic structure of [(NH₂)₂CS]₂Br₂·H₂O seen along its b_0 axis. The nitrogen atoms are the larger, the sulfur atoms the smaller black circles. The bromine atoms are the large dotted circles; the water molecules are somewhat smaller and dotted. The carbon atoms are fine-line shaded.

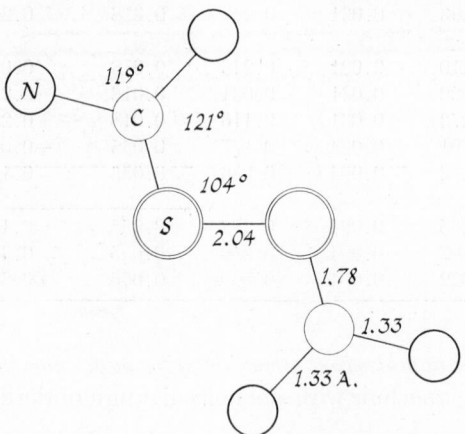

Fig. XIVA,176. Bond dimensions in the [(NH₂)₂CS]₂²⁺ cation in the bromide.

The other atoms are in the general positions:

$$(4g) \quad \pm(xyz;\ x,\bar{y},z+{}^1\!/_2)$$

with the parameters:

Atom	x	y	z
Br	0.2925	0.1307	0.0727
S	0.1200	0.6502	0.2550
C	0.186	0.410	0.356
N(1)	0.083	0.277	0.405
N(2)	0.334	0.332	0.368

The structure is shown in Figure XIVA,175. It is built up of $[(NH_2)_2$-CS$]_2$ cations, bromide anions and water molecules. The cations are planar and have the bond dimensions of Figure XIVA,176. The bromide ions are closest to nitrogen atoms of the cation with Br–N = 3.27 upwards, and to water molecules with the shortest Br–O = 3.34 A. These anions also approach to within 3.81 A. of sulfur atoms.

XIV,a140.   Unlike the bromide, *formamidinium disulfide diiodide monohydrate*, $[(NH_2)_2CS]_2I_2 \cdot H_2O$, is orthorhombic with a tetramolecular unit of the dimensions:

$$a_0 = 5.15 \text{ A.}; \quad b_0 = 16.52 \text{ A.}; \quad c_0 = 13.39 \text{ A.}$$

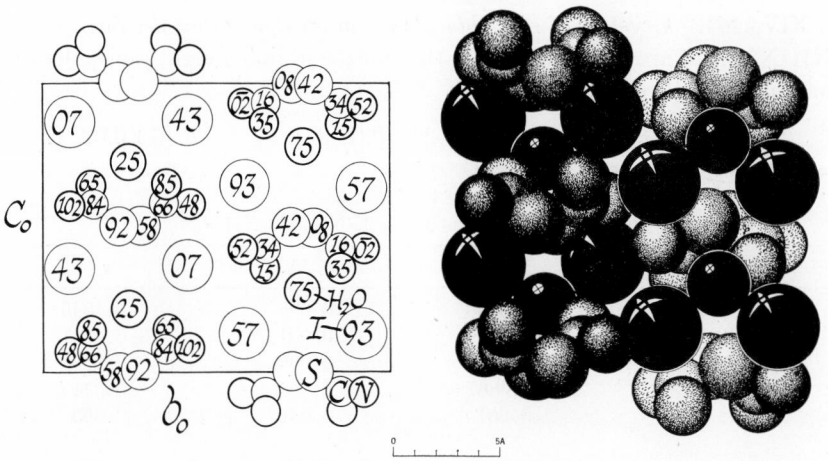

Fig. XIVA,177a (left). The orthorhombic structure of $[(NH_2)_2CS]_2I_2 \cdot H_2O$ projected along its a_0 axis. The different shape of the cation in this crystal and the bromide and their different relationships to the water molecules are apparent by comparing this figure with Figure XIVA,175a. Right-hand axes.

Fig. XIVA,177b (right). A packing drawing of the orthorhombic structure of $[(NH_2)_2CS]_2I_2 \cdot H_2O$ seen along its a_0 axis. The I⁻ anions are the large, the water molecules the smaller black circles. In the $[(NH_2)_2CS]_2^{2+}$ cations the sulfur atoms are the largest circles and the NH_2 radicals are heavily outlined.

The space group is V_h^{10} ($Pccn$) with water oxygen atoms in the positions:

$$(4c) \quad \pm (^1/_4 \; ^1/_4 \; u; \; ^1/_4, ^1/_4, u+^1/_2) \qquad \text{with } u = 0.214$$

All other atoms are in the general positions:

$$(8e) \quad \pm (xyz; \; x+^1/_2, y+^1/_2, \bar{z}; \; ^1/_2-x, y, z+^1/_2; \; x, ^1/_2-y, z+^1/_2)$$

with the following parameters:

Atom	x	y	z
I	0.428	0.0783	0.3547
S	0.578	0.2192	−0.008
C	0.665	0.139	0.071
N(1)	0.850	0.140	0.136
N(2)	0.481	0.080	0.076

The resulting structure is shown in Figure XIVA,177. The $[(NH_2)_2CS]_2$ cation in this crystal is of a somewhat different shape from that in the bromide (**XIV,a139**). It is probably not strictly planar, and there are different rotations about the S–S and S–C bonds. Thus the dihedral angle CSS/SSC is 105° for the iodide and 89° for the bromide.

XIV,a141. Crystals of *zinc chloride mono thiosemicarbazide*, $ZnCl_2 \cdot NH_2$-NH(NH$_2$)CS, are orthorhombic with a unit containing eight molecules and having the cell edges:

$$a_0 = 11.92 \pm 0.01 \text{ A.;} \quad b_0 = 7.28 \pm 0.02 \text{ A.;} \quad c_0 = 15.46 \pm 0.02 \text{ A.}$$

TABLE XIVA,64
Parameters of Atoms in $ZnCl_2 \cdot NH_2NH(NH_2)CS$

Atom	u	σ_u, A.	v
Zn(1)	0.3830	0.005	0.1945
Zn(2)	−0.3370	0.004	−0.4881
S(1)	0.2112	0.005	0.2603
S(2)	−0.3867	0.005	−0.3445
C(1)	0.261	0.031	0.365
C(2)	0.750	0.021	−0.308
N(1)	0.195	0.032	0.432
N(2)	0.367	0.029	0.378
N(3)	0.459	0.028	0.317
N(1′)	−0.229	0.026	−0.228
N(2′)	−0.172	0.019	−0.364
N(3′)	−0.168	0.031	−0.454

The space group was chosen as V_h^{16} ($Pnma$) with atoms in the positions:

$(4c)$ $\pm (u\ ^1/_4\ v;\ u+^1/_2,^1/_4,^1/_2-v)$

$(8d)$ $\pm (xyz;\ ^1/_2-x,y+^1/_2,z+^1/_2;\ x,^1/_2-y,z;\ x+^1/_2,y,^1/_2-z)$

All but the chlorine atoms are in $(4c)$. For them

Cl(1): $(8d)$ with $x = 0.4383,\ y = 0.0040,\ z = 0.1189$

Cl(2): $(8d)$ with $x = 0.3720,\ y = 0.4960,\ z = 0.5661$

$[\sigma_x = 0.003$ A., $\sigma_y = 0.015$ A., $\sigma_z = 0.004$ A. for both$]$

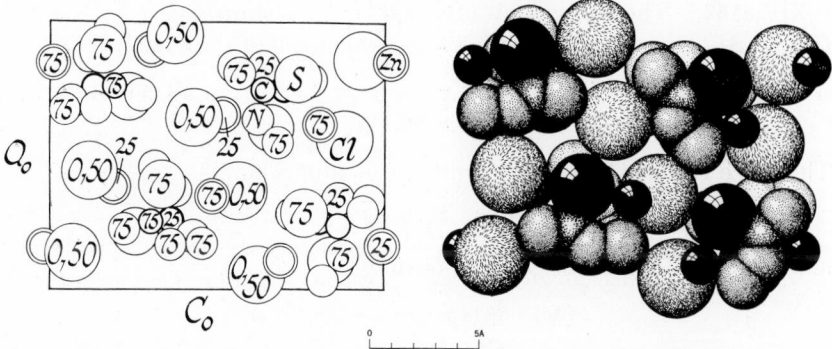

Fig. XIVA,178a (left). The orthorhombic structure of $ZnCl_2 \cdot NH_2NH(NH_2)CS$ projected along its b_0 axis. Right-hand axes.

Fig. XIVA,178b (right). A packing drawing of the orthorhombic $ZnCl_2 \cdot NH_2NH$-$(NH_2)CS$ arrangement viewed along its b_0 axis. The zinc atoms are the small, the sulfur atoms the large black circles. Chloride ions are the largest line-shaded circles. The nitrogen atoms are dotted; the atoms of carbon are triangular in shape.

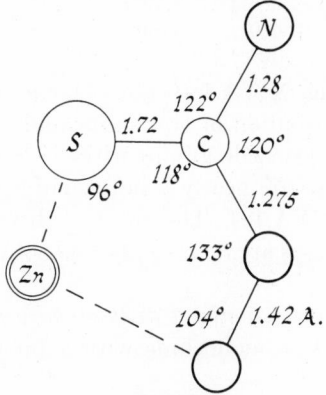

Fig. XIVA,179. Bond dimensions of the $NH_2NH(NH_2)CS$ molecule in $ZnCl_2 \cdot NH_2NH$-$(NH_2)CS$.

The parameters of the other atoms are those of Table XIVA,64.

The resulting structure is shown in Figure XIVA,178. The thiosemi-carbazide molecules it contains have the average bond dimensions of Figure XIVA,179. The zinc atoms are tetrahedrally coordinated with two chloride ions and one sulfur and one nitrogen belonging to a semicarbazide molecule (Zn–Cl = 2.24 A., Zn–S = 2.29 or 2.30 A., Zn–N = 2.08 or 2.11 A.). Between molecules the close separations are N–Cl = 3.27−3.72 A., and N–S = 3.23 A.

XIV,a142. The red crystals of *nickel bis-thiosemicarbazide*, Ni[NH$_2$NH-(NH$_2$)CS]$_2$, are monoclinic with a bimolecular cell of the edge lengths:

$$a_0 = 7.68 \text{ A.}; \quad b_0 = 4.32 \text{ A.}; \quad c_0 = 12.23 \text{ A.}, \quad \text{all } \pm 0.01 \text{ A.}$$
$$\beta = 93°48'$$

The space group is C_{2h}^5 ($P2_1/c$) with nickel atoms in

$$(2a) \quad 000; \, 0 \, ^1/_2 \, ^1/_2$$

and all other atoms in the general positions:

$$(4e) \quad \pm (xyz; \, x,^1/_2-y,z+^1/_2)$$

The determined parameters are the following:

Atom	x	$\sigma(x)$	y	$\sigma(y)$	z	$\sigma(z)$
S	0.2663	0.003	0.6587	0.005	0.5030	0.003
N(1)	0.4354	0.011	0.9505	0.022	0.3362	0.010
N(2)	0.1409	0.010	0.8758	0.016	0.3143	0.009
N(3)	−0.0171	0.011	0.7376	0.020	0.3679	0.010
C	0.2742	0.011	0.8490	0.018	0.3775	0.011

In this structure (Fig. XIVA,180) the nickel atoms have about them an essentially planar distribution of two sulfur and two nitrogen atoms sup-plied by two thiosemicarbazide groups with Ni–S = 2.15 A. and Ni–N(3) = 1.91 A. These groups are nearly if not exactly planar with the bond di-mensions of Figure XIVA,181. The shortest distances between atoms of different molecular groups are N(2)–N(3) = 2.83 A. and S–N(3) = 3.08 A.

XIV,a143. The α-form of *nickel di thiosemicarbazide sulfate trihydrate*, Ni(CH$_5$N$_3$S)$_2$SO$_4 \cdot$3H$_2$O, is monoclinic with a bimolecular unit of the di-mensions:

$$a_0 = 6.98 \text{ A.}; \quad b_0 = 16.43 \text{ A.}; \quad c_0 = 6.35 \text{ A.}, \quad \text{all } \pm 0.01 \text{ A.}$$
$$\beta = 99°12' \pm 6'$$

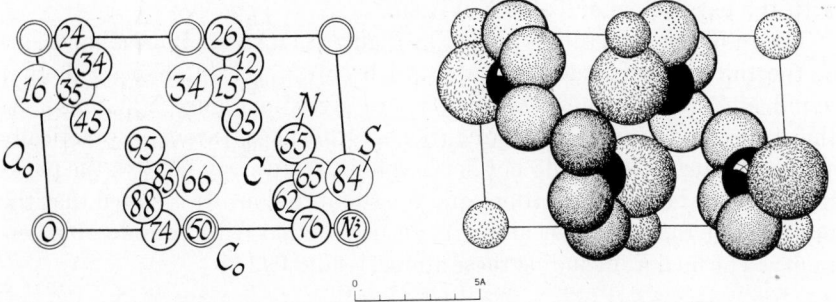

Fig. XIVA,180a (left). The monoclinic structure of Ni[NH₂NH(NH₂)CS]₂ projected along its b_0 axis. Right-hand axes.

Fig. XIVA,180b (right). A packing drawing of the monoclinic Ni[NH₂NH(NH₂)CS]₂ arrangement viewed along its b_0 axis. The carbon atoms are black, the nitrogen atoms dotted. The largest dashed circles are sulfur. Zinc atoms are the smallest hook-shaded circles.

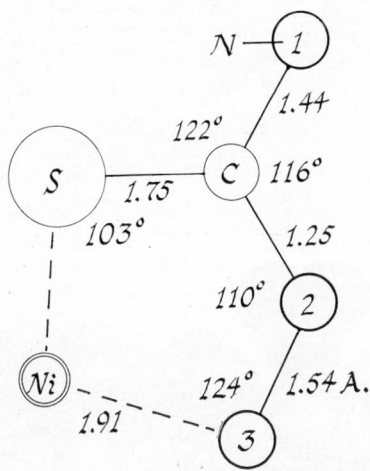

Fig. XIVA,181. Bond dimensions in NH₂NH(NH₂)CS in its compound with nickel.

The space group is C_{2h}^2 $(P2_1/m)$. Atoms of nickel are in the positions:

$$(2a) \quad 000; \; 0 \; {}^1\!/_2 \, 0$$

All other atoms are in the positions:

$$(2e) \quad \pm (u \; {}^1\!/_4 \, v)$$
$$(4f) \quad \pm (xyz; \; x,{}^1\!/_2 - y,z)$$

with the parameters of Table XIVA,65.

The resulting structure is shown in Figure XIVA,182. Bond dimensions of the thiosemicarbazide molecule, which agrees well in shape with that found in $Zn(CH_5N_3S)Cl_2$ (**XIV,a141**), are given in Figure XIVA,183. As this figure indicates, the coordination of the nickel atom is practically square. Water molecules do not lie close to it, but occupy holes in the packing between the complex cations and the sulfate anions. It is stated that the positions of the hydrogen atoms have been found with considerable certainty. The final R, including these atoms, is ca. 0.092.

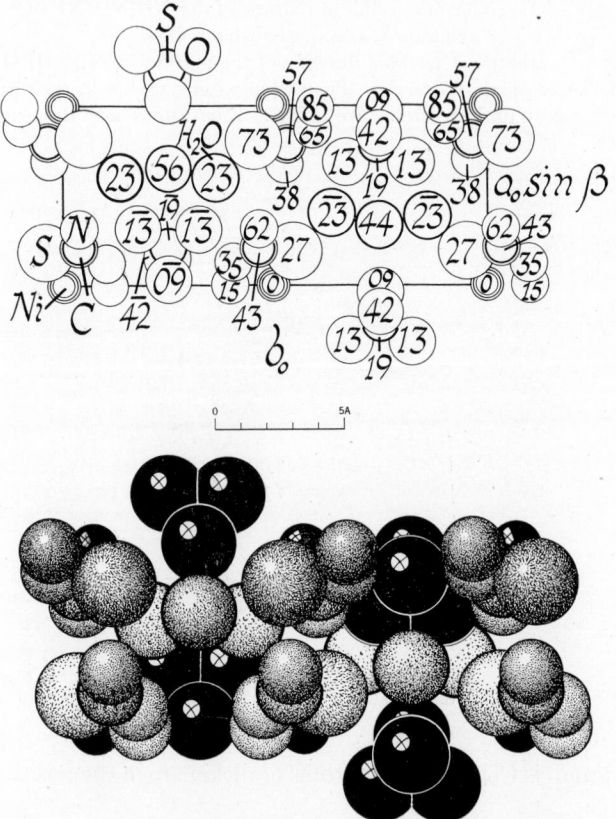

Fig. XIVA,182a (top). The monoclinic structure of $Ni(CH_5N_3S)_2SO_4 \cdot 3H_2O$ projected along its c_0 axis. Right-hand axes.

Fig. XIVA,182b (bottom). A packing drawing of the monoclinic $Ni(CH_5N_3S)_2SO_4 \cdot 3H_2O$ arrangement seen along its c_0 axis. Sulfate oxygens are the large black circles; the smaller black nickel atoms only partially show. The sulfur atoms are large and line shaded. Of the smaller dotted circles, the carbon atoms are more heavily ringed. The larger heavily ringed and dotted circles are the water molecules.

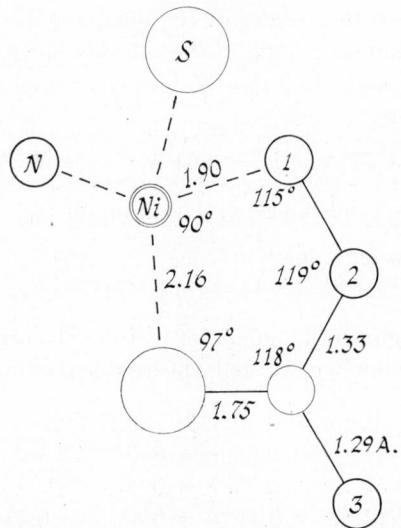

Fig. XIVA,183. Bond dimensions in the NH₂NH(NH₂)CS molecule in its hydrated compound with NiSO₄.

TABLE XIVA,65
Positions and Parameters of the Atoms in $Ni(CH_5N_3S)_2SO_4 \cdot 3H_2O$

Atom	Position	x	$\sigma(x)$	y	$\sigma(y)$	z	$\sigma(z)$
S(1)	(4f)	0.1915	0.006	0.0515	0.003	0.2685	0.005
S(2)	(2e)	−0.215	0.008	1/4	—	0.187	0.007
N(1)	(4f)	9.001	0.02	−0.1005	0.015	0.147	0.015
N(2)	(4f)	0.123	0.03	−0.1025	0.01	0.352	0.015
N(3)	(4f)	0.311	0.045	−0.034	0.02	0.619	0.02
C	(4f)	0.208	0.05	−0.0345	0.015	0.432	0.02
O(1)	(2e)	−0.041	0.025	1/4	—	0.086	0.05
O(2)	(2e)	−0.155	0.05	1/4	—	0.417	0.02
O(3)	(4f)	−0.324	0.02	0.1785	0.025	0.127	0.035
H₂O(1)	(2e)	0.342	0.02	1/4	0.015	0.438	0.02
H₂O(2)	(4f)	0.396	0.015	0.136	0.015	0.773	0.02
H(N,1)	(4f)	0.048	—	−0.145	—	0.060	—
H'(N,1)	(4f)	−0.130	—	−0.117	—	0.175	—
H(N,2)	(4f)	0.125	—	−0.150	—	0.445	—
H(N,3)	(4f)	0.375	—	−0.080	—	0.706	—
H'(N,3)	(4f)	0.344	—	0.013	—	0.719	—
H(H₂O)	(2e)	0.210	—	1/4	—	0.362	—
H'(H₂O)	(2e)	0.415	—	1/4	—	0.322	—
H''(H₂O)	(4f)	0.500	—	0.150	—	0.885	—
H'''(H₂O)	(4f)	0.380	—	0.172	—	0.655	—

XIV,a144.   Each of the halides of the guanidonium ion has a different crystal structure. That prevailing in the chloride has not been established, but *guanidonium bromide*, $C(NH_2)_3Br$, has a tetramolecular orthorhombic cell of the dimensions:

$$a_0 = 6.77 \text{ A.}; \quad b_0 = 8.64 \text{ A.}; \quad c_0 = 8.305 \text{ A.}$$

Its atoms are in the special and general positions of V_h^{16} (*Pmnb*):

(4c) $\pm(1/4\ u\ v;\ 1/4,u+1/2,1/2-v)$

(8d) $\pm(xyz;\ x+1/2,\bar{y},\bar{z};\ 1/2-x,y+1/2,1/2-z;\ \bar{x},1/2-y,z+1/2)$

Transferring the origin to the point ($1/4\ 1/4\ 0$) of the original description to agree with the coordinates just listed, the atomic parameters become

$$Br: u = 0.50, v = -0.179$$
$$C: u = 0.63, v = 0.36$$
$$N(1): u = 0.53, v = 0.26$$
$$N(2): x = 0.41, y = 0.68, z = 0.41$$

In the structure that results (Fig. XIVA,184), the guanidonium ion, $C(NH_2)_3{}^+$, is a centered triangle with $C-NH_2 = 1.20$ A. The NH_2-Br separations vary between 3.46 and 3.85 A.; Br–Br distances are 4.48 and 4.50 A.

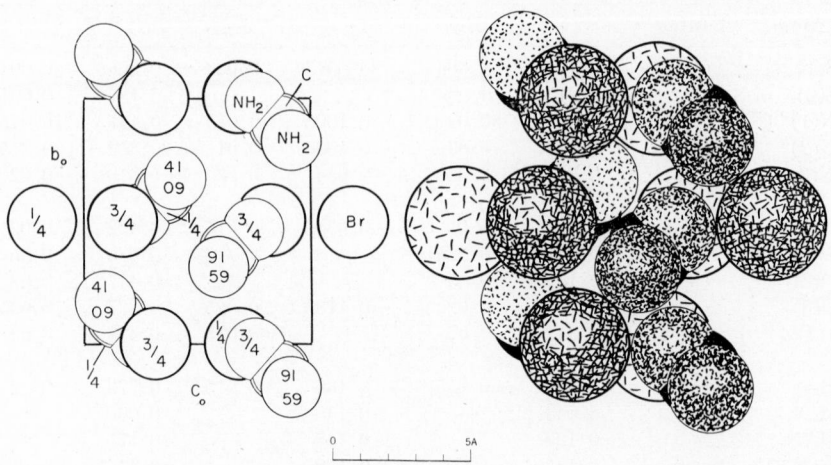

Fig. XIVA,184a (left). The orthorhombic structure of $C(NH_2)_3Br$ projected along its a_0 axis.   Left-hand axes.

Fig. XIVA,184b (right). A packing drawing of the orthorhombic structure of C-$(NH_2)_3Br$ seen along its a_0 axis.   The bromine atoms are large and line shaded; the NH_2 radicals are smaller and dotted.   Black carbon atoms only partly show.

XIV,a145. Crystals of *guanidonium iodide*, $C(NH_2)_3I$, have hexagonal rather than the orthorhombic symmetry that prevails for the bromide. Their tetramolecular unit has the dimensions:

$$a_0 = 7.19 \text{ A.}, \qquad c_0 = 12.30 \text{ A.}$$

The space group is C_{6v}^4 (*P6mc*) with atoms in the positions:

$$
\begin{array}{lll}
I(1): (2a) & 00u;\ 0,0,u+\tfrac{1}{2} & \text{with } u = 0.090 \\
I(2): (2b) & \tfrac{1}{3}\,\tfrac{2}{3}\,u;\ \tfrac{2}{3},\tfrac{1}{3},u+\tfrac{1}{2} & \text{with } u = -0.090 \\
C(1): (2b) & \text{with } u = 0.61 & \\
C(2): (2b) & \text{with } u = 0.324 & \\
NH_2(1): (6c) & u\bar{u}v; \qquad 2\bar{u}\ \bar{u}\ v; \qquad u\ 2u\ v; & \\
& \bar{u},u,v+\tfrac{1}{2};\ 2u,u,v+\tfrac{1}{2};\ \bar{u},2\bar{u},v+\tfrac{1}{2} & \\
& \text{with } u = -0.428,\ v = 0.11 & \\
NH_2(2): (6c) & \text{with } u = -0.238,\ v = -0.176 &
\end{array}
$$

This structure (Fig. XIVA,185) is built of alternate layers of I^- and $C(NH_2)_3^+$ ions parallel to the c_0 axis. As in the bromide, the carbon atoms

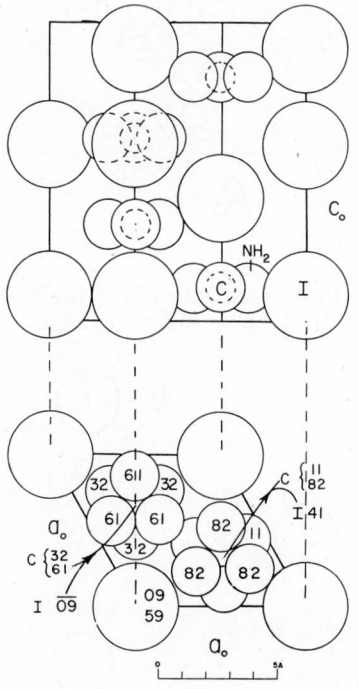

Fig. XIVA 185. Two projections of the hexagonal structure of $C(NH_2)_3I$. Left-hand axes.

are at the centers of triangles of NH_2 radicals, with $C–N = 1.18$ A. The closest approach of iodine atoms to these $C(NH_2)_3{}^+$ is through an $I–NH_2 = 3.71$ A.

XIV,a146. *Methylguanidinium nitrate*, $[(NH_2)_2CNHCH_3]NO_3$, is orthorhombic with a tetramolecular unit of the edge lengths:

$$a_0 = 13.243 \pm 0.005 \text{ A.}; \quad b_0 = 6.427 \pm 0.002 \text{ A.}; \quad c_0 = 7.289 \pm 0.005 \text{ A.}$$

The space group is $V_h{}^{16}$ (*Pnma*) with all atoms [except $H(6,7)$] in the special positions:

$$(4c) \quad \pm (u \; {}^1/_4 \; v; \; u+{}^1/_2, {}^1/_4, {}^1/_2-v)$$

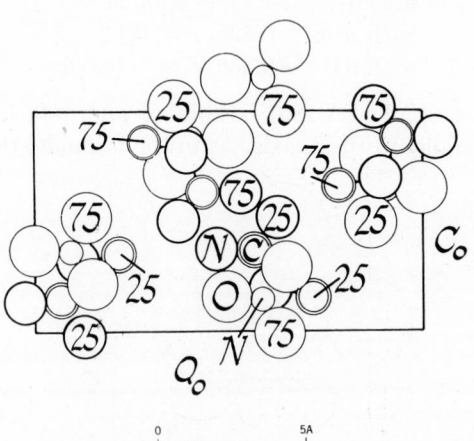

Fig. XIVA,186. The orthorhombic structure of $[(NH_2)_2CNHCH_3]NO_3$ projected along its b_0 axis. Right-hand axes.

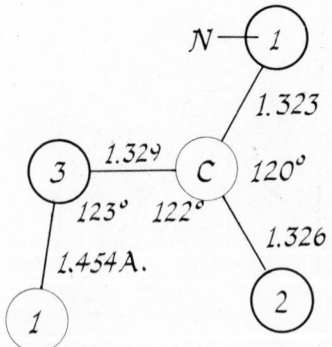

Fig. XIVA,187. The bond dimensions of the cation in $[(NH_2)_2CNHCH_3]NO_3$.

The determined parameters, including those assigned the hydrogen atoms, are listed in Table XIVA,66.

TABLE XIVA,66
Parameters of the Atoms in $[(NH_2)_2CNHCH_3]NO_3$

Atom	u	v
C(0)	0.4254	0.3582
C(1)	0.2801	0.1482
N(1)	0.0248	0.1231
N(2)	0.3660	0.5047
N(3)	0.3885	0.1886
N(4)	0.0919	0.6390
O(1)	0.4983	0.8484
O(2)	0.1451	0.7782
O(3)	0.1329	0.4840
H(1)	0.053	0.026
H(2)	0.062	0.209
H(3)	0.390	0.633
H(4)	0.306	0.498
H(5)	0.424	0.102
H(6,7)[a]	0.253	0.203, $y = 0.14$
H(8)	0.272	0.026

[a] These atoms are in the general positions (8d).

The resulting structure is shown in Figure XIVA,186. The bond dimensions of its molecular cations are those of Figure XIVA,187. Its nitrate ions are triangular, with N–O = 1.235, 1.243, and 1.254 A. Between the ions of opposite sign there are considered to be five different N–H–O bonds of lengths 2.875–3.091 A.

XIV,a147. *Tri aminoguanidinium chloride*, $C(NHNH_2)_3Cl$, is hexagonal with a bimolecular cell of the edge lengths:

$$a_0 = 7.528 \text{ A.}, \qquad c_0 = 6.253 \text{ A.}$$

The space group is C_{6h}^2 ($P6_3/m$) with atoms in the positions:

Cl: (2d) $^2/_3$ $^1/_3$ $^1/_4$; $^1/_3$ $^2/_3$ $^3/_4$
C: (2a) 0 0 $^1/_4$; 0 0 $^3/_4$
N(1): (6h) $\pm(u\ v\ ^1/_4;\ \bar{v},u-v,^1/_4;\ v-u,\bar{u},^1/_4)$
   with $u = 0.1845$, $v = 0.0205$
N(2): (6h) with $u = 0.2065$, $v = -0.1602$

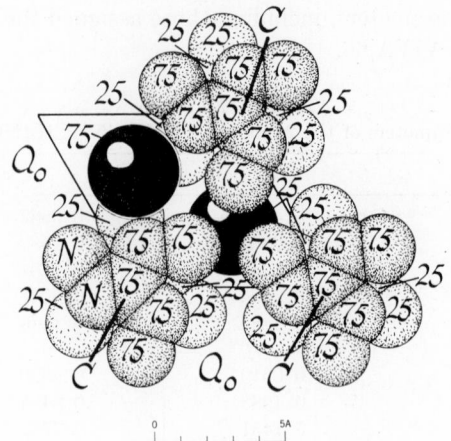

Fig. XIVA,188. The hexagonal structure of $C(NHNH_2)_3Cl$ projected along its c_0 axis. The chlorine atoms are black. Right-hand axes.

The structure is shown in Figure XIVA,188. Its $C(NHNH_2)_3$ cations are planar, with $C-N(1) = 1.318$ A. and $N(1)-N(2) = 1.450$ A.; the bond angles are 120°. Each chloride ion is hydrogen-bonded to three $N(1)$ atoms, with $N-Cl = 3.189$ A.; the $N(2)-Cl$ separations are 3.60 and 3.69 A.

XIV,a148. *Dicyandiamide*, $(NH_2)_2C(NCN)$, has been given a monoclinic unit containing eight molecules in a cell of the dimensions:

$$a_0 = 15.00 \text{ A.}; \quad b_0 = 4.44 \text{ A.}; \quad c_0 = 13.12 \text{ A.}; \quad \beta = 115°20'$$

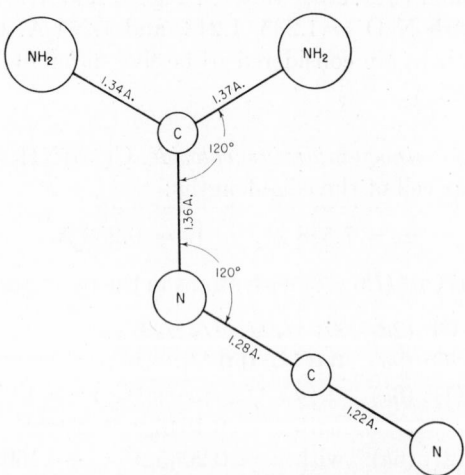

Fig. XIVA,189. Approximate bond dimensions in the molecule of $(NH_2)_2C(NCN)$.

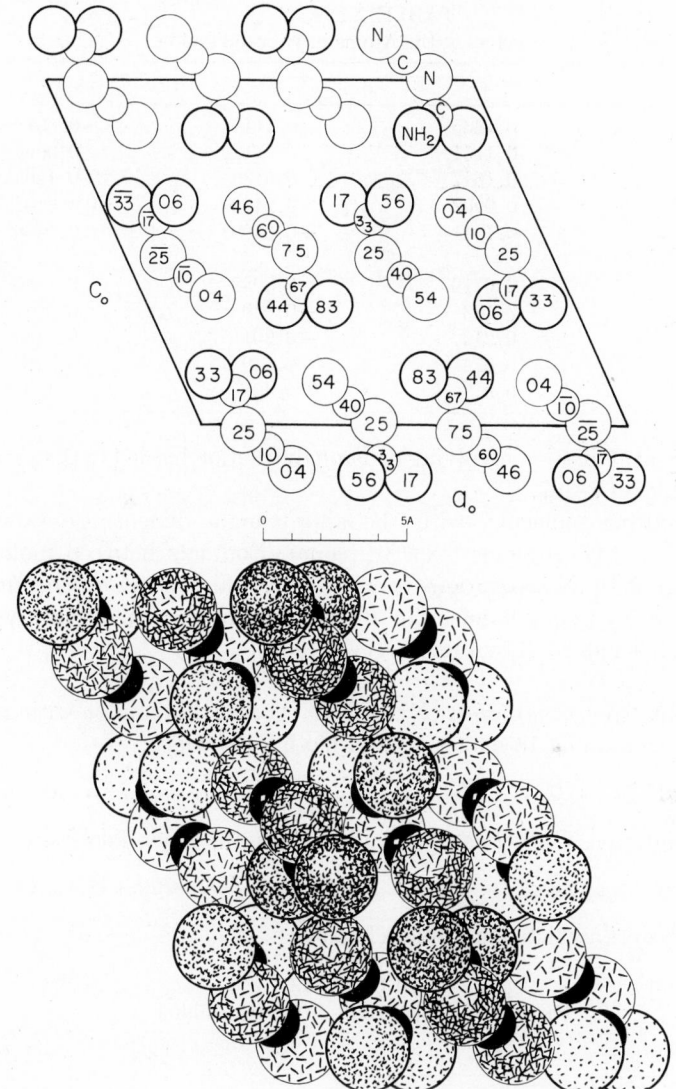

Fig. XIVA,190a (top). The monoclinic structure of $(NH_2)_2C(NCN)$ projected along its b_0 axis.   Left-hand axes.

Fig. XIVA,190b (bottom). A packing drawing of the monoclinic $(NH_2)_2C(NCN)$ arrangement seen along its b_0 axis.   The carbon atoms are black.   The NH_2 radicals are dotted; the other nitrogen atoms are line shaded.

All atoms have been put in general positions of C_{2h}^6 $(C2/c)$:

$(8f)$ $\pm (xyz;\ \bar{x},y,{}^1\!/_2-z;\ x+{}^1\!/_2,y+{}^1\!/_2,z;\ {}^1\!/_2-x,y+{}^1\!/_2,{}^1\!/_2-z)$

TABLE XIVA,67
Parameters of the Atoms in Dicyandiamide

Atom	x	y	z
C(1)	0.1335	0.1113	−0.0672
C(2)	0.1151	0.1717	0.0907
N(1)	0.1647	−0.0133	−0.1268
N(2)	0.0902	0.2488	−0.0162
N(3)	0.1811	−0.0400	0.1469
N(4)	0.0710	0.3088	0.1440
H(1)	0.183	−0.040	0.228
H(2)	0.212	−0.201	0.113
H(3)	0.074	0.149	0.216
H(4)	0.918	0.468	0.110

A recent re-examination involving neutron diffraction has led to the parameters of Table XIVA,67.

The resulting molecule, with the approximate dimensions shown in Figure XIVA,189, is almost exactly planar. Contacts between molecules are through NH_2–N separations of 2.94, 3.02, and 3.04 A. which presumably involve hydrogen bonds. The distribution of these molecules within the unit cell is indicated by Figure XIVA,190.

XIV,a149. *Nitroguanidine*, $(NH_2)_2CNNO_2$, is orthorhombic with a very large unit containing 16 molecules and having the edges:

$$a_0 = 17.58 \pm 0.09 \text{ A.}; \quad b_0 = 24.82 \pm 0.12 \text{ A.}; \quad c_0 = 3.58 \pm 0.02 \text{ A.}$$

The space group is C_{2v}^{19} ($Fdd2$) with all atoms in the general positions:

(16b) $xyz; \bar{x}\bar{y}z; 1/4-x,y+1/4,z+1/4; x+1/4,1/4-y,z+1/4;$ F.C.

The parameters are listed in Table XIVA,68.

TABLE XIVA,68
Parameters of the Atoms in Nitroguanidine

Atom	x	$\sigma(x)$	y	$\sigma(y)$	z
O(1)	0.1770	0.0002	0.0114	0.0002	0.208
O(2)	0.0743	0.0012	0.0256	0.0001	0.509
N(1)	0.2568	0.0008	0.0929	0.0002	0.000
N(2)	0.1953	0.0008	0.1719	0.0001	0.151
N(3)	0.1359	0.0004	0.0967	0.0005	0.325
N(4)	0.1306	0.0006	0.0425	0.0003	0.343
C	0.1980	0.0001	0.1180	0.0003	0.160

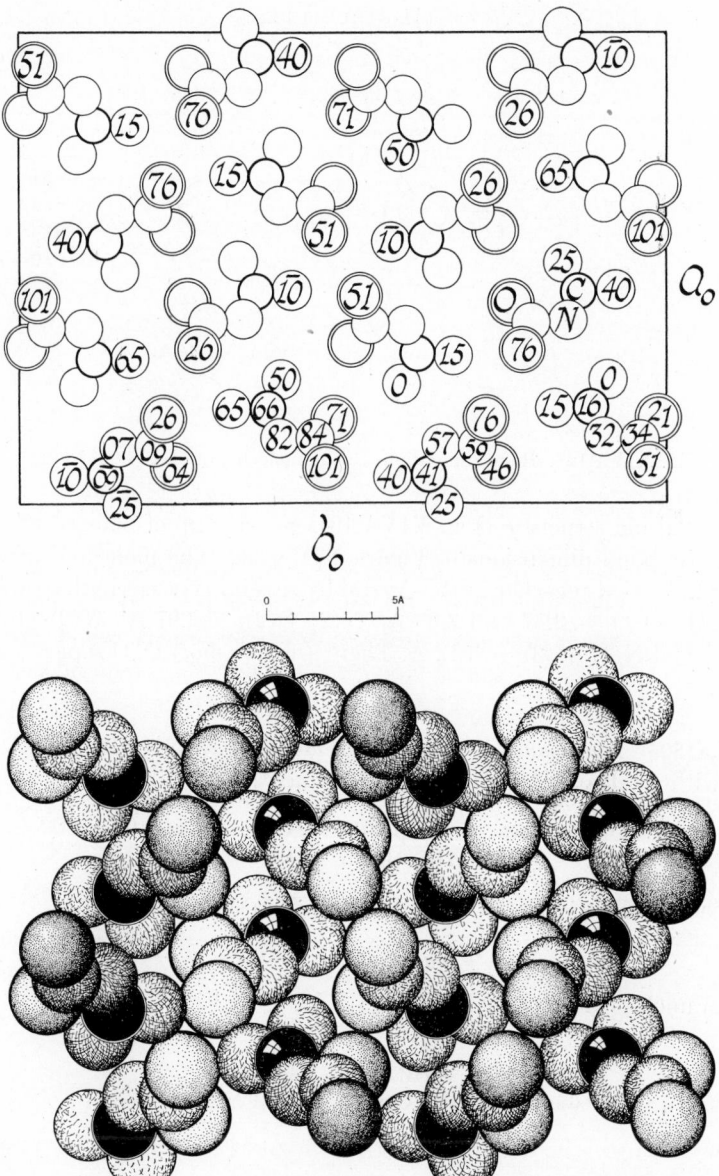

Fig. XIVA,191a (top). The orthorhombic structure of $(NH_2)_2CNNO_2$ projected along
its short c_0 axis.   Right-hand axes.
Fig. XIVA,191b (bottom). A packing drawing of the orthorhombic $(NH_2)_2CNNO_2$
structure viewed along its c_0 axis.   The carbon atoms are black, the oxygen atoms heavily
outlined and dotted.   Nitrogen atoms are line shaded.

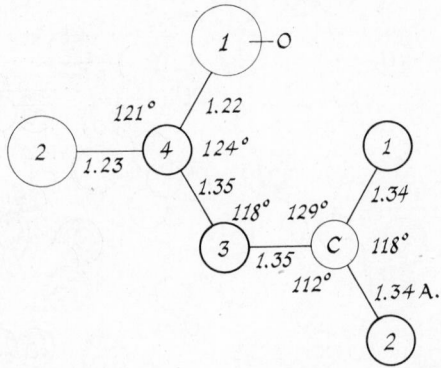

Fig. XIVA,192. Bond dimensions in the molecule of $(NH_2)_2CNNO_2$.

The resulting structure (Fig. XIVA,191) is made up of planar molecules having the bond dimensions of Figure XIVA,192. The molecules are considered to be tied together in the crystal by several types of hydrogen bond with $N(1)$–$O(1)$ = 2.57 and 3.03 A., $N(2)$–$O(2)$ = 2.97 A., $N(2)$–$O(3)$ = 3.28 A., and $N(1)$–$N(3)$ = 3.12 A. There are also short O–O separations of 2.90 A.

XIV,a150. *Aminoguanidine hydrochloride*, $(NH_2)_2CNHNH_2 \cdot Cl$, is monoclinic with a tetramolecular unit of the dimensions:

$$a_0 = 7.78 \pm 0.02 \text{ A.}; \quad b_0 = 11.07 \pm 0.04 \text{ A.}; \quad c_0 = 6.00 \pm 0.02 \text{ A.}$$
$$\beta = 103°5' \pm 10'$$

The space group is C_{2h}^5 $(P2_1/n)$ with all atoms in the positions:

$$(4e) \quad \pm (xyz; x+\tfrac{1}{2}, \tfrac{1}{2}-y, z+\tfrac{1}{2})$$

The parameters are those of Table XIVA,69.

TABLE XIVA,69
Parameters of the Atoms in $(NH_2)_2CNHNH_2 \cdot Cl$

Atom	x	σ_x, A.	y	σ_y, A.	z	σ_z, A.
Cl	0.5159	0.0053	0.3347	0.0058	0.1523	0.0066
N(1)	0.4294	0.019	−0.1134	0.019	0.1950	0.025
N(2)	0.2311	0.021	−0.0942	0.021	−0.1546	0.025
N(3)	0.2412	0.020	0.0543	0.021	0.1058	0.024
N(4)	0.3077	0.020	0.0969	0.019	0.3318	0.023
C	0.3038	0.024	−0.0534	0.023	0.0526	0.028

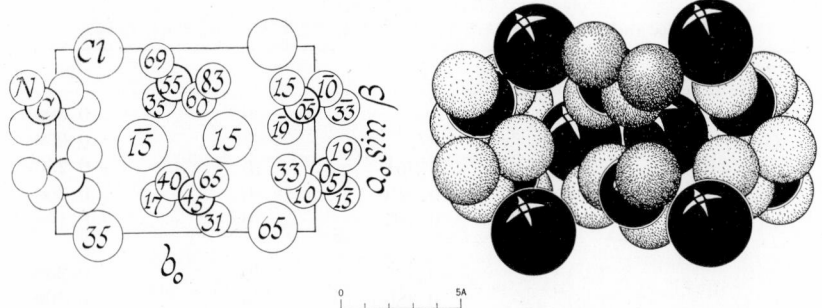

Fig. XIVA,193a (left). The monoclinic structure of $(NH_2)_2CNHNH_2Cl$ projected along its c_0 axis. Right-hand axes.
Fig. XIVA,193b (right). A packing drawing of the monoclinic $(NH_2)_2CNHNH_2Cl$, arrangement seen along its c_0 axis. Chlorine atoms are the large, carbon atoms the somewhat smaller black circles. Atoms of nitrogen are hook shaded.

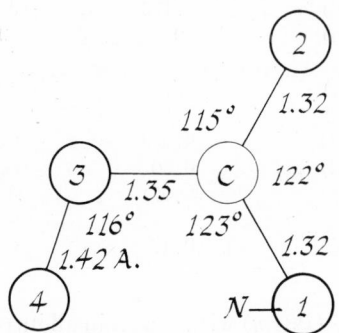

Fig. XIVA,194. Bond dimensions in the $[(NH_2)_2CNHNH_2]^+$ ion in its chloride.

The atomic arrangement is that of Figure XIVA,193. The molecules have the bond dimensions of Figure XIVA,194. The guanidinium part of the molecule is strictly planar and the amino group N(4) is 0.093 A. out of this plane. Between molecules the closest atomic approach, which probably corresponds to a hydrogen bond, is $N(3)$–$Cl = 3.13$ A. There are other close N–Cl separations of 3.31 and 3.39 A.

XIV,a151. *Guanidinium gallium sulfate hexahydrate*, $[C(NH_2)_3]Ga(SO_4)_2 \cdot 6H_2O$, is hexagonal with a trimolecular cell of the edge lengths:

$$a_0 = 11.82 \pm 0.03 \text{ A.}, \quad c_0 = 9.13 \pm 0.03 \text{ A.}$$

TABLE XIVA,70
Positions and Parameters of the Atoms in $[C(NH_2)_3]Ga(SO_4)_2 \cdot 6H_2O$

Atom	Position	x	y	z
Ga(1)	(1a)	0	0	0.000
Ga(2)	(2b)	$^1/_3$	$^2/_3$	0.059
S(1)	(3c)	0.351	0	−0.250
S(2)	(3c)	−0.318	0	0.306
O(1)	(3c)	0.455	0	−0.156
O(2)	(3c)	−0.433	0	0.238
O(3)	(3c)	0.386	0	−0.409
O(4)	(3c)	−0.328	0	0.470
O(5)	(6d)	0.342	0.120	−0.220
O(6)	(6d)	−0.314	−0.120	0.258
H₂O(1)	(3c)	0.140	0	0.130
H₂O(2)	(3c)	−0.144	0	−0.135
H₂O(3)	(6d)	−0.477	0.333	−0.066
H₂O(4)	(6d)	0.476	−0.336	0.174
C(1)	(1a)	0	0	0.50
C(2)	(2b)	$^1/_3$	$^2/_3$	0.52
N(1)	(3c)	0.113	0	0.50
N(2)	(6d)	0.553	0.333	0.52
H(1)	(6d)	0.20	0.08	0.50
H(2)	(6d)	0.55	0.42	0.52
H(3)	(6d)	0.47	0.25	0.52

The space group is C_{3v}^2 ($P31m$) with atoms in all its positions:

(1a) $00u$

(2b) $^1/_3\ ^2/_3\ u;\ ^2/_3\ ^1/_3\ u$

(3c) $u0v;\ 0uv;\ \bar{u}\bar{u}v$

(6d) $xyz;\ \bar{y},x-y,z;\ y-x,\bar{x},z;$
        $yxz;\ \bar{x},y-x,z;\ x-y,\bar{y},z$

The assigned parameters are those of Table XIVA,70.

In this structure the gallium atoms are octahedrally surrounded by water molecules, with Ga–OH₂ = 2.04–2.10 A., and these water molecules are hydrogen-bonded to sulfate oxygens with O–H–O = 2.51–2.64 A.

The corresponding *aluminum* compound, $[C(NH_2)_3]Al(SO_4)_2 \cdot 6H_2O$, is isostructural with

$$a_0 = 11.75 \pm 0.02 \text{ A.}, \qquad c_0 = 8.94 \pm 0.01 \text{ A.}$$

For this compound the structure could not be refined in satisfactory fashion and, as indicated in Table XIVA,71, there are important differences in some parameters in the two structures.

TABLE XIVA,71
Positions and Parameters given the Atoms in $[C(NH_2)_3]Al(SO_4)_2 \cdot 6H_2O$

Atom	Position	x	y	z
Al(1)	(1a)	0	0	0.0000
Al(2)	(2b)	$^1/_3$	$^2/_3$	0.4396
S(1)	(3c)	0.3469	0	−0.2502
S(2)	(3c)	−0.3200	0	0.3079
O(1)	(3c)	0.4688	0	−0.3351
O(2)	(3c)	−0.4246	0	0.2739
O(3)	(3c)	0.3651	0	−0.1185
O(4)	(3c)	−0.3180	0	0.4471
O(5)	(6d)	0.3512	0.1250	−0.3122
O(6)	(6d)	−0.3281	−0.1123	0.2025
H₂O(1)	(3c)	0.1367	0	0.1260
H₂O(2)	(3c)	−0.1375	0	−0.1073
H₂O(3)	(6d)	−0.4639	0.3327	0.3434
H₂O(4)	(6d)	0.4647	−0.3422	0.5294
C(1)	(1a)	0	0	0.4694
C(2)	(2b)	$^1/_3$	$^2/_3$	0.0110
N(1)	(6d)	0.0987	0	0.4124
N(2)	(6d)	0.2047	−0.3304	−0.0305

XIV,a152. Crystals of *creatine monohydrate*, $C_4N_3O_2H_9 \cdot H_2O$, are monoclinic with a tetramolecular cell of the dimensions:

$$a_0 = 12.50 \text{ A.}; \quad b_0 = 5.01 \text{ A.}; \quad c_0 = 12.16 \text{ A.}; \quad \beta = 109°$$

The space group is C_{2h}^5 ($P2_1/c$) with all atoms in the general positions:

$$(4e) \quad \pm (xyz; \ x, ^1/_2 - y, z + ^1/_2)$$

The selected parameters are listed in Table XIVA,72.

The molecular packing within the crystal is illustrated in Figure XIVA, 195. Bond lengths and angles in the molecule itself, which is a zwitter ion, are those of Figure XIVA,196. Evidently it is a guanidonium ion in which one hydrogen of an amino group is replaced by CH_3 and the other by ($-CH_2COO^-$). The important interionic hydrogen bonds have the values indicated in the figure.

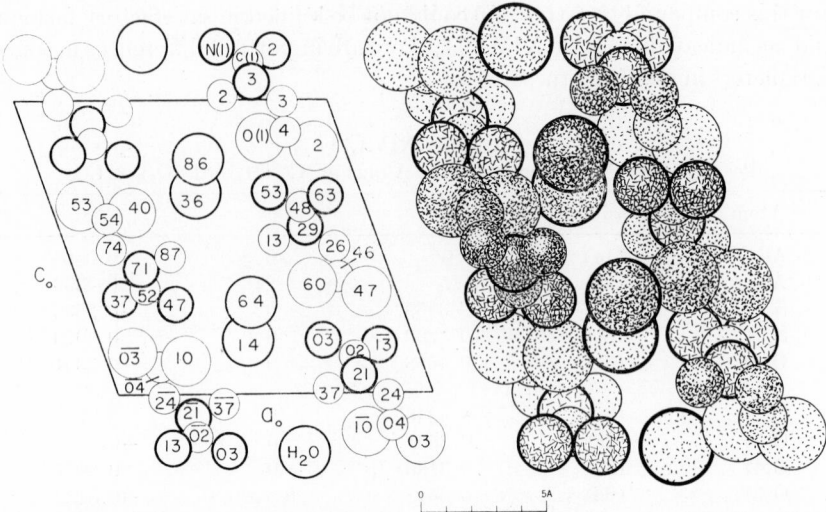

Fig. XIVA,195a (left). The monoclinic structure of creatine monohydrate projected along its b_0 axis. Left-hand axes.

Fig. XIVA,195b (right). A packing drawing of the monoclinic structure of creatine monohydrate seen along its b_0 axis. The oxygen atoms are the large, the carbon atoms the small, lightly outlined, and dotted circles. Atoms of nitrogen are line shaded. The water molecules are also heavily outlined, but larger and dotted.

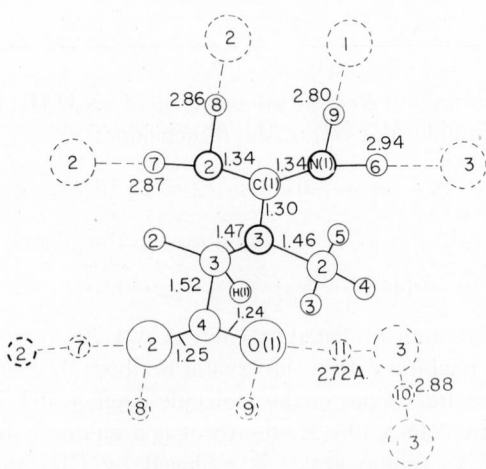

Fig. XIVA,196. Bond dimensions, including intermolecular hydrogen-bond lengths, in the molecules of creatine monohydrate.

TABLE XIVA,72T
Parameters of the Atoms in Creatine Monohydrate[a]

Atom	x	y	z
N(1)	0.289 (0.288)	0.472 (0.472)	0.306 (0.307)
N(2)	0.119 (0.120)	0.370 (0.366)	0.329 (0.327)
N(3)	0.224 (0.226)	0.715 (0.715)	0.434 (0.435)
C(1)	0.211 (0.211)	0.520 (0.522)	0.358 (0.360)
C(2)	0.331 (0.330)	0.874 (0.870)	0.472 (0.473)
C(3)	0.148 (0.149)	0.744 (0.741)	0.502 (0.504)
C(4)	0.170 (0.171)	0.542 (0.540)	0.600 (0.601)
O(1)	0.256 (0.256)	0.397 (0.396)	0.625 (0.626)
O(2)	0.098 (0.098)	0.529 (0.523)	0.651 (0.651)
$H_2O(3)$	0.484 (0.485)	0.356 (0.360)	0.689 (0.690)
H(1)	0.170	0.914	0.546
H(2)	0.073	0.734	0.457
H(3)	0.320	0.012	0.548
H(4)	0.396	0.754	0.516
H(5)	0.344	0.917	0.415
H(6)	0.352	0.527	0.317
H(7)	0.048	0.381	0.348
H(8)	0.116	0.214	0.272
H(9)	0.280	0.341	0.250
H(10)	0.500	0.206	0.750
H(11)	0.408	0.367	0.675

[a] The hydrogen parameters and those in parentheses for the other atoms are taken from 1955: J.

BIBLIOGRAPHY TABLE, CHAPTER XIVA

Compound	Paragraph	Literature
di-Acetonitrile dodecahydrodeca-borane $B_{10}H_{12}(CH_3CN)_2$	a29	1959: R&L
Aluminum hydride di trimethyl-amine $AlH_3 \cdot 2N(CH_3)_3$	a70	1963: H,N&P
Aluminum methyl dichloride $AlCH_3Cl_2$	a52	1962: A&P; 1963: A,P&I
Aminoguanidine hydrochloride $(NH_2)_2CNHNH_2 \cdot Cl$	a150	1957: B
tri-Aminoguanidinium chloride $C(NHNH_2)_3Cl$	a147	1957: O&P

(continued)

BIBLIOGRAPHY TABLE, CHAPTER XIVA (*continued*)

Compound	Paragraph	Literature
mono-Amyl ammonium bromide		
$NH_3 \cdot nC_5H_{11}Br$	a103	1928: H
mono-Amyl ammonium chloride		
$NH_3 \cdot nC_5H_{11}Cl$	a103	1928: H; 1930: H; 1933: S,M&H
mono-Amyl ammonium iodide		
$NH_3 \cdot nC_5H_{11}I$	a103	1928: H; 1930: W
Antimony pentachloride phos-		
phorus oxytrichloride		
$SbCl_5 \cdot POCl_3$	a66	1959: L&B
Antimony pentachloride phos-		
phorus oxytrimethyl		
$SbCl_5 \cdot PO(CH_3)_3$	a66	1961: B&L
Arsenomethane		
$(AsCH_3)_5$	a77	1957: B&W
Auric bromide trimethyl phosphine		
$AuBr_3 \cdot P(CH_3)_3$	a72	1946: P&W
Azochloramide		
$[ClNC(NH_2)N{=}]_2$	a24	1958: B
Azodicarbonamide		
$[NH_2C(O)N]_2$	a25	1961: B
di-Barium cupric formate		
tetrahydrate		
$CuBa_2(HCOO)_6 \cdot 4H_2O$	a100	1958: SR,S&SR
Barium formate		
$Ba(HCOO)_2$	a94	1928: N; 1949: N&S; 1951: S,K,-S&N
Biuret		
$NH_2C(O)NHC(O)NH_2$	a26	1961: H,Y&F
di-Bromo bis-(trimethyl arsine)		
μ-dibromodipalladium		
$[As(CH_3)_3 \cdot PdBr_2]_2$	a75	1938: M&W; W
mono-Butyl ammonium bromide		
$NH_3 \cdot nC_4H_9Br$	a103	1928: H
mono-Butyl ammonium chloride		
$NH_3 \cdot nC_4H_9Cl$	a103	1928: H
mono-Butyl ammonium iodide		
$NH_3 \cdot nC_4H_9I$	a103	1928: H; 1930: W
$(CH_3)_2NSO_2N(CH_3)_2$	a39	1962: J,S&L; 1963: J,S,L&L
Cacodyl cyanide		
$(CH_3)_2AsCN$	a74	1963: C&T
Cacodyl disulfide		
$[(CH_3)_2AsS]_2$	a78	1964: C&T
Cadmium chloride bis-biuret		
$CdCl_2[OC(NH_2)NHC(NH_2)O]_2$	a85	1960: C,N&F

(*continued*)

BIBLIOGRAPHY TABLE, CHAPTER XIVA (*continued*)

Compound	Paragraph	Literature
Cadmium chloride di thiourea		
CdCl$_2$·2(NH$_2$)$_2$CS	a131	1957: N,C&B
Cadmium chloride diurea		
CdCl$_2$·2(NH$_2$)$_2$CO	a125	1957: N,C&F
Cadmium chloride tetra thiourea		
CdCl$_2$·4(NH$_2$)$_2$CS	a134	1956: N,C&B
Cadmium formate		
Cd(HCOO)$_2$	a93	1961: H; 1963: H
Cadmium formate di thiourea		
Cd(HCOO)$_2$·2(NH$_2$)$_2$CS	a132	1962: N,F&B
Cadmium thiocyanate di thiourea		
Cd(NCS)$_2$·2(NH$_2$)$_2$CS	a135	1957: N,C&B
Calcium formate		
Ca(HCOO)$_2$	a93	1925: Y; 1928: N; 1948: N&O; 1963: H
Cerium formate		
Ce(HCOO)$_3$	a99	1962: M,S,F&G
di-Chloro bis-(trimethyl arsine)		
μ-dichlorodipalladium		
[As(CH$_3$)$_3$·PdCl$_2$]$_2$	a75	1938: M&W; W
hexa-Chloro dimethyl trisulfide		
(CCl$_3$)$_2$S$_3$	a32	1961: B
Cobaltous chloride tetra thiourea		
CoCl$_2$·4(NH$_2$)$_2$CS	a134	1956: N,C&B
Cobaltous thiocyanate di thiourea		
Co(NCS)$_2$·2(NH$_2$)$_2$CS	a135	1957: N,C&B
Copper formate		
Cu(HCOO)$_2$	a96	1961: B&K
Creatine monohydrate		
C$_4$N$_3$O$_2$H$_9$·H$_2$O	a152	1954: M&H; 1955: J
Cupric bromide tetra methylamine		
CuBr$_2$·4CH$_3$NH$_2$	a69	1954: L&M
Cupric formate tetrahydrate		
Cu(HCOO)$_2$·4H$_2$O	a97	1931: K; 1954: K,I&M
Cuprous chloride azomethane		
Cu$_2$Cl$_2$·N$_2$(CH$_3$)$_2$	a68	1960: B&D
Cuprous chloride tris-thiourea		
CuCl·3(NH$_2$)$_2$CS	a130	1959: K,O&P; 1964: O&K
Cyanamide		
NH$_2$C≡N	a14	1951: C; 1961: Z&K
di-Cyandiamide		
(NH$_2$)$_2$C(NCN)	a148	1940: H; 1964: R&O
di-Cyanotrisulfide		
NCS$_3$CN	a33	1964: F&L

(*continued*)

BIBLIOGRAPHY TABLE, CHAPTER XIVA (*continued*)

Compound	Paragraph	Literature
mono-Decyl ammonium iodide $NH_3 \cdot nC_{10}H_{21}I$	a103	1930: W
mono-Dodecyl ammonium iodide $NH_3 \cdot nC_{12}H_{25}I$	a103	1930: W
mono-Ethyl chlor mercury mercaptan C_2H_5SHgCl	a49	1939: J
tetra-Ethyl ammonium iodide $N(C_2H_5)_4I$	a108	1927: G; 1928: N; W; 1958: W&P
tetra-Ethyl ammonium plutonium hexachloride $[N(C_2H_5)_4]_2PuCl_6$	a116	1952: S&S
Ferrous chloride tetra thiourea $FeCl_2 \cdot 4(NH_2)_2CS$	a134	1956: N,C&B
Formamide $HCONH_2$	a89	1954: L&P
Formamidinium disulfide dibromide monohydrate $[(NH_2)_2CS]_2Br_2 \cdot H_2O$	a139	1957: F&J; 1958: F,J&T
Formamidinium disulfide diiodide monohydrate $[(NH_2)_2CS]_2I_2 \cdot H_2O$	a140	1957: F&J; 1958: F,J&T
Formamidoxime NH_2CHNOH	a90	1956: H&L
Formic acid $HCOOH$	a88	1952: H,P&F; 1953: H,P&F
di-Formyl hydrazine $OCHNHNHCHO$	a91	1958: T,K&N
Gadolinium formate $Gd(HCOO)_3$	a99	1943: P; 1962: M,S,F&G
Gallium hydride trimethylamine $GaH_3 \cdot N(CH_3)_3$	a71	1963: S&N
Guanidinium aluminum sulfate hexahydrate $[C(NH_2)_3]Al(SO_4)_2 \cdot 6H_2O$	a151	1958: V,Z&U; 1962: G&K
Guanidinium gallium sulfate hexahydrate $[C(NH_2)_3]Ga(SO_4)_2 \cdot 6H_2O$	a151	1959: G&B; 1960: G
Guanidonium bromide $C(NH_2)_3Br$	a144	1935: T
Guanidonium iodide $C(NH_2)_3I$	a145	1935: T

(*continued*)

BIBLIOGRAPHY TABLE, CHAPTER XIVA (*continued*)

Compound	Paragraph	Literature
mono-Heptyl ammonium chloride		
$NH_3 \cdot nC_7H_{15}Cl$	a103	1928: H
mono-Heptyl ammonium iodide		
$NH_3 \cdot nC_7H_{15}I$	a103	1928: H; 1930: W
mono-Hexyl ammonium bromide		
$NH_3 \cdot nC_6H_{13}Br$	a103	1928: H
mono-Hexyl ammonium chloride		
$NH_3 \cdot nC_6H_{13}Cl$	a103	1928: H
mono-Hexyl ammonium iodide		
$NH_3 \cdot nC_6H_{13}I$	a103	1928: H; 1930: W
Iodoform		
CHI_3	a2	1926: N; 1927: W; 1931: H&N; 1951: K,K&S; 1953: K,K&S
Iodoform sulfur		
$CHI_3 \cdot 3S_8$	a3	1931: H; 1937: W; 1962: B
di-Iodomethyl arsine		
CH_3AsI_2	a73	1963: C&T
Isocyanic acid		
HNCO	a13	1955: vD&C
Lead acetate thiourea		
$Pb(CH_3COO)_2 \cdot (NH_2)_2CS$	a138	1959: N&F; 1960: N,F&B
Lead formate		
$Pb(HCOO)_2$	a94	1928: N; 1932: H&Z; 1949: N&S; 1951: S,K,S&N
Lithium methoxide		
$LiOCH_3$	a46	1960: W; 1961: D&K
Magnesium formate dihydrate		
$Mg(HCOO)_2 \cdot 2H_2O$	a98	1963: O,N&W; 1964: O,N&W
Manganous chloride tetra thiourea		
$MnCl_2 \cdot 4(NH_2)_2CS$	a134	1956: N,C&B
Manganous formate dihydrate		
$Mn(HCOO)_2 \cdot 2H_2O$	a98	1963: O,N&W; 1964: O,N&W
Manganous thiocyanate di thiourea		
$Mn(NCS)_2 \cdot 2(NH_2)_2CS$	a135	1957: N,C&B
Mercuric chloride bis-biuret		
$HgCl_2[OC(NH_2)NHC(NH_2)O]_2$	a85	1960: N&C
Mercury methyl mercaptide		
$Hg(SCH_3)_2$	a50	1962: B&K; 1964: B&K
Methane		
CH_4	a1	1928: ML&P; 1931: M; 1939: M&S; S
Methyl alcohol		
CH_3OH	a5	1952: DA&D; T&L; 1959: KM

(*continued*)

BIBLIOGRAPHY TABLE, CHAPTER XIVA (*continued*)

Compound	Paragraph	Literature
Methyl alcohol bromine		
$2CH_3OH \cdot Br_2$	a6	1963: G&H; 1964: G&H
Methyl amine		
CH_3NH_2	a15	1953: A&L
mono-Methylamine–boron		
trifluoride		
$CH_3NH_2 \cdot BF_3$	a16	1950: G&H; 1951: H,G&O
mono-Methyl ammonium aluminum alum		
$(CH_3NH_3)Al(SO_4)_2 \cdot 12H_2O$	a105	1957: O,A,P&V; 1961: F&S; 1964: F&S
mono-Methyl ammonium bromide		
NH_3CH_3Br	a103	1928: H; 1961: G
mono-Methyl ammonium chloride		
NH_3CH_3Cl	a102	1928: H; 1946: H&L
mono-Methyl ammonium chloroplatinate		
$[NH_3CH_3]_2PtCl_6$	a104	1928: W
mono-Methyl ammonium chlorostannate		
$[NH_3CH_3]_2SnCl_6$	a104	1928: W
mono-Methyl ammonium iodide		
NH_3CH_3I	a103	1928: H
mono-Methyl chlor mercury mercaptan		
CH_3SHgCl	a49	1939: J
Methyl chloride		
CH_3Cl	a4	1953: B
Methyl guanidinium nitrate		
$[(NH_2)_2CNHCH_3]NO_3$	a146	1955: C&P
O-Methyl hydroxylamine hydrochloride		
$CH_3ONH_2 \cdot HCl$	a20	1964: L&R
S-Methyl isothiourea sulfate		
$[(NH_2)_2CSCH_3]_2SO_4$	a129	1962: S
Methyl lithium		
CH_3Li	a45	1964: W&L
Methyl meta dithiophosphonate		
CH_3PS_2	a31	1962: W
Methyl triethyl ammonium chlorostannate		
$[NCH_3(C_2H_5)_3]_2SnCl_6$	a116	1929: W&C

(*continued*)

BIBLIOGRAPHY TABLE, CHAPTER XIVA (*continued*)

Compound	Paragraph	Literature
Methyl triethyl phosphonium chlorostannate $[PCH_3(C_2H_5)_3]_2SnCl_6$	a116	1930: C&W
mono-Methyl urea $NH_2(NHCH_3)CO$	a119	1924: M; 1925: M; 1933: C&W
Methyl urea nitrate $(CH_3NH_2)NH_2CO \cdot NO_3$	a120	1957: B
di-Methyl amino borine $(BH_2)_3[N(CH_3)_2]_3$	a28	1959: T&L; 1961: T,M&L
di-Methyl amino boron dichloride $N(CH_3)_2BCl_2$	a27	1963: H
di-Methyl ammonium bromide–bromine $[(CH_3)_2NH_2Br]_2 \cdot Br_2$	a107	1959: S
di-Methyl ammonium chloride-iodine $[(CH_3)_2NH_2Cl]_2 \cdot I_2$	a107	1959: S
di-Methyl ammonium chloro-stannate $[NH_2(CH_3)_2]_2SnCl_6$	a106	1933: G; 1934: C&W
di-Methyl beryllium $(CH_3)_2Be$	a48	1950: R&S; 1951: S&R
di-Methyl ethyl sulfonium chloro-stannate $[S(CH_3)_2C_2H_5]_2SnCl_6$	a116	1930: C&W
di-Methyl gallium hydroxide $(CH_3)_2GaOH$	a53	1959: S&H
di-Methyl magnesium $(CH_3)_2Mg$	a48	1964: W
di-Methyl phosphinoborine $[(CH_3)_2PBH_2]_3$	a28	1955: H
di-Methyl sulfone $(CH_3)_2SO_2$	a35	1963: S
di-Methyl sulfonyl disulfide $(CH_3SO_2S)_2$	a36	1949: S&F; 1953: S
di-Methyl tellurium dichloride $(CH_3)_2TeCl_2$	a58	1958: C,S&MC
di-Methyl thallic bromide $(CH_3)_2TlBr$	a55	1934: P&C
di-Methyl thallic chloride $(CH_3)_2TlCl$	a55	1934: P&C
di-Methyl thallic iodide $(CH_3)_2TlI$	a55	1934: P&C

(*continued*)

BIBLIOGRAPHY TABLE, CHAPTER XIVA (continued)

Compound	Paragraph	Literature
tri-Methyl aluminum dimer		
$[Al(CH_3)_3]_2$	a51	1953: L&R
tri-Methyl amine–boron trifluoride		
$(CH_3)_3N \cdot BF_3$	a17	1951: G&H; G,H&H
tri-Methyl amine–boron tri- hydride		
$(CH_3)_3N \cdot BH_3$	a17	1951: G&H; H,G&O
tri-Methyl amine–iodine		
$(CH_3)_3N \cdot I_2$	a19	1959: S
tri-Methyl amine–iodine mono- chloride		
$(CH_3)_3N \cdot ICl$	a18	1960: H&H
tri-Methyl amine oxide		
$(CH_3)_3NO$	a21	1964: C,P,G&D
tri-Methyl amine oxide hydro- chloride		
$(CH_3)_3NO \cdot HCl$	a22	1957: R; 1960: R; 1962: C&D
tri-Methyl ammonium chloro- stannate		
$[NH(CH_3)_3]_2SnCl_6$	a116	1929: W&C
tri-Methyl antimony dibromide		
$(CH_3)_3SbBr_2$	a65	1938: W
tri-Methyl antimony dichloride		
$(CH_3)_3SbCl_2$	a65	1938: W
tri-Methyl antimony diiodide		
$(CH_3)_3SbI_2$	a65	1938: W
tri-Methyl ethyl ammonium chlorostannate		
$[N(CH_3)_3C_2H_5]_2SnCl_6$	a116	1929: W&C
tri-Methyl indium		
$(CH_3)_3In$	a54	1958: A&R
tri-Methyl oxosulfonium fluoborate		
$[(CH_3)_3SO]BF_4$	a38	1963: Z,B&MC
tri-Methyl oxosulfonium per- chlorate		
$[(CH_3)_3SO]ClO_4$	a37	1963: C,G&MC
bis(tri-Methyl phosphine oxide)- cobaltous nitrate		
$Co(NO_3)_2 \cdot 2[(CH_3)_3PO]$	a84	1963: C&S
tri-Methyl platinum chloride		
$(CH_3)_3PtCl$	a56	1935: C&W; 1947: R&S

(continued)

BIBLIOGRAPHY TABLE, CHAPTER XIVA (*continued*)

Compound	Paragraph	Literature
tri-Methyl sulfonium chloro-		
stannate		
[S(CH₃)₃]₂SnCl₆	a116	1930: C&W
tri-Methyl sulfonium iodide		
(CH₃)₃SI	a34	1941: M; 1959: Z&MC
tri-Methyl tin fluoride		
(CH₃)₃SnF	a57	1963: C,OB&T; 1964: C,OB&T
tetra-Methyl ammonium bromide		
N(CH₃)₄Br	III,b3	1927: V&S; 1928: W
tetra-Methyl ammonium cerium		
hexachloride		
[N(CH₃)₄]₂CeCl₆	a116	1954: C&K
tetra-Methyl ammonium chloride		
N(CH₃)₄Cl	III,b3	1927: V&S; 1928: W
tetra-Methyl ammonium chloro-		
platinate		
[N(CH₃)₄]₂PtCl₆	a116	1926: H
tetra-Methyl ammonium chloro-		
stannate		
[N(CH₃)₄]₂SnCl₆	a116	1929: W&C
tetra-Methyl ammonium cupric		
tetrachloride		
[N(CH₃)₄]₂CuCl₄	a115	1939: M; 1961: M&L
tetra-Methyl ammonium dichloro-		
iodide		
N(CH₃)₄ICl₂	a109	1939: M; 1964: V&V
tetra-Methyl ammonium ennea-		
iodide		
N(CH₃)₄I₉	a111	1955: J,H,F&R
tetra-Methyl ammonium		
fluosilicate		
[N(CH₃)₄]₂SiF₆	a117	1934: C
tetra-Methyl ammonium iodide		
N(CH₃)₄I	III,b3	1917: V; 1922: N; 1926: V&B; 1927:
		G; V&S; 1928: W; Z
tetra-Methyl ammonium mercury		
tribromide		
N(CH₃)₄HgBr₃	a113	1963: W
tetra-Methyl ammonium mercury		
trichloride		
N(CH₃)₄HgCl₃	a113	1963: W

(*continued*)

BIBLIOGRAPHY TABLE, CHAPTER XIVA (continued)

Compound	Paragraph	Literature
tetra-Methyl ammonium penta-iodide $N(CH_3)_4I_5$	a110	1951: H&R; 1957: B,H&W
tetra-Methyl ammonium perchlorate $N(CH_3)_4ClO_4$	a108	1929: H&I; 1964: MC
tetra-Methyl ammonium plutonium hexachloride $[N(CH_3)_4]_2PuCl_6$	a116	1952: S&S
tetra-Methyl ammonium silver iodide $N(CH_3)_4Ag_2I_3$	a112	1963: M
tetra-Methyl ammonium tetrachlorocobaltate $[N(CH_3)_4]_2CoCl_4$	a114	1959: M&L
tetra-Methyl ammonium tetrachlorozincate $[N(CH_3)_4]_2ZnCl_4$	a114	1959: M&L
tetra-Methyl ammonium uranium hexachloride $[N(CH_3)_4]_2UCl_6$	a116	1952: S&S
tetra-Methyl arsonium bromide $As(CH_3)_4Br$	a76	1963: C,S&M
bis tetra-Methyl disilanilene dioxide $[(CH_3)_4Si_2O]_2$	a59	1963: T,K&K
tetra-Methyl N,N'-bis-trimethyl silyl cyclodisilazane $[(CH_3)_2Si]_2[N_2Si_2(CH_3)_6]$	a63	1962: W
tetra-Methyl orthothiocarbonate $C(SCH_3)_4$	a10	1943: P&T; 1946: P&T
tetra-Methyl platinum $Pt(CH_3)_4$	a56	1947: R&S
tetra-Methyl stibonium tetrakis trimethyl siloxy aluminate $[(CH_3)_4Sb]Al[OSi(CH_3)_3]_4$	a64	1963: W
tetra-Methyl stibonium tetrakis trimethyl siloxy ferrate $[(CH_3)_4Sb]Fe[OSi(CH_3)_3]_4$	a64	1963: W
tetra-Methyl stibonium tetrakis trimethyl siloxy gallate $[(CH_3)_4Sb]Ga[OSi(CH_3)_3]_4$	a64	1963: W

(continued)

BIBLIOGRAPHY TABLE, CHAPTER XIVA (*continued*)

Compound	Paragraph	Literature
tetra-Methyl thiourea		
$[(CH_3)_2N]_2CS$	a127	1960: Z,A&G
tetra-Methyl thiuram mono-sulfide		
$[(CH_3)_2NC(S)S—]_2$	a40	1941: Y&M; 1960: T
octa-Methyl cyclo tetraphospho-nitrile		
$[PN(CH_3)_2]_4$	a43	1961: D
octa-Methyl cyclotetrasiloxane		
$(CH_3)_8Si_4O_4$	a60	1955: S,P&F
octa-(Methyl silsesquioxane)		
$(CH_3SiO_{1.5})_8$	a62	1960: L
octa-Methyl spiro-[5·5]penta-siloxane		
$(CH_3)_8Si_5O_6$	a61	1948: R&H
Neodymium formate		
$Nd(HCOO)_3$	a99	1962: M,S,F&G
Nickel chloride tetra thiourea		
$NiCl_2 \cdot 4(NH_2)_2CS$	a134	1956: C,N&B; N,C&B; 1963: LC&T
Nickel formate dihydrate		
$Ni(HCOO)_2 \cdot 2H_2O$	a98	1963: K&M
Nickel thiocyanate di thiourea		
$Ni(NCS)_2 \cdot 2(NH_2)_2CS$	a135	1957: N,B&F; N,C&B
Nickel bis-thiosemicarbazide		
$Ni[NH_2NH(NH_2)CS]_2$	a142	1962: C,N&F
Nickel di thiosemicarbazide sulfate trihydrate		
$Ni(CH_5N_3S)_2SO_4 \cdot 3H_2O$	a143	1962: G&R
Nitroguanidine		
$(NH_2)_2CNNO_2$	a149	1948: D&G; 1956: B,B,H&D
Nitroso(dimethyl dithiocarbonato)-cobalt		
$Co(NO)[S_2CN(CH_3)_2]_2$	a87	1962: A,O&R
cis-di-Nitrosomethane		
$(CH_3NO)_2$	a23	1963: G,P&vM
trans-di-Nitrosomethane		
$(CH_3NO)_2$	a12	1959: vM&G
mono-Octadecyl ammonium chloride		
$NH_3 \cdot nC_{3}H_{37}Cl$	a103	1932: B
mono-Octyl ammonium iodide		
$NH_3 \cdot nC_8H_{17}I$	a103	1930: W

(*continued*)

BIBLIOGRAPHY TABLE, CHAPTER XIVA (*continued*)

Compound	Paragraph	Literature
Palladous chloride tetra thiourea		
$PdCl_2 \cdot 4(NH_2)_2CS$	a136	1960: O,K,N&K
Pentaerythritol		
$C(CH_2OH)_4$	a7	1921: B&J; 1923: M&W; 1926: H&H; N; W&vM; 1927: H; M&W; S; S&S; W; 1928: E; H,S&S; M&vS; vM&S; M&W; N; S&H; 1937: L,C&G; 1938: N&W; 1958: H; S,C&C
Pentaerythritol tetraacetate		
$C(CH_2O \cdot COCH_3)_4$	a9	1928: G,M&R; K; M&R; vM&S; 1929: K; 1938: G&H
Pentaerythritol tetranitrate		
$C(CH_2ONO_2)_4$	a8	1925: K; 1928: G,M&R; 1947: B&L; 1963: T
Phosgene		
$COCl_2$	a11	1952: Z,A&L
Phosphonitrilic dimethylamide tetramer		
$P_4N_4[N(CH_3)_2]_8$	a44	1960: B; 1962: B
Phosphorus carbon trifluoride pentamer		
$(PCF_3)_5$	a42	1961: S&L
Phosphorus carbon trifluoride tetramer		
$(PCF_3)_4$	a41	1962: P&D
Plumbous chloride di thiourea		
$PbCl_2 \cdot 2(NH_2)_2CS$	a137	1958: N&F; 1959: N&F
Potassium copper(II) di biuret tetrahydrate		
$K_2Cu[NHC(O)NHC(O)NH_2]_2 \cdot 4H_2O$	a86	1961: F,S&T
Potassium methylate		
$KOCH_3$	a47	1963: W
Potassium methylene bis-nitroso-hydroxylamine		
$K_2[ONN(O)CH_2N(O)NO]$	a79	1959: B
Potassium methylene disulfonate		
$K_2[CH_2(SO_3)_2]$	a81	1962: T
Potassium O,O-dimethyl phosphor-dithionate		
$K[S_2P(OCH_3)_2]$	a83	1962: C,MG,H&D

(*continued*)

BIBLIOGRAPHY TABLE, CHAPTER XIVA (continued)

Compound	Paragraph	Literature
Praseodymium formate		
$Pr(HCOO)_3$	a99	1962: M,S,F&G
mono-Propyl ammonium bromide		
$NH_3C_3H_7Br$	a102	1928: H; 1950: K&L
mono-Propyl ammonium chloride		
$NH_3C_3H_7Cl$	a102, c3	1928: H; 1950: K&L
mono-Propyl ammonium iodide		
$NH_3C_3H_7I$	a102	1928: H
Samarium formate		
$Sm(HCOO)_3$	a99	1962: M,S,F&G
Selenium dimethane thiosulfonate		
$Se(S_2O_2CH_3)_2$	a67	1951: F,F&H
Sodium formate		
$Na(HCOO)$	a92	1938: Z; 1940: Z
Sodium hydroxymethane sulfinate dihydrate		
$Na(OHCH_2SO_2) \cdot 2H_2O$	a80	1955: T; 1962: T
Sodium iodide tri dimethyl-formamide		
$NaI \cdot 3(CH_3)_2NCHO$	a101	1962: G,P&vM
Sodium methanethiosulfonate monohydrate		
$NaS_2O_2CH_3 \cdot H_2O$	a82	1957: F&H; 1964: F&H
Strontium formate		
$Sr(HCOO)_2$	a94	1928: N; 1949: N&S
Strontium formate dihydrate		
$Sr(HCOO)_2 \cdot 2H_2O$	a95	1928: N; 1958: O; 1961: G&F; 1964: C
Sulfur dimethane thiosulfonate		
$S(S_2O_2CH_3)_2$	a67	1951: F,F&H; 1953: F
Tellurium dimethane thiosulfonate		
$Te(S_2O_2CH_3)_2$	a67	1951: F,F&H; 1954: F&V
di-Thiodimethyl dodecahydro-decaborane		
$B_{10}H_{12}[S(CH_3)_2]_2$	a30	1962: S&Z
Thiourea		
$(NH_2)_2CS$	a126	1928: D&N; H; 1932: W&C; 1958: K&T; Z&T; 1959: G&W; 1960: D&V; 1961: D&V

(continued)

BIBLIOGRAPHY TABLE, CHAPTER XIVA (*continued*)

Compound	Paragraph	Literature
Thiourea dioxide $(NH_2)_2CSO_2$	a128	1962: S&H
mono-Undecyl ammonium iodide $NH_3 \cdot nC_{11}H_{23}I$	a103	1930: W
Urea $(NH_2)_2CO$	a118	1921: B&J; 1923: M&W; 1928: H; 1930: W; 1932: W; 1934: W&C; 1941: L; 1952: V&D; 1956: GW; 1957: W,L&P; 1961: L&V; S,S&P; 1964: C&D
Urea ammonium bromide $(NH_2)_2CO \cdot NH_4Br$	a124	1963: C
Urea ammonium chloride $(NH_2)_2CO \cdot NH_4Cl$	a123	1960: R
Urea–peroxide $(NH_2)_2CO \cdot H_2O_2$	a122	1941: L,H&G
Urea phosphate $(NH_2)_2CO \cdot H_3PO_4$	a121	1957: SR,T&P
Zinc chloride di thiourea $ZnCl_2 \cdot 2(NH_2)_2CS$	a133	1958: K&T
Zinc chloride mono thiosemi- carbazide $ZnCl_2 \cdot NH_2NH(NH_2)CS$	a141	1960: C,N&B

B. DERIVATIVES OF ETHANE, ETHYLENE, AND ACETYLENE

1. Simple Ethane Derivatives

XIV,b1. Solid *ethane*, C_2H_6, is hexagonal with a bimolecular unit which at $-185°C$. has the dimensions:

$$a_0 = 4.46 \text{ A.}, \qquad c_0 = 8.19 \text{ A.}$$

The carbon atoms are in the following special positions derived from D_{6h}^4 $(P6_3/mmc)$:

$$(4f) \quad \pm(^1/_3\,{}^2/_3\,u;\, {}^2/_3,{}^1/_3,u+^1/_2)$$
with u between 0.15 and 0.16

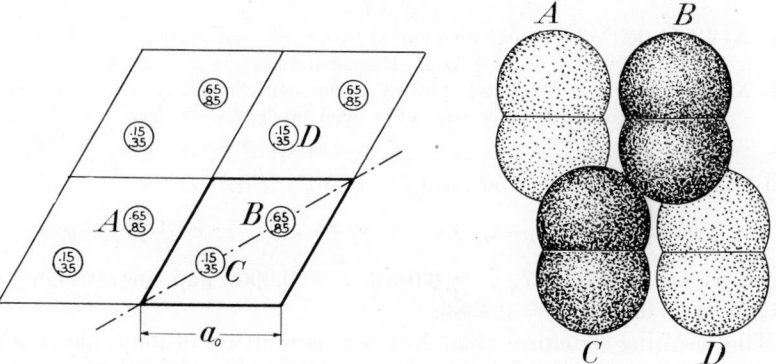

Fig. XIVB,1a (left).   A basal projection of the carbon atoms in four unit cells of the
hexagonal structure of ethane.

Fig. XIVB,1b (right).   A packing drawing of several molecules in the hexagonal
structure of ethane as projected on a vertical plane whose trace is the dot-and-dash line
of Fig. XIVB,1a.   Molecules are similarly lettered in the two drawings.

This structure (Fig. XIVB,1) gives a carbon to carbon distance within the molecule of 1.54 A. and a CH_3–CH_3 separation between adjacent molecules of ca. 3.6 A.

Solid *diborane*, B_2H_6, has the same arrangement with a cell of the dimensions:

$$a_0 = 4.54 \text{ A.}, \qquad c_0 = 8.69 \text{ A.}$$

XIV,b2.   Solid *cyanogen*, $(CN)_2$, at $-95°C$. is orthorhombic with a tetramolecular unit of the edge lengths:

$$a_0 = 6.31 \text{ A.}; \quad b_0 = 7.08 \text{ A.}; \quad c_0 = 6.19 \text{ A.}$$

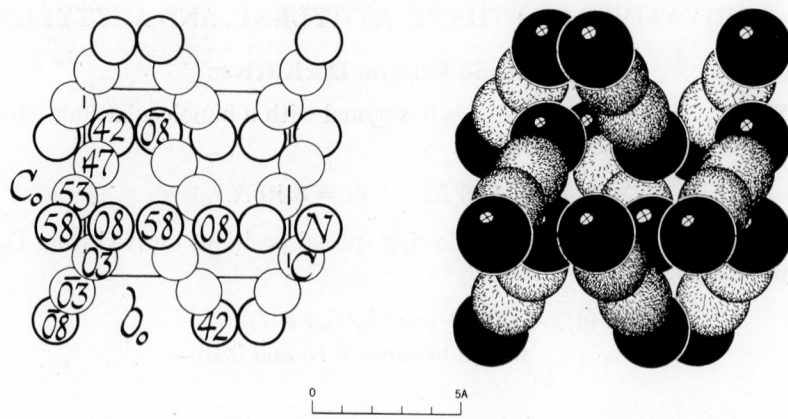

Fig. XIVB,2a (left).   The orthorhombic structure of solid cyanogen projected along
its a_0 axis.   Right-hand axes.
Fig. XIVB,2b (right).   A packing drawing of the orthorhombic structure of cyanogen
seen along its a_0 axis.   The black circles are nitrogen.

All atoms are in the general positions of V_h^{15} $(Pcab)$:

$$(8c) \quad \pm (xyz; \; {}^1/_2 - x, y, z + {}^1/_2; \; x, y + {}^1/_2, {}^1/_2 - z; \; x + {}^1/_2, {}^1/_2 - y, z)$$

For carbon, $x = 0.0327$, $y = 0.0483$, $z = 0.0900$ and for nitrogen, $x = 0.0857$, $y = 0.1271$, $z = 0.2386$.

The resulting structure (Fig. XIVB,2) is built up of molecules $N{\equiv}C{-}C{\equiv}N$ which are linear to within half a degree. In them $N{-}C = 1.13$ A. and $C{-}C = 1.37$ A.

XIV,b3.   *Ethane dithiocyanate*, $C_2H_4(SCN)_2$, is orthorhombic with a tetramolecular unit of the edge lengths:

$$a_0 = 7.39 \text{ A.}; \quad b_0 = 7.81 \text{ A.}; \quad c_0 = 11.48 \text{ A.}$$

The space group is V_h^{15} $(Pbca)$ with all atoms in the general positions:

$$(8c) \quad \pm (xyz; \; {}^1/_2 - x, y + {}^1/_2, z; \; x, {}^1/_2 - y, z + {}^1/_2; \; x + {}^1/_2, y, {}^1/_2 - z)$$

The parameters are:

Atom	x	y	z
S	0.268	0.027	0.0768
N	0.165	−0.088	0.298
C(1)	0.051	0.080	0.0165
C(2)	0.205	−0.029	0.207

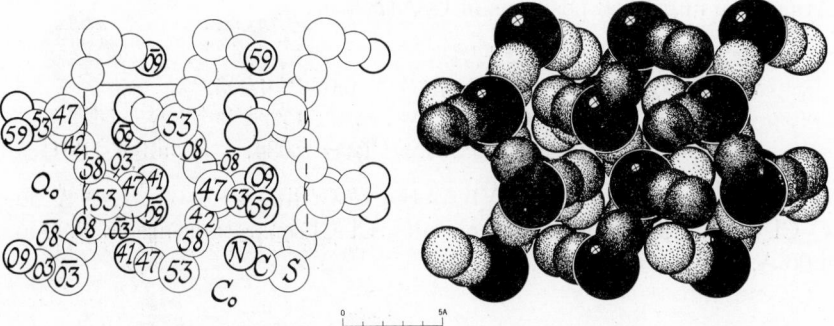

Fig. XIVB,3a (left).   The orthorhombic structure of $C_2H_4(SCN)_2$ projected along its b_0 axis.   Right-hand axes.

Fig. XIVB,3b (right).   A packing drawing of the orthorhombic structure of $C_2H_4(SCN)_2$ viewed along its b_0 axis.   The sulfur atoms are large and black; the nitrogen atoms are heavily outlined.

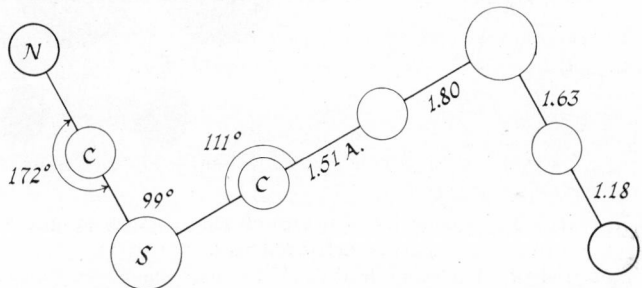

Fig. XIVB,4.   Bond dimensions in molecules of $C_2H_4(SCN)_2$.

The structure as shown in Figure XIVB,3 is built up of molecules having the bond dimensions of Figure XIVB,4. Approximate parameters for the hydrogen atoms have been given as

for H(1): $x = 0.056,\ y = 0.168,\ z = -0.056$
for H(2): $x = -0.019,\ y = 0.155,\ z = 0.082$

XIV,b4.   The structure of *1,2-dichloroethane*, $C_2H_4Cl_2$, below the broad transition centered around $-96°C.$, is monoclinic with a bimolecular cell of the dimensions:

$a_0 = 4.66$ A.; $b_0 = 5.42$ A.; $c_0 = 7.88$ A.; $\beta = 103°30'\ (-140°C.)$

Atoms are in general positions of C_{2h}^5 ($P2_1/c$):

(4e) $\pm(xyz;\ x,^1/_2-y,z+^1/_2)$
with $x(C) = 0.099\pm0.003$, $y(C) = 0.003\pm0.005$,
  $z(C) = 0.094\pm0.002$
$x(Cl) = 0.318$, $y(Cl) = 0.278$, $z(Cl) = 0.084$, all ±0.001

In this structure (Fig. XIVB,5) the C–C distance is ca. 1.49 A. and C–Cl = ca. 1.80 A. Between molecules, Cl–Cl separations are ca. 3.99 and 4.06 A.

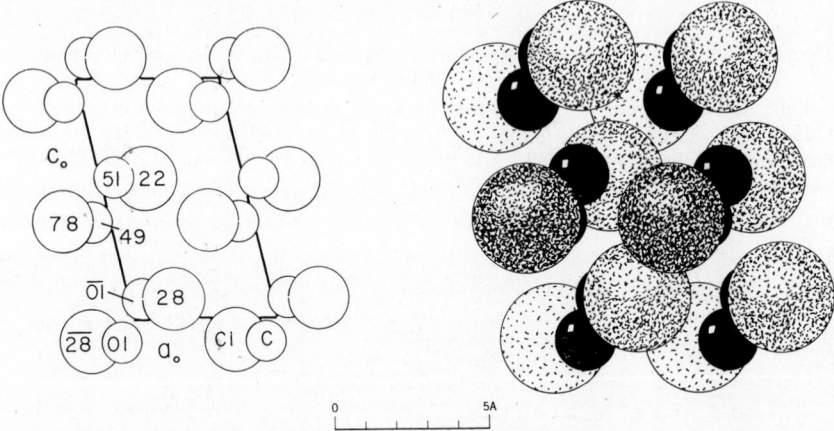

Fig. XIVB,5a (left). The monoclinic structure of solid $C_2H_4Cl_2$ projected along its
b_0 axis. Left-hand axes.
Fig. XIVB,5b (right). A packing drawing of the monoclinic structure of $C_2H_4Cl_2$
seen long its b_0 axis. The carbon atoms are black; the chlorine atoms are large and
dotted.

At 50°C. the symmetry remains monoclinic with a unit of the dimensions:

$$a_0 = 5.04 \text{ A.};\quad b_0 = 5.56 \text{ A.};\quad c_0 = 8.00 \text{ A.};\quad \beta = 109°30'$$

The space group is still C_{2h}^5 ($P2_1/c$) but there is strong evidence that the molecule is rotating about its Cl–Cl axis, and this is in accord with an absence of carbon peaks in the Fourier projections. The chlorine atoms were found to be in the general positions (4e) with $x = 0.303$, $y = 0.279$, $z = 0.074$.

XIV,b5. A *symmetrical diiodoethane*, $C_2H_4I_2$, and the isomorphous *symmetrical diiodoethylene*, $C_2H_2I_2$, are monoclinic pseudo-cubic; years ago, however, they were given approximate structures in terms of an orthorhombic space group. The bimolecular monoclinic cells have the dimensions:

$C_2H_4I_2$: $a_0 = 4.768$ A.; $b_0 = 12.897$ A.; $c_0 = 4.784$ A.
$\beta = 105°5'$

$C_2H_2I_2$: $a_0 = 4.58$ A.; $b_0 = 13.310$ A.; $c_0 = 4.58$ A.
$\beta = 105°20'$

The tetramolecular orthorhombic pseudo-units which were used to describe atomic positions are:

$C_2H_4I_2$: $a_0' = 7.582$ A.; $b_0' = 12.897$ A.; $c_0' = 5.810$ A.
$C_2H_2I_2$: $a_0' = 7.280$ A.; $b_0' = 13.310$ A.; $c_0' = 5.553$ A.

It was stated that the iodine atoms are in the special positions

$$(8f) \quad \pm(0uv;\ {}^1/_2,u+{}^1/_2,\bar{v};\ {}^1/_2,u,v+{}^1/_2;\ 0,u+{}^1/_2,{}^1/_2-v)$$

of $V_h{}^{18}$ (*Bmab*) with parameters that for both compounds are $u = 0.145$, $v = 0.27$. On other than x-ray considerations, it was concluded that the carbon atoms are in similar positions with $u(C) = -0.004$, $v(C) = 0.132$ for $C_2H_4I_2$ and the rather different $u(C) = 0.053$, $v(C) = -0.008$ for $C_2H_2I_2$. Especially in view of the lower symmetry actually possessed by these crystals, only a limited significance can be attached to these carbon positions and to the hydrogen positions that also have been proposed. A renewed study is obviously desirable.

XIV,b6. According to a recent study, the original structure assigned to *chloral hydrate*, $CCl_3CH(OH)_2$, is wrong. The symmetry is monoclinic with a tetramolecular cell of the dimensions:

$$a_0 = 11.50\text{ A.};\ b_0 = 6.04\text{ A.};\ c_0 = 9.60\text{ A.};\ \beta = 120°30'$$

The correct space group is $C_{2h}{}^5$ ($P2_1/c$) with all atoms in the positions:

$$(4e) \quad \pm(xyz;\ x,{}^1/_2-y,z+{}^1/_2)$$

The newly selected parameters are those of Table XIVB,1.

TABLE XIVB,1. Parameters of the Atoms in Chloral Hydrate

Atom	x	y	z
Cl(1)	0.2323	0.0378	−0.2289
Cl(2)	0.3139	−0.3547	−0.0267
Cl(3)	0.4051	0.0844	0.1146
O(1)	0.105	0.178	−0.035
O(2)	0.038	−0.200	−0.126
C(1)	0.256	−0.061	−0.048
C(2)	0.145	−0.065	−0.008

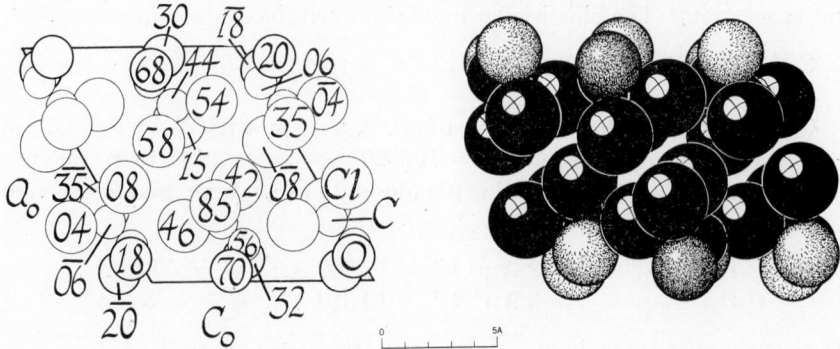

Fig. XIVB,6a (left). The monoclinic structure of chloral hydrate projected along its b_0 axis. Right-hand axes.

Fig. XIVB,6b (right). A packing drawing of the monoclinic structure of CCl_3CH-$(OH)_2$ viewed along its b_0 axis. The chlorine atoms are the large black circles; the smaller black carbon atoms only partially show. Hydroxyls are line shaded and heavily outlined.

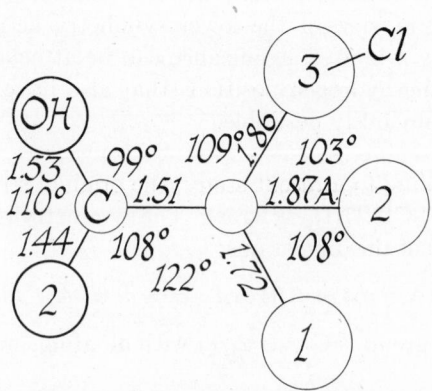

Fig. XIVB,7. Bond dimensions in the molecule of chloral hydrate.

The resulting structure is shown in Figure XIVB,6. It leads to molecules that have the bond dimensions of Figure XIVB,7. Between molecules, short atomic separations that could represent hydrogen bonds are O–O = 2.8 and 2.9 A., and O–Cl = 3.1 A.

XIV,b7. *Hexabrom ethane*, C_2Br_6, is orthorhombic with a tetramolecular unit of the edge lengths:

$$a_0 = 12.00 \text{ A.}; \quad b_0 = 10.64 \text{ A.}; \quad c_0 = 6.69 \text{ A.}$$

Atoms are in the following positions of V_h^{16} (*Pnma*):

(4c) $\pm (u\ ^1/_4\ v;\ u+^1/_2,^1/_4,^1/_2-v)$

(8d) $\pm (xyz;\ x+^1/_2,^1/_2-y,^1/_2-z;\ x,^1/_2-y,z;\ x+^1/_2,y,^1/_2-z)$

The bromine atoms were determined to be in two sets of (4c) and two of (8d) with the parameters of Table XIVB,2. Following indications from the x-ray data and using expected interatomic distances, the carbon atoms were placed in two sets of (4c) with the parameters stated in this table.

TABLE XIVB,2
Positions and Parameters of the Atoms in C_2Br_6

Atom	Position	x	y	z
C(1)	(4c)	0.111	$^1/_4$	-0.200
C(2)	(4c)	0.139	$^1/_4$	0.025
Br(1)	(4c)	0.001	$^1/_4$	0.180
Br(2)	(4c)	0.249	$^1/_4$	-0.355
Br(3)	(8d)	0.027	0.104	-0.265
Br(4)	(8d)	0.223	0.104	0.090

The molecule that results (Fig. XIVB,8) has a *"trans"* form in which C–C = 1.52 A. (partly by assumption), C–Br = 1.93 A., and C–C–Br = 113 or 108°. The centers of these molecules are in (4c), with $u = \ ^1/_8$, $v = -0.087$. Their closest approach to one another is through Br–Br separations of 3.7–3.8 A.

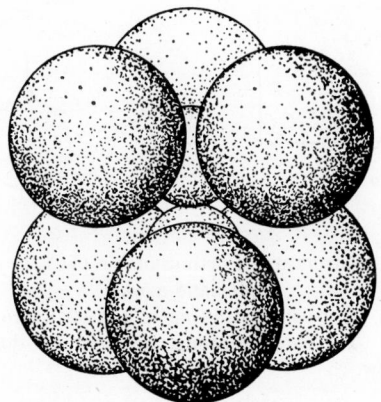

Fig. XIVB,8.   A packing drawing of the molecule of C_2Br_6.   The carbon atoms are the small, the bromine atoms the large dotted circles.

The form of *hexachlorethane*, C_2Cl_6, stable below 45°C., has this structure with a cell of the dimensions:

$$a_0 = 11.54 \text{ A.}; \quad b_0 = 10.165 \text{ A.}; \quad c_0 = 6.407 \text{ A.}$$

The chosen parameters, with carbon given positions to make C–C = 1.54 A. and C–Cl = 1.76 A., are those of Table XIVB,3.

TABLE XIVB,3
Positions and Parameters of the Atoms in Orthorhombic Hexachlorethane

Atom	Position	x	y	z
C(1)	(4c)	0.111	$1/4$	−0.212
C(2)	(4c)	0.139	$1/4$	0.023
Cl(1)	(4c)	0.240	$1/4$	−0.360
Cl(2)	(4c)	0.008	$1/4$	0.168
Cl(3)	(8d)	0.030	0.107	−0.278
Cl(4)	(8d)	0.221	0.108	0.086

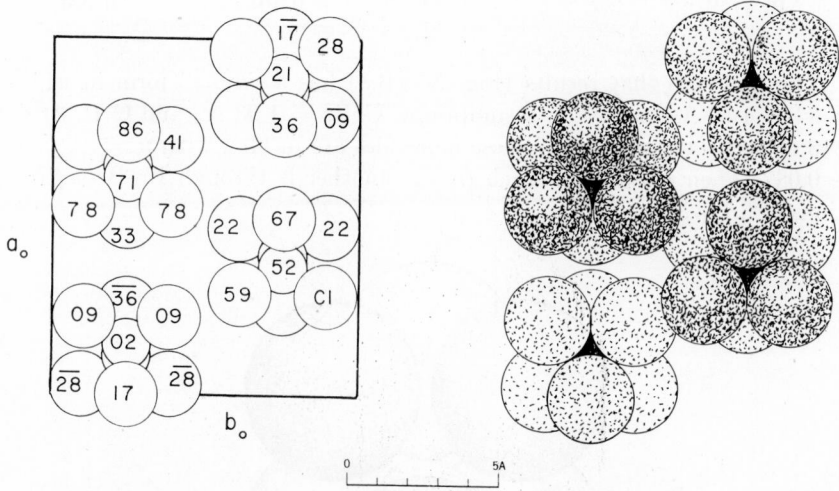

Fig. XIVB,9a (left). The structure of orthorhombic C_2Cl_6 projected along its c_0 axis. Left-hand axes.

Fig. XIVB,9b (right). A packing drawing of the structure of orthorhombic C_2Cl_6 viewed along its c_0 axis. The chlorine atoms are large and dotted; the black carbon atoms only partially show.

The resulting structure (valid also for the bromine compound) is shown in Figure XIVB,9. Between molecules, Cl–Cl ranges upwards from 3.64 A.

Many years ago the following seemingly isomorphous substances were found to have the cell dimensions:

Compound	a_0, A.	b_0, A.	c_0, A.
C_2Br_5F	11.84	10.75	6.55
$C_2Br_4(CH_3)_2$(II)	11.70	10.44	6.57
$C_2Br_3Cl_3$	11.77	10.44	6.54
BrC_2Cl_4Br	11.73	10.375	6.505
$Cl_3C \cdot CBr_2Cl$	11.61	10.35	6.51

XIV,b8. Crystals of *S-bis nitroamino ethane* (ethylene dinitroamine), $C_2H_4(NHNO_2)_2$, are orthorhombic with a tetramolecular unit of the dimensions:

$$a_0 = 10.67 \text{ A.}; \quad b_0 = 8.67 \text{ A.}; \quad c_0 = 6.16 \text{ A.}$$

All atoms are in general positions of V_h^{15} (*Pbca*):

(8c) $\pm (xyz; \; x+\frac{1}{2},\frac{1}{2}-y,\bar{z}; \; \bar{x},y+\frac{1}{2},\frac{1}{2}-z; \; \frac{1}{2}-x,\bar{y},z+\frac{1}{2})$

with the parameters:

Atom	x	y	z
C	0.035	0.005	0.107
NH	0.165	0.028	0.074
N	0.215	0.167	0.043
O(1)	0.327	0.173	0.014
O(2)	0.146	0.278	0.057

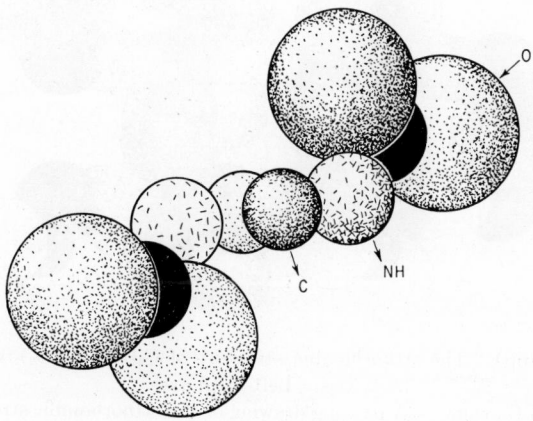

Fig. XIVB,10. The molecule of $C_2H_4(NHNO_2)_2$ as it occurs in its crystals. Atoms of nitrogen in the nitro radicals are black.

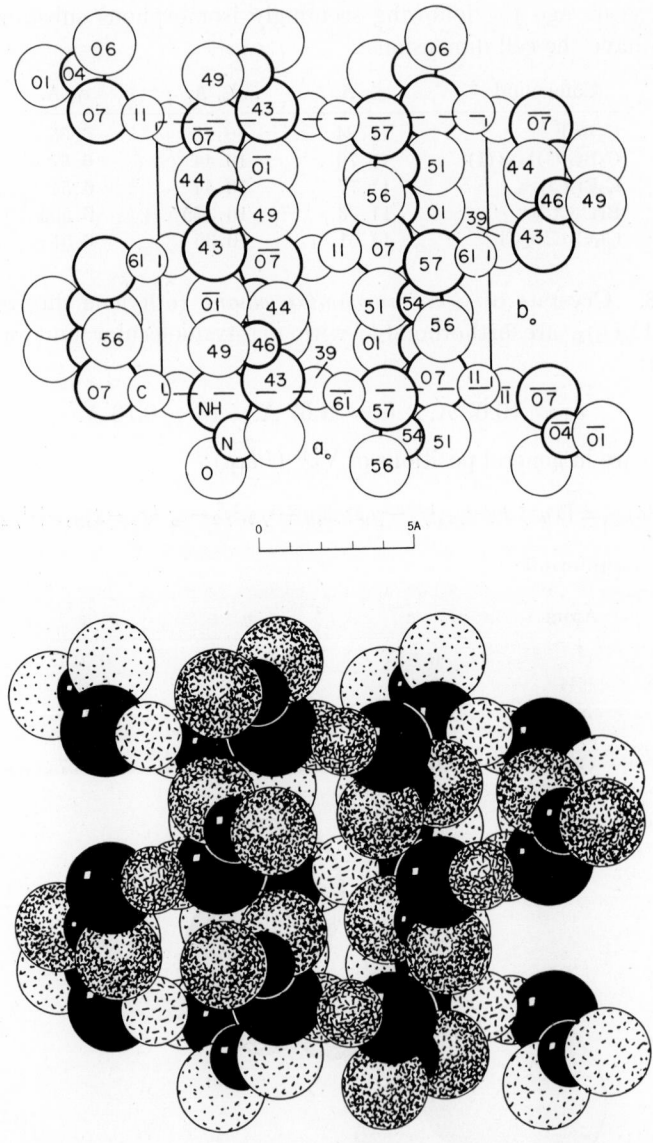

Fig. XIVB,11a (top). The orthorhombic structure of $C_2H_4(NHNO_2)_2$ projected along its c_0 axis. Left-hand axes.

Fig. XIVB,11b (bottom). A packing drawing of the orthorhombic structure of C_2H_4-$(NHNO_2)_2$ seen along its c_0 axis. The oxygen atoms are dotted, the carbon atoms smaller and line shaded. Atoms of nitrogen are black.

This leads to a centrosymmetrical molecule (Fig. XIVB,10) in which CH_2-CH_2 = 1.52 A., CH_2-NH = 1.41 A., $NH-N$ = 1.33 A., $N-O(1)$ = $N-O(2)$ = 1.21 A., CH_2-CH_2-NH = 111°26', CH_2-NH-N = 123°9', $NH-N-O(1)$ = 117°18', $NH-N-O(2)$ = 118°9', $O(1)-N-O(2)$ = 125°21'. Atoms of an $NHNO_2$ radical are coplanar. In the structure as a whole (Fig. XIVB,11), the molecular centers are in an approximately face-centered array $(4a)$ 000; F.C.; since all values of z lie close to zero, the sheet-like molecules are roughly normal to the c_0 axis.

XIV,b9. Crystals of *taurine*, $NH_2(C_2H_4)HSO_3$, are monoclinic with a tetramolecular unit of the dimensions:

$$a_0 = 5.30 \text{ A.}; \quad b_0 = 11.65 \text{ A.}; \quad c_0 = 7.94 \text{ A.}; \quad \beta = 94°17'$$

All atoms are in the following general positions of C_{2h}^5 $(P2_1/c)$:

$$(4e) \quad \pm(xyz; \ x,{}^1\!/_2-y,z+{}^1\!/_2)$$

with the parameters of Table XIVB,4. Hydrogen positions were obtained using the usual assumptions of a tetrahedral distribution and an $H-C$ = 1.08 A.

In the resulting structure (Fig. XIVB,12) the molecules have the bond dimensions of Figure XIVB,13. Between the terminal nitrogen atoms and oxygen atoms there are relatively short hydrogen bonds of 2.78 and 2.89 A. Other $N-O$ distances nearly as short are 2.96 and 3.03 A. The final R = 0.11.

TABLE XIVB,4
Parameters of the Atoms in Taurine

Atom	x	$\sigma(x)$	y	$\sigma(y)$	z	$\sigma(z)$
S	0.2961	0.0007	0.1515	0.0003	0.1488	0.0006
O(1)	0.5659	0.0013	0.1636	0.0006	0.2086	0.0011
O(2)	0.1564	0.0013	0.2589	0.0006	0.1434	0.0010
O(3)	0.2672	0.0012	0.0897	0.0006	−0.0119	0.0010
C(1)	0.1569	0.0018	0.0609	0.0009	0.3014	0.0014
C(2)	0.2869	0.0018	−0.0545	0.0009	0.3199	0.0014
N	0.2359	0.0014	−0.1295	0.0007	0.1674	0.0012
H(1)	0.192	—	0.109	—	0.419	—
H(2)	−0.005	—	0.065	—	0.246	—
H(3)	0.217	—	−0.096	—	0.415	—
H(4)	0.451	—	−0.048	—	0.335	—
H(5)	0.302	—	−0.216	—	0.190	—
H(6)	0.085	—	−0.133	—	0.162	—
H(7)	0.284	—	0.274	—	0.197	—

236 CRYSTAL STRUCTURES

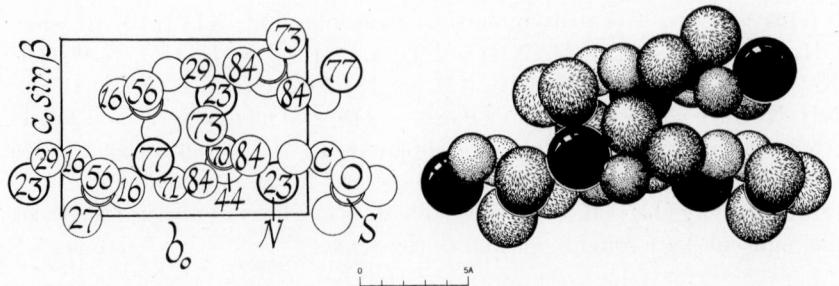

Fig. XIVB,12a (left).   The monoclinic structure of taurine projected along its a_0 axis.
Right-hand axes.
Fig. XIVB,12b (right).   A packing drawing of the monoclinic structure of NH₂-
(CH₂)₂HSO₃ viewed along its a_0 axis.   The nitrogen atoms are black; the oxygens are
line shaded.   Atoms of carbon are smaller and dotted.   The sulfur atoms barely show.

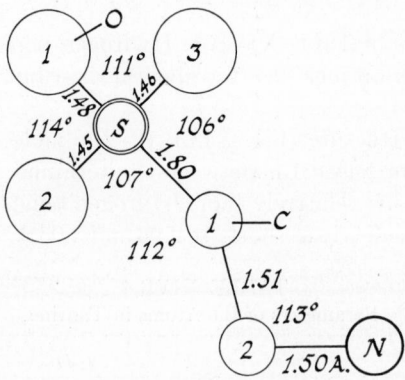

Fig. XIVB,13.   Bond dimensions in the molecule of taurine.

XIV,b10.   Crystals of *2,2′-di iodo diethyl trisulfide*, [I(C₂H₄)]₂S₃, are
tetragonal with an elongated cell of the edge lengths:

$$a_0 = 6.01 \text{ A.}, \qquad c_0 = 29.4 \text{ A.}$$

Atoms have been placed in the following positions of D_4^4 $(P4_12_1)$, the
origin of this description being displaced $(^1/_4\,^3/_4\,^1/_8)$ with respect to that
used in the International Tables:

(4a) $u\ \bar{u}\ ^1/_8$; $u+^1/_2,u,^3/_8$; $^1/_2-u,u+^1/_2,^5/_8$; $\bar{u},^1/_2-u,^7/_8$

(8b) xyz; $^1/_2-y,x,z+^1/_4$; $^1/_2-x,^1/_2-y,z+^1/_2$; $y,^1/_2-x,z+^3/_4$;
        $\bar{x},y+^1/_2,\bar{z}$; $\bar{y},\bar{x},^1/_4-z$; $y+^1/_2,x+^1/_2,^3/_4-z$; $x+^1/_2,\bar{y},^1/_2-z$

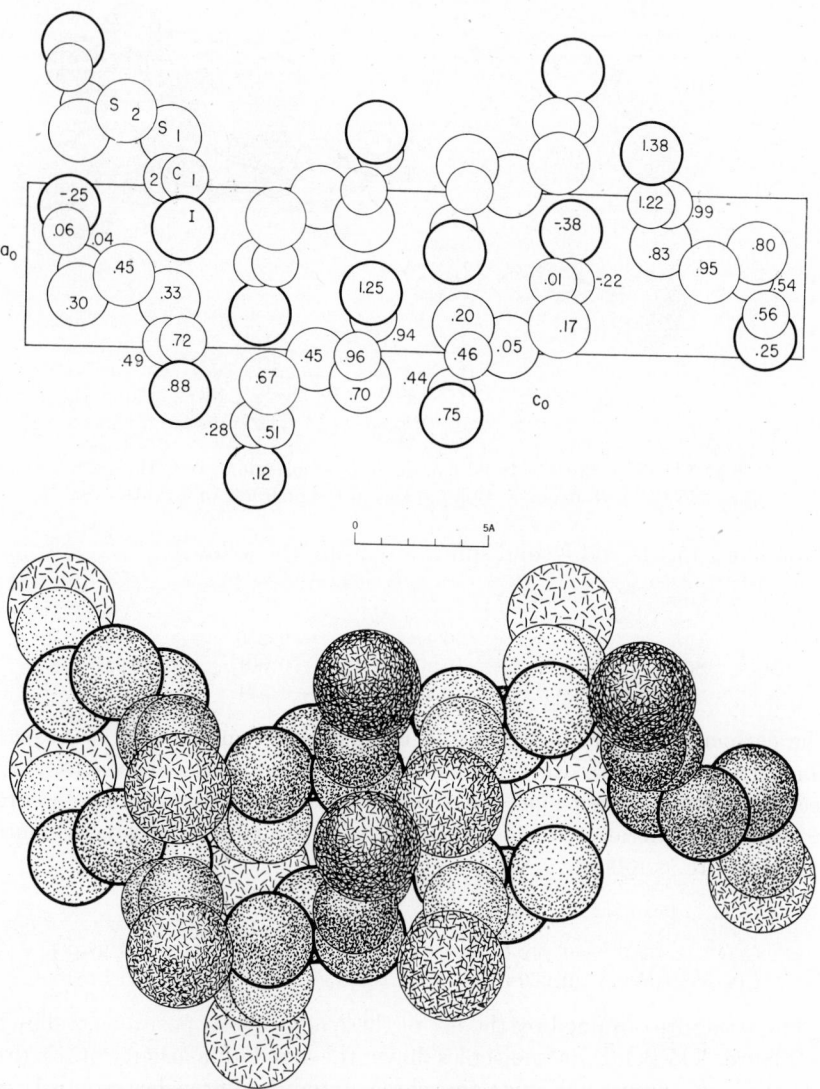

Fig. XIVB,14a (top). The tetragonal structure of $[I(C_2H_4)]_2S_3$ projected along an a_0 axis. Atoms of iodine are the large heavily ringed and sulfur the large light circles. The carbon atoms are the smaller circles. Left-hand axes.

Fig. XIVB,14b (bottom). A packing drawing of the tetragonal structure of $[I(C_2H_4)]_2S_3$ seen along an a_0 axis. The iodine atoms are line shaded; the sulfur atoms are dotted and heavily ringed.

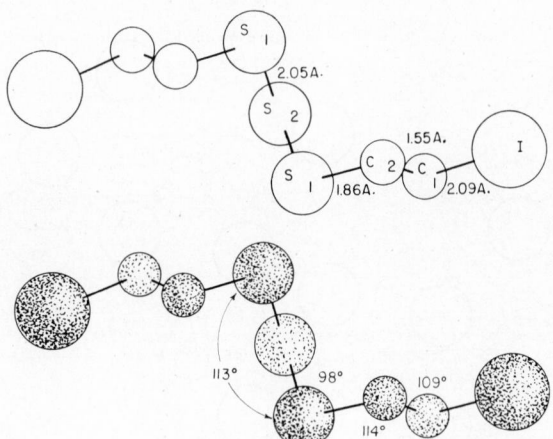

Fig. XIVB,15a (top).   Bond lengths in the molecule of $[I(C_2H_4)]_2S_3$.
Fig. XIVB,15b (bottom).   Bond angles in the molecule of $[I(C_2H_4)]_2S_3$.

Parameters for the iodine and sulfur atoms are the following:

Atom	Position	x	y	z
I	(8b)	0.885	0.250	0.948
S(1)	(8b)	0.674	0.301	0.067
S(2)	(4a)	0.554	−0.554	$^1/_8$

The carbon atoms did not give well defined peaks on the Fourier projections. In the original study the positions assigned them, chosen to yield acceptable bond dimensions, are given below in parentheses. From a subsequent reconsideration of the same data, new parameters, selected as more probable, are as follows:

Atom	Position	x	y	z
C(1)	(8b)	0.942 (0.600)	0.725 (0.233)	0.203 (0.986)
C(2)	(8b)	0.960 (0.617)	0.491 (0.447)	0.181 (0.015)

The structure obtained by the use of these new carbon positions is shown in Figure XIVB,14; its molecules have the bond dimensions of Figure XIVB,15. Between molecules the closest atomic approaches are I–I = 4.4 A. and I–S = 3.80 A.

XIV,b11. The *methyl cyanide–boron hydride*, $B_9H_{13}(CH_3CN)$, is monoclinic, pseudo-orthorhombic, with a bimolecular cell of the edge lengths:

$$a_0 = 5.639 \pm 0.007 \text{ A.}; \quad b_0 = 9.220 \pm 0.008 \text{ A.}; \quad c_0 = 9.813 \pm 0.020 \text{ A.}$$
$$\beta = 90° \pm 30'$$

The space group is C_{2h}^2 $(P2_1/m)$ with atoms in the positions:

$$(4f) \quad \pm (xyz; \; x,{}^1/_2-y,z)$$
$$(2e) \quad \pm (u \; {}^1/_4 v)$$

The determined parameters are those of Table XIVB,5.

In this structure the B–B bonds of the molecule have the dimensions of Figure XIVB,16; the B–H distances are 0.95–1.18 A. except for the two "bridge" hydrogens for which B–H = 1.21 and 1.35 A.

TABLE XIVB,5. Positions and Parameters of the Atoms in $B_9H_{13}(CH_3CN)$[a]

Atom	Position	x	y	z
C(1)	(2e)	0.987	$^1/_4$	0.169
C(2)	(2e)	0.111	$^1/_4$	0.296
N	(2e)	0.209	$^1/_4$	0.396
B(1)	(2e)	0.341	$^1/_4$	0.529
B(2)	(2e)	0.142	$^1/_4$	0.664
B(3)	(2e)	0.481	$^1/_4$	0.869
B(4)	(4f)	0.236	0.151	0.808
B(5)	(4f)	0.296	0.087	0.640
B(6)	(4f)	0.507	0.070	0.783
H(1)	(2e)	0.542	$^1/_4$	0.527
H(2)	(2e)	0.976	$^1/_4$	0.644
H(3)	(2e)	0.535	$^1/_4$	0.974
H(5)	(4f)	0.800	0.016	0.403
H(8)	(4f)	0.450	0.024	0.163
H(7)	(4f)	0.509	0.077	0.645
H(4)	(4f)	0.111	0.092	0.889
H(6)	(4f)	0.693	0.098	0.784
H(9)	(2e)	0.793	$^1/_4$	0.171
H(10)	(4f)	0.010	0.175	0.116

[a] The average σ for C, N, and B is ± 0.006 A.; for H it is ± 0.04 A.

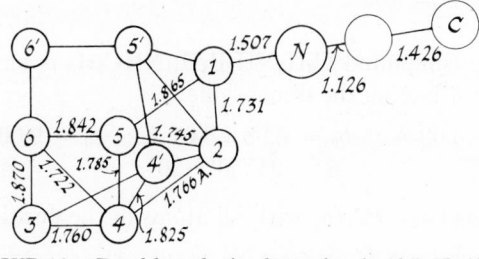

Fig. XIVB,16.   Bond lengths in the molecule of $B_9H_{13}(CH_3CN)$.

XIV,b12. The boron halides form complexes with methyl cyanide (acetonitrile) analogous to those formed with methyl amine (**XIV,a16**). Their orthorhombic tetramolecular cells have the dimensions:

$CH_3CN \cdot BF_3$: $a_0 = 7.76$ A.; $b_0 = 7.20$ A.; $c_0 = 8.34$ A.

$CH_3CN \cdot BCl_3$: $a_0 = 8.72$ A.; $b_0 = 7.30$ A.; $c_0 = 10.20$ A.

$CH_3CN \cdot BBr_3$: $a_0 = 8.91$ A.; $b_0 = 7.51$ A.; $c_0 = 10.94$ A.

A detailed analysis of structure has been made for *methyl cyanide–boron trifluoride*. Its space group is V_h^{16} (*Pnma*) with atoms in the special and general positions:

(4c) $\pm (u\ ^1/_4\ v;\ u+^1/_2,^1/_4,^1/_2-v)$

(8d) $\pm (xyz;\ x+^1/_2,^1/_2-y,^1/_2-z;\ \bar{x},y+^1/_2,\bar{z};\ ^1/_2-x,\bar{y},z+^1/_2)$

The chosen parameters are those of Table XIVB,6.

TABLE XIVB,6
Positions and Parameters of the Atoms in Methyl Cyanide–Boron Trifluoride

Atom	Position	x	y	z
F(1)	(8d)	0.398	0.097	0.273
F(2)	(4c)	0.270	$^1/_4$	0.471
B	(4c)	0.391	$^1/_4$	0.359
N	(4c)	0.573	$^1/_4$	0.460
C	(4c)	0.698	$^1/_4$	0.530
CH_3	(4c)	0.860	$^1/_4$	0.615
Proposed hydrogen positions				
H(1)	(8d)	0.87	0.13	0.69
H(2)	(4c)	0.96	$^1/_4$	0.53

In the resulting arrangement, as illustrated in Figure XIVB,17, B–F = 1.32 A. and F–B–F = 114°. As the table indicates, all atoms except those of fluorine in (8d) are coplanar. In the CH_3CN part of the molecule, CH_3–C = 1.44 A. and C–N = 1.13 A. Between the two parts of the molecule, N–B = 1.64 A. and N–B–F = 103°.

XIV,b13. The compound $C_2H_5NH_2B_8H_{11}NHC_2H_5$ is monoclinic with a tetramolecular unit having the dimensions:

$a_0 = 24.35 \pm 0.02$ A.; $b_0 = 5.98 \pm 0.01$ A.; $c_0 = 9.005 \pm 0.01$ A.

$\beta = 94°50' \pm 20'$

The space group is C_{2h}^5 ($P2_1/a$) with all atoms in the positions:

(4e) $\pm (xyz;\ x+^1/_2,^1/_2-y,z)$

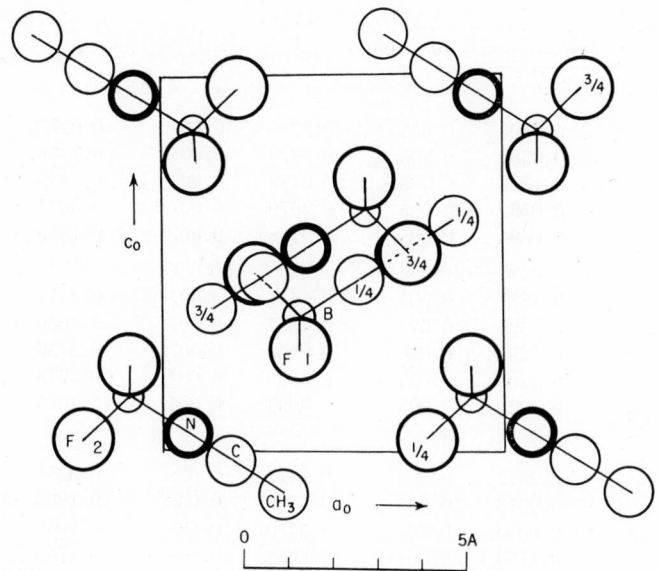

Fig. XIVB,17.   The orthorhombic structure of $CH_3CN \cdot BF_3$ projected along its b_0 axis.   Left-hand axes.

The determined parameters are those of Table XIVB,7.

In this structure (Fig. XIVB,18) the molecules have the bond lengths of Figure XIVB,19. The atoms are similarly numbered in the table and the figure. The terminal B–H = 1.04–1.31 A., and the bridge B–H = 1.32–1.37 A.

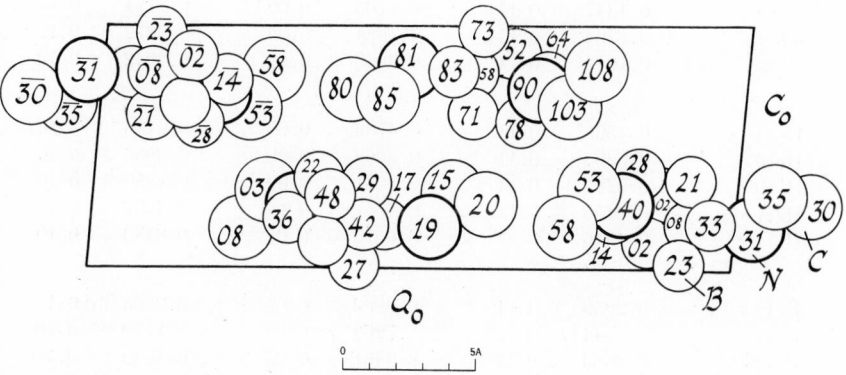

Fig. XIVB,18.   The monoclinic structure of $C_2H_5NH_2B_8H_{11}NHC_2H_5$ projected along its b_0 axis.   Right-hand axes.

TABLE XIVB,7
Positions of the Atoms in $C_2H_5NH_2B_8H_{11}NHC_2H_5$

Atom	x	σ_x, A.	y	σ_y, A.	z	σ_z, A.
B(1)	0.0803	0.005	0.2268	0.007	0.0277	0.006
B(2)	0.0388	0.005	0.3329	0.007	0.1853	0.006
B(3)	0.1323	0.005	0.0182	0.007	0.1153	0.006
B(4)	0.0762	0.005	0.2079	0.007	0.3622	0.006
B(5)	0.0694	0.005	0.0758	0.007	0.1859	0.006
B(6)	0.1877	0.005	0.1421	0.007	0.2377	0.006
B(7)	0.1496	0.005	0.2777	0.007	0.4026	0.006
B(8)	0.1293	0.005	0.0158	0.007	0.3133	0.006
N(9)	0.1803	0.003	0.3958	0.004	0.2773	0.004
N(10)	−0.0260	0.003	0.3093	0.004	0.1673	0.004
C(11)	0.2312	0.005	0.5275	0.008	0.3201	0.006
C(12)	0.2613	0.005	0.5760	0.008	0.1834	0.006
C(13)	−0.0554	0.005	0.3460	0.008	0.3049	0.006
C(14)	−0.1164	0.005	0.3031	0.008	0.2783	0.006
H(1)	0.1057	0.11	0.3755	0.09	0.0247	0.10
H(1′)	0.0615	0.11	0.1667	0.09	−0.0904	0.10
H(2)	0.0502	0.11	0.4992	0.09	0.2008	0.10
H(3)	0.1434	0.11	−0.1506	0.09	0.0591	0.10
H(4)	0.0488	0.11	0.1538	0.09	0.4459	0.10
H(5)	0.0294	0.11	−0.0704	0.09	0.1724	0.10
H(6)	0.2279	0.11	0.0481	0.09	0.2687	0.10
H(7)	0.1694	0.11	0.2831	0.09	0.5257	0.10
H(8)	0.1347	0.11	−0.1705	0.09	0.3774	0.10
H(1x)	0.1764	0.11	0.1384	0.09	0.0872	0.10
H(2x)	0.1016	0.11	0.3960	0.09	0.4042	0.10
H(9)	0.1595	0.11	0.4736	0.09	0.2136	0.10
H(10)	−0.0398	0.11	0.3959	0.09	0.0936	0.10
H(10′)	−0.0282	0.11	0.1667	0.09	0.1250	0.10
H(11)	0.2455	0.11	0.4041	0.09	0.4252	0.10
H(11′)	0.2246	0.11	0.6913	0.09	0.3489	0.10
H(12)	0.2959	0.11	0.6802	0.09	0.2164	0.10
H(12′)	0.2444	0.11	0.7262	0.09	0.1170	0.10
H(12″)	0.2683	0.11	0.4461	0.09	0.1348	0.10
H(13)	−0.0437	0.11	0.4888	0.09	0.3218	0.10
H(13′)	−0.0309	0.11	0.2160	0.09	0.3667	0.10

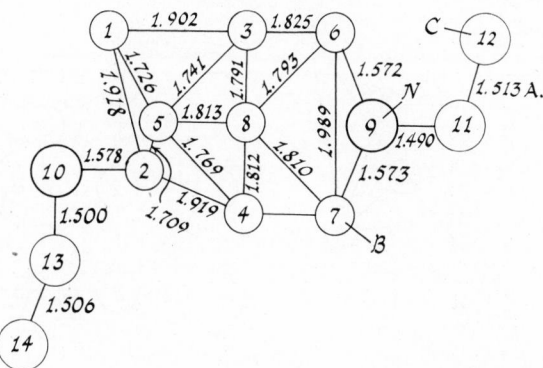

Fig. XIVB,19.   Bond lengths in the molecule of $C_2H_5NH_2B_8H_{11}NHC_2H_5$.

XIV,b14.   A structure determination has been made on a crystalline compound which proved to be *1-ethyl decaborane*, $B_{10}H_{13}C_2H_5$. It is ortho-rhombic with a tetramolecular unit of the edge lengths:

$$a_0 = 10.11 \text{ A.}; \quad b_0 = 14.40 \text{ A.}; \quad c_0 = 7.28 \text{ A.}$$

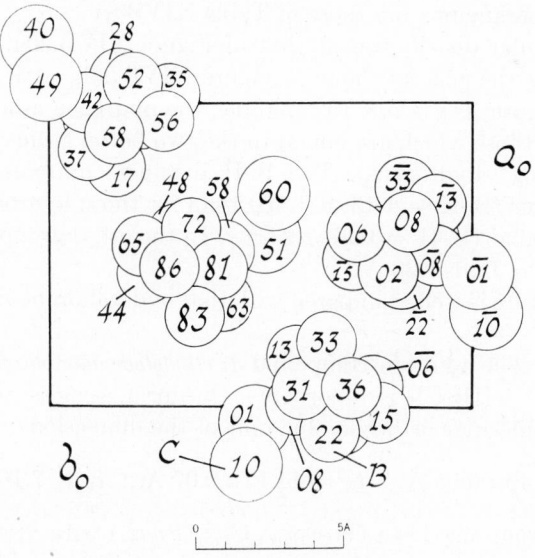

Fig. XIVB,20a.   The orthorhombic structure of $B_{10}H_{13}C_2H_5$ projected along its c_0 axis. Right-hand axes.

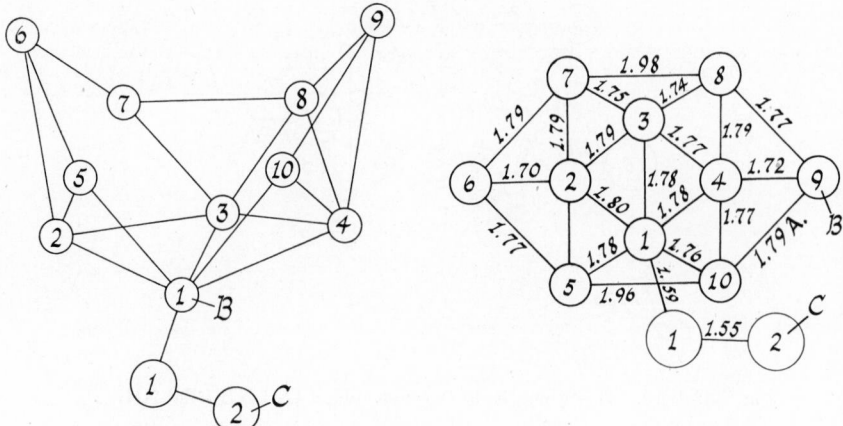

Fig. XIVB,20b (left).   The shape of the molecule of $B_{10}H_{13}C_2H_5$.
Fig. XIVB,20c (right).   Bond lengths in the molecule of $B_{10}H_{13}C_2H_5$.

The space group is V^4 ($P2_12_12_1$) and all atoms are in the positions:

$$(4a) \quad xyz; \; {}^1/_2-x,\bar{y},z+{}^1/_2; \; x+{}^1/_2,{}^1/_2-y,\bar{z}; \; \bar{x},y+{}^1/_2,{}^1/_2-z$$

The atomic parameters are those of Table XIVB,8.

The molecular distribution is that of Figure XIVB,20a. Each of these molecules has the general shape of Figure XIVB,20b, with the B–B separations of Figure XIVB,20c. In the table, the hydrogen atoms bearing two numbers are those which are bound to the two boron atoms thus identified (the "bridging" hydrogens). The B–H distances applying to them are usually greater (B–H = 1.13–1.36 A.) than for those hydrogen atoms that are singly bound (B–H = 0.97–1.23 A.). In the ethyl groups, C–C = 1.55 A. and C–H = 0.91–1.20 A.

This structure is to be compared with that of decaborane itself (**V,f30**).

XIV,b15.   The addition compound *di ethylether–monobromodichlorometh ane*, $(C_2H_5)_2O \cdot CHBrCl_2$, is dimorphous. One form, as measured at $-130°C$. has a tetramolecular orthorhombic unit of the dimensions:

$$a_0 = 9.49 \pm 0.05 \text{ A.}; \quad b_0 = 14.12 \pm 0.07 \text{ A.}; \quad c_0 = 7.16 \pm 0.06 \text{ A.}$$

The space group has been chosen as C_{2v}^9 ($Pna2_1$) with atoms in the positions:

$$(4a) \quad xyz; \; \bar{x},\bar{y},z+{}^1/_2; \; {}^1/_2-x,y+{}^1/_2,z+{}^1/_2; \; x+{}^1/_2,{}^1/_2-y,z$$

TABLE XIVB,8
Parameters of the Atoms in $B_{10}H_{13}C_2H_5$

Atom	x	y	z
C(1)	−0.057	0.510	0.011
C(2)	−0.191	0.533	0.104
B(1)	0.002	0.414	0.080
B(2)	−0.062	0.308	−0.018
B(3)	−0.072	0.330	0.224
B(4)	0.054	0.402	0.312
B(5)	0.097	0.351	−0.084
B(6)	0.061	0.231	−0.064
B(7)	−0.029	0.218	0.146
B(8)	0.049	0.281	0.364
B(9)	0.200	0.341	0.329
B(10)	0.172	0.413	0.132
H(2)	−0.15	0.32	−0.11
H(3)	−0.16	0.34	0.29
H(4)	0.04	0.46	0.41
H(5)	0.12	0.38	−0.23
H(6)	0.05	0.17	−0.16
H(7)	−0.10	0.15	0.16
H(8)	0.01	0.26	0.48
H(9)	0.28	0.35	0.43
H(10)	0.24	0.47	0.10
H(6–7)	0.08	0.19	0.09
H(5–6)	0.18	0.28	−0.07
H(9–10)	0.24	0.34	0.18
H(8–9)	0.15	0.25	0.32
H(C11)	0.02	0.57	0.03
H(C12)	−0.06	0.51	−0.14
H(C21)	−0.18	0.53	0.25
H(C22)	−0.22	0.61	0.05
H(C23)	−0.25	0.49	0.09

The chosen parameters and their standard deviations are those of Table XIVB,9.

The structure that results is shown in Figure XIVB,21. Both types of molecules are present; they are considered to be tied together by a C–H···O bond of 3.10 A.; other intermolecular bonds range upwards from 3.68 A. The final $R = 0.19$.

TABLE XIVB,9
Parameters of the Atoms in $(C_2H_5)_2O \cdot CHBrCl_2$

Atom	x	$\sigma(x)$	y	$\sigma(y)$	z	$\sigma(z)$
Br	0.1763	0.0006	0.2276	0.0004	0.000	—
Cl(1)	0.391	0.002	0.113	0.001	0.213	0.009
Cl(2)	0.391	0.002	0.112	0.001	−0.213	0.009
C(1)	0.273	0.010	0.115	0.006	0.000	—
C(2)	0.288	0.005	−0.152	0.004	0.000	—
C(3)	0.127	0.006	−0.152	0.004	0.050	—
O	0.084	0.006	−0.060	0.002	0.000	—
C(4)	−0.071	0.006	−0.057	0.004	0.050	—
C(5)	−0.119	0.005	0.055	0.004	0.000	—

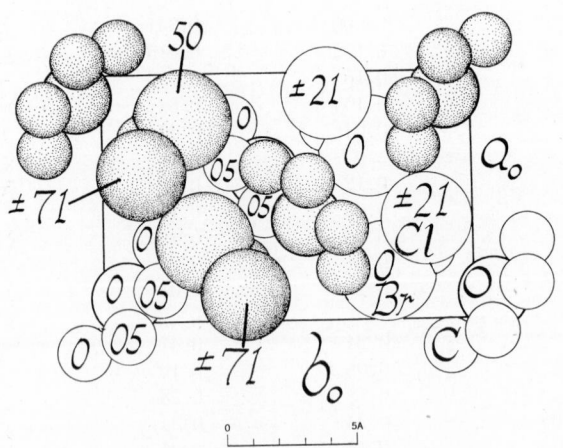

Fig. XIVB,21.   The orthorhombic structure of $(C_2H_5)_2O \cdot CHBrCl_2$ projected along its c_0 axis.   Right-hand axes.

XIV,b16.   Crystals of *thio triethyl phosphine*, $SP(C_2H_5)_3$, are hexagonal with a bimolecular unit of the edge lengths:

$$a_0 = 8.98 \pm 0.01 \text{ A.}, \qquad c_0 = 6.32 \pm 0.01 \text{ A.}$$

The x-ray data correspond to the requirements of the space group C_{6v}^4 $(P6_3mc)$. This would place the sulfur and phosphorus atoms in the positions:

$$(2b) \quad {}^1/_3\,{}^2/_3\,u; \; {}^2/_3, {}^1/_3, u + {}^1/_2$$

and it has been concluded that, taking $u(S) = {}^1/_2$, then $u(P) = -0.205$.

It is stated that the carbon atoms are not in the sixfold special positions (6c) of C_{6v}^4, as one might expect. Instead, it is thought that they are probably statistically distributed through half the positions of (12d):

$$(12d) \quad xyz; \quad \bar{y},x-y,z; \quad y-x,\bar{x},z;$$
$$\bar{y}\bar{x}z; \quad y-x,y,z; \quad x,x-y,z;$$
$$\bar{x},\bar{y},z+{}^1/_2; \; y,y-x,z+{}^1/_2; \; x-y,x,z+{}^1/_2;$$
$$y,x,z+{}^1/_2; \; x-y,\bar{y},z+{}^1/_2; \; \bar{x},y-x,z+{}^1/_2$$

Furthermore two sets of carbon parameters have been taken as possible and both are considered to contribute to the structure. They represent two enantiomorphous forms (A and B) of the molecule with the parameters listed below.

Atom	x	y	z
C(1,A)	0.453	0.169	0.406
C(1,B)	0.453	0.284	0.406
C(2,A)	0.317	0.183	0.335
C(2,B)	0.317	0.133	0.335

Several hypotheses dealing with the distribution of these isomers were discussed, but it was not possible to decide between disorder in their distribution or a kind of twinning of separate crystalline entities. In any case the significant interatomic distances arising from this determination are those of Figure XIVB,22.

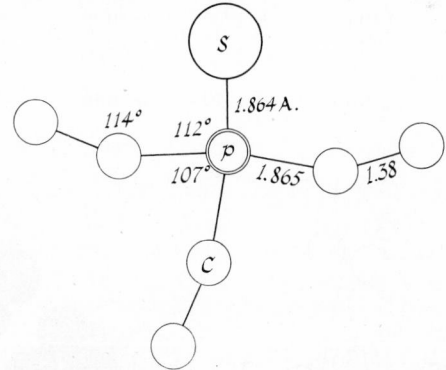

Fig. XIVB,22.   Bond dimensions in the molecule of $SP(C_2H_5)_3$.

The corresponding selenium compound, *seleno triethyl phosphine*, SeP-$(C_2H_5)_3$, is isostructural with

$$a_0 = 9.06 \text{ A.}, \qquad c_0 = 6.54 \text{ A.}$$

248 CRYSTAL STRUCTURES

The selenium and phosphorus atoms were found to be in (2b) with $u(Se) = 1/2$ and $u(P) = -0.200$. The same uncertainty exists as with the sulfur compound with respect to the distribution of the carbon atoms, but the parameters, to be used more or less statistically, are given as:

Atom	x	y	z
C(1,A)	0.455	0.169	0.420
C(1,B)	0.455	0.285	0.420
C(2,A)	0.318	0.185	0.345
C(2,B)	0.318	0.133	0.345

XIV,b17. *Tetraethyl diphosphine disulfide*, $(C_2H_5)_4P_2S_2$, is triclinic with a unimolecular cell of the dimensions:

$$a_0 = 8.98 \pm 0.03 \text{ A.}; \quad b_0 = 6.45 \pm 0.02 \text{ A.}; \quad c_0 = 6.15 \pm 0.02 \text{ A.}$$
$$\alpha = 113°0' \pm 30'; \quad \beta = 85°12' \pm 18'; \quad \gamma = 102°30' \pm 18'$$

The space group is C_i^1 ($P\bar{1}$) with all atoms in the general positions: (2i) $\pm(xyz)$. The determined parameters are those of Table XIVB,10.

TABLE XIVB,10
Parameters of the Atoms in $(C_2H_5)_4P_2S_2$

Atom	x	σ_x, A.	y	σ_y, A.	z	σ_z, A.
P	0.1197	0.0027	0.0484	0.0026	0.0614	0.0031
S	0.1832	0.0036	-0.1916	0.0033	0.1257	0.0037
C(1)	0.1193	0.0126	0.3168	0.0135	0.3217	0.0141
C(2)	0.2221	0.0117	0.1267	0.0129	-0.1725	0.0129
C(3)	0.2768	0.0144	0.4399	0.0148	0.4197	0.0154
C(4)	0.3915	0.0135	0.1312	0.0148	0.1734	0.0141

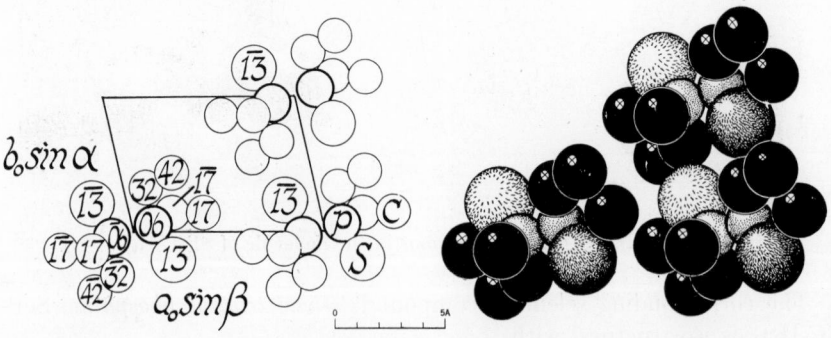

Fig. XIVB,23a (left). The triclinic structure of $(C_2H_5)_4P_2S_2$ projected along its c_0 axis. Right-hand axes.
Fig. XIVB,23b (right). A packing drawing of the triclinic structure of $(C_2H_5)_4P_2S_2$ seen along its c_0 axis. The carbon atoms are black, the phosphorus atoms smaller, dotted, and heavily outlined. Atoms of sulfur are line shaded.

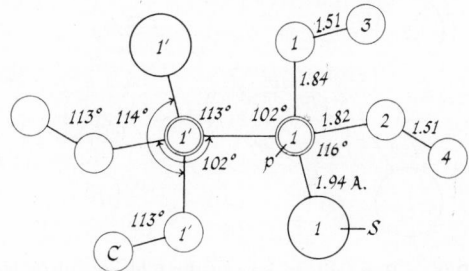

Fig. XIVB,24. Bond dimensions of the molecule of $(C_2H_5)_4P_2S_2$.

The structure is shown in Figure XIVB,23; its molecule has the bond dimensions of Figure XIVB,24. The shortest intermolecular atomic separations are P–C = 2.80 A. and C–C = 2.95 A.; others range upwards from a P–C = 3.15 A. and an S–C = 3.20 A.

XIV,b18. Crystals of a compound obtained by adding CS_2 to $P(C_2H_5)_3$ and designated as *triethylscarphane* are monoclinic. There are four molecules of the composition $(C_2H_5)_3P \cdot CS_2$ in a unit of the dimensions:

$$a_0 = 7.50 \pm 0.02 \text{ A.}; \quad b_0 = 11.97 \pm 0.02 \text{ A.}; \quad c_0 = 11.58 \pm 0.05 \text{ A.}$$
$$\beta = 90°12' \pm 12'$$

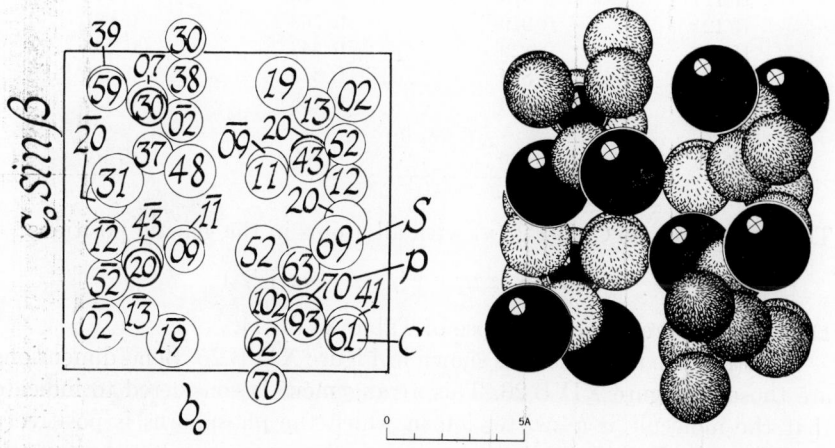

Fig. XIVB,25a (left). The monoclinic structure of $(C_2H_5)_3P \cdot CS_2$ projected along its a_0 axis. Right-hand axes.

Fig. XIVB,25b (right). A packing drawing of the monoclinic structure of $(C_2H_5)_3P \cdot$ CS_2 viewed along its a_0 axis. The sulfur atoms are the large, the phosphorus the small black circles. Atoms of carbon are line shaded.

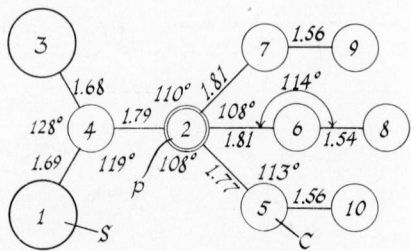

Fig. XIVB,26.   Bond dimensions in the molecule of $(C_2H_5)_3P \cdot CS_2$.

TABLE XIVB,11
Parameters of the Atoms in $(C_2H_5)_3P \cdot CS_2$

Atom	x	y	z
S(1)	0.9844	0.1145	0.1414
P(2)	0.8024	0.2441	0.3289
S(3)	0.6883	0.8398	0.4003
C(4)	0.8723	0.2308	0.1819
C(5)	0.9334	0.7507	0.1690
C(6)	0.8794	0.1282	0.4170
C(7)	0.8866	0.3725	0.3906
C(8)	0.3057	0.3723	0.0410
C(9)	0.0946	0.3793	0.3848
C(10)	0.0252	0.6453	0.2208
	Calculated hydrogen positions		
H(11)	0.978	0.825	0.216
H(12)	0.979	0.763	0.081
H(13)	0.829	0.443	0.345
H(14)	0.844	0.379	0.480
H(15)	0.025	0.128	0.420
H(16)	0.846	0.050	0.375

The space group is C_{2h}^5 $(P2_1/n)$ with all atoms in the general positions:

$$(4e) \quad \pm (xyz; \; x+^1/_2, ^1/_2-y, z+^1/_2)$$

Determined parameters are those of Table XIVB,11.

The structure that results is shown in Figure XIVB,25. Bond dimensions are those of Figure XIVB,26. This arrangement is considered to indicate that the molecule is a zwitter ion in which the phosphorus is positively charged and the corresponding negative charge is distributed over the sulfur atoms.

XIV,b19.   *2,2′,2″-tri-Amino-triethylamine trihydrochloride*, $N(CH_2CH_2-NH_2HCl)_3$, is cubic with a tetramolecular unit of the edge length:

$$a_0 = 10.870 \pm 0.001 \text{ A.}$$

The space group is T^4 ($P2_13$) with atoms in the positions:

(4a) $uuu; u+\frac{1}{2},\frac{1}{2}-u,\bar{u}; \frac{1}{2}-u,\bar{u},u+\frac{1}{2}; \bar{u},u+\frac{1}{2},\frac{1}{2}-u$

(12b) $xyz; x+\frac{1}{2},\frac{1}{2}-y,\bar{z}; \bar{x},y+\frac{1}{2},\frac{1}{2}-z; \frac{1}{2}-x,\bar{y},z+\frac{1}{2};$ tr

The following atoms are in (4a):

Cl(1):	$u = 0.2020$	$\sigma(u) = 0.0003$	
Cl(2):	$u = 0.4402$	$\sigma(u) = 0.00035$	
Cl(3):	$u = 0.9309$	$\sigma(u) = 0.0004$	
N(1):	$u = 0.7351$	$\sigma(u) = 0.0009$	

All other atoms, in (12b), have been given the parameters of Table XIVB,12.

In the molecular cations, $N(1)-C(1) = 1.52$ A., $C(1)-C(2) = 1.46$ A., $C(2)-N(2) = 1.47$ A.; $C(1)-N(1)-C(1) = 108°24'$, $N(1)-C(1)-C(2) = 111°36'$, and $C(1)-C(2)-N(2) = 112°36'$. Separations between Cl and N(2) lie between 3.14 and 3.27 A.; $N(1)-Cl = 3.68$ A.

TABLE XIVB,12
Parameters of Atoms in $N(CH_2CH_2NH_2HCl)_3$

Atom	x	y	z	σ
C(1)	0.6549	0.6515	0.8135	0.0013
C(2)	0.7183	0.6127	0.9255	0.0012
N	0.7527	0.7158	0.0058	0.0014
H(1)	0.578	0.695	0.830	—
H(2)	0.645	0.570	0.760	—
H(3)	0.970	0.675	0.537	—
H(4)	0.892	0.809	0.580	—
H(5)	0.470	0.680	0.225	—
H(6)	0.510	0.813	0.219	—

XIV,b20.   Crystals of *ethylene diammonium chloride*, $NH_3(CH_2)_2NH_3Cl_2$, are monoclinic with a bimolecular cell of the dimensions:

$$a_0 = 9.95 \text{ A.;} \quad b_0 = 6.89 \text{ A.;} \quad c_0 = 4.42 \text{ A.,} \qquad \text{all } \pm 0.01 \text{ A.}$$
$$\beta = 90°42' \pm 18'$$

The space group is C_{2h}^5 ($P2_1/a$) with all atoms in the general positions:

(4e) $\pm (xyz; x+\frac{1}{2},\frac{1}{2}-y,z)$

The established parameters are the following:

Atom	x	y	z
C	0.026 (0.022)	0.072 (0.078)	0.118 (0.115)
N	0.083 (0.079)	0.241 (0.245)	0.954 (0.953)
Cl	0.1703 (0.171)	0.5803 (0.582)	0.4058 (0.401)

The atomic arrangement is shown in Figure XIVB,27. The planar cation has the bond dimensions of Figure XIVB,28. Three chloride ions are closest to nitrogen atoms with N–Cl = 3.14, 3.20, and 3.22 A., and there is a fourth with N–Cl = 3.48 A.; they probably represent hydrogen bonds. The closest approach of chlorine atoms to one another is 3.89 A.

A second determination agreeing well with the foregoing was published at about the same time. Its parameters, after an exchange of the a_0 and c_0 axes and a translation of the origin to the point $^1/_2 b_0$, $^1/_2 c_0$, are given above in parentheses.

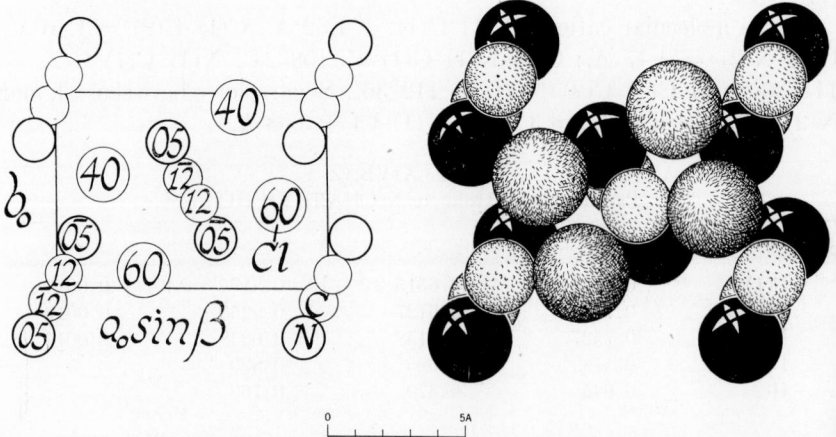

Fig. XIVB,27a (left).   The monoclinic structure of $NH_3(CH_2)_2NH_3Cl_2$ projected along its c_0 axis.   Right-hand axes.

Fig. XIVB,27b (right).   A packing drawing of the monoclinic structure of NH_3-$(CH_2)_2NH_3Cl_2$ viewed along its c_0 axis.   The nitrogen atoms are black.   Atoms of chlorine are the large line-shaded circles; the carbon atoms are smaller and hook shaded.

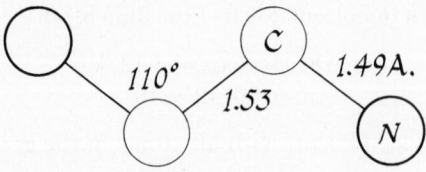

Fig. XIVB,28.   Bond dimensions in the molecular cation of $NH_3(CH_2)_2NH_3Cl_2$.

XIV,b21. *Ethylene diammonium sulfate*, $(NH_3CH_2-)_2SO_4$, forms tetragonal crystals which have a base-centered unit containing eight molecules and the cell edges:

$$a_0 = 8.47 \pm 0.02 \text{ A.,} \qquad c_0 = 18.03 \pm 0.04 \text{ A.}$$

The space group is one or the other of the enantiomorphous pair D_4^4 ($P4_122_1$) or D_4^8 ($P4_322_1$). Choosing the space group as D_4^4 and the unit as base-centered, sulfur atoms are in the positions:

(8a) $u00$; $\bar{u}\,0\,^1/_2$; $^1/_2\,u\,^3/_4$; $^1/_2\,\bar{u}\,^1/_4$;
$u+^1/_2,^1/_2,0$; $^1/_2-u,^1/_2,^1/_2$; $0,u+^1/_2,^3/_4$; $0,^1/_2-u,^1/_4$
with $u = 0.2291$

All other atoms are in the general positions:

(16b) xyz; $\bar{x},\bar{y},z+^1/_2$; $^1/_2-y,x,z+^3/_4$; $y+^1/_2,\bar{x},z+^1/_4$;
$x\bar{y}\bar{z}$; $\bar{x},y,^1/_2-z$; $^1/_2-y,\bar{x},^1/_4-z$; $y+^1/_2,x,^3/_4-z$;
$x+^1/_2,y+^1/_2,z$; $^1/_2-x,^1/_2-y,z+^1/_2$; $\bar{y},x+^1/_2,z+^3/_4$; $y,^1/_2-x,z+^1/_4$;
$x+^1/_2,^1/_2-y,\bar{z}$; $^1/_2-x,y+^1/_2,^1/_2-z$; $\bar{y},^1/_2-x,^1/_4-z$; $y,x+^1/_2,^3/_4-z$

For the heavier atoms the stated parameters are:

Atom	x	y	z
O(1)	0.123	0.125	0.0296
O(2)	0.329	0.063	0.0608
N	0.611	0.111	0.0643
C	0.744	0.005	0.0430

Reasonable positions have also been suggested for the hydrogen atoms. They are:

Atom	x	y	z
H(1)	0.64	0.12	0.114
H(2)	0.61	0.21	0.044
H(3)	0.53	0.07	0.055
H(4)	0.83	0.07	0.055
H(5)	0.76	0.10	0 055

There is an error in the original coordinate description, but it seems probable that the values given above represent the atomic arrangement.

The simpler diagonal unit containing four molecules would have the dimensions:

$$a_0 = 5.99 \text{ A.,} \qquad c_0 = 18.03 \text{ A.}$$

Sulfur atoms are in the positions:

(4a) $uu0$; $\bar{u}\,\bar{u}\,^1/_2$; $^1/_2-u,u+^1/_2,^1/_4$; $u+^1/_2,^1/_2-u,^3/_4$
with, as before, $u = 0.2291$

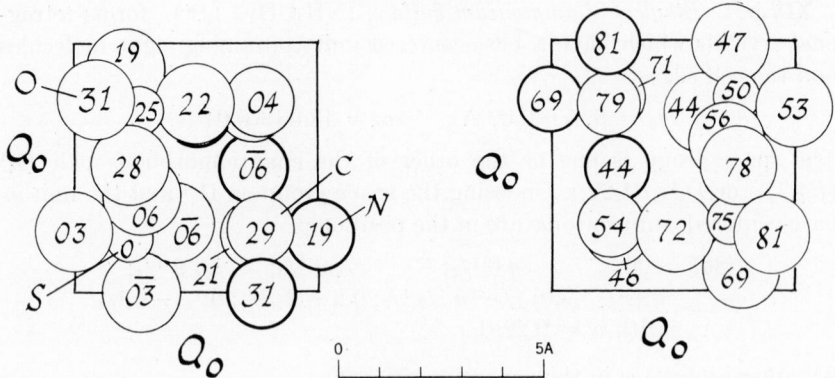

Fig. XIVB,29a (left). The contents of the lower half of the tetragonal unit of $(NH_3CH_2—)_2SO_4$ projected along its long c_0 axis. Right-hand axes.

Fig. XIVB,29b (right). The contents of the upper half of the tetragonal unit of $(NH_3CH_2—)_2SO_4$ projected along its c_0 axis.

The other heavy atoms, in the positions:

$$(8b) \quad xyz;\ \bar{x},\bar{y},z+{}^1/_2;\ {}^1/_2-y,x+{}^1/_2,z+{}^1/_4;\ y+{}^1/_2,{}^1/_2-x,z+{}^3/_4;$$
$$yx\bar{z};\ \bar{y},\bar{x},{}^1/_2-z;\ {}^1/_2-x,y+{}^1/_2,{}^1/_4-z;\ x+{}^1/_2,{}^1/_2-y,{}^3/_4-z$$

would have the parameters:

Atom	x	y	z
O(1)	−0.002	0.248	0.0296
O(2)	0.275	0.394	0.0608
N	0.500	0.722	0.0643
C	0.749	0.739	0.0430

Figure XIVB,29 indicates the distribution of the ions within this smaller cell. The $(CH_2NH_3)_2$ group has a *gauche* configuration (Fig. XIVB,30) with N–C = 1.49 A., C–C = 1.55 A., and the angle between the two planes N–C–C = 75°42′. Each nitrogen atom is closest to three oxygen atoms at distances between 2.80 and 2.82 A.; these can be considered to involve hydrogen bonds.

XIV,b22. *Choline chloride,* [(CH₃)₃NCH₂CH₂OH]Cl, is orthorhombic with a tetramolecular cell of the dimensions:

$$a_0 = 11.21\ A.;\quad b_0 = 11.59\ A.;\quad c_0 = 5.87\ A.,\qquad \text{all} \pm 0.02\ A.$$

The space group is V^4 ($P2_12_12_1$) with all atoms in the positions:

$$(4a) \quad xyz;\ {}^1/_2-x,\bar{y},z+{}^1/_2;\ x+{}^1/_2,{}^1/_2-y,\bar{z};\ \bar{x},y+{}^1/_2,{}^1/_2-z$$

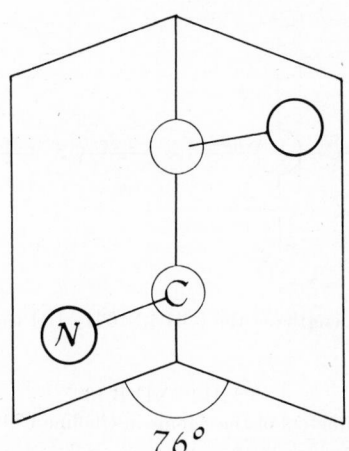

Fig. XIVB,30. The general shape of the $(NH_3CH_2—)_2{}^{2+}$ cation in its sulfate.

'he determined parameters are those of Table XIVB,13. The molecular arrangement in the crystal is shown in Figure XIVB,31. 'he molecules themselves have the bond lengths of Figure XIVB,32.

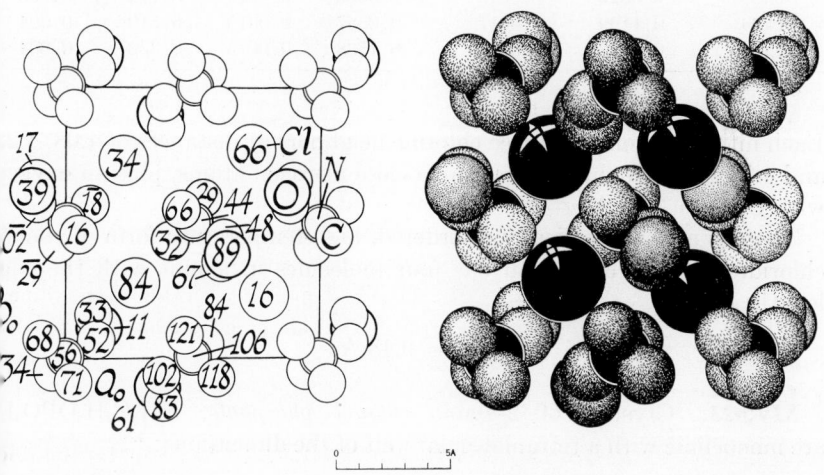

'ig. XIVB,31a (left). The orthorhombic structure of choline chloride projected along its c_0 axis. Right-hand axes.

Fig. XIVB,31b (right). A packing drawing of the orthorhombic structure of $[(CH_3)_3$-$CH_2CH_2OH]Cl$ seen along its c_0 axis. The Cl^- ions are the large, the nitrogen atoms he smaller black circles. The hydroxyls are heavily outlined and dotted; the carbon toms are line shaded.

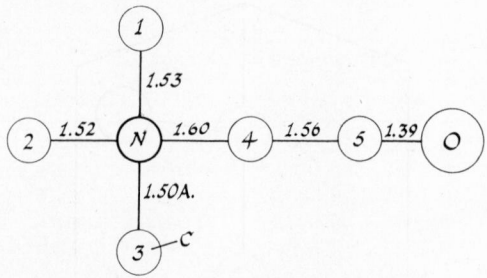

Fig. XIVB,32.   Bond lengths in the $(CH_3)_3N(CH_2)_2OH$ cation in choline chloride.

TABLE XIVB,13
Parameters of the Atoms in Choline Chloride

Atom	x	$\sigma(x)$	y	$\sigma(y)$	z	$\sigma(z)$
Cl	−0.2203	0.0003	0.2287	0.0003	0.1627	0.0007
O	0.1247	0.0010	0.1091	0.0009	0.114	0.002
N	−0.0059	0.0008	−0.0024	0.0009	0.558	0.002
C(1)	−0.0545	0.0014	−0.0560	0.0012	0.337	0.003
C(2)	−0.1011	0.0017	0.0678	0.0014	0.679	0.004
C(3)	0.0320	0.0013	−0.0983	0.0012	0.714	0.003
C(4)	0.1199	0.0017	0.0587	0.0015	0.519	0.004
C(5)	0.1122	0.0016	0.1538	0.0015	0.332	0.004

Each nitrogen atom has three chlorine neighbors at distances of 4.08, 4.2: and 4.28 A.; each chlorine has three close nitrogen atoms, plus an oxyge₁ with Cl–O = 3.03 A.

There is also a spherically disordered, high-temperature form of cholin₁ chloride which is cubic and has four molecules in a unit with the edg₁ length:

$$a_0 = 9.48 \text{ A.}$$

XIV,b23.   Crystals of *2-amino ethanol phosphate*, $NH_2C_2H_5OPO_3H$ are monoclinic with a tetramolecular unit of the dimensions:

$$a_0 = 9.04 \text{ A.;} \quad b_0 = 7.75 \text{ A.;} \quad c_0 = 8.86 \text{ A.;} \quad \beta = 102°27'$$

The space group is C_{2h}^5 $(P2_1/c)$ and all atoms are in the general position₁

$$(4e) \quad \pm(xyz; \ x,{}^1/_2-y,z+{}^1/_2)$$

with the parameters of Table XIVB,14.

TABLE XIVB,14
Parameters of Atoms in $NH_2C_2H_5OPO_3H$

Atom	x	$\sigma(x)$	y	$\sigma(y)$	z	$\sigma(z)$
P(1)	0.2332	0.0001	0.0231	0.0001	0.1130	0.0001
O(2)	0.3701	0.0003	−0.0766	0.0004	0.1899	0.0005
O(3)	0.1686	0.0003	−0.0136	0.0005	−0.0549	0.0004
O(4)	0.1063	0.0004	0.0023	0.0006	0.2058	0.0004
O(5)	0.2865	0.0003	0.2189	0.0004	0.1333	0.0004
C(6)	0.1797	0.0006	0.3514	0.0007	0.0742	0.0006
C(7)	0.2641	0.0006	0.5206	0.0007	0.0708	0.0006
N(8)	0.3478	0.0004	0.5715	0.0005	0.2284	0.0004
H(9)	0.281	0.006	0.552	0.007	0.288	0.006
H(10)	0.366	0.006	0.695	0.007	0.220	0.006
H(11)	0.439	0.007	0.533	0.008	0.277	0.007
H(12)	0.017	0.011	−0.018	0.012	0.152	0.010
H(13)	0.098	0.008	0.404	0.011	0.154	0.009
H(14)	0.123	0.006	0.314	0.008	−0.032	0.006
H(15)	0.168	0.007	0.597	0.009	0.028	0.007
H(16)	0.335	0.007	0.516	0.008	0.006	0.007

These lead to an arrangement having two of the four molecules as shown in Figure XIVB,33, the others being almost completely hidden beneath these. The individual molecules have the bond dimensions of Figure XIVB,34. The arrangement around the C(6)–C(7) bond is *gauche* rather than zigzag. The N–H distances lie between 0.90 and 0.98 A.; C–H =

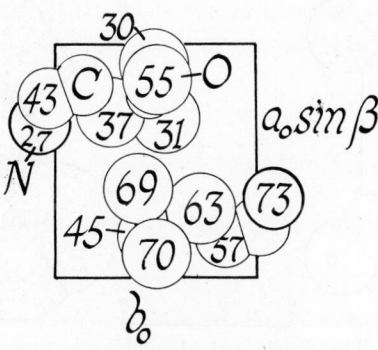

Fig. XIVB,33. A part of the monoclinic structure of 2-amino ethanol phosphate projected along its c_0 axis. Only two of the four molecules per cell are shown; they overlie the others. The phosphorus atoms scarcely show. Right-hand axes.

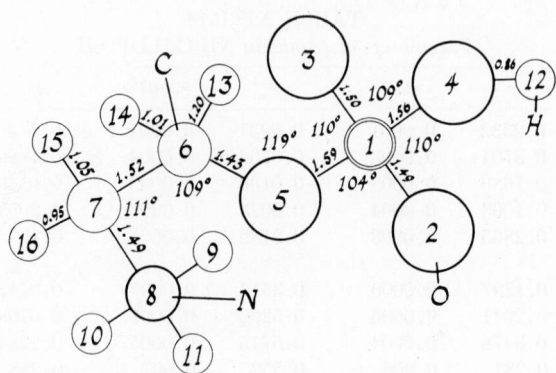

Fig. XIVB,34.   Bond dimensions in the molecule of 2-amino ethanol phosphate.

0.95–1.20 A. and O–H = 0.86 A. Four hydrogen bonds are noted: one between the phosphate oxygens of adjacent molecules with O–O = 2.56 A., and three between nitrogen and oxygens with N–O = 2.75, 2.76, and 2.80 A.

A second determination has been reported (1962: F,L&Y) which confirms the foregoing.

2. Simple Ethylene Derivatives

XIV,b24.   The structure of sylid *ethylene*, C_2H_4, differs from that for ethane. It is orthorhombic with a bimolecular cell having the edge lengths:

$$a_0 = 4.87 \text{ A.}; \quad b_0 = 6.46 \text{ A.}; \quad c_0 = 4.14 \text{ A.} \ (-175°\text{C.})$$

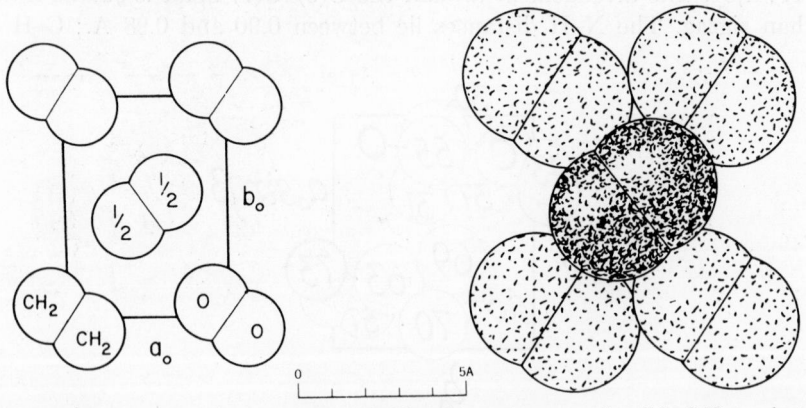

Fig. XIVB,35a (left).   The simple orthorhombic structure of solid C_2H_4 projected along its c_0 axis. Left-hand axes.

Fig. XIVB,35b (right).   A packing drawing of the orthorhombic structure of C_2H_4 seen along its c_0 axis.

Carbon atoms are in the following positions of V_h^{12} ($Pnnm$):

$$(4g) \quad \pm(uv0; \ u+\tfrac{1}{2}, \tfrac{1}{2}-v, \tfrac{1}{2})$$

$$\text{with } u = 0.11, \ v = 0.06$$

This leads to a body-centered packing, elongated along the b_0 axis, of molecules in which the double-bonded carbon separation is 1.33 A. (Fig. XIVB,35).

XIV,b25. Crystals of *tetracyanoethylene*, $(CN)_2C{=}C(CN)_2$, are monoclinic with a bimolecular unit of the dimensions:

$$a_0 = 7.51 \text{ A.}; \quad b_0 = 6.21 \text{ A.}; \quad c_0 = 7.00 \text{ A.}; \quad \beta = 97°10'$$

The space group is C_{2h}^5 ($P2_1/n$) with all atoms in the positions:

$$(4e) \quad \pm(xyz; \ x+\tfrac{1}{2}, \tfrac{1}{2}-y, z+\tfrac{1}{2})$$

The determined parameters are listed below:

Atom	x	$\sigma(x)$	y	$\sigma(y)$	z	$\sigma(z)$
C(1)	0.0028	0.0005	−0.0392	0.0006	−0.0866	0.0006
C(2)	0.0860	0.0006	0.0832	0.0006	−0.2279	0.0007
C(3)	−0.0756	0.0005	−0.2436	0.0006	−0.1465	0.0007
N(1)	0.1516	0.0006	0.1776	0.0007	−0.3379	0.0007
N(2)	−0.1377	0.0005	−0.4037	0.0006	−0.1933	0.0007

The structure is shown in Figure XIVB,36. It is built up of planar

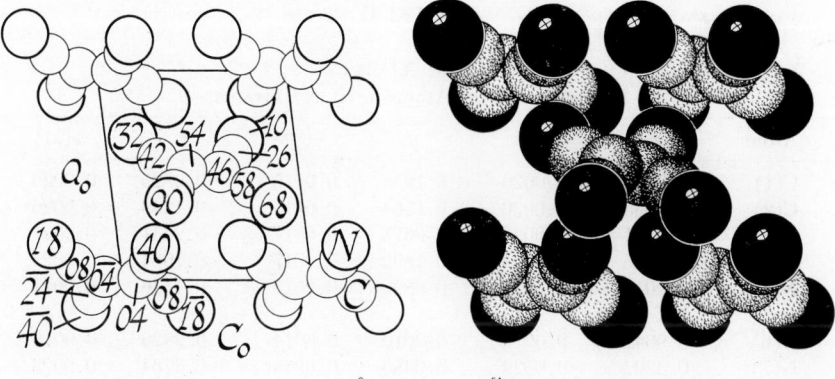

Fig. XIVB,36a (left). The monoclinic structure of $C_2(CN)_4$ projected along its b_0 axis. Right-hand axes.

Fig. XIVB,36b (right). A packing drawing of the monoclinic structure of $C_2(CN)_4$ viewed along its b_0 axis. The nitrogen atoms are black; the carbon atoms are curl shaded.

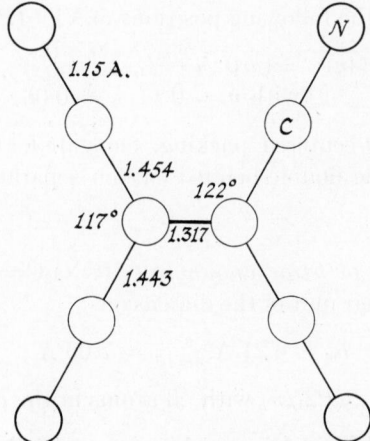

Fig. XIVB,37.   Bond dimensions in the planar molecules of $C_2(CN)_4$.

molecules, having the dimensions of Figure XIVB,37. Between molecules the C–N separations range upwards from 3.09 A.; the shortest N–N = 3.41 A. and the shortest C–C = 3.96 A.

XIV,b26. There is a tetramer of hydrocyanic acid which has been determined by x-ray analysis to be *diaminomaleonitrile*, $[NH_2C(CN)=]_2$. It is monoclinic with a tetramolecular unit of the dimensions:

$$a_0 = 6.44 \pm 0.01 \text{ A.}; \quad b_0 = 18.24 \pm 0.03 \text{ A.}; \quad c_0 = 5.22 \pm 0.01 \text{ A.}$$
$$\beta = 122°0'$$

TABLE XIVB,15
Parameters of the Atoms in HCN Tetramer

Atom	x	$\sigma(x)$	y	$\sigma(y)$	z	$\sigma(z)$
C(1)	0.7315	0.0020	0.1055	0.0015	−0.0825	0.0020
C(2)	0.7595	0.0020	0.1564	0.0015	0.1249	0.0020
C(3)	0.5031	0.0023	0.0675	0.0017	−0.2451	0.0021
C(4)	0.5626	0.0022	0.1692	0.0017	0.1684	0.0022
N(5)	0.9019	0.0021	0.0899	0.0017	−0.1546	0.0021
N(6)	0.9641	0.0019	0.2010	0.0014	0.2826	0.0019
N(7)	0.3202	0.0022	0.0380	0.0018	−0.3764	0.0023
N(8)	0.4048	0.0022	0.1794	0.0018	0.2028	0.0026
H(9)	0.046	0.039	0.110	0.029	−0.067	0.035
H(10)	0 878	0.034	0.050	0.027	−0.306	0.034
H(11)	0.115	0.037	0.183	0.025	0.320	0.037
H(12)	0.974	0.036	0.222	0.031	0.452	0.036

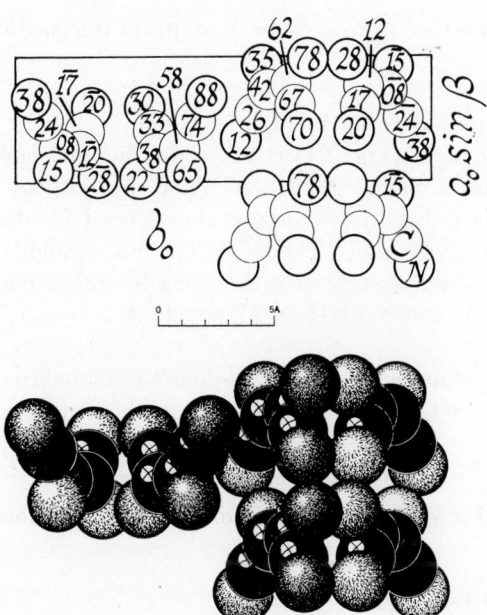

Fig. XIVB,38a (top).   The monoclinic structure of diaminomaleonitrile projected along
its c_0 axis.   Right-hand axes.

Fig. XIVB,38b (bottom).   A packing drawing of the monoclinic structure of diaminomaleonitrile seen along its c_0 axis.   The carbon atoms are black, the nitrogen atoms
line shaded.

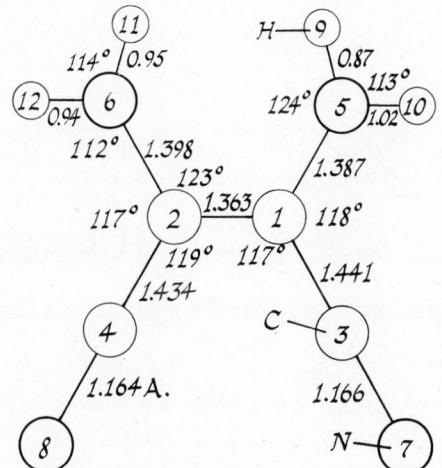

Fig. XIVB,39.   Bond dimensions in the molecule of diaminomaleonitrile.

The space group is C_{2h}^5 ($P2_1/c$) with all atoms in the positions:

$$(4e)\quad \pm(xyz;\ x,{}^1\!/_2-y,z+{}^1\!/_2)$$

and the parameters of Table XIVB,15.

In this arrangement (Fig. XIVB,38), the individual molecules have the bond dimensions of Figure XIVB,39. They are without symmetry, the two halves being twisted by 6° about the $C(1)\!=\!C(2)$ double bond. The bonds $C(2)$–$C(4)$–$N(8)$ and $C(1)$–$C(3)$–$N(7)$ are straight. The molecules are tied together by a system of rather long hydrogen bonds, with $N(5)$–H–$N(8)$ = 3.21 A. and $N(6)$–H–$N(8)$ = 3.12 A.

XIV,b27.   Crystals of *N-methyl 2,2-dimethyl sulfonyl vinylidene amine*, $(CH_3SO_2)_2C\!=\!C\!=\!NCH_3$, are orthorhombic with the tetramolecular cell:

$$a_0 = 12.082\ \text{A.};\quad b_0 = 8.656\ \text{A.};\quad c_0 = 8.627\ \text{A.}$$

The space group is V_h^{14} (*Pbcn*) and atoms are in the following positions of this group:

$$(4c)\quad \pm(0\ u\ {}^1\!/_4;\ {}^1\!/_2,u+{}^1\!/_2,{}^1\!/_4)$$
$$(8d)\quad \pm(xyz;\ x+{}^1\!/_2,y+{}^1\!/_2,{}^1\!/_2-z;\ {}^1\!/_2-x,y+{}^1\!/_2,z;\ \bar{x},y,{}^1\!/_2-z)$$

A detailed determination of parameters gave the values listed in Table XIVB,16.

These result in a molecule having the dimensions stated in Figure XIVB,40. It is noteworthy that the chain $C(2)$–$C(3)$–N–$C(4)$ is strictly

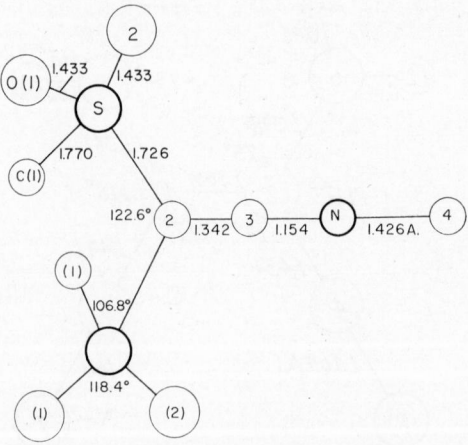

Fig. XIVB,40.   Bond dimensions in the molecule of $(CH_3SO_2)_2C\!=\!C\!=\!NCH_3$.

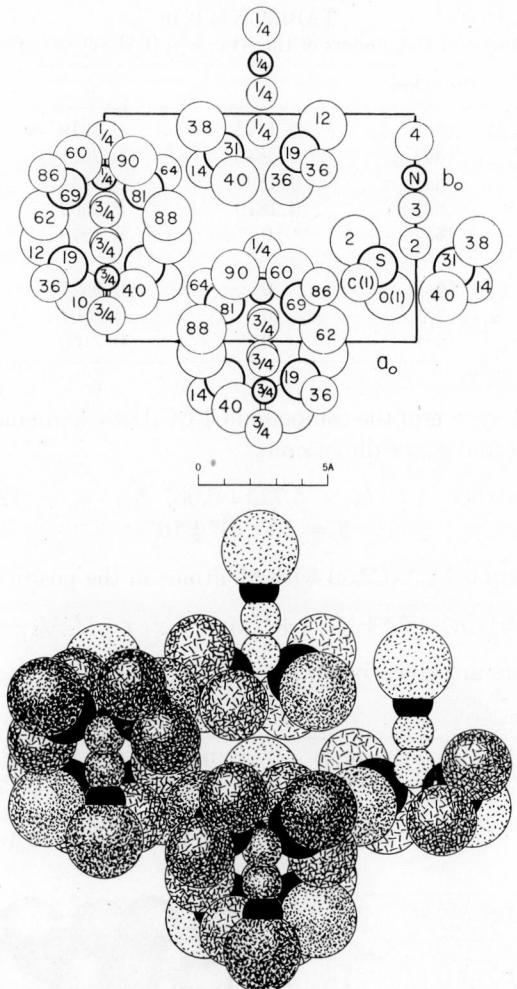

Fig. XIVB,41a (top).   The orthorhombic structure of $(CH_3SO_2)_2C{=}C{=}NCH_3$ projected along its c_0 axis.   Left-hand axes.

Fig. XIVB,41b (bottom).   A packing drawing of the orthorhombic structure of $(CH_3SO_2)_2C{=}C{=}NCH_3$ seen along its c_0 axis.   The sulfur atoms are the large, the nitrogen atoms the small black circles.   The carbon atoms are dotted, with the terminal CH_3 larger than the others.   Atoms of oxygen are line shaded.

linear. The packing of these molecules within the crystal is illustrated in Figure XIVB,41.

TABLE XIVB,16

Positions and Parameters of the Atoms in $(CH_3SO_2)_2C{=}C{=}NCH_3$

Atom	Position	x	y	z
S	(8d)	0.1175	0.3319	0.1889
O(1)	(8d)	0.0840	0.2021	0.0967
O(2)	(8d)	0.1893	0.4444	0.1207
C(1)	(8d)	0.1819	0.2562	0.3568
C(2)	(4c)	0	0.4276	$1/4$
C(3)	(4c)	0	0.5827	$1/4$
C(4)	(4c)	0	0.8807	$1/4$
N	(4c)	0	0.7160	$1/4$

XIV,b28. Crystals of the compound $Cl_2BC_2H_4BCl_2$ are monoclinic with a tetramolecular unit of the dimensions:

$$a_0 = 10.950\pm0.006 \text{ A.}; \quad b_0 = 5.723\pm0.007 \text{ A.}; \quad c_0 = 12.533\pm0.014 \text{ A.}$$
$$\beta = 100°31'\pm10'$$

The space group is C_{2h}^6 $(C2/c)$ with all atoms in the positions:

$$(8f) \quad \pm(xyz; x,\bar{y},z+1/2; x+1/2,y+1/2,z; x+1/2,1/2-y,z+1/2)$$

The parameters are given below.

Atom	x	y	z
Cl(1)	0.3825	0.6475	0.1215
Cl(2)	0.1275	0.639	0.175
C	0.2035	0.290	0.031
B	0.239	0.513	0.1045

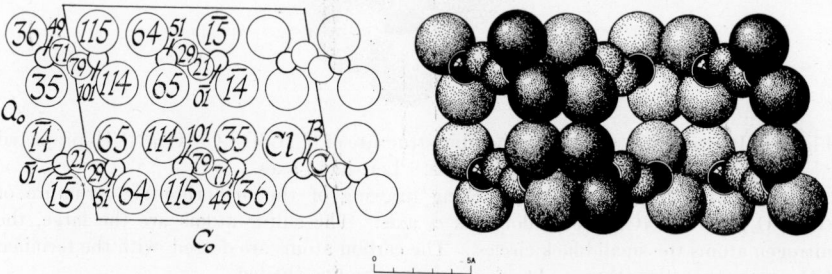

Fig. XIVB,42a (left). The monoclinic structure of $Cl_2BC_2H_4BCl_2$ projected along its b_0 axis. Right-hand axes.

Fig. XIVB,42b (right). A packing drawing of the monoclinic structure of $Cl_2BC_2H_4$-BCl_2 seen along its b_0 axis. The boron atoms are small and black; the carbon atoms are somewhat larger and line shaded. The large dotted circles are the chlorine atoms.

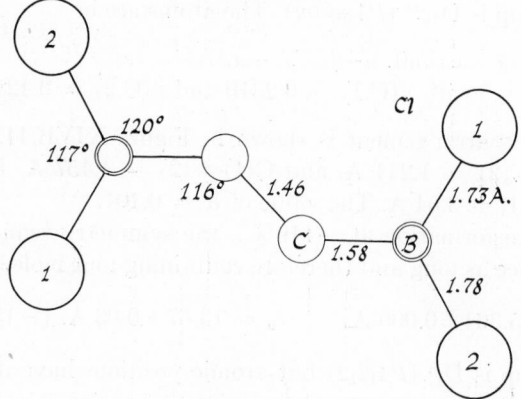

Fig. XIVB,43. Bond dimensions in the nearly planar molecule of $Cl_2BC_2H_4BCl_2$.

This structure (Fig. XIVB,42) is made up of molecules having the bond dimensions of Figure XIVB,43. They are nearly planar, with carbon 0.05 A. and boron 0.01 A. away from the plane set by the chlorine atoms and the midpoint between the carbon atoms. The final R for this structure is 0.105.

3. Simple Acetylene Derivatives

XIV,b29. The form of solid *acetylene*, C_2H_2, stable between a transition point at ca. $-140°C$. and a subliming temperature of $-84°C$., is cubic. Its tetramolecular cell has the edge length:

$$a_0 = 6.14 \text{ A. } (-117°C.)$$

Atoms are in the following special positions of T_h^6 ($Pa3$):

$$(8c) \quad \pm (uuu; u+^1/_2,^1/_2-u,\bar{u}; \bar{u},u+^1/_2,^1/_2-u; \, ^1/_2-u,\bar{u},u+^1/_2)$$

For carbon, $u = 0.056$. The Fourier projections contained an indication of probable hydrogen peaks and they lead to a hydrogen parameter of $u(H) = 0.155$.

In this structure there is within the linear molecule a C–C $= 1.20$ A. and a C–H $= 1.05$ A. Between molecules the shortest C–C $= 3.89$ A. and the shortest H–H $= 3.28$ A.

XIV,b30. *Dimethyl acetylene*, $CH_3C{\equiv}CCH_3$, is tetragonal. At $-50°C$. its bimolecular cell has the dimensions:

$$a_0 = 5.42 \text{ A.}, \qquad c_0 = 6.89 \text{ A.}$$

The space group is D_{4h}^{14} ($P4_2mnm$). The atoms are in

$$(4e) \quad \pm (uu0; \; u+{}^1/_2, {}^1/_2-u, {}^1/_2)$$
$$\text{with } u(C,1) = 0.2310 \text{ and } u(C,2) = 0.4210$$

The resulting arrangement is shown in Figure XIVB,44. In the molecules, $C(2){\equiv}C(2) = 1.211$ A. and $C(1)-C(2) = 1.457$ A. Between molecules, $CH_3-CH_3 = 3.54$ A. The value of $R = 0.101$.

Below a transformation at $-119°C.$, the symmetry remains tetragonal with a cell twice as long and therefore containing four molecules:

$$a_0 = 5.364 \pm 0.006 \text{ A.,} \quad c_0 = 13.67 \pm 0.03 \text{ A. } (-138°C.)$$

Its space group is D_4^4 ($P4_12_12$) but atomic positions have not been established for this form.

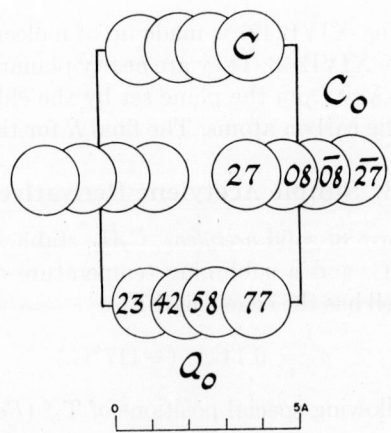

Fig. XIVB,44.   The tetragonal structure of dimethyl acetylene projected along an a_0 axis.   Right-hand axes.

XIV,b31.   Crystals of *cyanoacetylene*, $HC{\equiv}C(CN)$, are monoclinic with a bimolecular unit of the dimensions:

$$a_0 = 6.965 \pm 0.009 \text{ A.;} \quad b_0 = 6.300 \pm 0.010 \text{ A.;} \quad c_0 = 3.839 \pm 0.004 \text{ A.}$$
$$\beta = 110°30' \pm 6' \; (-25°C.)$$

The space group has been chosen as C_{2h}^2 ($P2_1/m$) with all atoms in the special positions:

$$(2e) \quad \pm (u \; {}^1/_4 \, v)$$

The determined parameters are as follows:

Atom	u	v
N	0.4009	0.2397
C(1)	0.5641	0.2399
C(2)	0.7630	0.2531
C(3)	0.9320	0.2581
H	0.067	0.258

The final R for these parameters is 0.049.

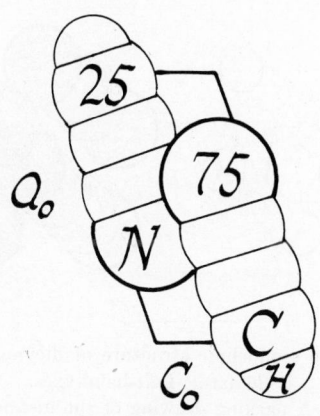

Fig. XIVB,45. The monoclinic structure of cyanoacetylene projected along its b_0 axis. Right-hand axes.

In this structure (Fig. XIVB,45), H–C = 0.95 A., C≡C = 1.18 A., C–C = 1.38 A., and C≡N = 1.14 A. The atoms of a molecule are collinear to within 0.015 A. and this line is straight to within the limit of error of the determination. The molecules appear tilted by ca. 1° to the a_0 axis; they can be thought of as tied together in chains by hydrogen bonds that have the length C–H$\cdots$N = 3.27 A.

XIV,b32. *Dicyanoacetylene*, N≡C—C≡C—C≡N, forms monoclinic crystals with a bimolecular unit having the dimensions:

$$a_0 = 8.93 \text{ A.}; \quad b_0 = 6.04 \text{ A.}; \quad c_0 = 3.86 \text{ A.}; \quad \beta = 99°20'$$

The space group is C_{2h}^5 $(P2_1/a)$ and all atoms of the centrosymmetric molecule are in the general positions:

$$(4e) \quad \pm(xyz; \ x+\tfrac{1}{2}, \tfrac{1}{2}-y, z)$$

The atomic parameters were found to be the following:

Atom	x	y	z
C(1)	0.032	0.082	0.056
C(2)	0.106	0.270	0.186
N	0.171	0.422	0.291

In the linear molecule that results, $C\equiv C$ = 1.14 A., $C–C$ = 1.37 A., and $C\equiv N$ = 1.19 A. An indication of the molecular packing can be had from Figure XIVB,46.

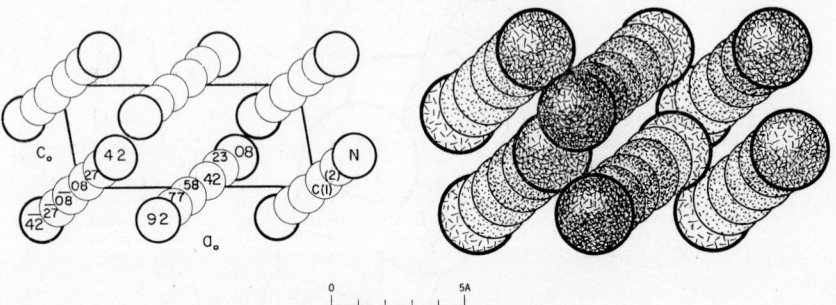

Fig. XIVB,46a (left). The monoclinic structure of dicyanoacetylene projected along its b_0 axis. Left-hand axes.

Fig. XIVB,46b (right). A packing drawing of the monoclinic structure of dicyano-acetylene viewed along its b_0 axis. The nitrogen atoms are heavily outlined and line shaded; the carbon atoms are dotted.

XIV,b33. *Dimethyl triacetylene*, $CH_3–C\equiv C–C\equiv C–C\equiv C–CH_3$, forms rhombohedral crystals with a unimolecular cell of the dimensions:

$$a_0 = 5.69 \text{ A.}, \qquad \alpha = 70°58'$$

The corresponding trimolecular hexagonal cell has the edge lengths:

$$a_0' = 6.60 \text{ A.}, \qquad c_0' = 12.66 \text{ A.}$$

The space group is D_{3d}^5 ($R\bar{3}m$) and all the carbon atoms are on the trigonal axes with the coordinates:

$$(6c) \quad \pm(00u; \; {}^1/_3,{}^2/_3,u+{}^2/_3; \; {}^2/_3,{}^1/_3,u+{}^1/_3)$$

Designating the methyl carbon as C(1) and, moving inward in the molecule, the others as C(2), C(3), and C(4), their parameters have been given the values $u(1) = 0.13352$, $u(2) = 0.24935$, $u(3) = 0.34405$, $u(4) = 0.45263$.

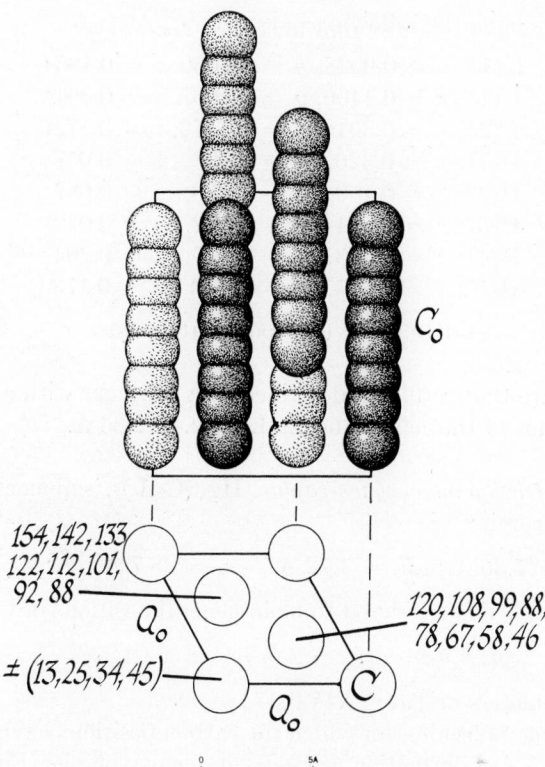

Fig. XIVB,47.   Two projections of the hexagonal structure of dimethyl triacetylene.

The resulting bond lengths are C(1)–C(2) = 1.466 A., C(2)≡C(3) = 1.196 A., C(3)–C(4) = 1.375 A , and C(4)≡C(4) = 1.199 A.
The resulting structure is shown in Figure XIVB,47.

4. Metal-Ethyl Compounds

XIV,b34.   The compound *ethyl lithium*, C_2H_5Li, forms orthorhombic crystals. Their large unit contains 16 molecules and has the edges:

$$a_0 = 7.24 \text{ A.}; \quad b_0 = 8.27 \text{ A.}; \quad c_0 = 18.11 \text{ A.}$$

The space group has been chosen as V_h^{14} (*Pcan*). Atoms are in the positions:

(8d) $\pm (xyz; x+\frac{1}{2},y+\frac{1}{2},\frac{1}{2}-z; x+\frac{1}{2},\frac{1}{2}-y,z; \bar{x},y,z+\frac{1}{2})$

Half the atoms have the following parameters:

$$Li(1): x = 0.0928, y = 0.1050, z = 0.1954$$
$$C(1): x = 0.4406, y = 0.1067, z = 0.0916$$
$$C(2): x = 0.3946, y = 0.1479, z = 0.1726$$
$$H(1): x = 0.320, y = 0.128, z = 0.057$$
$$H(2): x = 0.480, y = 0.980, z = 0.087$$
$$H(3): x = 0.554, y = 0.182, z = 0.072$$
$$H(4): x = 0.543, y = 0.147, z = 0.204$$
$$H(5): x = 0.315, y = 0.250, z = 0.173$$

The other half is said to have the coordinate values $x' = 0.4190-x$, $y' = y$, $z' = \frac{1}{2}-z$.

The structure that results is described as a tetramer with a tetrahedron of lithium atoms at the center and Li–Li = 2.42–2.63 A.

XIV,b35. *Diethyl mercury mercaptan*, $Hg(SC_2H_5)_2$, is monoclinic with the tetramolecular unit:

$$a_0 = 7.36 \text{ A.}; \quad b_0 = 4.72 \text{ A.}; \quad c_0 = 28.7 \text{ A.}; \quad \beta = 52°$$

All atoms have been considered to be in general positions of C_s^4 (Cc):

(4a) $xyz; x,\bar{y},z+\frac{1}{2}; x+\frac{1}{2},y+\frac{1}{2},z; x+\frac{1}{2},\frac{1}{2}-y,z+\frac{1}{2}$

with the parameters of Table XIVB,17.

The resulting molecules, for which the carbon positions have been chosen on other than x-ray diffraction grounds, are nearly planar (Fig. XIVB,48). In the structure, these molecules are aligned so that the b_0 axis is normal to S–Hg–S and the c_0 axis is approximately in the direction of the long axis of the molecule (Fig. XIVB,49).

TABLE XIVB,17. Parameters of Atoms in $Hg(SC_2H_5)_2$

Atom	x	y	z
Hg	0.00	0.22	0.00
S(1)	0.06	0.22	0.077
S(2)	−0.06	0.22	−0.077
$CH_2(1)$	−0.07	−0.09	0.122
$CH_3(1)$	−0.01	−0.09	0.165
$CH_2(2)$	0.07	0.53	−0.122
$CH_3(2)$	0.01	0.53	−0.165

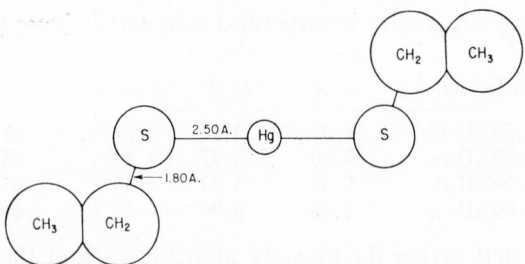

Fig. XIVB,48. Dimensions of the molecule of $Hg(SC_2H_5)_2$. The positions assigned the methyl groups puts them somewhat outside the plane of the other atoms.

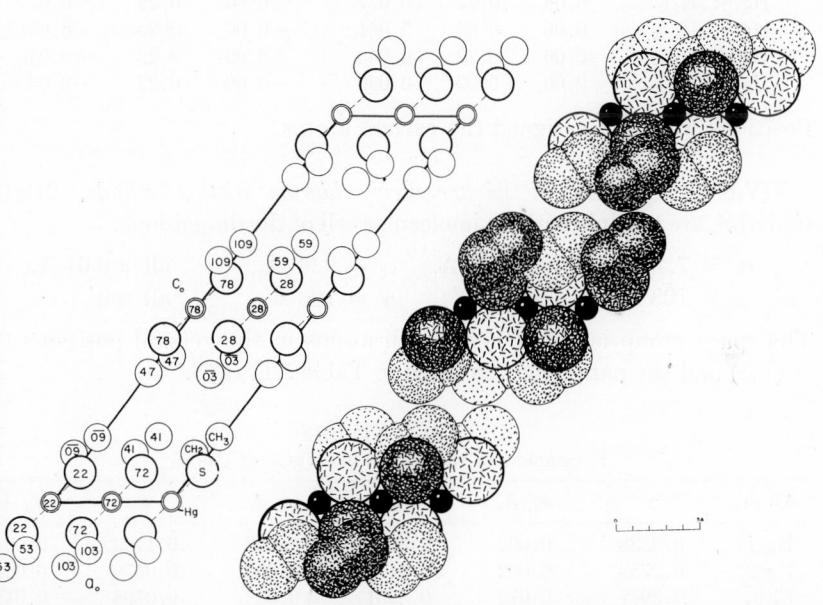

Fig. XIVB,49a (left). The monoclinic structure of $Hg(SC_2H_5)_2$ projected along its b_0 axis. It is to be noted that this structure has been described in terms of a cell having an unconventionally small, acute β angle. Left-hand axes.

Fig. XIVB,49b (right). A packing drawing of the monoclinic $Hg(SC_2H_5)_2$ structure viewed along its b_0 axis. The mercury atoms are small and black, the carbon atoms are dotted, and the atoms of sulfur line shaded and more heavily outlined.

The following apparently isostructural compounds have the cell dimen sions:

Compound	a_0, A.	b_0, A.	c_0, A.	β
$Hg(SC_3H_7)_2$	7.34	5.12	32.8	51°
$Hg(SC_5H_{11})_2$	7.36	5.27	43.1	53°
$Hg(SC_6H_{13})_2$	7.35	5.22	48.8	54°30′
$Hg(SC_7H_{15})_2$	7.45	5.30	56.1	50°30′

It was found that giving the mercury atoms in each of these crystals th parameters $x = z = 0$, $y = 0.22$, the sulfur parameters would have th following values:

Compound	S(1) Atom			S(2) Atom		
	x	y	z	x	y	z
$Hg(SC_3H_7)_2$	0.06	0.22	0.077	−0.06	0.22	−0.077
$Hg(SC_5H_{11})_2$	0.06	0.22	0.051	−0.06	0.22	−0.051
$Hg(SC_6H_{13})_2$	0.06	0.22	0.045	−0.06	0.22	−0.045
$Hg(SC_7H_{15})_2$	0.06	0.22	0.039	−0.06	0.22	−0.039

Positions were not assigned the carbon atoms.

XIV,b36. Crystals of *di mercuric chloride diethyl sulfide*, 2HgC $(C_2H_5)_2S$, are triclinic with a bimolecular cell of the dimensions:

$$a_0 = 7.43 \text{ A.}; \quad b_0 = 9.28 \text{ A.}; \quad c_0 = 9.92 \text{ A.}, \qquad \text{all} \pm 0.01 \text{ A.}$$
$$\alpha = 108°42'; \quad \beta = 99°0'; \quad \gamma = 103°6', \qquad \text{all} \pm 6'$$

The space group is C_i^1 ($P\bar{1}$) with all atoms in the general positions (2 $\pm (xyz)$ and the parameters as listed in Table XIVB,18.

TABLE XIVB,18
Parameters of the Atoms in $2HgCl_2 \cdot (C_2H_5)_2S$

Atom	x	σ_x, A.	y	σ_y, A.	z	σ_z, A
Hg(1)	0.8338	0.002	0.3924	0.003	0.1109	0.002
Hg(2)	0.2753	0.002	0.1179	0.003	0.0579	0.002
Cl(3)	0.3977	0.012	0.8201	0.018	0.0798	0.012
Cl(4)	0.9176	0.012	0.0531	0.018	0.1718	0.012
Cl(5)	0.4738	0.014	0.2548	0.021	0.2889	0.014
Cl(6)	0.1708	0.012	0.3930	0.019	0.0398	0.012
S(7)	0.9825	0.012	0.5814	0.019	0.3569	0.012
C(8)	0.7897	0.053	0.6553	0.077	0.4259	0.053
C(9)	0.6545	0.065	0.6810	0.087	0.3208	0.065
C(10)	0.1145	0.054	0.7400	0.077	0.3170	0.055
C(11)	0.2127	0.051	0.8912	0.074	0.4530	0.052

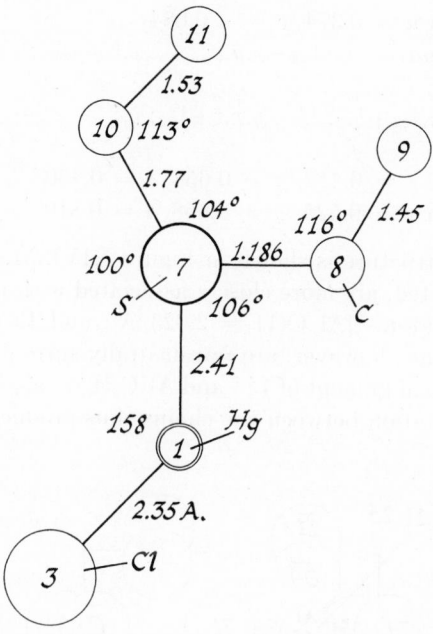

Fig. XIVB,50.   Bond dimensions in the complex cation in crystals of $2HgCl_2 \cdot (C_2H_5)_2S$.

This structure is made up of $[ClHgS(C_2H_5)_2]^+$, Cl^-, and $HgCl_2$ molecules. The cations have the bond dimensions of Figure XIVB,50. In the $HgCl_2$ molecules, $Hg(2)-Cl(4) = 2.33$ A., $Hg(2)-Cl(5) = 2.30$ A., and $Cl(4)-Hg(2)-Cl(5) = 171°48'$. The $Cl^-(6)$ anion is 2.70 and 2.85 A. distant from $Hg(1)$ and 2.88 A. from $Hg(2)$.

XIV,b37. *Lithium aluminum tetraethyl,* $LiAl(C_2H_5)_4$, is tetragonal with a bimolecular cell of the edge lengths:

$$a_0 = 9.99 \pm 0.01 \text{ A.}, \qquad c_0 = 5.411 \pm 0.005 \text{ A.}$$

The space group is D_{4h}^{15} $(P4_2/nmc)$ with atoms in the positions:

Li: (2a) $\pm (^1/_4 \, ^3/_4 \, ^1/_4)$
Al: (2b) $\pm (^3/_4 \, ^1/_4 \, ^1/_4)$
C(1): (8g) $\pm (^1/_4 \, u \, v; \; ^1/_4, ^1/_2-u,v; \; u,^1/_4,v+^1/_2; \; ^1/_2-u,^1/_4,v+^1/_2)$
        with $u = 0.5820$, $v = -0.0411$
        $[\sigma(u) = 0.0005, \sigma(v) = 0.0009]$
C(2): (8g) with $u = 0.4599$, $v = -0.2099$
        $[\sigma(u) = 0.0007, \sigma(v) = 0.0012]$

H(3): (8*g*)   with $u = 0.374, v = -0.084$
H(1): (16*h*) $\pm (xyz;$ $\frac{1}{2}-x,y,z;$ $x,\frac{1}{2}-y,z;$

$$\frac{1}{2}-x,\frac{1}{2}-y,z;$$

$$y,x,z+\frac{1}{2}; \; \frac{1}{2}-y,x,z+\frac{1}{2}; \; y,\frac{1}{2}-x,z+\frac{1}{2};$$

$$\frac{1}{2}-y,\frac{1}{2}-x,z+\frac{1}{2})$$

with $x = 0.412, y = 0.658, z = 0.450$
H(2): (16*h*)   with $x = 0.525, y = 0.658, z = 0.810$

The resulting structure is shown in Figure XIVB,51. The ethyl groups, as would be expected, are more closely associated with the aluminum than with the lithium atoms [Al–C(1) = 2.023 A., and Li–C(1) = 2.302 A.]. Both metallic atoms, however, are tetrahedrally surrounded by these C(1) atoms. There is an alignment of Li^+ and $Al(C_2H_5)_4{}^-$ along the c_0 axis with a considerable separation between the chains thus produced.

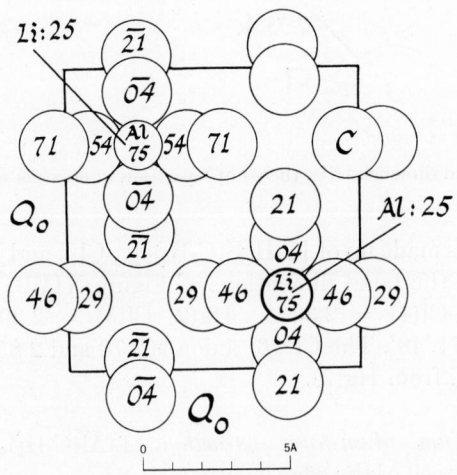

Fig. XIVB,51.   The tetragonal structure of $LiAl(C_2H_5)_4$ projected along its c_0 axis. Right-hand axes.

XIV,b38. The *aluminum triethyl potassium fluoride* complex, $2Al(C_2H_5)_3 \cdot$ KF, has hexagonal (rhombohedral) symmetry with a unimolecular rhombohedron of the dimensions:

$$a_0 = 8.95 \text{ A.}, \qquad \alpha = 55°20'$$

The corresponding trimolecular cell referred to hexagonal axes has the edges:

$$a_0 = 8.31 \text{ A.}, \qquad c_0 = 22.65 \text{ A.}$$

The space group has been chosen as C_{3i}^2 ($R\bar{3}$) with atoms in the positions:

K: (3a) 000; rh
F: (3b) 0 0 $^1/_2$; rh
Al: (6c) ±(00u); rh   with u = 0.419
C(1): (18f) ±(xyz; $\bar{y},x-y,z$; $y-x,\bar{x},z$); rh
          with x = 0.266, y = 0.180, z = 0.400
C(2): (18f) with x = 0.337, y = 0.368, z = 0.429

If the hydrogen atoms are placed so that H–C = 1.08 A., and the bond distribution around the carbon atoms is tetrahedral, they will be in (18f) with the following parameters:

Atom	x	y	z
H(1)	0.276	0.201	0.349
H(2)	0.353	0.119	0.414
H(3)	0.328	0.348	0.480
H(4)	0.250	0.432	0.416
H(5)	0.487	0.462	0.419

The resulting structure is shown in Figure XIVB,52. In it pairs of Al-$(C_2H_5)_3$ groups have between them a fluorine atom placed so that Al–F–Al is a straight line and Al–F = 1.82 A. In the groups, Al–C = 2.00 A. and C–C = 1.52 A. The closest atoms to potassium are ethyl carbons, with K–C = 3.28 A. The arrangement can accordingly be considered as an ionic one with $[Al(C_2H_5)_3FAl(C_2H_5)_3]^-$ functioning as the anion.

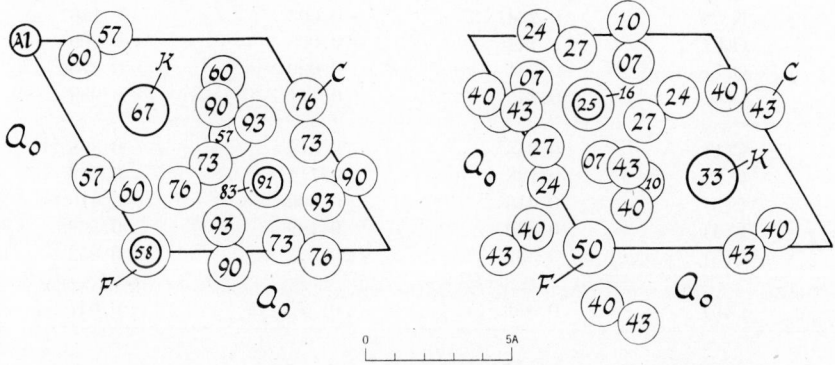

Fig. XIVB,52a (left). The contents of the upper half of the hexagonal unit of $2Al(C_2H_5)_3 \cdot KF$ projected along its c_0 axis. Right-hand axes.

Fig. XIVB,52b (right). The contents of the lower half of the hexagonal unit of $2Al$-$(C_2H_5)_3 \cdot KF$ projected along its c_0 axis. Right-hand axes.

XIV,b39. According to a preliminary report, crystals of the complex *uranyl nitrate–triethyl phosphate*, $UO_2(NO_3)_2 \cdot 2(C_2H_5)_3PO_4$, are triclinic with a unimolecular cell of the dimensions:

$$a_0 = 9.125 \text{ A.}; \quad b_0 = 8.655 \text{ A.}; \quad c_0 = 9.066 \text{ A.}$$
$$\alpha = 102°29'; \quad \beta = 98°56'; \quad \gamma = 97°55'$$

The space group is C_i^1 ($P\bar{1}$) with a uranium atom in the origin (1a) 000 and all other atoms in the general positions (2i) $\pm(xyz)$. The selected parameters are those listed in Table XIVB,19.

The environment of the uranium atom and the interatomic distances provided by this structure are shown in Figure XIVB,53. Some of the C–C and N–O separations are so short that the parameters cannot be more than approximately correct. The nitrate and phosphate oxygen neighbors as shown in this figure are practically coplanar with the uranium atom. Its two uranyl oxygen neighbors form a linear ion which makes an angle of 80° with this plane; in it, U–O = 1.71 A.

TABLE XIVB,19
Parameters of Atoms in $UO_2(NO_3)_2 \cdot 2(C_2H_5)_3PO_4$

Atom	x	y	z
O(1)	0.104	−0.042	0.146
P	0.191	0.424	0.231
O(2)	0.094	0.286	0.113
O(3)	0.290	0.534	0.169
O(4)	0.075	0.520	0.300
O(5)	0.287	0.390	0.371
N	0.234	−0.003	−0.186
O(6)	0.144	−0.098	−0.200
O(7)	0.331	−0.020	−0.275
O(8)	0.245	0.111	−0.089
C(1)	0.048	0.614	0.398
C(2)	0.128	0.714	0.462
C(3)	0.415	0.300	0.410
C(4)	0.399	0.199	0.500
C(5)	0.337	0.502	0.052
C(6)	0.371	0.575	−0.047

XIV,b40. Crystals of *trans, trans, trans, tris (2-chlorovinyl) dichlorostibine*, $(ClCH{=}CH)_3SbCl_2$, are monoclinic with a unit containing eight molecules and having the dimensions:

$$a_0 = 20.96 \text{ A.}; \quad b_0 = 7.00 \text{ A.}; \quad c_0 = 17.23 \text{ A.}; \quad \beta = 101°50'$$

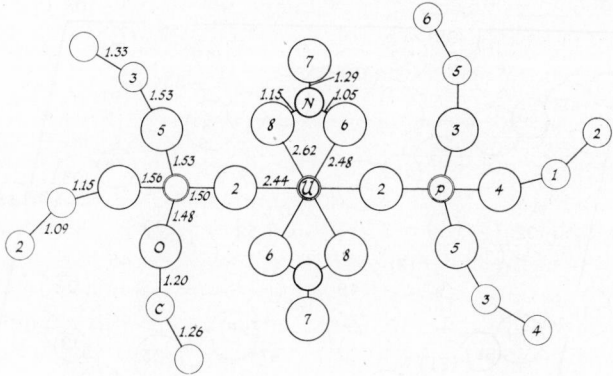

Fig. XIVB,53. Interatomic distances in the crystalline structure of $UO_2(NO_3)_2 \cdot 2(C_2H_5)_3PO_4$.

All atoms are in general positions of C_{2h}^6 $(C2/c)$:

(8*f*) $\pm(xyz;\ x,\bar{y},z+^1/_2;\ x+^1/_2,y+^1/_2,z;\ x+^1/_2,^1/_2-y,z+^1/_2)$

with the parameters of Table XIVB,20.

TABLE XIVB,20
Parameters of the Atoms in $(ClCH{=}CH)_3SbCl_2$

Atom	x	y	z
Sb	0.143	0.250	0.311
Cl(1)	0.083	−0.032	0.338
Cl(2)	0.203	0.534	0.284
Cl(3)	0.311	−0.051	0.484
Cl(4)	−0.023	0.563	0.398
Cl(5)	0.103	−0.018	0.060
C(1)	0.234	0.174	0.389
C(2)	0.240	0.008	0.424
C(3)	0.079	0.457	0.346
C(4)	0.028	0.398	0.371
C(5)	0.133	0.193	0.187
C(6)	0.112	0.027	0.158
H(1)	0.274	0.273	0.397
H(2)	0.200	−0.090	0.415
H(3)	0.088	0.607	0.341
H(4)	0.019	0.248	0.375
H(5)	0.144	0.300	0.148
H(6)	0.100	−0.080	0.197

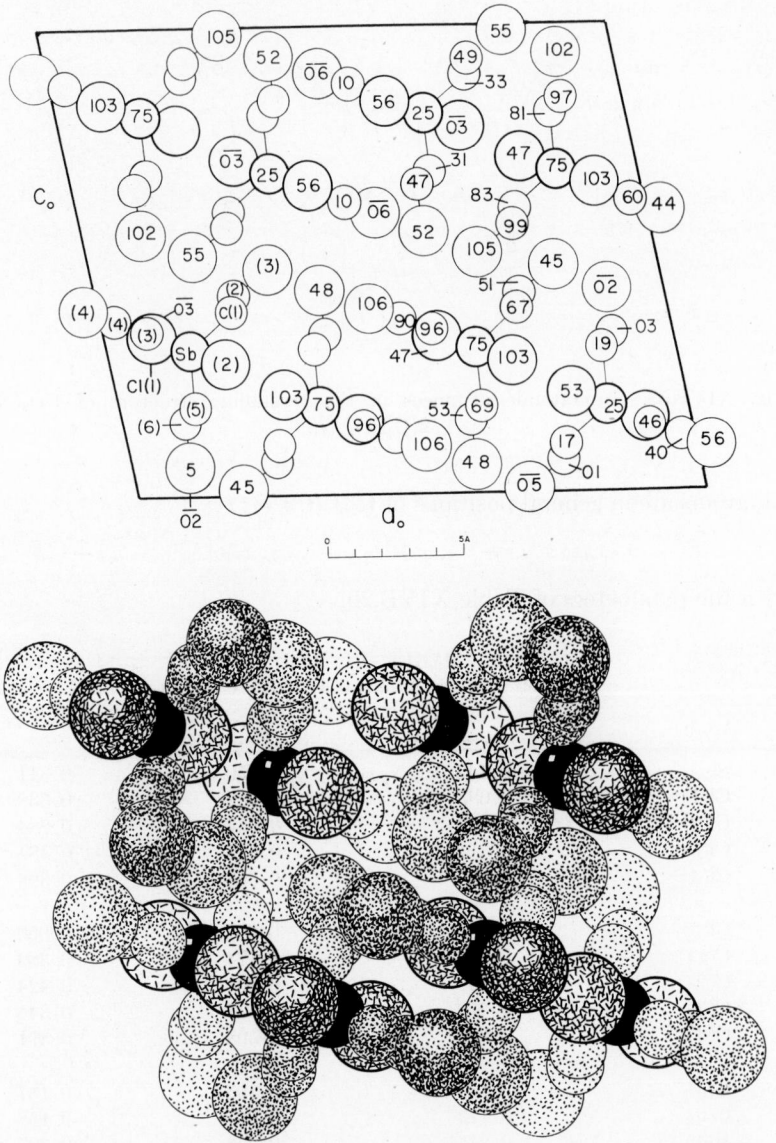

Fig. XIVB,54a (top). The monoclinic structure of (ClCH=CH)₃SbCl₂ projected along its b_0 axis. The two chlorine atoms directly bound to each antimony atom are more heavily ringed than the others. Left-hand axes.

Fig. XIVB,54b (bottom). A packing drawing of the monoclinic structure of (ClCH=CH)₃SbCl₂ viewed along its b_0 axis. The chlorine atoms in contact with the black antimony atoms are more heavily outlined and line shaded; the others are dotted. Atoms of carbon are the smaller dotted circles.

In the resulting structure (Fig. XIVB,54), each antimony atom is surrounded by two chlorines (Sb–Cl = 2.45 A.) and the carbon ends of three ClCH=CH radicals (Sb–C = 2.15 A.). In these radicals the carbon double bond has a length of 1.31 A. and C–Cl = 1.70 A. The proposed hydrogen positions are also listed in Table XIVB,20.

XIV,b41. Crystals of the *iron carbonyl ethyl mercaptide*, $[Fe(CO)_3C_2H_5S]_2$, are monoclinic with a unit that contains four of these dimeric molecules. Its cell dimensions are

$$a_0 = 8.98 \text{ A.}; \quad b_0 = 11.68 \text{ A.}; \quad c_0 = 15.65 \text{ A.}, \qquad \text{all} \pm 0.02 \text{ A.}$$
$$\beta = 107°15' \pm 15'$$

Atoms are in the general positions of the space group C_{2h}^5 $(P2_1/c)$:

$$(4e) \quad \pm (xyz; \ x, 1/2 - y, z + 1/2)$$

The determined parameters are listed in Table XIVB,21.

TABLE XIVB,21
Parameters of the Atoms in $[Fe(CO)_3C_2H_5S]_2$

Atom	x	$\sigma(x)$	y	$\sigma(y)$	z	$\sigma(z)$
Fe(1)	0.3472	0.0009	0.1717	0.0004	0.2300	0.0004
Fe(2)	0.1517	0.0009	0.2884	0.0004	0.2817	0.0004
S(3)	0.1170	0.0015	0.2338	0.0008	0.1370	0.0007
S(4)	0.1902	0.0015	0.0974	0.0007	0.3015	0.0006
C(5)	0.5274	0.0061	0.1656	0.0034	0.3292	0.0033
O(6)	0.6105	0.0037	0.1631	0.0021	0.3921	0.0020
C(7)	0.3761	0.0053	0.0527	0.0028	0.1730	0.0027
O(8)	0.4093	0.0037	−0.0319	0.0023	0.1317	0.0019
C(9)	−0.0463	0.0057	0.3042	0.0038	0.2860	0.0028
O(10)	−0.1642	0.0043	0.3185	0.0029	0.3032	0.0022
C(11)	0.2541	0.0056	0.3331	0.0036	0.3960	0.0031
O(12)	0.3330	0.0036	0.3466	0.0022	0.4693	0.0020
C(13)	0.2884	0.0050	0.0589	0.0027	0.4177	0.0022
C(14)	0.1531	0.0056	0.0426	0.0029	0.4618	0.0031
C(15)	0.4291	0.0062	0.2899	0.0038	0.1800	0.0031
O(16)	0.4803	0.0040	0.3552	0.0023	0.1462	0.0020
C(17)	0.1905	0.0060	0.4313	0.0037	0.2491	0.0032
O(18)	0.2150	0.0041	0.5218	0.0025	0.2260	0.0019
C(19)	−0.0198	0.0051	0.1196	0.0025	0.0987	0.0021
C(20)	−0.1956	0.0060	0.1490	0.0032	0.0619	0.0028

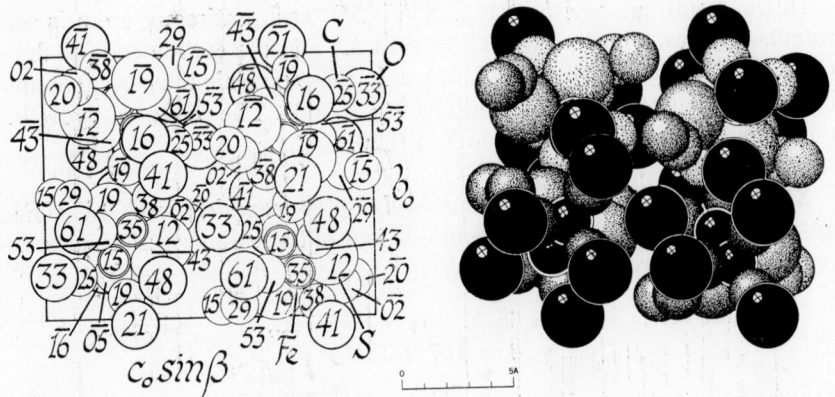

Fig. XIVB,55a (left). The monoclinic structure of $[Fe(CO)_3C_2H_5S]_2$ projected along its a_0 axis. Right-hand axes.

Fig. XIVB,55b (right). A packing drawing of the monoclinic structure of $[Fe(CO)_3C_2H_5S]_2$ viewed along its a_0 axis. The oxygen atoms are the large, the iron atoms the small black circles. The carbon atoms are small and dotted, the sulfur atoms large and line shaded.

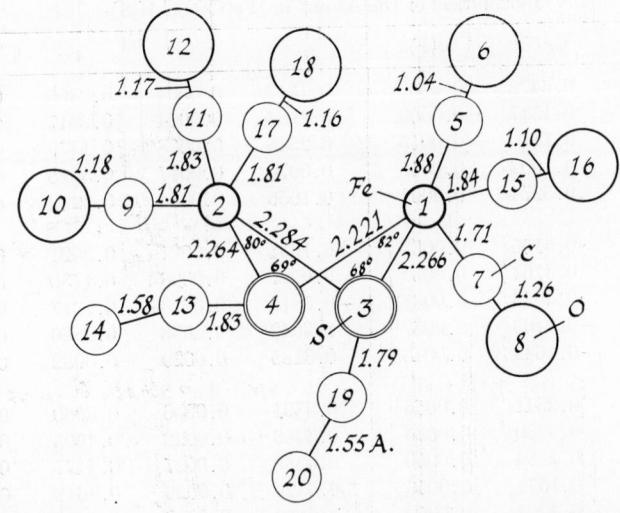

Fig. XIVB,56. Bond dimensions in the molecule of $[Fe(CO)_3C_2H_5S]_2$.

This is a molecular structure (Fig. XIVB,55) in which the separate molecules have the bond dimensions of Figure XIVB,56. Each iron atom is surrounded by two sulfur atoms and three carbonyl groups at the corners of a distorted tetragonal pyramid; the iron atoms are 2.537 A. apart. This structure is to be compared with that of Roussin's Red Ethyl Ester, $[Fe(NO)_2C_2H_5S]_2$, (**XIV,b43**) in which the iron atoms are tetrahedrally coordinated and the Fe_2S_2 association is a planar rhombus.

XIV,b42. The compound sometimes designated as *iron octacarbonyl dimethyl acetylene*, $CH_3CCCH_3 \cdot H_2Fe_2(CO)_8$, has been shown by x-ray analysis not to be a derivative of acetylene. It is monoclinic with a tetramolecular unit of the dimensions:

$$a_0 = 12.26 \text{ A.}; \quad b_0 = 7.47 \text{ A.}; \quad c_0 = 15.70 \text{ A.}; \quad \beta = 97°30'$$

The space group is C_{2h}^5 ($P2_1/c$) with all atoms in the positions:

$$(4e) \quad \pm (xyz; \ x,^1/_2-y,z+^1/_2)$$

The parameters have been determined to be those listed in Table XIVB,22.

TABLE XIVB,22
Parameters of the Atoms in $CH_3CCCH_3 \cdot H_2Fe_2(CO)_8$

Atom	x	σ_x, A.	y	σ_y, A.	z	σ_z, A.
Fe(1)	0.2178	0.0013	0.1697	0.0014	0.3273	0.0013
Fe(2)	0.2129	0.0013	0.1432	0.0014	0.1685	0.0013
O(5)	0.2334	0.0050	0.5301	0.0049	0.2231	0.0052
O(6)	0.3546	0.0056	−0.1456	0.0049	0.2558	0.0057
O(7)	0.1208	0.0069	−0.2037	0.0061	0.1013	0.0062
O(8)	0.0322	0.0072	0.3392	0.0072	0.0685	0.0070
O(9)	0.3726	0.0056	0.1995	0.0075	0.0528	0.0058
O(10)	−0.0012	0.0059	0.0471	0.0083	0.2632	0.0061
O(11)	0.1351	0.0071	0.4597	0.0076	0.4248	0.0065
O(12)	0.2471	0.0088	−0.0925	0.0089	0.4647	0.0079
C(1)	0.4124	0.0079	0.4795	0.0077	0.3522	0.0076
C(2)	0.3558	0.0066	0.3282	0.0067	0.3004	0.0069
C(3)	0.3870	0.0068	0.1439	0.0075	0.3084	0.0072
C(4)	0.4819	0.0069	0.0770	0.0081	0.3701	0.0078
C(5)	0.2639	0.0068	0.3528	0.0069	0.2360	0.0070
C(6)	0.3210	0.0069	0.0277	0.0069	0.2524	0.0066
C(7)	0.1580	0.0075	−0.0680	0.0078	0.1279	0.0070
C(8)	0.1014	0.0073	0.2667	0.0080	0.1084	0.0071
C(9)	0.3083	0.0069	0.1782	0.0073	0.0969	0.0072
C(10)	0.0898	0.0079	0.0958	0.0090	0.2802	0.0078
C(11)	0.1679	0.0080	0.3466	0.0083	0.3873	0.0073
C(12)	0.2354	0.0081	0.0060	0.0093	0.4106	0.0085

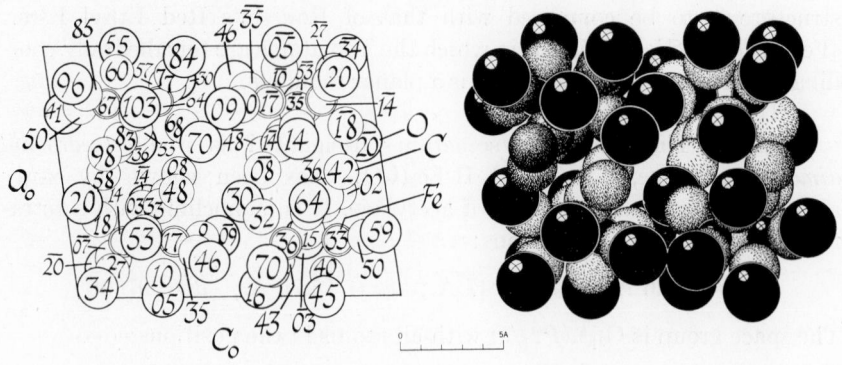

Fig. XIVB,57a (left).   The monoclinic structure of $CH_3CCCH_3 \cdot H_2Fe_2(CO)_8$ projected along its b_0 axis.   Right-hand axes.

Fig. XIVB,57b   (right).   A packing   drawing   of the   monoclinic   structure   of $CH_3CCCH_3 \cdot H_2Fe_2(CO)_8$ seen along its b_0 axis.   The oxygen atoms are the large, the iron atoms the smaller black circles.   Atoms of carbon are line shaded.

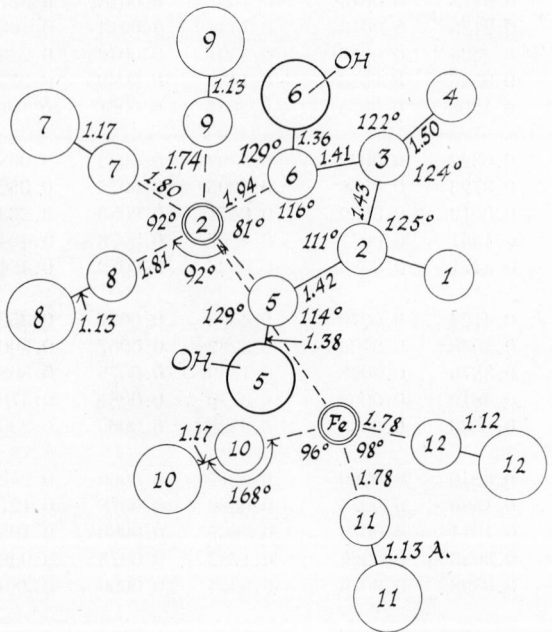

Fig. XIVB,58.   Bond dimensions in the molecule of $CH_3CCCH_3 \cdot H_2Fe_2(CO)_8$.

The structure is shown in Figure XIVB,57. Its complicated molecules have the bond dimensions of Figure XIVB,58. They consist of two Fe-(CO)$_3$ groups tied together by a direct Fe–Fe bond of 2.48 A. and having a C(OH)C(CH$_3$)C(CH$_3$)C(OH) group doubly associated with one of these iron atoms. The Fe(1) atom thus has a fourfold and the Fe(2) a sixfold coordination. Five of the six Fe–C–O groups are straight, but the angle Fe(1)–C(10)–O(10) is only 168°. The molecule has a plane of symmetry, though not one dictated by the crystal symmetry; it passes through the two iron atoms and C(9), C(10), O(9), and O(10), and bisects C(2)–C(3).

XIV,b43. *Roussin's Red Ethyl Ester*, [Fe(NO)$_2$C$_2$H$_5$S]$_2$, is monoclinic with a bimolecular cell of the dimensions:

$$a_0 = 7.81 \pm 0.02 \text{ A.}; \quad b_0 = 12.67 \pm 0.04 \text{ A.}; \quad c_0 = 7.01 \pm 0.02 \text{ A.}$$
$$\beta = 111°24'$$

The space group is C$_{2h}^5$ (P2$_1$/a) with all atoms in the positions:

$$(4e) \quad \pm (xyz; \ x + 1/2, 1/2 - y, z)$$

The determined parameters are those of Table XIVB,23.

The structure is shown in Figure XIVB,59. Its molecule, with the bond dimensions of Figure XIVB,60, has a shape suggesting that of the separate

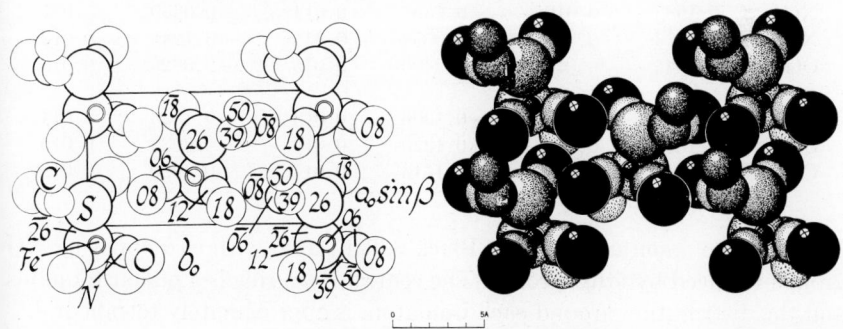

Fig. XIVB,59a (left). The monoclinic structure of [Fe(NO)$_2$C$_2$H$_5$S]$_2$ projected along its c_0 axis. Right-hand axes.

Fig. XIVB,59b (right). A packing drawing of the monoclinic structure of [Fe(NO)$_2$C$_2$H$_5$S]$_2$ seen along its c_0 axis. Oxygen atoms are the large, iron atoms the very small black circles. Atoms of nitrogen are heavily outlined and dotted. The carbon atoms are dotted but lightly outlined; the sulfur atoms are the largest line shaded circles.

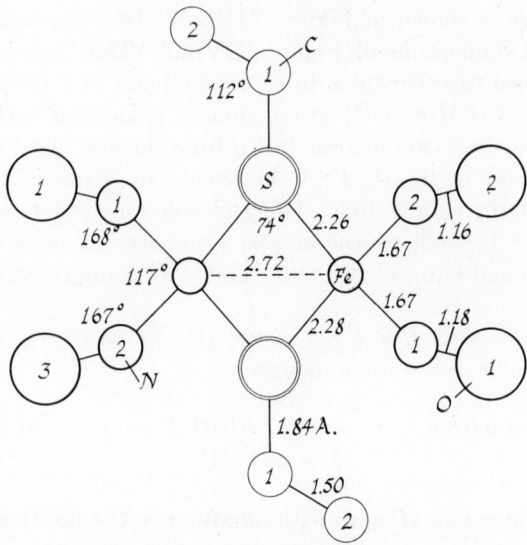

Fig. XIVB,60.   Bond dimensions in the molecule of Roussin's Red Ethyl Ester.

TABLE XIVB,23
Parameters of the Atoms in Roussin's Red Ethyl Ester

Atom	x	σ_x, A.	y	σ_y, A.	z	σ_z, A.
Fe	0.1412	0.002	0.0324	0.002	−0.0555	0.002
S	0.1562	0.003	−0.0051	0.003	0.2655	0.003
N(1)	0.1933	0.010	0.1597	0.014	−0.0639	0.011
N(2)	0.2673	0.011	−0.0558	0.014	−0.1245	0.012
O(1)	0.2545	0.012	0.2425	0.015	−0.0826	0.013
O(2)	0.3687	0.013	−0.1006	0.013	−0.1823	0.014
C(1)	0.1617	0.015	0.1220	0.017	0.3933	0.015
C(2)	0.3548	0.018	0.1616	0.017	0.4990	0.019

halves of the anion in Roussin's Black Salt (**X,a34**) with two opposing NO groups replaced by ethyl groups. The central Fe_2S_2 ring is a planar rhombus and the distribution around each iron atom is approximately tetrahedral.

5. Salts

XIV,b44.   Crystals of *palladous chloride ethylene*, $PdCl_2 \cdot C_2H_4$, are orthorhombic with a unit containing eight molecules and having the edge lengths:

$$a_0 = 15.41 \text{ A.}; \quad b_0 = 9.29 \text{ A.}; \quad c_0 = 7.23 \text{ A.}$$

TABLE XIVB,24
Positions and Parameters of the Atoms in $PdCl_2 \cdot C_2H_4$

Atom	Position	x	y	z
Pd(1)	(4g)	0.080	0.130	0
Pd(2)	(4h)	0.418	0.372	$^1/_2$
Cl(1)	(4g)	0.424	0.380	0
Cl(2)	(4h)	0.076	0.123	$^1/_2$
Cl(3)	(4g)	0.230	0.132	0
Cl(4)	(4h)	0.408	0.138	$^1/_2$
C(1)	(8i)	0.079	0.376	0.094
C(2)	(8i)	0.266	0.371	0.594

The space group is probably V_h^9 (*Pbam*) with atoms in the positions:

(4g)　$\pm (uv0; u+^1/_2, ^1/_2-v,0)$
(4h)　$\pm (u\,v\,^1/_2; u+^1/_2, ^1/_2-v, ^1/_2)$
(8i)　$\pm (xyz; \bar{x}\bar{y}z; ^1/_2-x, y+^1/_2, z; x+^1/_2, ^1/_2-y, z)$

The positions and parameters are those of Table XIVB,24.

The approximate structure that results is shown in Figure XIVB,61. Its palladium atoms are surrounded by squares of three chlorine atoms and one C_2H_4 group, with Pd–Cl = 2.18–2.43 A. and Pd–C_2H_4 = 2.28 or 2.34 A.

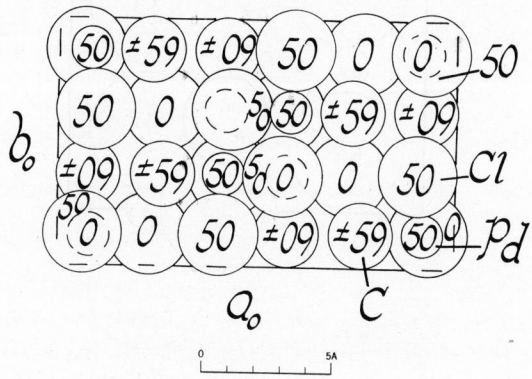

Fig. XIVB,61. The approximate orthorhombic structure of $PdCl_2 \cdot C_2H_4$ projected along its c_0 axis. Right-hand axes.

XIV,b45. A partial determination has been made of the structure of *Zeise's Salt*, $K(PtCl_3 \cdot C_2H_4) \cdot H_2O$. Its bimolecular monoclinic cell has the dimensions:

$$a_0 = 10.70 \text{ A.}; \quad b_0 = 8.42 \text{ A.}; \quad c_0 = 4.81 \text{ A.}; \quad \beta = 97°$$

The space group has been chosen as $C_2{}^2$ $(P2_1)$ with all atoms in the positions:

$$(2a) \quad xyz; \; \bar{x},y+{}^1/_2,\bar{z}$$

The parameters are given in Table XIVB,25.

TABLE XIVB,25
Parameters of the Atoms in Zeise's Salt

Atom	x	y	z
Pt	0.213	0.000	0.333
Cl(1)	0.428	0.000	0.529
Cl(2)	0.213	0.277	0.333
Cl(3)	0.213	0.723	0.333
K	0.438	0.313	0.933
C(1)	0.01	—	0.26
C(2)	0.07	—	0.99
H$_2$O	0.05	—	0.58

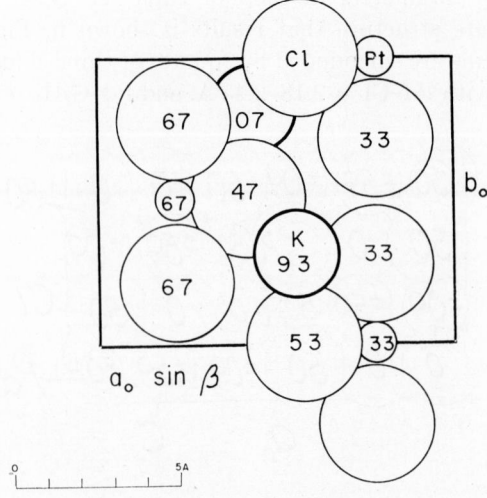

Fig. XIVB,62. Part of the monoclinic structure of $K(PtCl_3 \cdot C_2H_4) \cdot H_2O$ projected along its c_0 axis. Left-hand axes.

The $(PtCl_3C_2H_4)$ group in this structure (Fig. XIVB,62) is one in which platinum and the three chlorine atoms are coplanar. The Cl(1), which is *trans* to the C_2H_4 group, is farther (2.42 A.) from the platinum atom than the other two chlorines (Pt–Cl = 2.32 A.). The C–C bond of C_2H_4 is normal to the PtCl$_3$ plane, with C–C = 1.5 A. and Pt–C = ca. 2.2 A.

XIV,b46. *Potassium acetylene diolate*, $K_2O_2C_2$, is tetragonal with a bimolecular unit of the edge lengths:

$$a_0 = 3.927 \pm 0.02 \text{ A.}, \qquad c_0 = 12.75 \pm 0.05 \text{ A.}$$

The space group is D_{4h}^{17} $(I4/mmm)$ with atoms in the positions:

K: (4e) $\pm (00u; \; {}^1/_2, {}^1/_2, u + {}^1/_2)$ with $u = 0.143$
O: (4e)   with $u = 0.352$
C: (4e)   with $u = 0.452$

The structure is shown in Figure XIVB,63. Each potassium atom has five oxygen neighbors, with K–O = 2.66 or 2.78 A. In the (OC≡CO) anions, which are straight, O–C = 1.27 A. and C–C = 1.21 A.

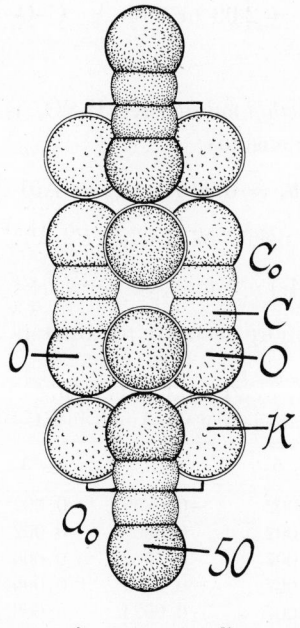

0   5A

Fig. XIVB,63. A packing drawing of the tetragonal structure of $K_2O_2C_2$ viewed along an a_0 axis. All atoms are at either 0 or $^1/_2$ along a_0. Right-hand axes.

Rubidium acetylene diolate, $Rb_2O_2C_2$, is isostructural, with

$$a_0 = 4.133 \pm 0.01 \text{ A.}, \qquad c_0 = 13.049 \pm 0.03 \text{ A.}$$

The atomic positions are

Rb: (4e)   with $u = 0.141$

$$O: (4e) \quad \text{with } u = 0.357$$
$$C: (4e) \quad \text{with } u = 0.454$$

These lead to the interatomic distances: Rb–O $= 2.82$ or 2.92 A., C–O $=$ 1.27 A., and C–C $= 1.20$ A. Rb–Rb along the c_0 axis $= 3.68$ A.

For *cesium acetylene diolate*, $Cs_2O_2C_2$, with the same structure:

$$a_0 = 4.371 \pm 0.01 \text{ A.}, \qquad c_0 = 13.523 \pm 0.03 \text{ A.}$$

$$Cs: (4e) \quad \text{with } u = 0.140$$
$$O: (4e) \quad \text{with } u = 0.361$$
$$C: (4e) \quad \text{with } u = 0.455$$

In these crystals Cs–O $= 2.99$ or 3.09 A., C–O $= 1.27$ A., and C–C $=$ 1.20 A.; Cs–Cs $= 3.79$ A.

XIV,b47. *Potassium ethyl sulfate*, $KC_2H_5SO_4$, is monoclinic with a tetramolecular unit of the dimensions:

$$a_0 = 11.62 \text{ A.}; \quad b_0 = 6.99 \text{ A.}; \quad c_0 = 7.51 \text{ A.}; \quad \beta = 100°18'$$

All atoms are in general positions of C_{2h}^5 $(P2_1/c)$:

$$(4e) \quad \pm (xyz; \ x, \tfrac{1}{2}-y, z+\tfrac{1}{2})$$

with the redetermined, more accurate parameters of Table XIVB,26.

TABLE XIVB,26
Parameters of the Atoms in $KC_2H_5SO_4$

Atom	x	σ_x, A.	y	σ_y, A.	z	σ_z, A.
K	0.1095	0.002	−0.0859	0.002	−0.1283	0.002
S	0.1597	0.002	0.0668	0.002	0.4211	0.002
O(1)	0.0704	0.007	0.1962	0.006	0.4659	0.007
O(2)	0.1232	0.007	−0.0374	0.006	0.2516	0.007
O(3)	0.2101	0.007	−0.0624	0.008	0.5645	0.006
O(4)	0.2610	0.006	0.2131	0.005	0.3956	0.007
C(1)	0.3606	0.009	0.1297	0.011	0.3335	0.011
C(2)	0.4501	0.012	0.2874	0.013	0.3381	0.012

This produces, as would be expected, a tetrahedral SO_4 to one atom of which a C_2H_5 radical is attached; the bonds within this $C_2H_5SO_4$ ion are shown in Figure XIVB,64. The S–O(4) distance is significantly longer than the others.

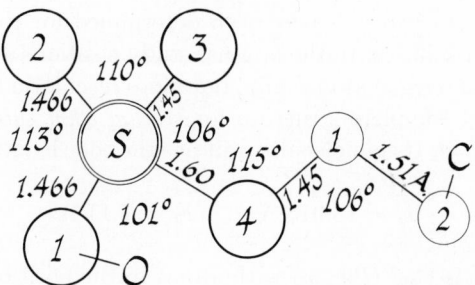

Fig. XIVB,64. Bond dimensions in the $(C_2H_5SO_4)^-$ anion of $KC_2H_5SO_4$.

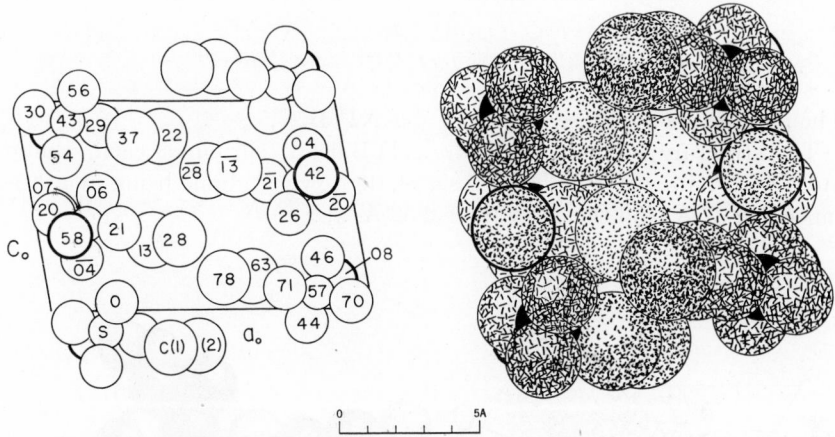

Fig. XIVB,65a (left). The monoclinic structure of $KC_2H_5SO_4$ projected along its b_0 axis. The potassium atom is heavily outlined. Left-hand axes.

Fig. XIVB,65b (right). A packing drawing of the monoclinic structure of $KC_2H_5SO_4$ seen along its b_0 axis. The black sulfur atoms only partly show. The oxygen atoms are line shaded; the atoms of carbon are large and dotted. The potassium atoms, also dotted, are heavily outlined.

The ionic packing in the crystal is illustrated in Figure XIVB,65. Double layers of anions have their sulfate ends facing a central layer of potassium ions, each sheet being parallel to (100). Bonding between sheets is therefore K–O on either side of the potassium ions and a van der Waals CH_3–CH_3 involving the other sides of the anions. These C–C separations are 3.80 A. within the same layer, and 3.90 A. or more between successive layers. Each potassium ion has eight oxygen neighbors at distances between 2.80 and 3.02 A.

XIV,b48. A structure has now been determined for the hexagonal hydrated rare earth sulfates. In the original study positions were found for the metal, sulfur, and oxygen atoms only, but more recently a further examination has led to a complete structure for *erbium ethyl sulfate nonahydrate*, $Er(C_2H_5SO_4)_3 \cdot 9H_2O$. Its bimolecular unit has the edge lengths:

$$a_0 = 13.915 \text{ A.}, \qquad c_0 = 7.11 \text{ A.}$$

The space group is C_{6h}^2 $(P6_3/m)$ with atoms in the positions:

(2c) $\pm (1/3 \ 2/3 \ 1/4)$
(6h) $\pm (u \ v \ 1/2; \ \bar{v}, u-v, 1/4; \ v-u, \bar{u}, 1/4)$
(12i) $\pm (xyz; \qquad \bar{y}, x-y, z; \qquad y-x, \bar{x}, z;$
         $\bar{x}, \bar{y}, z+1/2; \ y, y-x, z+1/2; \ x-y, x, z+1/2)$

The parameters are those listed in Table XIVB,27.

The structure is shown in Figure XIVB,66. It surrounds each erbium atom with nine water molecules, six of the oxygen atoms being at a distance of 2.37 A. and the other three 2.52 A. away.

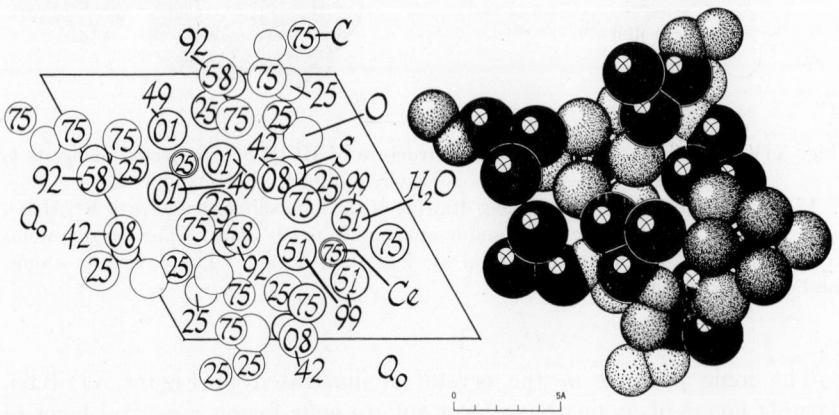

Fig. XIVB,66a (left). The hexagonal structure of $Ce(C_2H_5SO_4)_3 \cdot 9H_2O$ projected along its c_0 axis. Right-hand axes.

Fig. XIVB,66b (right). A packing drawing of the hexagonal structure of $Ce(C_2H_5SO_4)_3 \cdot 9H_2O$ seen along its c_0 axis. The oxygen atoms are large and black; the sulfur atoms do not show. The somewhat smaller black cerium atoms are partly hidden behind water molecules which are line shaded and heavily outlined. The carbon atoms are of medium size and dotted.

TABLE XIVB,27
Parameters of the Atoms in Er(C$_2$H$_5$SO$_4$)$_3$·9H$_2$O

Atom	Position	x	$\sigma(x)$	y	$\sigma(y)$	z	$\sigma(z)$
Er	(2c)	$^1/_3$	—	$^2/_3$	—	$^1/_4$	—
S	(6h)	−0.3191	0.0003	−0.3720	0.0003	$^1/_4$	—
O(1)	(6h)	−0.2489	0.0012	−0.4216	0.0012	$^1/_4$	—
O(2)	(6h)	−0.2289	0.0012	−0.2425	0.0012	$^1/_4$	—
O(3)	(12i)	0.3826	0.0009	0.3948	0.0009	0.5780	0.0019
H$_2$O(1)	(6h)	0.3509	0.0011	0.4947	0.0011	$^1/_4$	—
H$_2$O(2)	(12i)	0.2138	0.0008	0.5458	0.0008	0.4868	0.0017
C(1)	(6h)	−0.1772	0.0020	−0.0523	0.0020	$^1/_4$	—
C(2)	(6h)	−0.2727	0.0020	−0.1693	0.0020	$^1/_4$	—

For the isostructural *yttrium* salt, Y(C$_2$H$_5$SO$_4$)$_3$·9H$_2$O,

$$a_0 = 13.924 \pm 0.001 \text{ A.,} \qquad c_0 = 7.057 \pm 0.002 \text{ A.}$$

The parameters are those of Table XIVB,28. They lead to the separations
Y–6(OH$_2$) = 2.37 A. and Y–3(OH$_2$) = 2.55 A.

TABLE XIVB,28
Parameters of the Atoms in Y(C$_2$H$_5$SO$_4$)$_3$·9H$_2$O

Atom	Position	x	$\sigma(x)$	y	$\sigma(y)$	z	$\sigma(z)$
Y	(2c)	$^1/_3$	—	$^2/_3$	—	$^1/_4$	—
S	(6h)	−0.3187	0.0003	−0.3715	0.0003	$^1/_4$	—
O(1)	(6h)	−0.2485	0.0009	−0.4205	0.0009	$^1/_4$	—
O(2)	(6h)	−0.2298	0.0009	−0.2443	0.0009	$^1/_4$	—
O(3)	(12i)	0.3837	0.0006	0.3947	0.0006	0.5763	0.0014
H$_2$O(1)	(6h)	0.3525	0.0008	0.4936	0.0008	$^1/_4$	—
H$_2$O(2)	(12i)	0.2122	0.0005	0.5452	0.0005	0.4854	0.0014
C(1)	(6h)	−0.1793	0.0017	−0.0515	0.0017	$^1/_4$	—
C(2)	(6h)	−0.2751	0.0015	−0.1650	0.0015	$^1/_4$	—

A similar study was made of the *praseodymium* compound, Pr(C$_2$H$_5$-SO$_4$)$_3$·9H$_2$O. For it,

$$a_0 = 14.007 \text{ A.,} \qquad c_0 = 7.09 \text{ A.}$$

Its parameters are given in Table XIVB,29.

TABLE XIVB,29
Parameters of the Atoms in $Pr(C_2H_5SO_4)_3 \cdot 9H_2O$

Atom	Position	x	$\sigma(x)$	y	$\sigma(y)$	z
Pr	$(2c)$	$1/3$	—	$2/3$	—	$1/4$
S	$(6h)$	-0.3178	0.0008	-0.3710	0.0008	$1/4$
O(1)	$(6h)$	-0.2476	0.0024	-0.4207	0.0024	$1/4$
O(2)	$(6h)$	-0.2292	0.0025	-0.2422	0.0025	$1/4$
O(3)	$(12i)$	0.3819	0.0014	0.3958	0.0014	—
$H_2O(1)$	$(6h)$	0.3570	0.0026	0.4908	0.0025	$1/4$
$H_2O(2)$	$(12i)$	0.2081	0.0012	0.5418	0.0013	—
C(1)	$(6h)$	-0.1786	0.0041	-0.0566	0.0040	$1/4$
C(2)	$(6h)$	-0.2763	0.0044	-0.1748	0.0044	$1/4$

The original incomplete study was of the *cerium* compound, $Ce(C_2H_5SO_4)_3 \cdot 9H_2O$, which has a cell of the edge lengths:

$$a_0 = 14.048 \text{ A.}, \quad c_0 = 7.11 \text{ A.}$$

Parameters stated for this salt were:

Atom	Position	x	y	z
Ce	$(2c)$	$1/3$	$2/3$	$1/4$
S	$(6h)$	-0.33	-0.35	$1/4$
O(1)	$(6h)$	0.50	-0.40	$1/4$
O(2)	$(6h)$	-0.22	-0.23	$1/4$
O(3)	$(12i)$	0.33	0.41	0.58

They are in rough accord with those more recently determined for the other compounds except for the O(1) atom which has a very different value of x. Unit cells have been found for the following additional compounds:

Compound	a_0, A.	c_0, A.
$Dy(C_2H_5SO_4)_3 \cdot 9H_2O$	13.906	7.04
$Gd(C_2H_5SO_4)_3 \cdot 9H_2O$	13.931	7.06
$La(C_2H_5SO_4)_3 \cdot 9H_2O$	14.080	7.11
$Nd(C_2H_5SO_4)_3 \cdot 9H_2O$	13.992	7.07
$Sm(C_2H_5SO_4)_3 \cdot 9H_2O$	13.961	7.08

XIV,b49. Crystals of *disodium 1,2-ethane bis nitraminate*, $Na_2(NO_2-N-CH_2 \cdot CH_2-N-NO_2)$, are monoclinic with a bimolecular unit of the dimensions:

$$a_0 = 5.039 \text{ A.}; \quad b_0 = 18.39 \text{ A.}; \quad c_0 = 3.626 \text{ A.}; \quad \beta = 93°$$

Atoms are in the general positions of C_{2h}^5 $(P2_1/n)$:

$$(4e) \quad \pm (xyz; x+1/2, 1/2-y, z+1/2)$$

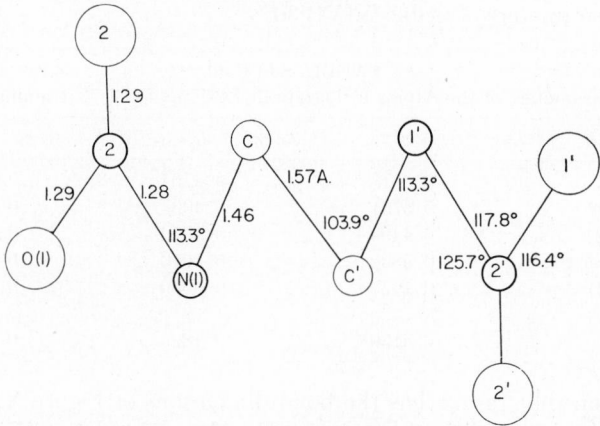

Fig. XIVB,67.   Bond dimensions in the anion of $Na_2[NO_2N(CH_2)_2NNO_2]$.

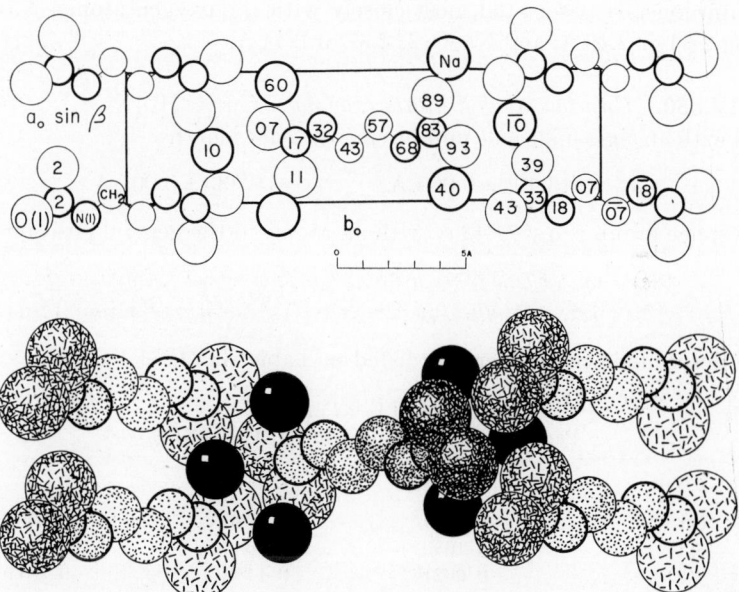

Fig. XIVB,68a (top).   The monoclinic structure of $Na_2[NO_2N(CH_2)_2NNO_2]$ projected along its c_0 axis.   Left-hand axes.

Fig. XIVB,68b (bottom).   A packing drawing of the monoclinic structure of Na_2-$[NO_2N(CH_2)_2NNO_2]$ seen along its c_0 axis.   The sodium atoms are black.   The oxygen atoms are large and line shaded; of the smaller dotted circles, the nitrogens are the more heavily ringed.

with the parameters of Table XIVB,30.

TABLE XIVB,30
Parameters of the Atoms in Disodium 1,2-Ethane Bis Nitraminate

Atom	x	y	z
C	0.4018	0.4704	0.4262
N(1)	0.5751	0.4097	0.3239
N(2)	0.4494	0.3572	0.171
O(1)	0.5830	0.3010	0.0708
O(2)	0.1970	0.3550	0.113
Na	0.4139	0.1825	0.1007

The anion that results has the bond dimensions of Figure XIVB,67. As would be anticipated, it closely resembles the molecule of bis-nitroamino-ethane itself (**XIV,b8**). Both are planar.

The packing in this ionic crystal is indicated in Figure XIVB,68. The sodium ions are associated most closely with the oxygen atoms, Na–O(1) being 2.34 or 2.40 A. and Na–O(2) 2.37 or 2.44 A.

XIV,b50. *Cuprous diethyl dithiocarbamate*, $Cu[(C_2H_5)_2NCS_2]$, is tetragonal with an eight-molecule unit having the edge lengths:

$$a_0 = 13.494 \pm 0.006 \text{ A.}, \qquad c_0 = 9.335 \pm 0.005 \text{ A.}$$

The space group is V_d^4 ($P\bar{4}2_1c$) with all atoms in the general positions:

$$(8e) \quad xyz; \; \bar{x}\bar{y}z; \; {}^1\!/_2-x,y+{}^1\!/_2,{}^1\!/_2-z; \; x+{}^1\!/_2,{}^1\!/_2-y,{}^1\!/_2-z;$$
$$\bar{y}x\bar{z}; \; y\bar{x}\bar{z}; \; y+{}^1\!/_2,x+{}^1\!/_2,z+{}^1\!/_2; \; {}^1\!/_2-y,{}^1\!/_2-x,z+{}^1\!/_2$$

The determined parameters are listed in Table XIVB,31.

TABLE XIVB,31
Parameters of the Atoms in $Cu[(C_2H_5)_2NCS_2]$

Atom	x	y	z
Cu	0.0704	0.0740	0.0967
S(1)	0.1680	0.0680	−0.1038
S(2)	−0.0021	0.1489	−0.2817
N	0.1885	0.1828	−0.3348
C(1)	0.1200	0.1350	−0.2445
C(2)	0.1583	0.2414	−0.4643
C(3)	0.2955	0.1779	−0.3098
C(4)	0.1454	0.3499	−0.4202
C(5)	0.3415	0.0877	−0.3869

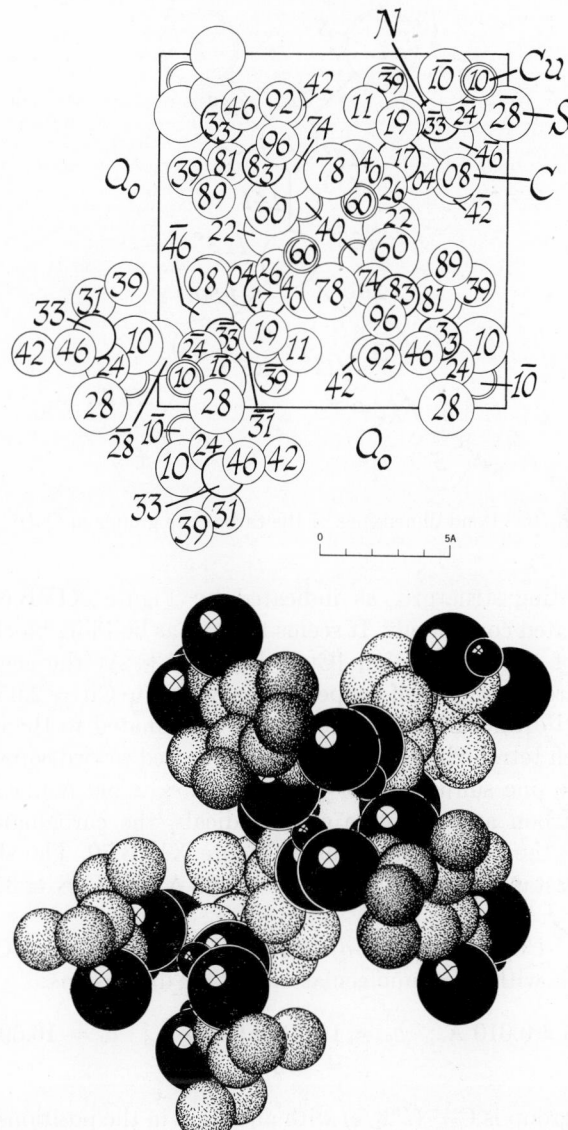

Fig. XIVB,69a (top). The tetragonal structure of Cu[(C$_2$H$_5$)$_2$NCS$_2$] projected along its c_0 axis. Right-hand axes.

Fig. XIVB,69b (bottom). A packing drawing of the tetragonal structure of copper diethyl dithiocarbamate viewed along its c_0 axis. The sulfur are the large, the copper the small black circles. Of the dotted circles the nitrogen atoms are the more heavily ringed.

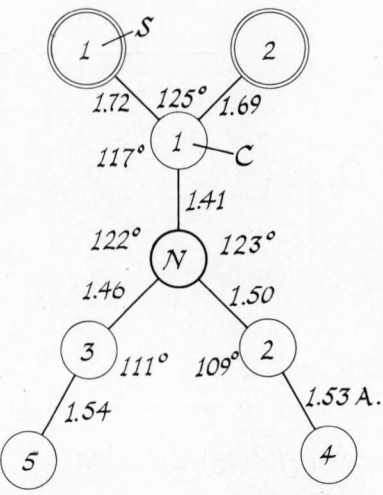

Fig. XIVB,70. Bond dimensions of the carbamate groups in $Cu[(C_2H_5)_2NCS_2]$.

The resulting structure, as indicated by Figure XIVB,69, is unique among chelated compounds. It seems to have as building blocks tetrameric molecules of the composition $[Cu(C_2H_5)_2NCS_2]_2$. At the centers of these tetramers are tetrahedra of copper atoms, with Cu–Cu = 2.658 or 2.757 A. Two diethyl dithiocarbamate groups are coordinated to the copper atoms forming each tetrahedral face with one sulfur tied to two copper atoms and the other to one sulfur (Cu–S = 2.246, 2.258, 2.290 A.). Except for the terminal carbon atoms of the ethyl radicals, the carbamate groups are planar with the bond dimensions of Figure XIVB,70. The shortest intermolecular distances are given as C–C = 3.63 A. and C–S = 3.69 A.

XIV,b51. *Cupric bis-N,N-diethyl dithiocarbamate,* $Cu[(C_2H_5)_2NCS_2]_2$, is monoclinic with a tetramolecular unit of the dimensions:

$$a_0 = 9.907 \pm 0.010 \text{ A.}; \quad b_0 = 10.627 \pm 0.005 \text{ A.}; \quad c_0 = 16.591 \pm 0.010 \text{ A.}$$
$$\beta = 113°52' \pm 5'$$

The space group is C_{2h}^5 ($P2_1/c$) with all atoms in the positions:

$$(4e) \quad \pm (xyz; \ x, \tfrac{1}{2} - y, z + \tfrac{1}{2})$$

The parameters are those of Table XIVB,32.

The resulting structure is shown in Figure XIVB,71. The bond dimensions in its ions are those of Figure XIVB,72. Each copper atom is at the

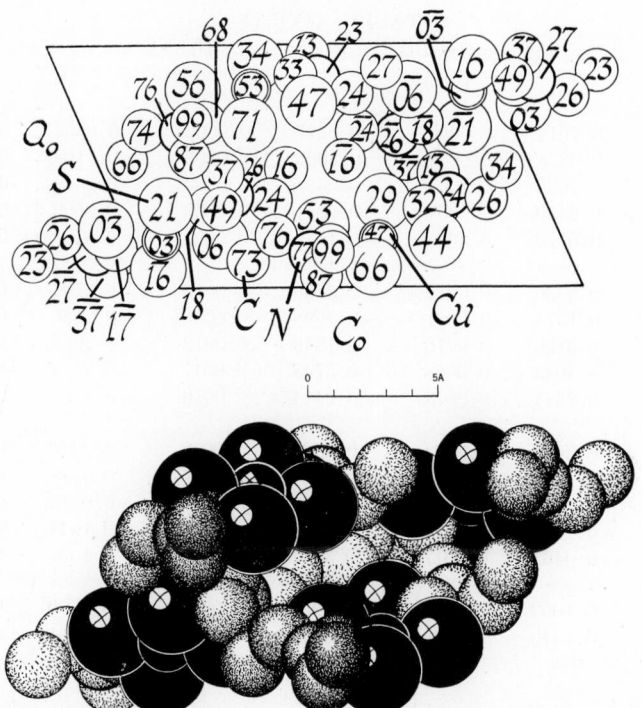

Fig. XIVB,71a (top).   The monoclinic structure of Cu[(C₂H₅)₂NCS₂]₂ projected along its b_0 axis.   Right-hand axes.

Fig. XIVB,71b (bottom).   A packing drawing of the monoclinic structure of Cu-[(C₂H₅)₂NCS₂]₂ seen along its b_0 axis.   The sulfur atoms are the large, the copper atoms the small black circles.   Of the dotted circles, the nitrogen are the more heavily outlined

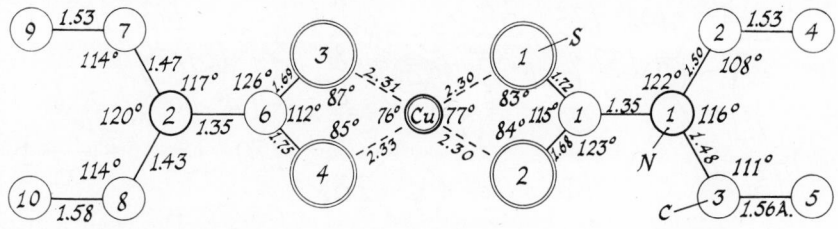

Fig. XIVB,72.   Bond dimensions in crystals of Cu[(C₂H₅)₂NCS₂]₂.

TABLE XIVB,32
Parameters of the Atoms in $Cu[(C_2H_5)_2NCS_2]_2$

Atom	x	$\sigma(x)$	y	$\sigma(y)$	z	$\sigma(z)$
Cu	0.1914	0.0002	0.0317	0.0002	0.0651	0.0001
S(1)	0.3338	0.0004	0.2098	0.0003	0.1102	0.0002
S(2)	0.2030	0.0004	0.0631	0.0003	0.2049	0.0002
S(3)	0.2459	0.0004	−0.0341	0.0003	−0.0509	0.0002
S(4)	0.0779	0.0003	−0.1645	0.0003	0.0283	0.0001
N(1)	0.3736	0.0012	0.2630	0.0010	0.2764	0.0006
N(2)	0.1374	0.0011	−0.2698	0.0008	−0.1026	0.0005
C(1)	0.3120	0.0016	0.1849	0.0012	0.2065	0.0006
C(2)	0.4643	0.0018	0.3748	0.0017	0.2744	0.0010
C(3)	0.3544	0.0018	0.2434	0.0016	0.3596	0.0008
C(4)	0.3604	0.0024	0.4862	0.0018	0.2361	0.0012
C(5)	0.4817	0.0029	0.1611	0.0026	0.4248	0.0011
C(6)	0.1522	0.0014	−0.1673	0.0010	−0.0512	0.0006
C(7)	0.2101	0.0016	−0.2654	0.0014	−0.1644	0.0008
C(8)	0.0465	0.0017	−0.3747	0.0013	−0.1037	0.0008
C(9)	0.1093	0.0025	−0.2260	0.0021	−0.2580	0.0008
C(10)	0.1349	0.0029	−0.4913	0.0019	−0.0491	0.0016

apex of a four-sided pyramid having sulfur atoms at the corners of the base; there is also a fifth, somewhat more distant, sulfur with $S(4')$–Cu = 2.86 A.

The *zinc* compound, $Zn[(C_2H_5)_2NCS_2]_2$, is isostructural with a cell of the dimensions:

$$a_0 = 10.015 \pm 0.010 \text{ A.}; \quad b_0 = 10.661 \pm 0.005 \text{ A.}; \quad c_0 = 16.357 \pm 0.010 \text{ A.}$$
$$\beta = 111°58' \pm 5'$$

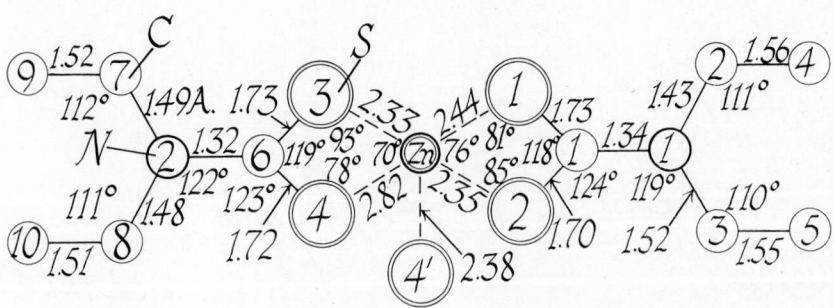

Fig. XIVB,73.   Bond dimensions in crystals of $Zn[(C_2H_5)_2NCS_2]_2$.

The parameters are those listed in Table XIVB,33. There is, as Figure XIVB,73 indicates, an important difference in the relation of the S(4) atoms to the metallic atoms in the two compounds, three different anions contributing closest neighbors to the zinc atom.

TABLE XIVB,33
Parameters of the Atoms in $Zn[(C_2H_5)_2\ NCS_2]_2$

Atom	x	$\sigma(x)$	y	$\sigma(y)$	z	$\sigma(z)$
Zn	0.1697	0.0002	0.0725	0.0002	0.0540	0.0001
S(1)	0.3403	0.0004	0.2441	0.0003	0.1123	0.0002
S(2)	0.2166	0.0004	0.0730	0.0003	0.2057	0.0002
S(3)	0.2439	0.0005	−0.0387	0.0003	−0.0445	0.0002
S(4)	0.0528	0.0004	−0.1727	0.0003	0.0286	0.0002
N(1)	0.3897	0.0015	0.2674	0.0010	0.2818	0.0007
N(2)	0.1345	0.0015	−0.2613	0.0009	−0.0985	0.0006
C(1)	0.3209	0.0017	0.2000	0.0013	0.2092	0.0007
C(2)	0.4726	0.0019	0.3780	0.0013	0.2868	0.0011
C(3)	0.3672	0.0022	0.2370	0.0016	0.3666	0.0010
C(4)	0.3739	0.0026	0.4960	0.0015	0.2575	0.0013
C(5)	0.4885	0.0034	0.1493	0.0028	0.4255	0.0013
C(6)	0.1409	0.0019	−0.1682	0.0011	−0.0438	0.0008
C(7)	0.2128	0.0022	−0.2618	0.0015	−0.1600	0.0012
C(8)	0.0436	0.0020	−0.3734	0.0012	−0.1051	0.0011
C(9)	0.1182	0.0031	−0.2198	0.0024	−0.2520	0.0011
C(10)	0.1322	0.0027	−0.4810	0.0015	−0.0515	0.0016

XIV,b52. *Nickel bis diethyl dithiocarbamate*, $Ni[(C_2H_5)_2NCS_2]_2$, though monoclinic has a structure different from that of the copper and zinc compounds just described (**XIV,b51**). It has a bimolecular unit of the dimensions:

$$a_0 = 6.23\ A.;\quad b_0 = 11.62\ A.;\quad c_0 = 11.55\ A.;\quad \beta = 95°$$

The space group is $C_{2h}^5\ (P2_1/c)$ with the nickel atoms in

$$(2a)\quad 000;\ 0\ ^1/_2\ ^1/_2$$

All other atoms are in the general positions:

$$(4e)\quad \pm(xyz;\ x,^1/_2-y,z+^1/_2)$$

The parameters have been stated to be those of Table XIVB,34.

TABLE XIVB,34
Parameters of Atoms in $Ni[(C_2H_5)_2NCS_2]_2$

Atom	x	y	z
S(1)	0.305	0.045	0.111
S(2)	0.114	0.171	0.067
C	0.128	0.147	0.120
N	0.200	0.223	0.192
CH$_2$(1)	0.083	0.329	0.192
CH$_2$(2)	0.360	0.176	0.278
CH$_3$(1)	0.069	0.338	0.284
CH$_3$(2)	0.240	0.121	0.371

In such a structure the nickel atoms would be surrounded by an approximate square of sulfur atoms with Ni–S = ca. 2.27 A. and S–Ni–S = 82°. Other important interatomic distances are C–N = 1.26 A., N–CH$_2$ = 1.45 A., and CH$_2$–CH$_3$ = 1.51 A. The C–S separation is reported to be 1.62 A., but the parameters of the C and S(2) atoms are such as to bring them closer than this to one another. Additional work seems called for.

XIV,b53. *Potassium xanthate*, $KS_2COC_2H_5$, is monoclinic with an eight-molecule unit of the dimensions:

$$a_0 = 4.40 \text{ A.}; \quad b_0 = 16.46 \text{ A.}; \quad c_0 = 18.89 \text{ A.}; \quad \beta = 92°7'$$

TABLE XIVB,35
Parameters of the Atoms in $K(S_2COC_2H_5)$

Atom	x	y	z
K(1)	0.739	0.176	0.282
K(2)	0.727	0.343	0.509
S(1)	0.216	0.475	0.571
S(2)	0.237	0.321	0.652
S(3)	0.241	0.090	0.387
S(4)	0.226	0.272	0.390
O(1)	0.464	0.453	0.700
O(2)	0.487	0.184	0.486
C(1)	0.316	0.419	0.643
C(2)	0.315	0.179	0.426
C(3)	0.424	0.463	0.310
C(4)	0.261	0.444	0.238
C(5)	0.591	0.112	0.524
C(6)	0.765	0.140	0.591

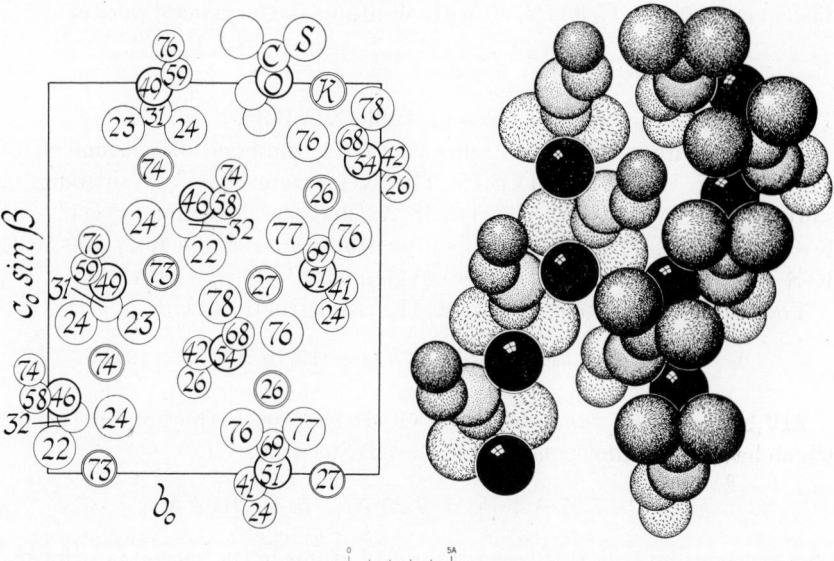

Fig. XIVB,74a (left).   The monoclinic structure of potassium xanthate projected
along its a_0 axis.   Right-hand axes.

Fig. XIVB,74b (right).   A packing drawing of the monoclinic $KS_2COC_2H_5$ structure
seen along its a_0 axis.   The sulfur atoms are large and line shaded; the oxygen atoms
are almost as big, but are dotted and heavily outlined.   The smaller carbon atoms are
hook shaded.   Atoms of potassium are black.

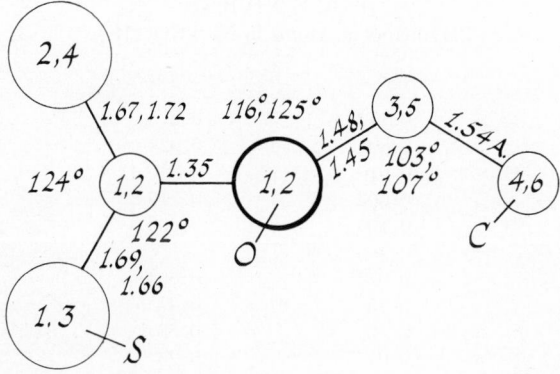

Fig. XIVB,75.   Bond dimensions of the xanthate anion in its potassium salt.   The
two sets of numbers apply to the two crystallographically different anions in the
crystal.

The space group is C_{2h}^5 $(P2_1/c)$ with all atoms in the general positions:

$$(4e) \quad \pm (xyz; \; x, 1/2 - y, z + 1/2)$$

The chosen parameters are those of Table XIVB,35.

The structure is shown in Figure XIVB,74, the bond dimensions of the xanthate ion in Figure XIVB,75. The K(1) atoms have six surrounding sulfur atoms, with K–S = 3.24–3.48 A., and a still nearer oxygen, with K–O = 2.86 A. There are seven sulfur atoms around the K(2) ions, with K–S = 3.25–3.54 A., and a nearer oxygen, with K–O = 2.84 A.

The *rubidium xanthate*, $RbS_2COC_2H_5$, is isostructural with

$$a_0 = 4.56 \text{ A.}; \quad b_0 = 16.65 \text{ A.}; \quad c_0 = 19.16 \text{ A.}; \quad \beta = 91°6'$$

XIV,b54. *Nickel xanthate*, $Ni(S_2COC_2H_5)_2$, forms orthorhombic crystals which have a tetramolecular unit of the edge lengths:

$$a_0 = 7.57 \text{ A.}; \quad b_0 = 7.23 \text{ A.}; \quad c_0 = 20.92 \text{ A.}$$

The space group is V_h^{15} $(Pbca)$ with nickel atoms in the special positions:

$$(4a) \quad 000; \; 1/2 \, 1/2 \, 0; \; 0 \, 1/2 \, 1/2; \; 1/2 \, 0 \, 1/2$$

and all other atoms in the general positions:

$$(8c) \quad \pm (xyz; \; 1/2 - x, y + 1/2, z; \; x, 1/2 - y, z + 1/2; \; x + 1/2, y, 1/2 - z)$$

The established parameters are given in Table XIVB,36.

TABLE XIVB,36
Parameters of Atoms in $Ni(S_2COC_2H_5)_2$

Atom	x	y	z
S(1)	0.112	0.275	0.0290
S(2)	0.891	0.024	0.0985
O	0.010	0.340	0.1497
C(1)	0.003	0.219	0.0990
C(2)	0.105	0.503	0.1483
C(3)	0.074	0.604	0.2089
H(2,1)	0.25	0.47	0.14
H(2,2)	0.06	0.58	0.10
H(3,1)	0.01	0.23	0.20
H(3,2)	0.20	0.14	0.23
H(3,3)	0.00	0.01	0.24

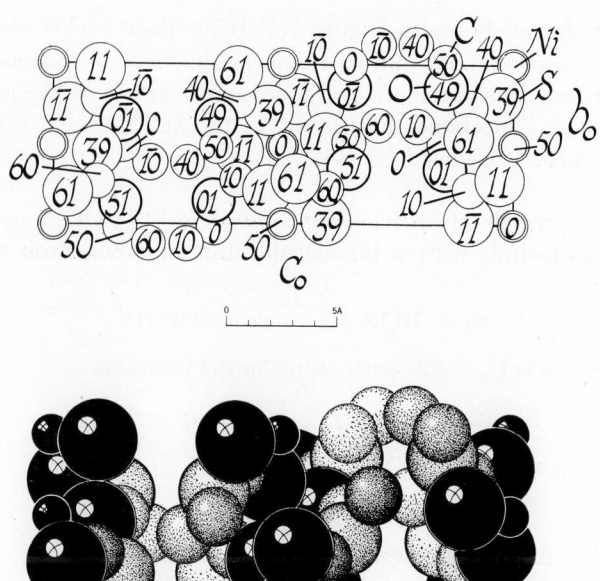

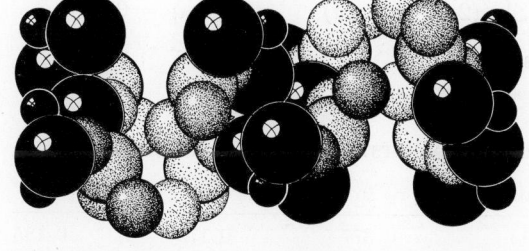

Fig. XIVB,76a (top).   The orthorhombic structure of nickel xanthate projected along
its a_0 axis.   Right-hand axes.

Fig. XIVB,76b (bottom).   A packing drawing of the orthorhombic nickel xanthate
structure viewed along its a_0 axis.   The sulfur atoms are the large, the nickel atoms the
small black circles.   Atoms of carbon are short-line shaded; those of oxygen are larger,
heavily outlined, and dotted.

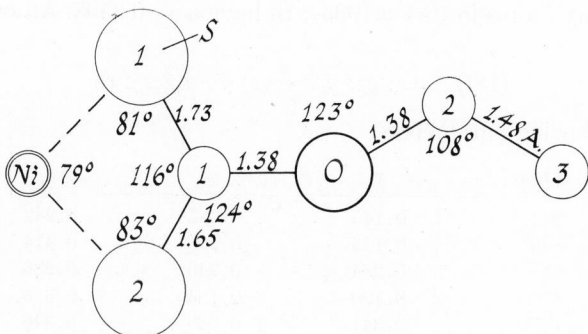

Fig. XIVB,77.   Bond dimensions of the xanthate anion in its nickel salt.

The structure is shown in Figure XIVB,76. Each nickel atom is surrounded by four coplanar sulfur atoms, with Ni–S = 2.23 or 2.24 A.; above and below this plane are two more sulfur atoms at the greater distance of 3.41 A. The bond dimensions of the xanthate ion are given in Figure XIVB,77. The final $R = 0.17$.

XIV,b55. Crystals of *antimonious xanthate*, $Sb(S_2COC_2H_5)_3$, are hexagonal, rhombohedral, with a bimolecular unit rhombohedron having the dimensions:

$$a_0 = 10.13 \text{ A.}, \qquad \alpha = 103°31'$$

The space group is C_{3i}^2 $(R\overline{3})$ with atoms in the positions:

$$Sb: (2c) \quad \pm(uuu) \qquad \text{with } u = 0.3508$$

The other atoms are in

$$(6f) \quad \pm(xyz; zxy; yzx)$$

with the following parameters:

Atom	x	y	z
S(1)	0.377	0.175	0.145
S(2)	0.557	0.475	0.209
O	0.570	0.249	0.040
C(1)	0.505	0.308	0.131
C(2)	0.687	0.342	0.008
C(3)	0.764	0.254	0.936

The six-molecule cell referred to hexagonal axes has the dimensions:

$$a_0' = 15.91 \text{ A.}, \qquad c_0' = 12.81 \text{ A.}$$

The antimony atoms in $(6c)$ $\pm(00u)$; rh have $u = 0.3508$. All other atoms are in

$$(18f) \quad \pm(xyz; \bar{y},x-y,z; y-x,\bar{x},z); \text{ rh}$$

with the following parameters:

Atom	x	y	z
S(1)	0.145	0.087	0.232
S(2)	0.143	0.205	0.414
O	0.284	0.246	0.286
C(1)	0.190	0.184	0.315
C(2)	0.341	0.338	0.346
C(3)	0.446	0.382	0.318

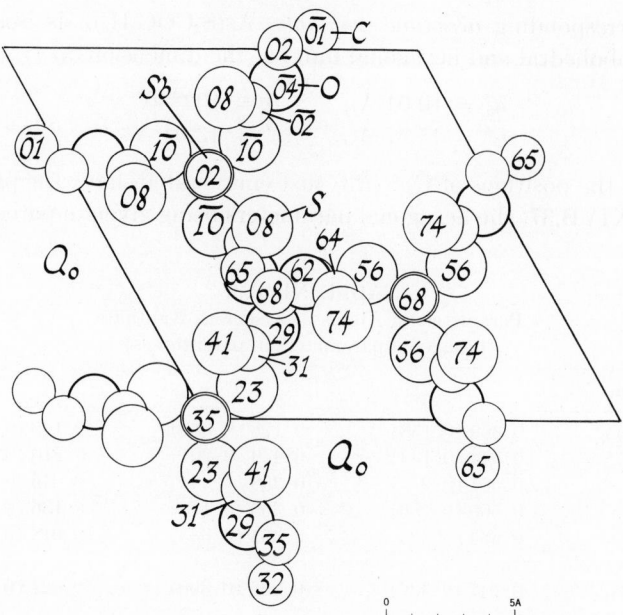

Fig. XIVB,78. The contents of the lower half of the hexagonal structure of $Sb(S_2COC_2H_5)_3$ projected along its c_0 axis. Right-hand axes.

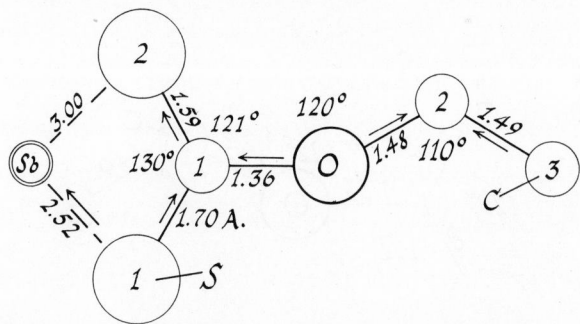

Fig. XIVB,79. Bond dimensions of the xanthate group in $Sb(S_2COC_2H_5)_3$

The structure as a whole is shown in Figure XIVB,78. It is made up of molecules in which each antimony atom is surrounded by three sulfur atoms of three xanthate groups at a distance of 2.52 A., and by the other three sulfur atoms of these groups with Sb–S(2) = 3.00 A. The bond dimensions of the xanthate group are those of Figure XIVB,79. In the group, the S(1), C(1), S(2), O part is planar.

The corresponding *arsenious xanthate*, $As(S_2COC_2H_5)_3$, is isostructural with rhombohedral and hexagonal units of the dimensions:

$$a_0 = 10.04 \text{ A.,} \qquad \alpha = 102°40'$$
$$a_0' = 15.67 \text{ A.,} \qquad c_0' = 13.04 \text{ A.}$$

Atoms, in the positions of $C_{3i}{}^2$ ($R\overline{3}$) just enumerated, have the parameters of Table XIVB,37, the hexagonal parameters being given in parentheses.

TABLE XIVB,37
Parameters of Atoms in Arsenious Xanthate
(hexagonal parameters in parentheses)

Atom	x	y	z
S(1)	0.369 (0.138)	0.178 (0.086)	0.145 (0.231)
S(2)	0.544 (0.133)	0.479 (0.201)	0.210 (0.411)
O	0.567 (0.277)	0.258 (0.244)	0.046 (0.290)
C(1)	0.500 (0.186)	0.305 (0.178)	0.136 (0.314)
C(2)	0.684 (0.333)	0.351 (0.333)	0.018 (0.351)
C(3)	0.747 (0.436)	0.265 (0.390)	0.921 (0.311)

The bond dimensions thus found for this crystal are those of Figure XIVB,80.

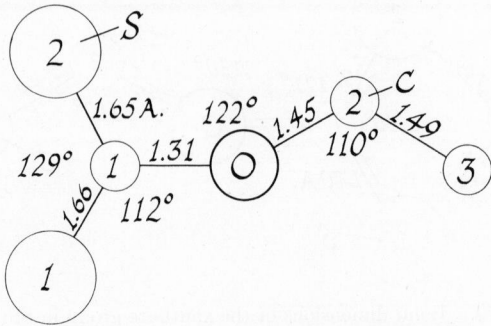

Fig. XIVB,80.   Bond dimensions of the xanthate group in $As(S_2COC_2H_5)_3$.

XIV,b56.   *Oxomolybdenum xanthate*, $Mo_2O_3(C_2H_5OCS_2)_2$, is monoclinic with a bimolecular cell of the dimensions:

$$a_0 = 10.72 \text{ A.;} \quad b_0 = 13.57 \text{ A.;} \quad c_0 = 10.86 \text{ A.,} \qquad \text{all} \pm 0.03 \text{ A.}$$
$$\beta = 123°30'$$

TABLE XIVB,38
Parameters of the Atoms in $Mo_2O_3(C_2H_5OCS_2)_2$

Atom	x	$\sigma(x)$	y	$\sigma(y)$	z	$\sigma(z)$
Mo(1)	0.23642	0.00035	0.00000	—	0.08156	0.0004
Mo(2)	0.28520	0.00034	0.26739	0.0003	0.17356	0.0003
O(1)	0.2584	0.0031	0.1327	0.0024	0.1285	0.0035
O(2)	0.0972	0.0026	−0.0062	0.0023	−0.0929	0.0026
O(3)	0.1635	0.0027	0.3179	0.0020	0.0121	0.0027
S(1)	0.1684	0.0012	−0.0362	0.0008	0.2591	0.0012
S(2)	0.2402	0.0012	−0.1844	0.0008	0.1181	0.0012
C(1)	0.1836	0.0044	−0.1543	0.0029	0.2336	0.0044
O(4)	0.1557	0.0025	−0.2296	0.0019	0.3024	0.0026
C(5)	0.1676	0.0053	−0.3311	0.0034	0.2723	0.0053
C(6)	0.1512	0.0055	−0.3865	0.0044	0.3774	0.0063
S(3)	0.41985	0.0010	−0.0039	0.0009	0.0067	0.0011
S(4)	0.5212	0.0011	−0.0182	0.0009	0.3134	0.0011
C(2)	0.5555	0.0046	−0.0159	0.0038	0.1910	0.0045
O(5)	0.6998	0.0032	−0.0091	0.0027	0.2231	0.0032
C(7)	0.7310	0.0063	0.0003	0.0052	0.1072	0.0068
C(8)	0.7802	0.0079	0.1028	0.0059	0.0983	0.0086
S(5)	0.5343	0.0011	0.2586	0.0009	0.2117	0.0011
S(6)	0.43325	0.0011	0.4207	0.0009	0.3104	0.0011
C(3)	0.5692	0.0043	0.3663	0.0028	0.2953	0.0042
O(6)	0.7064	0.0027	0.4099	0.0020	0.3612	0.0026
C(9)	0.8311	0.0060	0.3670	0.0039	0.3605	0.0056
C(10)	0.9655	0.0048	0.4337	0.0033	0.4597	0.0048
S(7)	0.11805	0.0012	0.2844	0.0010	0.2668	0.0012
S(8)	0.4156	0.0011	0.2011	0.0008	0.4547	0.0011
C(4)	0.2631	0.0046	0.2332	0.0033	0.4284	0.0047
O(7)	0.2389	0.0028	0.2268	0.0019	0.5427	0.0029
C(11)	0.3573	0.0070	0.2019	0.0047	0.6946	0.0072
C(12)	0.2977	0.0068	0.2320	0.0049	0.7859	0.0070

The space group has been chosen as C_2^2 ($P2_1$) with all atoms in the positions:

$$(2a) \quad xyz; \ \bar{x}, y + \frac{1}{2}, \bar{z}$$

The given parameters are listed in Table XIVB,38.

In the resulting structure (Fig. XIVB,81) pairs of molybdenum atoms are joined by an O(1) atom, with Mo–O = 1.851 or 1.872 A.; a second oxygen [O(2) or O(3)] is equally bound to each, with Mo–O = 1.644 or

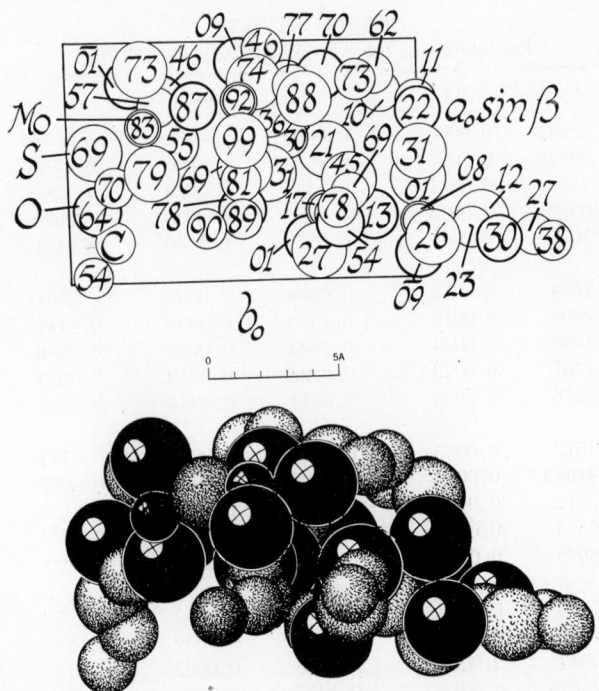

Fig. XIVB,81a (top). The monoclinic structure of $Mo_2O_3(C_2H_5OCS_2)_2$ projected along its c_0 axis. Right-hand axes.

Fig. XIVB,81b (bottom). A packing drawing of the monoclinic $Mo_2O_3(C_2H_5OCS_2)_2$ arrangement viewed along its c_0 axis. The molybdenum atoms are the small, the sulfur atoms the large black circles. The small line shaded circles are the carbon atoms; the oxygen atoms are somewhat larger, hook shaded, and more heavily ringed.

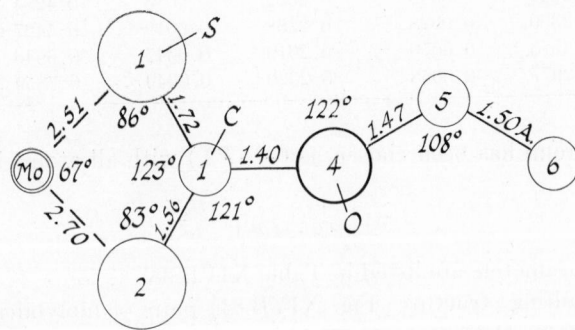

Fig. XIVB,82. Bond dimensions in crystals of $Mo_2O_3(C_2H_5OCS_2)_2$

1.649 A. Each molybdenum atom has also about it four sulfur atoms from two xanthate groups, with Mo–S lying between 2.458 and 2.715 A. The average bond lengths of the two crystallographically different xanthate ions in the crystal are given in Figure XIVB,82.

XIV,b57. Crystals of *sodium iodide triacetonate*, $NaI \cdot 3(CH_3)_2CO$, are hexagonal with a bimolecular cell of the edge lengths:

$$a_0 = 11.39 \pm 0.03 \text{ A.}, \qquad c_0 = 6.611 \pm 0.01 \text{ A.}$$

The space group has been chosen as C_6^6 ($P6_3$) with atoms in the positions:

Na: (2a) $00u;\ 0,0,u+^1/_2$ with $u = 0.000$ (arbitrary)
 I: (2b) $^1/_3\,^2/_3\,u;\ ^2/_3,\,^1/_2,u+^1/_2$ with $u = 0.250$
C(0): (6c) $xyz;$ $\bar{y},x-y,z;$ $y-x,\bar{x},z;$
    $\bar{x},\bar{y},z+^1/_2;\ y,y-x,z+^1/_2;\ x-y,x,z+^1/_2$
    with $x = 0.272,\ y = 0.260,\ z = 0.295$
C(1): (6c)   with $x = 0.375,\ y = 0.338,\ z = 0.120$
C(2): (6c)   with $x = 0.306,\ y = 0.303,\ z = 0.518$
C(3): (6c)   with $x = 0.165,\ y = 0.156,\ z = 0.250$

The structure is shown in Figure XIVB,83. As it indicates, each atom of sodium is octahedrally surrounded by six oxygen atoms at a distance of 2.46 A. The iodine ions are closest to methyl groups, with I–C lying between 4.09 and 4.37 A. As the figure and the parameter values indicate, the molecules of acetone lie almost exactly in the plane $x = y$.

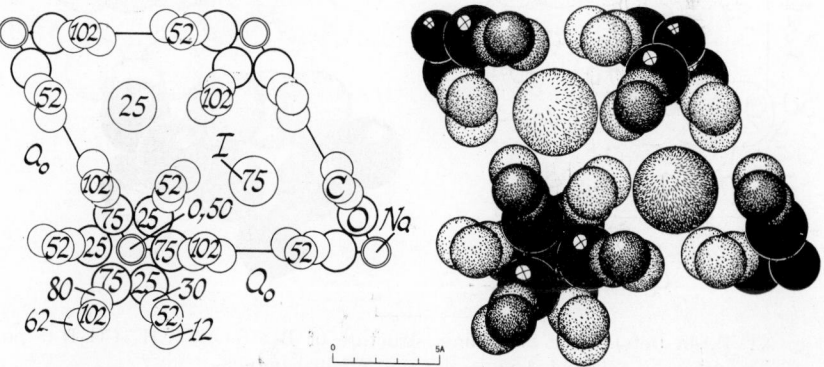

Fig. XIVB,83a (left).   The hexagonal structure of $NaI \cdot 3(CH_3)_2CO$ projected along its c_0 axis. Right-hand axes.
Fig. XIVB,83b (right).   A packing drawing of the hexagonal $NaI \cdot 3(CH_3)_2CO$ structure seen along its c_0 axis. The oxygen atoms are the large, the sodium atoms the slightly smaller black circles. The atoms of carbon are small and dotted, those of iodine large and line shaded.

XIV,b58.   *Barium pentathionate* forms an addition compound with *acetone* which crystallizes with water and has the composition $BaS(S_2O_3)_2 \cdot (CH_3)_2CO \cdot H_2O$. The symmetry is monoclinic with a bimolecular unit of the dimensions:

$$a_0 = 5.04 \text{ A.}; \quad b_0 = 10.47 \text{ A.}; \quad c_0 = 13.61 \text{ A.}; \quad \beta = 104°$$

The space group is C_{2h}^2 ($P2_1/m$) and atoms are in the positions:

$$(2e) \quad \pm (u\ ^1/_4\ v)$$
$$(4f) \quad \pm (xyz;\ x,^1/_2-y,z)$$

The determined positions and parameters are those listed in Table XIVB,39.

The layered structure that results is shown in Figure XIVB,84. Its pentathionate anion has the bond dimensions of Figure XIVB,85; they are to be compared with those found for this ion in **X,b55.**

The corresponding seleno thionate, $BaSe(S_2O_3)_2 \cdot (CH_3)_2CO \cdot H_2O$, is isomorphous with a cell of the dimensions:

$$a_0 = 5.02 \text{ A.}; \quad b_0 = 10.56 \text{ A.}; \quad c_0 = 13.78 \text{ A.}; \quad \beta = 105°$$

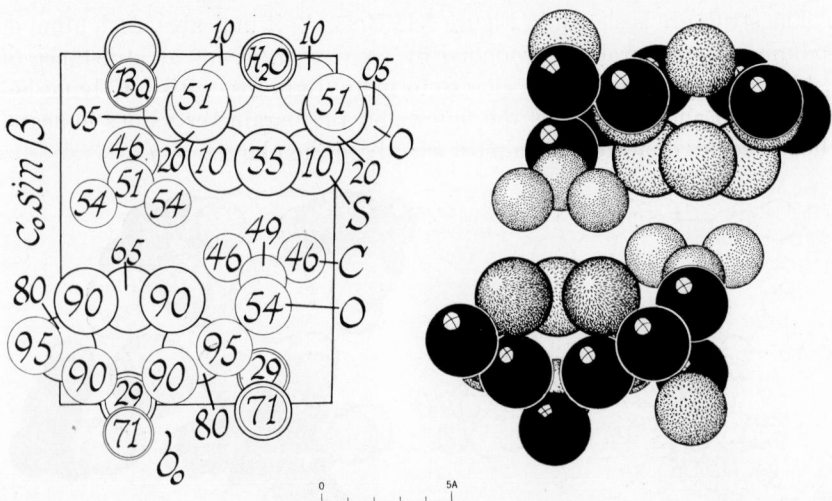

Fig. XIVB,84a (left).   The monoclinic structure of $BaS(S_2O_3)_2 \cdot (CH_3)_2CO \cdot H_2O$ projected along its a_0 axis.   Right-hand axes.

Fig. XIVB,84b (right).   A packing drawing of the monoclinic structure of $BaS(S_2O_3)_2 \cdot (CH_3)_2CO \cdot H_2O$ seen along its a_0 axis.   Both the barium and the oxygen atoms are black, the "windows" of the oxygen atoms being round.   The carbon atoms are small and dotted; the water molecules are hook shaded.   Atoms of sulfur are more heavily ringed and are line shaded.

Fig. XIVB,85. Bond dimensions of the pentathionate ions in $BaS(S_2O_3)_2 \cdot (CH_3)_2CO \cdot H_2O$.

TABLE XIVB,39
Positions and Parameters of the Atoms in Barium Pentathionate–Acetone

Atom	Position	x	y	z
Ba	(2e)	0.711	$1/4$	−0.088
S(1)	(4f)	0.798	0.015	0.153
S(2)	(4f)	0.903	0.097	0.299
S(3)	(2e)	0.655	$1/4$	0.294
O(1)	(4f)	0.495	0.010	0.127
O(2)	(4f)	0.953	−0.105	0.169
O(3)	(4f)	0.897	0.102	0.085
H_2O	(2e)	0.292	$1/4$	0.010
O(4)	(2e)	0.459	$1/4$	−0.285
C(1)	(2e)	0.510	$1/4$	−0.372
C(2)	(4f)	0.539	0.125	−0.430

The following three additional compounds with tetrahydrofuran also have the similar cells:

Compound	a_0, A.	b_0, A.	c_0, A.	β
$BaS(S_2O_3)_2 \cdot (CH_2)_4O \cdot H_2O$	5.03	10.56	13.81	104°
$BaSe(S_2O_3)_2 \cdot (CH_2)_4O \cdot H_2O$	5.03	10.69	13.97	105°
$BaTe(S_2O_3)_2 \cdot (CH_2)_4O \cdot H_2O$	5.00	10.82	14.21	106°

6. Acetic Acid Derivatives and Salts

XIV,b59. Crystalline *acetic acid*, CH_3COOH, is orthorhombic with a tetramolecular unit of the edge lengths:

$a_0 = 13.32 \pm 0.02$ A.; $b_0 = 4.08 \pm 0.01$ A.; $c_0 = 5.77 \pm 0.01$ A. (5°C.)

The space group has been chosen as C_{2v}^9 ($Pna2_1$) with all atoms in the positions:

(4a) $xyz; \bar{x},\bar{y},z+1/2; 1/2-x,y+1/2,z+1/2; x+1/2,1/2-y,z$

312   CRYSTAL STRUCTURES

The parameters follow:

Atom	x	$\sigma(x)$	y	$\sigma(y)$	z	$\sigma(z)$
O(1)	0.1295	0.0003	0.103	0.002	0.000	0.001
O(2)	0.2526	0.0004	0.369	0.002	0.178	0.001
C(1)	0.1641	0.0006	0.276	0.003	0.170	0.002
C(2)	0.0868	0.0006	0.372	0.003	0.357	0.002

The structure is shown in Figure XIVB,86. In the molecules, C–C = 1.54 A., and C–O = 1.24 and 1.29 A. These molecules are tied together by O–H–O bonds of length 2.61 A. into chains that run roughly parallel to the (011) plane. In the molecules, C–C–O = 122° and O–C–O = 116°.

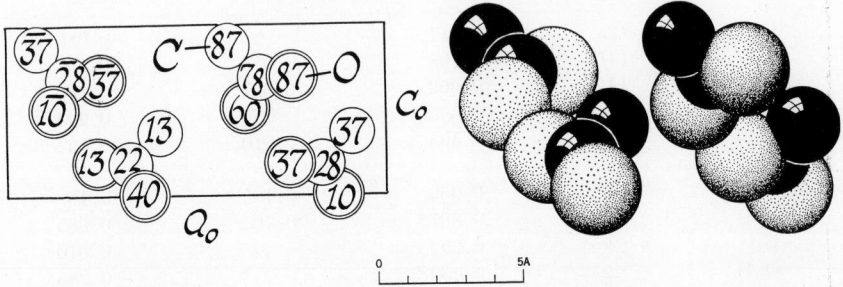

Fig. XIVB,86a (left). The orthorhombic structure of acetic acid projected along its b_0 axis. Right-hand axes.

Fig. XIVB,86b (right). A packing drawing of the orthorhombic acetic acid arrangement viewed along its b_0 axis. The oxygen atoms are large and dotted; the carbon atoms are black.

XIV,b60. Crystals of *acetamide*, $CH_3C(O)NH_2$, are hexagonal (rhombohedral) with a rhombohedral unit containing six molecules and having the dimensions:

$$a_0 = 7.99 \text{ A.}, \qquad \alpha = 91°25'$$

Its hexagonal cell contains 18 molecules and has the dimensions:

$$a_0' = 11.44 \text{ A.}, \qquad c_0' = 13.49 \text{ A.}$$

All atoms are in general positions of C_{3v}^6 ($R3c$) which have, in terms of hexagonal axes, the coordinates:

(18b) $xyz;\quad \bar{y},x-y,z;\quad y-x,\bar{x},z;$
$\bar{y},\bar{x},z+{}^1/_2;\ x,x-y,z+{}^1/_2;\ y-x,y,z+{}^1/_2;$ rh

Their parameters are:

Atom	x	y	z
C	0.333	0.005	0.034
O	0.225	−0.004	0.000
NH_2	0.333	−0.090	0.096
CH_3	0.477	0.123	0.033

This rather complicated structure is built up of planar molecules having the dimensions of Figure XIVB,87. In the crystal, six of these form a ring in which intermolecular contacts are maintained through NH_2–O separations of 2.86 A.

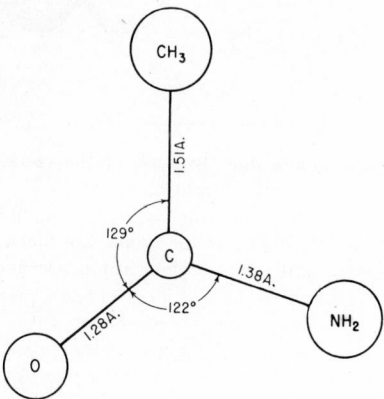

Fig. XIVB,87.   Bond dimensions in the planar molecule of acetamide.

XIV,b61. *Thioacetamide*, $CH_3C(S)NH_2$, is monoclinic with an eight-molecule unit of the dimensions:

$$a_0 = 11.062 \text{ A.}; \quad b_0 = 10.005 \text{ A.}; \quad c_0 = 7.170 \text{ A.}, \qquad \text{all } \pm 0.005 \text{ A.}$$
$$\beta = 99°30' \pm 12'$$

The space group is C_{2h}^5 $(P2_1/a)$ with atoms in the positions:

$$(4e) \quad \pm (xyz; \; x+{}^1\!/_2, {}^1\!/_2 - y, z)$$

The established parameters, including those for hydrogen, are listed in Table XIVB,40.

In this structure (Fig. XIVB,88), both types of molecule are planar, except for the hydrogen atoms, and have the agreeing bond dimensions of

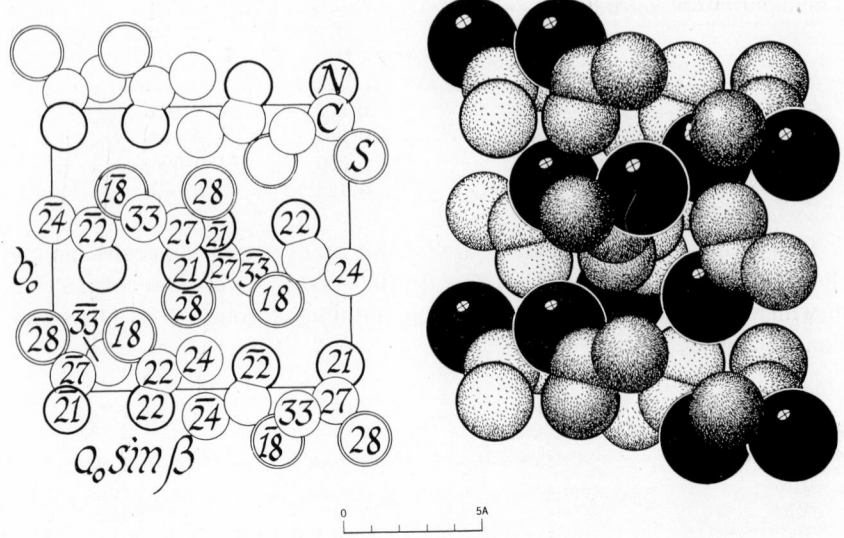

Fig. XIVB,88a (left).   The monoclinic structure of thioacetamide projected along its
c_0 axis.   Right-hand axes.

Fig. XIVB,88b (right).   A packing drawing of the monoclinic structure of thioacetamide seen along its c_0 axis.   The sulfur atoms are black; the atoms of nitrogen
are hook shaded and heavily outlined.   Carbon atoms are fine-line shaded.

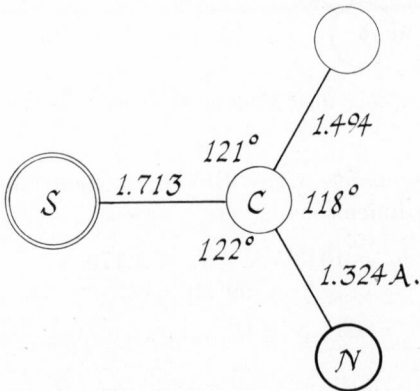

Fig. XIVB,89.   Bond dimensions of the molecule of thioacetamide.

Figure XIVB,89. The shortest interatomic distances between molecules are N–N = 3.45 A. and N–S = 3.396–3.504 A.

TABLE XIVB,40.   Parameters of the Atoms in $CH_3C(S)NH_2$

Atom	x	y	z
S(1)	0.2531	0.1850	0.1804
C(1)	0.3565	0.0590	0.2163
C(2)	0.4889	0.0896	0.2386
N(1)	0.3224	−0.0659	0.2228
H(1)	0.384	−0.134	0.200
H(2)	0.189	−0.091	0.209
H(5)	0.515	−0.012	0.254
H(6)	0.515	0.147	0.345
H(7)	0.520	0.128	0.102
S(2)	−0.0365	0.1878	−0.2756
C(3)	0.0644	0.0597	−0.2678
C(4)	0.1851	0.0766	−0.3284
N(2)	0.0393	−0.0596	−0.2080
H(3)	0.122	−0.137	−0.161
H(4)	−0.081	−0.071	−0.139
H(8)	0.207	0.138	−0.446
H(9)	0.236	0.138	−0.219
H(10)	0.211	0.002	−0.369

XIV,b62.   Crystals of *monofluoroacetamide*, CH_2FCONH_2, are triclinic with a bimolecular unit having the dimensions:

$$a_0 = 5.103 \text{ A.}; \quad b_0 = 5.343 \text{ A.}; \quad c_0 = 6.901 \text{ A.}$$
$$\alpha = 104°46'; \quad \beta = 100°26'; \quad \gamma = 100°7'$$

TABLE XIVB,41.   Parameters of the Atoms in CH_2FCONH_2

Atom	x	$\sigma(x)$	y	$\sigma(y)$	z	$\sigma(z)$
C(1)	0.2658	0.0041	0.6666	0.0043	0.3421	0.0045
C(2)	0.1563	0.0036	0.3874	0.0040	0.1947	0.0039
N	0.3274	0.0032	0.2639	0.0035	0.1134	0.0036
O	0.9070	0.0026	0.2898	0.0031	0.1602	0.0033
F	0.5464	0.0027	0.7441	0.0029	0.3662	0.0032
H(1)	0.146	—	0.792	—	0.271	—
H(2)	0.208	—	0.646	—	0.479	—
H(3)	0.271	—	0.083	—	0.000	—
H(4)	0.521	—	0.354	—	0.125	—

The space group is C_i^1 $(P\bar{1})$ with all atoms in the positions $(2i)$ $\pm(xyz)$. The selected parameters are given in Table XIVB,41.

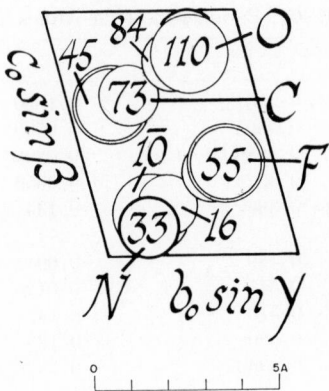

Fig. XIVB,90. The triclinic structure of CH_2FCONH_2 projected along its a_0 axis.
Right-hand axes.

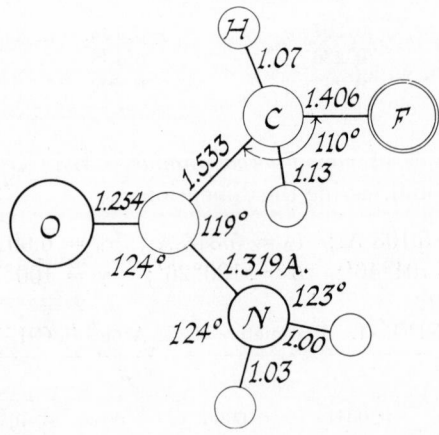

Fig. XIVB,91. Bond dimensions in the CH_2FCONH_2 molecule.

The molecules in this structure (Fig. XIVB,90) have the bond dimensions of Figure XIVB,91.   Molecules are tied together by hydrogen bonds
having the lengths N–O = 2.955 and 2.878 A.   The nearest fluorine neighbor is an oxygen atom 3.47 A. away.   All the atoms except the hydrogens
are coplanar to within 0.01 A.; the amino hydrogens depart from this plane
by less than 0.2 A.

XIV,b63. Crystals of *bromacetamide*, $BrCH_2CONH_2$, are monoclinic with a tetramolecular cell of the edge lengths:

$$a_0 = 10.44 \text{ A.}; \quad b_0 = 5.22 \text{ A.}; \quad c_0 = 7.77 \text{ A.}; \quad \beta = 99°34'$$

The space group is C_{2h}^5 $(P2_1/c)$ with atoms in the positions:

$$(4e) \quad \pm (xyz; \ x, \tfrac{1}{2}-y, z+\tfrac{1}{2})$$

The parameters are:

Atom	x	y	z
Br	0.111	0.316	0.180
C(1)	0.203	−0.007	0.178
C(2)	0.325	0.014	0.096
O	0.375	−0.201	0.067
N	0.383	0.237	0.075

The molecules that make up this structure (Fig. XIVB,92) have the bond dimensions of Figure XIVB,93, with a σ that is ca. 0.04 A. These molecules are planar within the limit of error of the determination except for the bromine atom which is ca. 0.3 A. out of the best plane through the other atoms.

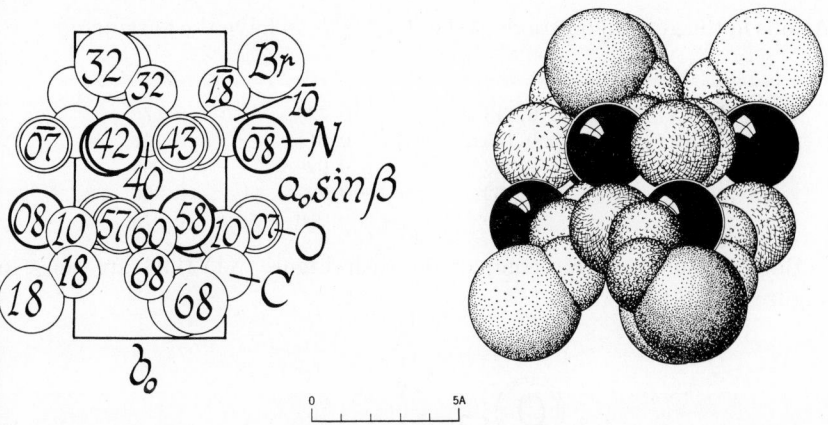

Fig. XIVB,92a (left). The monoclinic structure of $CH_2BrCONH_2$ projected along its c_0 axis. Right-hand axes.

Fig. XIVB,92b (right). A packing drawing of the monoclinic structure of CH_2-$BrCONH_2$ viewed along its c_0 axis. The bromine atoms are large and dotted; the nitrogen atoms are black. Oxygen atoms are line shaded; the smaller carbon atoms are hook shaded.

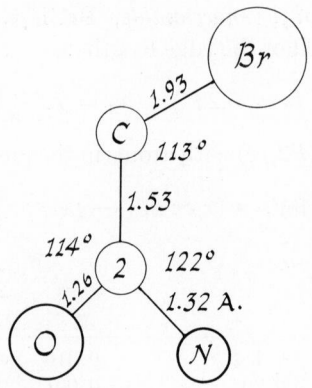

Fig. XIVB,93.   Bond dimensions of the $CH_2BrCONH_2$ molecule.

Chloracetamide, $ClCH_2CONH_2$, would appear to be isostructural, though there are differences between the two determinations that have been made. In the one that gives the two compounds the same structure, the chlorine compound has a unit of the dimensions:

$$a_0 = 10.281 \text{ A.}; \quad b_0 = 5.145 \text{ A.}; \quad c_0 = 7.429 \text{ A.}, \qquad \text{all} \pm 0.01 \text{ A.}$$
$$\beta = 98°49'$$

Atoms in the general positions $(4e)$ of C_{2h}^5 $(P2_1/c)$ have the parameters:

Atom	x	y	z
Cl	0.113	0.320	0.175
C(1)	0.195	0.011	0.162
C(2)	0.323	0.020	0.096
O	0.362	-0.201	0.066
N	0.379	0.250	0.075

Bond dimensions, for comparison with Figure XIVB,93 are those of Figure XIVB,94.

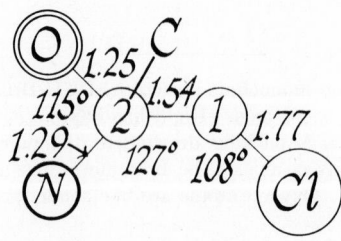

Fig. XIVB,94.   Bond dimensions of the molecule in α-chloracetamide.

XIV,b64.   There seems to be a second, perhaps unstable, modification of
chloracetamide, $ClCH_2CONH_2$, which also is monoclinic. Its unit is tetramolecular and of nearly the same dimensions as those of the stable form
(**XIV,b63**) [with a_0 and c_0 interchanged].   They are

$$a_0 = 7.45 \text{ A.}; \quad b_0 = 5.15 \text{ A.}; \quad c_0 = 10.27 \text{ A.}; \quad \beta = 102°30'$$

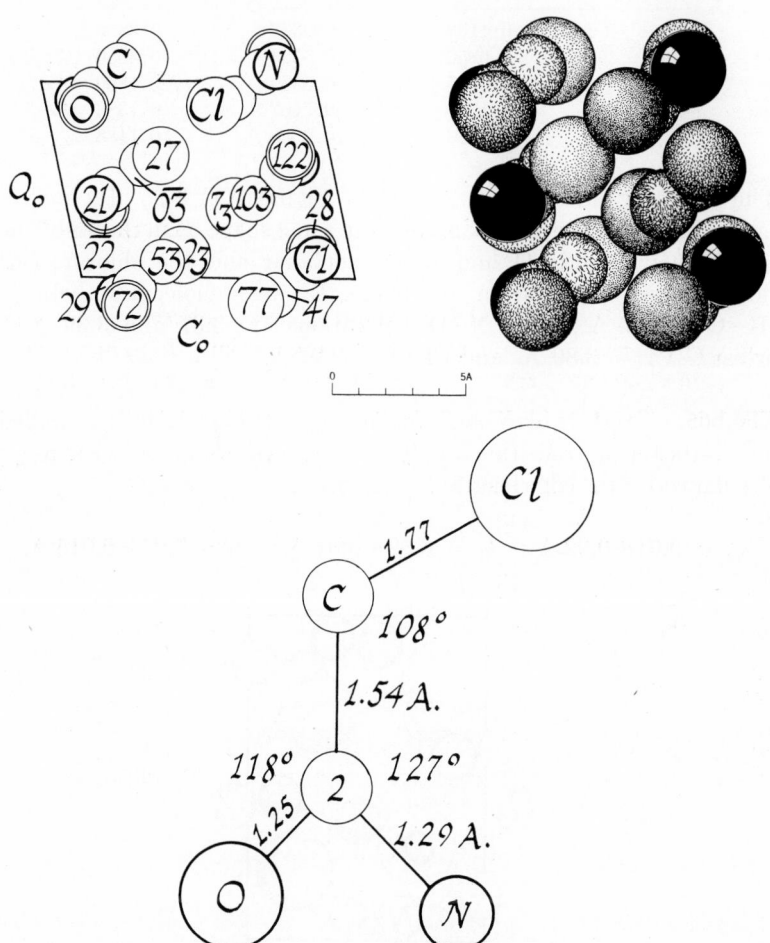

Fig. XIVB,95a (top left).   The β form of chloracetamide projected along the b_0 axis of
its monoclinic structure.   Right-hand axes.

Fig. XIVB, 95b (top right).   A packing drawing of the β form of chloracetamide seen
along its monoclinic b_0 axis.   The nitrogen atoms are black; the chlorine atoms are
large and dotted.   The oxygen atoms, almost as big, are hook shaded; the smaller
carbon atoms are line shaded.

Fig. XIVB,95c (bottom).   Bond dimensions in the molecule of the β form of chloracetamide.

The space group is C_{2h}^5 ($P2_1/a$) with all atoms in the positions:

$$(4e) \quad \pm (xyz; \ x+^1/_2, ^1/_2-y,z)$$

The parameters are the following:

Atom	x	y	z
Cl	0.1133	0.2333	0.3850
C(1)	0.0850	0.5333	0.3050
C(2)	−0.0517	0.5333	0.1800
O	−0.1367	0.7167	0.1383
N	−0.0983	0.2950	0.1183

The resulting structure is shown in Figure XIVB,95. Its molecules (Fig. XIVB,95c) have approximately the same shape as in the stable form. The atoms C(1), C(2), O, and N are coplanar and the chlorine–carbon bond is twisted by 21° from this plane. Between molecules there is an N–H···O = 3.05 A.; other N···O separations are 3.37 and 3.39 A. The shortest C–Cl = 3.80 A. and Cl–Cl = 3.97 A.

XIV,b65. Crystals of *N-methylacetamide*, $CH_3C(O)NHCH_3$, studied at −35°C. (below a transition at 10°C.) are orthorhombic with a tetramolecular cell of the edge lengths:

$$a_0 = 9.61 \pm 0.02 \text{ A.}; \quad b_0 = 6.52 \pm 0.01 \text{ A.}; \quad c_0 = 7.24 \pm 0.015 \text{ A.}$$

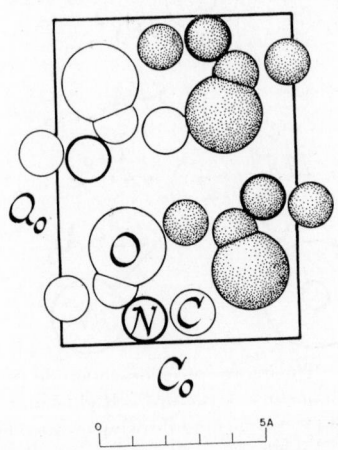

Fig. XIVB,96. The orthorhombic structure of methylacetamide projected along its b_0 axis. Right-hand axes.

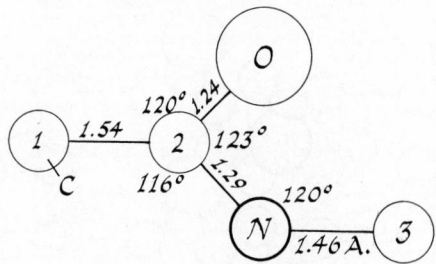

Fig. XIVB,97.   Bond dimensions in the molecule of methylacetamide.

The space group has been chosen as V_h^{16} (*Pnma*) with all the heavy atoms in the special positions:

$$(4c) \quad \pm (u\ ^1/_4\ v;\ u+^1/_2,^1/_4,^1/_2-v)$$

The parameters are:

Atom	u	v	σ, A.
C(1)	0.1343	0.0306	0.011
C(2)	0.1697	0.2394	0.011
C(3)	0.0877	0.5520	0.011
N	0.0660	0.3529	0.007
O	0.2917	0.2906	0.005

The atomic arrangement is shown in Figure XIVB,96. The molecules it contains have the bond dimensions of Figure XIVB,97. The shortest intermolecular distances are C(1)–O = 3.50 A. and C(1)–C(3) = 3.51 A.

XIV,b66. *Diacetyl hydrazine,* $(-NHCOCH_3)_2$, is orthorhombic with a tetramolecular unit having the edge lengths:

$$a_0 = 18.30 \pm 0.05\ \text{A.}; \quad b_0 = 6.51 \pm 0.02\ \text{A.}; \quad c_0 = 4.79 \pm 0.02\ \text{A.}$$

The space group has been chosen as V_h^{18} (*Ccma*) with all the heavy atoms in the special positions:

$$(8f) \quad \pm (u0v;\ u,^1/_2,^1/_2-v;\ u+^1/_2,^1/_2,v;\ u+^1/_2,0,^1/_2-v)$$

The parameters are:

Atom	u	σ_u, A.	v	σ_v, A.
N	0.0352	0.005	0.0560	0.004
C(1)	0.0921	0.007	-0.1203	0.005
O	0.0843	0.005	-0.3734	0.003
C(2)	0.1656	0.007	0.0204	0.006

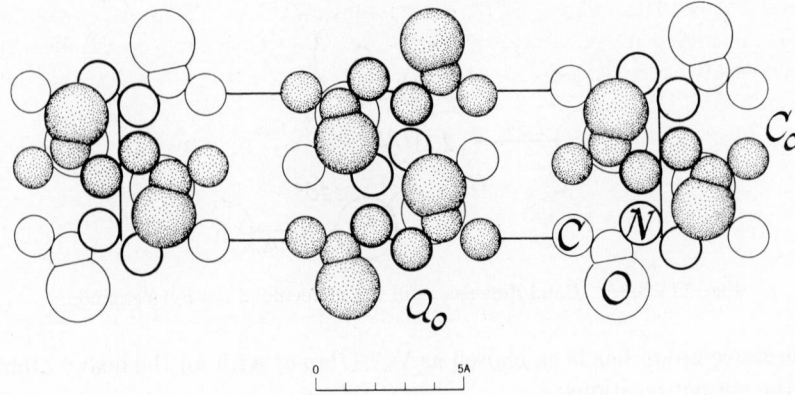

Fig. XIVB,98. The orthorhombic structure of diacetyl hydrazine projected along its b_0 axis. The shaded molecules are at a height of half the b_0 axis; the others are at $b_0 = 0$. Right-hand axes.

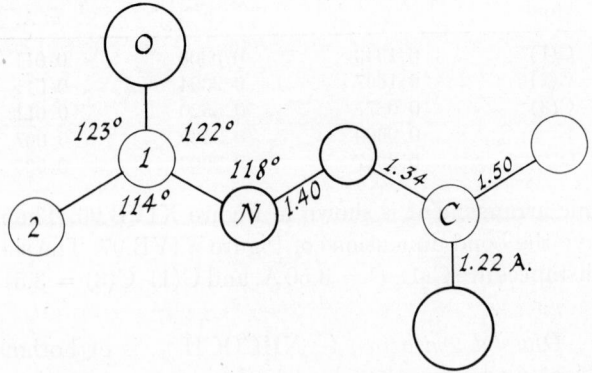

Fig. XIVB,99. Bond dimensions of the diacetyl hydrazine molecule.

Two of the three kinds of hydrogen atom are also in (8*f*), the third, belonging to the methyl groups, is in the general positions:

(16*g*) $\pm (xyz; \; ^1/_2-x,y,z+^1/_2; \; x+^1/_2,y+^1/_2,z; \; \bar{x},y+^1/_2,z+^1/_2;$
           $\bar{x}y\bar{z}; \; x+^1/_2,y,^1/_2-z; \; ^1/_2-x,y+^1/_2,\bar{z}; \; x,y+^1/_2,^1/_2-z)$

The parameters of these hydrogen atoms are given as follows:

Atom	x	y	z
H(1)	0.038	0	0.247
H(2)	0.156	0	0.227
H(3)	0.188	0.108	−0.038

The resulting structure is shown in Figure XIVB,98. It is composed of molecules that have the bond lengths of Figure XIVB,99. These molecules are tied together into chains running along the c_0 axis by hydrogen bonds between nitrogen and oxygen, with N–H–O = 2.877 A.

This structure is to be compared to that of the analogous diformyl hydrazine (**XIV,a91**).

XIV,b67. *Ammonium trifluoroacetate*, $NH_4(CF_3COO)$, is monoclinic with a tetramolecular cell of the dimensions:

$$a_0 = 12.65 \text{ A.}; \quad b_0 = 8.20 \text{ A.}; \quad c_0 = 4.83 \text{ A.}; \quad \beta = 100°36'$$

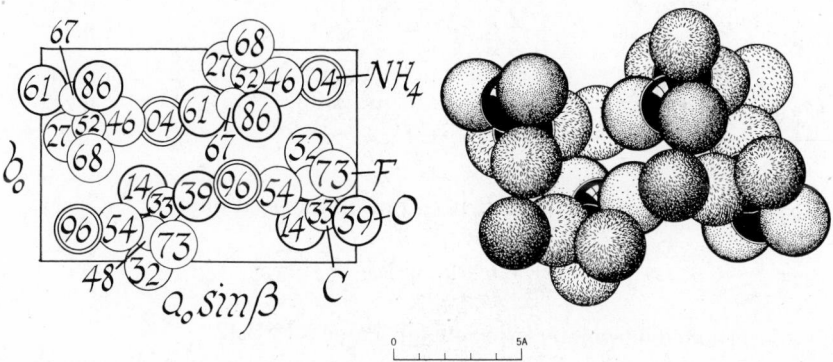

Fig. XIVB,100a (left).   The monoclinic structure of NH_4CF_3COO projected along its c_0 axis.   Right-hand axes.

Fig. XIVB,100b (right).   A packing drawing of the monoclinic structure of NH_4CF_3COO viewed along its c_0 axis.   The carbon atoms are black; the oxygen atoms are heavily ringed and dotted.   The fluorine atoms are line and the NH_4 ions are hook shaded.

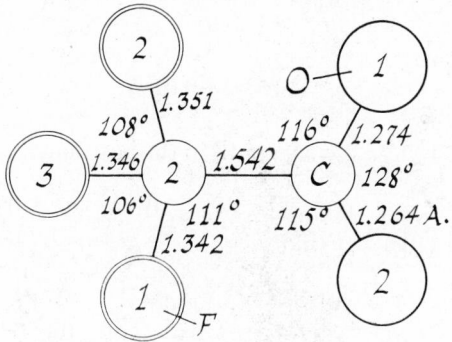

Fig. XIVB,101.   Bond dimensions of the CF_3COO anion in NH_4CF_3COO.

TABLE XIVB,42
Parameters of the Atoms in $NH_4(CF_3COO)$

Atom	x	σ_x, A.	y	σ_y, A.	z
F(1)	−0.2444	0.0037	0.3490	0.0043	0.5447
F(2)	−0.0797	0.0040	0.4132	0.0040	0.7302
F(3)	−0.1563	0.0046	0.5112	0.0033	0.3216
O(1)	−0.0084	0.0036	0.1960	0.0040	0.3934
O(2)	−0.1784	0.0040	0.1697	0.0036	0.1380
C(1)	−0.1077	0.0047	0.2325	0.0044	0.3271
C(2)	−0.1464	0.0051	0.3775	0.0051	0.4859
N	−0.1169	0.0040	−0.1621	0.0041	0.0413
H(1)	−0.143	—	−0.079	—	0.052
H(2)	−0.069	—	−0.168	—	−0.102
H(3)	−0.068	—	−0.191	—	0.231
H(4)	−0.181	—	−0.217	—	0.006

The space group is C_{2h}^5 ($P2_1/a$) with atoms in the positions:

$$(4e) \quad \pm (xyz; \ x+{}^1/_2, {}^1/_2-y, z)$$

The determined parameters are listed in Table XIVB,42.

In this crystal (Fig. XIVB,100) the nonrotating ammonium ions have N–H = 0.76–1.04 A. The anions have the bond dimensions of Figure XIVB,101; in the crystal they are tied together by four kinds of N–H–O bond having lengths between 2.87 and 2.92 A.

XIV,b68. *Sodium hydrogen acetate*, $NaH(C_2H_3O_2)_2$, though cubic, has a complicated structure with a unit which, containing 24 molecules, has the edge length:

$$a_0 = 15.92 \pm 0.01 \text{ A.}$$

The space group is T_h^7 ($Ia3$) with atoms in the positions:

Na(1): (8a) $000; 0 \ {}^1/_2 \ {}^1/_2; \ {}^1/_2 \ 0 \ {}^1/_2; \ {}^1/_2 \ {}^1/_2 \ 0;$ B.C.
Na(2): (16c) $\pm (uuu; \qquad\qquad u+{}^1/_2, {}^1/_2-u, \bar{u};$
$\bar{u}, u+{}^1/_2, {}^1/_2-u; \ {}^1/_2-u, \bar{u}, u+{}^1/_2);$ B.C.
with $u = 0.11956$
H(1): (24d) $\pm (0 \ {}^1/_4 \ u; \ {}^1/_2 \ {}^3/_4 \ u);$ tr; B.C.
with $u = 0.137$

All other atoms are in the general positions:

$(48e)$ $\pm(xyz;\ x,\bar{y},{}^1/_2-z;\ z,\bar{x},{}^1/_2-y;\ y,\bar{z},{}^1/_2-x)$; tr; B.C.

with the following parameters, those for the methyl hydrogens being considered approximate only:

Atom	x	y	z
O(1)	0.03213	0.39144	0.09994
O(2)	0.07525	0.26521	0.13970
C(1)	0.08964	0.33971	0.11036
C(2)	0.17885	0.35908	0.08904
H(2)	0.204	0.402	0.136
H(3)	0.181	0.388	0.027
H(4)	0.215	0.301	0.089

In this structure, which is too complicated to be shown in the usual projection, each sodium atom is surrounded by six oxygen atoms. For Na(1), they are at the corners of a regular, trigonal antiprism, with Na(1)–O = 2.404 A. For Na(2), the antiprism is not regular and Na(2)–O = 2.441 or 2.445 A. In the acetate groups, C(1)–O(1) = 1.243 A., C(1)–O(2) = 1.295 A., and C(1)–C(2) = 1.492 A., all ±0.01 A. These radicals are essentially planar with C(1) being only 0.005 A. from the plane defined by CO_2.

XIV,b69. *Lithium acetate dihydrate*, $Li(CH_3COO)\cdot 2H_2O$, is orthorhombic with a tetramolecular cell of the edge lengths:

$$a_0 = 6.86\ \text{A.};\quad b_0 = 11.49\ \text{A.};\quad c_0 = 6.59\ \text{A.}$$

The space group has been chosen as $C_{2v}{}^{11}$ ($Cmm2$) with atoms in the positions:

C(1): (4e) $0uv;\ 0\bar{u}v;\ {}^1/_2,u+{}^1/_2,v;\ {}^1/_2,{}^1/_2-u,v$
            with $u = 0.325,\ v = -0.248$
C(2): (4e)   with $u = 0.272,\ v = -0.118$
O(1): (4e)   with $u = 0.334,\ v = 0.118$
O(2): (4e)   with $u = 0.156,\ v = 0.005$
 Li: (4d) $u0v;\ \bar{u}0v;\ u+{}^1/_2,{}^1/_2,v;\ {}^1/_2-u,{}^1/_2,v$
            with $u = 0.162,\ v = 0.147$
H_2O: (8f) $xyz;$ $\bar{x}\bar{y}z;$ $\bar{x}yz;$ $x\bar{y}z;$
            $x+{}^1/_2,y+{}^1/_2,z;\ {}^1/_2-x,{}^1/_2-y,z;\ {}^1/_2-x,y+{}^1/_2,z;\ x+{}^1/_2,{}^1/_2-y,z$
            with $x = 0.156,\ y = 0.140,\ z = 0.452$

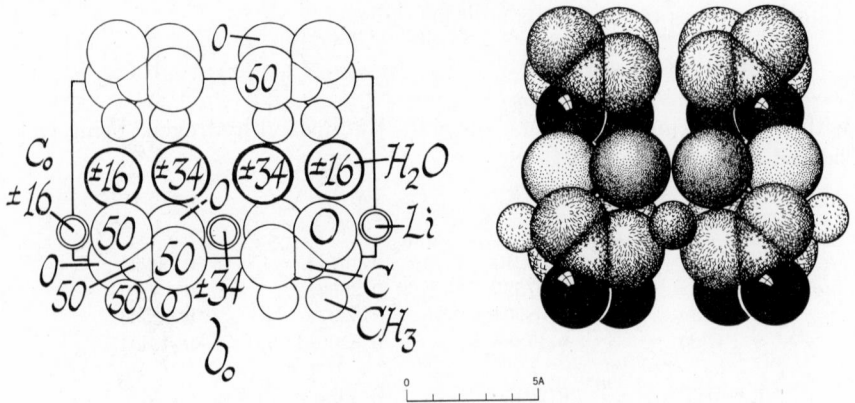

Fig. XIVB,102a (left).   The orthorhombic structure of $LiCH_3COO \cdot 2H_2O$ projected
along its a_0 axis.   Right-hand axes.

Fig. XIVB,102b (right).   A packing drawing of the orthorhombic structure of
$LiCH_3COO \cdot 2H_2O$ seen along its a_0 axis.   The methyl groups are black, the lithium
atoms small and hook shaded.   Line shaded carboxyl oxygen atoms are on either side
of the pie-shaped and cross shaded carbon atoms.   Water molecules are heavily ringed
and dotted.

The resulting structure is shown in Figure XIVB,102. The lithium atoms
have six oxygen neighbors, four of which belong to the acetate ions, with
$Li-O = 2.27$ A. The two $Li-H_2O$ distances are 2.57 A. In the acetate ions,
$C(1)-C(2) = 1.55$ A., $C(2)-O(1) = 1.22$ A., and $C(2)-O(2) = 1.33$ A.
There are thought to be weak hydrogen bonds of lengths 3.08 and 3.14 A.
between these anions. This is a layer-like structure in which lithium acetate layers are separated along the c_0 axis by layers of water.

XIV,b70.   Crystals of *cupric acetate monohydrate*, $Cu(CH_3COO)_2 \cdot H_2O$,
and of the corresponding *chromous* salt, $Cr(CH_3COO)_2 \cdot H_2O$, have the same
structure. They are monoclinic with eight molecules in cells of the dimensions:

For $Cu(CH_3COO)_2 \cdot H_2O$:
$a_0 = 13.15$ A.; $b_0 = 8.52$ A.; $c_0 = 13.90$ A.; $\beta = 117°0'$
For $Cr(CH_3COO)_2 \cdot H_2O$:
$a_0 = 13.15$ A.; $b_0 = 8.55$ A.; $c_0 = 13.94$ A.; $\beta = 117°0'$

All atoms are in general positions of C_{2h}^6 $(C2/c)$:

$(8f)$ $\pm (xyz; \ x,\bar{y},z+{}^1/_2; \ x+{}^1/_2,y+{}^1/_2,z; \ x+{}^1/_2,{}^1/_2-y,z+{}^1/_2)$

TABLE XIVB,43
Parameters of the Atoms in Cupric Acetate Monohydrate

Atom	x	y	z
Cu	0.450	0.088	0.044
O(1)	0.677	−0.100	0.101
O(2)	0.598	0.050	0.175
O(3)	0.492	−0.250	0.017
O(4)	0.400	−0.108	0.087
H₂O	0.367	0.200	0.132
C(1)	0.697	−0.025	0.192
C(2)	0.807	−0.008	0.296
C(3)	0.427	−0.250	0.063
C(4)	0.385	−0.400	0.095

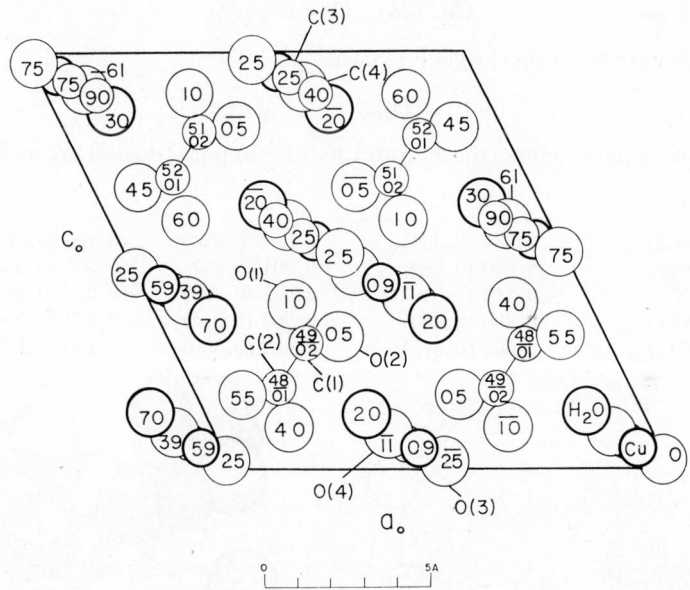

Fig. XIVB,103. The monoclinic structure of $Cu(CH_3COO)_2 \cdot H_2O$ projected along its
b_0 axis.   Left-hand axes.

The parameters of Table XIVB,43 have been established for the cupric salt
and it has been shown that the same values serve to explain the data from
the chromous compound.

These lead to a structure (Fig. XIVB,103) in which pairs of metal atoms
are in relatively close contact (Cu–Cu = 2.64 A.). It is accordingly considered that the molecule is more appropriately written in the doubled form:

$R_2(CH_3COO)_4 \cdot 2H_2O$. Besides its near metal neighbor, each copper atom, for instance, has one water molecule at a distance of 2.20 A. and four carboxyl oxygens ca. 1.97 A. away. The separate acetate groups are planar, with C–C = 1.52 A. and C–O between 1.29 and 1.36 A. Between the carboxyl and the water oxygen atoms, O–O = 2.82 or 2.89 A.

XIV,b71. Two moderately well agreeing structures have been established for crystals of *copper diacetato diammine*, $Cu(CH_3COO)_2 \cdot 2NH_3$. They are monoclinic with a bimolecular cell of the dimensions:

$$a_0 = 5.51 \pm 0.02 \text{ A.}; \quad b_0 = 10.42 \pm 0.04 \text{ A.}; \quad c_0 = 7.44 \pm 0.03 \text{ A.}$$
$$\beta = 106°30'$$

The space group is C_{2h}^5 ($P2_1/c$) with atoms in the positions:

$$Cu: (2a) \quad 000; 0\ {}^1/_2\ {}^1/_2$$

The other atoms are in the general positions:

$$(4e) \quad \pm(xyz;\ x,{}^1/_2-y,z+{}^1/_2)$$

The chosen parameters (those from 1963: BS in parentheses) are as follows:

Atom	x	y	z
O(1)	0.335 (0.335)	0.097 (0.093)	0.109 (0.117)
O(2)	0.099 (0.082)	0.261 (0.257)	0.002 (0.017)
N	−0.112 (−0.077)	0.025 (0.020)	0.227 (0.243)
C(1)	0.290 (0.302)	0.209 (0.214)	0.073 (0.081)
C(2)	0.531 (0.513)	0.302 (0.309)	0.120 (0.133)

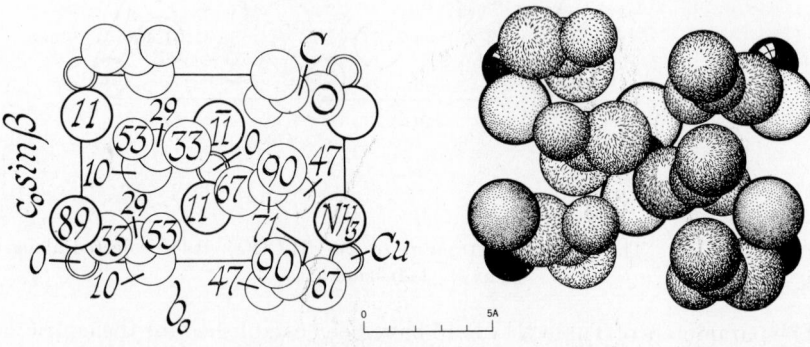

Fig. XIVB,104a (left).   The monoclinic structure of $Cu(CH_3COO)_2 \cdot 2NH_3$ projected along its a_0 axis.   Right-hand axes.

Fig. XIVB,104b (right).   A packing drawing of the monoclinic structure of $Cu(CH_3COO)_2 \cdot 2NH_3$ viewed along its a_0 axis.   The copper atoms are black.   The carboxyl oxygen atoms are line shaded; the carbon atoms are slightly smaller.   The ammonia molecules are dotted and heavily ringed.

The atomic arrangement is shown in Figure XIVB,104. Atoms of copper are surrounded by two O(1) atoms and the nitrogen atoms of the two ammonia molecules at the corners of a slightly distorted square, with Cu–O(1) = 2.07 A., Cu–N = 1.97 A., and O(1)–Cu–N = 92°. The O(2) atoms are considerably more distant, with Cu–O(2) = 2.77 A. In the acetate groups, C(1)–C(2) = 1.59 A., C(1)–O(1) = 1.21 A., and C(1)–O(2) = 1.17 A. (according to 1963: S,A&M).

XIV,b72. Crystals of *zinc acetate dihydrate*, Zn(CH$_3$COO)$_2$·2H$_2$O, are monoclinic with a tetramolecular unit of the dimensions:

$$a_0 = 14.50 \text{ A.}; \quad b_0 = 5.32 \text{ A.}; \quad c_0 = 11.02 \text{ A.}; \quad \beta = 100°0'$$

TABLE XIVB,44
Positions and Parameters of the Atoms in Zinc Acetate Dihydrate

Atom	Position	x	y	z
Zn	(4e)	0	0.125	$^1/_4$
O(1)	(8f)	0.071	0.200	0.095
O(2)	(8f)	0.103	0.423	0.267
H$_2$O	(8f)	0.088	−0.150	0.354
C(1)	(8f)	0.113	0.400	0.145
C(2)	(8f)	0.183	0.527	0.083

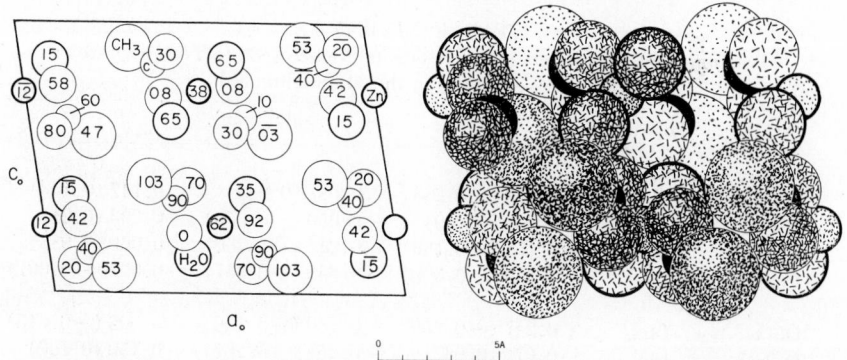

Fig. XIVB,105a (left).   The monoclinic structure of Zn(CH$_3$COO)$_2$·2H$_2$O projected along its b_0 axis. Left-hand axes.

Fig. XIVB,105b (right).   A packing drawing of the monoclinic structure of Zn-(CH$_3$COO)$_2$·2H$_2$O seen along its b_0 axis. The carboxyl carbon atoms are black; the methyl groups are large and dotted. Zinc atoms are small, dotted, and heavily outlined. Of the line shaded oxygen atoms, those belonging to the water molecules are heavily ringed.

Atoms are in the following positions of C_{2h}^6 $(C2/c)$:

(4e) $\pm (0\ u\ ^1/_4;\ ^1/_2, u + ^1/_2, ^1/_4)$

(8f) $\pm (xyz;\ x, \bar{y}, z + ^1/_2;\ x + ^1/_2, y + ^1/_2, z;\ x + ^1/_2, ^1/_2 - y, z + ^1/_2)$

with the parameters of Table XIVB,44.

The structure that results is shown in Figure XIVB,105. Each zinc atom has four carboxyl oxygen neighbors at distances of 2.17 and 2.18 A., and two water oxygens 2.14 A. away. The close approach to one another of the carboxyl oxygens of an acetate ion (2.21 A.) prevents this sixfold coordination from being near to that of a regular octahedron. Within the planar acetate ions, C(1)–C(2) = 1.48 A., C(1)–O(1) = 1.30 A., and C(1)–O(2) = 1.38 A.

XIV,b73.   The *tetrahydrate* of *nickel acetate*, $Ni(CH_3COO)_2 \cdot 4H_2O$, has a bimolecular monoclinic cell with the dimensions:

$$a_0 = 4.75\ \text{A.};\quad b_0 = 11.77\ \text{A.};\quad c_0 = 8.44\ \text{A.};\quad \beta = 93°36'$$

Atoms are in the following positions of C_{2h}^5 $(P2_1/c)$:

(2a) $000;\ 0\ ^1/_2\ ^1/_2$

(4e) $\pm (xyz;\ x, ^1/_2 - y, z + ^1/_2)$

The determined parameters are those of Table XIVB,45.

TABLE XIVB,45
Positions and Parameters of the Atoms in Nickel Acetate Tetrahydrate
and (in parentheses) in the Magnesium Salt

Atom	Position	x	y	z
Ni(Mg)	(2a)	0	0	0
H₂O(1)	(4e)	−0.273 (−0.283)	0.087 (0.083)	0.147 (0.133)
H₂O(2)	(4e)	0.233 (0.200)	−0.050 (−0.050)	0.203 (0.217)
C(1)	(4e)	−0.158 (−0.183)	−0.247 (−0.250)	0.060 (0.083)
C(2)	(4e)	−0.333 (−0.300)	−0.350 (−0.367)	−0.007 (−0.001)
O(1)	(4e)	−0.247 (−0.250)	−0.150 (−0.150)	0.008 (−0.015)
O(2)	(4e)	0.075 (0.067)	−0.258 (−0.267)	0.150 (0.150)

In this structure (Fig. XIVB,106) each metal atom is surrounded by an octahedron of oxygen atoms, two of which are carboxylic, the other four being from water. The separations are: Ni–H₂O = 2.06 and 2.11 A., and Ni–O(1) = 2.12 A. In the planar acetate ions, C(1)–C(2) = 1.56 A., C(1)–

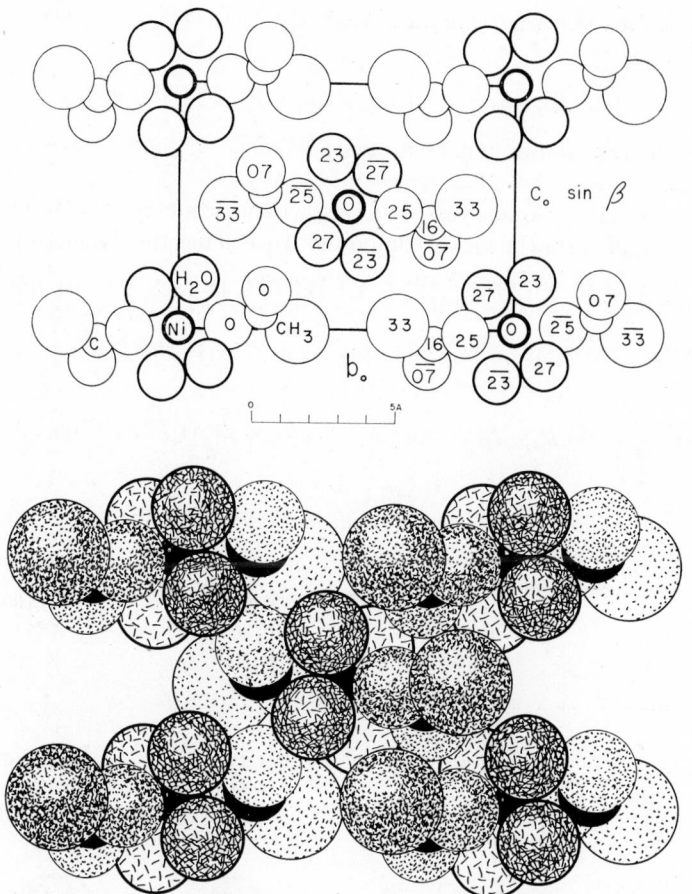

Fig. XIVB,106a (top). The monoclinic structure of $Ni(CH_3COO)_2 \cdot 4H_2O$ projected along its a_0 axis. Left-hand axes.

Fig. XIVB,106b (bottom). A packing drawing of the monoclinic structure of $Ni(CH_3COO)_2 \cdot 4H_2O$ viewed along its a_0 axis. Carboxyl carbon atoms are the larger, nickel atoms the smaller black circles; but both show only incompletely. Methyl groups are the large, oxygen atoms the smaller dotted circles. Water molecules are heavily outlined and line shaded.

$O(1) = 1.29$ A., and $C(1)$–$O(2) = 1.31$ A. There are considered to be hydrogen bonds between the second carboxyl oxygen $O(2)$ and $H_2O(2)$ [2.59 A.], between $O(2)$ and $H_2O(1)$ [2.64 A.], and between $O(1)$ and $H_2O(1)$ [2.66 A.].

Magnesium acetate tetrahydrate, $Mg(CH_3COO)_2 \cdot 4H_2O$, is isostructural, with the unit:

$$a_0 = 4.75 \text{ A.}; \quad b_0 = 11.79 \text{ A.}; \quad c_0 = 8.52 \text{ A.}; \quad \beta = 94°54'$$

The parameters found for its atoms are those in parentheses in Table XIVB,45.

The cobalt salt also has this structure with parameters said to be about the same as those for the nickel compound. Its cell has the dimensions:

$$a_0 = 4.77 \text{ A.}; \quad b_0 = 11.85 \text{ A.}; \quad c_0 = 8.42 \text{ A.}; \quad \beta = 94°30'$$

XIV,b74. Crystals of *uranium tetraacetate*, $U(CH_3COO)_4$, are monoclinic with a tetramolecular cell of the dimensions:

$$a_0 = 17.80 \text{ A.}; \quad b_0 = 8.35 \text{ A.}; \quad c_0 = 8.33 \text{ A.}; \quad \beta = 106°50'$$

TABLE XIVB,46
Parameters of Atoms in $U(CH_3COO)_4$

Atom	x	y	z
O(1)	0.033	0.267	0.100
O(2)	0.075	0.208	−0.117
O(3)	0.083	−0.100	0.083
O(4)	0.135	−0.136	−0.120
C(1)	0.067	0.309	−0.010
C(2)	0.100	0.473	0.007
C(3)	0.145	−0.117	0.038
C(4)	0.222	−0.124	0.173

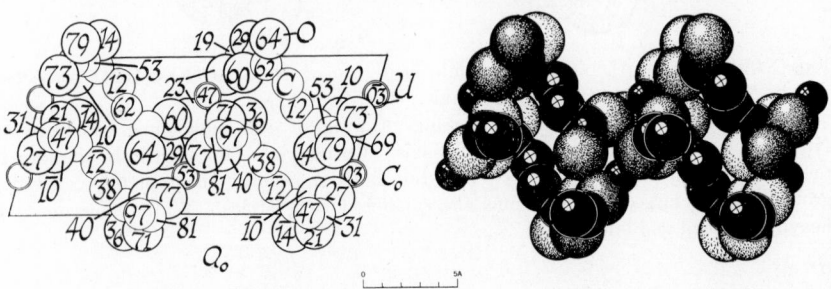

Fig. XIVB,107a (left).   The monoclinic structure of $U(CH_3COO)_4$ projected along its b_0 axis.   Right-hand axes.

Fig. XIVB,107b (right).   A packing drawing of the monoclinic $U(CH_3COO)_4$ arrangement seen along its b_0 axis.   The uranium atoms are the small, the carbon atoms the somewhat larger black circles.   The atoms of oxygen are large and line shaded.

The space group has been chosen as C_{2h}^6 ($C2/c$). Atoms of uranium are in the special positions:

$$(4e) \quad \pm (0\ u\ ^1/_4;\ ^1/_2, u+^1/_2, ^1/_4) \qquad \text{with } u = 0.032$$

All other atoms are in the general positions:

$$(8f) \quad \pm (xyz;\ x, \bar{y}, z+^1/_2;\ x+^1/_2, y+^1/_2, z;\ x+^1/_2, ^1/_2-y, z+^1/_2)$$

with the parameters of Table XIVB,46.

This arrangement is shown in Figure XIVB,107. Each acetate ion is near to two uranium atoms while a uranium atom has 10 oxygen neighbors in the form of an antiprism with centered square faces. Two acetate neighbors have both their oxygens near to a uranium, with U–O = 2.52 and 2.80 A.; the other six oxygens are furnished by six acetate ions, with U–O = 2.52 A.

XIV,b75. Crystals of *sodium uranyl acetate*, $Na(UO_2)(CH_3COO)_3$, are cubic with a unit that contains four molecules and has the edge length:

$$a_0 = 10.670\ A.$$

Atoms are in the following positions of T^4 ($P2_13$):

$(4a) \quad uuu;\ u+^1/_2, ^1/_2-u, \bar{u};\ ^1/_2-u, \bar{u}, u+^1/_2;\ \bar{u}, u+^1/_2, ^1/_2-u$

$(12b) \quad xyz;\ x+^1/_2, ^1/_2-y, \bar{z};\ y+^1/_2, ^1/_2-z, \bar{x};\ z+^1/_2, ^1/_2-x, \bar{y};\ \text{tr}$

with the positions and recently redetermined parameters of Table XIVB,47.

The resulting UO_2 groups are linear, with U–O = 2.18 and 2.25 A., but each uranium also has about it six acetate oxygen atoms not much more

TABLE XIVB,47

Positions and Parameters of the Atoms in Sodium Uranyl Acetate

Atom	Position	x	$\sigma(x)$	y	$\sigma(y)$	z	$\sigma(z)$
U	(4a)	0.4292	0.0003	0.4292	0.0003	0.4292	0.0003
Na	(4a)	0.8289	0.0006	0.8289	0.0006	0.8289	0.0006
O(1)	(4a)	0.5210	0.0020	0.5210	0.0020	0.5210	0.0020
O(2)	(4a)	0.3360	0.0020	0.3360	0.0020	0.3360	0.0020
O(3)	(12b)	0.3820	0.0020	0.2910	0.0010	0.6080	0.0010
O(4)	(12b)	0.5510	0.0010	0.2410	0.0010	0.5000	0.0020
C(1)	(12b)	0.4820	0.0010	0.2290	0.0020	0.5890	0.0010
C(2)	(12b)	0.5090	0.0020	0.1180	0.0020	0.6830	0.0010
H(1)	(12b)	0.43	—	0.06	—	0.67	—
H(2)	(12b)	0.52	—	0.15	—	0.77	—
H(3)	(12b)	0.58	—	0.07	—	0.65	—

distant, with U–O = 2.47–2.51 A. Each Na$^+$ ion likewise is surrounded by six acetate oxygen atoms at 2.38 and 2.40 A. Within the planar acetate ions C–C = 1.52 A. and C–O = 1.26 and 1.29 A.

The following two isomorphous compounds have the cube edges:

$$Na(NpO_2)(CH_3COO)_3: \ a_0 = 10.680 \ A.$$
$$Na(PuO_2)(CH_3COO)_3: \ a_0 = 10.664 \ A.$$

There are also cubic propionates that seem to have this type of structure. Two of these have the cube edges:

$$K(UO_2)(C_2H_5COO)_3: \ a_0 = 11.52 \ A.$$
$$NH_4(UO_2)(C_2H_5COO)_3: \ a_0 = 11.64 \ A.$$

XIV,b76. *Silver uranyl acetate*, $Ag(UO_2)(CH_3COO)_3 \cdot xH_2O$, has a structure which in some ways resembles that of the sodium salt (**XIV,b75**) in spite of the fact that it appears to contain some water of crystallization. Its crystals are tetragonal rather than cubic and their big cell, which contains sixteen molecules, has the edges:

$$a_0 = 12.98 \ A., \qquad c_0 = 28.10 \ A.$$

All atoms have been placed in general positions of C_{4h}^6 ($I4_3/a$):

$$(16f) \ \pm (xyz; \ y+\tfrac{1}{4}, \tfrac{1}{4}-x, z+\tfrac{3}{4}; \ \tfrac{1}{2}-x, \bar{y}, z+\tfrac{1}{2}; \ \tfrac{1}{4}-y, x+\tfrac{3}{4}, z+\tfrac{1}{4}); B.C$$

TABLE XIVB,48
Parameters of the Atoms in Silver Uranyl Acetate[a]

Atom	x	y	z
U	0.210	0.025	0.18
Ag	0.175	0.225	0.04
O(1)	0.17	0.89	0.15
O(2)	0.25	0.16	0.21
C(1)	0.27	0.33	0.19
C(2)	0.50	0.44	0.17
C(3)	0.48	0.38	0.28
O(1a)	0.31	0.52	0.16
O(1b)	0.32	0.48	0.23
O(2a)	0.46	0.62	0.12
O(2b)	0.61	0.56	0.15
O(3a)	0.48	0.54	0.29
O(3b)	0.62	0.50	0.25

[a] The origin of the description used here is displaced (0 $\tfrac{1}{2}$ $\tfrac{1}{8}$) with respect to that employed in the International Tables.

with the parameters of Table XIVB,48. Positions for the light atoms, insofar as they have been selected, have been based on packing considerations. The amount of water in these crystals was not established and it is apparent that more work is called for.

XIV,b77. At room temperature *basic beryllium acetate*, $Be_4O(CH_3COO)_6$, is cubic with a unit containing eight molecules and having the edge length:

$$a_0 = 15.72 \text{ A.}$$

Atoms are in the following positions of T_h^4 ($Fd3$) with the recently redetermined parameters stated below:

O(1): (8a)   000; $^1/_4\,^1/_4\,^1/_4$; F.C.
Be: (32e) $uuu;$ $u\bar{u}\bar{u};$
             $\bar{u}u\bar{u};$ $\bar{u}\bar{u}u;$
             $^1/_4-u,^1/_4-u,^1/_4-u;\ ^1/_4-u,u+^1/_4,u+^1/_4;$
             $u+^1/_4,^1/_4-u,u+^1/_4;\ u+^1/_4,u+^1/_4,^1/_4-u;$ F.C.
             with $u = -0.0611$
CH₃: (48f) $v00; \bar{v}00; v+^1/_4,^1/_4,^1/_4; ^1/_4-v,^1/_4,^1/_4;$ tr; F.C.
             with $v = 0.2950$
C: (48f)   with $v = 0.1997$
O(2): (96g) $xyz; x\bar{y}\bar{z};\ ^1/_4-x,^1/_4-y,^1/_4-z;\ ^1/_4-x,y+^1/_4,z+^1/_4;$
             $\bar{x}\bar{y}z; \bar{x}y\bar{z};\ x+^1/_4,^1/_4-y,z+^1/_4; x+^1/_4,y+^1/_4,^1/_4-z;$ tr; F.C
             with $x = -0.1617, y = -0.0576, z = -0.0386$

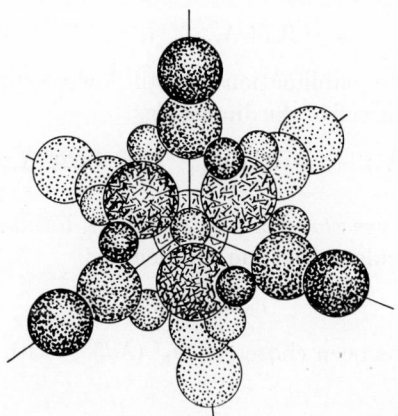

Fig. XIVB,108.   A packing drawing showing the approximate shape of the molecule of $Be_4O(CH_3COO)_6$ in its crystals.   In this representation the oxygen atoms are the small, the carbon atoms the large dotted circles; the beryllium atoms are large and line shaded.

Positions given the hydrogen atoms correspond to a rotational disorder in the orientation of the methyl radicals. It is expressed by placing half atoms in $(96g)$ with the three sets of parameters:

$$^1/_2\,H(1)\colon x = -0.038,\; y = -0.052,\; z = -0.317$$
$$^1/_2\,H(2)\colon x = 0.025,\; y = 0.061,\; z = 0.317$$
$$^1/_2\,H(3)\colon x = 0.065,\; y = 0.008,\; z = 0.317$$

This structure is a diamond arrangement of $Be_4O(CH_3COO)_6$ molecules having the shape shown in Figure XIVB,108. Within this molecule the central oxygen atom is surrounded by a tetrahedron of beryllium atoms with Be–O(1) = 1.666 A., and beyond these by twelve carboxyl oxygen atoms with O(1)–O(2) = 2.82 A. Each beryllium atom, too, is tetrahedrally surrounded by four oxygen atoms, three of which are carboxyl O(2), with Be–O(2) = 1.624 A. In the acetate radical, CH_3–C = 1.500 A. and C–O = 1.264 A.

As the temperature is raised to 40°C., $(hk0)$ and $(kh0)$ reflections gradually become equal, with the resulting greater symmetry of x-ray pattern which this implies. It has been found that this corresponds to a slight shift in the positions of the carboxyl oxygen atoms which move from the positions just stated to ones defined by the slightly different parameters: $x = -0.17, y = z = -0.05$.

At higher temperatures the cubic form transforms to one of low symmetry. Above 160°C. it is hexagonal, the unimolecular unit cell being rhombohedral with

$$a_0 = 9.74 \text{ A.}, \qquad \alpha = 54°20'$$

Material obtained by sublimation has still lower symmetry, with a tetramolecular monoclinic cell of the dimensions:

$$a_0 = 13.78 \text{ A.;} \quad b_0 = 9.26 \text{ A.;} \quad c_0 = 16.24 \text{ A.;} \quad \beta = 98°55'$$

XIV,b78. *Zinc oxyacetate*, $Zn_4O(CH_3COO)_6$, forms cubic crystals which have an eight-molecule unit of the edge length:

$$a_0 = 16.45 \text{ A.}$$

The space group has been chosen as O_h^7 ($Fd3m$) with atoms placed in the positions:

Zn: (32e) $uuu;$ $u\bar{u}\bar{u};$
               $^1/_4-u,^1/_4-u,^1/_4-u;\; ^1/_4-u,u+^1/_4,u+^1/_4;$ tr; F.C.
               with $u = -0.069$

O(1): (8a) $000;\; ^1/_4\,^1/_4\,^1/_4;$ F.C.

C(1): (48f) $u00; \bar{u}00; u+{}^1/_4,{}^1/_4,{}^1/_4; {}^1/_4-u,{}^1/_4,{}^1/_4;$ tr; F.C.
with $u = 0.220$

C(2): (48f)   with $u = 0.314$

O(2): (96g) $uuv;$ $u\bar{u}\bar{v};$

                     $\bar{u}u\bar{v};$ $\bar{u}\bar{u}v;$

                     ${}^1/_4-u,{}^1/_4-u,{}^1/_4-v; {}^1/_4-u,u+{}^1/_4,v+{}^1/_4;$

                     $u+{}^1/_4,{}^1/_4-u,v+{}^1/_4; u+{}^1/_4,u+{}^1/_4,v+{}^1/_4;$ tr; F.C.
with $u = -0.047, v = -0.185$

In this structure the central O(1) atom is tetrahedrally surrounded by four zinc atoms, each of which is in turn tetrahedrally enveloped by one O(1) and three O(2) atoms [Zn–O(1) = 1.96 A. and Zn–O(2) = 1.98 A.]. In the acetate groups, C(1)–C(2) = 1.55 A., C(2)–O(2) = 1.24 A., and O(2)–C(2)–O(2) = 125°.

This is very nearly the structure of $Be_4O(CH_3COO)_6$ (**XIV,b77**). In spite of the higher symmetry of the zinc salt, atomic coordinates are the same for all but the acetate oxygens, and the parameters are not very different in the two cases. These oxygen atoms also are not very differently placed; the beryllium compound departs from holohedral symmetry only to the degree that $y \neq z$ for oxygen.

XIV,b79. *Dipotassium nitroacetate*, $K_2(NO_2CHCOO)$, forms orthorhombic crystals that have a bimolecular unit of the edge lengths:

$$a_0 = 10.28 \text{ A.}; \quad b_0 = 7.49 \text{ A.}; \quad c_0 = 3.54 \text{ A.}$$

The space group is C_{2v}^2 (*Pb2m*) with all atoms except those of potassium in special positions:

$$(2a) \quad uv0; \bar{u},v+{}^1/_2,0$$

TABLE XIVB,49. Positions and Parameters of the Atoms in Dipotassium Nitroacetate

Atom	Position	x	y	z
K(1)	(2b)	0.033	0.579	$^1/_2$
K(2)	(2b)	0.369	0.275	$^1/_2$
N	(2a)	0.333	0.645	0
C(1)	(2a)	0.200	0.921	0
C(2)	(2a)	0.316	0.829	0
O(1)	(2a)	0.084	0.842	0
O(2)	(2a)	0.201	0.090	0
O(3)	(2a)	0.242	0.528	0
O(4)	(2a)	0.451	0.598	0
H (assumed)	(2a)	0.386	0.903	0

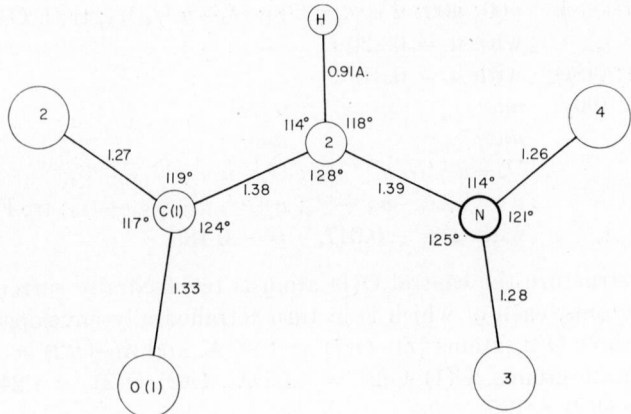

Fig. XIVB,109.　Bond dimensions in the nitroacetate anion of $K_2(NO_2CHCOO)$.

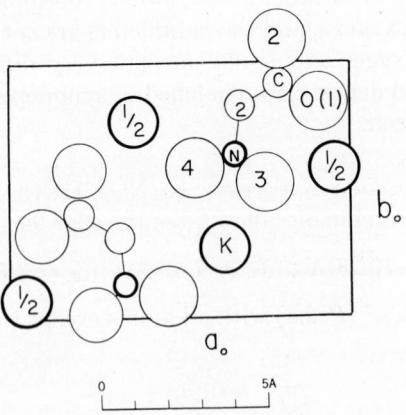

Fig. XIVB,110.　The orthorhombic structure of $K_2(NO_2CHCOO)$ projected along its c_0 axis.　Left-hand axes.

The potassium atoms are in

$$(2b) \quad u\,v\,{}^1\!/_2; \; \bar{u},v+{}^1\!/_2,{}^1\!/_2$$

The chosen parameters are listed in Table XIVB,49.

This structure gives a strictly planar bivalent anion with the bond angles and lengths indicated in Figure XIVB,109. The layered packing of these and the potassium ions is shown in Figure XIVB,110. Shortest interionic distances, between potassium and one or another of the four oxygen atoms, range from 2.70 to 3.11 A.; each potassium ion has eight such oxygen neighbors.

XIV,b80. The compound *manganese dichlorophosphate diethylacetate,* $Mn(PO_2Cl_2)_2(CH_3COOC_2H_5)_2$, is monoclinic with a tetramolecular unit of the dimensions:

$$a_0 = 14.27 \text{ A.}; \quad b_0 = 13.87 \text{ A.}; \quad c_0 = 10.04 \text{ A.}; \quad \beta = 95°56'$$

The space group is C_{2h}^5 $(P2_1/a)$ with all atoms in the positions:

$$(4e) \quad \pm (xyz; \; x+^1/_2, ^1/_2-y, z)$$

The determined parameters are listed in Table XIVB,50.

TABLE XIVB,50
Parameters of the Atoms in $Mn(PO_2Cl_2)_2(CH_3COOC_2H_5)_2$

Atom	x	σ_x, A.	y	σ_y, A.	z	σ_z, A.
Mn	0.5091	0.0033	0.4367	0.0037	0.2466	0.0029
Cl(1)	0.6718	0.0100	0.6970	0.0101	0.1582	0.0088
Cl(2)	0.4661	0.0091	0.7651	0.0097	0.0675	0.0081
Cl(3)	0.2818	0.0130	0.6313	0.0130	0.3937	0.0110
Cl(4)	0.2346	0.0110	0.4266	0.0120	0.4930	0.0100
P(1)	0.5481	0.0068	0.6458	0.0070	0.0725	0.0056
P(2)	0.3449	0.0070	0.5070	0.0071	0.4520	0.0058
O(1)	0.5128	0.0190	0.5774	0.0200	0.1651	0.0170
O(2)	0.5639	0.0170	0.6186	0.0180	−0.0653	0.0150
O(3)	0.3813	0.0180	0.4655	0.0190	0.3351	0.0160
O(4)	0.4059	0.0180	0.5230	0.0190	0.5774	0.0160
O(5)	0.5067	0.0190	0.2846	0.0210	0.3208	0.0170
O(6)	0.4779	0.0200	0.1402	0.0210	0.4032	0.0180
O(7)	0.6376	0.0210	0.4050	0.0220	0.1568	0.0190
O(8)	0.7752	0.0220	0.3804	0.0230	0.0695	0.0200
C(1)	0.4679	0.0450	0.2737	0.0480	0.5470	0.0400
C(2)	0.4890	0.0300	0.2388	0.0300	0.4145	0.0270
C(3)	0.4918	0.0350	0.0977	0.0370	0.2778	0.0310
C(4)	0.4689	0.0450	−0.0060	0.0460	0.2933	0.0400
C(5)	0.7419	0.0430	0.2841	0.0460	0.2411	0.0390
C(6)	0.7112	0.0300	0.3629	0.0300	0.1562	0.0250
C(7)	0.7403	0.0390	0.4567	0.0410	−0.0347	0.0360
C(8)	0.8253	0.0490	0.4932	0.0500	−0.0839	0.0440

In this structure, of which half the molecules are shown in Figure XIVB,-111, the $PO_2Cl_2^-$ anions are tetrahedral, with P–O = 1.46 A. and P–Cl = 2.01 A. The manganese atoms are octahedrally surrounded by four phosphate and two acetate oxygens, with Mn–O = 2.11–2.24 A. In the acetate molecules, C–C = 1.42–1.54 A., and C–O = 1.18–1.38 A.

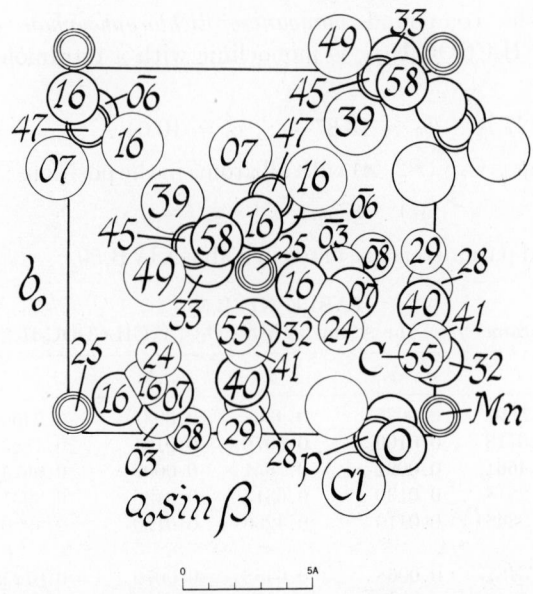

Fig. XIVB,111. A projection along its c_0 axis of half the contents of the monoclinic unit of $Mn(PO_2Cl_2)_2(CH_3COOC_2H_5)_2$. The rest of the atoms are to be obtained by inversion through the origin. Right-hand axes.

XIV,b81. Crystals of *acetyl choline bromide*, $CH_3CO \cdot O \cdot (CH_2)_2 \cdot N(CH_3)_3$-Br, are monoclinic with a tetramolecular unit of the dimensions:

$$a_0 = 11.10 \text{ A.}; \quad b_0 = 13.67 \text{ A.}; \quad c_0 = 7.18 \text{ A.}; \quad \beta = 110°0'$$

The space group has been chosen as C_2^2 ($P2_1$) though, according to a recent paper, it has the higher symmetry C_{2h}^5 ($P2_1/a$). In terms of the lower symmetry group all atoms are in the general positions:

$$(2a) \quad xyz; \ \bar{x}, y + {}^1/_2, \bar{z}$$

In the structure that gives best agreement with the intensity data and satisfactory interatomic distances, the two molecular ions that are present have different shapes. The parameters are those of Table XIVB,51.

As can be seen from Figure XIVB,112, the two ions differ from one another by rotation of the substituted ammonium part around the C(3)–C(4) bond. As a result, one is more extended than the other which has a quasiring shape. The structure as a whole is shown in Figure XIVB,113. There are rather short C(3)–O(2) separations of ca. 2.95 A. involving adjacent

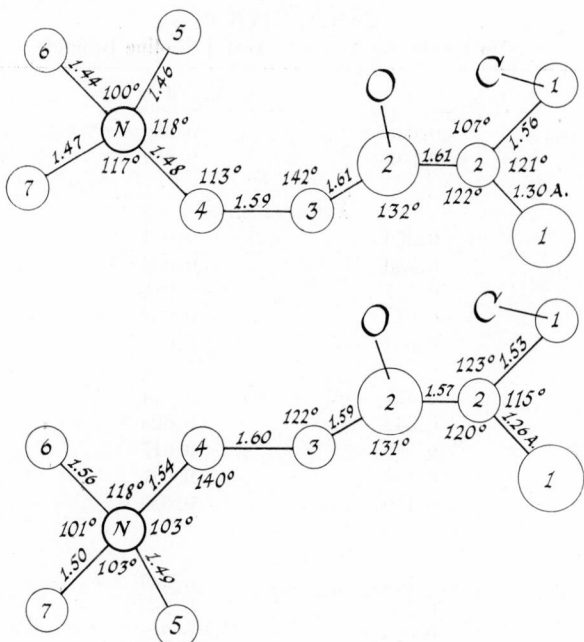

Fig. XIVB,112a (top).   Bond dimensions of the molecular anion of acetyl choline bromide in its "ring form."

Fig. XIVB,112b (bottom).   Bond dimensions of the "extended" molecular anion of
acetyl choline bromide.

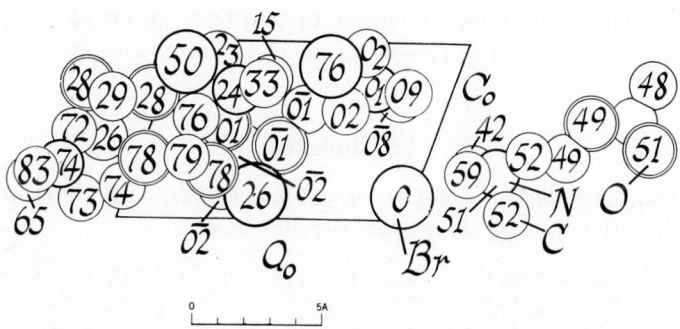

Fig. XIVB,113.   The monoclinic structure of acetyl choline bromide projected along
its b_0 axis.   Right-hand axes.

TABLE XIVB,51
Parameters of the Atoms in Acetyl Choline Bromide

Atom	x	y	z
Br(1)	0.0367	0.000	0.142
Br(2)	0.535	0.264	0.142
		Extended ion	
N	0.250	0.014	0.731
O(1)	0.750	0.008	0.533
O(2)	0.517	−0.008	0.417
C(1)	0.672	−0.017	0.194
C(2)	0.653	−0.017	0.394
C(3)	0.478	−0.014	0.611
C(4)	0.342	0.025	0.611
C(5)	0.339	0.017	0.939
C(6)	0.167	−0.081	0.700
C(7)	0.150	0.089	0.733
		Quasi-ring ion	
N	0.733	0.239	0.742
O(1)	0.269	0.283	0.792
O(2)	0.025	0.283	0.667
C(1)	0.150	0.261	0.431
C(2)	0.164	0.294	0.644
C(3)	0.978	0.244	0.842
C(4)	0.856	0.228	0.906
C(5)	0.744	0.222	0.536
C(6)	0.658	0.153	0.739
C(7)	0.667	0.333	0.722

molecules. The Br(1) ions are closest to two O(2) atoms at distances of 3.36 and 3.43 A.; Br(2) is closest to an O(1) atom, with Br(2)–O(1) = 3.22 A.

7. Chelates

XIV,b82. *Bis ethylene diamine cupric thiocyanate,* $Cu(C_2N_2H_8)_2(SCN)_2$, is triclinic with a unimolecular cell of the dimensions:

$$a_0 = 7.352 \text{ A.}; \quad b_0 = 9.364 \text{ A.}; \quad c_0 = 6.585 \text{ A.}$$
$$\alpha = 86°56'; \quad \beta = 113°23'; \quad \gamma = 125°8'$$

Taking the space group as C_i^1 $(P\bar{1})$ the copper atom is in the origin $(1a)$ 000 and all other atoms are in the general positions $(2i)$ $\pm(xyz)$. The determined parameters are those of Table XIVB,52, the choice of hydrogen positions involving the assumption that they are tetrahedrally distributed around carbon and nitrogen with the distances C–H = 1.075 A. and N–H = 1.005 A.

TABLE XIVB,52
Parameters of Atoms in $Cu(C_2N_2H_8)_2(SCN)_2$

Atom	x	y	z
S	0.5945	0.2303	0.3451
N(1)	0.4142	0.2138	0.6521
N(2)	0.0765	0.1994	−0.1661
N(3)	−0.0403	0.1316	0.1859
C(1)	0.4959	0.2226	0.5294
C(2)	0.1582	0.3588	−0.0149
C(3)	−0.0260	0.2801	0.0922
H(1)	−0.077	0.159	−0.310
H(2)	0.213	0.232	−0.205
H(3)	0.146	0.452	−0.112
H(4)	0.346	0.426	0.115
H(5)	0.039	0.383	0.226
H(6)	−0.207	0.233	−0.037
H(7)	0.096	0.180	0.346
H(8)	−0.207	0.047	0.186

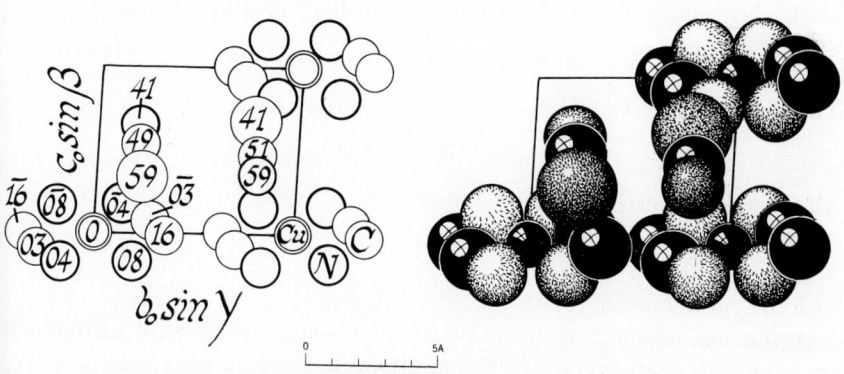

Fig. XIVB,114a (left).   The triclinic structure of $Cu(C_2N_2H_8)_2(SCN)_2$ projected along its a_0 axis.   Right-hand axes.

Fig. XIVB,114b (right).   A packing drawing of the triclinic structure of $Cu(C_2N_2H_8)_2$-$(SCN)_2$ viewed along its a_0 axis.   The copper atoms are the small, the carbon atoms the larger black circles.   The nitrogen atoms are short-line shaded and heavily ringed; the large sulfur circles are dotted.

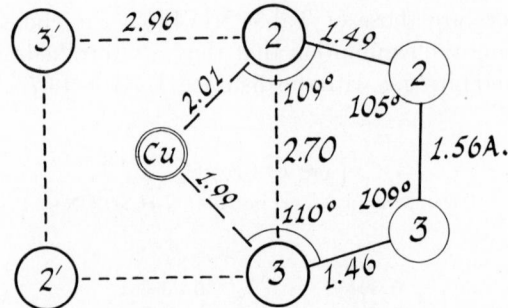

Fig. XVIB,115. Bond dimensions in crystals of $Cu(C_2N_2H_8)_2(SCN)_2$.

In this structure (Fig. XIVB,114) the copper atom is surrounded by four almost equally distant nitrogen atoms, with Cu–N = 2.01 or 1.99 A. Other interatomic distances are shown in Figure XIVB,115. Two sulfur atoms of thiocyanate ions are on a line roughly normal (89°) to the Cu–4N plane, with Cu–S = 3.27 A. These thiocyanate ions are linear, with C–N = 1.16 A. and C–S = 1.62 A. The shortest distance between a thiocyanate and an ethylene diamine nitrogen is 3.09 A.; all other distances between molecules or ions exceed this. The final reliability index for this determination is $R = 0.061$.

XIV,b83. Crystals of *bis ethylene diamine cupric nitrate*, $Cu(C_2N_2H_8)_2$-$(NO_3)_2$, are monoclinic with a bimolecular cell of the dimensions:

$$a_0 = 8.302 \text{ A.}; \quad b_0 = 10.052 \text{ A.}; \quad c_0 = 8.065 \text{ A.}; \quad \beta = 111°6'$$

The space group is C_{2h}^5 $(P2_1/c)$. The copper atoms are in

$$(2a) \quad 000; \ 0 \ ^1/_2 \ ^1/_2$$

All other atoms are in the general positions:

$$(4e) \quad \pm (xyz; \ x, ^1/_2-y, z+^1/_2)$$

with the parameters of XIVB,53.

In this arrangement (Fig. XIVB,116) the copper atoms are surrounded by an approximate square of amino nitrogen atoms, with Cu–N = 2.012 and 2.044 A., and N–Cu–N = 86°. There are two, more distant, nitrate oxygen atoms nearly normal to this plane; for them Cu–O = 2.593 A. In the nitrate ion, N–O = 1.248, 1.259, and 1.267 A.; the other bond dimensions are those of Figure XIVB,117. Between molecules there are considered to be N–H–O bonds with values between 3.01 and 3.07 A.

TABLE XIVB,53
Parameters of Atoms in $Cu(C_2N_2H_8)_2(NO_3)_2$

Atom	x	y	z
C(1)	0.3271	0.1318	0.1770
C(2)	0.3601	−0.0195	0.2007
N(1)	0.1639	0.1567	0.0245
N(2)	0.2000	−0.0904	0.1865
N(3)	0.1687	0.5571	0.1658
O(1)	0.2282	0.6279	0.0743
O(2)	0.2023	0.4338	0.1817
O(3)	0.0960	0.6117	0.2608

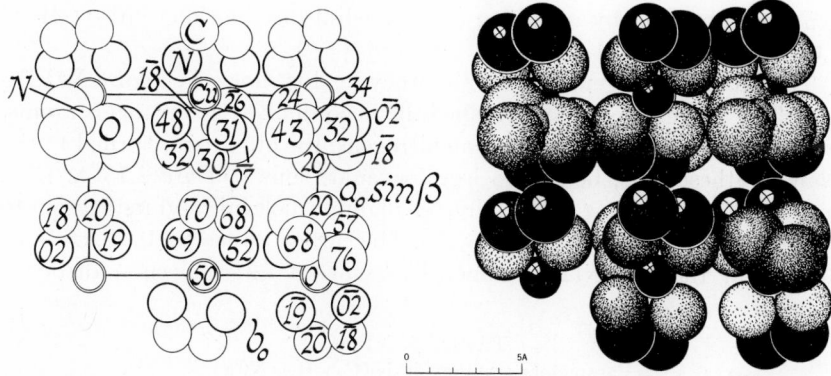

Fig. XIVB,116a (left). The monoclinic structure of $Cu(C_2N_2H_8)_2(NO_3)_2$ projected along its c_0 axis. Right-hand axes.

Fig. XIVB,116b (right). A packing drawing of the monoclinic structure of $Cu(C_2N_2H_8)_2(NO_3)_2$ viewed along its c_0 axis. The carbon atoms are the large, the copper atoms the smaller black circles. Atoms of diamine nitrogen are hook shaded and heavily outlined; the nitrate nitrogen atoms scarcely show within the groups of large line shaded oxygens.

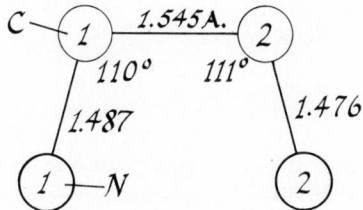

Fig. XIVB,117. Bond dimensions of the ethylene diamine molecule in $Cu(C_2N_2H_8)$-$(NO_3)_2$.

XIV,b84. *trans-bis Ethylene diamine nickel isocyanate,* $Ni(C_2N_2H_8)_2$-$(NCS)_2$, forms monoclinic crystals which have a bimolecular cell of the dimensions:

$$a_0 = 10.28 \text{ A.}; \quad b_0 = 8.26 \text{ A.}; \quad c_0 = 8.88 \text{ A.}; \quad \beta = 121°3'$$

The space group is C_{2h}^5 $(P2_1/a)$ with the nickel atoms in:

$$(2a) \quad 000; \; ^1/_2 \, ^1/_2 \, 0$$

and all others in the general positions:

$$(4e) \quad \pm (xyz; \; x+^1/_2, ^1/_2-y, z)$$

The determined parameters are given in Table XIVB,54. Those for hydrogen were obtained by assuming a tetrahedral distribution, with C–H = 1.075 A. and N–H = 1.005 A.

The atomic arrangement to which they lead is shown in Figure XIVB,-118. The nickel atoms are octahedrally coordinated to nitrogen atoms. Four of these are ethylene diamine nitrogens, the other two, normal to the plane of these four, belong to isothiocyanate ions and are 2.15 A. from nickel. The dimensions of the ethylene diamine molecule and its relation to nickel are shown in Figure XIVB,119. The carbon atoms are 0.34 A. above and below the Ni–4N(2,3) plane. The shortest intermolecular distance,

TABLE XIVB,54
Parameters of Atoms in $Ni(C_2N_2H_8)_2(NCS)_2$

Atom	x	y	z
S	0.0427	−0.4607	0.3318
C(1)	0.0708	−0.3256	0.2155
C(2)	−0.1321	0.0973	0.2148
C(3)	0.0365	0.0846	0.3367
N(1)	0.0874	−0.2291	0.1265
N(2)	−0.1693	−0.0128	0.0639
N(3)	0.1204	0.1155	0.2476
H(1)	−0.177	−0.127	0.097
H(2)	−0.269	0.021	−0.041
H(3)	−0.164	0.220	0.170
H(4)	−0.190	0.059	0.280
H(5)	0.063	−0.035	0.391
H(6)	0.071	0.172	0.441
H(7)	0.226	0.070	0.318
H(8)	0.127	0.235	0.232

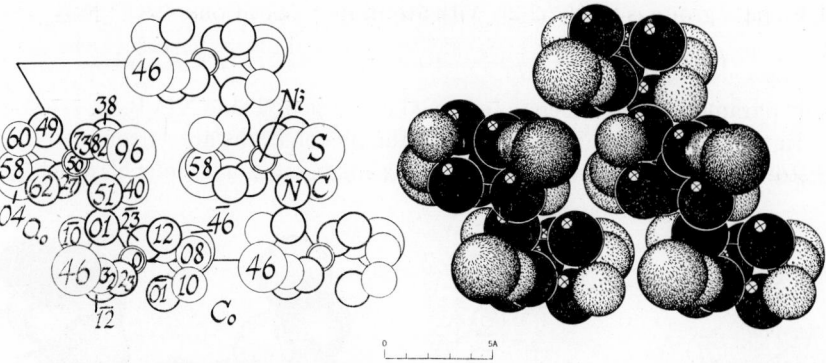

Fig. XIVB,118a (left). The monoclinic structure of $Ni(C_2N_2H_8)_2(NCS)_2$ projected along its b_0 axis. Right-hand axes.

Fig. XIVB,118b (right). A packing drawing of the monoclinic structure of $Ni(C_2N_2H_8)_2(NCS)_2$ viewed along its b_0 axis. The nitrogens are the large black circles; the slightly smaller, black nickel atoms show only poorly. The sulfur atoms are large and line shaded; the smaller carbon atoms are hook shaded.

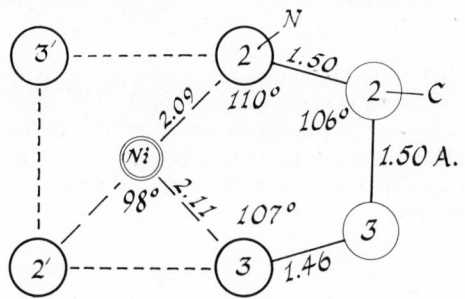

Fig. XIVB,119. Bond dimensions of the $C_2N_2H_8$ molecules and the enclosed nickel atoms in crystals of $Ni(C_2N_2H_8)_2(NCS)_2$.

between sulfur and an $N(2)$ atom, is 3.34 A.; other separations, between an $N(1)$ and $N(2)$ of neighboring molecules, range upwards from 3.54 A. The NCS ion is nearly straight $(N–C–S = 178°)$ with $N–C = 1.20$ A. and $C–S = 1.64$ A.

XIV,b85. Crystals of *2,2',2''-triamino triethylamine–nickel thiocyanate*, $Ni(SCN)_2 \cdot N(C_2H_4NH_2)_3$, are orthorhombic with a tetramolecular unit of the edge lengths:

$$a_0 = 10.821 \text{ A.}; \quad b_0 = 14.719 \text{ A.}; \quad c_0 = 8.620 \text{ A.}$$

The space group is V^4 ($P2_12_12_1$) with atoms in the positions:

(4a) $xyz;$ $^1/_2-x,\bar{y},z+^1/_2;$ $x+^1/_2,^1/_2-y,\bar{z};$ $\bar{x},y+^1/_2,^1/_2-z$

The parameters have been determined as those of Table XIVB,55.

In this structure (Fig. XIVB,120), the nickel atoms are surrounded by a distorted octahedral arrangement of six nitrogen atoms, four of which are

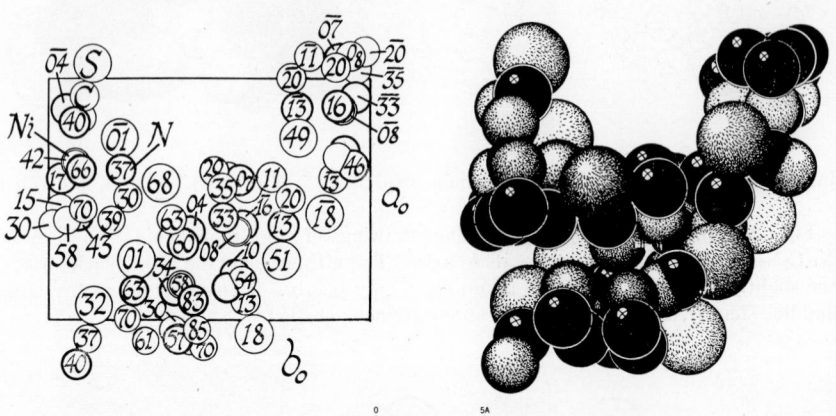

Fig. XIVB,120a (left). The orthorhombic structure of Ni(SCN)$_2$·N(C$_2$H$_4$NH$_2$)$_3$ projected along its c_0 axis. Right-hand axes.

Fig. XIVB,120b (right). A packing drawing of the orthorhombic Ni(SCN)$_2$·N(C$_2$H$_4$NH$_2$)$_3$ arrangement seen along its c_0 axis. The carbon atoms are black; the nickel atoms, which also are black, scarcely show. The atoms of nitrogen are hook shaded and heavily outlined; the sulfur atoms are large and line shaded.

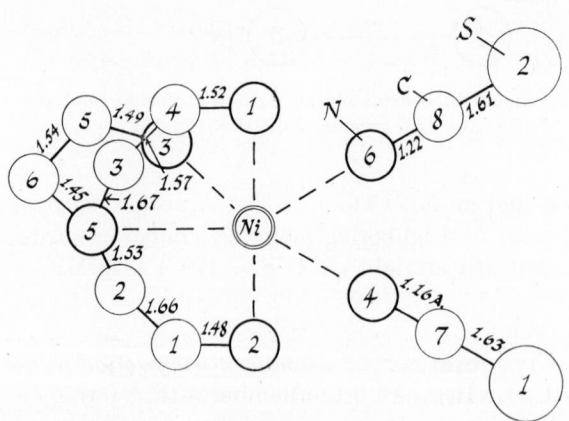

Fig. XIVB,121.   Bond dimensions and atomic associations in crystals of Ni(SCN)$_2$·N(C$_2$H$_4$NH$_2$)$_3$.

TABLE XIVB,55
Parameters of the Atoms in $Ni(SCN)_2 \cdot N(C_2H_4NH_2)_3$

Atom	x	y	z
C(8)	0.0811	0.3852	0.1298
C(7)	0.3266	0.6305	0.0375
C(6)	0.6036	0.3052	0.1137
C(5)	0.5062	0.2464	0.1963
C(4)	0.5477	0.4532	0.3509
C(3)	0.6106	0.4795	0.1962
C(2)	0.6005	0.4393	0.9182
C(1)	0.5189	0.3762	0.7964
N(6)	0.1862	0.4107	0.1041
N(5)	0.5664	0.3971	0.0742
N(4)	0.3674	0.5573	0.0374
N(3)	0.3816	0.2676	0.1325
N(2)	0.3883	0.3945	0.8379
N(1)	0.4077	0.4507	0.3293
S(2)	0.4424	0.1395	0.8186
S(1)	0.2423	0.2724	0.5107
Ni	0.3720	0.4172	0.0853

from the amine molecule and two from thiocyanate groups; their separations Ni–N lie between 2.020 and 2.243 A., with N(5) and the NCS distances being shorter than the others. Other interatomic distances are those of Figure XIVB,121. The final $R = 0.136$.

XIV,b86. *Nickel nitrate tris ethylene diamine*, $Ni(NO_3)_2 \cdot (NH_2CH_2CH_2\text{-}NH_2)_3$, is hexagonal with a bimolecular cell of the edge lengths:

$$a_0 = 8.87 \pm 0.01 \text{ A.}, \quad c_0 = 11.41 \pm 0.02 \text{ A.}$$

The space group is D_6^6 ($P6_322$) with atoms in the positions:

Ni: $(2d)$ $^2/_3\,^1/_3\,^1/_4$; $^1/_3\,^2/_3\,^3/_4$
N(NO$_3$): $(4f)$ $\pm(^1/_3\,^2/_3\,u;\ ^2/_3,\ ^1/_3,\ u+^1/_2)$
with $u = 0.1110$, $\sigma(u) = 0.0015$

All other atoms are in the general positions:

$(12i)$ $xyz;$ $\bar{y},x-y,z;$ $y-x,\bar{x},z;$
$yx\bar{z};$ $\bar{x},y-x,\bar{z};$ $x-y,\bar{y},\bar{z};$
$\bar{x},\bar{y},z+^1/_2;\ y,y-x,z+^1/_2;\ x-y,x,z+^1/_2;$
$\bar{y},\bar{x},^1/_2-z;\ x,x-y,^1/_2-z;\ y-x,y,^1/_2-z$

with the parameters of Table XIVB,56.

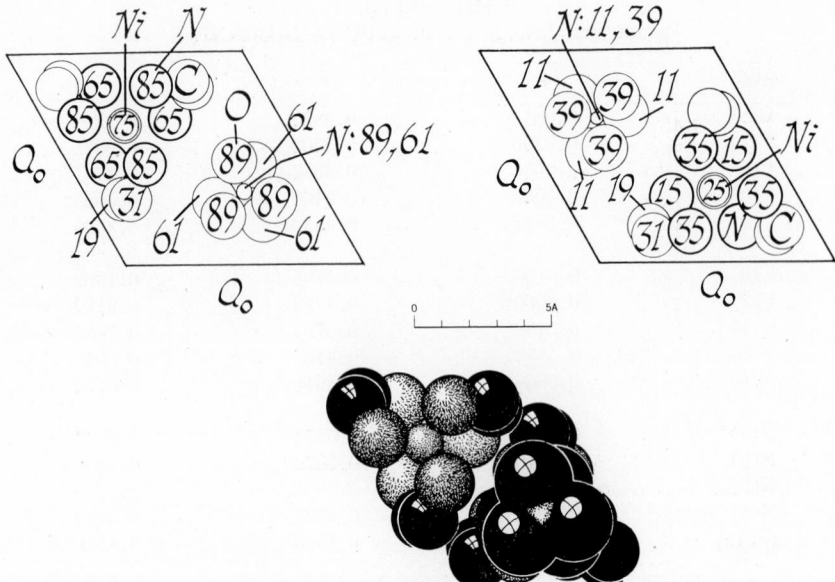

Fig. XIVB,122a (top left).   Half the molecules in the hexagonal unit of $Ni(NO_3)_2 \cdot (C_2N_2H_8)_3$ projected along its c_0 axis.   Right-hand axes.

Fig. XIVB,122b (top right).   The second half of the molecules in the hexagonal unit of $Ni(NO_3)_2 \cdot (C_2N_2H_8)_3$ projected along its c_0 axis.   Right-hand axes.

Fig. XIVB,122c (bottom).   A packing drawing of the hexagonal structure of $Ni(NO_3)_2 \cdot (C_2N_2H_8)_3$ viewed along its c_0 axis.   Atoms of oxygen are the large, of carbon, the smaller black circles.   Nickel atoms are small and dotted; those of nitrogen are line shaded.   One triangular nitrate nitrogen atom is shown.

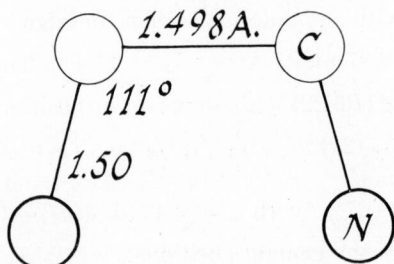

Fig. XIVB,123.   Bond dimensions of the $C_2N_2H_8$ molecule in $Ni(NO_3)_2 \cdot (C_2N_2H_8)_3$.

In the structure (Fig. XIVB,122) each nickel atom has around it six nitrogen atoms, two from each ethylene diamine molecule, with Ni–N = 2.12 A. Bond lengths in the molecule are those of Figure XIVB,123. The

TABLE XIVB,56
Parameters of Atoms in $Ni(NO_3)_2 \cdot (NH_2CH_2CH_2NH_2)_3$

Atom	x	$\sigma(x)$	y	$\sigma(y)$	z	$\sigma(z)$
$N(NH_2)$	0.4588	0.0017	0.1402	0.0015	0.3507	0.0016
C	0.2911	0.0018	0.1290	0.0023	0.3140	0.0017
O	0.2117	0.0016	0.5189	0.0016	0.1110	0.0018
H(1)	0.186	—	−0.003	—	0.333	—
H(2)	0.274	—	0.223	—	0.364	—
H(3)	0.451	—	0.024	—	0.339	—
H(4)	0.479	—	0.173	—	0.437	—

nitrate ions have their usual triangular shape, with N–O = 1.21 A. Each oxygen atom has two close amino hydrogen atoms, with H–O = 2.25 and 2.31 A., and three more distant (2.51, 2.64, and 2.69 A.); the two short separations may represent hydrogen bonds.

XIV,b87. Crystals of the complex *mercury tetrathiocyanate–copper diethylene diamine*, $Hg(SCN)_4 \cdot Cu[(NH_2CH_2)_2]_2$, are monoclinic. Their tetramolecular unit has the dimensions:

$$a_0 = 7.51 \text{ A.}; \quad b_0 = 18.03 \text{ A.}; \quad c_0 = 14.20 \text{ A.}; \quad \beta = 96°32'$$

All atoms, except copper, are in general positions of C_{2h}^5 $(P2_1/c)$:

$$(4e) \quad \pm (xyz; \; x, ^1/_2 - y, z + ^1/_2)$$

Half the copper atoms, Cu(1), are in

$$(2b) \quad ^1/_2 \, 0 \, 0; \; ^1/_2 \, ^1/_2 \, ^1/_2$$

the other half, Cu(2), in

$$(2d) \quad ^1/_2 \, 0 \, ^1/_2; \; ^1/_2 \, ^1/_2 \, 0$$

Parameters found for all the atoms in general positions are given in Table XIVB,57.

The resulting structure (Fig. XIVB,124) is one in which $Hg(SCN)_4$ groups and two ethylene diamine molecules chelated to a copper atom form a pair of linked five-membered rings (Fig. XIVB,125). The Hg–S distances lie between 2.49 and 2.61 A. This analysis suggests that the SCN ion is not linear: the angle S–C–N lies between 136 and 167°. The angle Hg–S–C ranges between 86 and 104°. The $Cu[(NH_2CH_2)_2]_2$ group, namely the copper atom and the four nitrogen atoms with which it is squarely coordinated, is planar; the carbon atoms lie ca. 0.35 and 0.55 A. above and below this

TABLE XIVB,57
Parameters of the Generally Placed Atoms in $Hg(SCN)_4 \cdot Cu[(NH_2CH_2)_2]_2$

Atom	x	y	z
Hg	0.133	0.228	0.287
S(1)	−0.012	0.116	0.203
S(2)	0.049	0.355	0.230
S(3)	0.046	0.206	0.456
S(4)	0.479	0.208	0.298
C(1)	−0.038	0.066	0.289
C(2)	0.160	0.410	0.310
C(3)	0.135	0.277	0.490
C(4)	0.439	0.123	0.302
N(1)	−0.057	0.036	0.377
N(2)	0.286	0.446	0.362
N(3)	0.291	0.296	0.537
N(4)	0.469	0.051	0.318

Ethylene diamine group

Atom	x	y	z
N(5)	0.633	0.097	0.047
N(6)	0.717	0.018	−0.081
C(5)	0.741	0.130	−0.018
C(6)	0.863	0.068	−0.047
N(7)	0.704	0.072	0.521
N(8)	0.358	0.079	0.569
C(7)	0.598	0.133	0.545
C(8)	0.502	0.112	0.629

Alternate positions around $1/2\ 0\ 0$; $1/2\ 1/2\ 1/2$

Atom	x	y	z
N(5′)	0.590	0.088	0.068
N(6′)	0.380	−0.026	0.128
C(5′)	0.474	0.083	0.135
C(6′)	0.511	0.008	0.187

plane. There are two thiocyanate nitrogens that come fairly close to each copper atom, with Cu(1)–N(2) = 2.58 A. and Cu(2)–N(4) = 2.81 A. These two additional coordinations about each copper atom are in a line about normal to the plane of the other four.

It should be noted that there is a second possible orientation of the ethylene diamine groups about the copper atoms at $1/2\ 0\ 0$; $1/2\ 1/2\ 1/2$. These have the primed values at the end of the table.

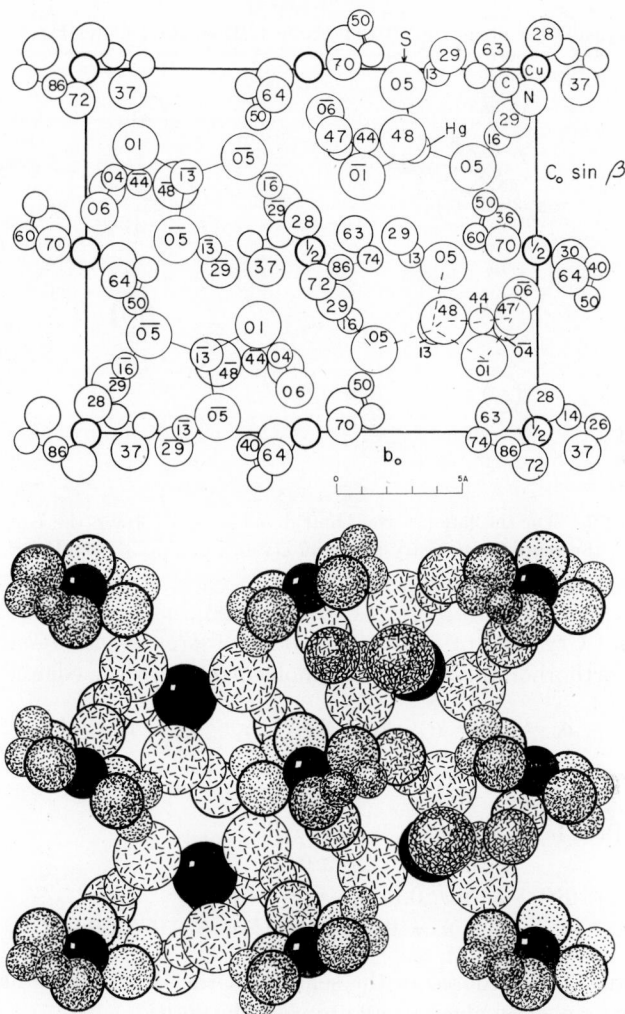

Fig. XIVB,124a (top).   The monoclinic structure of $Cu(C_2N_2H_8)_2Hg(SCN)_4$ projected
along its a_0 axis.   Left-hand axes.

Fig. XIVB,124b (bottom).   A packing drawing of the monoclinic structure of
$Cu(C_2N_2H_8)_2Hg(SCN)_4$ seen along its a_0 axis.   Copper atoms are the small, mercury
atoms the larger black circles.   Atoms of the thiocyanate groups are line shaded, with
sulfur the largest and carbon the smallest circles.   In the ethylene diamine molecules
the carbon atoms are small and dotted;   the nitrogens are larger, heavily ringed, and also
dotted.

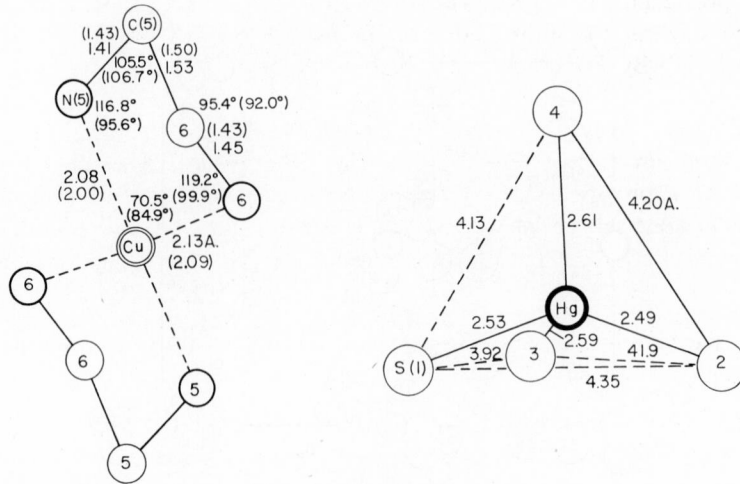

Fig. XIVB,125. On the left are the bond dimensions involving the copper atoms, and on the right the mercury atoms in crystals of $Cu(C_2N_2H_8)_2Hg(SCN)_4$.

XIV,b88. Crystals of *ethylene diamine–tribromo platinum*, $C_2N_2H_8 \cdot$ $PtBr_3$, are orthorhombic with a tetramolecular cell of the edge lengths:

$$a_0 = 5.60 \text{ A.}; \quad b_0 = 14.77 \text{ A.}; \quad c_0 = 10.12 \text{ A.}$$

The space group is V_h^{17} (*Cmcm*) with atoms in the positions:

Pt: (4c) $\pm (0 \ u \ ^1/_4; \ ^1/_2, u + ^1/_2, ^1/_4)$
with $u = 0.232 \pm 0.002$

Br(1): (8f) $\pm (0uv; \ 0, u, ^1/_2 - v; \ ^1/_2, u + ^1/_2, v; \ ^1/_2, u + ^1/_2, ^1/_2 - v)$
with $u = 0.112 \pm 0.002, \ v = 0.075 \pm 0.002$

The structure is disordered in the sense that the Br(2) atoms have a randomness expressed by placing half atoms in the positions:

(8g) $\pm (u \ v \ ^1/_4; \ \bar{u} \ v \ ^1/_4; \ u + ^1/_2, v + ^1/_2, ^1/_4; \ ^1/_2 - u, v + ^1/_2, ^1/_4)$
with $u = 0.442 \pm 0.002, \ v = 0.239 \pm 0.002$

Positions for the ethylene diamine atoms were not established.

In this incompletely determined structure the molecular groups, arranged in chains, are of two sorts centered around Pt(II) and Pt(IV) atoms. Normal to the chain, each platinum atom has four neighbors at the corners of a square, two being bromine atoms with Pt–Br = 2.51 A. and the other two presumably NH$_2$ groups of a C$_2$N$_2$H$_8$ molecule. Normal to and between these squares are additional bromine atoms with Pt(IV)–Br = 2.48 A. and Pt(II)–Br = 3.125 A.

XIV,b89. Wolfram's Red Salt, *platinous platinic tetra ethylamine chloride tetrahydrate*, Pt$_2$(C$_2$H$_4$NH$_2$)$_8$Cl$_6 \cdot$ 4H$_2$O, has a disordered tetragonal structure. An approximate atomic arrangement has been described in terms of a unimolecular subcell having the edge lengths:

$$a_0 = 13.28 \pm 0.05 \text{ A.,} \qquad c_0 = 5.39 \pm 0.05 \text{ A.}$$

The space group was chosen as C$_{4v}^9$ (*I4mm*) with atoms in the positions:

Pt: (2a)   00u; $^1/_2,^1/_2,u+^1/_2$
with u = 0 (arbitrary)
Cl(1): (2a)   with u = 0.418 or 0.582

observed disorder being explained by saying that half an atom is in each position.

Cl(2): (8c) uuv; $\bar{u}\bar{u}v$; $\bar{u}uv$; $u\bar{u}v$; B.C.
with u = 0.213, v = 0.745

Since there are but four Cl(2) atoms in the subcell, there must be some unspecified kind of disorder among these eight positions.

N: (8d) $u0v$; $\bar{u}0v$; $0uv$; $0\bar{u}v$; B.C.
with u = 0.148, v = 0.000
C(1): (8d)   with u = 0.190, v = 0.240
C(2): (8d)   with u = 0.308, v = 0.240

The positions of the water molecules were not found and it is considered that, as with the chlorine atoms, there is disorder in the way they are distributed.

XIV,b90. Crystals of *trans di ethylene diamine cobaltic chloride*, CoCl$_3 \cdot$ 2C$_2$H$_4$(NH$_2$)$_2$, are monoclinic with a bimolecular cell of the dimensions:

$$a_0 = 9.49 \text{ A.;} \quad b_0 = 8.96 \text{ A.;} \quad c_0 = 6.26 \text{ A.;} \quad \beta = 109°16'$$

The space group is C$_{2h}^5$ (*P2$_1$/a*) with atoms in the positions:

Co: (2a)   000; $^1/_2 \, ^1/_2 \, 0$
Cl(1): (2b) $^1/_2 \, 0 \, 0$; $0 \, ^1/_2 \, 0$

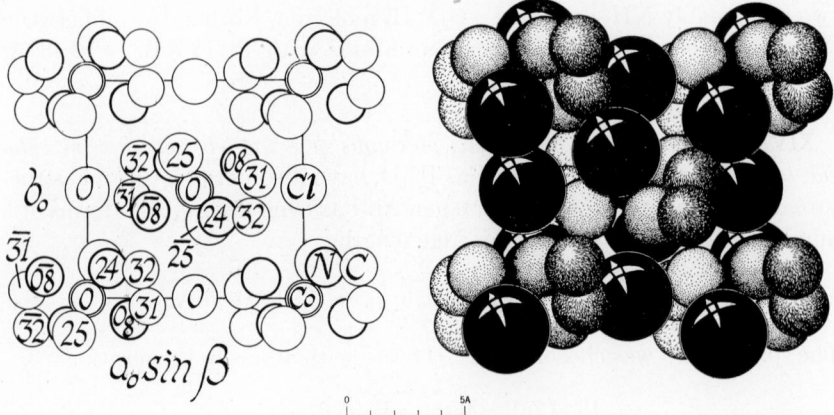

Fig. XIVB,126a (left).   The monoclinic structure of $CoCl_3(C_2N_2H_8)_2$ projected along its c_0 axis.   Right-hand axes.

Fig. XIVB,126b (right).   A packing drawing of the monoclinic $CoCl_3(C_2N_2H_8)_2$ structure viewed along its c_0 axis.   The large black circles are chlorine; the small black cobalt atoms scarcely show.   The dotted nitrogen atoms are heavily ringed; the carbon circles, of equal size, are line shaded.

and all other atoms in the general positions:

$$(4e) \quad \pm (xyz; \ x+^1/_2, ^1/_2-y, z)$$

Established parameters are as follows:

Atom	x	y	z
Cl(2)	0.060	0.136	−0.254
N(1)	0.195	−0.104	0.083
N(2)	0.091	0.156	0.236
C(1)	0.280	−0.042	0.307
C(2)	0.252	0.126	0.318

The structure that results is shown in Figure XIVB,126. In it each cobalt atom is surrounded by an approximate square of nitrogen atoms, with Co–N = 1.98 and 2.01 A., and by two chlorine atoms forming a line normal to this square, with Co–Cl(2) = 2.22 A. In the ethylene diamine molecules, N–C = 1.47 A. and C–C = 1.54 A.

XIV,b91.   Crystals of the *hydrochloride dihydrate* of *trans dichloro di ethylene diamine cobalti chloride*, $[Co(NH_2CH_2CH_2NH_2)_2Cl_2]Cl \cdot HCl \cdot 2H_2O$, are monoclinic with a bimolecular cell of the dimensions:

$$a_0 = 10.68 \ A.; \quad b_0 = 7.89 \ A.; \quad c_0 = 9.09 \ A.; \quad \beta = 110°26'$$

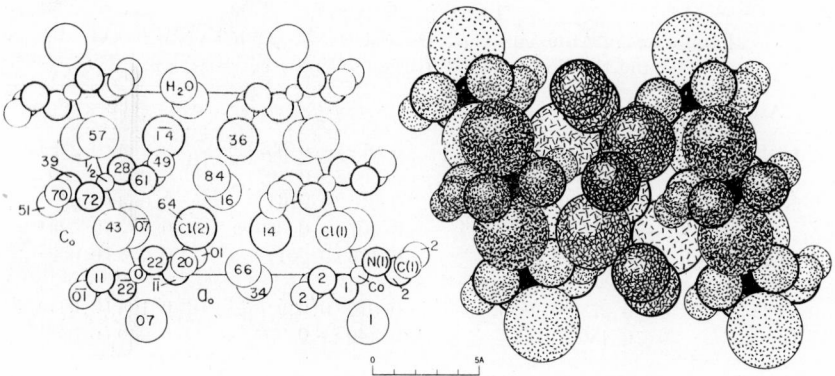

Fig. XIVB,127a (left). The monoclinic structure of $CoCl_3(C_2N_2H_8)_2 \cdot HCl \cdot 2H_2O$ projected along its b_0 axis. In this drawing the chlorine atoms closely associated with cobalt are less heavily outlined than the others. Left-hand axes.

Fig. XIVB,127b (right). A packing drawing of the monoclinic $CoCl_3(C_2N_2H_8)_2 \cdot HCl \cdot 2H_2O$ arrangement seen along its b_0 axis. The cobalt atoms are black. Of the chlorine atoms, which are the largest circles, those coordinated with cobalt are dotted. The water molecules are the smaller line shaded circles. The atoms of the ethylene diamine groups are dotted, the nitrogens being larger and more heavily ringed than the carbon.

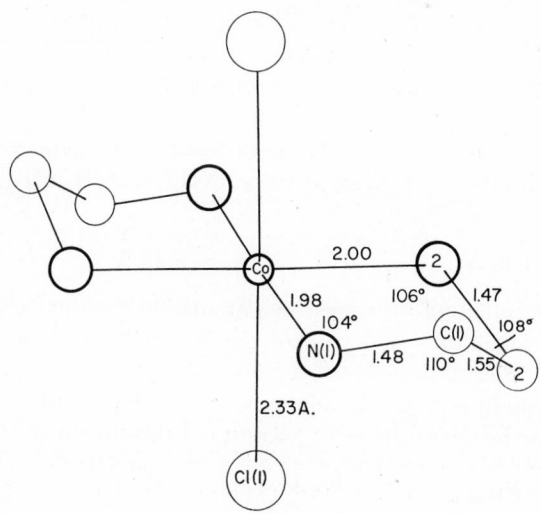

Fig. XIVB,128. The dimensions of the $CoCl_2(C_2N_2H_8)_2$ complex in $CoCl_3(C_2N_2H_8)_2 \cdot HCl \cdot 2H_2O$.

TABLE XIVB,58
Parameters of Atoms in [Co($NH_2CH_2CH_2NH_2$)$_2Cl_2$]Cl·HCl·$2H_2O$
and the Corresponding Bromide (in parentheses)

Atom	x	y	z
Cl[Br](1)	0.036 (0.060)	−0.072 (−0.037)	0.260 (0.275)
Cl[Br](2)	0.328 (0.344)	−0.363 (−0.397)	0.260 (0.269)
N(1)	0.092 (0.092)	0.219 (0.216)	0.069 (0.045)
N(2)	0.175 (0.175)	−0.107 (−0.099)	0.019 (0.024)
C(1)	0.225 (0.225)	0.196 (0.201)	0.056 (0.032)
C(2)	0.278 (0.281)	0.015 (0.029)	0.108 (0.091)
O	0.480 (0.506)	−0.340 (−0.350)	0.030 (0.048)

Atoms have been placed in the following positions of C_{2h}^5 ($P2_1/c$):
The cobalt atoms are in ($2a$) 000; 0 $^1/_2$ $^1/_2$
All other atoms are in

$$(4e) \quad \pm (xyz; \; x, ^1/_2 - y, z + ^1/_2)$$

with the parameters of Table XIVB,58.

In the resulting structure (Fig. XIVB,127) the two chlorine atoms and the two $NH_2CH_2CH_2NH_2$ groups are *trans* to one another, with Co–Cl = 2.33 A., and Co–N = 1.98 or 2.00 A. Thus the coordination of the cobalt atoms is octahedral. The complexes around cobalt (Fig. XIVB,128) are in sheets, centered in b_0c_0, separated from one another by the remaining chloride ions and H_2O. The water molecules occur in pairs with O–O = 2.66 A., and these pairs are surrounded by Cl⁻ ions at distances of 2.90 and 3.06 A.

Crystals of the isostructural bromide *trans cobalti bromide bis ethylene diamine hydrobromide dihydrate*, CoBr₃·(NH_2CH_2·CH_2NH_2)$_2$·HBr·$2H_2O$, have the cell:

$$a_0 = 10.98 \text{ A.}; \quad b_0 = 8.18 \text{ A.}; \quad c_0 = 9.46 \text{ A.}; \quad \beta = 113°12'$$

The atomic parameters are those in parentheses in the table. The bond dimensions to which they lead are shown in Figure XIVB,129.

The corresponding *chromium* compound, [Cr($NH_2CH_2CH_2NH_2$)$_2Cl_2$]Cl·HCl·$2H_2O$, also has this structure. Its unit has the dimensions:

$$a_0 = 10.97 \text{ A.}; \quad b_0 = 7.88 \text{ A.}; \quad c_0 = 9.12 \text{ A.}; \quad \beta = 111°30'$$

Choosing the axes to agree with those employed above, the parameters were found to be as listed in Table XIVB,59.

Fig. XIVB,129. Bond dimensions of the $CoBr_2(C_2N_2H_8)_2$ complex in $CoBr_3(C_2N_2H_8)_2 \cdot HBr \cdot 2H_2O$.

TABLE XIVB,59
Parameters of Atoms in $[Cr(NH_2CH_2CH_2NH_2)_2Cl_2]Cl \cdot HCl \cdot 2H_2O$

Atom	x	y	z
Cl(1)	0.050	−0.038	0.269
Cl(2)	0.345	−0.387	0.272
N(1)	0.100	0.235	0.042
N(2)	0.185	−0.106	0.023
C(1)	0.227	0.200	0.037
C(2)	0.288	0.022	0.100
O	0.478	−0.350	0.038

The four nitrogen atoms that are in a square coordination about each chromium atom are at the distances Cr–N = 2.11 or 2.13 A.; the two chlorine atoms normal to their plane are 2.33 A. from chromium. In the ethylene diamine group, C–N = 1.44 or 1.48 A. and C–C = 1.57 A.

XIV,b92. Crystals of DL-*tris-ethylene diamine cobaltic chloride trihydrate*, $[Co(NH_2CH_2CH_2NH_2)_3]Cl_3 \cdot 3H_2O$, are hexagonal with a tetramolecular unit of the edge lengths:

$$a_0 = 11.50 \text{ A.}, \qquad c_0 = 15.52 \text{ A.}$$

The space group is D_{3d}^4 ($P\bar{3}c1$) with atoms in the positions:

$$Co: (4d) \quad \pm(^1/_3\ ^2/_3\ u;\ ^1/_3,^2/_3,u+^1/_2)$$
$$\text{with } u = 0.125$$

All other atoms have been placed in the general positions:

$$(12g) \quad \pm(xyz; \qquad \bar{y},x-y,z; \qquad y-x,\bar{x},z;$$
$$y,x,^1/_2-z;\ \bar{x},y-x,^1/_2-z;\ x-y,\bar{y},^1/_2-z)$$

with the parameters of Table XIVB,60.

The resulting structure is shown in Figure XIVB,130. Around each Co^{3+} atom are six nitrogen atoms from three ethylene diamine groups dis-

TABLE XIVB,60
Parameters of Atoms in $[Co(NH_2CH_2CH_2NH_2)_3]Cl_3 \cdot 3H_2O$

Atom	x	y	z
Cl	0.097	0.493	0.372
H_2O	0.087	0.225	0.873
C(1)	0.180	0.397	0.077
C(2)	0.215	0.397	0.173
N(1)	0.192	0.525	0.050
N(2)	0.333	0.525	0.200

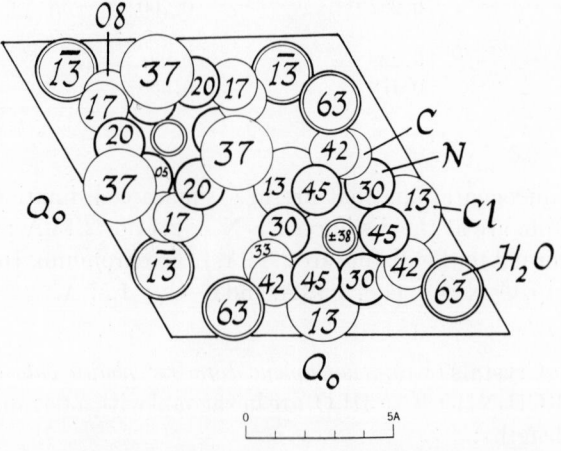

Fig. XIVB,130.   The contents of the lower half of the hexagonal cell of $CoCl_3(C_2N_2H_8)_3 \cdot 3H_2O$ projected along its c_0 axis.   Right-hand axes.

tributed at the corners of a slightly distorted octahedron (Co–N = 2.00 A.). Each [Co(NH₂CH₂CH₂NH₂)₃] group is surrounded by nine chloride ions, with Cl–N = 3.13–3.41 A. and Cl–C = 3.30 and 3.67 A. The dimensions of the ethylene diamine molecule are normal. The water molecules lie rather loosely in channels that run along the c_0 axis and in accordance with this the water content is variable; with careful heating the crystals can be dehydrated without a change in cell dimensions.

XIV,b93. The crystals of D-*tris ethylene diamine cobaltic bromide monohydrate*, D-[Co(NH₂CH₂CH₂NH₂)₃]Br₃·H₂O, are tetragonal with a tetramolecular unit of the dimensions:

$$a_0 = 9.95 \pm 0.03 \text{ A.}, \qquad c_0 = 16.73 \pm 0.05 \text{ A.}$$

The space group could be $D_4{}^4$ ($P4_12_12$), or the enantiomorphic $D_4{}^8$, but adopting the absolute configuration previously determined for the cationic complex, it is chosen as $D_4{}^8$ ($P4_32_12$). Atoms of cobalt, bromine, and oxygen are in the special positions:

$$(4a) \quad uu0; \, \bar{u}\,\bar{u}\,{}^1\!/_2; \, {}^1\!/_2-u,u+{}^1\!/_2,{}^3\!/_4; \, u+{}^1\!/_2,{}^1\!/_2-u,{}^1\!/_4$$

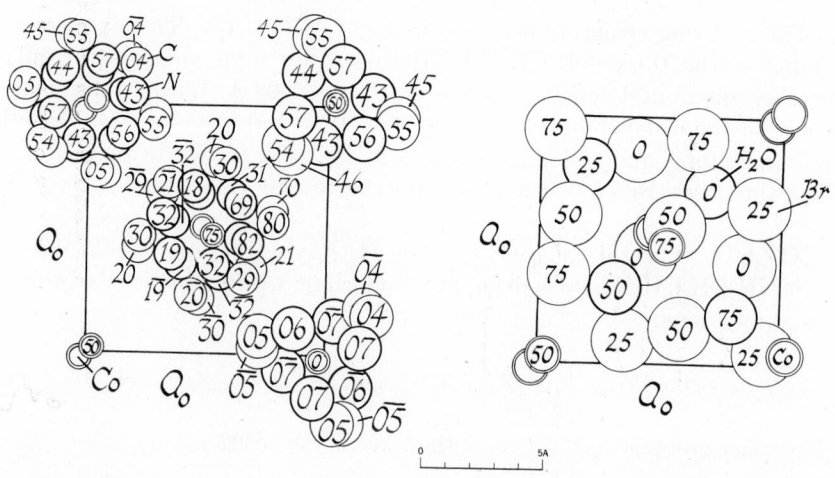

Fig. XIVB,131a (left). Part of the tetragonal structure of CoBr₃(C₂N₂H₈)₃·H₂O projected along its c_0 axis. Only the Co(C₂H₂N₈)₃ groups are shown in this drawing. Right-hand axes.

Fig. XIVB,131b (right). The distribution of the bromine atoms and water molecules in the tetragonal crystals of CoBr₃(C₂N₂H₈)₃·H₂O as projected along the c_0 axis. Right-hand axes.

with the parameters: $u(Br,1) = 0.450$, $u(O) = -0.295$ and $u(Co) = -0.020$. All other atoms are in the general positions:

(8b) $xyz;\ \bar{x},\bar{y},z+{}^1/_2;\ {}^1/_2-y,x+{}^1/_2,z+{}^3/_4;\ y+{}^1/_2,{}^1/_2-x,z+{}^1/_4;$
       $yx\bar{z};\ \bar{y},\bar{x},{}^1/_2-z;\ {}^1/_2-x,y+{}^1/_2,{}^3/_4-z;\ x+{}^1/_2,{}^1/_2-y,{}^1/_4-z$

with the parameters of Table XIVB,61.

TABLE XIVB,61
Parameters of Atoms in D-[Co(NH₂CH₂CH₂NH₂)₃]Br₃·H₂O

Atom	x	y	z
Br(2)	0.139	0.590	0.500
C(1)	0.663	0.313	0.295
C(2)	0.218	0.433	0.205
C(3)	0.217	0.463	0.295
N(1)	0.532	0.360	0.319
N(2)	0.362	0.395	0.190
N(3)	0.324	0.558	0.322

The resulting structure is shown in Figure XIVB,131. The six nitrogen atoms of the three NH₂CH₂CH₂NH₂ molecules form an almost regular octahedron around cobalt, with Co–N = 1.98–2.03 A. Within these molecules, of the usual shape, C–N = 1.44–1.50 A. and C–C = 1.54 A. Both bromine atoms and water molecules are also close to the nitrogen atoms, with the closest Br–N = 3.20 A. and the shortest Br–O = 2.90 A.

XIV,b94. Crystals of *trans dichloro bis ethylene diamine cobalti nitrate*, [Co(NH₂CH₂CH₂NH₂)₂Cl₂]NO₃, are monoclinic with a bimolecular unit of the dimensions:

$$a_0 = 6.33 \text{ A.};\quad b_0 = 9.25 \text{ A.};\quad c_0 = 10.62 \text{ A.};\quad \beta = 114°30'$$

The space group is C_{2h}^5 ($P2_1/c$) with atoms in the positions:

Co: (2a)   000; $0\ {}^1/_2\ {}^1/_2$

Other atoms of the cation are in the general positions:

(4e) $\pm(xyz;\ x,{}^1/_2-y,z+{}^1/_2)$

The assigned parameters are listed below.

Atom	x	y	z
Cl	-0.235	-0.149	0.052
N(1)	0.260	-0.133	0.100
N(2)	0.083	0.096	0.183
C(1)	0.358	-0.112	0.250
C(2)	0.322	0.055	0.270

It is considered that the nitrate ions must be disordered since they do not have centers of symmetry and the nitrogen atoms must lie close to such centers. Positions have been proposed which displace the nitrogen atoms by small amounts from (2c) 0 $^1/_2$ 0; 0 0 $^1/_2$, but further work would be required to place the atoms of these anions with certainty.

XIV,b95. D-*tris Ethylene diamine cobaltic chloride sodium chloride hexahydrate*, $2[Co(NH_2CH_2CH_2NH_2)_3Cl_3] \cdot NaCl \cdot 6H_2O$, is hexagonal with a unimolecular cell of the edge lengths:

$$a_0 = 11.47 \pm 0.03 \text{ A.}, \quad c_0 = 8.06 \pm 0.02 \text{ A.}$$

The space group is C_3^1 ($P3$) but the structure has been described in terms of C_6^6 ($P6_3$) with sodium and chlorine showing some disorder. Atoms are in the positions:

$$\text{Co: } (2b) \quad ^1/_3\, ^2/_3\, u; \, ^2/_3, ^1/_3, u+^1/_2 \quad \text{with } u = ^1/_2$$

The atoms of chlorine associated with cobalt and those of the ethylene diamine molecules are in general positions:

$$(6c) \quad xyz; \quad \bar{y},x-y,z; \quad y-x,\bar{x},z;$$
$$\bar{x},\bar{y},z+^1/_2; \, y,y-x,z+^1/_2; \, x-y,x,z+^1/_2$$

with the following parameters:

Atom	x	y	z
N(1)	0.783	0.295	0.854
N(2)	0.647	0.183	0.146
C(1)	0.773	0.167	0.907
C(2)	0.742	0.140	0.093
Cl(1)	0.617	0.100	0.528

It is thought that best agreement with the experimental data is provided by assuming that these crystals are made up of submicroscopic twins of about equal amounts of two structures in which the NaCl and H_2O have the following coordinates:

$$(2a) \quad 00u; \; 0,0,u+{}^1\!/_2$$

The parameters assigned these atoms are:

Atom	Position	x	y	z
For structure A:				
Na	$(2a)$	0	0	0.770
Cl	$(2a)$	0	0	0.270
$H_2O(1)$	$(6c)$	0.897	0.098	0.560
$H_2O(2)$	$(6c)$	0.098	0.201	0.980
For structure B:				
Na	$(2a)$	0	0	0.230
Cl	$(2a)$	0	0	0.730
$H_2O(1)$	$(6c)$	0.897	0.098	0.440
$H_2O(2)$	$(6c)$	0.098	0.201	0.020

In this arrangement six nitrogen atoms are octahedrally distributed around the cobalt atoms, with Co–N = 1.98 or 2.00 A. There are also six water molecules around the sodium ions, with Na–OH$_2$ = 2.62 A. The ethylene diamine molecules have their usual size, with N–C = 1.47 or 1.48 A., and C–C = 1.54 A. Each complex cation has nine Cl⁻ neighbors, with N–Cl = 3.17–3.19 A., and these are considered to be indicative of hydrogen bondings.

XIV,b96. *Ammonium cobaltic ethylene diamine tetraacetate dihydrate,* $NH_4Co[-CH_2NH(CH_2CO_2)_2]_2 \cdot 2H_2O$, is orthorhombic with a tetramolecular unit of the dimensions:

$$a_0 = 6.46 \text{ A.}; \quad b_0 = 23.16 \text{ A.}; \quad c_0 = 10.09 \text{ A.}$$

The space group is V⁴ ($P2_12_12_1$) with all atoms in the positions:

$$(4a) \quad xyz; \; {}^1\!/_2-x,\bar{y},z+{}^1\!/_2; \; x+{}^1\!/_2,{}^1\!/_2-y,\bar{z}; \; \bar{x},y+{}^1\!/_2,{}^1\!/_2-z$$

The determined parameters are those listed in Table XIVB,62.

The resulting structure is shown in Figure XIVB,132. Each cobalt atom is octahedrally surrounded by the two nitrogen and four of the oxygen atoms of the ethylene diamine tetraacetate ion, with Co–N = 1.925±0.005 A., Co–O = 1.900±0.015 A. The coordination about the metal atom is, except for the replacement of a carboxyl oxygen by a water oxygen, similar to that prevailing in $NiH_2[-CH_2NH(CH_2CO_2)_2]_2 \cdot H_2O$ (**XIV,b99**).

The *rubidium* salt, $RbCo[-CH_2NH(CH_2CO_2)_2]_2 \cdot 2H_2O$, is isostructural with the dimensions:

$$a_0 = 6.43 \text{ A.}; \quad b_0 = 23.08 \text{ A.}; \quad c_0 = 10.18 \text{ A.}$$

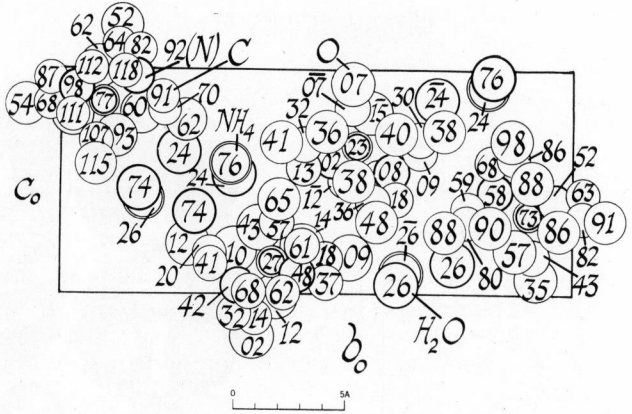

Fig. XIVB,132. The orthorhombic structure of $NH_4Co[-CH_2NH(CH_2CO_2)_2]_2 \cdot 2H_2O$ projected along its a_0 axis. Right-hand axes.

TABLE XIVB,62
Parameters of the Atoms in $NH_4Co[-CH_2NH(CH_2CO_2)_2]_2 \cdot 2H_2O$

Atom	x	y	z
Co	0.2748	0.4087	0.1401
NH_4	0.764	0.3365	-0.4218
N(1)	0.424	0.3510	0.0420
N(2)	0.482	0.4650	0.0940
C(1)	0.638	0.3726	0.0182
C(2)	0.617	0.4376	-0.0086
C(3)	0.411	0.2980	0.1262
C(4)	0.196	0.2962	0.1880
C(5)	0.610	0.4744	0.2140
C(6)	0.574	0.4273	0.3159
C(7)	0.370	0.5199	0.0560
C(8)	0.181	0.5227	0.1489
C(9)	0.321	0.3424	-0.0855
C(10)	0.139	0.3828	-0.1040
O(1)	0.116	0.2510	0.2231
O(2)	0.655	0.4310	0.4236
O(3)	0.089	0.5672	0.1666
O(4)	0.019	0.3797	-0.1967
O(5)	0.102	0.3448	0.1902
O(6)	0.426	0.3924	0.2968
O(7)	0.139	0.4745	0.2145
O(8)	0.119	0.4230	-0.0125
$H_2O(1)$	0.743	0.2659	0.3643
$H_2O(2)$	0.236	0.3428	-0.4726

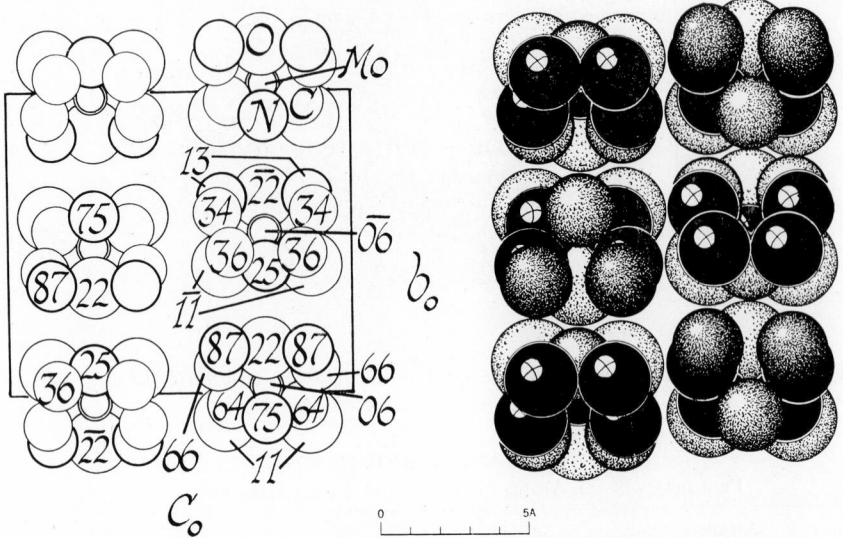

Fig. XIVB,133a (left).   The orthorhombic structure of $MoO_3[NH_2(CH_2)_2NH(CH_2)_2-$
$NH_2]$ projected along its a_0 axis.   Right-hand axes.

Fig. XIVB,133b (right).   A packing drawing of the orthorhombic structure of
$MoO_3[NH_2(CH_2)_2NH(CH_2)_2NH_2]$ seen along its a_0 axis.   The carbon atoms are the
large black circles; the small black atoms of molybdenum scarcely show.   The nitrogen
atoms are line shaded and more heavily outlined; the oxygen atoms are somewhat
larger and dotted.

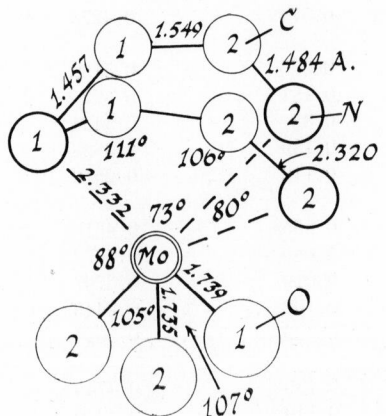

Fig. XIVB,134.   Bond dimensions in crystals of $MoO_3[NH_2(CH_2)_2NH(CH_2)_2NH_2]$.

Its heavy atoms have the parameters: for Co, $x = 0.2638$, $y = 0.4092$, $z = 0.1339$; for Rb, $x = 0.7475$, $y = 0.3402$, $z = -0.4099$.

XIV,b97. *Trioxo(diethylene triamine) molybdenum* $MoO_3[NH_2(CH_2)_2-NH(CH_2)_2NH_2]$ is orthorhombic with a tetramolecular cell of the edge lengths:

$$a_0 = 6.863 \pm 0.008 \text{ A.}; \quad b_0 = 10.250 \pm 0.007 \text{ A.}; \quad c_0 = 11.705 \pm 0.005 \text{ A.}$$

The space group has been chosen as V_h^{11} (*Pbcm*) with atoms in the positions:

$$(4d) \quad \pm (u \ v \ ^1/_4; \ \bar{u}, v + ^1/_2, ^1/_4)$$
$$(8e) \quad \pm (xyz; \ x,y,^1/_2 - z; \ \bar{x}, y + ^1/_2, z; \ x, ^1/_2 - y, z + ^1/_2)$$

The determined positions and parameters are those of Table XIVB,63.

TABLE XIVB,63
Parameters of the Atoms in $MoO_3[NH_2(CH_2)_2]_2NH$

Atom	Position	x	$\sigma(x)$	y	$\sigma(y)$	z	$\sigma(z)$
Mo	(4d)	0.0566	0.0002	0.0198	0.0001	$^1/_4$	—
O(1)	(4d)	0.2227	0.0015	0.1476	0.0009	$^1/_4$	—
O(2)	(8e)	0.1106	0.0010	0.9220	0.0007	0.3686	0.0006
N(1)	(4d)	0.7465	0.0017	0.9274	0.0011	$^1/_4$	—
N(2)	(8e)	0.8696	0.0012	0.1396	0.0008	0.3733	0.0007
C(1)	(8e)	0.6420	0.0016	0.9582	0.0010	0.3548	0.0009
C(2)	(8e)	0.6610	0.0015	0.1041	0.0012	0.3877	0.0011
H(N 1)	(4d)	0.250	—	0.333	—	$^1/_4$	—
H(1,N 2)	(8e)	0.100	—	0.603	—	0.033	—
H(2,N 2)	(8e)	0.160	—	0.743	—	0.103	—
H(1,C 1)	(8e)	0.283	—	0.413	—	0.080	—
H(2,C 1)	(8e)	0.417	—	0.083	—	0.427	—
H(1,C 2)	(8e)	0.310	—	0.107	—	0.000	—
H(2,C 2)	(8e)	0.400	—	0.160	—	0.183	—

The structure is shown in Figure XIVB,133. The molecule that is present has the bond dimensions of Figure XIVB,134. These molecules are associated together by hydrogen bonds to oxygen involving each amine hydrogen, with N(1)–H–O(1) = 2.875 A., N(2)–H–O(2) = 2.943 and 3.043 A. Nonbonded interatomic distances range upwards from N(2)–C(1) = 2.44 A. The final $R = 0.058$.

CRYSTAL STRUCTURES

TABLE XIVB,64

Parameters of the Atoms in $[(CH_3)_2NCH_2\text{—}]_2AlH_3$

Atom	x	$\sigma(x)$	y	$\sigma(y)$	z	$\sigma(z)$
Al(1)	0.0236	0.0001	0.2144	0.0001	0.2457	0.0001
Al(2)	−0.0043	0.0001	0.0124	0.0001	−0.1630	0.0001
N(1)	0.2106	0.0004	0.2235	0.0002	0.1396	0.0003
N(2)	0.1696	0.0004	0.0429	0.0002	−0.0510	0.0003
N(3)	−0.1636	0.0004	0.2024	0.0002	0.3577	0.0003
N(4)	−0.3294	0.0004	0.0237	0.0002	0.2200	0.0003
C(1)	0.3311	0.0006	0.2422	0.0003	0.2128	0.0005
C(2)	0.1916	0.0006	0.2873	0.0003	0.0578	0.0005
C(3)	0.1830	0.0006	−0.0191	0.0003	0.0336	0.0005
C(4)	0.2979	0.0006	0.0471	0.0003	−0.1205	0.0005
C(5)	0.2500	0.0006	0.1498	0.0003	0.0797	0.0005
C(6)	0.1355	0.0006	0.1186	0.0003	0.0021	0.0005
C(7)	−0.1149	0.0006	0.2117	0.0003	0.4738	0.0005
C(8)	−0.2642	0.0006	0.2641	0.0003	0.3308	0.0005
C(9)	−0.2071	0.0006	−0.0290	0.0003	0.2270	0.0005
C(10)	−0.3890	0.0006	0.0189	0.0003	0.1059	0.0005
C(11)	−0.2293	0.0006	0.1243	0.0003	0.3510	0.0005
C(12)	−0.2850	0.0006	0.1065	0.0003	0.2350	0.0005
H(1,Al 1)	−0.068	—	0.229	—	0.144	—
H(2,Al 1)	0.088	—	0.279	—	0.329	—
H(3,Al 1)	0.123	—	0.156	—	0.310	—
H(4,Al 2)	0.021	—	0.094	—	−0.239	—
H(5,Al 2)	0.080	—	−0.068	—	−0.197	—
H(6,Al 2)	−0.120	—	0.028	—	−0.066	—
H(7,C1)	0.428	—	0.235	—	0.176	—
H(8,C 1)	0.322	—	0.289	—	0.258	—
H(9,C 1)	0.333	—	0.206	—	0.263	—
H(10,C 1)	0.271	—	0.299	—	0.004	—
H(11,C 1)	0.101	—	0.291	—	0.016	—
H(12,C 2)	0.165	—	0.329	—	0.078	—
H(13,C 2)	0.257	—	−0.014	—	0.086	—
H(14,C 2)	0.120	—	−0.035	—	0.096	—
H(15,C 3)	0.201	—	−0.066	—	−0.008	—
H(16,C 3)	0.294	—	0.082	—	−0.180	—
H(17,C 3)	0.314	—	0.005	—	−0.152	—

(continued)

TABLE XIVB,64 (*continued*)

Atom	x	$\sigma(x)$	y	$\sigma(y)$	z	$\sigma(z)$
H(18,C 4)	0.366	—	0.057	—	−0.078	—
H(19,C 4)	0.275	—	0.120	—	0.159	—
H(20,C 4)	0.338	—	0.163	—	0.040	—
H(21,C 5)	0.104	—	0.149	—	−0.051	—
H(22,C 5)	0.044	—	0.120	—	0.029	—
H(23,C 6)	−0.049	—	0.183	—	0.487	—
H(24,C 6)	−0.171	—	0.204	—	0.517	—
H(25,C 7)	−0.057	—	0.248	—	0.519	—
H(26,C 8)	−0.209	—	0.314	—	0.348	—
H(27,C 8)	−0.276	—	0.273	—	0.252	—
H(28,C 8)	−0.345	—	0.256	—	0.396	—
H(29,C 9)	−0.252	—	−0.072	—	0.202	—
H(30,C 9)	−0.170	—	−0.019	—	0.298	—
H(31,C 9)	−0.143	—	−0.021	—	0.173	—
H(32,C 10)	−0.463	—	0.048	—	0.111	—
H(33,C 10)	−0.431	—	−0.020	—	0.086	—
H(34,C 10)	−0.331	—	0.046	—	0.045	—
H(35,C 11)	−0.284	—	0.110	—	0.402	—
H(36,C 11)	−0.150	—	0.090	—	0.371	—
H(37,C 12)	−0.223	—	0.127	—	0.187	—
H(38,C 12)	−0.369	—	0.133	—	0.226	—

XIV,b98. *N,N,N′,N′-Tetramethyl ethylene diamine, aluminum hydride,* $[(CH_3)_2NCH_2-]_2AlH_3$, is orthorhombic with a unit containing eight molecules and having the dimensions:

$$a_0 = 9.554 \pm 0.004 \text{ A.}; \quad b_0 = 17.241 \pm 0.009 \text{ A.}; \quad c_0 = 11.866 \pm 0.006 \text{ A.}$$

The space group is V^4 ($P2_12_12_1$) with all atoms in the positions:

(4a) $xyz; \frac{1}{2}-x,\bar{y},z+\frac{1}{2}; x+\frac{1}{2},\frac{1}{2}-y,\bar{z}; \bar{x},y+\frac{1}{2},\frac{1}{2}-z$

Their parameters are listed in Table XIVB,64. In placing the hydrogen atoms, the usual assumptions of C–H = 1.0 A. and of tetrahedral bonding around carbon were made to supplement observations on difference Fourier maps.

In the resulting structure (Fig. XIVB,135) the tetramethyl ethylene diamine groups are bound together in chains by aluminum atoms which make contact with their nitrogens.

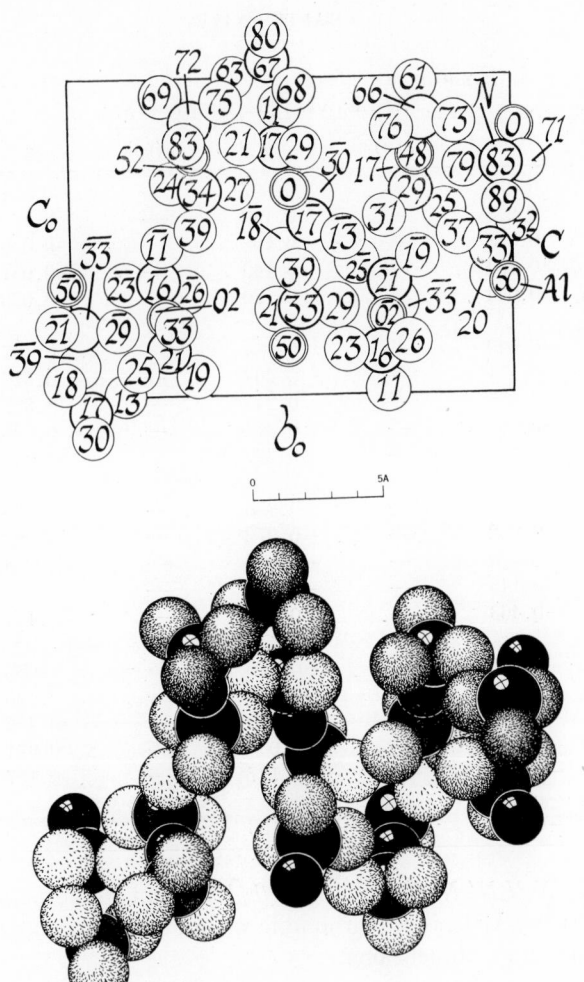

Fig. XIVB,135a (top). The orthorhombic structure of [(CH₃)₂NCH₂—]₂AlH₃ projected along its a_0 axis. Right-hand axes.

Fig. XIVB,135b (bottom). A packing drawing of the orthorhombic structure of [(CH₃)₂NCH₂—]₂AlH₃ viewed along its a_0 axis. The nitrogen atoms are the large, the aluminum atoms the slightly smaller black circles. Atoms of carbon are line shaded.

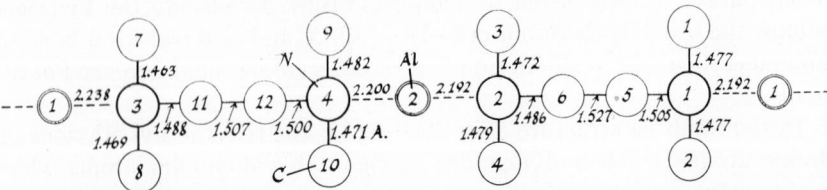

Fig. XIVB,136. Bond dimensions in crystals of [(CH₃)₂NCH₂—]₂AlH₃.

The bond dimensions are those of Figure XIVB,136. Three hydrogen atoms are arranged about each aluminum atom in a plane roughly normal to the N–Al–N direction to give aluminum a five fold coordination; these Al–H separations lie between 1.50 and 1.68 A. The final R for observed reflections is 0.064.

XIV,b99. Crystals of *nickel dihydrogen ethylene diamine tetraacetate monohydrate*, $NiH_2[—CH_2NH(CH_2CO_2)_2]_2 \cdot H_2O$, are monoclinic with a tetramolecular unit of the dimensions:

$$a_0 = 11.71 \text{ A.}; \quad b_0 = 6.94 \text{ A.}; \quad c_0 = 16.65 \text{ A.}; \quad \beta = 91°12'$$

The space group is C_{2h}^5 ($P2_1/c$) with all atoms in the general positions:

$$(4e) \quad \pm (xyz; \ x,^1/_2-y,z+^1/_2)$$

The chosen parameters are those of Table XIVB,65.

TABLE XIVB,65
Parameters of the Atoms in $NiH_2[—CH_2NH(CH_2CO_2)_2]_2 \cdot H_2O$

Atom	x	y	z
Ni	0.1712	0.3212	0.4242
N(1)	0.2518	0.364	0.3122
N(2)	0.3048	0.135	0.4523
C(1)	0.1832	0.267	0.2485
C(2)	0.0980	0.120	0.2822
C(3)	0.3777	0.279	0.3252
C(4)	0.3597	0.093	0.3732
C(5)	0.3857	0.232	0.5093
C(6)	0.3462	0.443	0.5297
C(7)	0.2493	−0.035	0.4858
C(8)	0.1520	0.018	0.5377
C(9)	0.2617	0.579	0.2968
C(10)	0.3192	0.634	0.2203
O(1)	0.1192	−0.115	0.5863
O(2)	0.0422	0.022	0.2357
O(3)	0.3217	0.820	0.2085
O(4)	0.3635	0.521	0.1710
O(5)	0.4040	0.524	0.5818
$H_2O(6)$	0.0390	0.519	0.4123
O(7)	0.0820	0.123	0.3585
O(8)	0.2633	0.511	0.4928
O(9)	0.1037	0.179	0.5283

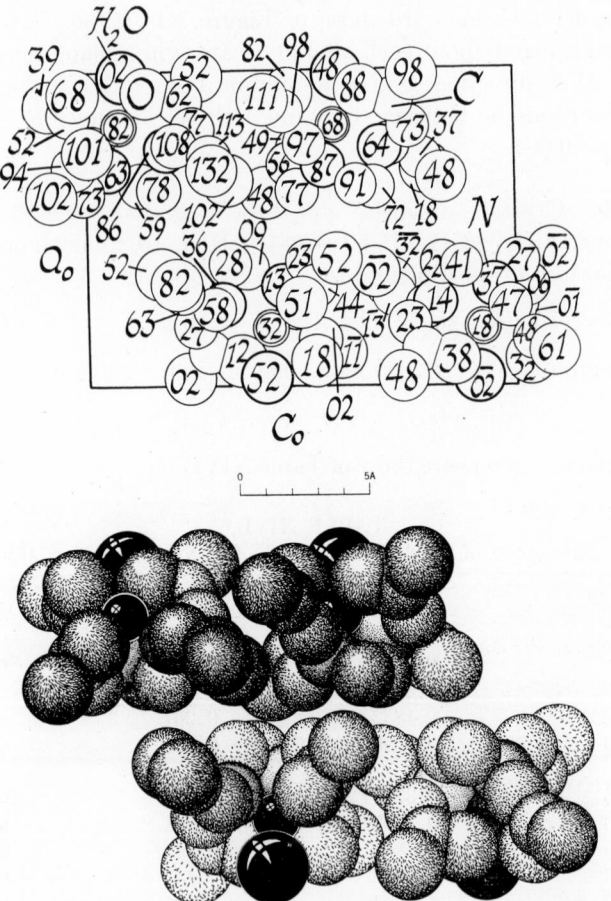

Fig. XIVB,137a (top). The monoclinic structure of $NiH_2[-CH_2NH(CH_2CO_2)_2]_2 \cdot H_2O$ projected along its b_0 axis. The nickel atoms are small and doubly ringed. Right-hand axes.

Fig. XIVB,137b (bottom). A packing drawing of the monoclinic structure of $NiH_2-[-CH_2NH(CH_2CO_2)_2]_2 \cdot H_2O$ seen along its b_0 axis. The water molecules are the large, the nickel atoms the small black circles. The atoms of oxygen are line shaded; those of carbon are smaller and hook shaded. Nitrogen atoms are dotted and heavily outlined.

The structure is shown in Figure XIVB,137. The nickel atom is octahedrally surrounded by two nitrogen and four oxygen atoms, one of which is from a water molecule. The general shape of this complex is shown in Figure XIVB,138.

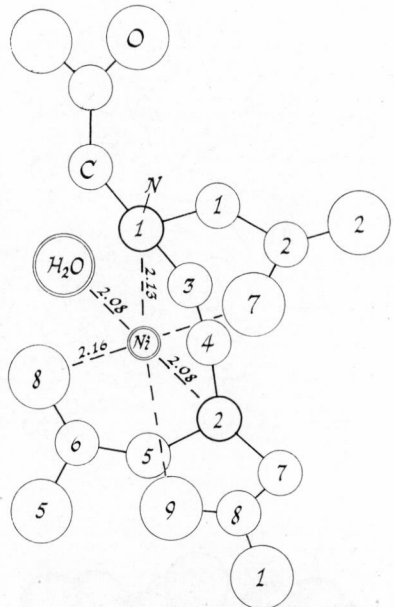

Fig. XIVB,138.   The environment of the nickel atoms in crystals of NiH₂[—CH₂NH-(CH₂CO₂)₂]₂·H₂O.

The corresponding *copper* compound is isostructural with

a_0 = 11.61 A.; b_0 = 7.00 A.; c_0 = 16.50 A.; β = 92°0′

XIV,b100.   *Platinous chloride diammine bis(acetamidine) monohydrate,* PtCl₂(NH₃)₂[CH₃C(NH₂)NH]₂·H₂O, is monoclinic with a tetramolecular cell of the dimensions:

a_0 = 15.940 A.; b_0 = 6.250 A.; c_0 = 17.246 A.; β = 130°23′

The space group is C_{2h}^6 ($A2/a$) with

Pt: (4a)   000; ¹/₂ 0 0; 0 ¹/₂ ¹/₂; ¹/₂ ¹/₂ ¹/₂
H₂O: (4e)   ± (¹/₄ u 0; ¹/₄,u+¹/₂,¹/₂)
with u = 0.8440

All other atoms are in the general positions:

(8f)   ± (xyz; x+¹/₂,ȳ,z; x,y+¹/₂,z+¹/₂; x+¹/₂,¹/₂−y,z+¹/₂)

The determined parameters are listed in Table XIVB,66.

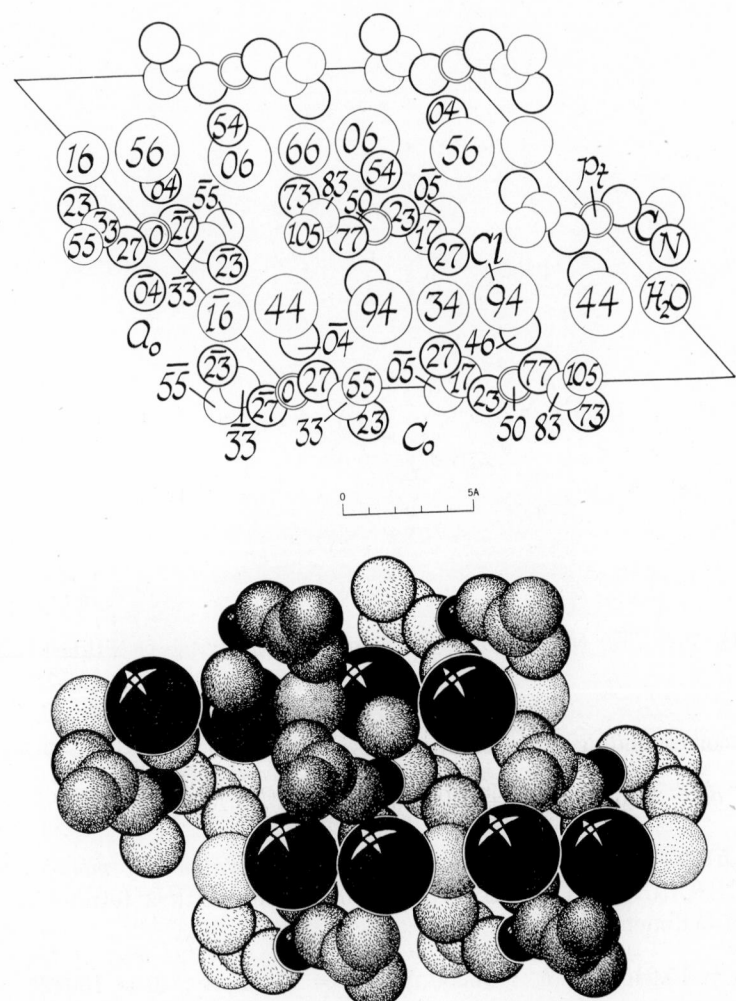

Fig. X1VB,139a (top). The monoclinic structure of $Pt(NH_3)_2[CH_3C(NH_2)NH]_2Cl_2 \cdot$ H_2O projected along its b_0 axis. Right-hand axes.

Fig. XIVB,139b (bottom). A packing drawing of the monoclinic structure of Pt- $(NH_3)_2[CH_3C(NH_2)NH]_2Cl_2\cdot H_2O$ seen along its b_0 axis. The large black circles are chlorine; the small black platinum atoms only partially show. Heavily ringed nitrogen atoms are line shaded; the hook-shaded carbon atoms are of similar size. Water molecules are slightly larger and dotted.

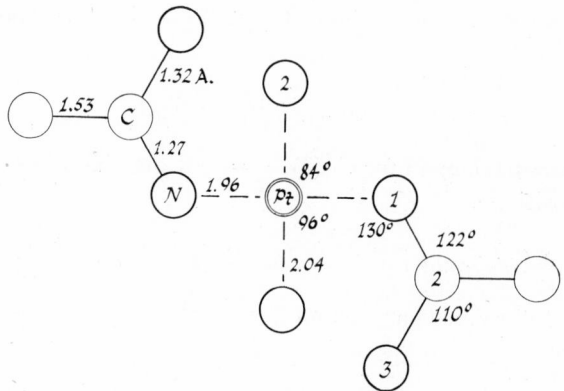

Fig. XIVB,140.   Bond dimensions in crystals of $Pt(NH_3)_2[CH_3C(NH_2)NH]_2Cl_2 \cdot H_2O$

TABLE XIVB,66

Parameters of Atoms in $PtCl_2(NH_3)_2[CH_3C(NH_2)NH]_2 \cdot H_2O$

Atom	x	y	z
C(1)	0.0095	0.5474	0.1626
C(2)	0.9778	0.3341	0.1059
N(1)	0.0225	0.2669	0.0707
N(2)	0.1641	0.9573	0.1175
N(3)	0.9049	0.2330	0.1054
Cl	0.2634	0.4408	0.1485

In this structure (Fig. XIVB,139), the environment of the fourfold co-ordinated platinum atoms and the bond dimensions of the acetonitrile groups are given in Figure XIVB,140. Platinum and its four coordinated nitrogen atoms are coplanar, but the acetonitrile is not planar. The water molecules are hydrogen-bonded to two chloride ions, with Cl–O = 3.01 A., and to two nitrogen atoms with O–N = 3.08 A. Chloride ions also come close to other atoms, with Cl–N = 3.40 A., Cl–O = 3.21 A., and Cl–C = 3.81 A.

XIV,b101.   Crystals of *cuprous iodide–methyl isocyanide*, $CuI \cdot CH_3NC$, are monoclinic with a unit that contains eight molecules and has the dimensions:

$$a_0 = 13.88 \text{ A.}; \quad b_0 = 13.20 \text{ A.}; \quad c_0 = 5.765 \text{ A.}; \quad \beta = 105°$$

The space group has been chosen as C_{2h}^6 $(C2/c)$ with the copper atoms in two sets of

$$(4e)   \pm (0 \; u \; {}^1/_4; \; {}^1/_2, u + {}^1/_2, {}^1/_4)$$

For Cu(1), $u = 0.005$, and for Cu(2), $u = -0.254$. All other atoms are in the general positions:

$$(8f)   \pm (xyz; \; x, \bar{y}, z + {}^1/_2; \; x + {}^1/_2, y + {}^1/_2, z; \; x + {}^1/_2, {}^1/_2 - y, z + {}^1/_2)$$

For them the determined parameters are:

Atom	x	y	z
I	0.1175	0.119	0.056
C(1)	0.072	0.314	0.567
N	0.124	0.351	0.444
C(2)	0.184	0.381	0.317

The resulting structure is shown in Figure XIVB,141. The cuprous ion has its usual tetrahedral coordination. In this case the Cu(1) atom is surrounded by four iodine atoms (Cu–I = 2.668 or 2.638 A.), and the Cu(2) atom by two iodine atoms (Cu–I = 2.731 A.) and two isocyanide carbon atoms (Cu–C = 1.81 A.). Within the methyl isocyanide groups, C(1)–N = 1.24 A. and N–CH$_3$ = 1.30 A.

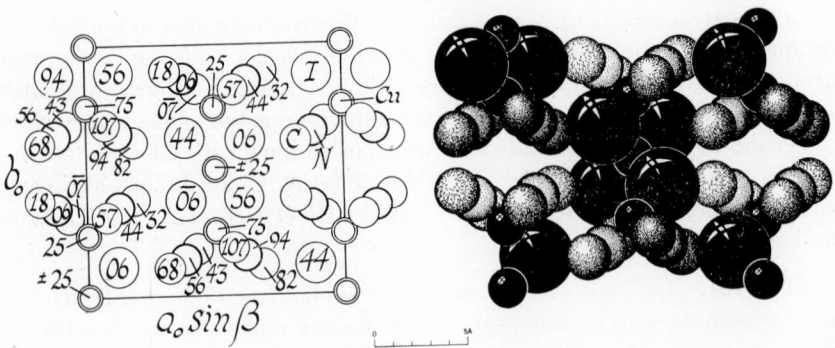

Fig. XIVB,141a (left).   The monoclinic structure of CuI·CH$_3$CN projected along its c_0 axis.   Right-hand axes.

Fig. XIVB,141b (right).   A packing drawing of the monoclinic CuI·CH$_3$CN arrangement seen along its c_0 axis.   Iodine atoms are the large, copper atoms the small black circles.   The nitrogen atoms are heavily outlined and dotted; the carbon atoms are line shaded.

XIV,b102.   *Zinc chloride di acetonitrile,* $ZnCl_2 \cdot 2CH_3CN$, is orthorhombic with a tetramolecular unit of the edge lengths:

$$a_0 = 12.83 \text{ A.}; \quad b_0 = 10.04 \text{ A.}; \quad c_0 = 6.65 \text{ A.}$$

The space group is V_h^{16} (*Pnma*) with atoms in the positions:

$(4c)$ $\pm (u\ ^1/_4\ v;\ u+^1/_2,^1/_4,^1/_2-v)$

$(8d)$ $\pm (xyz;\ ^1/_2-x,y+^1/_2,z+^1/_2;\ x,^1/_2-y,z;\ x+^1/_2,y,^1/_2-z)$

The selected positions and parameters are those of Table XIVB,67.

In this structure, (Fig. XIVB,142) the zinc atoms are tetrahedrally surrounded by two chlorine and two nitrogen atoms, with $Zn-Cl = 2.17$ A. and $Zn-N = 2.0$ A.

TABLE XIVB,67
Positions and Parameters of the Atoms in $ZnCl_2 \cdot 2CH_3CN$

Atom	Position	x	y	z
Zn	$(4c)$	0.997	$^1/_4$	0.250
Cl(1)	$(4c)$	0.924	$^1/_4$	0.547
Cl(2)	$(4c)$	0.161	$^1/_4$	0.164
N	$(8d)$	0.925	0.100	0.125
C(1)	$(8d)$	0.889	0.035	0.243
C(2)	$(8d)$	0.825	0.955	0.391

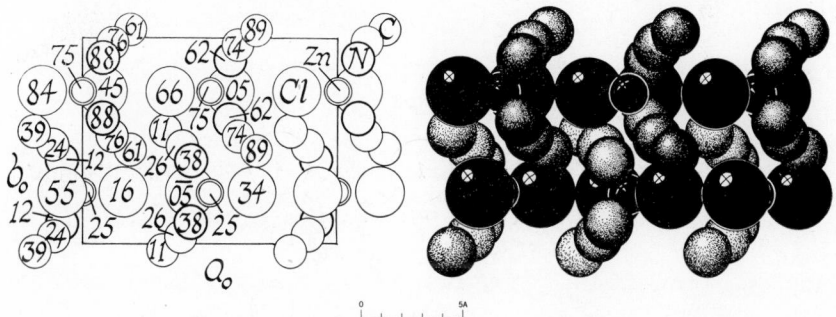

Fig. XIVB,142a (left).   The orthorhombic structure of $ZnCl_2 \cdot 2CH_3CN$ projected along its c_0 axis.   Right-hand axes.

Fig. XIVB,142b (right).   A packing drawing of the orthorhombic $ZnCl_2 \cdot 2CH_3CN$ structure seen along its c_0 axis.   The chlorine atoms are the large, the zinc atoms the small black circles.   The nitrogen atoms are heavily outlined and dotted; the carbon atoms are line shaded.

XIV,b103.   Structures have been determined for two compounds of $CuCl_2$ with CH_3CN. One of these, the yellow-brown *cupric chloride mono acetonitrile*, $CuCl_2 \cdot CH_3CN$, has a tetramolecular monoclinic unit of the dimensions:

$$a_0 = 3.84 \pm 0.01 \text{ A.}; \quad b_0 = 7.91 \pm 0.01 \text{ A.}; \quad c_0 = 18.35 \pm 0.02 \text{ A.}$$
$$\beta = 91°6' \pm 6'$$

The space group is C_{2h}^5 ($P2_1/c$) with atoms in the positions:

$$(4e) \quad \pm (xyz; \; x, {}^1\!/_2 - y, z + {}^1\!/_2)$$

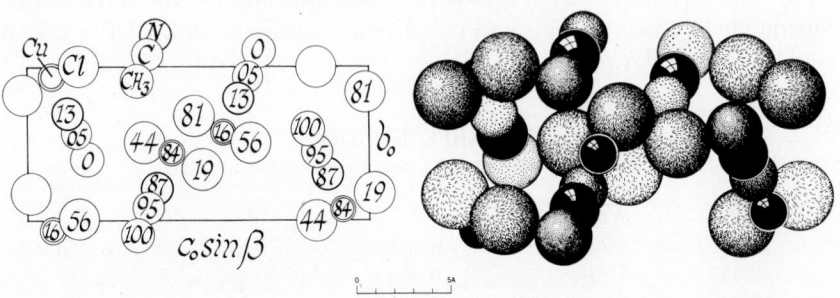

Fig. XIVB,143a (left).   The monoclinic structure of $CuCl_2 \cdot CH_3CN$ projected along its a_0 axis.   Right-hand axes.

Fig. XIVB,143b (right).   A packing drawing of the monoclinic $CuCl_2 \cdot CH_3CN$ structure viewed along its a_0 axis.   The carbon atoms are black, the methyl groups slightly larger and dotted.   Atoms of nitrogen are heavily ringed and hook shaded.   The largest, line shaded circles are chlorine; the small black circles are copper.

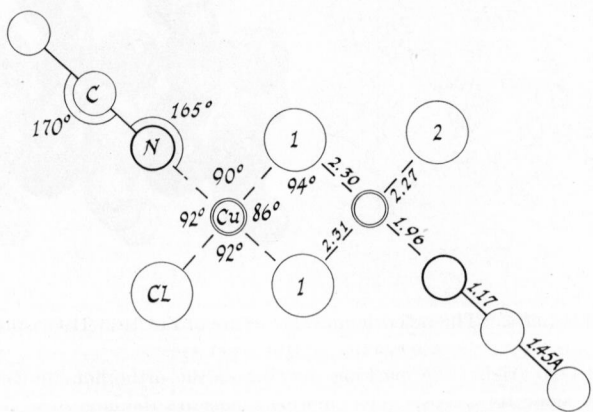

Fig. XIVB,144.   Bond dimensions in crystals of $CuCl_2 \cdot CH_3CN$.

TABLE XIVB,68
Parameters of Atoms in CuCl$_2$·CH$_3$CN

Atom	x	$\sigma(x)$	y	$\sigma(y)$	z (average)
Cu	0.8386	0.0011	0.0766	0.0006	0.0783
Cl(1)	0.1885	0.0020	0.1720	0.0010	−0.0145
Cl(2)	0.4380	0.0022	−0.0143	0.0010	0.1582
N	0.8700	0.0093	0.3018	0.0034	0.1220
C(1)	0.9465	0.0189	0.4396	0.0033	0.1520
C(2)	0.0023	0.0186	0.5994	0.0040	0.1771

The atomic parameters are given in Table XIVB,68.
The structure is shown in Figure XIVB,143; its molecules have the bond dimensions of Figure XIVB,144. They are stacked one above another along the a_0 axis of the crystal in such a way that there are two chlorine atoms of adjacent molecules distant 2.79 or 3.08 A. from each copper atom.

XIV,b104. The dark red *tri cupric chloride diacetonitrile*, 3CuCl$_2$·2CH$_3$CN, is monoclinic with a bimolecular unit of the dimensions:

$$a_0 = 6.78 \pm 0.01 \text{ A.}; \quad b_0 = 6.13 \pm 0.01 \text{ A.}; \quad c_0 = 16.51 \pm 0.02 \text{ A.}$$
$$\beta = 105°42' \pm 6'$$

The space group is C$_{2h}^5$ (*P*2$_1$/*c*) with one set of copper atoms in the positions:

$$\text{Cu(1): } (2a) \quad 000; 0\,^1/_2\,^1/_2$$

All other atoms are in the positions:

$$(4e) \quad \pm (xyz; x,^1/_2-y,z+^1/_2)$$

with the parameters of Table XIVB,69.

TABLE XIVB,69
Parameters of Atoms in 3CuCl$_2$·2CH$_3$CN

Atom	x	$\sigma(x)$	y	$\sigma(y)$	z	$\sigma(z)$
Cu(2)	0.7581	0.0003	0.4434	0.0002	0.04262	0.00010
Cl(1)	0.1726	0.0007	−0.2855	0.0005	0.07324	0.00018
Cl(2)	0.9384	0.0007	0.1556	0.0005	0.11657	0.00018
Cl(3)	0.3865	0.0007	−0.7343	0.0005	0.03442	0.00020
N	0.7484	0.0025	0.6008	0.0021	0.1457	0.0007
C(1)	0.7041	0.0031	0.7303	0.0019	0.1867	0.0009
C(2)	0.6392	0.0033	0.9007	0.0024	0.2377	0.0011

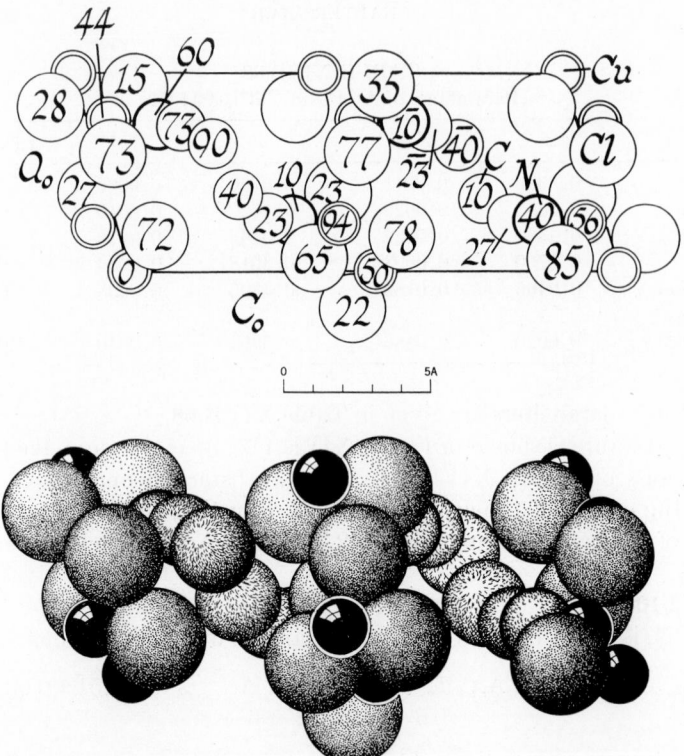

Fig. XIVB,145a (top). The monoclinic structure of $3CuCl_2 \cdot 2CH_3CN$ projected along its b_0 axis. Right-hand axes.

Fig. XIVB,145b (bottom). A packing drawing of the monoclinic structure of $3CuCl_2 \cdot 2CH_3CN$ projected along its b_0 axis. The copper atoms are black; the large chlorine atoms are dotted. Carbon atoms are line shaded; the atoms of nitrogen are heavily ringed and hook shaded.

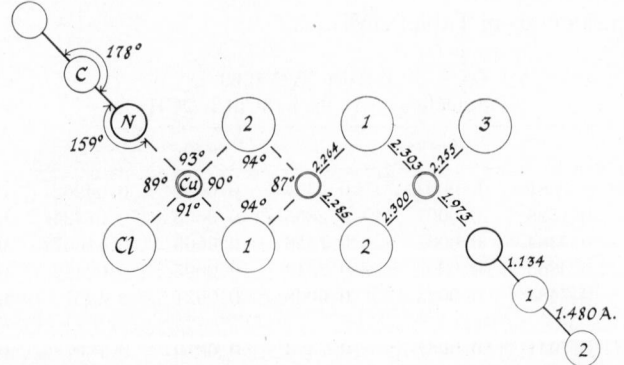

Fig. XIVB,146. Bond dimensions in crystals of $3CuCl_2 \cdot 2CH_3CN$.

The structure as a whole is illustrated in Figure XIVB,145. The molecules have the dimensions of Figure XIVB,146. The molecular stacking along the b_0 axis resembles that in $CuCl_2 \cdot CH_3CN$ (**XIV,b103**) in putting chlorine atoms of adjacent molecules above and below each copper atom, with Cu–Cl = 2.71, 3.01, or 3.18 A.

XIV,b105. Crystals of *hexa methylisocyanide, ferrous chloride trihydrate*, $Fe(CH_3NC)_6Cl_2 \cdot 3H_2O$, are hexagonal with a unimolecular cell of the dimensions:

$$a_0 = 10.47 \text{ A.}, \qquad c_0 = 5.315 \text{ A.}$$

Atoms have been placed in the following special positions of D_{3d}^3 ($P\bar{3}m1$):

Fe: $(1a)$ 000
$H_2O(1)$: $(2d)$ $\pm(^1/_3 \, ^2/_3 \, u)$ with $u = -0.1$
$2Cl+1H_2O(2)$: $(3e)$ $^1/_2 \, 0 \, 0; \, 0 \, ^1/_2 \, 0; \, ^1/_2 \, ^1/_2 \, 0$
C: $(6i)$ $\pm(u\bar{u}v; \, u \, 2u \, v; \, 2\bar{u} \, \bar{u} \, v)$
        with $u = 0.125$, $v = 0.200$
N: $(6i)$ with $u = 0.205$, $v = 0.330$
CH_3: $(6i)$ with $u = 0.313$, $v = 0.460$

The molecule that results has (CH_3NC) radicals octahedrally distributed about the central iron atom (Fig. XIVB,147) with the significant atomic separations: Fe–C = 1.85 A., C–N = 1.18 A., N–CH₃ = 1.47 A. The angle

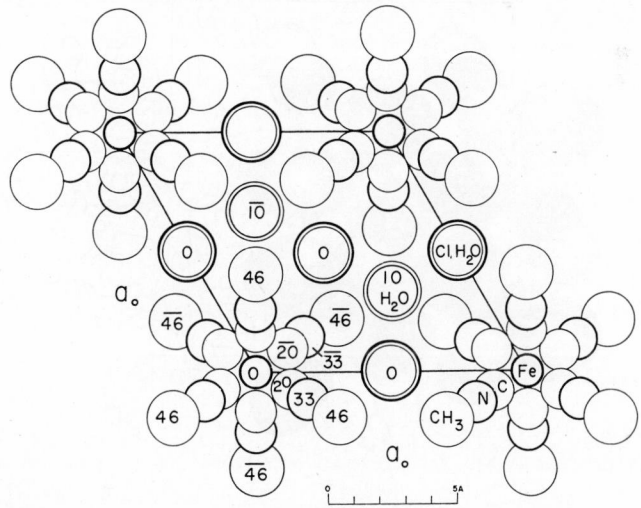

Fig. XIVB,147.   The hexagonal structure of $Fe(CH_3NC)_6Cl_2 \cdot 3H_2O$ projected along its c_0 axis. Right-hand axes.

C–N–CH$_3$ = 173°. These groups are closely packed in the structure with the three water molecules and two Cl$^-$ ions lying in holes left between. The shortest separations involving them and atoms of the Fe(CH$_3$NC)$_6^{2+}$ ion are Cl$^-$–H$_2$O(1) = 3.06 A., CH$_3$–H$_2$O(1) = 3.39 A., CH$_3$–Cl$^-$ = 3.60 A.

XIV,b106. The β form of *ferrous cyanide tetra methyl isonitrile*, Fe(CH$_3$NC)$_4$(CN)$_2$, is orthorhombic with a tetramolecular unit of the edge lengths:

$$a_0 = 8.45 \pm 0.02 \text{ A.;} \quad b_0 = 13.24 \pm 0.01 \text{ A.;} \quad c_0 = 11.62 \pm 0.02 \text{ A.}$$

The space group is V$_h^{15}$ (*Pbca*) with the iron atoms in the special positions:

$$(4a) \quad 000; \; ^1/_2 \, ^1/_2 \, 0; \; 0 \, ^1/_2 \, ^1/_2; \; ^1/_2 \, 0 \, ^1/_2$$

All other atoms are in the general positions:

$$(8c) \quad \pm (xyz; \; ^1/_2 - x, y + ^1/_2, z; \; x, ^1/_2 - y, z + ^1/_2; \; x + ^1/_2, y, ^1/_2 - z)$$

with the parameters of Table XIVB,70.

The resulting structure (Fig. XIVB,148) is a packing of neutral molecules which have a *trans* configuration. The Fe–C distances are ca. 1.8 A.;

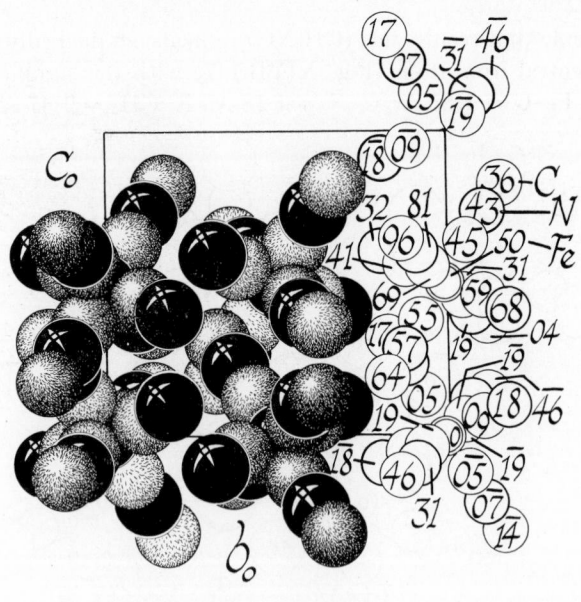

Fig. XIVB,148. The orthorhombic structure of Fe(CH$_3$NC)$_4$(CN)$_2$ projected along its a_0 axis. Right-hand axes.

TABLE XIVB,70
Parameters of the Atoms in Fe(CH₃NC)₄(CN)₂

Atom	x	y	z
C(1)	0.047	−0.066	0.128
N(1)	0.072	−0.115	0.220
CH₃(1)	0.140	−0.168	0.312
C(2)	0.190	−0.052	−0.054
N(2)	0.308	−0.081	−0.104
CH₃(2)	0.456	−0.124	−0.129
C(3)	0.093	0.112	0.054
N(3)	0.181	0.183	0.095

the other atomic distances agree, within the limit of accuracy of the determination, with those to be expected.

XIV,b107. The red form of *cobaltous perchlorate penta methyl isonitrile,*
Co(CH₃NC)₅·(ClO₄)₂ is orthorhombic with a tetramolecular unit of the
dimensions:

$$a_0 = 13.17 \text{ A.}; \quad b_0 = 12.47 \text{ A.}; \quad c_0 = 12.49 \text{ A.}, \quad \text{all } \pm 0.02 \text{ A.}$$

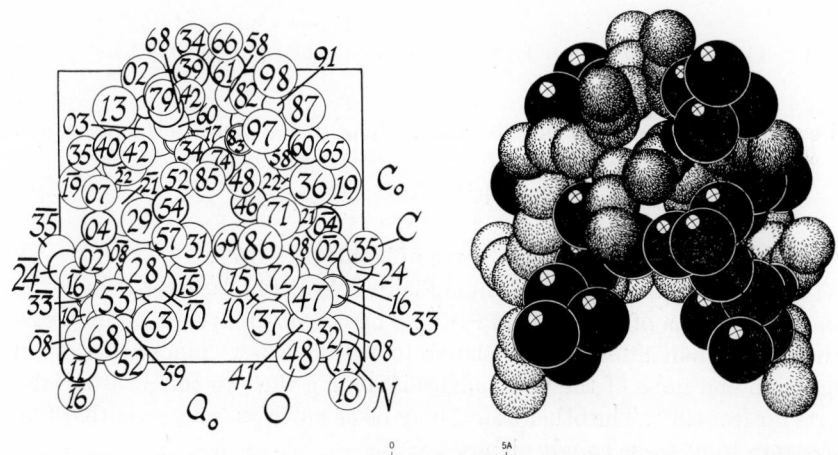

Fig. XIVB,149a (left). The orthorhombic structure of the red form of Co(CH₃NC)₅-
(ClO₄)₂ projected along its b_0 axis. The chlorine atoms are the small single-ringed, the
cobalt atoms the double-ringed circles. Right-hand axes.

Fig. XIVB,149b (right). A packing drawing of the orthorhombic structure of
Co(CH₃NC)₅(ClO₄)₂ viewed along its b_0 axis. The oxygen are the large, the cobalt the
small black circles. The chlorine atoms do not show. The nitrogen atoms are hook
shaded and heavily outlined; the atoms of carbon, of the same size, are line shaded.

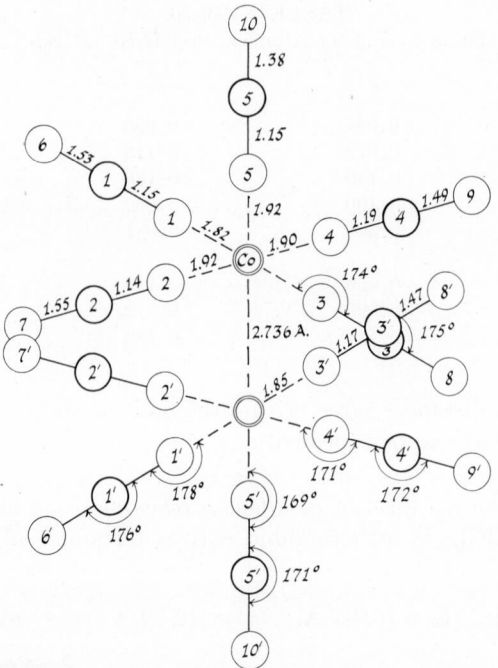

Fig. XIVB,150. Bond dimensions in crystals of Co(CH₃NC)₅(ClO₄)₂.

The space group has been chosen as V^3 ($P2_12_12$) with atoms in the positions:

(4c) $xyz;\ \bar{x}\bar{y}z;\ x+^1/_2,^1/_2-y,\bar{z};\ ^1/_2-x,y+^1/_2,\bar{z}$

The determined parameters are those of Table XIVB,71.

The resulting structure is shown in Figure XIVB,149. Its cations have the bond dimensions of Figure XIVB,150. The cobalt atom and its four sur-rounding carbon atoms are coplanar to within a few hundredths of an angstrom and none of the bond angles involving the two cobalt atoms de-parts far from 90°. The other atoms may be as much as several tenths of an angstrom from these cobalt planes.

XIV,b108. *Cuprous chloride tetrakis thioacetamide*, CuCl·4SC(CH₃)NH₂, was one of the early chelated compounds to be studied in detail with x-rays. Its tetragonal crystals have a bimolecular unit of the dimensions:

$a_0 = 12.449$ A., $c_0 = 5.469$ A.,   both ±0.005 A.

TABLE XIVB,71
Parameters of the Atoms in $Co(CH_3NC)_5 \cdot (ClO_4)_2$

Atom	x	$\sigma(x)$	y	$\sigma(y)$	z	$\sigma(z)$
Co	0.0991	0.0003	0.3286	0.0002	0.2371	0.0002
C(1)	0.128	0.002	−0.097	0.001	0.178	0.002
C(2)	0.079	0.002	0.082	0.001	0.093	0.002
C(3)	0.057	0.002	0.163	0.001	0.292	0.002
C(4)	0.109	0.002	−0.017	0.001	0.380	0.001
C(5)	0.239	0.003	0.076	0.002	0.243	0.002
C(6)	0.157	0.003	−0.290	0.002	0.089	0.002
C(7)	0.056	0.004	0.157	0.003	−0.107	0.003
C(8)	−0.013	0.003	0.348	0.003	0.364	0.002
C(9)	0.138	0.003	−0.072	0.002	0.585	0.002
C(10)	0.414	0.006	0.148	0.004	0.285	0.004
N(1)	0.144	0.002	−0.179	0.002	0.141	0.002
N(2)	0.063	0.002	0.109	0.002	0.007	0.002
N(3)	0.023	0.002	0.243	0.002	0.326	0.002
N(4)	0.129	0.002	−0.042	0.002	0.470	0.002
N(5)	0.321	0.003	0.101	0.002	0.263	0.002
Cl(1)	0.2082	0.0008	0.4081	0.0007	0.1254	0.0007
Cl(2)	0.3337	0.0009	0.7469	0.0007	0.3991	0.0007
O(1)	0.309	0.003	0.371	0.003	0.141	0.003
O(2)	0.147	0.006	0.319	0.004	0.093	0.004
O(3)	0.172	0.003	0.468	0.002	0.213	0.002
O(4)	0.219	0.004	0.476	0.003	0.031	0.003
O(5)	0.427	0.003	0.687	0.002	0.396	0.002
O(6)	0.346	0.005	0.859	0.004	0.410	0.004
O(7)	0.282	0.006	0.724	0.005	0.305	0.005
O(8)	0.277	0.004	0.714	0.003	0.490	0.003

The space group is $S_4{}^2(I\overline{4})$ and according to a recent redetermination, the copper and chlorine atoms are distributed as follows:

Cu: (2a)   000; $^1/_2\,^1/_2\,^1/_2$
Cl: (2d) $0\,^1/_2\,^3/_4$; $^1/_2\,0\,^1/_4$

The other atoms are in the general positions:

(8g) $xyz;\ \bar{x}\bar{y}z;\ y\bar{x}\bar{z};\ \bar{y}x\bar{z};$ B.C.

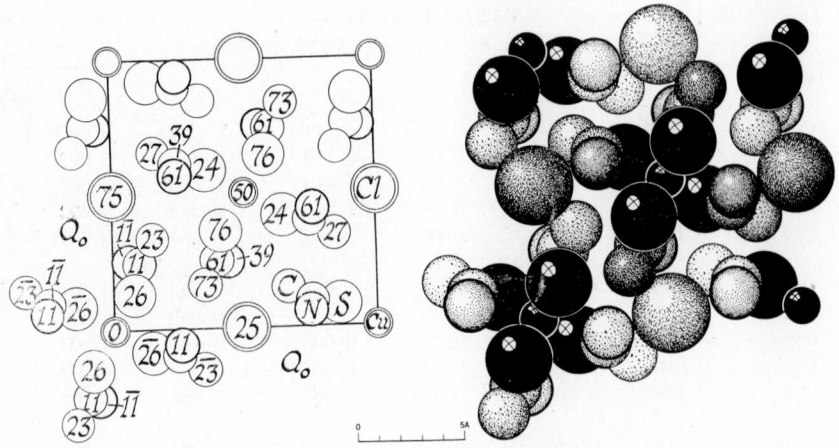

Fig. XIVB,151a (left).   The tetragonal structure of $CuCl \cdot 4SC(CH_3)NH_2$ projected along its c_0 axis.   Right-hand axes.

Fig. XIVB,151b (right).   A packing drawing of the tetragonal $CuCl \cdot 4SC(CH_3)NH_2$ arrangement viewed along its c_0 axis.   The sulfur atoms are the large, the copper atoms the small black circles.   The chlorine atoms are the largest, line shaded circles.   The nitrogen atoms are heavily outlined and dotted.

with the parameters:

Atom	x	y	z
S	0.0758	0.1273	0.2651
C(1)	0.0979	0.2442	0.1137
C(2)	0.1506	0.3373	0.2348
N	0.0657	0.2575	−0.1094

The atomic positions thus defined are significantly different from those found in the earlier study.

The structure resulting from the redetermination is shown in Figure XIVB,151. As is the case with other cuprous compounds, the fourfold coordination of the copper atom is tetrahedral, with $Cu-S = 2.343$ A. The planar thioacetamide groups have the same bond dimensions as in crystals of thioacetamide itself (**XIV,b61**) with $C-C = 1.494$ A., $C-N = 1.324$ A., and $C-S = 1.713$ A. Each Cl^- ion has four NH_2 neighbors at the corners of a flattened tetrahedron, with $Cl^--N = 3.22$ A.; these are thought to represent hydrogen bonds.

The corresponding *silver* salt, $AgCl \cdot 4SC(CH_3)NH_2$, is presumably isostructural with a cell of the dimensions:

$$a_0 = 12.59 \text{ A.}, \qquad c_0 = 5.37 \text{ A.}$$

XIV,b109. Crystals of *trans platinum dichloride ethylene dimethylamine*, $PtCl_2 \cdot C_2H_4 \cdot NH(CH_3)_2$, are monoclinic with a bimolecular cell of the dimensions:

$$a_0 = 7.77 \pm 0.02 \text{ A.}; \quad b_0 = 8.67 \pm 0.03 \text{ A.}; \quad c_0 = 6.65 \pm 0.02 \text{ A.}$$
$$\beta = 102°$$

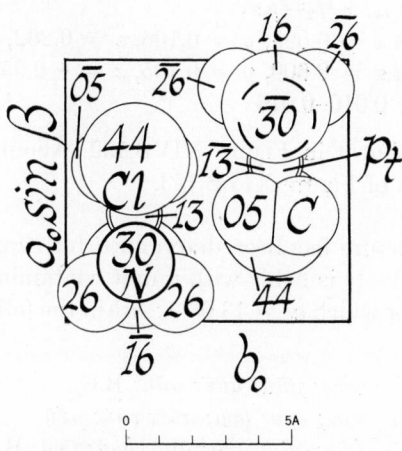

Fig. XIVB,152. The monoclinic structure of $PtCl_2 \cdot C_2H_4 \cdot NH(CH_3)_2$ projected along its c_0 axis. Right-hand axes.

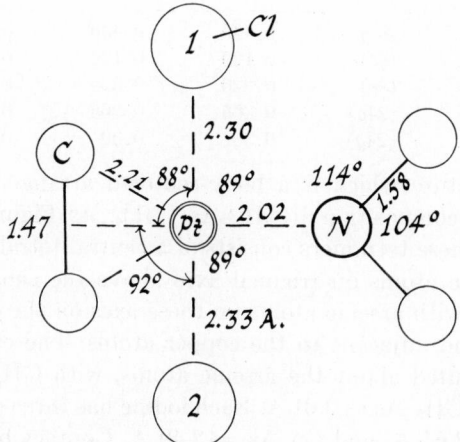

Fig. XIVB,153. Bond dimensions in crystals of $PtCl_2 \cdot C_2H_4 \cdot NH(CH_3)_2$.

The space group has been taken as C_{2h}^2 ($P2_1/m$) with atoms in the positions:

Pt: (2e) $\pm(u\;^1/_4\;v)$
          with $u = 0.4112$, $v = 0.1323$, $\sigma = 0.0008$
Cl(1): (2e) with $u = 0.197$, $v = -0.164$, $\sigma = 0.004$
Cl(2): (2e) with $u = 0.625$, $v = 0.436$, $\sigma = 0.004$
N: (2e)   with $u = 0.222$, $v = 0.300$, $\sigma = $ ca. 0.019
CH$_3$: (4f) $\pm(xyz;\;x,^1/_2-y,z)$
          with $x = 0.095$, $y = 0.106$, $z = 0.262$, $\sigma = $ ca. 0.010
CH$_2$: (4f) with $x = 0.600$, $y = 0.165$, $z = -0.050$,
          $\sigma = 0.010\text{--}0.015$

The structure is shown in Figure XIVB,152. About platinum, atoms have the distribution of Figure XIVB,153.

XIV,b110. A structure has been determined for *cuprous iodide triethyl arsine*, $CuI \cdot As(C_2H_5)_3$. It is cubic with a unit containing eight molecules. Atoms for this cell, for which $a_0 = 13.08$ A., are in the following special positions of T_d^3 ($I\bar{4}3m$):

(8c) $uuu;\;u\bar{u}\bar{u};\;\bar{u}u\bar{u};\;\bar{u}\bar{u}u;$ B.C.
(24g) $uuv;\;vuu;\;uvu;\;u\bar{u}\bar{v};\;v\bar{u}\bar{u};\;u\bar{v}\bar{u};$
       $\bar{u}\bar{u}v;\;\bar{v}\bar{u}u;\;\bar{u}\bar{v}u;\;\bar{u}u\bar{v};\;\bar{v}uu;\;\bar{u}v\bar{u};$ B.C.

The parameters, those for carbon atoms being chosen from considerations of available space rather than from x-ray data, are listed below.

Atom	Position	x	y	z
Cu	(8c)	0.430	0.430	0.430
I	(8c)	0.120	0.120	0.120
As	(8c)	0.320	0.320	0.320
CH$_2$	(24g)	0.365	0.365	0.18
CH$_3$	(24g)	0.30	0.30	0.10

This is a structure which is a body-centered arrangement of two big molecules of the composition $[(C_2H_5)_3As \cdot CuI]_4$. As Figure XIVB,154 indicates, each of these tetramers consists of a central tetrahedron of copper atoms with iodine atoms on trigonal axes above the center of the tetrahedral faces and with arsenic atoms on these axes on the opposite sides of the tetrahedra and adjacent to the copper atoms. The ethyl radicals are trigonally distributed about the arsenic atoms, with $CH_3\text{--}CH_2$ chosen as 1.59 A. and with $CH_2\text{--}As$ as 2.01 A. Each iodine has three copper neighbors at a distance of 2.65 A. and $Cu\text{--}As = 2.49$ A. Contact between adjacent tetrahedra is obtained through $I\text{--}CH_3$ separations of 3.34 A.

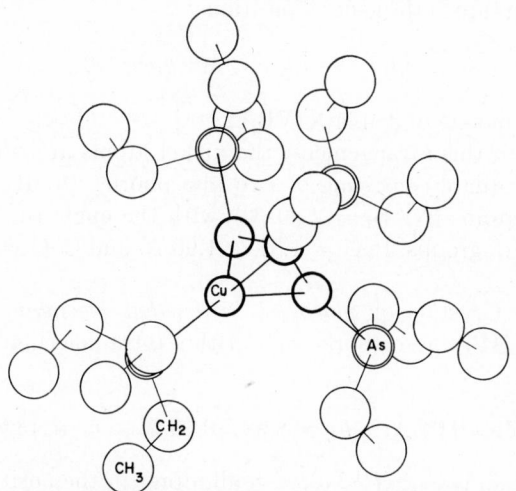

Fig. XIVB,154. The general atomic distribution in the molecule of $(C_2H_5)_3AsCuI$.

The corresponding *phosphine*, $CuI \cdot P(C_2H_5)_3$, is isostructural with

$$a_0 = 13.05 \text{ A.}$$

XIV,b111. Crystals of the addition compound *nickel bromide bis(triethyl phosphine)*, $NiBr_2 \cdot 2P(C_2H_5)_3$, are monoclinic with a bimolecular unit of the dimensions:

$$a_0 = 7.60 \text{ A.}; \quad b_0 = 11.5 \text{ A.}; \quad c_0 = 13.5 \text{ A.}; \quad \beta = 124°40'$$

The space group is C_{2h}^5 $(P2_1/c)$ and according to a preliminary announcement, atoms are in the positions:

$$\text{Ni: } (2a) \quad 000; 0 \, ^1/_2 \, ^1/_2$$

TABLE XIVB,72
Parameters of the Atoms in $NiBr_2 \cdot 2P(C_2H_5)_3$

Atom	x	y	z
Br	0.274	0.041	0.193
P	0.033	0.172	−0.079
C(1)	0.110	0.134	−0.181
C(2)	0.288	0.108	−0.200
C(3)	−0.200	0.260	−0.118
C(4)	0.275	0.307	0.019
C(5)	0.223	0.297	0.014
C(6)	0.056	0.391	−0.015

All other atoms are in the general positions:

$$(4e) \quad \pm (xyz; \; x,^1/_2 - y, z + ^1/_2)$$

with the parameters of Table XIVB,72.

According to this arrangement, the nickel atoms are surrounded by an approximate square consisting of two phosphorus (Ni–P = 2.26 A.) and two bromine atoms (Ni–Br = 2.30 A.), with the angle Br–Ni–P = 90°30′. In the $P(C_2H_5)_3$ groups, P–C = 1.87 or 1.90 A. and C–C = 1.53–1.56 A.

XIV,b112. Crystals of *platinum bis(triethyl phosphine) hydrobromide*, $Pt[(C_2H_5)_3P]_2HBr$, are orthorhombic with a tetramolecular cell of the edge lengths:

$$a_0 = 14.76 \pm 0.04 \text{ A.}; \quad b_0 = 8.92 \pm 0.03 \text{ A.}; \quad c_0 = 13.87 \pm 0.04 \text{ A.}$$

The space group is $C_{2v}{}^9$ ($Pn2_1a$) with all atoms in the positions:

$$(4a) \quad xyz; \; \bar{x}, y + ^1/_2, \bar{z}; \; ^1/_2 - x, y + ^1/_2, z + ^1/_2; \; x + ^1/_2, y, ^1/_2 - z$$

The determined parameters are those of Table XIVB,73.

TABLE XIVB,73
Parameters of the Atoms in $Pt[(C_2H_5)_3P]_2HBr$

Atom	x	y	z
Pt	0.1420	0.2500	0.0602
Br	0.2255	0.2500	0.2221
P(1)	0.003	0.250	0.128
P(2)	0.271	0.250	−0.027
C(1)	−0.094	0.233	0.040
C(2)	−0.162	0.333	0.083
C(3)	0.017	0.133	0.242
C(4)	0.008	−0.017	0.225
C(5)	−0.033	0.450	0.167
C(6)	0.025	0.516	0.258
C(7)	0.269	0.350	−0.154
C(8)	0.200	0.250	−0.217
C(9)	0.355	0.383	0.020
C(10)	0.355	0.533	0.020
C(11)	0.325	0.067	−0.058
C(12)	0.393	0.017	0.025

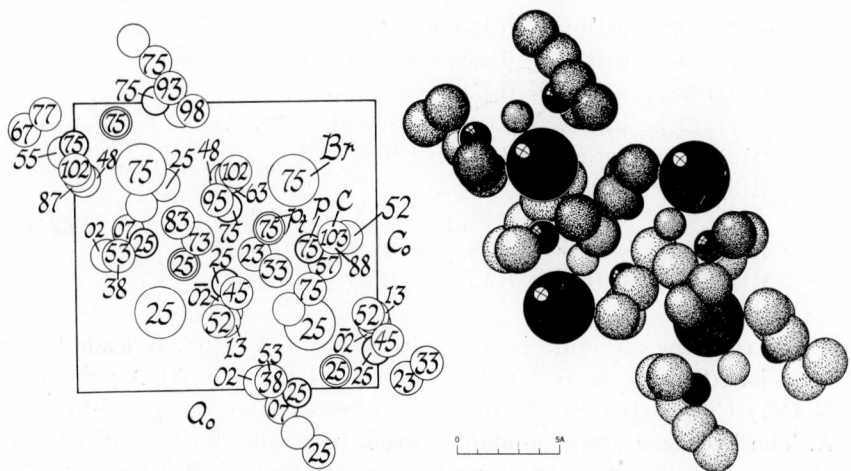

Fig. XIVB,155a (left). The orthorhombic structure of $Pt[(C_2H_5)_3P]_2HBr$ projected along its b_0 axis. Right-hand axes.

Fig. XIVB,155b (right). A packing drawing of the orthorhombic $Pt[(C_2H_5)_3P]_2HBr$ structure viewed along its b_0 axis. The phosphorus atoms are the small, the bromine atoms the large black circles. The platinum atoms are small and dotted. Carbon atoms are line shaded.

The resulting structure is shown in Figure XIVB,155. In its molecules the platinum, phosphorus and bromine atoms are coplanar, with Pt–P = 2.26 A. and Pt–Br = 2.56 A. The angles P–Pt–Br = 93°42′ and 94°6′, and therefore it is concluded that the coordination about the platinum atoms is square with hydrogen occupying the fourth corner. The positions of the carbon atoms are stated to be too approximate to give significance to their apparent separations.

8. Ethyl Substituted Ammonium Halides

XIV,b113. *Monoethyl ammonium bromide,* $C_2H_5NH_3Br$, at room temperature is monoclinic with a bimolecular cell of the dimensions:

$$a_0 = 8.361 \pm 0.010 \text{ A.}; \quad b_0 = 6.261 \pm 0.005 \text{ A.}; \quad c_0 = 4.630 \pm 0.005 \text{ A.}$$
$$\beta = 93°0′ \pm 12′$$

In a recent redetermination the space group was found to be C_{2h}^2 $(P2_1/m)$ with atoms in the positions:

Br: (2e) $\pm (u \; ^1/_4 \, v)$ with $u = -0.1619$, $v = -0.3968$

$\text{N}: (2e)$ with $u = 0.135, v = 0.125$
$\text{C}(1): (2e)$ with $u = 0.281, v = -0.050$
$\text{C}(2): (2e)$ with $u = 0.430, v = 0.153$
$\text{H}(1): (2e)$ with $u = 0.04, v = -0.01$
$\text{H}(2): (2e)$ with $u = 0.53, v = 0.04$
$\text{H}(3): (4f)$ $\pm (xyz; x, {}^1\!/_2 - y, z)$
                   with $x = 0.14, y = 0.12, z = 0.25$
$\text{H}(4): (4f)$ with $x = 0.28, y = 0.12, z = -0.17$
$\text{H}(5): (4f)$ with $x = 0.43, y = 0.12, z = 0.28$

The resulting structure is shown in Figure XIVB,156. It leads to the bond lengths $\text{N–C}(1) = 1.499$ A. and $\text{C}(1)–\text{C}(2) = 1.521$ A.; the angle $\text{N–C}(1)–\text{C}(2) = 109°16'$. N–H and C–H distances were assumed to be 1.0 A. The hydrogen atoms around nitrogen lie in the direction of bromine atoms with which they are considered to establish N–H–Br $= 3.37$ and 3.38 A. bridges. Contacts between molecules are between hydrogens, with H–H $= 2.8$–2.9 A.

The corresponding *chloride*, $\text{C}_2\text{H}_5\text{NH}_3\text{Cl}$, has this structure with

$$a_0 = 8.18 \text{ A.}; \quad b_0 = 5.95 \text{ A.}; \quad c_0 = 4.51 \text{ A.}; \quad \beta = 93°12'$$

The *iodide* also is isostructural with

$$a_0 = 8.69 \text{ A.}; \quad b_0 = 6.64 \text{ A.}; \quad c_0 = 4.82 \text{ A.}; \quad \beta = 92°6'$$

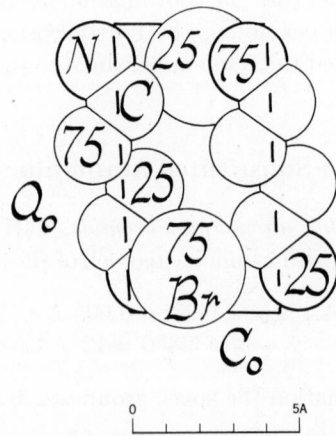

Fig. XIVB,156.   The monoclinic $\text{C}_2\text{H}_5\text{NH}_3\text{Br}$ structure projected along its b_0 axis. Right-hand axes.

XIV,b114. The *triethyl ammonium halides* have hexagonal symmetry and a structure which is somewhat related to that of $LiI \cdot 3H_2O$ (**X,c2**). Their bimolecular units have the dimensions:

$(C_2H_5)_3NHBr$: $a_0 = 8.56$ A., $c_0 = 7.49$ A.
$(C_2H_5)_3NHCl$: $a_0 = 8.38$ A., $c_0 = 7.08$ A.
$(C_2H_5)_3NHI$: $a_0 = 8.78$ A., $c_0 = 7.74$ A.

In a study made long ago ionic centers were placed in the following special positions of C_{6v}^4 ($P6_3mc$):

$(2b)$ $^1/_3\ ^2/_3\ u;\ ^2/_3,^1/_3,u+^1/_2$

For the halogen atoms u has been taken as 0, and for nitrogen atoms it has been given a value which, for the chloride, lies between 0.58 and 0.42. One set of carbon atoms (CH_2) may be in

$(6c)$ $u\bar{u}v;$ $u\ 2u\ v;$ $2\bar{u}\ \bar{u}\ v;$
$\bar{u},u,v+^1/_2;\ \bar{u},2\bar{u},v+^1/_2;\ 2u,u,v+^1/_2$
with $u = $ ca. 0.27, $v = $ ca. 0.50

In view of the fact that the ethyl groups are undoubtedly "rotating," nothing definite was considered known about their CH_3 carbons.

A projection of this structure is shown in Figure XIVB,157. This is the ionic distribution originally proposed for $(Li \cdot 3H_2O)I$, but in the more recent study the cation of the lithium salt was placed in $(2a)$ $00u;\ 0,0,u+^1/_2$.

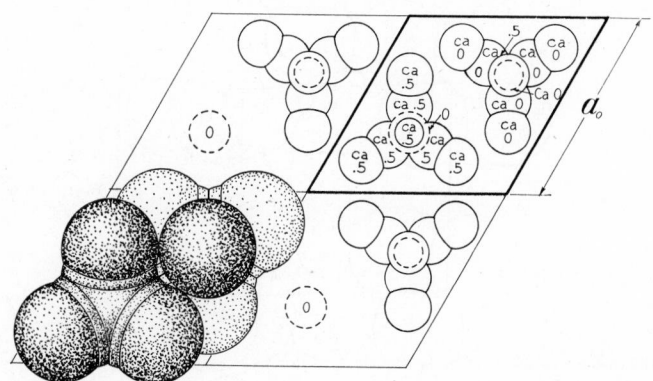

Fig. XIB,157.   Four units of the hexagonal structure of $(C_2H_5)_3NHCl$ projected along its c_0 axis.   Contents of the unit are shown in the upper right, a packing in the lower left corner.

If this is correct, the analogy to the present structure is restricted to halogen positions.

The *triethyl sulfonium iodide*, $(C_2H_5)_3SI$, has an orthorhombic structure which may be a slight distortion of this hexagonal arrangement. Its tetramolecular unit has the dimensions:

$$a_0 = 15.81 \text{ A.}; \quad b_0 = 8.69 \text{ A.}; \quad c_0 = 7.35 \text{ A.}$$

The space group probably is V_h^{16} (*Pnma*) or C_{2v}^9 (*Pna*). The relation to the ammonium salt becomes clear by expressing the hexagonal unit of $(C_2H_5)_3$-NHI in orthohexagonal axes to yield the dimensions:

$$a_0' = \sqrt{3}\, a_0 = 15.21 \text{ A.}; \quad b_0 = 8.78 \text{ A.}; \quad c_0 = 7.74 \text{ A.}$$

XIV,b115   *Monoethyl ammonium chlorostannate*, $(NH_3C_2H_5)_2SnCl_6$, and $(NH_3C_2H_5)_2PtCl_6$ have hexagonal symmetry. Their unimolecular cells have the edges:

$(NH_3C_2H_5)_2SnCl_6$: $a_0 = 7.24$ A., $c_0 = 8.41$ A.
$(NH_3C_2H_5)_2PtCl_6$: $a_0 = 7.13$ A., $c_0 = 8.53$ A.

Atoms of the chlorostannate ion have been found to be in the following positions of D_{3d}^3 (*P$\bar{3}$m*):

$$\text{Sn(Pt)}: (1a) \quad 000$$
$$\text{Cl}: (6i) \quad \pm(u\bar{u}v;\ 2\bar{u}\ \bar{u}\ v;\ u\ 2u\ v)$$
$$\text{with } u = 0.16, v = 0.17$$

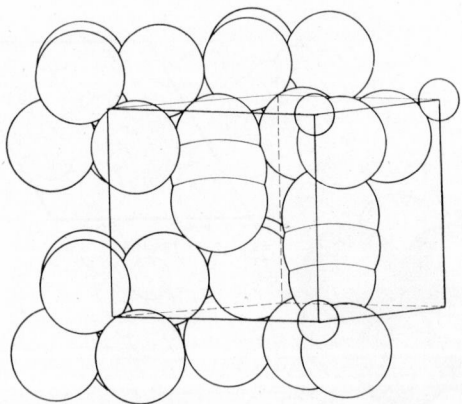

Fig. XIVB,158.   A perspective view of the hexagonal unit of the $(NH_3C_2H_5)_2SnCl_6$ structure indicating how some of its atoms pack.   The $SnCl_6$ ions, at corners of the cell, are separated in vertical layers by $(NH_3C_2H_5)^+$ cations.

Positions of the carbon atoms were not established since undoubtedly the ethyl radicals are "rotating," but the NH_3 ends of the $NH_3C_2H_5$ ions probably are in $(2d)$ $\pm(^1/_3\ ^2/_3\ u)$, with u = ca. 0.184.

This is obviously a layer structure (Fig. XIVB,158) which, except for the different axial ratio demanded by the elongated $(NH_3C_2H_5)$ ions, is like that for the β form of $(NH_4)_2SiF_6$ **(IX,c14)**. As the figure indicates, layers of closely packed $SnCl_6$ ions normal to the c_0 axis are separated by these $(NH_3C_2H_5)$ groups.

XIV,b116. *Tetraethyl ammonium heptaiodide*, $N(C_2H_5)_4I_7$, has been studied at room temperature, and also at $-175°C.$ to see if more information about the atomic positions could be obtained at the lower temperature. The symmetry is orthorhombic with a unit that contains four molecules and has the edge lengths:

$$a_0 = 11.502\ A.;\quad b_0 = 15.641\ A.;\quad c_0 = 12.357\ A.\ (20°C.)$$
$$a_0 = 11.236\ A.;\quad b_0 = 15.365\ A.;\quad c_0 = 12.345\ A.\ (-175°C.)$$

Uncertainty as to the true space group arises from disorder in the orientation of the ethyl groups, but the iodine atoms have been found to be in positions which can be described in terms of the following coordinates of V_h^{18} $(Acam)$:

$I(1)$: $(4a)$ $000;\ ^1/_2\ ^1/_2\ 0;\ 0\ ^1/_2\ ^1/_2;\ ^1/_2\ 0\ ^1/_2$

$I(2)$: $(8f)$ $\pm(uv0;\ ^1/_2-u,v,^1/_2;\ u,v+^1/_2,^1/_2;\ ^1/_2-u,v+^1/_2,0)$
with $u = 0.9233$, $v = 0.1805$ $(\sigma = 0.0015\ A.)$

$I(3)$: $(16g)$ $\pm(xyz;\qquad\qquad x+^1/_2,^1/_2-y,z;$
$\qquad\qquad\ xy\bar{z};\qquad\qquad x+^1/_2,^1/_2-y,\bar{z};$
$\qquad\qquad\ x,y+^1/_2,z+^1/_2;\ x+^1/_2,\bar{y},z+^1/_2;$
$\qquad\qquad\ x,y+^1/_2,^1/_2-z;\ x+^1/_2,\bar{y},^1/_2-z)$
with $x = 0.1561$, $y = 0.1947$, $z = 0.1795$

In V_h^{18} there is no way to distribute the ethyl groups tetrahedrally about the nitrogen atoms which are in

$$(4b)\quad 0\ 0\ ^1/_2;\ ^1/_2\ ^1/_2\ ^1/_2;\ 0\ ^1/_2\ 0;\ ^1/_2\ 0\ 0$$

but half atoms, corresponding to two orientations of the tetrahedra, could be put in four sets of $(16g)$ with the following parameters:

Atom	x	y	z
C(1)	0.100	0.047	0.442
C(2)	0.933	0.074	0.425
C(3)	0.858	0.050	0.328
C(4)	0.864	0.022	0.350

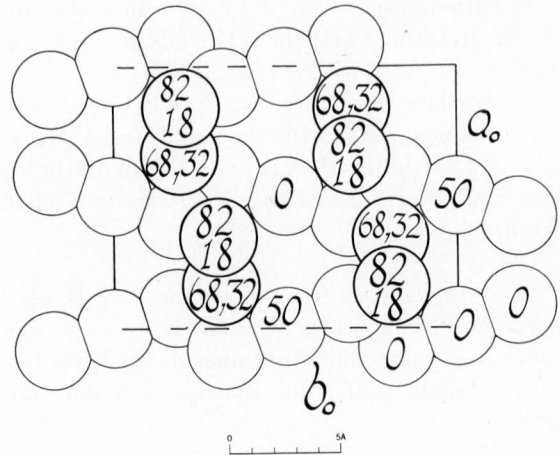

Fig. XIVB,159. The distribution of the iodine atoms in crystals of $N(C_2H_5)_4I_7$ as projected along the c_0 axis of its orthorhombic unit. Right-hand axes.

There are low peaks in the Fourier syntheses which are compatible with this interpretation. It is also possible that the space group is the lower symmetry C_{2v}^{17} (*Aba2*) in which the carbon atoms could be distributed without disorder if the extra Fourier peaks are considered as ghosts. The contributions of these light atoms to the entire structure, however, are not great enough to distinguish between these two possibilities.

In any event, the iodine atoms are distributed as shown in Figure XIVB,-159 and the structure consists of I_3^- and I_2 groups together with the $N(C_2H_5)_4$ cations. In the I_3 anion at $-175°C$., $I-I = 2.904$ A.; in the I_2 molecules, it is 2.735 A. There is no evidence requiring disorder in the distribution of the iodine atoms.

XIV,b117. Tetragonal *dimethyl diethyl ammonium chlorostannate*, $[N(CH_3)_2(C_2H_5)_2]_2SnCl_6$, has a bimolecular unit of the dimensions:

$$a_0 = 9.065 \text{ A.,} \qquad c_0 = 14.12 \text{ A.}$$

The ionic centers are distributed as in $[N(CH_3)_4]_2SiF_6$ (**XIV,a117**), but a different distribution was chosen for the halogen atoms. This gives the customary Sn-Cl separations, but turns the $SnCl_6$ ions through different orientations with respect to one another. Expressed in terms of special positions of the holohedral space group D_{4h}^6 (*P4/mnc*), the assigned atomic coordinates are

Sn: (2*a*)   000; B.C.

N: (4d) $0 \; ^1/_2 \; ^1/_4$; $^1/_2 \; 0 \; ^1/_4$; B.C.

Cl(1): (4e) $\pm (00u)$; B.C.   with $u = 0.17$

Cl(2): (8h) $\pm (uv0; v\bar{u}0; v+^1/_2,u+^1/_2,^1/_2; u+^1/_2,^1/_2-v,^1/_2)$
with $u = 0.23$, $v = 0.13$

The superscribed tetramolecular pseudocube has the cell edges:

$$a_0' = \sqrt{2}\, a_0 = 12.82 \text{ A.}, \qquad c_0' = c_0 = 14.12 \text{ A.}$$

9. Oxalic Acid and Derivatives

XIV,b118.   Anhydrous *oxalic acid*, $(COOH)_2$, is dimorphous with an orthorhombic and a monoclinic modification. For the α, orthorhombic, form, the tetramolecular unit has the edge lengths:

$$a_0 = 6.546 \text{ A.}; \quad b_0 = 7.847 \text{ A.}; \quad c_0 = 6.086 \text{ A.}, \qquad \text{all } \pm 0.005 \text{ A.}$$

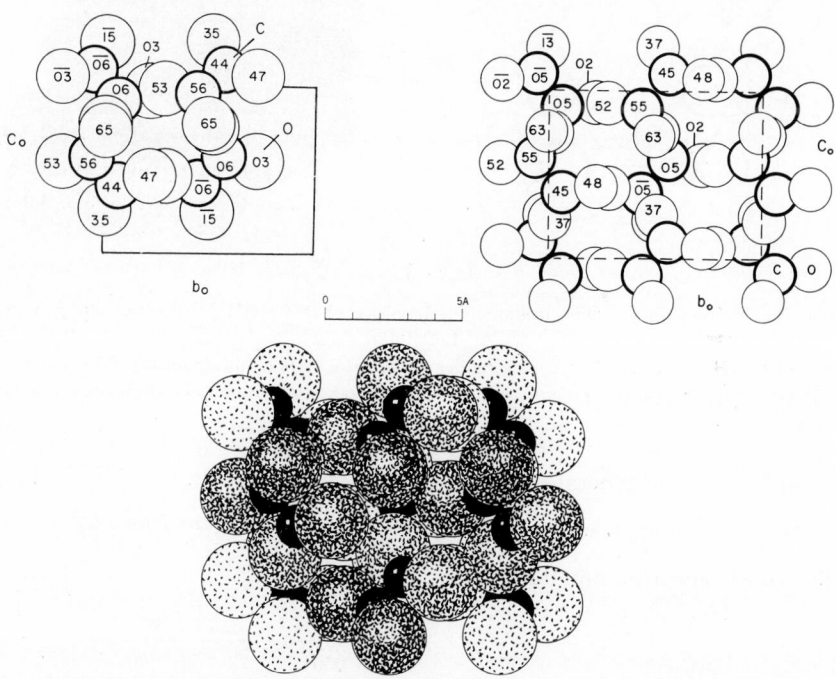

Fig. XIVB,160a (top left).   The orthorhombic structure of α-$(COOH)_2$ projected along its a_0 axis according to its most recent study.   Right-hand axes.

Fig. XIVB,160b (top right).   The a_0 projection of the α-$(COOH)_2$ structure according to its original parameters.

Fig. XIVB,160c (bottom).   A packing drawing of the orthorhombic α-oxalic acid structure viewed along its a_0 axis.   The atoms of carbon are black.

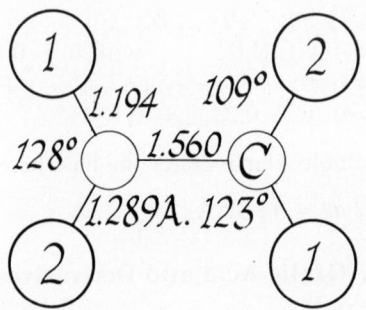

Fig. XIVB,161.   Bond dimensions of the oxalic acid molecule in α-(COOH)₂.

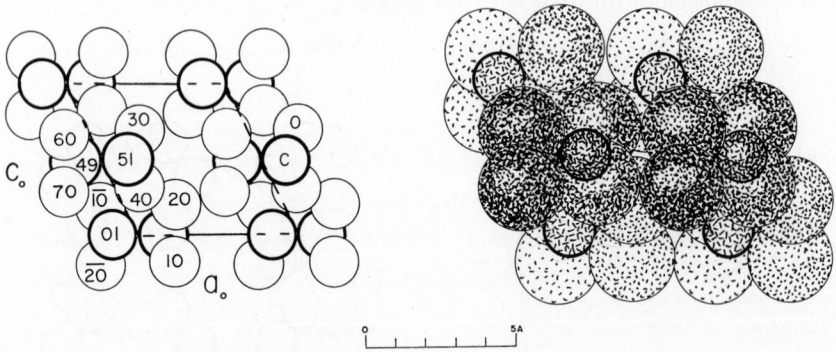

Fig. XIVB,162a (left).   The monoclinic structure of β-(COOH)₂ projected along its b_0
axis.   Left-hand axes.
  Fig. XIVB,162b (right).   A packing drawing of the monoclinic structure of β-(COOH)₂
viewed along its b_0 axis.   One of the line shaded, heavily outlined carbon atoms of each
molecule does not show.

Atoms are in the general positions of V_h^{15} (*Pcab*):

(8c) $\pm(xyz;\ x,y+{}^1/_2,{}^1/_2-z;\ x+{}^1/_2,{}^1/_2-y,z;\ {}^1/_2-x,y,z+{}^1/_2)$

The most recent parameters are:

Atom	x	y	z
C	0.0600	0.0551	0.9150
O(1)	0.1553	0.9941	0.7674
O(2)	0.0283	0.2141	0.9561

As can be seen by comparing Figures XIVB,160a and b, the new parameters represent only a minor shift in those originally assigned. The molecule,
according to this refinement, has the dimensions of Figure XIVB,161.

For the less thoroughly studied monoclinic, β, modification, there are two molecules in a cell of the dimensions:

$$a_0 = 5.30 \text{ A.}; \quad b_0 = 6.09 \text{ A.}; \quad c_0 = 5.51 \text{ A.}; \quad \beta = 115°30'$$

Its atoms are in the general positions of C_{2h}^5 $(P2_1/c)$:

(4e)　$\pm (xyz; x,{}^1/_2-y,z+{}^1/_2)$
For C: $x = 0.15$, $y = -0.01$, $z = 0.00$
For O(1): $x = 0.17$, $y = 0.10$, $z = -0.17$
For O(2): $x = 0.32$, $y = -0.13$, $z = 0.20$

The interatomic distances in the molecules of this structure (Fig. XIVB,-162) are substantially those in the α form.

XIV,b119. *Oxalic acid dihydrate*, $(COOH)_2 \cdot 2H_2O$, is monoclinic with a bimolecular unit of the dimensions:

$$a_0 = 6.119 \pm 0.004 \text{ A.}; \quad b_0 = 3.604 \pm 0.003 \text{ A.}; \quad c_0 = 12.051 \pm 0.005 \text{ A.}$$
$$\beta = 106°16' \pm 6'$$

Atoms are in the general positions of C_{2h}^5 $(P2_1/n)$:

(4e)　$\pm (xyz; x+{}^1/_2,{}^1/_2-y,z+{}^1/_2)$

Several especially careful studies of the parameters have been made using both x-rays and neutrons; the most accurate values thus obtained are listed in Table XIVB,74, the results of two determinations employing x-rays being in parentheses and in brackets.

The structure as a whole is shown in Figure XIVB,163. Bond dimensions in the molecule, as defined by the neutron study, are shown in Figure

TABLE XIVB,74
Parameters[a] of the Atoms in $(COOH)_2 \cdot 2H_2O$

Atom	x	y	z
C	-0.0450 [-0.0454] (-0.0455)	0.054 (0.0543)	0.0501 [0.0514] (0.0507)
O(1)	0.0817 [0.0852] (0.0854)	-0.057 (-0.0523)	0.1467 [0.1492] (0.1485)
O(2)	-0.2235 [-0.2187] (-0.2204)	0.227 (0.2156)	0.0375 [0.0373] (0.0364)
O(H$_2$O)	-0.0452 [-0.0467] (-0.0473)	0.114 (0.1028)	0.3222 [0.3207] (0.3204)
H(1)	-0.1361 [-0.1255] —	-0.062 —	0.3506 [0.3355] —
H(2)	0.0729 [0.048] —	0.118 —	0.3870 [0.383] —
H(3)	0.0227 [0.010] —	0.015 —	0.2180 [0.210] —

[a] The neutron results are given first, followed by two sets of x-ray parameters (in square brackets and parentheses).

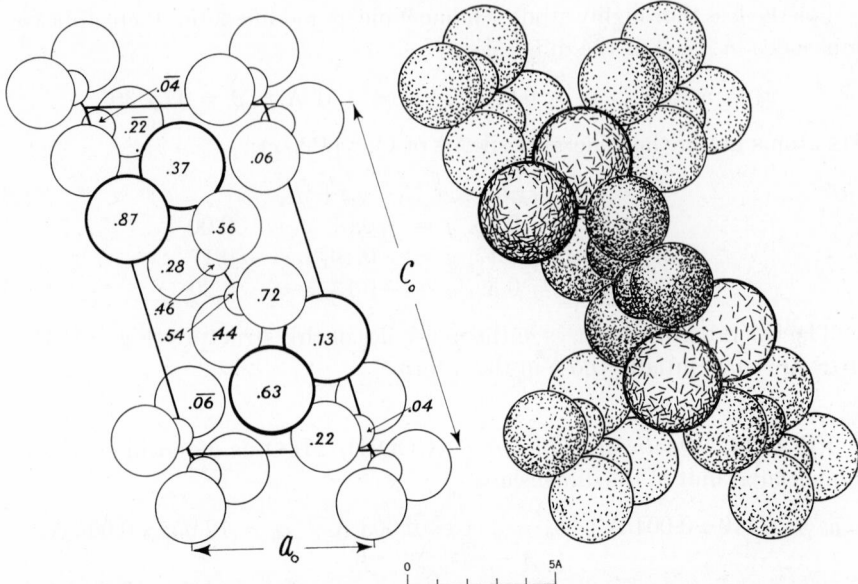

Fig. XIVB,163a (left).   A projection of the monoclinic structure of $(COOH)_2 \cdot 2H_2O$ along its b_0 axis.   The smallest circles are carbon;   the largest, more heavily outlined, are water molecules.   Left-hand axes.

Fig. XIVB,163b (right).   A packing drawing of the monoclinic structure of $(COOH)_2 \cdot 2H_2O$ seen along its b_0 axis.   The carbons are the smaller of the dotted circles.   The water molecules are more heavily ringed and line shaded.

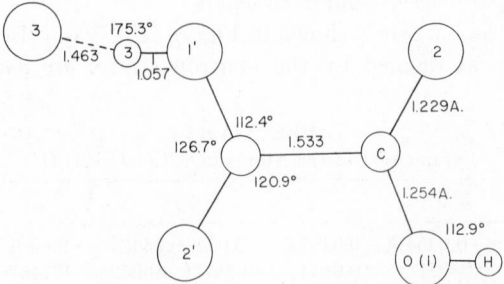

Fig. XIVB,164.   Bond dimensions found by neutron diffraction for the molecules in crystals of $(COOH) \cdot 2H_2O$.

XIVB,164. The molecule is planar even to the hydrogen atoms which are within 0.02 A. of the plane of the rest. This determination gives a particularly clear picture of the positions of the water hydrogens, H(1) and H(2).

In the well-defined H_2O molecules, $H(1)-O(3) = 0.968$ A., $H(2)-O(3) = 0.945$ A., and the angle $H(1)-O(3)-H(2) = 105°41'$. The carboxyl hydrogen $H(3)$ is 1.463 A. from the water oxygen $O(3)$.

It has been noted that the choice of different axes for this dihydrate brings out a close relationship between its structure and those found for the chemically analogous acetylene dicarboxylic and diacetylene dicarboxylic acid dihydrates. The pseudo-cells that apply to these axes and the parameters that refer to them are stated in **XIV,c68** and **69**.

Measurements made of the cell dimensions of the isostructural form of deuterium oxalic acid dideuterate show a clearly measurable effect of replacement by heavy hydrogen. Using the original axes, values found for the a_0 and c_0 axes of $(COOD)_2 \cdot 2D_2O$ are:

$$a_0 = 6.149 \text{ A.}; \quad c_0 = 12.071 \text{ A.}; \quad \beta = 106°33'$$

XIV,b120. *Deuterium oxalic acid dideuterate*, $(COOD)_2 \cdot 2D_2O$, has a second, β, modification which is also monoclinic with a bimolecular unit of the dimensions:

$$a_0 = 10.04 \text{ A.}; \quad b_0 = 5.06 \text{ A.}; \quad c_0 = 5.16 \text{ A.}, \qquad \text{all } \pm 0.01 \text{ A.}$$
$$\beta = 99°12' \pm 6'$$

The space group is C_{2h}^5 $(P2_1/a)$ with all atoms in the positions:

$$(4e) \quad \pm (xyz; \ x+{}^1/_2, {}^1/_2-y, z)$$

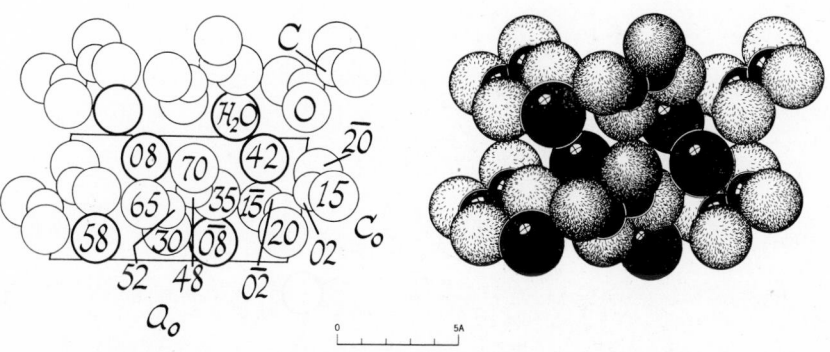

Fig. XIVB,165a (left). The monoclinic structure of β-$(COOD)_2 \cdot 2D_2O$ projected along its b_0 axis. Right-hand axes.

Fig. XIVB,165b (right). A packing drawing of the β form of $(COOD)_2 \cdot 2D_2O$ seen along its b_0 axis. The water molecules are the large, the carbon atoms the small black circles. Carboxyl oxygens are line shaded.

The parameters have been given as:

Atom	x	y	z
C	0.050	0.025	0.400
O(1)	0.039	0.199	0.249
O(2)	0.147	−0.153	0.440
H_2O	0.331	−0.077	0.151

The structure is shown in Figure XIVB,165. Like the α-form, it is built up of planar $(COOD)_2$ molecules tied together with deuterium bonds involving the water molecules; in this crystal they have the lengths 2.58, 2.82, and 2.89 A.

XIV,b121. The triclinic *oxamide*, $(CONH_2)_2$ has a unimolecular cell of the dimensions:

$$a_0 = 3.625 \text{ A.}; \quad b_0 = 5.188 \text{ A.}; \quad c_0 = 5.658 \text{ A.}$$
$$\alpha = 83°42'; \quad \beta = 114°6'; \quad \gamma = 115°10'$$

Atoms are in general positions of C_i^1 ($P\bar{1}$): $\pm(xyz)$ with the parameters:

Atom	x	y	z
C	−0.0025	0.4940	0.3638
N	−0.0032	0.2770	−0.2371
O	−0.0061	0.2792	0.2832

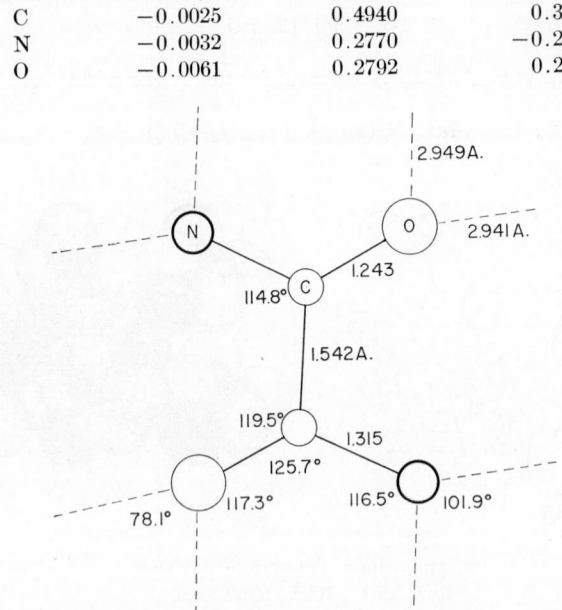

Fig. XIVB,166.   The bond dimensions in crystals of $(CONH_2)_2$.   Dashed lines correspond to intermolecular hydrogen bonds.

The bond lengths and angles of the resulting planar molecule are given in Figure XIVB,166. Within the molecular sheets the hydrogen bond lengths are 2.941 and 2.949 A.

XIV,b122. Crystals of *oxalyl bromide*, $(COBr)_2$, are monoclinic with a bimolecular unit of the dimensions:

$$a_0 = 6.18 \text{ A.}; \quad b_0 = 5.46 \text{ A.}; \quad c_0 = 7.80 \text{ A.}; \quad \beta = 112°24'$$

The space group is C_{2h}^5 ($P2_1/c$) with all atoms in the general positions:

$$(4e) \quad \pm(xyz; x,{}^1/_2-y,z+{}^1/_2)$$

Determined parameters are as follows:

Atom	x	y	z
Br	0.321	0.228	0.162
O	0.883	0.191	0.377
C	0.020	0.119	0.906

The structure that results is shown in Figure XIVB,167. Its molecules are planar within the limits of experimental error, with C–C = 1.56 A., C–O = 1.17 A., O–Br = 1.84 A., and C–C–O = 122°, O–C–Br = 128°. Between molecules the shortest O–Br = 3.27 A., this short separation giving rise to sheets of molecules; the shortest Br–O between the molecules of different sheets is 3.90 A.

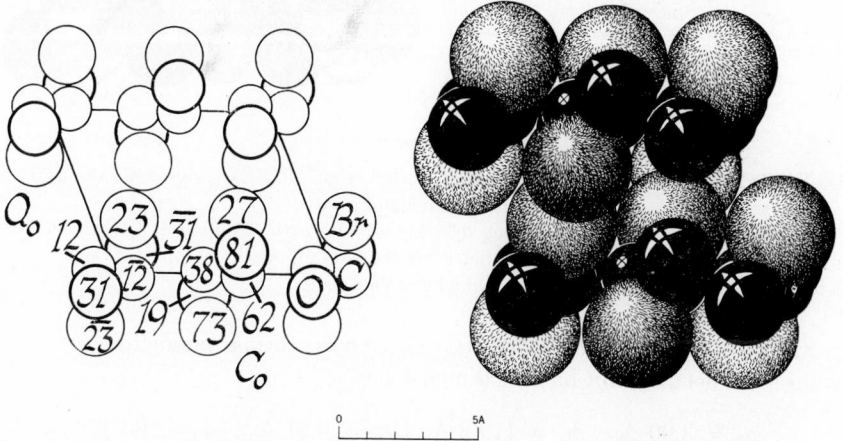

Fig. XIVB,167a (left). The monoclinic structure of $(COBr)_2$ projected along its b_0 axis. Right-hand axes.

Fig. XIVB,167b (right). A packing drawing of the monoclinic $(COBr)_2$ arrangement seen along its b_0 axis. Oxygen atoms are the large, carbon atoms the smaller black circles. Bromine atoms are still larger and fine-line shaded.

XIV,b123. Unlike the corresponding bromide, crystals of *oxalyl chloride*, $(COCl)_2$, are orthorhombic with a tetramolecular unit having

$$a_0 = 6.44 \text{ A.}; \quad b_0 = 6.08 \text{ A.}; \quad c_0 = 11.93 \text{ A.}$$

The space group is V_h^{15} (*Pbca*) with all atoms in the general positions:

(8c)	$\pm (xyz; \ x+^1/_2,^1/_2-y,\bar{z}; \ \bar{x},y+^1/_2,^1/_2-z; \ ^1/_2-x,\bar{y},z+^1/_2)$

Approximate parameters have been chosen as:

Atom	x	y	z
Cl	−0.084	−0.083	0.167
O	0.118	0.230	0.050
C	−0.006	0.062	0.053

The resulting arrangement, very different from that found for the bromide, is shown in Figure XIVB,168. Between molecules the shortest O–Cl is 3.50 A. and there is no evidence for the molecular sheets found for the other compound.

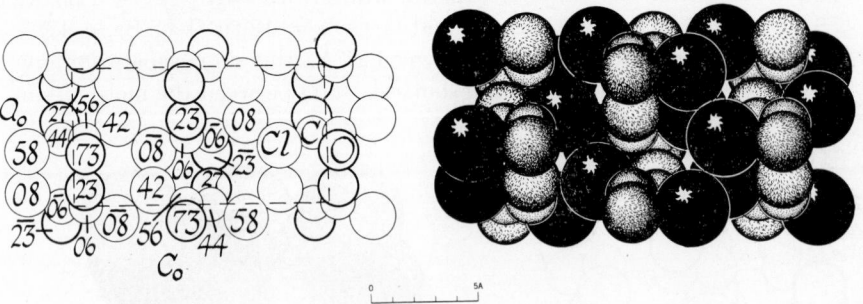

Fig. XIVB,168a (left). The orthorhombic structure of $(COCl)_2$ projected along its b_0 axis. Right-hand axes.

Fig. XIVB,168b (right). A packing drawing of the orthorhombic $(COCl)_2$ structure viewed along its b_0 axis. The large black circles are chlorine. Atoms of carbon are dotted; those of oxygen are heavily outlined and line shaded.

XIV,b124. *Dimethyl oxalate*, $(CH_3)_2(COO)_2$, forms monoclinic crystals whose bimolecular unit has the dimensions:

$$a_0 = 3.90 \text{ A.}; \quad b_0 = 11.88 \text{ A.}; \quad c_0 = 6.21 \text{ A.}; \quad \beta = 103°6'$$

All atoms are in general positions of C_{2h}^5 (*P2_1/n*):

(4e)	$\pm (xyz; \ x+^1/_2,^1/_2-y,z+^1/_2)$

with the parameters:

Atom	x	y	z
C(1)	0.095	0.052	0.056
C(2)	0.272	0.143	0.405
O(1)	0.244	0.119	−0.031
O(2)	0.095	0.051	0.267

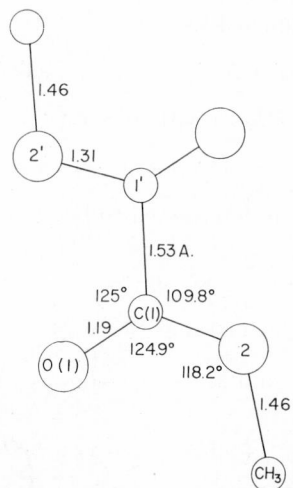

Fig. XIVB,169. Bond dimensions in the molecule of $(COOCH_3)_2$.

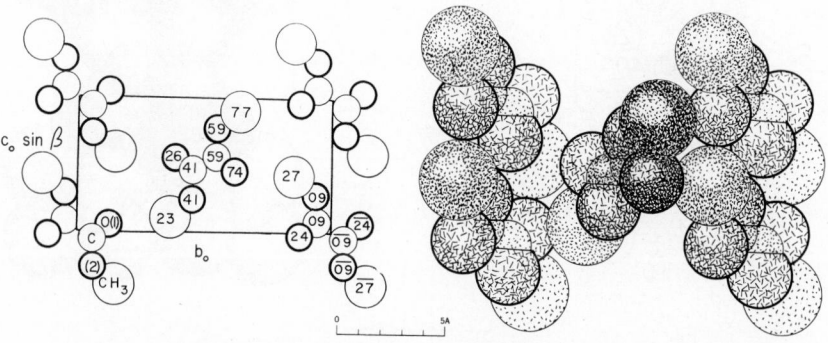

Fig. XIVB,170a (left). The monoclinic structure of $(COOCH_3)_2$ projected along its a_0 axis. Left-hand axes.

Fig. XIVB,170b (right). A packing drawing of the monoclinic $(COOCH_3)_2$ arrangement seen along its a_0 axis. The carbon atoms are dotted, methyl groups being the larger. Atoms of oxygen are line shaded and heavily ringed.

These lead to a molecule which is planar within the limits of experimental error (±0.05 A.) and has the dimensions of Figure XIVB,169. The molecular packing within the crystal is illustrated in Figure XIVB,170. It is a zigzag, layer-like assembly of the type shown, for instance, by naphthalene and anthracene. The shortest intermolecular distances are a C(2)–C(1) = 3.35 A. and an O(2)–O(2) = 3.37 A.

XIV,b125. The monoclinic crystals of *lithium oxalate*, $Li_2C_2O_4$, have a bimolecular unit of the dimensions:

$$a_0 = 3.400 \text{ A.}; \quad b_0 = 5.156 \text{ A.}; \quad c_0 = 9.055 \text{ A.}; \quad \beta = 95°36'$$

The space group is C_{2h}^5 $(P2_1/n)$ with all atoms in the general positions:

$$(4e) \quad \pm(xyz; \; x+^1/_2,^1/_2-y,z+^1/_2)$$

The parameters have been determined to be:

Atom	x	y	z
C	0.1058	0.1342	0.0074
O(1)	0.1331	0.2370	0.1338
O(2)	0.2385	0.2203	−0.1065
Li	−0.0567	−0.0134	0.2918

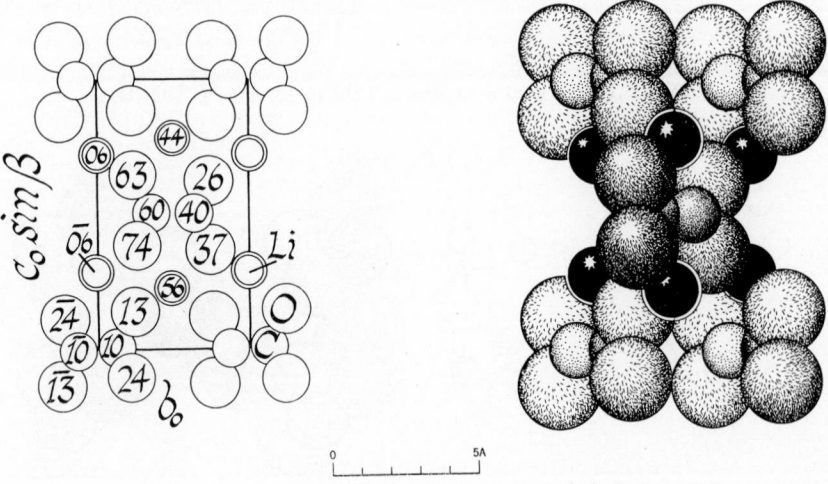

Fig. XIVB,171a (left).   The monoclinic structure of $Li_2C_2O_4$ projected along its a_0 axis. Right-hand axes.

Fig. XIVB,171b (right).   A packing drawing of the monoclinic $Li_2C_2O_4$ structure viewed along its a_0 axis.   The lithium atoms are the small black, the carbon atoms the small dotted circles. Atoms of oxygen are larger and fine-line shaded.

The resulting structure is shown in Figure XIVB,171. The oxalate ion is planar to within a few thousandths of an angstrom, with C–C = 1.561 A., C–O = 1.264 and 1.252 A. The lithium ions are surrounded by somewhat irregular tetrahedra of oxygen atoms, with Li–O = 1.931–2.071 A. The two O–C–C angles are 116° and O(1)–C–O(2) = 127°. The final R = 0.060.

XIV,b126. *Sodium oxalate*, $Na_2C_2O_4$, forms monoclinic crystals with a bimolecular cell of the dimensions:

$$a_0 = 10.35 \text{ A.}; \quad b_0 = 5.26 \text{ A.}; \quad c_0 = 3.46 \text{ A.}; \quad \beta = 92°54'$$

The space group is C_{2h}^5 $(P2_1/a)$ and all atoms are in the general positions:

$$(4e) \quad \pm(xyz;\ x+^1/_2, ^1/_2-y, z)$$

The parameters as determined are:

Atom	x	y	z
Na	0.353	0.053	0.307
O(1)	0.152	−0.114	0.163
O(2)	0.067	0.260	0.228
C	0.064	0.040	0.107

The resulting structure (Fig. XIVB,172) contains sodium and strictly planar oxalate ions. Each sodium has six oxygen neighbors at distances between 2.29 and 2.64 A. In the oxalate ion, C–O(1) = C–O(2) = 1.23 A. and C–C = 1.54 A.; the two angles C–C–O are, however, different, with C–C–O(1) = 121° and C–C–O(2) = 115°.

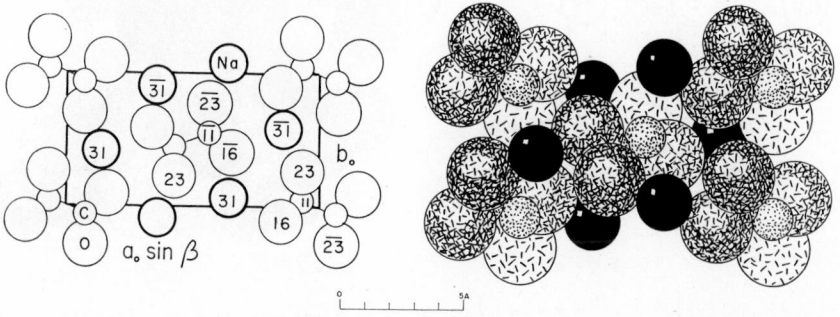

Fig. XIVB,172a (left). The monoclinic structure of $Na_2C_2O_4$ projected along its c_0 axis. Left-hand axes.

Fig. XIVB,172b (right). A packing drawing of the monoclinic $Na_2C_2O_4$ arrangement seen along its c_0 axis. The sodium atoms are black, the carbon atoms small and dotted. Oxygen atoms are line shaded.

XIV,b127.   *Potassium oxalate monohydrate*, $K_2(COO)_2 \cdot H_2O$, is monoclinic with the tetramolecular unit:

$$a_0 = 9.32 \text{ A.}; \quad b_0 = 6.17 \text{ A.}; \quad c_0 = 10.65 \text{ A.}; \quad \beta = 110°58'$$

All its atoms are in the general positions of C_{2h}^6 $(C2/c)$:

$$(8f) \quad \pm (xyz; \bar{x},y,{}^1\!/_2 - z; x + {}^1\!/_2, y + {}^1\!/_2, z; {}^1\!/_2 - x, y + {}^1\!/_2, {}^1\!/_2 - z)$$

with the parameters:

Atom	x	y	z
K	0.125	0.17	0.135
C	0.24	−0.32	0.06
O(1)	0.15	−0.25	0.12
O(2)	0.33	−0.45	0.10
H_2O	0.00	−0.40	0.25

The ions in this structure are distributed as shown in Figure XIVB,173. The C–C separation in an oxalate ion is 1.60 A. and one of the oxygen atoms is only 1.14 A. distant from carbon; the other C–O distance is 1.30 A. Each potassium ion has five oxygen neighbors at distances between 2.65 and 2.85 A. and a sixth at 3.05 A. The water molecules are tetrahedrally surrounded by two oxygen and two potassium atoms at a separation of 3.30 A.

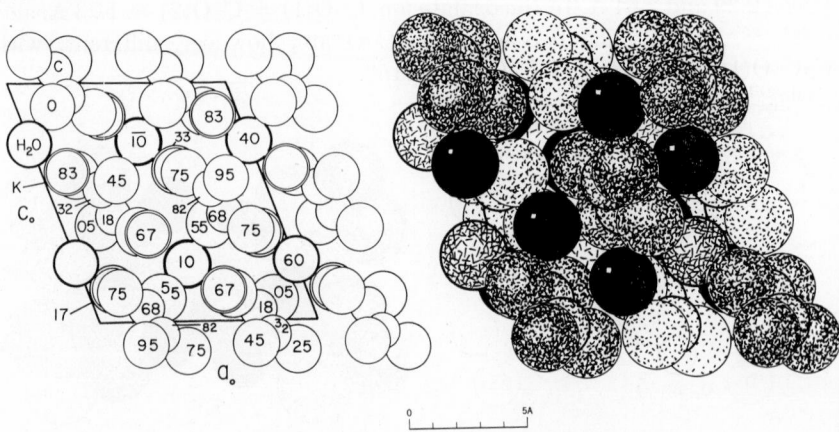

Fig. XIVB,173a (left).   The monoclinic structure of $K_2C_2O_4 \cdot H_2O$ projected along its b_0 axis.   Left-hand axes.

Fig. XIVB,173b (right).   A packing drawing of the monoclinic $K_2C_2O_4 \cdot H_2O$ arrangement viewed along its b_0 axis.   The potassium atoms are black, the water molecules heavily outlined and line shaded.   Both carbon and carboxyl oxygens are dotted, the latter being the larger.

The corresponding *rubidium* salt, $Rb_2(COO)_2 \cdot H_2O$, has the same structure with a cell of the dimensions:

$$a_0 = 9.66 \text{ A.;} \quad b_0 = 6.38 \text{ A.;} \quad c_0 = 11.20 \text{ A.;} \quad \beta = 110°30'$$

Atomic parameters, however, were not determined.

XIV,b128. *Potassium acid oxalate*, $KH(COO)_2$, and the isomorphous rubidium salt are monoclinic with tetramolecular cells of the dimensions:

$KH(COO)_2$:
$$a_0 = 4.32 \text{ A.;} \quad b_0 = 12.88 \text{ A.;} \quad c_0 = 10.32 \text{ A.;} \quad \beta = 133°29'$$
$RbH(COO)_2$:
$$a_0 = 4.30 \text{ A.;} \quad b_0 = 13.63 \text{ A.;} \quad c_0 = 10.39 \text{ A.;} \quad \beta = 133°15'$$

TABLE XIVB,75
Parameters of the Atoms in $KH(COO)_2$

Atom	x	y	z
K	0.35	0.069	0.275
C(1)	0.13	0.185	−0.135
C(2)	0.13	0.155	−0.285
O(1)	0.13	0.285	−0.107
O(2)	0.13	0.055	−0.313
O(3)	0.13	0.11	−0.05
O(4)	0.13	0.23	−0.37
H[a]	0.13	0.23 or 0.29	−0.485

[a] These are positions proposed for hydrogen atoms.

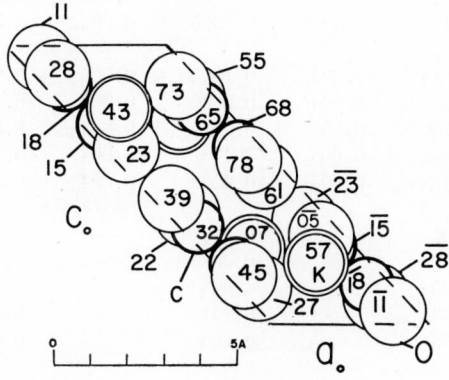

Fig. XIVB,174.   The monoclinic structure of KHC_2O_4 projected along its b_0 axis.   Left-hand axes.

For the potassium salt which has been studied in detail, all atoms have been found to be in general positions of C_{2h}^5 ($P2_1/c$):

$$(4e) \quad \pm(xyz; \; x,{}^1/_2-y,z+{}^1/_2)$$

with the parameters of Table XIVB,75.

In this structure (Fig. XIVB,174), C–C = 1.59 A., C–O = 1.30 or 1.32 A. Each potassium ion has six oxygen neighbors at distances between 2.55 and 2.85 A., as well as several somewhat more distant. There has been speculation as to the positions of the hydrogen atoms.

XIV,b129. *Potassium tetroxalate dihydrate,* $KHC_2O_4 \cdot H_2C_2O_4 \cdot 2H_2O$, is triclinic with a bimolecular unit of the dimensions:

$$a_0 = 7.047 \text{ A.}; \quad b_0 = 10.595 \text{ A.}; \quad c_0 = 6.355 \text{ A.}$$
$$\alpha = 101°24'; \quad \beta = 100°13'; \quad \gamma = 94°5'$$

TABLE XIVB,76

Parameters of the Atoms in $KHC_2O_4 \cdot H_2C_2O_4 \cdot 2H_2O$

Atom	x	$\sigma(x)$	y	$\sigma(y)$	z	$\sigma(z)$
K	0.1484	0.001	0.2729	0.001	0.1660	0.001
O(1)	0.3337	0.004	0.0679	0.003	0.3283	0.003
O(2)	0.9278	0.004	0.1537	0.003	0.5050	0.003
O(3)	0.5187	0.004	0.1570	0.003	0.6538	0.003
O(4)	0.1101	0.004	0.0765	0.003	0.7660	0.003
O(5)	0.5600	0.004	0.3085	0.003	0.1821	0.004
O(6)	0.8442	0.003	0.4306	0.003	0.2769	0.004
O(7)	0.3470	0.004	0.3555	0.003	0.6432	0.004
O(8)	0.6260	0.003	0.4824	0.003	0.7756	0.004
$H_2O(1)$	0.9486	0.003	0.3565	0.003	0.7855	0.003
$H_2O(2)$	0.7614	0.004	0.1305	0.003	0.0567	0.004
C(1)	0.4533	0.004	0.0632	0.004	0.4884	0.005
C(2)	0.0150	0.005	0.0649	0.004	0.5840	0.005
C(3)	0.6688	0.005	0.4189	0.004	0.2465	0.005
C(4)	0.4493	0.005	0.4632	0.004	0.7217	0.005
H(1)	0.627	—	0.242	—	0.150	—
H(2)	0.837	—	0.388	—	0.757	—
H(3)	0.800	—	0.075	—	0.133	—
H(4)	0.933	—	0.241	—	0.617	—
H(5)	0.473	—	0.225	—	0.617	—
H(6)	0.700	—	0.092	—	0.923	—
H(7)	0.000	—	0.417	—	0.773	—

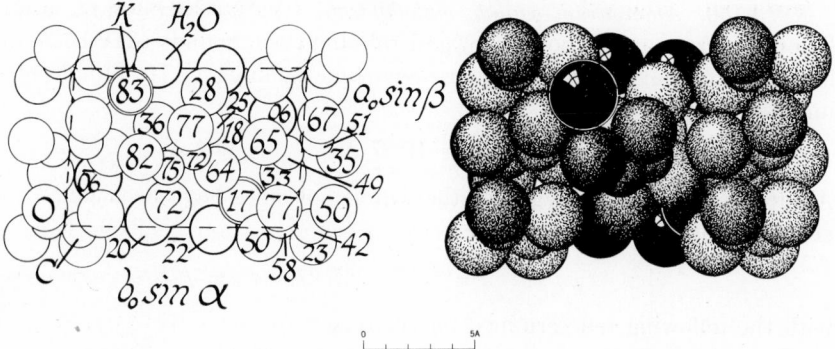

Fig. XIVB,175a (left). The triclinic structure of $KHC_2O_4 \cdot H_2C_2O_4 \cdot 2H_2O$ projected along its c_0 axis. Right-hand axes.

Fig. XIVB,175b (right). A packing drawing of the triclinic structure of $KHC_2O_4 \cdot H_2C_2O_4 \cdot 2H_2O$ seen along its c_0 axis. The potassium atoms are the larger black circles. The black water molecules, almost as big, have round highlights. Carboxyl oxygens are line shaded; the somewhat smaller carbon atoms are dotted.

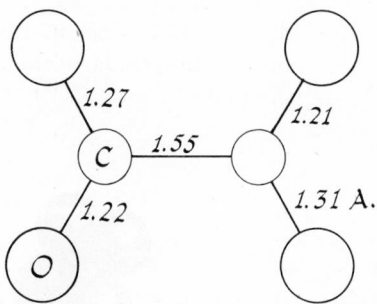

Fig. XIVB,176. Bond lengths in the oxalate ions of $KHC_2O_4 \cdot H_2C_2O_4 \cdot 2H_2O$.

The space group was chosen as $C_i{}^1$ ($P\bar{1}$) with all atoms in the positions $(2i)$ $\pm(xyz)$. The determined parameters including those for hydrogen are listed in Table XIVB,76.

The resulting structure is shown in Figure XIVB,175. There are two crystallographically different oxalate ions and an acid oxalate ion in the structure; within the limit of error, all have the bond dimensions shown in Figure XIVB,176. Each potassium atom has eight oxygen neighbors at distances between 2.87 and 3.04 A.; two of these are water oxygens. A ninth oxygen is 3.25 A. away. A total of six hydrogen bonds (O–H–O = 2.50–2.85 A.) can be recognized.

412 CRYSTAL STRUCTURES

XIV,b130. *Ammonium oxalate monohydrate*, $(NH_4)_2(COO)_2 \cdot H_2O$, is unlike the corresponding potassium and rubidium compounds (**XIV,b127**) in having orthorhombic rather than monoclinic symmetry. Its bimolecular unit has the edge lengths:

$$a_0 = 8.04 \text{ A.;} \quad b_0 = 10.27 \text{ A.;} \quad c_0 = 3.82 \text{ A.}$$

All atoms except the oxygen of the water molecules have been placed in general positions of V^3 $(P2_12_12)$:

$$(4c) \quad xyz; \ \bar{x}\bar{y}z; \ x+\tfrac{1}{2},\tfrac{1}{2}-y,\bar{z}; \ \tfrac{1}{2}-x,y+\tfrac{1}{2},\bar{z}$$

with the following redetermined parameters:

$$C: x = 0.092, y = 0.027, z = 0.066$$
$$O(1): x = 0.200, y = -0.056, z = 0.140$$
$$O(2): x = 0.118, y = 0.142, z = 0.001$$
$$NH_4: x = 0.386, y = 0.228, z = 0.424$$

The water molecules were put in

$$(2b) \quad 0\,\tfrac{1}{2}\,u; \ \tfrac{1}{2}\,0\,\bar{u} \quad \text{with } u = 0.192$$

The resulting structure (Fig. XIVB,177) has interatomic distances of the expected order but, unlike certain other oxalates that have been studied, it makes this ion nonplanar; the plane of one (COO) is turned through 28° with respect to the other.

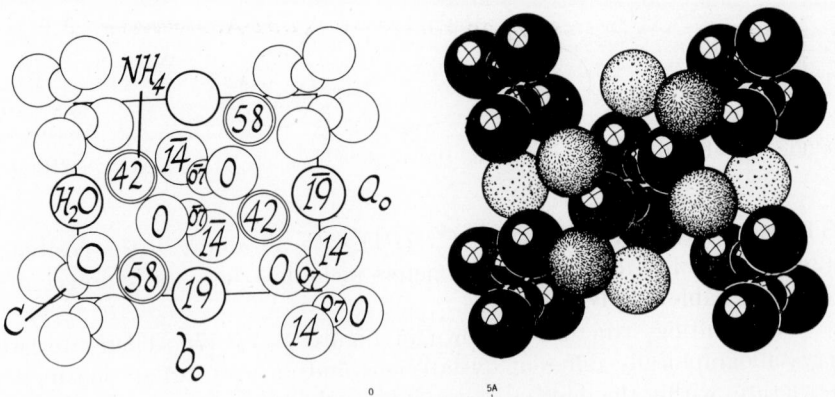

Fig. XIVB,177a (left). The orthorhombic structure of $(NH_4)_2C_2O_4 \cdot H_2O$ projected along its c_0 axis. Right-hand axes.

Fig. XIVB,177b (right). A packing drawing of the orthorhombic structure of $(NH_4)_2C_2O_4 \cdot H_2O$ seen along its c_0 axis. Both the larger oxygen and the carbon atoms of the oxalate ions are black. Water molecules are hook shaded; NH_4 cations are heavily ringed and line shaded.

XIV,b131. Crystals of *ammonium oxamate*, $NH_2CO \cdot COONH_4$, are monoclinic with a tetramolecular cell of the dimensions:

$$a_0 = 3.607 \pm 0.002 \text{ A.}; \quad b_0 = 10.007 \pm 0.004 \text{ A.}; \quad c_0 = 11.930 \pm 0.005 \text{ A.}$$
$$\beta = 95°36' \pm 5'$$

The space group is C_{2h}^5 ($P2_1/n$) with all atoms in the positions:

$$(4e) \quad \pm (xyz; \ x+^1/_2, ^1/_2-y, z+^1/_2)$$

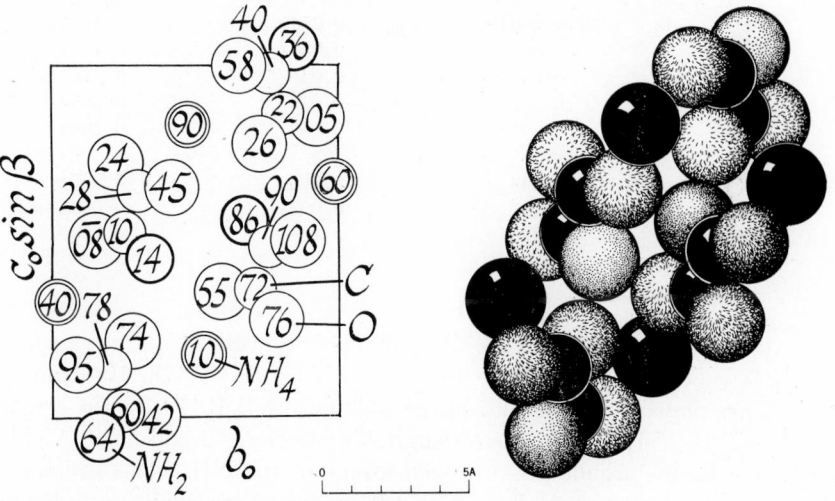

Fig. XIVB,178a (left). The monoclinic structure of ammonium oxamate projected along its a_0 axis. Right-hand axes.

Fig. XIVB,178b (right). A packing drawing of the monoclinic structure of NH_2-$CO \cdot COONH_4$ seen along its a_0 axis. The black NH_4 cations are somewhat larger than the black carbon atoms. The NH_2 groups are heavily outlined and dotted; the oxygen atoms are fine-line shaded.

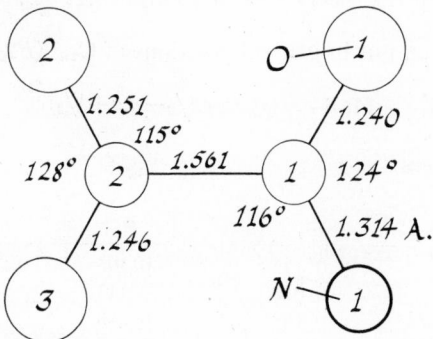

Fig. XIVB,179. Bond dimensions of the oxamate ions in $NH_2CO \cdot COONH_4$.

TABLE XIVB,77
Parameters of the Atoms in $NH_2CO \cdot COONH_4$

Atom	x	y	z
C(1)	−0.4047	0.2424	0.0172
C(2)	−0.2200	0.1967	0.1350
N(1)	−0.3601	0.1614	−0.0671
N(2)	0.4020	0.0186	0.3353
O(1)	−0.5783	0.3494	0.0071
O(2)	−0.2606	0.2745	0.2150
O(3)	−0.0505	0.0881	0.1384
H(1)	0.553	−0.046	0.3114
H(2)	0.257	−0.029	0.3882
H(3)	0.232	0.056	0.2688
H(4)	0.546	0.086	0.3674
H(5)	−0.227	0.092	−0.0583
H(6)	−0.473	0.179	−0.1331

The determined parameters are those of Table XIVB,77.

The structure that results is shown in Figure XIVB,178. Its oxamate anions are planar with the dimensions of Figure XIVB,179. In the crystal there are six N–H $\cdots$ O bonds of lengths between 2.812 and 3.043 A. The hydrogen atoms around the nitrogen atoms of the NH_4^+ ions are tetrahedrally arranged, with N–H = 0.91–1.02 A.; these are significantly longer than the amidic N(1)–H separations of 0.86 A.

XIV,b132. A structure has been assigned the monoclinic crystals of *silver oxalate*, $Ag_2(COO)_2$. The bimolecular unit has the dimensions:

$$a_0 = 3.46 \text{ A.}; \quad b_0 = 6.16 \text{ A.}; \quad c_0 = 9.47 \text{ A.}; \quad \beta = 76°$$

All atoms have been put in general positions of C_{2h}^5 $(P2_1/c)$:

$$(4e) \quad \pm (xyz; \ x, {}^1/_2 - y, z + {}^1/_2)$$

with the parameters:

Atom	x	y	z
Ag	0.145	0.495	0.156
C	−0.070	0.115	−0.012
O(1)	0.163	0.216	−0.110
O(2)	−0.163	0.187	0.100

The positions of the carbon atoms were suggested by Fourier projections and assigned according to the need for acceptable interatomic distances. It is to be noted that in the resulting $(COO)_2$ ions the carbon atoms are displaced somewhat from the plane of the oxygen atoms. Each silver atom is much closer to one oxygen atom of two $(COO)_2$ ions (2.17 A.) than to other oxalate oxygens (2.30 A.).

XIV,b133. *Calcium oxalate monohydrate,* $CaC_2O_4 \cdot H_2O$, (whewhellite) is monoclinic with a unit containing eight molecules and having the dimensions:

$$a_0 = 6.24 \text{ A.}; \quad b_0 = 14.58 \text{ A.}; \quad c_0 = 9.89 \text{ A.}; \quad \beta = 107°$$

The space group is C_{2h}^5 ($P2_1/c$) with all atoms in the general positions:

$$(4e) \quad \pm(xyz; x,^1/_2-y,z+^1/_2)$$

A refined series of parameters is listed in Table XIVB,78.

The resulting structure is shown in Figure XIVB,180. The coordination is different around the two calcium atoms. It is sevenfold with one neigh-

TABLE XIVB,78
Parameters of the Atoms in $CaC_2O_4 \cdot H_2O$

Atom	x	y	z
C(1)	0.0079	0.3220	0.2513
C(2)	0.0012	0.4296	0.2540
C(3)	0.5421	0.1103	0.1852
C(4)	0.4822	0.1279	0.3200
O(1)	0.0279	0.2811	0.1486
O(2)	0.9981	0.4671	0.1389
O(3)	0.0301	0.2844	0.3727
O(4)	0.9911	0.4669	0.3605
O(5)	0.3827	0.1212	0.0727
O(6)	0.7546	0.1261	0.1995
O(7)	0.2690	0.1275	0.3017
O(8)	0.6398	0.1170	0.4376
H₂O(1)	0.4032	0.4013	0.1059
H₂O(2)	0.6084	0.3461	0.3977
Ca(1)	0.0049	0.1227	0.0639
Ca(2)	0.0305	0.1263	0.4438

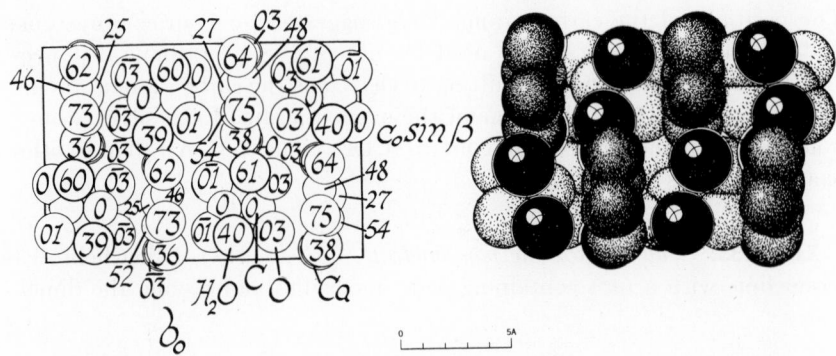

Fig. XIVB,180a (left).   The monoclinic structure of $CaC_2O_4 \cdot H_2O$ projected along its a_0 axis. Right-hand axes.

Fig. XIVB,180b (right).   A packing drawing of the monoclinic structure of $CaC_2O_4 \cdot H_2O$ seen along its a_0 axis. The water molecules are black. The calcium atoms also are black but with broad white borders. The large carboxyl oxygens are line shaded; the carbon atoms are almost as big and are similarly shaded.

bor a water oxygen (Ca–O = 2.33–2.58 A.). The oxalate ions have their expected dimensions.

XIV,b134. *Ferrous oxalate dihydrate* (humboldine), $FeC_2O_4 \cdot 2H_2O$ has monoclinic symmetry with a tetramolecular cell of the dimensions:

$$a_0 = 12.060 \text{ A.}; \quad b_0 = 5.550 \text{ A.}; \quad c_0 = 9.804 \text{ A.}; \quad \beta = 127°58'$$

The space group has been chosen as C_{2h}^6 ($C2/c$) with iron atoms in the positions:

$$(4e) \quad \pm(0 \ u \ ^1/_4; \ ^1/_2, u + ^1/_2, ^1/_4) \qquad \text{with } u = 0.171$$

All other atoms are in the general positions:

$$(8f) \quad \pm(xyz; \ x, \bar{y}, z + ^1/_2; \ x + ^1/_2, y + ^1/_2, z; \ x + ^1/_2, ^1/_2 - y, z + ^1/_2)$$

with the following parameters:

Atom	x	y	z
C	0.050	0.671	0.350
O(1)	0.086	0.474	0.422
O(2)	0.086	0.868	0.422
H_2O	0.172	0.171	0.250

In this structure (Fig. XIVB,181) the iron atoms are octahedrally surrounded by four carboxyl oxygens (Fe–O = 2.14 A.) and by two water oxygens (Fe–OH_2 = 2.07 A.). In the oxalate ions, C–C = 1.546 A., and C–O = 1.223 A. The angle O–C–O = 126°44' and C–C–O = 116°38'. Of

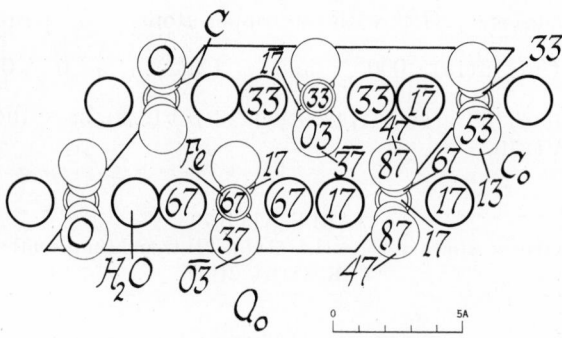

Fig. XIVB,181. The monoclinic structure of $FeC_2O_4 \cdot 2H_2O$ projected along its b_0 axis. Right-hand axes.

the angles involving iron, $O–Fe–O = 76°18'$ and $Fe–O–C = 115°13'$. The final R for data from different zones has lain between 0.128 and 0.117.

XIV,b135. Crystals of *cupric ammonium oxalate dihydrate*, $Cu(NH_4)_2$-$(C_2O_4)_2 \cdot 2H_2O$, are triclinic with a bimolecular unit of the dimensions:

$$a_0 = 8.91 \pm 0.02 \text{ A.}; \quad b_0 = 10.65 \pm 0.02 \text{ A.}; \quad c_0 = 6.95 \pm 0.03 \text{ A.}$$
$$\alpha = 122°35'; \quad \beta = 83°52'; \quad \gamma = 109°7'$$

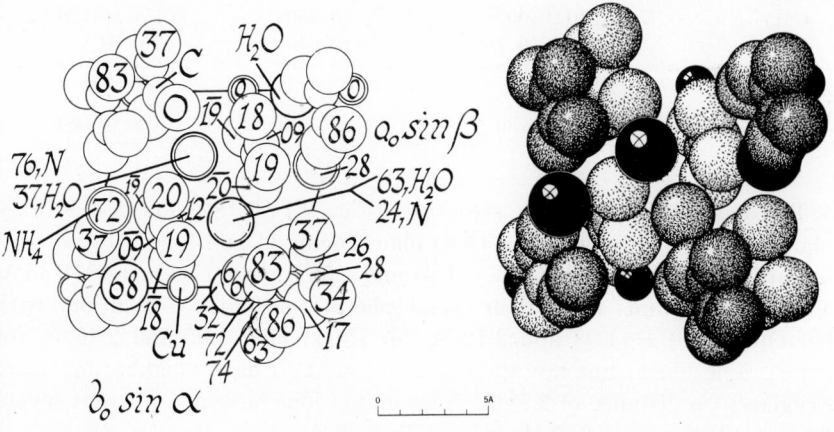

Fig. XIVB,182a (left). The triclinic structure of $Cu(NH_4)_2(C_2O_4)_2 \cdot 2H_2O$ projected along its c_0 axis. Right-hand axes.

Fig. XIVB,182b (right). A packing drawing of the triclinic structure of $Cu(NH_4)_2$-$(C_2O_4)_2 \cdot 2H_2O$ seen along its c_0 axis. The copper atoms are the small, the NH_4^+ ions the large black circles. The water molecules are hook shaded and heavily outlined. Car boxyl oxygens, of equal size, are line shaded; the smaller carbon atoms are dotted.

The space group is C_i^1 ($P\overline{1}$) with the copper atoms in the positions:

$$Cu(1): (1a) \quad 000 \quad \text{and} \quad Cu(2): (1c) \quad 0\,{}^1/_2\,0$$

The other atoms, in the general positions $(2i) \pm (xyz)$, have the parameters of Table XIVB,79.

TABLE XIVB,79
Parameters of Atoms in $Cu(NH_4)_2(C_2O_4)_2\cdot2H_2O$ and (in parentheses) in $CuK_2(C_2O_4)_2\cdot2H_2O$

Atom	x	y	z
$NH_4[K](1)$	0.337 (0.337)	0.367 (0.367)	0.240 (0.240)
$NH_4[K](2)$	0.614 (0.617)	0.044 (0.033)	0.283 (0.283)
O(1)	0.048 (0.037)	0.225 (0.234)	0.660 (0.630)
O(2)	0.100 (0.100)	0.120 (0.124)	0.830 (0.820)
O(3)	0.713 (0.705)	0.035 (0.022)	0.630 (0.614)
O(4)	0.198 (0.204)	0.025 (0.024)	0.140 (0.150)
O(5)	0.782 (0.780)	0.420 (0.425)	−0.192 (−0.192)
O(6)	0.867 (0.874)	0.420 (0.405)	0.180 (0.150)
O(7)	0.614 (0.610)	0.280 (0.242)	0.193 (0.180)
O(8)	0.534 (0.530)	0.300 (0.278)	−0.200 (−0.180)
$H_2O(1)$	0.335 (0.340)	0.358 (0.354)	0.633 (0.633)
$H_2O(2)$	0.980 (0.970)	0.245 (0.233)	0.320 (0.290)
C(1)	0.000 (0.000)	0.130 (0.150)	0.720 (0.730)
C(2)	−0.180 (−0.180)	0.040 (0.040)	0.740 (0.720)
C(3)	0.684 (0.680)	0.350 (0.340)	−0.120 (−0.100)
C(4)	0.714 (0.730)	0.340 (0.310)	0.090 (0.083)

The oxalate ions of this structure (Fig. XIVB,182) are not strictly planar, the angles between the COO planes being 11 and 15° in the two different ions. In these ions, C–C = 1.61 and 1.58 A. and C–O = 1.20–1.35 A. The copper atoms have their usual planar fourfold coordination, with Cu–O(oxalate) = 1.98 and 2.12 A. for Cu(1), and 1.97 and 2.13 A. for Cu(2). The Cu(1) but not the Cu(2) atoms also have neighboring water oxygens at a distance of 2.74 A. Ammonium ions have an apparent sevenfold coordination, with NH_4–O = 2.71–3.32 A.

The isostructural *potassium* salt, $CuK_2(C_2O_4)_2\cdot2H_2O$, has the cell:

$$a_0 = 8.66 \text{ A.;} \quad b_0 = 10.19 \text{ A.;} \quad c_0 = 6.86 \text{ A.}$$
$$\alpha = 120°41'; \quad \beta = 83°52'; \quad \gamma = 110°18'$$

Copper atoms are placed as in the ammonium compound, and the other atoms have the parameters in parentheses of Table XIVB,79. In this case the angles between the COO planes in the two nonequivalent anions are 12 and 24°. Atomic relationships are similar to those in the ammonium salt but probably less accurately known.

XIV,b136. Crystals of the coordination compound *trans potassium dioxalato diaquo chromiate trihydrate*, $K[Cr(C_2O_4)_2(H_2O)_2] \cdot 3H_2O$, are monoclinic with the bimolecular unit:

$$a_0 = 7.85 \text{ A.}; \quad b_0 = 5.72 \text{ A.}; \quad c_0 = 13.88 \text{ A.}; \quad \beta = 109°30'$$

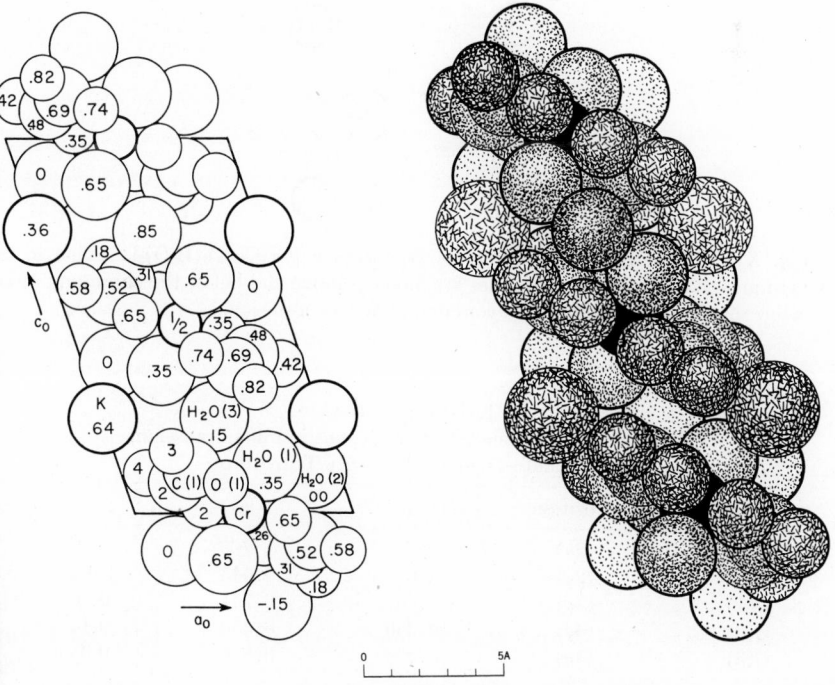

Fig. XIVB,183a (left). The monoclinic structure of $K[Cr(C_2O_4)_2(H_2O)_2] \cdot 3H_2O$ projected along its b_0 axis. Left-hand axes.

Fig. XIVB,183b (right). A packing drawing of the monoclinic structure of $KCr(C_2O_4)_2 \cdot 5H_2O$ viewed along its b_0 axis. The potassium atoms are large and line shaded; the water molecules are dotted and heavily outlined. Carbon atoms are large, dotted, and lightly outlined; the smaller carboxyl oxygen atoms are line shaded and more heavily outlined.

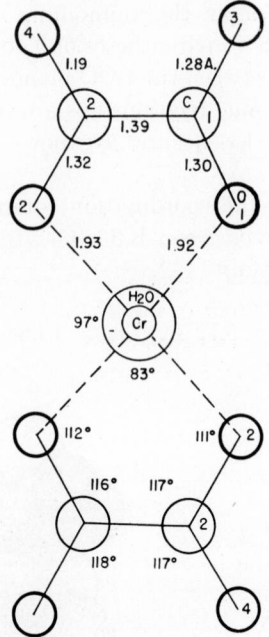

Fig. XIVB,184. Bond dimensions in the complex $[Cr(C_2O_4)_2(H_2O)_2]^-$ anion in its potassium salt. The water molecules are directly above and below the chromium atom on a line that is normal to the plane containing the rest of the ion.

TABLE XIVB,80
Positions and Parameters of the Atoms in *trans* Potassium
Dioxalato Diaquo Chromiate Trihydrate

Atom	Position	x	y	z
Cr	(2b)	$1/2$	$1/2$	0
K	(2e)	0	0.645	$1/4$
O(1)	(4g)	0.451	0.741	0.083
O(2)	(4g)	0.296	0.350	0.021
O(3)	(4g)	0.256	0.816	0.171
O(4)	(4g)	0.109	0.423	0.109
C(1)	(4g)	0.318	0.687	0.115
C(2)	(4g)	0.232	0.476	0.080
H₂O(1)	(4g)	0.667	0.350	0.128
H₂O(2)	(4g)	0.867	0.000	0.112
H₂O(3)	(2f)	$1/2$	0.147	$1/4$

The space group is C_{2h}^4 ($P2/c$) with atoms in the special and general positions:

(2b) $^1/_2\ ^1/_2\ 0;\ ^1/_2\ ^1/_2\ ^1/_2$
(2e) $0\ u\ ^1/_4;\ 0\ \bar{u}\ ^3/_4$
(2f) $^1/_2\ u\ ^1/_4;\ ^1/_2\ \bar{u}\ ^3/_4$
(4g) $\pm(xyz;\ x,\bar{y},z+^1/_2)$

The established parameters are those of Table XIVB,80.

This structure is illustrated in Figure XIVB,183. Its dioxalato diaquo chromiate ion has the dimensions of Figure XIVB,184. This ion is planar except for the H_2O groups which are situated 2.02 A. above and below the chromium atom on a line normal to the ionic plane and passing through the chromium.

XIV,b137. *Ammonium trioxalato chromiate dihydrate,* $(NH_4)_3[Cr(C_2O_4)_3]\cdot 2H_2O$, crystallizes with less water than the corresponding potassium salt (**XIV,b136**) and has a different structure. Its bimolecular triclinic cell has the dimensions:

$$a_0 = 7.79\ \text{A.};\quad b_0 = 10.90\ \text{A.};\quad c_0 = 10.73\ \text{A.}$$
$$\alpha = 98°10';\quad \beta = 112°36';\quad \gamma = 67°22'$$

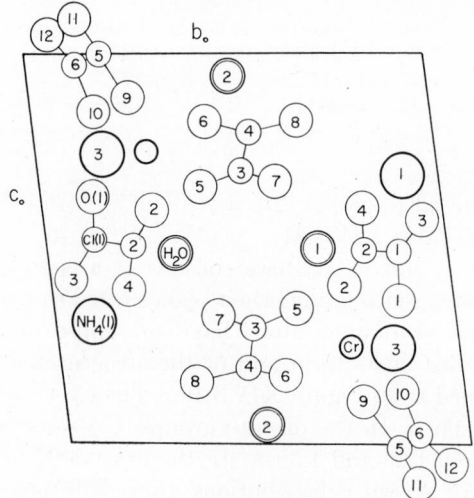

Fig. XIVB,185.   The triclinic structure of $(NH_4)_3[Cr(C_2O_4)_3]\cdot 2H_2O$ projected on its b_0c_0 plane.   Numbers refer to the atoms similarly labeled in Table XIVB,81.

TABLE XIVB,81
Parameters of the Atoms in $(NH_4)_3[Cr(C_2O_4)_3] \cdot 2H_2O$

Atom	x	y	z
O(1)	0.300	0.137	−0.370
O(2)	0.467	0.295	−0.405
O(3)	0.250	0.058	−0.578
O(4)	0.467	0.195	−0.602
O(5)	0.127	0.398	−0.343
O(6)	0.442	0.443	−0.175
O(7)	−0.060	0.610	−0.320
O(8)	0.283	0.660	−0.168
O(9)	0.307	0.250	−0.100
O(10)	0.627	0.167	−0.143
O(11)	0.450	0.140	0.097
O(12)	0.800	0.060	0.057
H$_2$O(1)	0.750	0.328	−0.500
H$_2$O(2)	0.825	0.500	−0.057
NH$_4$(1)	0.817	0.090	−0.685
NH$_4$(2)	0.133	0.150	−0.838
NH$_4$(3)	0.883	0.170	−0.257
Cr	0.385	0.273	−0.256
C(1)	0.330	0.128	−0.483
C(2)	0.413	0.218	−0.497
C(3)	0.117	0.510	−0.300
C(4)	0.273	0.540	−0.200
C(5)	0.477	0.177	−0.007
C(6)	0.653	0.128	−0.033

The atoms, in general positions $(2i)$ $\pm(xyz)$ of C_i^1 $(P\bar{1})$, have been given the parameters of Table XIVB,81.

The trioxalato chromiate ion here consists of a central chromium and three planar oxalate groups having their planes tilted so that the six oxygen atoms nearest the chromium atom have an approximately octahedral distribution. An idea of its form and of the structure of the crystal as a whole can be gained from Figure XIVB,185. The Cr–O distances vary between 1.89 and 2.06 A. In the oxalato groups, C–C = 1.40±0.01 A. and C–O varies between 1.23 and 1.35 A. In the crystal, N–O distances range upwards from 2.70 A. with distributions around nitrogen which are different for the three kinds of ammonium group. One water molecule has two ammonium neighbors, the other none.

XIV,b138. Racemic *tripotassium ferric oxalate trihydrate*, $K_3Fe(C_2O_4)_3 \cdot$ $3H_2O$, is monoclinic with a tetramolecular unit of the dimensions:

$$a_0 = 7.66 \text{ A.}; \quad b_0 = 19.87 \text{ A.}; \quad c_0 = 10.27 \text{ A.}; \quad \beta = 105°6'$$

The space group is C_{2h}^5 $(P2_1/c)$ with all atoms in the positions:

$$(4e) \quad \pm(xyz; x, 1/2 - y, z + 1/2)$$

The determined parameters are those of Table XIVB,82.

TABLE XIVB,82
Parameters of the Atoms in $K_3Fe(C_2O_4)_3 \cdot 3H_2O$

Atom	x	y	z
Fe	0.250	0.132	0.250
K(1)	−0.015	0.078	0.666
K(2)	0.510	0.079	0.838
K(3)	0.750	0.083	0.252
C(1)	0.253	0.068	0.485
C(2)	0.435	0.098	0.498
C(3)	0.205	0.261	0.288
C(1′)	0.245	0.056	0.015
C(2′)	0.050	0.088	0.003
C(3′)	0.338	0.259[a]	0.210
O(1)	0.161	0.066	0.368
O(2)	0.490	0.125	0.392
O(3)	0.220	0.035	0.582
O(4)	0.555	0.080	0.600
O(5)	0.132	0.213	0.318
O(6)	0.180	0.314	0.315
O(1′)	0.330	0.060	0.138
O(2′)	0.022	0.122	0.103
O(3′)	0.260	0.030	0.925
O(4′)	−0.068	0.072	0.898
O(5′)	0.370	0.208	0.178
O(6′)	0.410	0.308	0.164
$H_2O(1)$	0.600	0.184	0.048
$H_2O(2)$	−0.070	0.175	0.460
$H_2O(3)$	0.248	−0.009	0.254

[a] According to a private communication, the y parameter for C(3′) should be 0.259 (as given here) rather than the 0.059 of the original article. The improbably close approach of $H_2O(3)$ to O(1′) is to be taken as an indication that these water molecules have rather indeterminate positions.

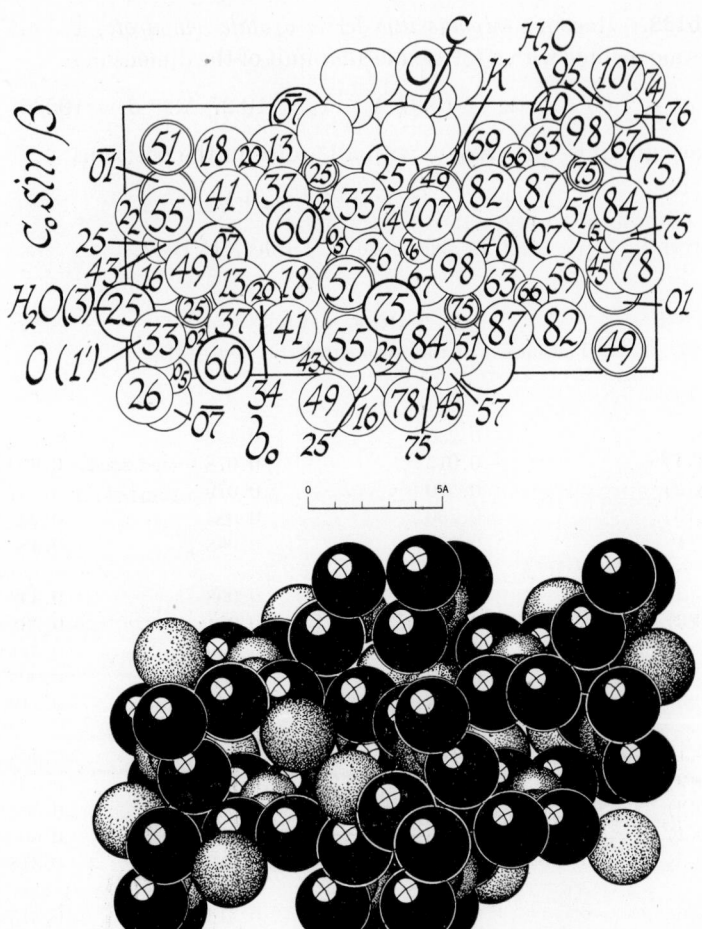

Fig. XIVB,186a (top).   The monoclinic structure of $K_3Fe(C_2O_4)_3 \cdot 3H_2O$ projected along its a_0 axis.   Right-hand axes.

Fig. XIVB,186b (bottom).   A packing drawing of the monoclinic structure of K_3Fe-$(C_2O_4)_3 \cdot 3H_2O$ viewed along its a_0 axis.   The carboxyl oxygens are black; the carbon atoms are dotted.   The potassium atoms are hook shaded;   the water molecules are heavily outlined and line shaded.   The small, black iron atoms scarcely show.

In the resulting structure (Fig. XIVB,186) the iron atoms are octa-hedrally surrounded by six carboxyl oxygens, with Fe–O = 2.01–2.06 A. Each atom of potassium is surrounded by eight oxygens, with K–O = 2.64–3.05 A. A water molecule has about it two potassium atoms and two water oxygens at especially short distances (2.35 and 2.48 A.) attributed to hydrogen bonding. In the oxalate ions, C–C = 1.56 A. and C–O = 1.19–1.29 A. For this structure R = ca. 0.15.

Salts of other trivalent metals are isostructural. Several years ago it was shown that for $K_3Cr(C_2O_4)_3 \cdot 3H_2O$ the unit has the dimensions:

$$a_0 = 7.71 \text{ A.}; \quad b_0 = 19.74 \text{ A.}; \quad c_0 = 10.40 \text{ A.}; \quad \beta = 108°0'$$

At the time a partial structure was proposed using C_{2h}^6 instead of C_{2h}^5 as space group.

The aluminum compound $K_3Al(C_2O_4)_3 \cdot 3H_2O$ has the cell:

$$a_0 = 7.65 \text{ A.}; \quad b_0 = 19.78 \text{ A.}; \quad c_0 = 10.32 \text{ A.}; \quad \beta = 106°2'$$

Another salt which seems to have this arrangement is $Rb_3Cr(C_2O_4)_3 \cdot xH_2O$ for which

$$a_0 = 7.81 \text{ A.}; \quad b_0 = 19.69 \text{ A.}; \quad c_0 = 10.40 \text{ A.}; \quad \beta = 108°0'$$

XIV,b139. *Tetrasodium zirconium tetroxalate trihydrate*, $Na_4Zr(C_2O_4)_4 \cdot 3H_2O$, forms orthorhombic crystals having a tetramolecular unit of the edge lengths:

$$a_0 = 7.42 \pm 0.01 \text{ A.}; \quad b_0 = 11.81 \pm 0.02 \text{ A.}; \quad c_0 = 19.74 \pm 0.02 \text{ A.}$$

The space group is V^5 in the axial orientation $B22_12$. The following atoms are in special positions:

Na(2): (4a) $u00; \bar{u} \, {}^1/_2 \, 0; u+{}^1/_2,0,{}^1/_2; {}^1/_2-u,{}^1/_2,{}^1/_2$
          with $u = 0.3392 \pm 0.0010$
Na(3): (4b) ${}^1/_2 \, {}^1/_4 \, u; {}^1/_2 \, {}^3/_4 \, \bar{u}; 0,{}^1/_4,u+{}^1/_2; 0,{}^3/_4,{}^1/_2-u$
          with $u = 0.4117 \pm 0.0003$
H$_2$O(1): (4b)   with $u = 0.2489 \pm 0.0007$
     Zr: (4b)   with $u = 0.60658 \pm 0.0001$

All other atoms are in the general positions:

(8c) $xyz; \quad x\bar{y}\bar{z}; \quad x+{}^1/_2,y,z+{}^1/_2; \quad x+{}^1/_2,\bar{y},{}^1/_2-z;$
$\bar{x},y+{}^1/_2,\bar{z}; \bar{x},{}^1/_2-y,z; {}^1/_2-x,y+{}^1/_2,{}^1/_2-z; {}^1/_2-x,{}^1/_2-y,z+{}^1/_2$

The determined parameters are those of Table XIVB,83.

In this structure (Fig. XIVB,187) each zirconium atom has eight oxygen neighbors, two from each of the four oxalate radicals that surround it. Their coordination is, however, that of a bisphenoid of $\overline{4}2m$ symmetry, rather than of the square antiprism that prevails for the analogous acetylacetonate (**XIV,c48**). The averaged bond dimensions for the two kinds of oxalate group around a zirconium atom are shown in Figure XIVB,188; the differences between the two Zr–O distances for a C_2O_4 radical are significant. Presumably the zirconium and its surrounding oxalate radicals are to be considered as forming a complex anion of the type $[Zr(C_2O_4)_4]^{4-}$. Each of the three kinds of Na$^+$ ion has a different environment: for Na(2) the coor-

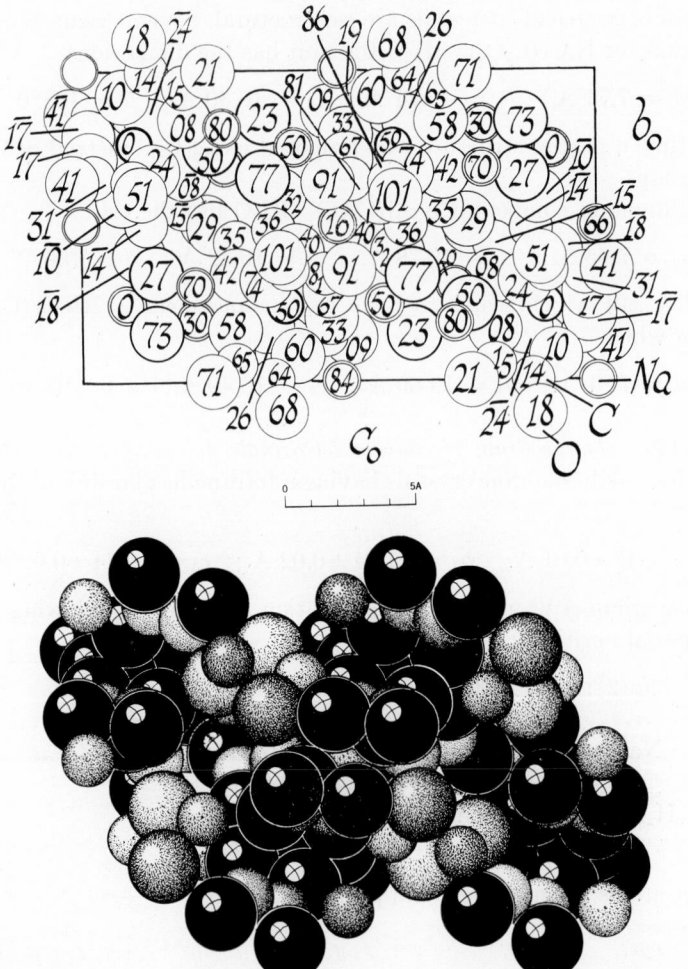

Fig. XIVB,187a (top).   The orthorhombic structure of $Na_4Zr(C_2O_4)_4 \cdot 3H_2O$ projected
along its a_0 axis.   Right-hand axes.

Fig. XIVB,187b (bottom).   A packing drawing of the orthorhombic $Na_4Zr(C_2O_4)_4 \cdot$
$3H_2O$ structure seen along its a_0 axis.   The carboxyl oxygens are black; the smaller
carbon atoms are dotted.   Atoms of sodium also are small, but hook shaded.   Water
molecules are line shaded.   Atoms of zirconium cannot be distinguished.

dination is eightfold, with Na–O = 2.41–2.76 A., for Na(3) it is sixfold,
with Na–O = 2.38–2.53 A., and for Na(1) it is sevenfold with Na–O =
2.34–2.72 A. For Na(1) and Na(3), two of these oxygens belong to water
molecules which, in this crystal, seal hollows in the structure.

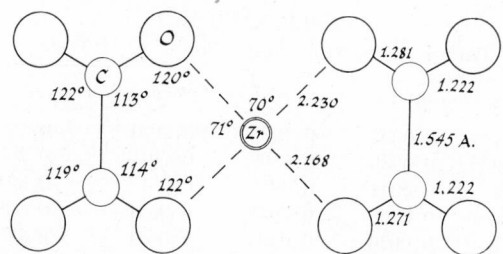

Fig. XIVB,188. Bond dimensions in crystals of $Na_4Zr(C_2O_4)_4 \cdot 3H_2O$.

TABLE XIVB,83
Parameters of Atoms in $Na_4Zr(C_2O_4)_4 \cdot 3H_2O$

Atom	x	$\sigma(x)$	y	$\sigma(y)$	z	$\sigma(z)$
O(1)	0.1670	0.0012	0.2875	0.0006	0.0163	0.0004
O(2)	0.0967	0.0011	0.0810	0.0005	0.0785	0.0003
O(3)	0.2451	0.0014	0.3290	0.0007	0.1394	0.0004
O(4)	0.0810	0.0011	0.1568	0.0005	0.2009	0.0003
O(5)	0.4138	0.0017	0.3722	0.0010	−0.0224	0.0005
O(6)	0.1833	0.0011	−0.0905	0.0007	0.1147	0.0005
O(7)	0.5102	0.0021	0.4038	0.0017	0.1117	0.0007
O(8)	0.2104	0.0021	0.0025	0.0009	0.2441	0.0005
C(1)	0.3147	0.0012	0.3419	0.0007	0.0229	0.0004
C(2)	0.1414	0.0012	0.0088	0.0007	0.1231	0.0003
C(3)	0.3651	0.0013	0.3613	0.0008	0.0977	0.0004
C(4)	0.1460	0.0012	0.0562	0.0006	0.1964	0.0003
Na(1)	0.2013	0.0012	0.3153	0.0006	0.2760	0.0004
H₂O(2)	0.2338	0.0063	0.1565	0.0031	0.3541	0.0012

The corresponding *hafnium* compound, $Na_4Hf(C_2O_4)_4 \cdot 3H_2O$, has the same structure with

$$a_0 = 7.43 \pm 0.01 \text{ A.}; \quad b_0 = 11.85 \pm 0.02 \text{ A.}; \quad c_0 = 19.76 \pm 0.02 \text{ A.}$$

XIV,b140. Binuclear *potassium molybdenum(VI) oxalate hydrate*, $K_2(MoO_2C_2O_4 \cdot H_2O)_2O$, is monoclinic with a bimolecular unit of the dimensions:

$$a_0 = 7.508 \text{ A.}; \quad b_0 = 14.235 \text{ A.}; \quad c_0 = 6.602 \text{ A.}; \quad \beta = 94°33'$$

The space group is C_{2h}^5 $(P2_1/c)$ with one oxygen atom [O(8)] in the origin $(2a)$ $000; 0\ ^1/_2\ ^1/_2$ and all other atoms in the positions:

$$(4e) \quad \pm(xyz; x,^1/_2-y,z+^1/_2)$$

TABLE XIVB,84
Parameters of Atoms in $K_2(MoO_2C_2O_4 \cdot H_2O)_2O$

Atom	x	$\sigma(x)$	y	$\sigma(y)$	z	$\sigma(z)$
Mo	0.1011	0.0003	0.1077	0.0001	0.1254	0.0003
C(1)	0.4603	0.0036	0.1800	0.0016	0.2637	0.0034
C(2)	0.4842	0.0041	0.0673	0.0017	0.2546	0.0037
O(1)	0.3053	0.0026	0.2051	0.0011	0.1993	0.0024
O(2)	0.3478	0.0026	0.0269	0.0011	0.1784	0.0023
O(3)	0.5812	0.0028	0.2271	0.0012	0.3225	0.0026
O(4)	0.6260	0.0031	0.0353	0.0013	0.3179	0.0028
O(5)	0.0360	0.0030	0.1959	0.0013	0.0369	0.0028
O(6)	0.0573	0.0030	0.0893	0.0013	0.3697	0.0028
H_2O	0.2228	0.0028	0.1172	0.0012	−0.1878	0.0026
K	0.8061	0.0008	0.3260	0.0004	0.1392	0.0008

The determined parameters are given in Table XIVB,84.
In this structure (Fig. XIVB,189) the complex anion consists of two
$(MoO_2C_2O_4)$ groups tied together by the oxygen atom O(8). Its bond dimensions are shown in Figure XIVB,190. As is usual, the oxalate groups are
planar. The water oxygens [O(7)] complete the sixfold coordination of the
molybdenum atoms, with $Mo–OH_2 = 2.330$ A. Potassium atoms are surrounded by eight oxygen atoms at distances between 2.67 and 3.25 A.

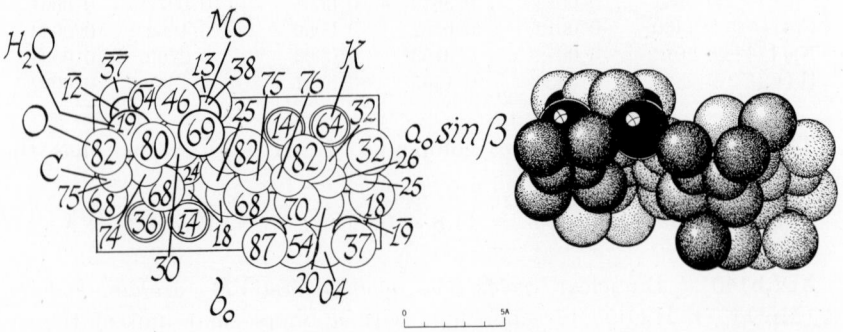

Fig. XIVB,189a (left).   The monoclinic structure of $K_2[MoO_2(C_2O_4) \cdot H_2O]_2O$ projected
along its c_0 axis.   Right-hand axes.
Fig. XIVB,189b (right).   A packing drawing of the monoclinic structure of $K_2[MoO_2-$
$(C_2O_4) \cdot H_2O]_2O$ seen along its c_0 axis.   The potassium atoms are heavily outlined and
dotted; the water molecules are black.   Carboxyl oxygen atoms are line shaded; the
smaller carbon atoms are dotted.   Two of the small black molybdenum atoms partly
show in the upper left.

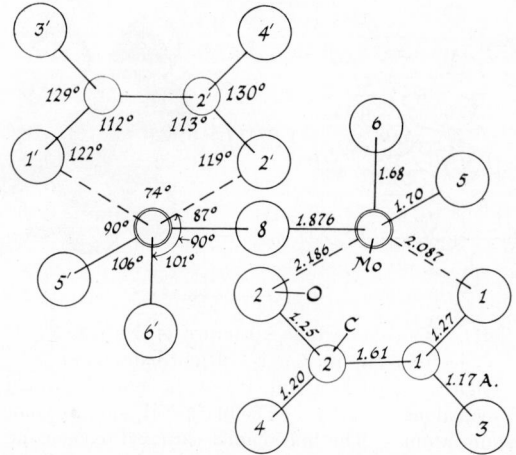

Fig. XIVB,190. Bond dimensions in crystals of $K_2[MoO_2(C_2O_4)\cdot H_2O]_2O$.

XIV,b141. Crystals of the compound *sodium ammonium oxalato-molybdate dihydrate*, $NaNH_4(MoO_3C_2O_4)\cdot 2H_2O$, are monoclinic with a tetramolecular unit of the dimensions:

$$a_0 = 9.30\pm 0.03 \text{ A.}; \quad b_0 = 13.43\pm 0.01 \text{ A.}; \quad c_0 = 7.85\pm 0.03 \text{ A.}$$

$$\beta = 98°$$

TABLE XIVB,85
Parameters of the Atoms in $NaNH_4(MoO_3C_2O_4)\cdot 2H_2O$

Atom	x	y	z
Mo	0.479	0.260	0.519
Na	0.098	0.391	0.017
NH$_4$	0.369	0.016	0.801
O(1)	0.613	0.124	0.504
O(2)	0.844	0.065	0.494
O(3)	0.946	0.257	0.543
O(4)	0.704	0.308	0.552
O(5)	0.418	0.381	0.586
O(6)	0.319	0.187	0.562
O(7)	0.503	0.276	0.289
H$_2$O(1)	0.127	0.113	0.228
H$_2$O(2)	0.044	0.067	0.829
C(1)	0.757	0.132	0.514
C(2)	0.809	0.237	0.521

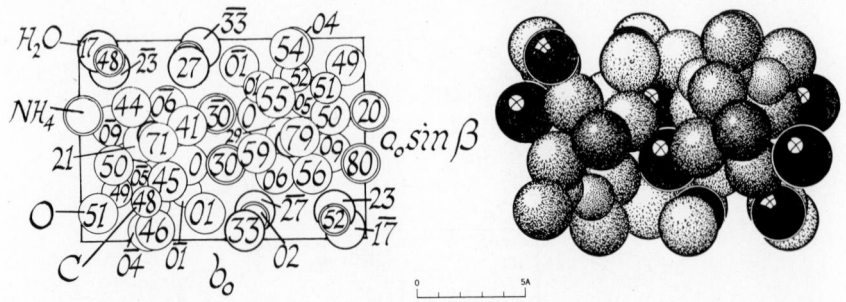

Fig. XIVB,191a (left).   The monoclinic structure of NaNH₄(MoO₃C₂O₄)·2H₂O projected along its c_0 axis.   Right-hand axes.

Fig. XIVB,191b (right).   A packing drawing of the monoclinic structure of NaNH₄-
(MoO₃C₂O₄)·2H₂O seen along its c_0 axis.   The black NH₄ ions are somewhat larger than
the black molybdenum atoms.   The line shaded carboxyl oxygens are larger than the
dotted carbon atoms.   Water molecules are heavily outlined and hook shaded.

The space group is C_{2h}^5 ($P2_1/c$) with all atoms in the positions:

$$(4e) \quad \pm(xyz;\ x,^1/_2-y,z+^1/_2)$$

The chosen parameters are those of Table XIVB,85.

The structure is shown in Figure XIVB,191. In it each molybdenum
atom is surrounded by a distorted octahedron of oxygen atoms, with
Mo–O = 2.230–2.42 A. for the oxalate oxygens and for the O(7) atoms
that are shared by adjacent MoO₆ octahedra; for the other oxygens, Mo–O
= 1.815, 1.850, and 1.878 A. The sodium ions also are surrounded by
six oxygen atoms (two from water molecules) with Na–O = 2.34–2.48 A.
The coordination of the NH₄⁺ ions is less definite, with NH₄–O ranging
upwards from 2.59 A.

XIV,b142.   A structure has been given *potassium platinum thiooxalate*,
K₂Pt(COS)₄. It and the two other isomorphous salts that have been investigated have tertamolecular monoclinic cells of the dimensions:

K₂Pt(COS)₄:
$a_0 = 11.16$ A.; $b_0 = 7.63$ A.; $c_0 = 22.62$ A.; $\beta = 144°4'$

$K_2Pd(COS)_4$:

$a_0 = 11.13$ A.; $b_0 = 7.78$ A.; $c_0 = 22.56$ A.; $\beta = 144°5'$

$K_2Ni(COS)_4$:

$a_0 = 10.99$ A.; $b_0 = 7.80$ A.; $c_0 = 22.23$ A.; $\beta = 144°1'$

Atoms have been placed in the following positions of C_{2h}^6 $(A2/a)$:

(4e) $\pm(^1/_4\, u\, 0;\ ^1/_4, u+^1/_2, ^1/_2)$

(8f) $\pm(xyz;\ x+^1/_2, \bar{y}, z;\ x, y+^1/_2, z+^1/_2;\ x+^1/_2, ^1/_2-y, z+^1/_2)$

with, for the three compounds, the parameters of Table XIVB,86. The resulting structure is molecular rather than ionic with all atoms of a molecule, including those of potassium, lying in a plane parallel to the a face of the crystal.

TABLE XIVB,86
Positions and Parameters of the Atoms in $K_2M(COS)_4$

Atom	Position	x	y	z
		$K_2Pt(COS)_4$ *and* $K_2Pd(COS)_4$		
Pt(Pd)	(4e)	$^1/_4$	0.125	0
K	(8f)	0.250	0.125	0.288
S(1)	(8f)	0.250	0.363	0.072
S(2)	(8f)	0.250	−0.113	−0.072
C(1)	(8f)	0.250	0.225	0.137
C(2)	(8f)	0.250	0.025	−0.137
O(1)	(8f)	0.250	0.294	0.182
O(2)	(8f)	0.250	−0.044	−0.182
		$K_2Ni(COS)_4$		
Ni	(4e)	$^1/_4$	0.125	0
K	(8f)	0.250	0.125	0.288
S(1)	(8f)	0.250	0.360	0.063
S(2)	(8f)	0.250	−0.110	−0.063
C(1)	(8f)	0.250	0.224	0.130
C(2)	(8f)	0.250	0.026	−0.130
O(1)	(8f)	0.250	0.292	0.175
O(2)	(8f)	0.250	−0.042	−0.175

BIBLIOGRAPHY TABLE, CHAPTER XIVB

Compound	Paragraph	Literature
Acetamide		
$CH_3C(O)NH_2$	b60	1929: H&L; 1940: S&H
Acetic acid		
CH_3COOH	b59	1956: J; 1958: J&T
Acetyl choline bromide		
$CH_3 \cdot CO \cdot O \cdot (CH_2)_2 \cdot N(CH_3)_3Br$	b81	1959: S; 1963: D
di-Acetyl hydrazine		
$(-NHCOCH_3)_2$	b66	1960: S
Acetylene		
C_2H_2	b29	1952: S&K
Aluminum triethyl potassium		
fluoride		
$2Al(C_2H_5)_3 \cdot KF$	b38	1961: N,A,P&Z; 1963: A&P
2-Amino ethanol phosphate		
$NH_2C_2H_5OPO_3H$	b23	1961: K; 1962: F,L&Y
di-Aminomaleonitrile		
$[NH_2C(CN){=}]_2$	b26	1961: P&L
2,2′,2″-tri-Amino triethylamine–		
nickel thiocyanate		
$Ni(SCN)_2 \cdot N(C_2H_4NH_2)_3$	b85	1935: C&W; 1958: H&W; R; 1959: R
2,2′,2″-tri-Amino-triethylamine		
trihydrochloride		
$N(CH_2CH_2NH_2HCl)_3$	b19	1935: J&B; 1963: R&G
Ammonium cobaltic ethylene		
diamine tetraacetate		
$NH_4Co[-CH_2NH(CH_2CO_2)_2]_2 \cdot$		
$2H_2O$	b96	1959: W&H
Ammonium oxalate monohydrate		
$(NH_4)_2(COO)_2 \cdot H_2O$	b130	1926: W; 1936: H&J; 1952: J&P
Ammonium oxamate		
$NH_2CO \cdot COONH_4$	b131	1963: B&S
Ammonium trifluoroacetate		
$NH_4(CF_3COO)$	b67	1964: C,J&W
Ammonium trioxalatochromiate		
dihydrate		
$(NH_4)_3[Cr(C_2O_4)_3] \cdot 2H_2O$	b137	1951: vN&S; 1952: vN&S
Ammonium uranyl propioniate		
$NH_4(UO_2)(C_2H_5COO)_3$	b75	1957: F,N&T
di-Amyl mercury mercaptan		
$Hg(SC_5H_{11})_2$	b35	1937: W
Antimonious xanthate		
$Sb(S_2COC_2H_5)_3$	b55	1938: D&R; 1961: G

(continued)

BIBLIOGRAPHY TABLE, CHAPTER XIVB (*continued*)

Compound	Paragraph	Literature
Arsenious xanthate		
$As(S_2COC_2H_5)_3$	**b55**	1938: B&D; 1960: C&G
Barium pentathionate–acetone		
monohydrate		
$BaS(S_2O_3)_2 \cdot (CH_3)_2CO \cdot H_2O$	**b58**	1956: F&T; 1958: F&T
Barium pentathionate tetrahydro-		
furan monohydrate		
$BaS(S_2O_3)_2 \cdot (CH_2)_4O \cdot H_2O$	**b58**	1956: F&T
Barium selenopentathionate–		
acetone monohydrate		
$BaSe(S_2O_3)_2 \cdot H_2O \cdot (CH_3)_2CO$	**b58**	1956: F&T
Barium selenopentathionate		
tetrahydrofuran monohydrate		
$BaSe(S_2O_3)_2 \cdot (CH_2)_4O \cdot H_2O$	**b58**	1956: F&T
Barium telluropentathionate		
tetrahydrofuran monohydrate		
$BaTe(S_2O_3)_2 \cdot (CH_2)_4O \cdot H_2O$	**b58**	1956: F&T
basic Beryllium acetate		
$Be_4O(CH_3COO)_6$	**b77**	1923: B; B&M; 1926: M&A; 1934: P&S; 1943: B; 1949: W&S; W,S&K; 1951: S; 1952: H&H; S; 1954: S,S&N; 1959: T; T&W; T,W&P; 1960: S
di-Borane		
B_2H_6	**b1**	1925: M,B&P; M&P
Bromacetamide		
$BrCH_2CONH_2$	**b63**	1959: D
hexa-Bromethane		
C_2Br_6	**b7**	1942: S&W
trans di-Bromo di ethylene		
diamine cobalti bromide		
hydrobromide dihydrate		
$[Co(NH_2CH_2CH_2NH_2)_2Br_2]Br \cdot$		
$HBr \cdot 2H_2O$	**b91**	1959: O,K,S&K
di-Bromo tetrachloro ethane		
BrC_2Cl_4Br		
$Cl_3C \cdot CBr_2Cl$	**b7**	1928: Y
tri-Bromo trichloro ethane		
$C_2Br_3Cl_3$	**b7**	1928: Y
tetra-Bromo dimethyl ethane		
$C_2Br_4(CH_3)_2(II)$	**b7**	1928: Y
penta-Bromo fluoroethane		
C_2Br_5F	**b7**	1928: Y

(*continued*)

BIBLIOGRAPHY TABLE, CHAPTER XIVB (*continued*)

Compound	Paragraph	Literature
Calcium oxalate monohydrate		
$CaC_2O_4 \cdot H_2O$	b133	1961: C; H; 1962: C&S
Cerium ethyl sulfate nonahydrate		
$Ce(C_2H_5SO_4)_3 \cdot 9H_2O$	b48	1937: K
Cesium acetylene diolate		
$Cs_2O_2C_2$	b46	1964: W&B
Chloracetamide		
$ClCH_2CONH_2$	b63, b64	1955: D; 1956: K; P&S; 1957: D; K
Chloral hydrate		
$CCl_3CH(OH)_2$	b6	1937: E; 1950: K&N; 1963: O
1,2-di-Chlorethane		
$C_2H_4Cl_2$	b4	1951: M&L; 1953: R&L; 1961: L,W,M&L
hexa-Chlorethane		
C_2Cl_6	b7	1928: Y; 1934: W; 1953: A,O&W; S&A; 1959: S&A
1,2-bis-(di-Chloro borane)–ethane		
$Cl_2BC_2H_4BCl_2$	b28	1956: M&L
trans di-Chloro di ethylene diamine chromic chloride hydrochloride dihydrate		
$[Cr(NH_2CH_2CH_2NH_2)_2Cl_2]Cl \cdot HCl \cdot 2H_2O$	b91	1960: O,K&K
trans di-Chloro di ethylene diamine cobaltic chloride hydrochloride dihydrate		
$[Co(NH_2CH_2CH_2NH_2)_2Cl_2]Cl \cdot HCl \cdot 2H_2O$	b91	1952: N,S&K
trans di-Chloro di ethylene diamine cobalti nitrate		
$[Co(NH_2CH_2CH_2NH_2)_2Cl_2]NO_3$	b94	1963: O&K
trans, trans, trans, tris (2-Chloro-vinyl) dichlorostibine		
$(ClCH{=}CH)_3SbCl_2$	b40	1952: S,K&K; 1953: S&K
Choline chloride		
$[(CH_3)_3NCH_2CH_2OH]Cl$	b22	1956: S; 1960: S&T; 1961: S&C
Chromous acetate monohydrate		
$Cr(CH_3COO)_2 \cdot H_2O$	b70	1953: vN,S&dW; vN&S
Cobalt acetate tetrahydrate		
$Co(CH_3COO)_2 \cdot 4H_2O$	b73	1953: vN&S
Cobaltous perchlorate penta methyl isonitrile		
$Co(CH_3NC)_5 \cdot (ClO_4)_2$	b107	1964: C,D&W

(*continued*)

BIBLIOGRAPHY TABLE, CHAPTER XIVB (continued)

Compound	Paragraph	Literature
Copper diacetato diammine $Cu(CH_3COO)_2 \cdot 2NH_3$	b71	1963: BS; S,A&M
Cupric acetate monohydrate $Cu(CH_3COO)_2 \cdot H_2O$	b70	1953: A&S; vN&S; vN,S&dW
Cupric ammonium oxalate dihydrate $Cu(NH_4)_2(C_2O_4)_2 \cdot 2H_2O$	b135	1962: V
Cupric chloride mono acetonitrile $CuCl_2 \cdot CH_3CN$	b103	1964: W&R
Cupric bis-N,N-diethyl dithiocarbamate $Cu[(C_2H_5)_2NCS_2]_2$	b51	1963: B; B,D,M,M,V&Z
Cupric dihydrogen ethylene diamine tetraacetate monohydrate $CuH_2[—CH_2NH(CH_2CO_2)_2]_2 \cdot H_2O$	b99	1959: S&H
Cupric potassium oxalate dihydrate $CuK_2(C_2O_4)_2 \cdot 2H_2O$	b135	1962: V
tri-Cupric chloride di acetonitrile $3CuCl_2 \cdot 2CH_3CN$	b104	1964: W&R
Cuprous chloride tetrakis thioacetamide $CuCl \cdot 4SC(CH_3)NH_2$	b108	1936: C,S,W&W; C,W&W; 1962: T&R
Cuprous diethyl dithiocarbamate $Cu[(C_2H_5)_2NCS_2]$	b50	1963: H
Cuprous iodide–methyl isocyanide $CuI \cdot CH_3NC$	b101	1960: F,T&H
Cuprous iodide–triethyl arsine $CuI \cdot As(C_2H_5)_3$	b110	1936: M,P&W; W
Cuprous iodide triethyl phosphine $CuI \cdot P(C_2H_5)_3$	b110	1936: M,P&W; W
Cyanoacetylene $HC{\equiv}C(CN)$	b31	1958: S&C
di-Cyanoacetylene $N{\equiv}C—C{\equiv}C—C{\equiv}N$	b32	1953: H&C
tetra-Cyanoethylene $(CN)_2C{=}C(CN)_2$	b25	1960: B&T
Cyanogen $(CN)_2$	b2	1963: P&H
Deuterium oxalic acid dideuterate $(COOD)_2 \cdot 2D_2O$	b119, b120	1939: R&U; 1964: F,I&S

(continued)

BIBLIOGRAPHY TABLE, CHAPTER XIVB (*continued*)

Compound	Paragraph	Literature
Dysprosium ethyl sulfate nonahydrate $Dy(C_2H_5SO_4)_3 \cdot 9H_2O$	**b48**	1937: K
Erbium ethyl sulfate nonahydrate $Er(C_2H_5SO_4)_3 \cdot 9H_2O$	**b48**	1959: F&R
Ethane C_2H_6	**b1**	1925: M,B&P; M&P
Ethane dithiocyanate $C_2H_4(SCN)_2$	**b3**	1958: B&F
mono-Ethyl ammonium bromide $C_2H_5NH_3Br$	**b113**	1928: H; 1958: J
mono-Ethyl ammonium chloride $C_2H_5NH_3Cl$	**b113**	1958: J
mono-Ethyl ammonium iodide $C_2H_5NH_3I$	**b113**	1928: H
mono-Ethyl ammonium chloro-platinate $(NH_3C_2H_5)_2PtCl_6$	**b115**	1928: W
mono-Ethyl ammonium chloro-stannate $(NH_3C_2H_5)_2SnCl_6$	**b115**	1928: W
1-Ethyl decaborane $B_{10}H_{13}C_2H_5$	**b14**	1964: P
Ethyl lithium C_2H_5Li	**b34**	1959: D; 1963: D
di-Ethyl amino borane $C_2H_5NH_2B_8H_{11}NHC_2H_5$	**b13**	1963: L,S&L
di-Ethyl ether–monobromodi-chloromethane $(C_2H_5)_2O \cdot CHBrCl_2$	**b15**	1964: A&TM
di-Ethyl mercury mercaptan $Hg(SC_2H_5)_2$	**b35**	1937: W
tri-Ethyl ammonium bromide $(C_2H_5)_3NHBr$	**b114**	1928: H
tri-Ethyl ammonium chloride $(C_2H_5)_3NHCl$	**b114**	1928: H
tri-Ethyl ammonium iodide $(C_2H_5)_3NHI$	**b114**	1928: H
tri-Ethylscarphane $(C_2H_5)_3P \cdot CS_2$	**b18**	1962: M&T
tri-Ethyl sulfonium iodide $(C_2H_5)_3SI$	**b114**	1940: M

(*continued*)

BIBLIOGRAPHY TABLE, CHAPTER XIVB (*continued*)

Compound	Paragraph	Literature
tetra-Ethyl ammonium heptaiodide $N(C_2H_5)_4I_7$	b116	1955: H&W; 1958: H&W
tetra-Ethyl diphosphine disulfide $(C_2H_5)_4P_2S_2$	b17	1961: D&W
Ethylene C_2H_4	b24	1935: K&T; 1938: T; 1944: B
Ethylene diamine–tribromo platinum $C_2N_2H_8 \cdot PtBr_3$	b88	1961: R&R
trans di-Ethylene diamine– cobaltic chloride $CoCl_3 \cdot 2C_2H_4(NH_2)_2$	b90	1955: B,G&P; 1959: B,G&P
bis Ethylene diamine cupric nitrate $Cu(C_2N_2H_8)_2(NO_3)_2$	b83	1964: K&L
bis Ethylene diamine cupric thiocyanate $Cu(C_2N_2H_8)_2(SCN)_2$	b82	1964: B&L
trans-bis Ethylene diamine nickel isocyanate $Ni(C_2N_2H_8)_2(NCS)_2$	b84	1963: B&L
D-tris-Ethylene diamine cobaltic bromide monohydrate D-$[Co(NH_2CH_2CH_2NH_2)_3]Br_3 \cdot H_2O$	b93	1955: S,N,S&K; 1962: N
D-tris-Ethylene diamine cobaltic chloride sodium chloride hexahydrate $2[Co(NH_2CH_2CH_2NH_2)_3Cl_3] \cdot NaCl \cdot 6H_2O$	b95	1955: S,N,S&K; 1957: N,S,S&K; S,N,S&K
DL-tris-Ethylene diamine cobaltic chloride trihydrate $[Co(NH_2CH_2CH_2NH_2)_3]Cl_3 \cdot 3H_2O$	b92	1956: N,S&K
Ethylene diammonium chloride $NH_3(CH_2)_2NH_3Cl_2$	b20	1963: A&H; K,K&Y
Ethylene diammonium sulfate $[NH_3CH_2—]_2SO_4$	b21	1927: B; 1961: S
Ferrous cyanide tetra methyl isonitrile $Fe(CH_3NC)_4(CN)_2$	b106	1957: H&P
Ferrous oxalate dihydrate $FeC_2O_4 \cdot 2H_2O$	b134	1957: M&G; 1959: C

(*continued*)

438

CRYSTAL STRUCTURES

BIBLIOGRAPHY TABLE, CHAPTER XIVB (*continued*)

Compound	Paragraph	Literature
mono-Fluoroacetamide		
CH_2FCONH_2	b62	1962: H&S
Gadolinium ethyl sulfate		
nonahydrate		
$Gd(C_2H_5SO_4)_3 \cdot 9H_2O$	b48	1937: K
di-Heptyl mercury mercaptan		
$Hg(SC_7H_{15})_2$	b35	1937: W
di-Hexyl mercury mercaptan		
$Hg(SC_6H_{13})_2$	b35	1937: W
sym-di-Iodoethane		
$C_2H_4I_2$	b5	1935: K
sym-di-Iodoethylene		
$C_2H_2I_2$	b5	1935: K
2,2'-di-Iododiethyl trisulfide		
$[I(C_2H_4)]_2S_3$	b10	1948: D,M&R; D&R; 1950: D
Iron carbonyl ethyl mercaptide		
$[Fe(CO)_3C_2H_5S]_2$	b41	1963: D&W
Iron octacarbonyl dimethyl		
acetylene		
$CH_3CCCH_3 \cdot H_2Fe_2(CO)_8$	b42	1961: H&M
Lanthanum ethyl sulfate		
nonahydrate		
$La(C_2H_5SO_4)_3 \cdot 9H_2O$	b48	1937: K
Lithium acetate dihydrate		
$Li(CH_3COO) \cdot 2H_2O$	b69	1958: A&P
Lithium aluminum tetraethyl		
$LiAl(C_2H_5)_4$	b37	1964: G,D&B
Lithium oxalate		
$Li_2C_2O_4$	b125	1921: B&J; 1964: B&S
Magnesium acetate tetrahydrate		
$Mg(CH_3COO)_2 \cdot 4H_2O$	b73	1952: P; 1957: S,K&P
Manganese dichlorophosphate		
diethylacetate		
$Mn(PO_2Cl_2)_2(CH_3COOC_2H_5)_2$	b80	1963: D&R
di-Mercuric chloride diethyl sulfide		
$2HgCl_2 \cdot (C_2H_5)_2S$	b36	1964: B
Mercury tetrathiocyanate–copper		
diethylene diamine		
$Hg(SCN)_4 \cdot Cu[(NH_2CH_2)_2]_2$	b87	1950: S&C; 1953: S
N-Methylacetamide		
$CH_3C(O)NHCH_3$	b65	1960: K&P

(*continued*)

BIBLIOGRAPHY TABLE, CHAPTER XIVB (*continued*)

Compound	Paragraph	Literature
Methyl cyanide–boron hydride $B_9H_{13}(CH_3CN)$	b11	1961: W,S&L
Methyl cyanide–boron tribromide $CH_3CN \cdot BBr_3$	b12	1951: G&S
Methyl cyanide–boron trichloride $CH_3CN \cdot BCl_3$	b12	1951: G&S
Methyl cyanide–boron trifluoride $CH_3CN \cdot BF_3$	b12	1950: H,O,B&S; 1951: G&S; H,G&O
N-Methyl 2,2-dimethyl sulfonyl vinylidine amine $(CH_3SO_2)_2C{=}C{=}NCH_3$	b27	1954: W
di-Methyl acetylene $CH_3C{\equiv}CCH_3$	b30	1955: P&P; 1959: M,S&P
di-Methyl diethyl ammonium chlorostannate $[N(CH_3)_2(C_2H_5)_2]_2SnCl_6$	b117	1929: W&C
di-Methyl oxalate $(CH_3)_2(COO)_2$	b124	1953: D&J
di-Methyl triacetylene $CH_3{-}C{\equiv}C{-}C{\equiv}C{-}C{\equiv}C{-}CH_3$	b33	1952: J&R
N,N,N',N'-tetra-Methyl ethylene diamine, aluminum hydride $[(CH_3)_2NCH_2{-}]_2AlH_3$	b98	1964: P
hexa-Methylisocyanide, ferrous chloride trihydrate $Fe(CH_3NC)_6Cl_2 \cdot 3H_2O$	b105	1945: P&B
Neodymium ethyl sulfate nonahydrate $Nd(C_2H_5SO_4)_3 \cdot 9H_2O$	b48	1937: K
Nickel acetate tetrahydrate $Ni(CH_3COO)_2 \cdot 4H_2O$	b73	1934: H; 1953: vN&S
Nickel bromide bis(triethyl phosphine) $NiBr_2 \cdot 2P(C_2H_5)_3$	b111	1958: S&T
Nickel bis diethyl dithiocarbamate $Ni[(C_2H_5)_2NCS_2]_2$	b52	1958: S&S; 1960: S&L; 1963: F&S
Nickel dihydrogen ethylene diamine tetraacetate monohydrate $NiH_2[{-}CH_2NH(CH_2CO_2)_2]_2 \cdot H_2O$	b99	1959: S&H

(*continued*)

BIBLIOGRAPHY TABLE, CHAPTER XIVB (*continued*)

Compound	Paragraph	Literature
Nickel nitrate tris ethylene diamine $Ni(NO_3)_2 \cdot (NH_2CH_2CH_2NH_2)_3$	b86	1951: W&A; 1960: S&A
Nickel xanthate $Ni(S_2COC_2H_5)_2$	b54	1938: B; 1963: F
S-bis (Nitroamino) ethane $C_2H_4(NHNO_2)_2$	b8	1948: L&W
Oxalic acid $(COOH)_2$	b118	1924: H&M; 1935: H; 1947: B; 1952: C,D&J
Oxalic acid dihydrate $(COOH)_2 \cdot 2H_2O$	b119	1924: H&M; 1926: W; 1934: Z; 1936: R&W; 1939: B,H&P; R&U; 1942: B,H&P; 1947: D&R; 1953: A&C; 1954: G; P
Oxamide $(CONH_2)_2$	b121	1938: M&vdW; 1953: R; 1954: A&D
Oxalyl bromide $(COBr)_2$	b122	1961: G&H; 1962: G&H
Oxalyl chloride $(COCl)_2$	b123	1962: G&H
tri-Oxo(diethylene triamine) molybdenum $MoO_3[NH_2(CH_2)_2NH(CH_2)_2NH_2]$	b97	1964: C&E
Oxomolybdenum xanthate $Mo_2O_3(C_2H_5OCS_2)_2$	b56	1964: B,C&W
Palladous chloride ethylene $PdCl_2 \cdot C_2H_4$	b44	1955: D&B
Platinous chloride diammine bis(acetamidine) monohydrate $PtCl_2(NH_3)_2[CH_3C(NH_2)NH]_2 \cdot H_2O$	b100	1962: S
Platinous platinic tetraethylamine chloride tetrahydrate $Pt_2(C_2H_4NH_2)_8Cl_6 \cdot 4H_2O$	b89	1961: C&H
Platinum bis(triethyl phosphine) hydrobromide $Pt[(C_2H_5)_3P]_2HBr$	b112	1960: O,P&R
trans-Platinum dichloride ethylene dimethylamine $PtCl_2 \cdot C_2H_4 \cdot NH(CH_3)_2$	b109	1960: A,O&R
Potassium acetylene diolate $K_2O_2C_2$	b46	1963: W&B

(*continued*)

BIBLIOGRAPHY TABLE, CHAPTER XIVB (*continued*)

Compound	Paragraph	Literature
Potassium acid oxalate		
$KH(COO)_2$	**b128**	1935: H
Tri-Potassium aluminum oxalate		
trihydrate		
$K_3Al(C_2O_4)_3 \cdot 3H_2O$	**b138**	1958: H
Tri-Potassium chromic oxalate		
trihydrate		
$K_3Cr(C_2O_4)_3 \cdot 3H_2O$	**b138**	1952: vN&S
trans-Potassium dioxalato diaquo		
chromiate trihydrate		
$K[Cr(C_2O_4)_2(H_2O)_2] \cdot 3H_2O$	**b136**	1950: vN&S; 1951: vN&S
Potassium ethyl sulfate		
$KC_2H_5SO_4$	**b47**	1953: J; 1958: T
Tri-Potassium ferric oxalate		
trihydrate		
$K_3Fe(C_2O_4)_3 \cdot 3H_2O$	**b138**	1958: H
Potassium molybdenum(VI)		
oxalate hydrate		
$K_2(MoO_2C_2O_4 \cdot H_2O)_2O$	**b140**	1964: C,M&W
Potassium nickel thiooxalate		
$K_2Ni(COS)_4$	**b142**	1935: C,W&W
Potassium nitroacetate		
$K_2(NO_2CHCOO)$	**b79**	1954: S,L&M
Potassium oxalate monohydrate		
$K_2(COO)_2 \cdot H_2O$	**b127**	1935: H
Potassium palladium thiooxalate		
$K_2Pd(COS)_4$	**b142**	1935: C,W&W
Potassium platinum thiooxalate		
$K_2Pt(COS)_4$	**b142**	1935: C,W&W
Potassium tetroxalate dihydrate		
$KHC_2O_4 \cdot H_2C_2O_4 \cdot 2H_2O$	**b129**	1964: H
Potassium uranyl propionate		
$K(UO_2)(C_2H_5COO)_3$	**b75**	1957: F,N&T
Potassium xanthate		
$KS_2COC_2H_5$	**b53**	1963: M&T
Praseodymium ethyl sulfate		
nonahydrate		
$Pr(C_2H_5SO_4)_3 \cdot 9H_2O$	**b48**	1937: K; 1959: F&R
n-Propyl mercury mercaptan		
$Hg(SC_3H_7)_2$	**b35**	1937: W
Roussin's Red Ethyl Ester		
$[Fe(NO)_2C_2H_5S]_2$	**b43**	1958: T,R&C
Rubidium acetylene diolate		
$Rb_2O_2C_2$	**b46**	1964: W&B

(*continued*)

442   CRYSTAL STRUCTURES

BIBLIOGRAPHY TABLE, CHAPTER XIVB (*continued*)

Compound	Paragraph	Literature
Rubidium acid oxalate		
$RbH(COO)_2$	b128	1935: H
tri-Rubidium chromic oxalate		
hydrate		
$Rb_3Cr(C_2O_4)_3 \cdot xH_2O$	b138	1952: vN&S
Rubidium cobaltic ethylene		
diamine tetraacetate dihydrate		
$RbCo[-CH_2NH(CH_2CO_2)_2]_2 \cdot$	b96	1959: W&H
$2H_2O$		
Rubidium oxalate monohydrate		
$Rb_2(COO)_2 \cdot H_2O$	b127	1935: H
Rubidium xanthate		
$RbS_2COC_2H_5$	b53	1963: M&T
Samarium ethyl sulfate		
nonahydrate		
$Sm(C_2H_5SO_4)_3 \cdot 9H_2O$	b48	1937: K
Seleno triethyl phosphine		
$SeP(C_2H_5)_3$	b16	1959: vM&L; 1960: vM&L
Silver chloride tetrakis		
thioacetamide		
$AgCl \cdot 4SC(CH_3)NH_2$	b108	1936: C,W&W
Silver oxalate		
$Ag_2(COO)_2$	b132	1943: G
Silver uranyl acetate hydrate		
$Ag(UO_2)(CH_3COO)_3 \cdot xH_2O$	b76	1936: F
Sodium ammonium oxalato-		
molybdate dihydrate		
$NaNH_4(MoO_3C_2O_4) \cdot 2H_2O$	b141	1962: A&B; 1963: A&B
Sodium hydrogen acetate		
$NaH(C_2H_3O_2)_2$	b68	1922: W; 1959: S; 1961: S&M
Sodium iodide triacetonate		
$NaI \cdot 3(CH_3)_2CO$	b57	1963: P,G&vM
Sodium neptunyl acetate		
$Na(NpO_2)(CH_3COO)_3$	b75	1949: Z
Sodium oxalate		
$Na_2C_2O_4$	b126	1954: J&P
Sodium plutonyl acetate		
$Na(PuO_2)(CH_3COO)_3$	b75	1949: Z
Sodium uranyl acetate		
$Na(UO_2)(CH_3COO)_3$	b75	1930: dJ; 1935: F; 1959: Z&P

(*continu*

BIBLIOGRAPHY TABLE, CHAPTER XIVB (*continued*)

Compound	Paragraph	Literature
di-Sodium 1,2-ethane bis nitraminate $Na_2(NO_2\!\!-\!\!N\!\!-\!\!CH_2\cdot CH_2\!\!-$ $N\!\!-\!\!NO_2)$	b49	1953: A&W
tetra-Sodium hafnium tetroxalate trihydrate $Na_4Hf(C_2O_4)_4\cdot 3H_2O$	b139	1963: G,S&H
tetra-Sodium zirconium tetroxalate trihydrate $Na_4Zr(C_2O_4)_4\cdot 3H_2O$	b139	1961: H,G&S; 1963: G,S&H
Taurine $NH_2(C_2H_4)HSO_3$	b9	1963: S&Y; 1964: D
Thioacetamide $CH_3C(S)NH_2$	b61	1960: T
Thio triethyl phosphine $SP(C_2H_5)_3$	b16	1959: vM&L
Uranium tetraacetate $U(CH_3COO)_4$	b74	1964: J,G&B
Uranyl nitrate–triethyl phosphate $UO_2(NO_3)_2\cdot 2(C_2H_5)_3PO_4$	b39	1959: F&L; 1960: F&L
Yttrium ethyl sulfate nonahydrate $Y(C_2H_5SO_4)_3\cdot 9H_2O$	b48	1937: K; 1959: F&R
Zeise's Salt $K(PtCl_3\cdot C_2H_4)\cdot H_2O$	b45	1954: W&M; 1955: W&M
Zinc acetate dihydrate $Zn(CH_3COO)_2\cdot 2H_2O$	b72	1953: vN,S&T
Zinc bis-N,N-diethyl dithio-carbamate $Zn[(C_2H_5)_2NCS_2]_2$	b51	1953: S&H; 1963: B,D,M,M,V&Z
Zinc chloride di acetonitrile $ZnCl_2\cdot 2CH_3CN$	b102	1962: I&Z
Zinc oxyacetate $Zn_4O(CH_3COO)_6$	b78	1954: K&S

C. LONGER-CHAIN COMPOUNDS

1. Three-Carbon Chain Compounds

XIV,c1. Crystals of *β-nitropropionic acid*, $NO_2 \cdot CH_2 \cdot CH_2 \cdot COOH$, are monoclinic with a unit cell that contains eight molecules and has the dimensions:

$$a_0 = 20.02 \text{ A.}; \quad b_0 = 5.20 \text{ A.}; \quad c_0 = 9.57 \text{ A.}; \quad \beta = 92°26'$$

TABLE XIVC,1
Parameters of the Atoms in β-Nitropropionic Acid

Atom	x	y	z
C(1)	0.066	0.292	0.403
C(2)	0.116	0.109	0.330
C(3)	0.146	0.255	0.209
N	0.181	0.494	0.246
O(1)	0.050	0.217	0.531
O(2)	0.040	0.485	0.352
O(3)	0.192	0.558	0.371
O(4)	0.205	0.612	0.148

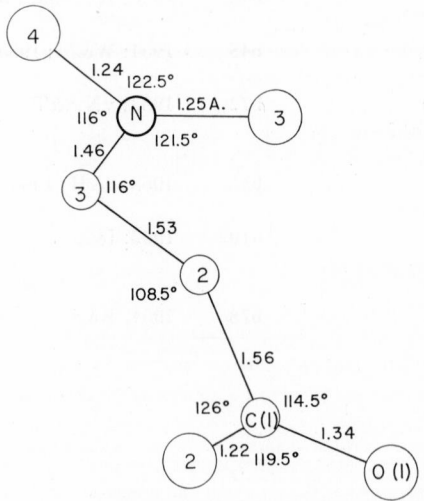

Fig. XIVC,1. Bond dimensions in the molecule of $NO_2(CH_2)_2COOH$.

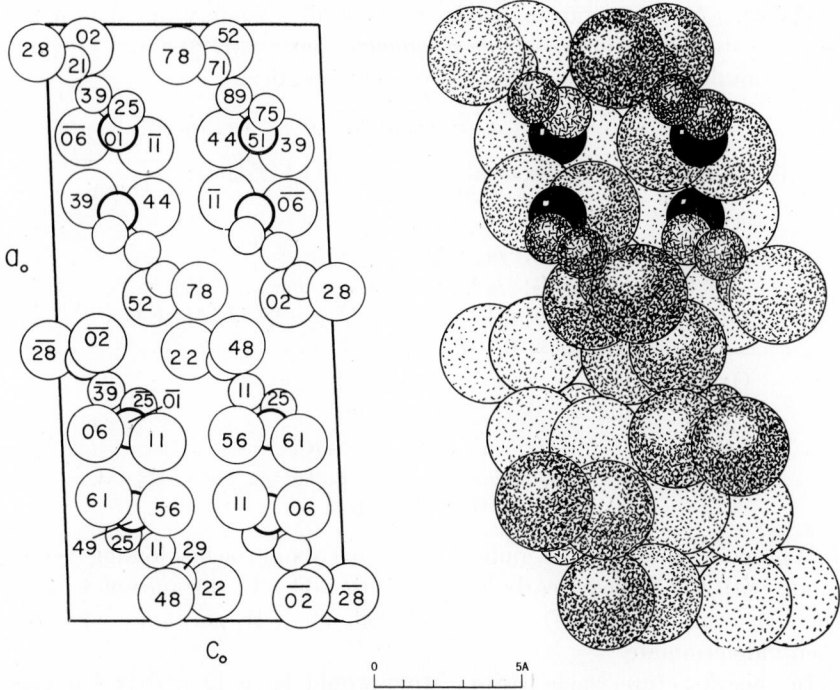

Fig. XIVC,2a (left). The monoclinic structure of $NO_2(CH_2)_2COOH$ projected along its b_0 axis. Of the smaller circles the nitrogens are heavily, the carbons lightly outlined. The larger circles are oxygen. Right-hand axes.

Fig. XIVC,2b (right). A packing drawing of the monoclinic structure of $NO_2(CH_2)_2$-COOH seen along its b_0 axis. The carbon atoms are the small, the oxygen atoms the large dotted circles. Atoms of nitrogen are black.

Atoms are in the general positions:

$$(8f) \quad \pm(xyz;\ x+1/2,\bar{y},z;\ x,y+1/2,z+1/2;\ x+1/2,1/2-y,z+1/2)$$

of C_{2h}^6 $(A2/a)$ with the parameters of Table XIVC,1.

This leads to a molecule having the bond lengths and angles of Figure XIVC,1. As in many of the other structures studied, the two carbon–oxygen distances in the carboxyl group are unequal; in this case, C–O(1) = 1.34 A. and C–O(2) = 1.22 A. The molecular distribution within the crystal is indicated in Figure XIVC,2. Hydrogen bonding between adjacent molecules leads to carboxyl O–O distances as short as 2.66 A.

XIV,c2.   An atomic arrangement was described many years ago for the cubic crystals of *barium dicalcium propionate*, $BaCa(C_2H_5COO)_6$. The unit cube, containing eight molecules, has the edge length:

$$a_0 = 18.20 \text{ A.}$$

Atoms are in the following special positions of O_h^7 ($Fd3m$):

Ba: (8a) $000; {}^1/_4\,{}^1/_4\,{}^1/_4$; F.C.

Ca: (16c) ${}^1/_8\,{}^1/_8\,{}^1/_8; {}^1/_8\,{}^3/_8\,{}^3/_8$;

  ${}^3/_8\,{}^1/_8\,{}^3/_8; {}^3/_8\,{}^3/_8\,{}^1/_8$; F.C.

C: (48f) $u00; \bar{u}00; u+{}^1/_4,{}^1/_4,{}^1/_4; {}^1/_4-u,{}^1/_4,{}^1/_4$; tr; F.C.

  with $u = 0.194$

O: (96g) $uuv;$ $u\bar{u}\bar{v};$

  $\bar{u}u\bar{v};$ $\bar{u}\bar{u}v;$

  ${}^1/_4-u,{}^1/_4-u,{}^1/_4-v; {}^1/_4-u,u+{}^1/_4,v+{}^1/_4;$

  $u+{}^1/_4,{}^1/_4-u,v+{}^1/_4; u+{}^1/_4,u+{}^1/_4,{}^1/_4-v;$ tr; F.C.

  with $u = 0.042, v = 0.160$.

It is proposed that the C_2H_5 radical is rotating about the line joining its CH_2 group with the carboxyl carbon atom; with a $C-CH_2$ distance of 1.54 A., the CH_2 would be in (48f) with $u = 0.280$. The CH_3 position is of course then indeterminate.

In this structure, each barium atom would have 12 carboxyl oxygen neighbors at a distance of 3.11 A., while each calcium atom would be equidistant from six of these oxygen atoms (Ca–O = 2.23 A.). For the chosen parameters, C–O = 1.24 A.

XIV,c3.   In paragraph **XIV,a102** it was pointed out that at room temperature the propyl ammonium halides have the same structure as methyl ammonium chloride. At reduced temperatures, *propyl ammonium chloride*, $NH_3C_3H_7Cl$, passes by a second-order transition to a low-temperature form. From photographs made at somewhat below $-100°C$. it was concluded that this low $NH_3C_3H_7Cl$ has a tetramolecular monoclinic cell of the dimensions:

$$a_0 = 9.06 \text{ A.}; \quad b_0 = 8.58 \text{ A.}; \quad c_0 = 7.34 \text{ A.}; \quad \beta = 98°$$

The space group was chosen as C_{2h}^3 ($C2/m$) with all atoms except chlorine in

  (4i) $\pm (u0v; u+{}^1/_2,{}^1/_2,v)$

The chloride ions are in

  (4e) $\pm ({}^1/_4\,{}^1/_4\,0; {}^1/_4\,{}^3/_4\,0)$

and the following approximate positions, corresponding to a nonrotating cation, have been proposed for the carbon and nitrogen atoms:

$C(1)$: $u = v = 0.3$ $C(2)$: $u = 0.2, v = 0.4$
$C(3)$: $u = 0.25, v = 0.6$ N: $u = 0.26, v = 0.09$

XIV,c4. *Tetra-N-propyl ammonium bromide*, $(C_3H_7)_4NBr$, is tetragonal with a bimolecular unit of the edge lengths:

$$a_0 = 8.24 \pm 0.01 \text{ A.}, \qquad c_0 = 10.92 \pm 0.01 \text{ A.}$$

The space group is $S_4{}^2$ ($I\bar{4}$) with atoms in the positions:

Br: (2a)   000; $1/2\ 1/2\ 1/2$
N: (2c) $1/2\ 0\ 1/4; 0\ 1/2\ 3/4$
C(1): (8g) $xyz; \bar{x}\bar{y}z; y\bar{x}\bar{z}; \bar{y}x\bar{z};$ B.C.
        with $x = 0.413 \pm 0.004, y = 0.121 \pm 0.004, z = 0.163 \pm 0.003$
C(2): (8g)   with $x = 0.311 \pm 0.004, y = 0.249 \pm 0.005, z = 0.231 \pm 0.004$
C(3): (8g)   with $x = 0.246 \pm 0.006, y = 0.367 \pm 0.006, z = 0.131 \pm 0.004$

The resulting structure is shown in Figure XIVC,3. It has a ZnS type arrangement of ions with each bromine surrounded by four $(C_3H_7)_4N^+$ cations and vice versa; all three carbon atoms of a propyl radical are about equally distant from bromine (3.91–3.96 A.). In the cations, all bond

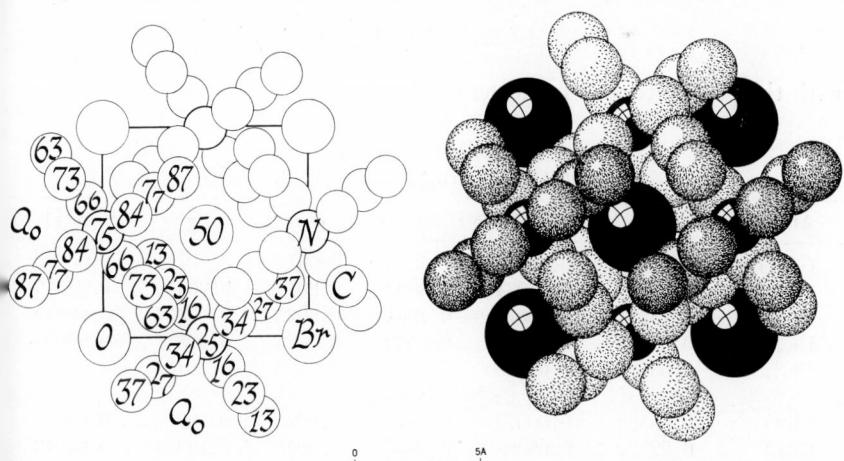

Fig. XIVC,3a (left). The tetragonal structure of $(C_3H_7)_4NBr$ projected along its c_0 axis. Right-hand axes.
Fig. XIVC,3b (right). A packing drawing of the tetragonal $(C_3H_7)_4NBr$ arrangement seen along its c_0 axis. The bromine atoms are the large, the nitrogens the small black circles. Atoms of carbon are line shaded.

lengths are the same (1.55 A.) to within the limit of error.

Positions have been assigned the hydrogen atoms attached to C(1) and C(2) on the assumptions that C–H = 1.10 A. and that the angles are tetrahedral. Their parameters thus would be:

Atom	x	y	z
H(1),C1	0.332	0.051	0.012
H(2),C1	0.506	0.184	0.108
H(3),C2	0.209	0.191	0.279
H(4),C2	0.387	0.316	0.297

XIV,c5. The red compound between *cupric chloride* and *N-propyl alcohol*, $5CuCl_2 \cdot 2C_3H_7OH$, is monoclinic with a bimolecular unit of the edge lengths:

$$a_0 = 10.17 \pm 0.02 \text{ A.}; \quad b_0 = 6.04 \pm 0.01 \text{ A.}; \quad c_0 = 18.30 \pm 0.02 \text{ A.}$$
$$\beta = 94°36' \pm 6'$$

The space group is C_{2h}^5 in the orientation $P2_1/n$ with one set of copper atoms in the positions:

$$\text{Cu(1): } (2a) \quad 000; \; {}^1/_2 \, {}^1/_2 \, {}^1/_2$$

All other atoms are in the positions:

$$(4e) \quad \pm (xyz; \; x+{}^1/_2, {}^1/_2-y, z+{}^1/_2)$$

with the parameters of Table XIVC,2.

TABLE XIVC,2
Parameters of the Atoms in $5CuCl_2 \cdot 2C_3H_7OH$

Atom	x	$\sigma(x)$	y	$\sigma(y)$	z	$\sigma(z)$
Cu(2)	0.05721	0.00039	0.4333	0.0011	0.10711	0.00020
Cu(3)	0.10731	0.00034	0.8892	0.0011	0.20455	0.00019
Cl(1)	0.13291	0.00078	0.2951	0.0022	0.00364	0.00039
Cl(2)	−0.09236	0.00077	0.1540	0.0022	0.09827	0.00039
Cl(3)	−0.03805	0.00075	0.5997	0.0025	0.20101	0.00042
Cl(4)	0.18884	0.00075	0.7443	0.0021	0.10309	0.00038
Cl(5)	0.22789	0.00083	0.2053	0.0023	0.19354	0.00043
O	0.4340	0.0020	0.5432	0.0057	0.2181	0.0011
C(1)	0.5837	0.0052	0.4989	0.0121	0.1831	0.0029
C(2)	0.5821	0.0059	0.5819	0.0171	0.1068	0.0034
C(3)	0.5631	0.0053	0.8452	0.0167	0.0936	0.0030

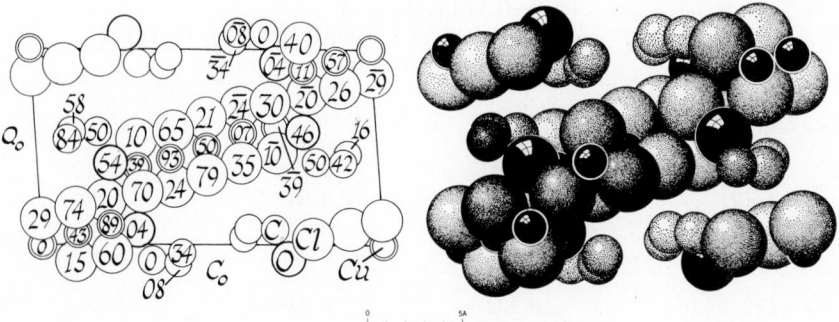

Fig. XIVC,4a (left). The monoclinic structure of $5CuCl_2 \cdot 2C_3K_7OH$ projected along its b_0 axis. Right-hand axes.

Fig. XIVC,4b (right). A packing drawing of the monoclinic structure of $5CuCl_2 \cdot 2C_3H_7OH$ viewed along its b_0 axis. The oxygen atoms are the large, the copper atoms the small black circles. The large dotted circles are chlorine; the smaller are carbon.

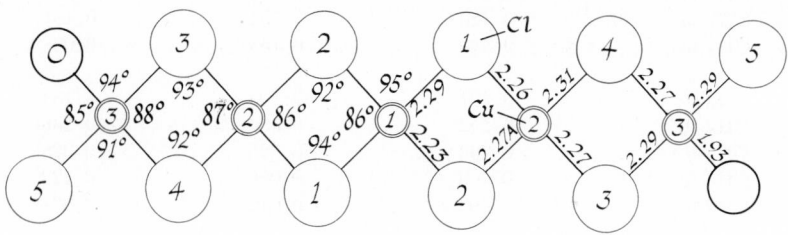

Fig. XIVC,5. Bond dimensions in crystals of $5CuCl_2 \cdot 2C_3H_7OH$.

The structure is shown in Figure XIVC,4. The square coordination around each of the copper atoms gives rise to the bond dimensions of Figure XIVC,5.

XIV,c6. A partial structure has been found for the orthorhombic *di-N-propyl gold cyanide*, $(C_3H_7)_2AuCN$, which has been assigned the following large cell containing 16 molecules:

$$a_0 = 17.06 \text{ A.}; \quad b_0 = 22.36 \text{ A.}; \quad c_0 = 10.0 \text{ A.}$$

Gold atoms have been put in four sets of the general positions of C_{2v}^5 (*Pca*):

(4a) $xyz; \; \bar{x},\bar{y},z+\frac{1}{2}; \; \frac{1}{2}-x,y,z+\frac{1}{2}; \; x+\frac{1}{2},\bar{y},z$

with the parameters listed in Table XIVC,3.

TABLE XIVC,3
Parameters Found for the Gold Atoms and Proposed for the Other
Atoms in $(C_3H_7)_2AuCN$

Atom	x	y	z
Au(1)	−0.024	0.14	0.00
Au(2)	0.058	0.326	0.307
Au(3)	0.242	0.362	0.295
Au(4)	0.277	0.177	−0.012
C(1)	−0.044	0.252	0.185
C(2)	0.122	0.348	0.300
C(3)	0.263	0.250	0.110
C(4)	0.096	0.154	−0.005
N(1)	−0.037	0.211	0.118
N(2)	0.057	0.339	0.302
N(3)	0.255	0.291	0.177
N(4)	0.161	0.163	−0.007
$CH_2(4a)$	0.396	0.191	−0.017
$CH_2(4a')$	0.426	0.152	0.101
$CH_3(4a)$	0.516	0.163	0.098
$CH_2(4b)$	0.291	0.103	−0.134
$CH_2(4b')$	0.252	0.080	−0.236
$CH_3(4b)$	0.263	0.025	−0.328
$CH_2(1a)$	−0.009	0.066	−0.122
$CH_2(1a')$	0.002	0.010	−0.035
$CH_3(1a)$	0.013	−0.045	−0.126
$CH_2(1b)$	−0.142	0.125	0.005
$CH_2(1b')$	−0.195	0.181	−0.054
$CH_3(1b)$	−0.285	0.170	−0.050
$CH_2(2a)$	−0.178	0.310	0.313
$CH_2(2a')$	−0.207	0.349	0.194
$CH_3(2a)$	−0.297	0.338	0.198
$CH_2(2b)$	−0.072	0.399	0.429
$CH_2(2b')$	−0.033	0.422	0.530
$CH_3(2b)$	−0.044	0.476	0.622
$CH_2(3a)$	0.227	0.436	0.417
$CH_2(3a')$	0.216	0.491	0.330
$CH_3(3a)$	0.205	0.546	0.421
$CH_2(3b)$	0.361	0.376	0.290
$CH_2(3b')$	0.414	0.320	0.349
$CH_3(3b)$	0.503	0.331	0.345

The positions ascribed to the light carbon and nitrogen atoms yield etrameric molecules in which the four gold atoms at the corners of a quare are joined by $N{\equiv}C$ radicals. Two propyl radicals associated with ach gold atom are located along prolongations of the sides of the square. The possible parameters thus given the numerous light atoms have also een listed in Table XIVC,3.

XIV,c7. A structure has been described for the monoclinic crystals of *ilver thiocyanate tripropyl phosphine*, $AgSCN \cdot P(C_3H_7)_3$. Their tetramolecu-*ir unit has the dimensions:

$$a_0 = 6.32 \text{ A.}; \quad b_0 = 14.00 \text{ A.}; \quad c_0 = 18.24 \text{ A.}; \quad \beta = 114^0$$

The space group has been chosen as C_{2h}^5 ($P2_1/c$) with all atoms in the gen-*ral positions:

$$(4e) \quad \pm (xyz; \; x, {}^1/_2 - y, z + {}^1/_2)$$

The reported parameters are those of Table XIVC,4.

TABLE XIVC,4
Parameters of the Atoms in $AgSCN \cdot P(C_3H_7)_3$

Atom	x	y	z
Ag	0.0260	0.5700	0.4195
S	0.2730	−0.0805	0.0890
P	0.2110	0.2140	0.1380
N	0.309	−0.510	0.399
C(SCN)	0.470	−0.015	0.080
C(1)	0.234	0.281	0.062
C(2)	0.342	0.241	0.003
C(3)	0.419	0.208	0.443
C(4)	0.490	0.686	0.292
C(5)	0.443	0.627	0.232
C(6)	0.333	0.606	0.161
C(7)	0.249	0.312	0.183
C(8)	0.081	0.367	0.202
C(9)	0.122	−0.067	0.315

In the resulting structure each silver atom is reported to have four close eighbors, with Ag–S = 2.83 or 2.88 A., Ag–P = 2.48 A., and Ag–N = .10 A. In the thiocyanate radicals, S–C = 1.81 A. and C–N = 1.42 A. The three carbon atoms around the phosphorus atoms are at distances of

1.71, 1.73, and 1.84 A.; within the propyl radicals, C–C = 1.45–1.59 A. The stated parameters, however, make P–C(4) in the neighborhood of 2.40 A.; it would appear that the phosphorus parameters are not as given.

XIV,c8. The *α-platinum (tripropyl phosphine) chloride thiocyanate*, α-Pt$[P(C_3H_7)_3]$Cl(SCN), is monoclinic with a tetramolecular cell of the dimensions:

$$a_0 = 7.54 \pm 0.02 \text{ A.}; \quad b_0 = 13.62 \pm 0.03 \text{ A.}; \quad c_0 = 15.09 \pm 0.03 \text{ A.}$$
$$\beta = 95°0'$$

The space group has been chosen as C_{2h}^5 $(P2_1/n)$ with all atoms in the positions:

$$(4e) \quad \pm (xyz; \, x+^1/_2, ^1/_2-y, z+^1/_2)$$

The parameters are those of Table XIVC,5.

TABLE XIVC,5
Parameters of the Atoms in α-Pt$[P(C_3H_7)_3]$Cl(SCN)[a]

Atom	x	σ_x, A.	y	σ_y, A.	z	σ_z, A.
Pt	0.1553	0.0047	0.0540	0.0066	0.1688	0.0038
Cl	0.048	0.024	−0.010	0.041	0.300	0.023
S	0.250	0.027	0.134	0.042	0.036	0.024
P	0.352	0.027	0.138	0.043	0.249	0.027
N	−0.023	0.148	(−0.047)	0.201	0.113	0.123
C(1)	−0.125	0.148	(−0.075)	0.190	0.043	0.113
C(2)	0.717	—	0.196	—	0.073	—
C(3)	0.637	—	0.128	—	0.138	—
C(4)	0.510	—	0.209	—	0.170	—
C(5)	−0.060	—	0.299	—	0.300	—
C(6)	0.073	—	0.247	—	0.258	—
C(7)	0.210	—	0.198	—	0.315	—
C(8)	0.720	—	0.053	—	0.453	—
C(9)	0.587	—	0.125	—	0.400	—
C(10)	0.497	—	0.057	—	0.327	—

[a] In the original paper the two y parameters in parentheses were given as $y(N) = 0.45$ and $y(C) = 0.425$. These do not result in recognizable thiocyanate groups and it is presumed they should be increased by $^1/_2$ to yield the values stated here.

The structure apparently is that shown in Figure XIVC,6. It is built up of trimeric molecules in which $[Pt(SCN)]_2$ form a kind of planar, eight membered ring with the attached phosphorus and chlorine atoms also lying in the plane. The coordination of the platinum atoms is therefor

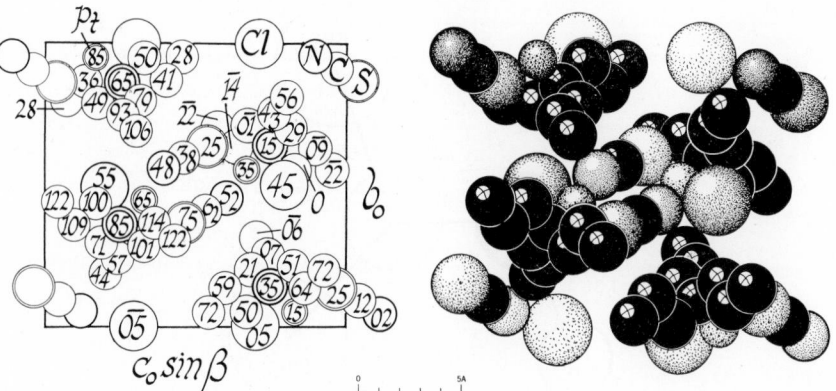

Fig. XIVC,6a (left). The monoclinic structure given α-Pt[P(C₃H₇)₃]Cl(SCN) projected along its a_0 axis. Right-hand axes.

Fig. XIVC,6b (right). A packing drawing of the monoclinic structure assigned α-Pt[P(C₃H₇)₃]Cl(SCN) as seen along its a_0 axis. Both the carbon and the phosphorus atoms are black, the carbons having circular highlights. Atoms of nitrogen are dotted and heavily outlined; the sulfur atoms are larger and short-line shaded. The smallest dot-and-cross shaded circles are platinum.

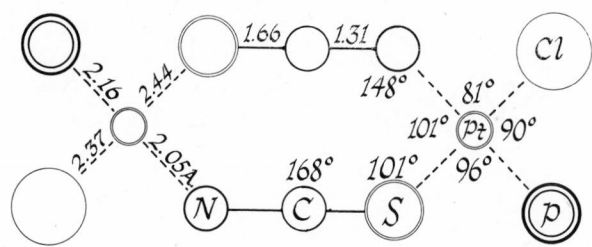

Fig. XIVC,7. Bond dimensions in crystal of α-Pt[P(C₃H₇)₃]Cl(SCN).

fourfold and approximately square, with the dimensions of Figure XIVC,7. The positions of the propyl carbon atoms are only approximate.

XIV,c9. Crystals of *acetoxime*, $(CH_3)_2C{=}NOH$, are hexagonal with a cell containing six of these molecules and having the edges:

$$a_0 = 10.61 \text{ A.,} \qquad c_0 = 7.02 \text{ A.}$$

The space group has been chosen as C_{6h}^2 ($P6_3/m$) with all atoms in the special positions:

$$(6h) \quad \pm(u\,v\,{}^1/_4;\ \bar{v},u{-}v,{}^1/_4;\ v{-}u,\bar{u},{}^1/_4)$$

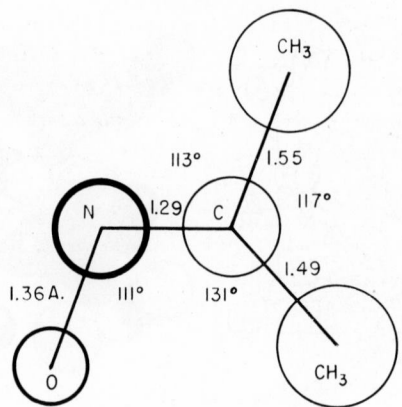

Fig. XIVC,8. Bond dimensions in the planar molecule of $(CH_3)_2C{=}NOH$.

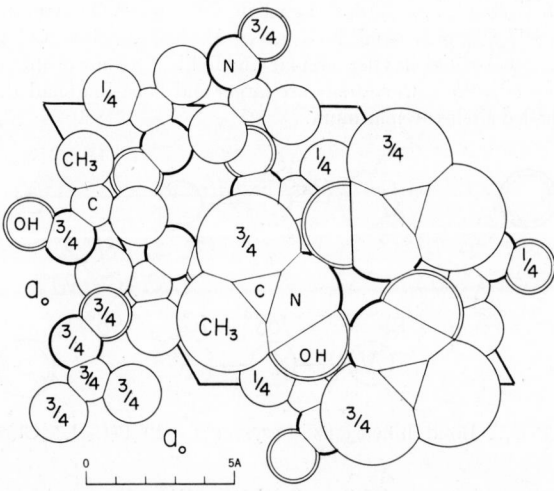

Fig. XIVC,9. The hexagonal structure of acetoxime projected along its c_0 axis. Molecules at the right of the figure are drawn with larger atoms to suggest the type of packing that prevails.

The chosen parameters are given below.

Atom	x	y	z
C(1)	0.362	0.013	$^1/_4$
C(2)	0.203	0.963	$^1/_4$
C(3)	0.403	0.892	$^1/_4$
O	0.432	0.245	$^1/_4$
N	0.474	0.143	$^1/_4$

The molecules of this arrangement are necessarily planar; their bond lengths and angles are those of Figure XIVC,8. Between molecules the shortest atomic separation is O–N = 2.78 A. As Figure XIVC,9 shows, presumptive hydrogen bonds link the molecules into trimers lying completely in planes normal to the c_0 axis; the separation between these molecular layers thus is $1/2$ c_0.

XIV,c10. *Sodium pyruvate*, $CH_3C(O)COONa$, is monoclinic with a tetramolecular cell of the dimensions:

$$a_0 = 22.25 \text{ A.}; \quad b_0 = 5.31 \text{ A.}; \quad c_0 = 3.71 \text{ A.}; \quad \beta = 98°12'$$

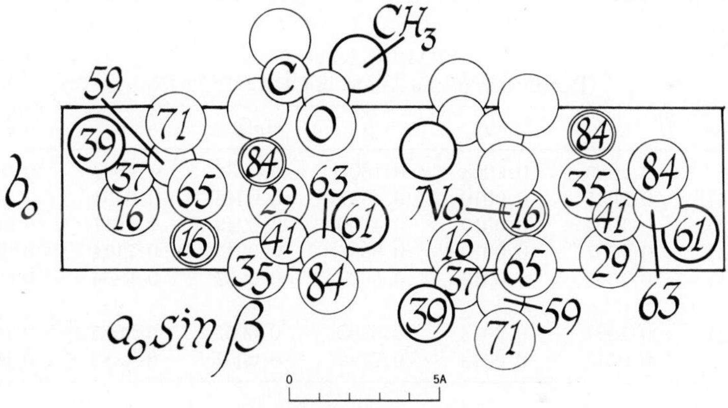

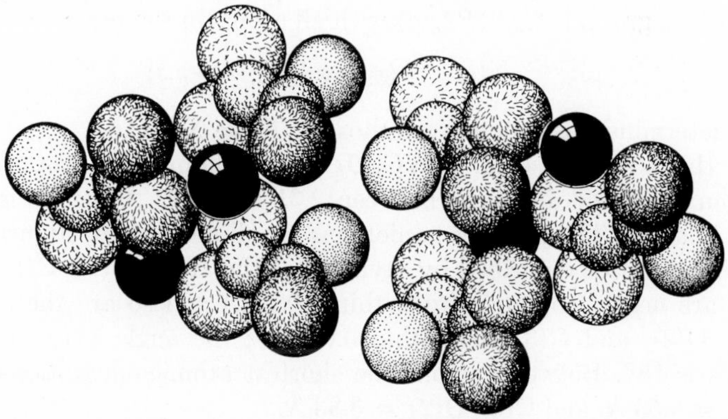

Fig. XIVC,10a (top). The monoclinic structure of sodium pyruvate projected along its c_0 axis. Right-hand axes.

Fig. XIVC,10b (bottom). A packing drawing of the monoclinic $CH_3C(O)COONa$ arrangement viewed along its c_0 axis. The sodium atoms are black. Methyl groups are dotted and heavily ringed; the other carbon atoms are small and hook shaded. Atoms of oxygen are line shaded.

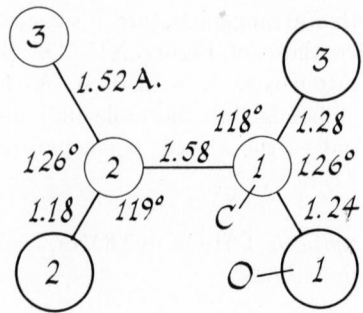

Fig. XIVC,11. Bond dimensions in the pyruvate anion in its sodium salt.

TABLE XIVC,6
Parameters of the Atoms in $CH_3C(O)COONa$

Atom	x	$\sigma(x)$	y	$\sigma(y)$	z	$\sigma(z)$
Na	0.1989	0.0008	0.1802	0.0032	0.1638	0.0054
O(1)	0.2035	0.0010	0.4887	0.0043	0.6491	0.0073
O(2)	0.1004	0.0010	0.4100	0.0043	0.1617	0.0073
O(3)	0.1707	0.0010	0.8843	0.0043	0.7144	0.0073
C(1)	0.1657	0.0013	0.6608	0.0057	0.5854	0.0097
C(2)	0.1024	0.0013	0.5803	0.0057	0.3685	0.0097
C(3)	0.0523	0.0013	0.7722	0.0057	0.3882	0.0097

The space group is C_{2h}^5 $(P2_1/a)$ with all atoms in the positions:

$$(4e) \quad \pm(xyz; x+^1/_2,{}^1/_2-y,z)$$

The determined parameters are those of Table XIVC,6.

In this structure (Fig. XIVC,10) each sodium atom has around it six oxygen atoms at distances between 2.31 and 2.72 A.; there is a seventh 3.07 A. away. The pyruvate anion has the dimensions of Figure XIVC,11. Its shape can be described by saying that the atoms C(2), C(1), O(1), and O(3) are nearly coplanar (to within 0.03 A.) and so are the atoms C(1), C(2), O(2), and C(3) (to within 0.05 A.); the angle between these two planes is 18°. Between anions the shortest atomic separations are C(3)–C(3) = 3.53 A. and C(3)–O(2) = 3.83 A.

XIV,c11. The symmetry of crystals of *acrylic acid*, CH_2=CH—COOH, is orthorhombic. Its unit, containing eight molecules, has the edges:

$$a_0 = 9.966 \text{ A.}; \quad b_0 = 11.744 \text{ A.}; \quad c_0 = 6.306 \text{ A.}$$

The space group has been determined as $V_h{}^{26}$ (*Ibam*) with all atoms in the special positions:

$$(8j) \quad \pm (uv0;\ u\,\bar{v}\,{}^1\!/_2)\,;\ \mathrm{B.C.}$$

The parameters are those of Table XIVC,7.

In this arrangement (Fig. XIVC,12) the planar molecules are tied together in pairs by hydrogen bonds between oxygen atoms, with O–H–O =

TABLE XIVC,7
Parameters of the Atoms in Acrylic Acid

Atom	u	v	σ
O(1)	0.1655	0.0635	0.0007
O(2)	−0.0402	0.1362	0.0007
C(1)	0.0854	0.1487	0.0009
C(2)	0.1492	0.2615	0.0010
C(3)	0.0765	0.3538	0.0010
H(1)	0.113	−0.004	—
H(2)	0.256	0.258	—
H(3)	0.130	0.442	—
H(4)	−0.014	0.347	—

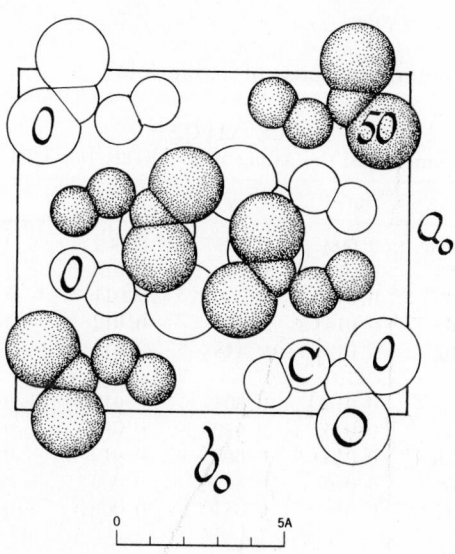

0   5A

Fig. XIVC,12. The orthorhombic structure of acrylic acid projected along its c_0 axis. Molecules lie in two layers along c_0, those at $c_0 = 0.50$ being shaded. Right-hand axes.

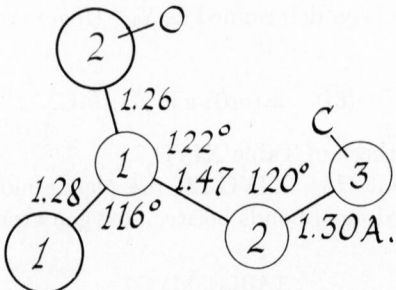

Fig. XIVC,13. Bond dimensions in the planar molecule of CH₂=CHCOOH.

2.66 A. The bond dimensions of the molecule are shown in Figure XIVC,13. The shortest intermolecular distances other than the oxygen contacts through hydrogen bonds are C–O = 3.19 A. The final R = 0.102.

XIV,c12. Crystals of *tetracarbonyl (acrylonitrile) iron*, Fe(CO)₄(CH₂= CHCN), are monoclinic with a tetramolecular cell of the dimensions:

$$a_0 = 12.09 \pm 0.03 \text{ A.}; \quad b_0 = 11.45 \pm 0.02 \text{ A.}; \quad c_0 = 6.585 \pm 0.01 \text{ A.}$$
$$\beta = 110°24' (120°K.)$$

The space group is C_{2h}^5 ($P2_1/a$) with all atoms in the general positions:

$$(4e) \quad \pm (xyz; x+^1/_2, ^1/_2-y,z)$$

Determined parameters are those of Table XIVC,8.

TABLE XIVC,8
Parameters of the Atoms in Fe(CO)₄(CH₂=CHCN)

Atom	x	$\sigma(x)$	y	$\sigma(y)$	z	$\sigma(z)$
Fe	−0.2396	0.002	0.4247	0.002	0.2053	0.003
C(1)	−0.373	0.014	0.373	0.013	0.025	0.014
C(2)	−0.160	0.013	0.399	0.011	−0.007	0.014
C(3)	−0.160	0.014	0.304	0.012	0.355	0.015
C(4)	−0.322	0.013	0.448	0.011	0.418	0.016
C(5)	−0.251	0.015	0.604	0.012	0.136	0.016
C(6)	−0.141	0.015	0.579	0.013	0.292	0.016
C(7)	−0.128	0.015	0.608	0.012	0.513	0.016
O(1)	−0.459	0.010	0.342	0.009	−0.090	0.010
O(2)	−0.117	0.010	0.381	0.009	−0.136	0.010
O(3)	−0.110	0.009	0.227	0.008	0.450	0.010
O(4)	−0.369	0.012	0.456	0.010	0.541	0.012
N	−0.114	0.012	0.634	0.011	0.697	0.013

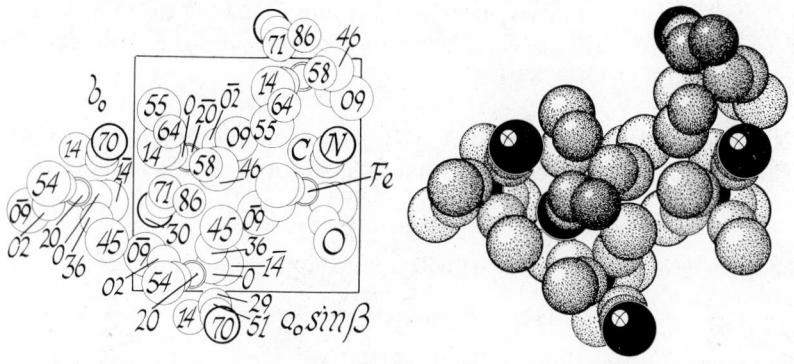

Fig. XIVC,14a (left). The monoclinic structure of Fe(CO)$_4$(CH$_2$=CHCN) projected along its c_0 axis. Right-hand axes.

Fig. XIVC,14b (right). A packing drawing of the monoclinic structure of Fe(CO)$_4$-(CH$_2$=CHCN) seen along its c_0 axis. The nitrogen atoms are black; those of oxygen are a little larger and short-line shaded. Atoms of carbon are slightly smaller and hook shaded. Only parts of the small black iron atoms show.

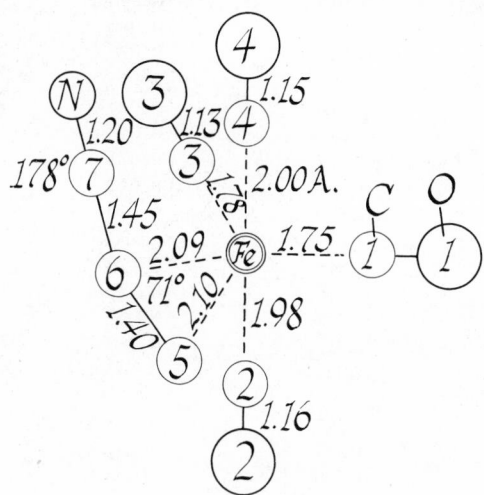

Fig. XIVC,15. Bond dimensions in crystals of Fe(CO)$_4$(CH$_2$=CHCN).

The resulting atomic arrangement is shown in Figure XIVC,14. The significant bond lengths in the monomeric molecules that exist are those of Figure XIVC,15. As they indicate, the iron atom is equally close to the two double-bond carbons of the acrylonitrile group. The lines between iron and the carbonyl atoms are essentially straight; angles between the Fe–

C(2) and Fe–C(4) bonds are nearly right, those between Fe–C(3) and both Fe–C(1) and Fe–C(6) are ca. 109°. Between molecules the shortest atomic separations are O–O = 3.04 A.

XIV,c13.   Crystals of *copper-N,N-di propyl dithiocarbamate*, $Cu[(C_3H_7)_2-NCS_2]_2$, are monoclinic with a tetramolecular unit of the dimensions:

$$a_0 = 13.25 \text{ A.}; \quad b_0 = 18.60 \text{ A.}; \quad c_0 = 8.27 \text{ A.}; \quad \beta = 99°48'$$

The space group is C_{2h}^5 ($P2_1/a$) with atoms in the positions:

$$(4e) \quad \pm(xyz; \ x+\tfrac{1}{2},\tfrac{1}{2}-y,z)$$

The parameters are those of Table XIVC,9.

TABLE XIVC,9
Parameters of the Atoms in $Cu[(C_3H_7)_2NCS_2]_2$

Atom	x	y	z
Cu	0.055	0.072	0.925
S(1)	0.897	0.094	0.763
S(3)	0.013	0.963	0.794
N(1)	0.850	0.980	0.570
C(1)	0.910	0.010	0.697
C(3)	0.775	0.025	0.464
C(5)	0.677	0.018	0.540
C(7)	0.592	0.071	0.463
C(9)	0.867	0.912	0.488
C(11)	0.809	0.850	0.552
C(13)	0.816	0.776	0.467
S(2)	0.116	0.189	0.976
S(4)	0.231	0.057	0.997
N(2)	0.319	0.186	0.030
C(2)	0.234	0.147	0.995
C(4)	0.317	0.264	0.007
C(6)	0.311	0.304	0.169
C(8)	0.400	0.279	0.307
C(10)	0.418	0.156	0.015
C(12)	0.443	0.136	0.846
C(14)	0.548	0.100	0.888

The resulting structure is shown in Figure XIVC,16. Each Cu^{2+} atom is surrounded by four sulfur atoms from two thiocarbamate groups situated at the corners of a trapezoid, with Cu–S = 2.32 A. and S–Cu–S = ca. 76

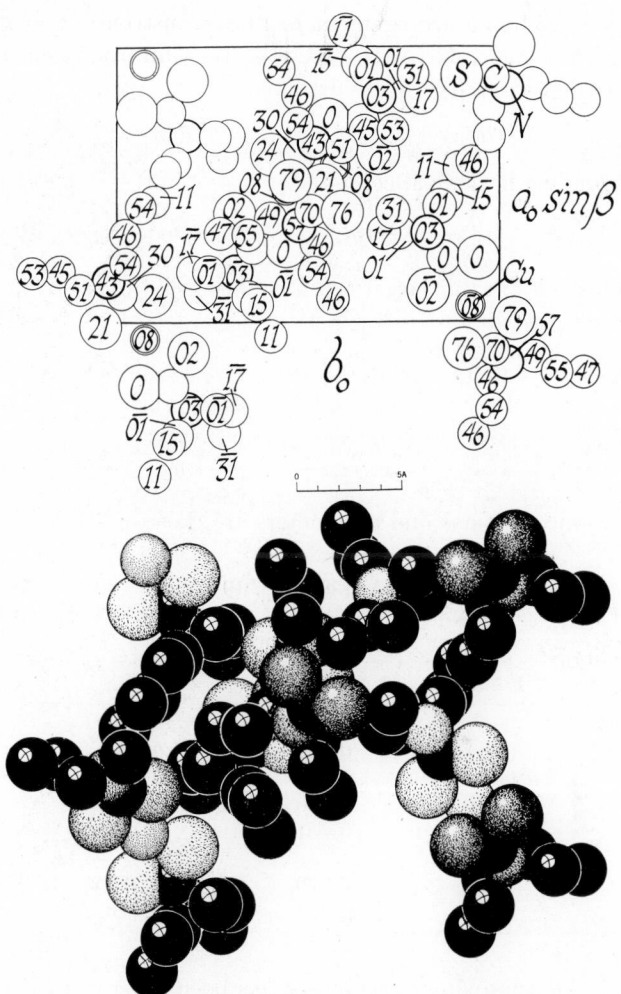

Fig. XIVC,16a (top). The monoclinic structure of $Cu[(C_3H_7)_2NCS_2]_2$ projected along its c_0 axis. Right-hand axes.

Fig. XIVC,16b (bottom). A packing drawing of the monoclinic structure of $Cu[(C_3H_7)_2NCS_2]_2$ viewed along its c_0 axis. The carbon atoms are black; the nitrogen atoms of similar size are hook shaded. Oxygen atoms are larger and dot-and-line shaded. Where they show, the copper atoms are dotted.

and 100°. There is also a more distant sulfur atom with its bond roughly normal to the plane of the trapezoid, and Cu–S = 2.71 A. In the thiocarbamates, C–S = 1.673–1.731 A., N–C = 1.327–1.470 A., and C–C = 1.529–1.566 A.

XIV,c14. An atomic arrangement has been reported for *nickel-N,N-di propyl dithiocarbamate*, $Ni[(C_3H_7)_2NCS_2]_2$. Its unit rhombohedron, containing three molecules, has the dimensions:

$$a_0 = 14.81 \text{ A.}, \qquad \alpha = 116°35'$$

The corresponding hexagonal cell is

$$a_0' = 25.20 \text{ A.}, \qquad c_0' = 8.30 \text{ A.}$$

The nine molecules in this larger cell have been said to be in the following positions of D_{3d}^5 ($R\overline{3}m$):

$$(9e) \quad \tfrac{1}{2}\,0\,0;\ 0\,\tfrac{1}{2}\,0;\ \tfrac{1}{2}\,\tfrac{1}{2}\,0;\ \text{rh}$$
$$(18h) \quad \pm(u\bar{u}v;\ u\,2u\,v;\ 2\bar{u}\,\bar{u}\,v);\ \text{rh}$$
$$(36i) \quad \pm(xyz;\ \bar{y},x-y,z;\ y-x,\bar{x},z;$$
$$\bar{y}\bar{x}z;\ x,x-y,z;\ y-x,y,z);\ \text{rh}$$

Assigned atomic positions and parameters are those of Table XIVC,10.

TABLE XIVC,10
Positions and Parameters of the Atoms in Nickel-*N,N*-di Propyl
Dithiocarbamate

Atom	Position	x	y	z
Ni	(9e)	$\tfrac{1}{2}$	0	0
C	(18h)	0.552	−0.552	0.916
N	(18h)	0.586	−0.586	0.862
S	(36i)	0.470	0.077	0.940
C(1)	(36i)	0.551	0.170	0.000
C(2)	(36i)	0.520	0.212	0.979
C(3)	(36i)	0.484	0.211	0.120

XIV,c15. An approximate structure has been described for the *diallyl-silanediol*, $(CH_2=CHCH_2SiOH)_2$. It is monoclinic with a tetramolecular unit of the dimensions:

$$a_0 = 14.62 \text{ A.}; \quad b_0 = 4.95 \text{ A.}; \quad c_0 = 13.40 \text{ A.}; \quad \beta = 114°30'$$

The space group is C_{2h}^5 ($P2_1/a$) with atoms in the positions:

$$(4e) \quad \pm(xyz;\ x+\tfrac{1}{2},\tfrac{1}{2}-y,z)$$

The assigned parameters are listed in Table XIVC,11.
The molecules that result have the bond dimensions of Figure XIVC,17.

TABLE XIVC,11
Parameters of the Atoms in $(C_3H_5SiOH)_2$

Atom	x	y	z
O(1)	0.084	0.257	0.463
O(2)	0.138	−0.257	0.463
Si	0.082	0.000	0.386
C(1)	−0.054	0.091	0.296
C(2)	0.146	−0.091	0.296
C(3)	−0.115	0.071	0.259
C(4)	0.182	−0.071	0.259
C(5)	−0.152	0.244	0.153
C(6)	0.138	−0.244	0.153

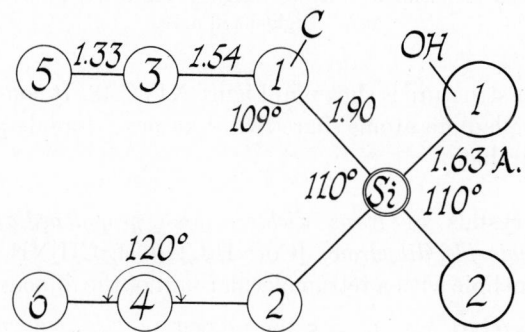

Fig. XIVC,17. Bond dimensions given the molecules in $(CH_2{=}CHCH_2SiOH)_2$.

XIV,c16. According to a preliminary note, crystals of $PdClC_3H_5$ are monoclinic with a tetramolecular unit of the dimensions:

$$a_0 = 7.50 \text{ A.}; \quad b_0 = 7.42 \text{ A.}; \quad c_0 = 8.60 \text{ A.}; \quad \beta = 93°34'$$

The space group is C_{2h}^5 ($P2_1/n$) with all atoms in the positions:

$$(4e) \quad \pm (xyz; x{+}^1/_2, {}^1/_2{-}y, z{+}^1/_2)$$

The selected parameters are as follows:

Atom	x	y	z
Pd	0.126	0.183	−0.050
Cl	−0.172	0.068	−0.113
C(1)	0.375	0.310	−0.044
C(2)	0.244	0.450	−0.050
C(3)	0.140	0.430	−0.165

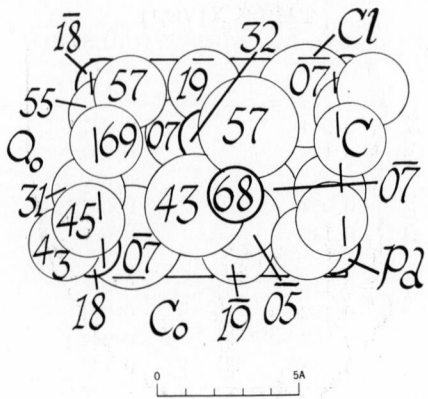

Fig. XIVC,18. The monoclinic structure assigned PdClC₃H₅ projected along its b_0 axis. Right-hand axes.

The resulting structure is shown in Figure XIVC,18. It consists of dimers in which two palladium atoms share two chlorines. More detailed information seems desirable.

XIV,c17. Crystals of *trans dichloro bis-*L-*propylenediamine cobaltic chloride hydrochloride dihydrate*, [Co(NH₂CH(CH₃)CH₂NH₂)₂Cl₂]Cl·HCl· 2H₂O, are monoclinic with a tetramolecular unit of the dimensions:

$$a_0 = 22.092 \pm 0.004 \text{ A.}; \quad b_0 = 8.406 \pm 0.002 \text{ A.}; \quad c_0 = 9.373 \pm 0.004 \text{ A.}$$
$$\beta = 99°39' \pm 7'$$

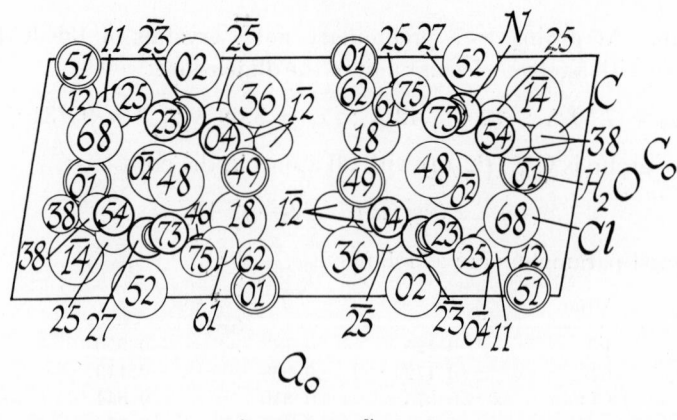

Fig. XIVC,19a. The monoclinic structure of [Co(NH₂CH(CH₃)CH₂NH₂)₂Cl₂]Cl· HCl·2H₂O projected along its b_0 axis. Right-hand axes.

The space group of these optically active crystals is $C_2{}^3$ ($C2$) with all atoms in the general positions:

$$(4c) \quad xyz; \; \bar{x}y\bar{z}; \; x+{}^1/_2,y+{}^1/_2,z; \; {}^1/_2-x,y+{}^1/_2,\bar{z}$$

TABLE XIVC,12
Parameters of the Atoms in $Co[C_3H_6(NH_2)_2]_2Cl_3\cdot HCl\cdot 2H_2O$

Atom	x	y	z
Co	0.250	0.000	0.250
Cl(1)	0.392	0.365	0.171
Cl(2)	0.109	0.681	0.327
Cl(3)	0.267	0.022	0.017
Cl(4)	0.232	−0.022	0.483
N(1)	0.268	−0.232	0.251
N(2)	0.338	0.036	0.320
N(1′)	0.231	0.235	0.255
N(2′)	0.164	−0.039	0.175
C(1)	0.334	−0.250	0.261
C(2)	0.365	−0.119	0.355
C(3)	0.434	−0.117	0.345
C(1′)	0.168	0.246	0.167
C(2′)	0.136	0.109	0.211
C(3′)	0.067	0.117	0.154
O(1)	0.441	0.488	0.473
O(2)	0.059	0.509	0.029

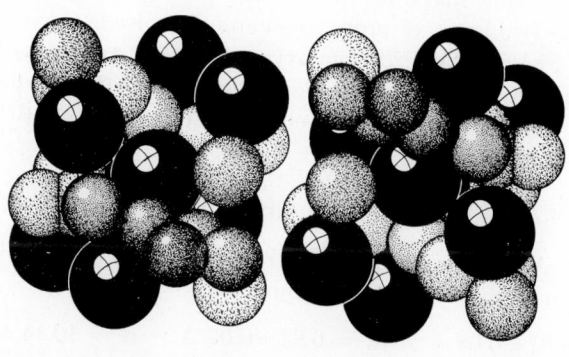

Fig. XIVC,19b. A packing drawing of the monoclinic structure of $[Co(NH_2CH(CH_3)CH_2NH_2)_2Cl_2]Cl\cdot HCl\cdot 2H_2O$ seen along its b_0 axis. The large black circles are chlorine. Atoms of nitrogen are heavily outlined and dotted; those of carbon, of equal size, are hook shaded. Molecules of water are somewhat larger and short-line shaded. Cobalt atoms do not show.

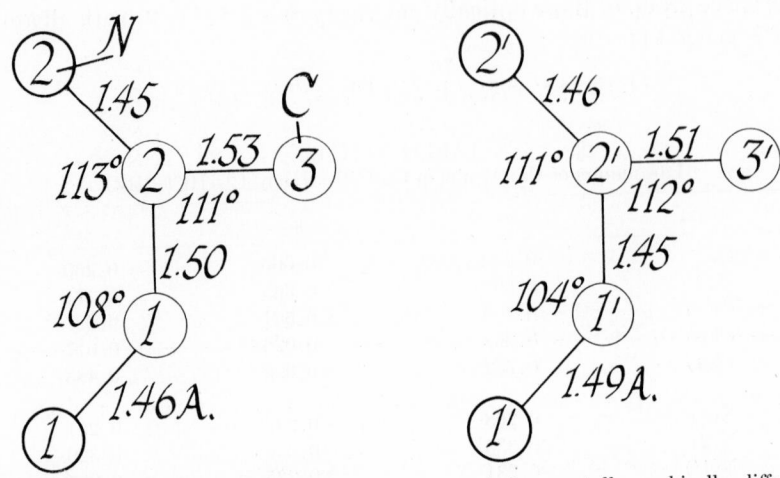

Fig. XIVC,20a (left). Bond dimensions in one of the crystallographically different propylene diamine molecules found in crystals of [Co(NH₂CH(CH₃)CH₂NH₂)₂Cl₂]-Cl·HCl·2H₂O.

Fig. XIVC,20b (right). Bond dimensions in the other propylene diamine molecule in the cobaltic hydrochloride.

The parameters are listed in Table XIVC,12.

The structure is shown in Figure XIVC,19. The cobalt atoms are octahedrally surrounded by the four nitrogen atoms of two propylenediamine groups, with Co–N = 1.94–2.02 A., and by two chlorine atoms normal to the Co–4N plane (Co–Cl = 2.29 A.). This complex cation thus has substantially the same shape as that of the corresponding ethylenediamine complex. The two kinds of propylenediamine molecule have the bond dimensions of Figure XIVC,20. The oxygen atoms of two molecules of water are 2.57 A. apart and this is considered to point to the presence of (H₂O·H·OH₂)⁺ cations, each of which is closest to Cl⁻ anions, with Cl⁻–O ranging upwards from 3.04 A.

2. Four-Carbon Chain Compounds

XIV,c18. *N-Butyric acid*, $CH_3(CH_2)_2COOH$, at −43°C. is monoclinic with a tetramolecular unit of the dimensions:

$$a_0 = 8.01 \pm 0.08 \text{ A.}; \quad b_0 = 6.82 \pm 0.02 \text{ A.}; \quad c_0 = 10.14 \pm 0.03 \text{ A.}$$
$$\beta = 111°27'$$

The space group has been chosen as C_{2h}^3 ($C2/m$) with all the atoms in the special positions:

$$(4i) \quad \pm (u0v; \ u + ^1/_2, ^1/_2, v)$$

The determined parameters are the following:

Atom	u	$\sigma(u)$	v	$\sigma(v)$
O(1)	0.2359	0.0006	0.0379	0.0004
O(2)	0.9987	0.0006	0.8382	0.0004
C(3)	0.1582	0.0010	0.8986	0.0006
C(4)	0.2932	0.0010	0.8248	0.0007
C(5)	0.2072	0.0013	0.6647	0.0008
C(6)	0.3471	0.0012	0.5968	0.0008

The molecular arrangement, as shown in Figure XIVC,21, is, like that of fatty acids, a parallel stacking of the molecules with the carboxyl groups facing one another across a center of symmetry and bound together by hydrogen bonds. The dimensions of these molecules are shown in Figure XIVC,22.

At $-55°C$. there is a transition to a lower-temperature form whose structure has not yet been studied.

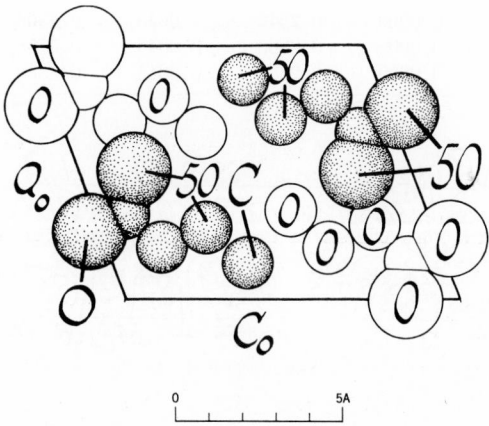

Fig. XIVC,21. The monoclinic structure of $CH_3(CH_2)_2COOH$ projected along its b_0 axis. Molecules in the $b_0 = 0.50$ layer are shaded. Right-hand axes.

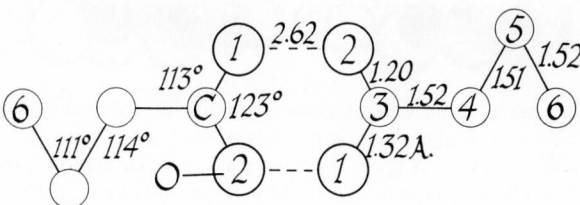

Fig. XIVC,22. Bond dimensions in crystals of butyric acid.

468   CRYSTAL STRUCTURES

XIV,c19. *Sodium α-ketobutyrate*, $C_2H_5C(O)COONa$, is orthorhombic with an eight-molecule cell of the edge lengths:

$$a_0 = 29.28 \text{ A.}; \quad b_0 = 6.045 \text{ A.}; \quad c_0 = 5.90 \text{ A.}$$

The space group is V_h^{14} (*Pbcn*) with all atoms in the general positions:

$$(8d) \quad \pm(xyz; \; x+{}^1/_2,y+{}^1/_2,{}^1/_2-z; \; {}^1/_2-x,y+{}^1/_2,z; \; x,\bar{y},z+{}^1/_2)$$

The established parameters are those of Table XIVC,13.

TABLE XIVC,13
Parameters of the Atoms in $C_2H_5C(O)COONa$

Atom	x	$\sigma(x)$	y	$\sigma(y)$	z	$\sigma(z)$
Na	0.2111	0.0005	0.2450	0.0025	0.0762	0.0025
O(1)	0.2136	0.0007	0.1578	0.0033	0.4599	0.0034
O(2)	0.1330	0.0007	0.1642	0.0033	0.2584	0.0034
O(3)	0.1855	0.0007	0.3769	0.0033	0.7312	0.0034
C(1)	0.1838	0.0009	0.2381	0.0045	0.5566	0.0046
C(2)	0.1336	0.0009	0.2542	0.0045	0.4564	0.0046
C(3)	0.0938	0.0009	0.2865	0.0045	0.5758	0.0046
C(4)	0.0488	0.0009	0.2481	0.0045	0.4568	0.0046

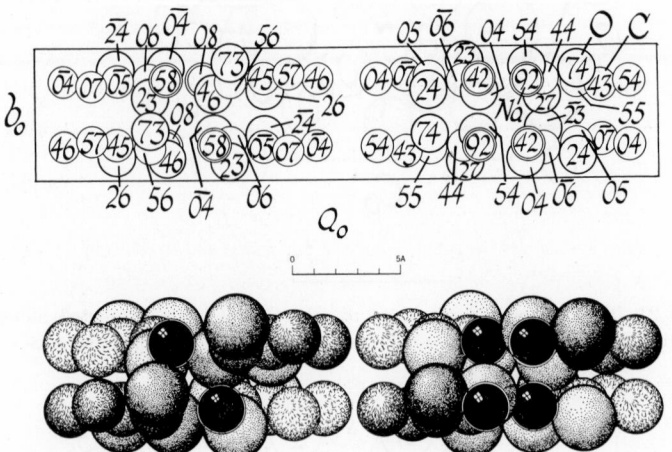

Fig. XIVC,23a (top). The orthorhombic structure of $C_2H_5C(O)COONa$ projected along its c_0 axis. Right-hand axes.
Fig. XIVC,23b (bottom). A packing drawing of the orthorhombic $C_2H_5C(O)COONa$ arrangement seen along its c_0 axis. The sodium atoms are black. Atoms of carbon are fine-line shaded; those of oxygen are larger, dotted, and heavily outlined.

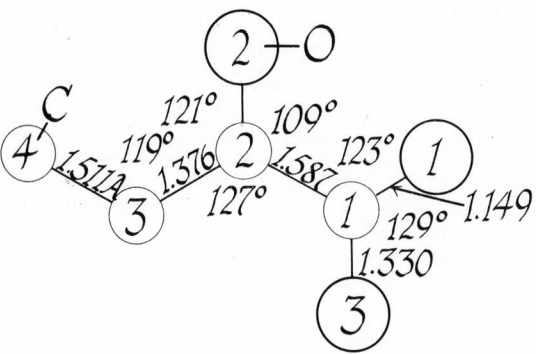

Fig. XIVC,24. Bond dimensions in the ketobutyrate anion in $C_2H_5C(O)COONa$.

The resulting arrangement is shown in Figure XIVC,23. Bond dimensions in the anion are those of Figure XIVC,24, the C(1), C(2), O(1), and O(3) atoms being nearly coplanar. Each sodium atom has six oxygen neighbors at a distance of 2.5 ± 0.2 A.; five of these are carboxyl oxygens, the sixth is the keto oxygen, O(2).

This is a structure closely related to that of sodium pyruvate (**XIV,c10**); it is practically isostructural with sodium-2-oxocaprylate (**XIV,c119**).

XIV,c20. *Tetra-N-butyl ammonium fluoride hydrate*, $(C_4H_9)_4NF \cdot 32.8$-H_2O, is one of the complex clathrate compounds that has a structure based on an $H_{40}O_{20}$ pentagonal dodecahedral framework. The symmetry is tetragonal with a unit that contains five molecules and has the edge lengths:

$$a_0 = 23.52 \pm 0.01 \text{ A.,} \qquad c_0 = 12.30 \pm 0.01 \text{ A. } (-25°C.)$$

The space group was chosen as C_{4h}^2 ($P4_2/m$) and a structure has been assigned which, however, involves a certain measure of disorder. Two oxygens of the water framework occupy half the requisite equivalent positions. There is similar disorder in the arrangement of the carbon atoms. Nitrogen and fluorine atoms together occupy the two special positions $(2f)$. Parameters, as stated in the original paper, apply to an origin displaced by $^1/_2 b_0$ with respect to the usual description.

XIV,c21. *Tri-N-butyl sulfonium fluoride hydrate*, $(C_4H_9)_3SF \cdot 20H_2O$, is another clathrate compound which is built upon $H_{40}O_{20}$ as a pentagonal dodecahedral framework. The symmetry is in this case cubic with a bimolecular cell of the edge length:

$$a_0 = 12.26 \text{ A. } (-80°C.)$$

The space group is $O_h{}^3$ ($Pm3n$) and a highly disordered structure based on it has been described. This places atoms in the following positions with the fractional occupancy noted in the third column:

Atom	Position	Occupancy	x	y	z
O(1)	(16i)	1	0.1849	0.1849	0.1849
O(2)	(24k)	$^5/_6$	0	0.3042	0.1215
O(3)	(24k)	$^1/_6$	0	0.2192	0.0714
S	(12g)	$^1/_6$	0	$^1/_2$	0.2601
F	(12f)	$^1/_6$	0	0	0.033
C	(24k)	$^1/_4$	0	0.3481	0.1603

It should be noted that this is basically the same arrangement proposed for the hydrate of chlorine [Pauling, L., and Marsh, R. E., *Proc. Natl. Acad. Sci. U.S.*, **38**, 112 (1952)].

XIV,c22. Crystals of *pentamethylethyl alcohol* (2,3,3-trimethyl butanol) *hemihydrate*, $C(CH_3)_3C(CH_3)_2OH \cdot {}^1/_2H_2O$, are orthorhombic with a unit containing eight molecules and having the edges:

$$a_0 = 21.42 \text{ A.}; \quad b_0 = 10.55 \text{ A.}; \quad c_0 = 7.62 \text{ A.} \ (-15°\text{C.})$$

The space group has been considered to be $C_{2v}{}^{21}$ (*Iba2*). Water molecules are in the positions:

(4a) $00u; \ 0,0,u+{}^1/_2$; B.C.   with $u = 0$ (arbitrary)

All other atoms are in the general positions:

(8c) $xyz; \ \bar{x}\bar{y}z; \ x,\bar{y},z+{}^1/_2; \ \bar{x},y,z+{}^1/_2$; B.C.

The parameters are said to be those listed in Table XIVC,14.

TABLE XIVC,14
Parameters of the Atoms in $C(CH_3)_3C(CH_3)_2OH \cdot {}^1/_2H_2O$

Atom	x	y	z
O	0.0639	0.1336	0.2493
C(1)	0.1279	0.2844	0.4147
C(2)	0.0980	0.2521	0.2362
C(3)	0.1498	0.2389	0.0984
C(4)	0.1200	0.2076	−0.0814
C(5)	0.1858	0.3659	0.0827
C(6)	0.1947	0.1336	0.1522
C(7)	0.0532	0.3583	0.1811

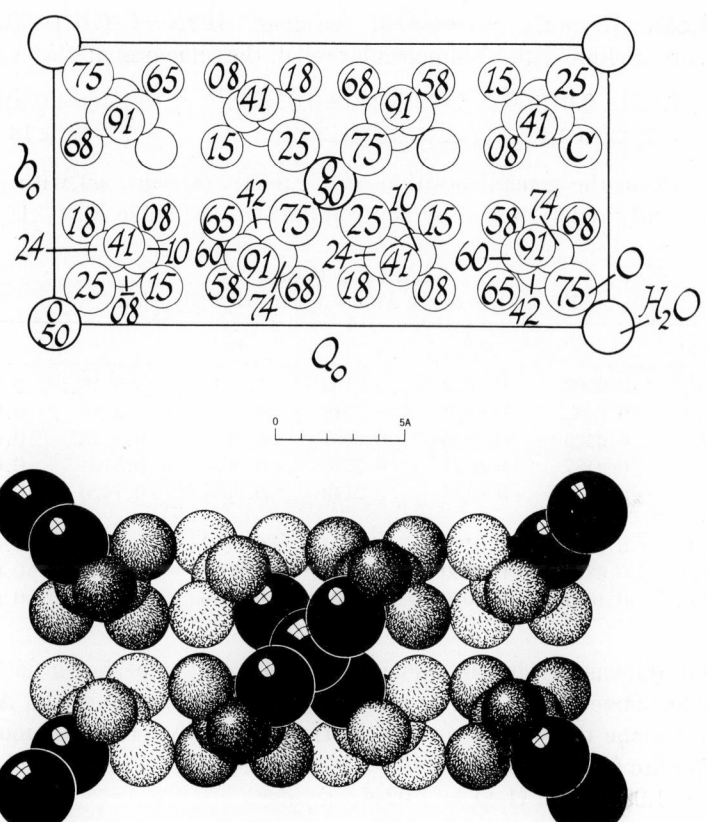

Fig. XIVC,25a (top). The orthorhombic structure of $C(CH_3)_3C(CH_3)_2OH \cdot 1/2H_2O$ projected along its c_0 axis. Right-hand axes.

Fig. XIVC,25b (bottom). A packing drawing of the orthorhombic structure of $C(CH_3)_3C(CH_3)_2OH \cdot 1/2H_2O$ viewed along its c_0 axis. Both the water and the hydroxyl oxygens are black, the hydroxyls having round highlights. Atoms of carbon are line shaded.

The resulting structure is shown in Figure XIVC,25. The molecules which are present are considered to be bound together along the c_0 axis by hydrogen bonds involving both the water molecules and the hydroxyl oxygens.

The compound 1,1,1-trichloro-2-methylpropyl alcohol hemihydrate, $CCl_3C(CH_3)_2OH \cdot 1/2H_2O$, is said to be isostructural with a cell of the dimensions:

$$a_0 = 21.20 \text{ A.}; \quad b_0 = 10.40 \text{ A.}; \quad c_0 = 7.57 \text{ A.} \ (-15°C.)$$

XIV,c23. Crystals of *dimethyl glyoxime*, $HON=C(CH_3) \cdot C(CH_3)=NOH$, are triclinic with a unimolecular cell of the dimensions:

$$a_0 = 6.053 \pm 0.015 \text{ A.}; \quad b_0 = 6.292 \pm 0.015 \text{ A.}; \quad c_0 = 4.468 \pm 0.010 \text{ A.}$$
$$\alpha = 122°22' \pm 12'; \quad \beta = 91°34' \pm 6'; \quad \gamma = 77°38' \pm 18'$$

Atoms are in the general positions of C_i^1 $(P\bar{1})$: $(2i)$ $\pm(xyz)$ with parameters recently redetermined by neutron diffraction (Table XIVC,15).

TABLE XIVC,15
Parameters of the Atoms in Dimethyl Glyoxime

Atom	x	$\sigma(x)$	y	$\sigma(y)$	z	$\sigma(z)$
C(1)	0.0907	0.0013	0.0196	0.0014	0.9016	0.0029
C(2)	0.7062	0.0027	0.1918	0.0022	0.3086	0.0031
N	0.0526	0.0013	0.2437	0.0016	0.9623	0.0033
O	0.2187	0.0027	0.2718	0.0035	0.8104	0.0076
H(1)	0.5881	0.0156	0.2149	0.0097	0.1420	0.0116
H(2)	0.7415	0.0110	0.3764	0.0082	0.3849	0.0087
H(3)	0.6374	0.0100	0.1647	0.0075	0.5098	0.0330
H(4)	0.1631	0.0046	0.4304	0.0046	0.7952	0.0151

Bond dimensions in the molecules that result are shown in Figure XIVC,26; none of the heavier atoms departs by more than 0.03 A. from the best plane through them. The N–H–O bonds that tie these molecules together are far from collinear since the angle N–H–O is 140°; in this bond, N–H = 1.90 A. and H–O = 1.02 A.

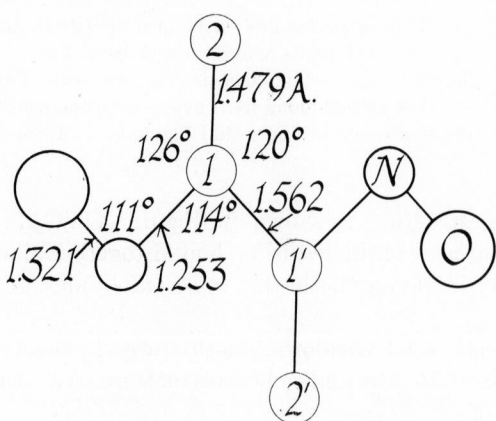

Fig. XIVC,26. Bond dimensions in molecules of dimethyl glyoxime.

XIV,c24. Crystals of *copper dimethyl glyoxime*, $Cu[(CH_3)_2C_2N_2O_2H]_2$, are monoclinic with a tetramolecular cell of the dimensions:

$$a_0 = 9.80 \pm 0.04 \text{ A.}; \quad b_0 = 17.10 \pm 0.06 \text{ A.}; \quad c_0 = 7.12 \pm 0.03 \text{ A.}$$
$$\beta = 107°20' \pm 20' \ (20° \text{ C.})$$

$$a_0 = 9.71 \pm 0.04 \text{ A.}; \quad b_0 = 16.88 \pm 0.06 \text{ A.}; \quad c_0 = 7.08 \pm 0.03 \text{ A.}$$
$$\beta = 108°26' \pm 20' \ (-140° \text{ C.})$$

TABLE XIVC,16

Parameters of the Atoms in Copper Dimethyl Glyoxime

Atom	x	y	z
Cu	0.034	0.097	0.382
C(1)	0.231	0.018	0.248
C(2)	0.636	0.006	0.783
C(3)	0.897	0.040	0.772
C(4)	0.893	0.118	0.856
N(1)	0.216	0.086	0.312
N(2)	0.989	0.011	0.711
O(1)	0.314	0.146	0.345
O(2)	0.095	0.059	0.711
H(1)	0.927	0.159	0.806
H(2)	0.868	0.157	0.955
H(3)	0.837	0.127	0.906
H(4)	0.625	0.034	0.787
H(5)	0.597	0.014	0.761
H(6)	0.399	0.012	0.707
C(5)	0.797	0.191	0.352
C(6)	0.658	0.217	0.365
C(7)	0.923	0.246	0.386
C(8)	0.401	0.163	0.876
N(3)	0.835	0.121	0.339
N(4)	0.042	0.212	0.400
O(3)	0.655	0.246	0.912
O(4)	0.728	0.061	0.316
H(7)	0.667	0.250	0.318
H(8)	0.635	0.233	0.422
H(9)	0.625	0.185	0.422
H(10)	0.465	0.139	0.955
H(11)	0.403	0.133	0.904
H(12)	0.365	0.137	0.806

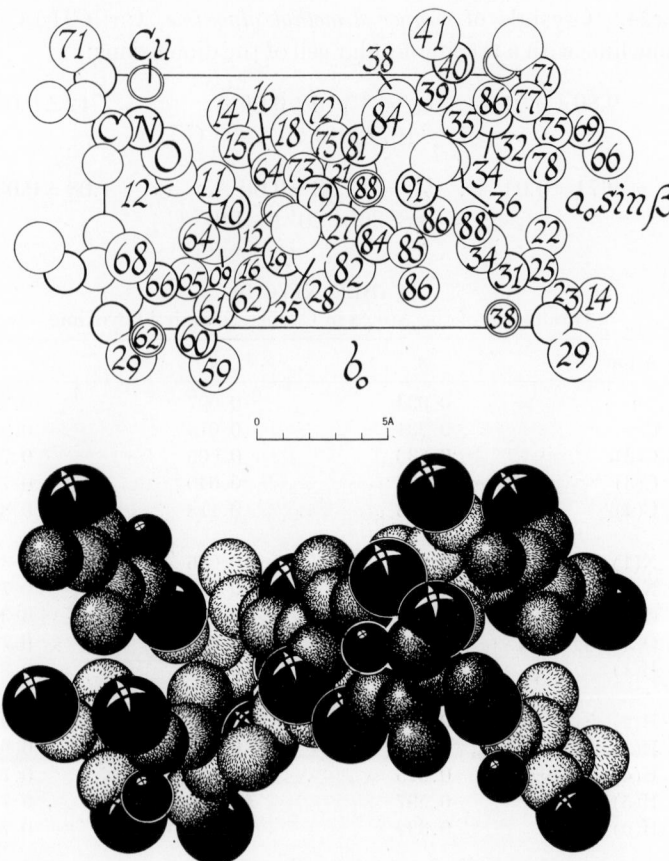

Fig. XIVC,27a (top). The monoclinic structure of copper dimethyl glyoxime projected along its c_0 axis. Right-hand axes.

Fig. XIVC,27b (bottom). A packing drawing of the monoclinic structure of Cu[(CH₃)₂C₂N₂O₂H]₂ seen along its c_0 axis. Atoms of oxygen are the large, of copper the small black circles. The carbon atoms are line shaded; the nitrogens are hook shaded and heavily outlined.

There is no transition between these temperatures and the space group is C_{2h}^5 $(P2_1/n)$. All atoms are in the general positions:

$$(4e) \quad \pm (xyz; \; x+{}^1/_2, {}^1/_2-y, z+{}^1/_2)$$

with the parameters listed in Table XIVC,16.

The atomic arrangement that results is shown in Figure XIVC,27. It molecules have the bond dimensions of Figure XIVC,28. The coordination about copper is unusual. Though there are four nitrogen atoms nearly

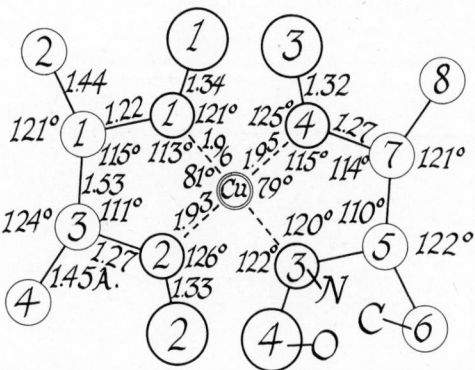

Fig. XIVC,28. Bond dimensions in crystals of copper dimethyl glyoxime.

equidistant from the metal atom, it does not lie in their plane but together with them makes a very flat pyramid. Above the copper there is also an oxygen atom of another molecule, with Cu–O = 2.43 A. The extension of this line in the opposite direction encounters, not another oxygen, but a methyl group 3.76 A. away and hence the total coordination of the copper is five. There are hydrogen bonds joining the oxygen atoms that belong to the two dimethyl glyoxime groups associated with a copper atom; for them, O(1)–O(3) = 2.53 A. and O(2)–O(4) = 2.70 A.

XIV,c25. The orthorhombic crystals of *nickel dimethyl glyoxime*, $Ni[(CH_3)_2C_2N_2O_2H]_2$, contain four molecules in a unit of the edge lengths:

$$a_0 = 16.68 \text{ A.}; \quad b_0 = 10.44 \text{ A.}; \quad c_0 = 6.49 \text{ A.}$$

The space group is V_h^{26} (*Ibam*) with the nickel atoms in

$$(4c) \quad 000; \, 0\,0\,^1/_2; \, \text{B.C.}$$

TABLE XIVC,17
Parameters of the Atoms in Nickel Dimethyl Glyoxime

Atom	u	$\sigma(u)$	v	$\sigma(v)$
O(1)	0.1577	0.0009	0.0991	0.0014
O(2)	0.0543	0.0009	0.2589	0.0015
N(1)	0.1111	0.0009	0.9931	0.0019
N(2)	0.9909	0.0011	0.1765	0.0014
C(1)	0.0808	0.0013	0.7741	0.0022
C(2)	0.1440	0.0013	0.8813	0.0020
C(3)	0.0988	0.0015	0.6334	0.0024
C(4)	0.2348	0.0015	0.8575	0.0023

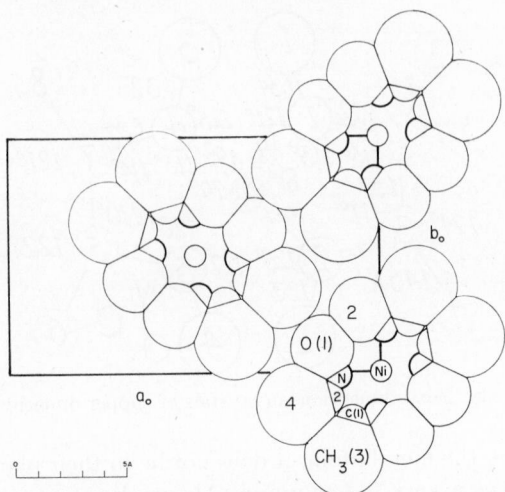

Fig. XIVC,29. Half the molecules in the orthorhombic structure of nickel dimethyl glyoxime projected along its c_0 axis. Those shown here are in the plane $c_0 = 0$. Left-hand axes.

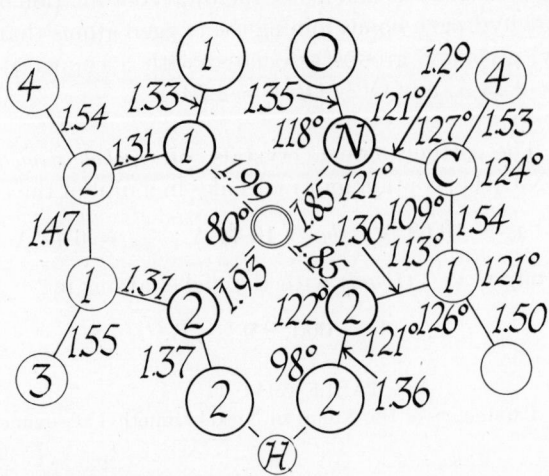

Fig. XIVC,30. Bond dimensions in crystals of nickel dimethyl glyoxime and the isostructural palladous compound. Bond dimensions on the right apply to the nickel, those on the left to the palladium compound.

and all other atoms in

$$(8j) \quad \pm (uv0; \bar{u}\,v\,{}^1/_2); \text{ B.C.}$$

The recently redetermined parameters are listed in Table XIVC,17.

They lead to a structure (Fig. XIVC,29) which is a stacking of planar molecules along the c_0 axis. The dimensions of these molecules are those given in the right-hand half of Figure XIVC,30. Of particular interest is the short O–H–O bond of 2.44 A. The molecules stacked one above another are turned through 90° about the c_0 axis, thus avoiding contact between the large methyl groups. With the new parameters, $R = 0.124$.

The analogous *palladium dimethyl glyoxime*, $Pd[(CH_3)_2C_2N_2O_2H]_2$, has this structure with a cell of the edge lengths:

$$a_0 = 16.85 \text{ A.}; \quad b_0 = 10.49 \text{ A.}; \quad c_0 = 6.52 \text{ A.}$$

The recently established parameters of Table XIVC,18 lead to the molecular bond dimensions of the left half of Figure XIVC,30. For this determination $R = 0.065$.

TABLE XIVC,18. Parameters of the Atoms in Palladium Dimethyl Glyoxime

Atom	u	$\sigma(u)$	v	$\sigma(v)$
O(1)	0.1618	0.0012	0.0960	0.0020
O(2)	0.0500	0.0012	0.2654	0.0021
N(1)	0.1179	0.0013	0.9901	0.0025
N(2)	0.9868	0.0014	0.1829	0.0022
C(1)	0.0864	0.0017	0.7744	0.0028
C(2)	0.1471	0.0018	0.8748	0.0028
C(3)	0.1076	0.0019	0.6305	0.0031
C(4)	0.2373	0.0020	0.8522	0.0031

The *platinum dimethyl glyoxime*, $Pt[(CH_3)_2C_2N_2O_2H]_2$ also has been shown to have this atomic arrangement. Its cell dimensions are

$$a_0 = 16.73 \pm 0.06 \text{ A.}; \quad b_0 = 10.59 \pm 0.05 \text{ A.}; \quad c_0 = 6.47 \pm 0.02 \text{ A.}$$

TABLE XIVC,19. Parameters of the Atoms in Platinum Dimethyl Glyoxime

Atom	u	$\sigma(u)$	v	$\sigma(v)$
O(1)	0.172	0.0002	0.073	0.0003
O(2)	0.042	0.0002	0.268	0.0003
N(1)	0.115	0.0002	0.988	0.0003
N(2)	0.993	0.0002	0.184	0.0003
C(1)	0.084	0.0003	0.776	0.0005
C(2)	0.148	0.0003	0.880	0.0005
C(3)	0.097	0.0003	0.639	0.0005
C(4)	0.234	0.0003	0.856	0.0005

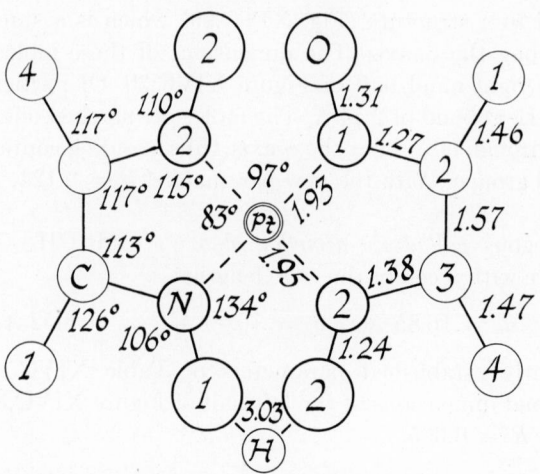

Fig. XIVC,31. Bond dimensions in crystals of platinous dimethyl glyoxime.

Choosing its axes to conform to those employed for the preceding two compounds, the less accurately established parameters are those of Table XIVC,19.

The resulting bond dimensions are shown in Figure XIVC,31.

XIV,c26.   *Cobalti bis dimethyl glyoximino diammine nitrate,*
$Co[(-C(CH_3)NOH)_2]_2NO_3 \cdot 2NH_3$, is monoclinic with a bimolecular cell of the dimensions:

$$a = 9.80 \pm 0.03 \text{ A.}; \quad b_0 = 6.32 \pm 0.02 \text{ A.}; \quad c_0 = 12.47 \pm 0.04 \text{ A.}$$
$$\beta = 97°$$

The space group is C_{2h}^5 ($P2_1/c$) with atoms in the positions:

$$Co: (2a) \quad 000; 0\,^1/_2\,^1/_2$$
$$N(NO_3): (2b) \quad ^1/_2\,0\,0; ^1/_2\,^1/_2\,^1/_2$$

All other atoms are in the general positions:

$$(4e) \quad \pm (xyz; x,^1/_2-y,z+^1/_2)$$

with the parameters of Table XIVC,20.

TABLE XIVC,20. Parameters of Atoms in $Co[(-C(CH_3)NOH)_2]_2NO_3 \cdot 2NH_3$

Atom	x	y	z
N(1)	0.109	0.943	0.140
N(2)	0.142	0.214	0.012
N(3)	0.883	0.198	0.067
C(1)	0.212	0.233	0.098
C(2)	0.216	0.040	0.172
C(3)	0.314	0.003	0.274
C(4)	0.316	0.405	0.103
O(1)	0.101	0.748	0.180
O(2)	0.135	0.366	0.933

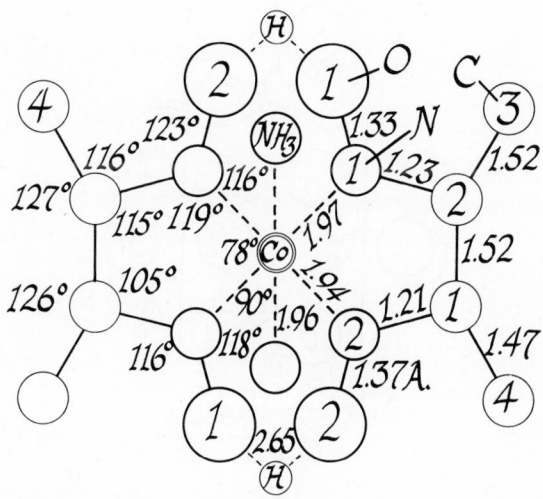

Fig. XIVC,32. Bond dimensions in crystals of $Co[(-C(CH_3)NOH)_2]_2NO_3 \cdot 2NH_3$

Each cobalt atom is octahedrally surrounded by six nitrogen atoms, four from the two organic parts of the complex and two from the ammonia molecules. Their distances and the rest of the bond dimensions are shown in Figure XIVC,32. Positions were not found for the nitrate oxygens which are thought to be freely rotating about an axis through the nitrogen atom slightly turned to the b_0 axis. The short O(1)–O(2) separation (2.65 A.) between the two organic groups associated with a cobalt atom is attributed to the presence of a hydrogen bond. Between different molecules, the shortest distances are O–C(3) = 2.92 A., NH_3–O = 3.14, and 3.18 A.

480 CRYSTAL STRUCTURES

XIV,c27. The *trans* dimer of *nitrosoisobutane*, $[(CH_3)_2CHCH_2NO—]_2$, forms monoclinic crystals. A unit containing two molecules of the dimer has the dimensions:

$$a_0 = 8.84 \text{ A.}; \quad b_0 = 9.93 \text{ A.}; \quad c_0 = 6.14 \text{ A.}; \quad \beta = 96°10'$$

The space group is C_{2h}^5 ($P2_1/c$) with all atoms in the general positions:

$$(4e) \quad \pm(xyz; \ x,{}^1\!/_2-y,z+{}^1\!/_2)$$

The chosen parameters are given in Table XIVC,21, those for hydrogen having been set by chemical assumptions.

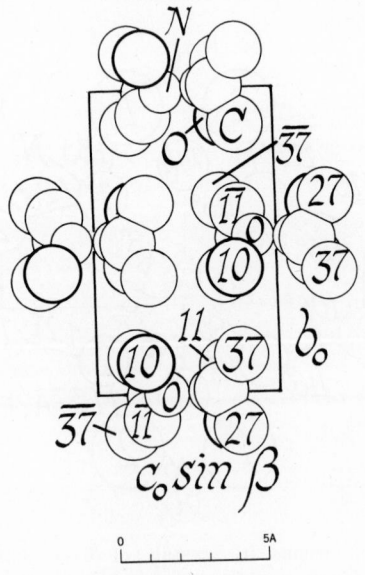

Fig. XIVC,33. The monoclinic structure of $[(CH_3)_2CHCH_2NO—]_2$ projected along its a_0 axis. Right-hand axes.

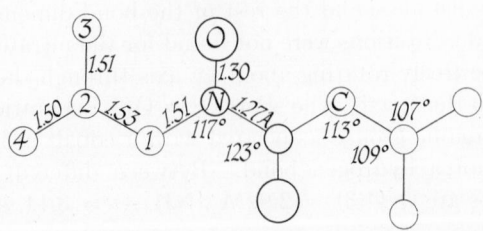

Fig. XIVC,34. Bond dimensions in the molecule of $[(CH_3)_2CHCH_2NO—]_2$.

TABLE XIVC,21
Parameters of the Atoms in the *trans* Dimer of Nitrosoisobutane

Atom	x	y	z
O	−0.095	−0.078	0.275
N	0.000	−0.003	0.397
C(1)	0.113	0.083	0.295
C(2)	0.273	0.023	0.333
C(3)	0.267	−0.117	0.220
C(4)	0.373	0.117	0.208
H(1)	0.118	0.182	0.365
H(2)	0.067	0.093	0.122
H(3)	0.320	0.012	0.487
H(4)	0.195	−0.183	0.295
H(5)	0.220	−0.107	0.048
H(6)	0.382	−0.160	0.237
H(7)	0.378	0.217	0.330
H(8)	0.327	0.127	0.035
H(9)	0.488	0.075	0.230

The atomic arrangement is shown in Figure XIVC,33. The bond dimensions are given in Figure XIVC,34. In the molecule, the N_2O_2 portion and the attached C(1) atoms are planar. The final $R = 0.198$.

XIV, c28. Crystals of *silver perfluorobutyrate*, $CF_3(CF_2)_2COOAg$, are monoclinic with a tetramolecular unit of the dimensions:

$$a_0 = 6.46 \pm 0.01 \text{ A.}; \quad b_0 = 9.01 \pm 0.02 \text{ A.}; \quad c_0 = 13.11 \pm 0.04 \text{ A.}$$
$$\beta = 100°12' \pm 18'$$

TABLE XIVC,22
Parameters of the Atoms in C_3F_7COOAg

Atom	x	y	z
O(1)	0.325	−0.120	0.083
O(2)	0.325	0.110	0.089
F(1)	0.650	−0.140	0.190
F(2)	0.740	0.040	0.155
F(3)	0.410	0.035	0.325
F(4)	0.550	0.205	0.260
F(5)	0.740	−0.135	0.395
F(6)	0.930	0.030	0.360
F(7)	0.740	0.080	0.440

The space group is $C_2{}^3$ ($C2$). Silver atoms are in

$$(2a)\quad 0u0;\ {}^1/_2,u+{}^1/_2,0 \qquad \text{with } u = 0.161$$

The fluorine and oxygen atoms are in the general positions:

$$(4c)\quad xyz;\ \bar{x}y\bar{z};\ x+{}^1/_2,y+{}^1/_2,z;\ {}^1/_2-x,y+{}^1/_2,\bar{z}$$

with the parameters of Table XIVC,22. Positions for the carbon atoms were not established.

XIV,c29. Crystals of *erythritol*, $CH_2OH(CHOH)_2CH_2OH$, are tetragonal with an eight-molecule unit of the edge lengths:

$$a_0 = 12.802 \pm 0.004 \text{ A.,} \qquad c_0 = 6.837 \pm 0.003 \text{ A.}$$

The space group is $C_{4h}{}^6$ ($I4_1/a$) with atoms in the positions:

$$(16f)\quad \pm(xyz;\ x,y+{}^1/_2,\bar{z};\ {}^3/_4-y,x+{}^1/_4,z+{}^1/_4;\ y+{}^1/_4,{}^1/_4-x,z+{}^1/_4);\quad \text{B.C.}$$

The parameters from two agreeing determinations (those from 1959: B&P in parentheses) are as follows:

Atom	x	y	z
C(1)	0.019 (0.018)	0.001 (0.003)	0.108 (0.105)
C(2)	−0.039 (−0.035)	−0.080 (−0.075)	0.236 (0.237)
O(1)	0.011 (0.012)	0.108 (0.105)	0.185 (0.175)
O(2)	−0.148 (−0.146)	−0.049 (−0.048)	0.243 (0.242)

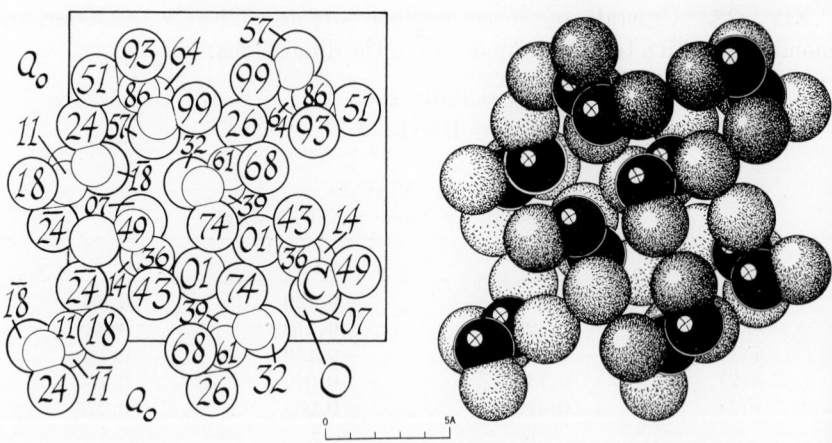

Fig. XIVC,35a (left). The tetragonal structure of erythritol projected along its c_0 axis. Right-hand axes.

Fig. XIVC,35b (right). A packing drawing of the tetragonal structure of $CH_2OH(CHOH)_2CH_2OH$ seen along its c_0 axis. The carbon atoms are black, the oxygens short-line shaded.

Fig. XIVC,36. Bond dimensions in the molecule of erythritol.

The structure is shown in Figure XIVC,35. Its molecules are planar with he bond dimensions of Figure XIVC,36. The molecules are held together n the crystal by the hydrogen bonds $O(1)-H-O(1) = 2.68$ A. and $O(2)-$ $I-O(2) = 2.72$ A.

XIV,c30. *N-tert-Butyl propylamine hydrochloride,* $C_3H_7NHC(CH_3)$· ICl, is orthorhombic with a tetramolecular unit of the edge lengths:

$$a_0 = 14.15 \text{ A.}; \quad b_0 = 8.90 \text{ A.}; \quad c_0 = 7.73 \text{ A.}$$

ll atoms have been placed in the general positions of $C_{2v}{}^9$ $(Pna2_1)$:

$$(4a) \quad xyz; \; \bar{x},\bar{y},z+{}^1/_2; \; {}^1/_2-x,y+{}^1/_2,z+{}^1/_2; \; x+{}^1/_2,{}^1/_2-y,z$$

TABLE XIVC,23
Parameters of the Atoms in $C_3H_7NHC(CH_3)_3 \cdot HCl$

Atom	x	y	z
Cl	0.406	0.117	0.255
N	0.100	0.537	0.160
C(1)	0.037	0.943	0.255
C(2)	0.022	0.780	0.202
C(3)	0.108	0.683	0.259
C(4)	0.185	0.436	0.176
C(5)	0.172	0.284	0.069
C(6)	0.279	0.506	0.139
C(7)	0.178	0.378	0.366

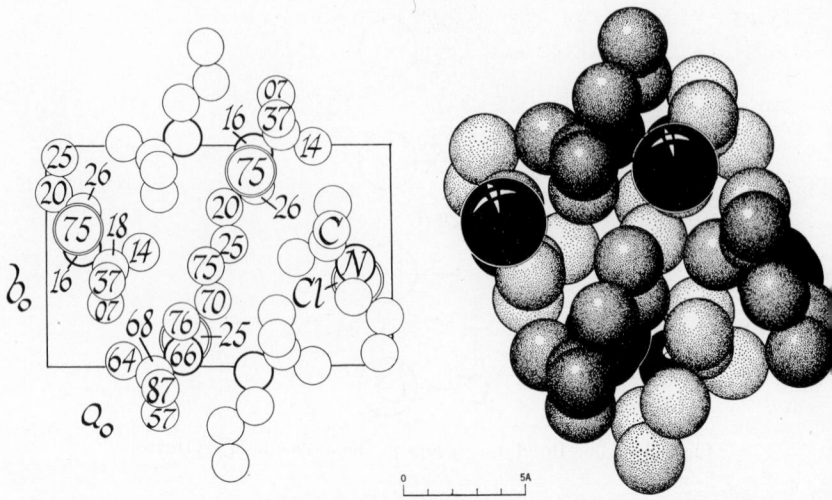

Fig. XIVC,37a (left). The orthorhombic structure of $C_3H_7NHC(CH_3)_3 \cdot HCl$ pro
jected along its c_0 axis. Right hand-axes.

Fig. XIVC,37b (right). A packing drawing of the orthorhombic structure of C_3H
$NHC(CH_3)_3 \cdot HCl$ seen along its c_0 axis. The chlorine atoms are black, as are the poorl
shown nitrogens. Carbon atoms are dotted.

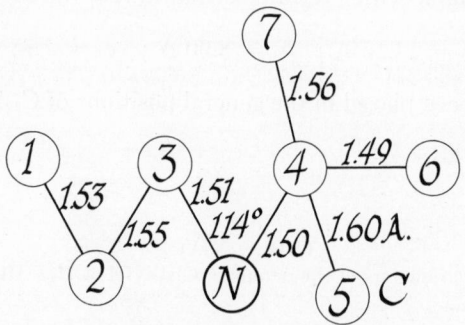

Fig. XIVC,38. Bond lengths in the complex cation of $C_3H_7NHC(CH_3)_3 \cdot HCl$.

with the parameters of Table XIVC,23.

The structure is shown in Figure XIVC,37. The bond dimensions in th
cation are those of Figure XIVC,38. Chloride ions are 3.16 and 3.21 A
away from nitrogen atoms and 3.87 A. and more from methyl group
The final $R = 0.15.$

XIV,c31. Crystals of *tetramethylene diammonium chloride*, $NH_2(CH_2)_4$-$NH_2 \cdot 2HCl$, are monoclinic with a bimolecular unit of the dimensions:

$$a_0 = 10.73 \text{ A.}; \quad b_0 = 8.35 \text{ A.}; \quad c_0 = 4.60 \text{ A.}, \quad \text{all } \pm 0.01 \text{ A.}$$
$$\beta = 92°30' \pm 18'$$

The space group is C_{2h}^5 $(P2_1/a)$ with all atoms in the general positions:

$$(4e) \quad \pm (xyz; \ x+\tfrac{1}{2}, \tfrac{1}{2}-y, z)$$

Determined parameters are as follows:

Atom	x	y	z
C(1)	0.076	0.204	0.079
C(2)	0.025	0.063	0.903
N	0.126	0.331	0.885
Cl	0.158	0.600	0.375

The structure is shown in Figure XIVC,39, the planar organic cations having the bond dimensions of Figure XIVC,40. Each nitrogen atom has four nearest Cl⁻ ions, with one N–Cl = 3.19 A. and three N–Cl = 3.28 A.;

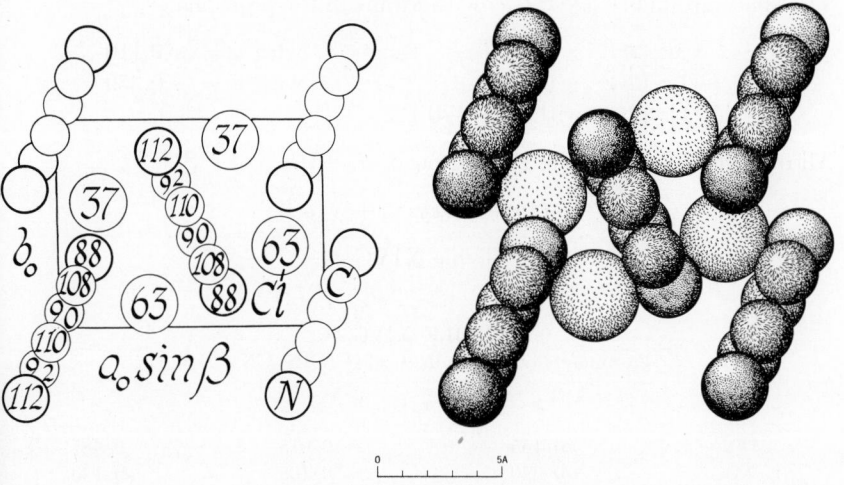

Fig. XIVC,39a (left). The monoclinic structure of $NH_2(CH_2)_4NH_2 \cdot 2HCl$ projected along its c_0 axis. Right-hand axes.

Fig. XIVC,39b (right). A packing drawing of the monoclinic $NH_2(CH_2)_4NH_2 \cdot 2HCl$ arrangement viewed along its c_0 axis. The chlorine atoms are large and hook shaded. Of the smaller atoms the carbons are line shaded, the nitrogens heavily outlined and dotted.

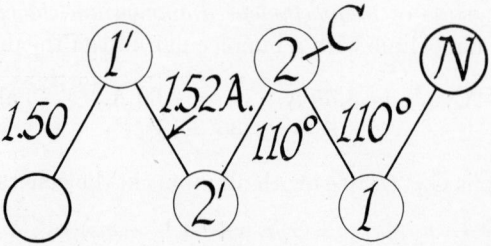

Fig. XIVC,40. Bond dimensions in the cation of $NH_2(CH_2)_4NH_2 \cdot 2HCl$.

three of these may involve hydrogen bonds. There are close relationships between this structure and that of the analogous ethylene diamine chloride (**XIV,b20**), the ionic distributions being similar in the two crystals.

XIV,c32. Crystals of *cuprous nitrate bis succinonitrile*, $[Cu(NC(CH_2)_2CN)_2]NO_3$, are monoclinic with a bimolecular unit of the dimensions:

$$a_0 = 11.62 \pm 0.03 \text{ A.}; \quad b_0 = 5.31 \pm 0.01 \text{ A.}; \quad c_0 = 9.53 \pm 0.03 \text{ A.}$$
$$\beta = 98°48' \pm 12'$$

The space group is C_{2h}^4 $(P2/a)$ with atoms in the positions:

Cu: (2e) $\pm (\frac{1}{4}\, u\, 0)$ with $u = -0.110$
O(1): (2f) $\pm (\frac{1}{4}\, u\, \frac{1}{2})$ with $u = -0.359$
N(1): (2f)   with $u = -0.123$

All other atoms are in the general positions:

$$(4g) \quad \pm (xyz;\ x+\frac{1}{2}, \bar{y}, z)$$

with the parameters listed in Table XIVC,24.

TABLE XIVC,24
Parameters of Atoms in $[Cu(NC(CH_2)_2CN)_2]NO_3$

Atom	x	y	z
O(2)	0.155	−0.015	0.487
N(2)	0.340	0.108	0.143
N(3)	0.153	−0.331	0.108
C(1)	0.385	0.247	0.225
C(2)	0.455	0.427	0.330
C(3)	0.028	−0.607	0.262
C(4)	0.100	−0.450	0.175

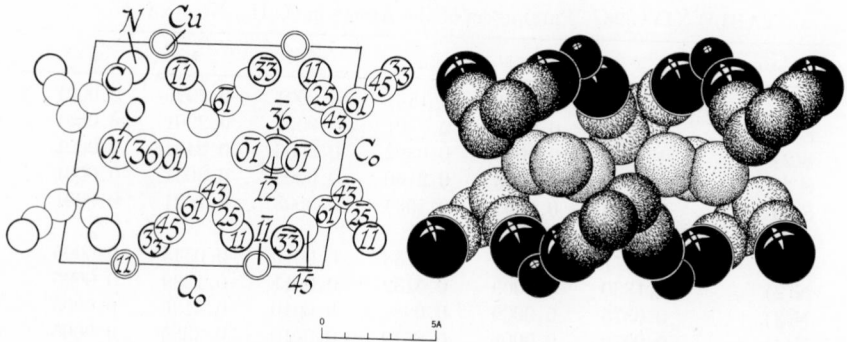

Fig. XIVC,41a (left). The monoclinic structure of $[Cu(NC(CH_2)_2CN)_2]NO_3$ projected along its b_0 axis. Right-hand axes.

Fig. XIVC,41b (right). A packing drawing of the monoclinic $[Cu(NC(CH_2)_2CN)_2]NO_3$ arrangement seen along its b_0 axis. Both the NH_2 groups and the smaller copper atoms are black. Atoms of carbon are line shaded. The dotted oxygen atoms obscure the nitrogen atoms they surround.

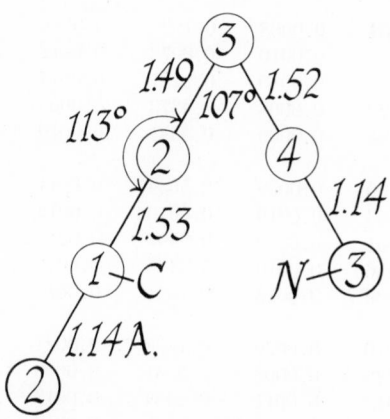

Fig. XIVC,42. Bond dimensions in the succinonitrile molecule in its compound with cuprous nitrate.

The structure that results is shown in Figure XIVC,41. Each cuprous atom is, as usual, tetrahedrally surrounded, in this case by the nitrogen atoms of four different succinonitrile groups (Cu–N = 1.96 or 2.02 A.). Within these groups the bond dimensions are those of Figure XIVC,42. The nitrate anions are situated between chains of copper succinonitrile complexes that run parallel to the a_0 axis; they have their usual triangular shape, with N–O = 1.23 or 1.25 A.

488   CRYSTAL STRUCTURES

TABLE XIVC,25.   Parameters of the Atoms in $(C_4H_9)_4NCu(S_2C_4N_2)_2$

Atom	x	$\sigma(x)$	y	$\sigma(y)$	z	$\sigma(z)$
Cu	0.1366	0.0001	0.1851	0.0001	0.2326	0.0001
S(1)	0.1412	0.0003	0.3401	0.0003	0.2201	0.0001
S(2)	0.1366	0.0003	0.0302	0.0003	0.2451	0.0001
S(3)	0.0916	0.0003	0.2160	0.0003	0.3031	0.0001
S(4)	0.1780	0.0003	0.1534	0.0003	0.1621	0.0001
N(1)	0.2477	0.0009	0.2738	0.0010	0.0545	0.0006
N(2)	0.1830	0.0009	0.5132	0.0011	0.1239	0.0005
N(3)	0.0375	0.0009	0.0986	0.0010	0.4136	0.0005
N(4)	0.0979	0.0008	0.8519	0.0010	0.3385	0.0005
N(5)	0.3307	0.0006	0.3446	0.0007	0.9111	0.0003
C(1)	0.1719	0.0008	0.3466	0.0010	0.1618	0.0005
C(2)	0.1903	0.0008	0.2659	0.0010	0.1392	0.0005
C(3)	0.2217	0.0010	0.2701	0.0012	0.0928	0.0007
C(4)	0.1799	0.0010	0.4402	0.0013	0.1412	0.0006
C(5)	0.1043	0.0008	0.0247	0.0010	0.3035	0.0005
C(6)	0.0864	0.0008	0.1029	0.0010	0.3274	0.0005
C(7)	0.0584	0.0010	0.0973	0.0012	0.3752	0.0007
C(8)	0.1011	0.0010	0.9287	0.0012	0.3235	0.0005
C(9)	0.3670	0.0008	0.4373	0.0009	0.9327	0.0004
C(10)	0.3306	0.0009	0.4681	0.0010	0.9789	0.0005
C(11)	0.3730	0.0009	0.5634	0.0011	0.9966	0.0005
C(12)	0.3343	0.0010	0.6019	0.0012	0.0392	0.0006
C(13)	0.3531	0.0009	0.2592	0.0010	0.9436	0.0005
C(14)	0.4456	0.0010	0.2435	0.0012	0.9582	0.0006
C(15)	0.4466	0.0013	0.1561	0.0013	0.9935	0.0007
C(16)	0.5316	0.0020	0.1486	0.0021	0.0195	0.0010
C(17)	0.3748	0.0008	0.3331	0.0009	0.8637	0.0005
C(18)	0.3412	0.0011	0.2454	0.0013	0.8349	0.0006
C(19)	0.3944	0.0010	0.2409	0.0012	0.7873	0.0006
C(20)	0.3717	0.0012	0.1512	0.0013	0.7598	0.0006
C(21)	0.2335	0.0008	0.3475	0.0009	0.9031	0.0004
C(22)	0.1982	0.0009	0.4306	0.0010	0.8725	0.0005
C(23)	0.1004	0.0011	0.4318	0.0012	0.8731	0.0006
C(24)	0.0611	—	0.5162	—	0.8431	—
H(1,C 9)	0.37	—	0.49	—	0.91	—
H(2,C 9)	0.43	—	0.43	—	0.95	—
H(1,C 10)	0.32	—	0.42	—	0.00	—
H(2,C 10)	0.29	—	0.48	—	0.97	—
H(1,C 11)	0.43	—	0.54	—	0.01	—
H(2,C 11)	0.38	—	0.61	—	0.97	—

(continued)

TABLE XIVC,25 (continued)

Atom	x	$\sigma(x)$	y	$\sigma(y)$	z	$\sigma(z)$
H(1,C 12)	0.36	—	0.65	—	0.04	—
H(2,C 12)	0.27	—	0.58	—	0.03	—
H(3,C 12)	0.34	—	0.56	—	0.07	—
H(1,C 13)	0.33	—	0.20	—	0.92	—
H(2,C 13)	0.32	—	0.27	—	0.97	—
H(1,C 14)	0.48	—	0.24	—	0.93	—
H(2,C 14)	0.45	—	0.28	—	0.97	—
H(1,C 15)	0.43	—	0.15	—	0.03	—
H(2,C 15)	0.45	—	0.12	—	0.97	—
H(1,C 16)	0.49	—	0.10	—	0.04	—
H(2,C 16)	0.54	—	0.11	—	0.99	—
H(3,C 16)	0.57	—	0.19	—	0.03	—
H(1,C 17)	0.44	—	0.32	—	0.88	—
H(2,C 17)	0.35	—	0.38	—	0.85	—
H(1,C 18)	0.38	—	0.25	—	0.82	—
H(2,C 18)	0.34	—	0.20	—	0.85	—
H(1,C 19)	0.38	—	0.29	—	0.77	—
H(2,C 19)	0.47	—	0.24	—	0.80	—
H(1,C 20)	0.32	—	0.18	—	0.74	—
H(2,C 20)	0.40	—	0.10	—	0.77	—
H(3,C 20)	0.31	—	0.11	—	0.76	—
H(1,C 21)	0.22	—	0.29	—	0.88	—
H(2,C 21)	0.21	—	0.35	—	0.93	—
H(1,C 22)	0.21	—	0.49	—	0.89	—
H(2,C 22)	0.22	—	0.43	—	0.84	—
H(1,C 23)	0.08	—	0.38	—	0.86	—
H(2,C 23)	0.10	—	0.46	—	0.90	—
H(1,C 24)	0.08	—	0.57	—	0.86	—
H(2,C 24)	0.00	—	0.53	—	0.87	—
H(3,C 24)	0.05	—	0.49	—	0.81	—

XIV,c33. *Tetra-N-butyl ammonium copper bis (maleonitrile dithiolate)*, $(C_4H_9)_4NCu(S_2C_4N_2)_2$, is monoclinic with a unit containing eight molecules and having the dimensions:

$$a_0 = 15.59 \pm 0.02 \text{ A.}; \quad b_0 = 13.83 \pm 0.01 \text{ A.}; \quad c_0 = 27.94 \pm 0.03 \text{ A.}$$
$$\beta = 93°52' \pm 2'$$

The space group is C_{2h}^6 ($I2/c$) with all atoms in the general positions:

$$(8f) \quad \pm (xyz; x,\bar{y},z+\frac{1}{2}); \text{ B.C.}$$

The determined parameters are those of Table XIVC,25.

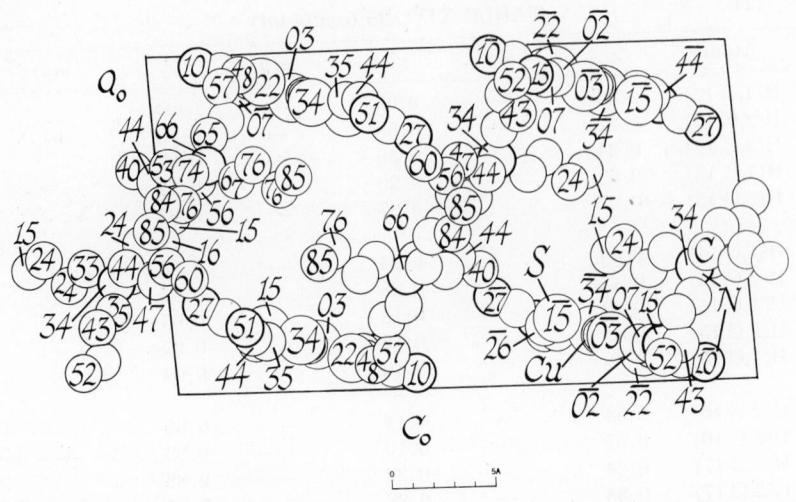

Fig. XIVC,43. Half the molecules in the monoclinic structure of $(C_4H_9)_4NCu(S_2C_4N_2)_2$ projected along its b_0 axis. Right-hand axes.

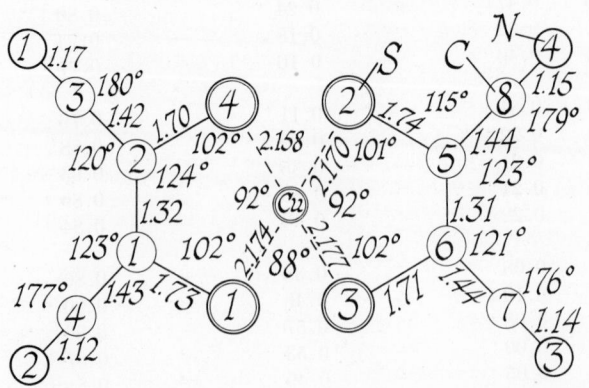

Fig. XIVC,44. Bond dimensions in crystals of $(C_4H_9)_4NCu(S_2C_4N_2)_2$.

Part of the structure is shown in Figure XIVC,43. Its complex anions, which are nearly planar, have the bond dimensions of Figure XIVC,44. In the cation, the four butyl radicals are tetrahedrally arranged around the central nitrogen atom, with N–C = 1.52 A. and C–C = 1.52±0.02 A.

XIV,c34.   Crystals of *di(tetra-N-butyl ammonium) cobalt(II) bis (maleonitrile dithiolate)*, $[(C_4H_9)_4N]_2Co(S_2C_4N_2)_2$, are triclinic with a unimolecular cell of the dimensions:

$$a_0 = 10.77 \text{ A.}; \quad b_0 = 12.35 \text{ A.}; \quad c_0 = 9.81 \text{ A.}, \qquad \text{all } \pm 0.01 \text{ A.}$$
$$\alpha = 88°30'; \quad \beta = 114°48'; \quad \gamma = 93°30', \qquad \text{all } \pm 6'$$

The space group is C_i^1 $(P\bar{1})$ with the cobalt atom in $(1a)$ 000 and all other atoms in $(2i)$ $\pm(xyz)$. The determined parameters are those of Table XIVC,26. Positions for hydrogen were given by the experimental data.

TABLE XIVC,26
Parameters of the Atoms in $[(C_4H_9)_4N]_2Co(S_2C_4N_2)_2$

Atom	x	$\sigma(x)$	y	$\sigma(y)$	z	$\sigma(z)$
S(1)	0.0401	0.0002	0.1381	0.0002	0.1462	0.0002
S(2)	−0.2156	0.0002	0.0303	0.0002	−0.1093	0.0002
C(1)	−0.1165	0.0007	0.1930	0.0005	0.0876	0.0007
C(2)	−0.2283	0.0007	0.1463	0.0006	−0.0214	0.0008
C(3)	−0.1243	0.0008	0.2883	0.0007	0.1575	0.0009
C(4)	−0.3583	0.0010	0.1883	0.0007	−0.0717	0.0010
N(1)	−0.1273	0.0007	0.3662	0.0006	0.2172	0.0008
N(2)	−0.4643	0.0010	0.2247	0.0007	−0.1113	0.0010
N(3)	0.0867	0.0005	0.2783	0.0004	−0.2929	0.0006
C(5)	0.1928	0.0007	0.2589	0.0005	−0.1328	0.0007
C(6)	0.2939	0.0007	0.3510	0.0006	−0.0626	0.0008
C(7)	0.3797	0.0008	0.3242	0.0007	0.1021	0.0009
C(8)	0.4976	0.0010	0.4089	0.0008	0.1789	0.0011
C(9)	−0.0023	0.0007	0.1740	0.0006	−0.3418	0.0008
C(10)	−0.1167	0.0008	0.1779	0.0007	−0.5001	0.0009
C(11)	−0.2078	0.0009	0.0756	0.0008	0.4697	0.0010
C(12)	−0.3248	0.0010	0.0689	0.0008	0.3116	0.0011
C(13)	0.1538	0.0007	0.3116	0.0005	−0.3963	0.0007
C(14)	0.2348	0.0007	0.2286	0.0006	−0.4230	0.0008
C(15)	0.3224	0.0008	0.2751	0.0007	−0.5004	0.0009
C(16)	0.4075	0.0009	0.1952	0.0007	0.4737	0.0010
C(17)	0.0004	0.0007	0.3718	0.0006	−0.2930	0.0008
C(18)	−0.0671	0.0008	0.3599	0.0006	−0.1848	0.0009
C(19)	−0.1712	0.0009	0.4468	0.0007	−0.2126	0.0010
C(20)	−0.2991	0.0011	0.4290	0.0009	−0.3551	0.0012

(continued)

CRYSTAL STRUCTURES

TABLE XIVC,26 (*continued*)

Atom	x	$\sigma(x)$	y	$\sigma(y)$	z	$\sigma(z)$
H(1,C 5)	0.13	—	0.23	—	−0.08	—
H(2,C 5)	0.24	—	0.20	—	−0.15	—
H(1,C 6)	0.35	—	0.37	—	−0.12	—
H(2,C 6)	0.25	—	0.42	—	−0.06	—
H(1,C 7)	0.32	—	0.32	—	0.16	—
H(2,C 7)	0.41	—	0.26	—	0.11	—
H(1,C 8)	−0.43	—	0.42	—	0.12	—
H(2,C 8)	−0.47	—	0.35	—	0.21	—
H(3,C 8)	−0.53	—	0.48	—	0.16	—
H(1,C 9)	−0.04	—	0.15	—	−0.25	—
H(2,C 9)	0.06	—	0.12	—	−0.34	—
H(1,C 10)	−0.18	—	0.21	—	−0.48	—
H(2,C 10)	−0.07	—	0.18	—	−0.57	—
H(1,C 11)	−0.25	—	0.08	—	0.55	—
H(2,C 11)	−0.14	—	0.05	—	0.45	—
H(1,C 12)	−0.38	—	0.11	—	0.33	—
H(2,C 12)	−0.27	—	0.04	—	0.26	—
H(3,C 12)	−0.33	—	0.14	—	0.25	—
H(1,C 13)	0.22	—	0.39	—	−0.35	—
H(2,C 13)	0.08	—	0.33	—	−0.47	—
H(1,C 14)	0.29	—	0.20	—	−0.34	—
H(2,C 14)	0.18	—	0.17	—	−0.49	—
H(1,C 15)	0.38	—	0.33	—	−0.44	—
H(2,C 15)	0.26	—	0.31	—	−0.58	—
H(1,C 16)	0.47	—	0.16	—	0.56	—
H(2,C 16)	0.37	—	0.13	—	0.43	—
H(3,C 16)	0.46	—	0.24	—	0.43	—
H(1,C 17)	−0.06	—	0.37	—	−0.40	—
H(2,C 17)	0.06	—	0.44	—	−0.28	—
H(1,C 18)	0.01	—	0.36	—	−0.07	—
H(2,C 18)	−0.12	—	0.29	—	−0.20	—
H(1,C 19)	−0.21	—	0.44	—	−0.12	—
H(2,C 19)	−0.14	—	0.52	—	−0.23	—
H(1,C 20)	−0.28	—	0.44	—	−0.44	—
H(2,C 20)	−0.33	—	0.37	—	−0.32	—
H(3,C 20)	−0.36	—	0.49	—	−0.33	—

Fig. XIVC,45. Bond dimensions in crystals of $[(C_4H_9)_4N]_2Co(S_2C_4N_2)_2$.

The complex anion that is present in this structure has the bond dimensions of Figure XIVC,45. It has a center of symmetry and is essentially planar. The tetrabutyl ammonium cation has its four butyl radicals tetrahedrally distributed around the nitrogen atom, with N–C = 1.51–1.53 A.; C–C distances range between 1.48 and 1.54 A. The final R = 0.091.

XIV,c35. *Butadiene iron tricarbonyl*, $Fe(CO)_3 \cdot C_4H_6$, forms orthorhombic crystals which have a tetramolecular unit of the edge lengths:

$$a_0 = 11.6 \text{ A.}; \quad b_0 = 11.1 \text{ A.}; \quad c_0 = 6.2 \text{ A.}$$

The space group has been chosen as V_h^{16} (*Pnma*) with atoms in the positions:

(4c) $\pm (u\ ^1/_4\ v;\ u+^1/_2,^1/_4,^1/_2-v)$

(8d) $\pm (xyz;\ ^1/_2-x,y+^1/_2,z+^1/_2;\ x,^1/_2-y,z;\ x+^1/_2,y,^1/_2-z)$

The determined parameters are those of Table XIVC,27.

TABLE XIVC,27
Positions and Parameters of the Atoms in $Fe(CO)_3 \cdot C_4H_6$

Atom	Position	x	y	z
Fe	(4c)	0.0788	$^1/_4$	0.1000
C(1)	(4c)	−0.0434	$^1/_4$	−0.0634
O(1)	(4c)	−0.1271	$^1/_4$	−0.1717
C(2)	(8d)	0.1630	0.1345	−0.0194
O(2)	(8d)	0.2144	0.0585	−0.0928
C(3)	(8d)	0.0012	0.1226	0.3154
C(4)	(8d)	0.0997	0.1848	0.4082

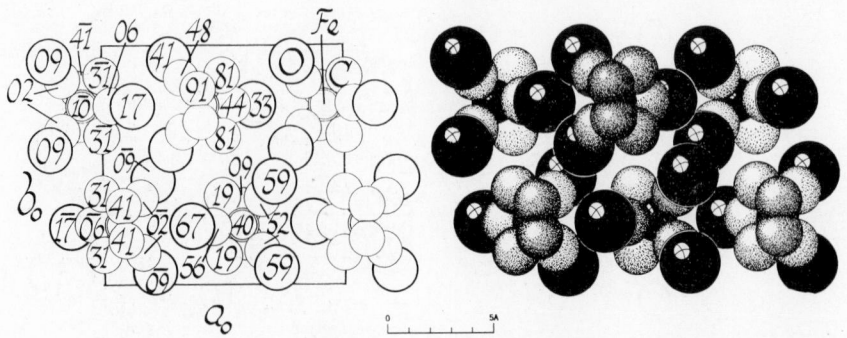

Fig. XIVC,46a (left). The orthorhombic structure of $Fe(CO)_3 \cdot C_4H_6$ projected along its c_0 axis. Right-hand axes.

Fig. XIVC,46b (right). A packing drawing of the orthorhombic $Fe(CO)_3 \cdot C_4H_6$ structure seen along its c_0 axis. The large black circles are oxygen; only parts of the smaller black iron atoms show. Atoms of carbon are short-line shaded.

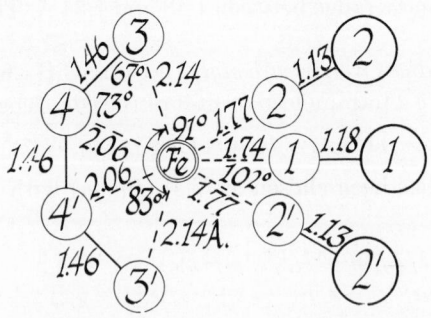

Fig. XIVC,47. Bond dimensions in crystals of $Fe(CO)_3 \cdot C_4H_6$.

The structure as a whole is shown in Figure XIVC,46. Interatomic distances are those of Figure XIVC,47. The butadiene molecule is strictly planar and this plane is inclined 61° to the trigonal axis of the $Fe(CO)_3$ group. In this group the Fe–C–O angles are 178 or 179°. Positions were not specified for the hydrogen atoms. The final $R = 0.077$.

3. Five-Carbon Chain Compounds

XIV,c36. Crystalline *pentane*, C_5H_{12}, is orthorhombic with a tetramolecular unit which at $-145°C$. has the edge lengths:

$$a_0 = 4.10 \text{ A.}; \quad b_0 = 9.04 \text{ A.}; \quad c_0 = 14.70 \text{ A.}, \qquad \text{all } \pm 0.02 \text{ A.}$$

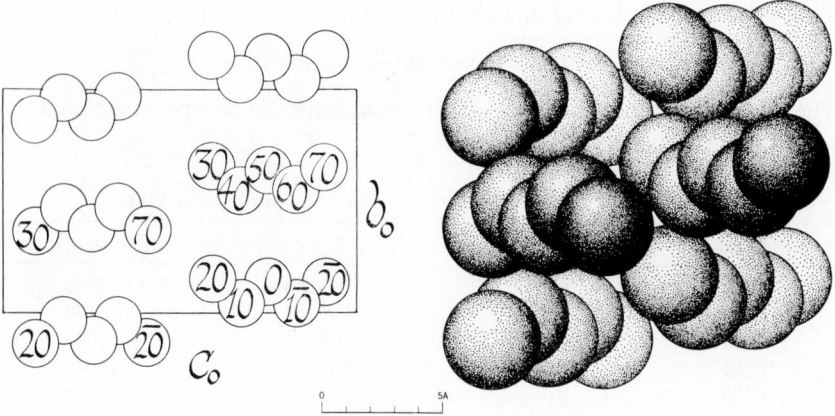

Fig. XIVC,48a (left). The orthorhombic structure of C_5H_{12} projected along its a_0 axis. Right-hand axes.
Fig. XIVC,48b (right). A packing drawing of the orthorhombic structure of pentane seen along its a_0 axis.

The space group has been selected as V_h^{14} (*Pbcn*) with atoms in the positions:

C(1): (8d) $\pm (xyz; \ x+\frac{1}{2},y+\frac{1}{2},\frac{1}{2}-z; \ \frac{1}{2}-x,y+\frac{1}{2},z; \ x,\bar{y},z+\frac{1}{2})$
            with $x = -0.1993$, $y = 0.1313$, $z = 0.0870$
C(2): (8d)   with $x = -0.1003$, $y = 0.0376$, $z = 0.1683$
C(3): (4c) $\pm (0 \ u \ \frac{1}{4}; \frac{1}{2},u+\frac{1}{2},\frac{1}{4})$ with $u = 0.1319$

The resulting structure is shown in Figure XIVC,48. In the molecule, C(1)–C(2) = 1.520 A. and C(2)–C(3) = 1.529 A.; the bond angles are 112.2°. Between molecules the van der Waals contacts are ca. 3.9 A. As the figure indicates, this chain is too short to give a structure closely resembling those of the long-chain hydrocarbons and their derivatives.

Two other arrangements based on C_{2v}^5 (*Pbc*2_1) were considered, but they give poorer agreement with the data than the above. The final $R = 0.064$ though this was not decisive for the chosen structure.

XIV,c37. *Valeric acid*, C_4H_9COOH, has a structure different from that of the shorter chain fatty acids. It is monoclinic with a tetramolecular cell of the dimensions:

$$a_0 = 5.55 \pm 0.01 \ \text{A.}; \quad b_0 = 9.664 \pm 0.007 \ \text{A.}; \quad c_0 = 11.341 \pm 0.011 \ \text{A.}$$
$$\beta = 101°49' \ (-135°C.)$$

The space group is C_{2h}^5 ($P2_1/c$) with all atoms in the general positions:

$$(4e) \quad \pm (xyz;\ x,{}^1\!/_2-y,z+{}^1\!/_2)$$

The determined parameters, involving anisotropic temperature factors, are listed in Table XIVC,28.

TABLE XIVC,28. Parameters of the Atoms in Valeric Acid

Atom	x	y	z
O(1)	0.0978	0.3280	0.4689
O(2)	0.1957	0.5403	0.4146
C(1)	0.2208	0.4112	0.4065
C(2)	0.3862	0.3400	0.3324
C(3)	0.5174	0.4414	0.2628
C(4)	0.6607	0.3582	0.1805
C(5)	0.8196	0.4553	0.1136

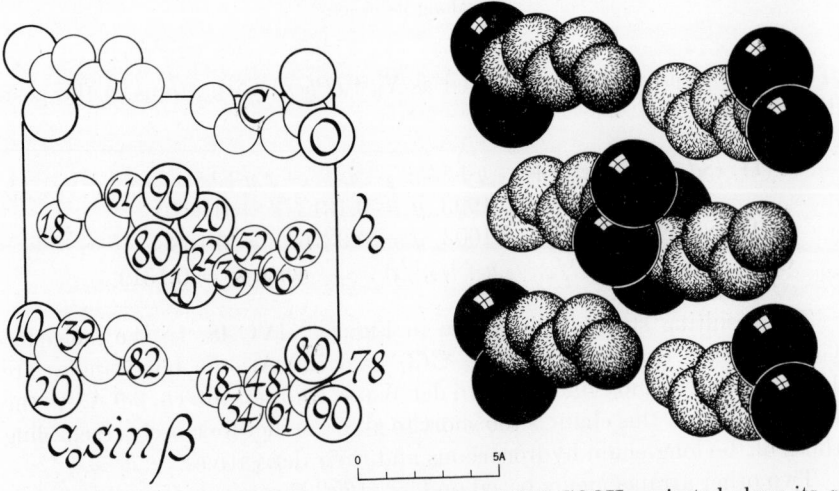

Fig. XIVC,49a (left). The monoclinic structure of C_4H_9COOH projected along its a_0 axis. Right-hand axes.

Fig. XIVC,49b (right). A packing drawing of the monoclinic C_4H_9COOH structure seen along its a_0 axis. The oxygen atoms are black, the carbon atoms line shaded.

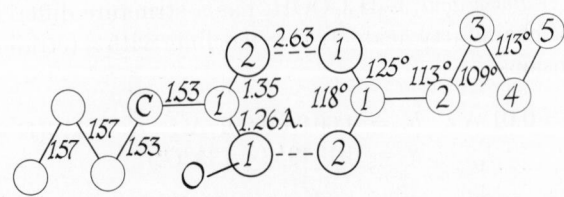

Fig. XIVC,50. Bond dimensions in crystals of valeric acid.

The resulting structure is shown in Figure XIVC,49; the molecular bond dimensions are those of Figure XIVC,50. As with other acids, the carboxyl groups of adjacent molecules are tied together by hydrogen bonds. The C(2), C(1), O(1), and O(2) atoms are coplanar, but the planes of adjacent molecules, though parallel, are displaced by 0.12 A. The C(3), C(4), and C(5) atoms are -0.026, -0.159, and -0.071 A. out of the plane of the other atoms of a molecule. The final R of this structure is 0.11.

XIV,c38. *Tetra isoamyl ammonium fluoride hydrate*, $(C_5H_{11})_4NF \cdot 38H_2O$, when crystallized with this large amount of water forms clathrate-type crystals. The symmetry is orthorhombic with a bimolecular unit of the dimensions:

$$a_0 = 12.08 \pm 0.01 \text{ A.}; \quad b_0 = 21.61 \pm 0.02 \text{ A.}; \quad c_0 = 12.82 \pm 0.01 \text{ A.}$$

TABLE XIVC,29
Positions and Parameters of the Atoms in $(C_5H_{11})_4NF \cdot 38H_2O$

Atom	Position	x	y	z
N	(2f)	0.0135	1/4	1/2
F	(2f)	0.5128	1/4	1/2
C(1)	(8l)	0.0853	0.1895	0.4576
C(2)	(8l)	0.1642	0.1744	0.5408
C(3)	(8l)	0.1998	0.1079	0.5068
C(4)	(8l)	0.2626	0.0744	0.5949
C(5)	(8l)	0.2840	0.1072	0.4032
C(6)	(8l)	0.9396	0.2739	0.4063
C(7)	(8l)	0.8559	0.2262	0.3715
C(8)	(8l)	0.8014	0.2606	0.2681
C(9)	(8l)	0.7263	0.1987	0.2142
C(10)	(8l)	0.7203	0.3095	0.2757
O(1)	(8l)	0.0563	0.1396	0.1725
O(2)	(8l)	0.4412	0.3571	0.1794
O(3)	(8l)	0.0624	0.9445	0.2982
O(4)	(8l)	0.4425	0.9431	0.3028
O(5)	(8l)	0.2517	0.0841	0.1086
O(6)	(8l)	0.2543	0.9636	0.1782
O(7)	(4i)	0.0642	0.8162	0
O(8)	(4i)	0.4514	0.8147	0
O(9)	(4i)	0.2588	0.8900	0
O(10)	(4k)	0.1681	1/4	0.2158
O(11)	(4k)	0.3901	1/4	0.2926
O(12)	(4j)	0.1507	0.9204	1/2
O(13)	(4j)	0.3894	0.8885	1/2

The space group is V_h^5 (*Pbmm*) with atoms in the positions:

(2*f*)　　$\pm (u\ ^1/_4\ ^1/_2)$
(4*i*)　　$\pm (uv0;\ u,^1/_2-v,0)$
(4*j*)　　$\pm (u\ v\ ^1/_2;\ u,^1/_2-v,^1/_2)$
(4*k*)　　$\pm (u\ ^1/_4\ v;\ u\ ^1/_4\ \bar{v})$
(8*l*)　　$\pm (xyz;\ xy\bar{z};\ x,^1/_2-y,z;\ x,^1/_2-y,\bar{z})$

The distribution among these positions, and the chosen parameters, are those of Table XIVC,29.

This structure is a complex of water molecules and F^- anions associated through hydrogen bondings to form polyhedra in which voids exist large enough to accommodate the $(C_5H_{11})_4N^+$ cations.

XIV,c39. *Calcium arabonate pentahydrate*, $Ca(C_5O_6H_9)_2 \cdot 5H_2O$, forms optically active monoclinic crystals which have a bimolecular unit of the dimensions:

$$a_0 = 14.74\ \text{A.};\quad b_0 = 5.82\ \text{A.};\quad c_0 = 11.40\ \text{A.};\quad \beta = 99°$$

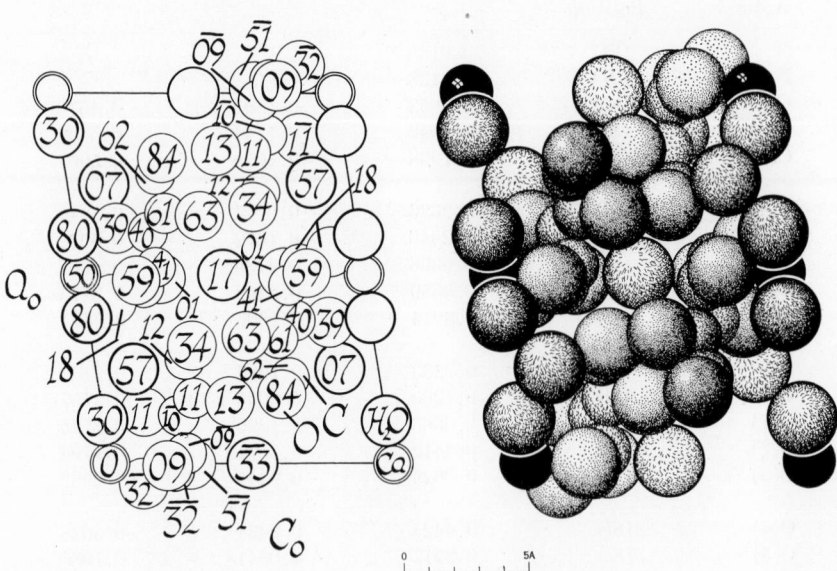

Fig. XIVC,51a (left). The monoclinic structure of $Ca(C_5O_6H_9)_2 \cdot 5H_2O$ projected along its b_0 axis. Right-hand axes.

Fig. XIVC,51b (right) A packing drawing of the monoclinic structure of $Ca(C_5O_6H_9)_2 \cdot 5H_2O$ viewed along its b_0 axis. The calcium atoms are black, the water molecules line shaded and heavily outlined. Carboxyl oxygen atoms are dotted; the smaller atoms of carbon are hook shaded.

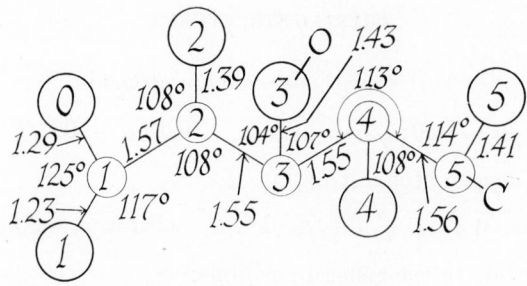

Fig. XIVC,52. Bond dimensions in the arabonate anion in $Ca(C_5O_6H_9)_2 \cdot 5H_2O$.

TABLE XIVC,30

Parameters of Atoms in $Ca(C_5O_6H_9)_2 \cdot 5H_2O$ and $Sr(C_5O_6H_9)_2 \cdot 5H_2O$ [in parentheses]

Atom	x	y	z
C(1)	−0.0248 (−0.023)	−0.322 (−0.315)	0.218 (0.219)
C(2)	0.0195 (0.018)	−0.094 (−0.090)	0.2741 (0.280)
C(3)	0.1249 (0.124)	−0.110 (−0.119)	0.2765 (0.283)
C(4)	0.1718 (0.170)	0.111 (0.094)	0.3335 (0.333)
C(5)	0.2763 (0.277)	0.120 (0.108)	0.3261 (0.331)
O(0)	−0.0550 (−0.059)	−0.321 (−0.323)	0.1117 (0.121)
O(1)	−0.0096 (−0.009)	−0.506 (−0.499)	0.2825 (0.287)
O(2)	−0.0170 (−0.019)	0.091 (0.082)	0.2051 (0.210)
O(3)	0.1355 (0.139)	−0.110 (−0.120)	0.1536 (0.163)
O(4)	0.1620 (0.160)	0.132 (0.128)	0.4552 (0.456)
O(5)	0.3169 (0.315)	0.338 (0.337)	0.3495 (0.352)
O(6)	0.1145 (0.122)	0.303 (0.313)	0.0012 (−0.006)
O(7)	0.2618 (0.260)	−0.427 (−0.429)	0.1414 (0.147)
O(8)	0.0000 (0.000)	−0.333 (−0.344)	0.5000 (0.500)
	Proposed hydrogen positions		
H(1)	0.010	−0.05	0.362
H(2)	0.138	−0.25	0.293
H(3)	0.158	0.27	0.317
H(4)	0.303	0.02	0.387
H(5)	0.285	0.02	0.235
H(6)	0.495	−0.40	0.415
H(7)	0.483	0.22	0.234
H(8)	0.405	0.45	0.035
H(9)	0.163	0.40	0.047
H(10)	0.278	0.43	0.208
H(11)	0.300	−0.38	0.095
H(12)	0.175	−0.22	0.150
H(13)	0.330	0.43	0.480
H(14)	0.378	0.28	0.400

The space group is $C_2{}^3$ $(C2)$ with the calcium atoms in

$$(2a) \quad 0u0; \; {}^1/_2,u+{}^1/_2,0 \qquad \text{with } u = 0.000$$

and one set of oxygen atoms [O(8)] in

$$(2b) \quad 0 \; u \; {}^1/_2; \; {}^1/_2,u+{}^1/_2,{}^1/_2 \qquad \text{with } u = -0.333$$

All other atoms are in the general positions:

$$(4c) \quad xyz; \; \bar{x}y\bar{z}; \; x+{}^1/_2,y+{}^1/_2,z; \; {}^1/_2-x,y+{}^1/_2,\bar{z}$$

with the parameters of Table XIVC,30.

In this structure (Fig. XIVC,51) each calcium atom has around it eight oxygen atoms of which three come from each of two anions; the others are provided by water molecules. The coordination is that of a distorted square Archimedes antiprism, with Ca–O = 2.44–2.52 A. Bond dimensions of the arabonate anion are those of Figure XIVC,52; its carbon atoms are practically coplanar, with the oxygen atoms lying on either side of this plane. The proposed hydrogen positions are based on the assumption that the O–O separations shorter than 3.2 A. involve hydrogen bonds.

The *strontium* compound is isostructural with a unit having

$$a_0 = 14.79 \text{ A.}; \quad b_0 = 5.83 \text{ A.}; \quad c_0 = 11.78 \text{ A.}; \quad \beta = 99° \; (\pm 0.5\%)$$

The assigned atomic parameters are given in parentheses in Table XIVC,30.

XIV,c40. Crystals of 2-*methyl trans but-2-enoic acid, tiglic acid,* $CH_3CH:C(CH_3)COOH$, are triclinic with a bimolecular unit of the dimensions:

$$a_0 = 7.70 \text{ A.}; \quad b_0 = 5.23 \text{ A.}; \quad c_0 = 7.42 \text{ A.}, \qquad \text{all } \pm 0.02 \text{ A.}$$
$$\alpha = 96°36'; \quad \beta = 86°30'; \quad \gamma = 106°0', \qquad \text{all } \pm 30'$$

The space group is $C_i{}^1$ $(P\bar{1})$ with all atoms in the general positions: $(2i)$ $\pm(xyz)$. The determined parameters are those of Table XIVC,31.

TABLE XIVC,31
Parameters of the Atoms in Tiglic Acid

Atom	x	y	z
C(1)	0.7000	0.3254	0.5557
C(2)	0.7863	0.5333	0.4298
C(3)	0.7667	0.5095	0.2523
C(4)	0.8707	0.7155	0.1450
C(5)	0.6440	0.2710	0.1506
O(6)	0.9773	0.9290	0.2310
O(7)	0.8530	0.6980	-0.0217

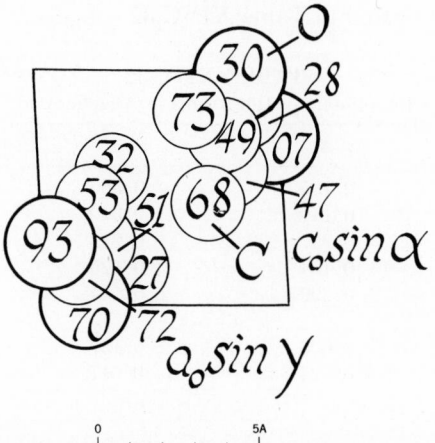

Fig. XIVC,53. The triclinic structure of tiglic acid projected along its b_0 axis. Right-hand axes.

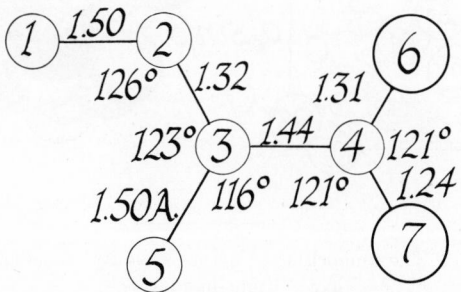

Fig. XIVC,54. Bond dimensions in molecules of tiglic acid.

The resulting structure is shown in Figure XIVC,53. The molecules it contains have the bond dimensions of Figure XIVC,54. These molecules are tied together in the crystal by hydrogen bonds between carboxyl oxygens, with O–H–O = 2.64 A.

XIV,c41. Crystals of *angelic acid*, the *cis* form of *2-methyl but-2-enoic acid*, $CH_3CH:C(CH_3)COOH$, are monoclinic with a tetramolecular unit of the dimensions:

$$a_0 = 7.66 \text{ A.}; \quad b_0 = 11.60 \text{ A.}; \quad c_0 = 6.67 \text{ A.}, \quad \text{all} \pm 0.02 \text{ A.}$$
$$\beta = 100°0' \pm 30'$$

The space group is C_{2h}^5 $(P2_1/a)$ with all atoms in the general positions:

$$(4e) \quad \pm(xyz; \ x+{}^1/_2, {}^1/_2-y, z)$$

The parameters are listed in Table XIVC,32.

TABLE XIVC,32
Parameters of the Atoms in Angelic Acid

Atom	x	y	z
C(1)	0.105	0.360	0.117
C(2)	0.181	0.310	0.318
C(3)	0.207	0.213	0.361
C(4)	0.134	0.126	0.170
C(5)	0.299	0.154	0.530
O(6)	0.006	0.137	0.015
O(7)	0.124	0.017	0.231

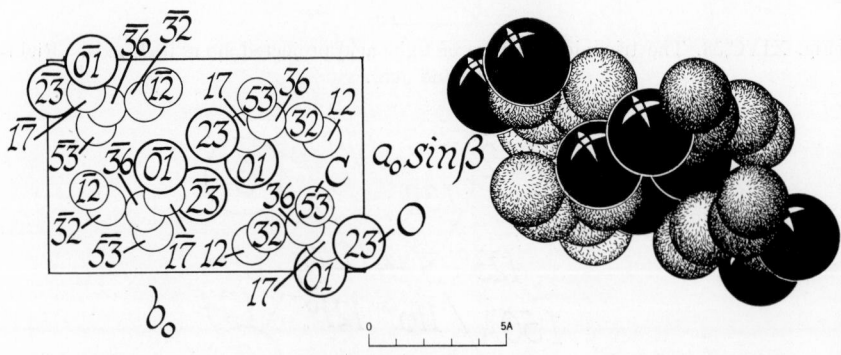

Fig. XIVC,55a (left). The monoclinic structure of angelic acid projected along its c_0 axis. Right-hand axes.

Fig. XIVC,55b (right). A packing drawing of the monoclinic structure of angelic acid seen along its c_0 axis. The oxygen atoms are black, the carbon atoms line shaded.

The resulting atomic arrangement is shown in Figure XIVC,55. As with the *trans* form (**XIV,c40**), the molecules are held together by hydrogen bonds between carboxyl oxygen atoms. The planes of the molecules are practically coincident with the ($20\bar{1}$) plane of the crystal which has a pronounced cleavage in this direction.

XIV,c42. Crystals of a *hexamer* of *acrylonitrile*, more accurately described as *1,1,4,4-tetra-(2-cyanoethyl)-1,4-dicyano-trans-2-butene*, [NC-$(CH_2)_2$]$_2$C(CN)CH=CHC(CN)[NC$(CH_2)_2$]$_2$, are monoclinic with a bimolecular cell of the dimensions:

$$a_0 = 15.56 \text{ A.}; \quad b_0 = 6.42 \text{ A.}; \quad c_0 = 9.24 \text{ A.}; \quad \beta = 108°18'$$

The space group is C_{2h}^5 $(P2_1/a)$ with all atoms in the positions:

$$(4e) \quad \pm (xyz;\ x+{}^1/_2, {}^1/_2-y, z)$$

The determined parameters are those of Table XIVC,33.

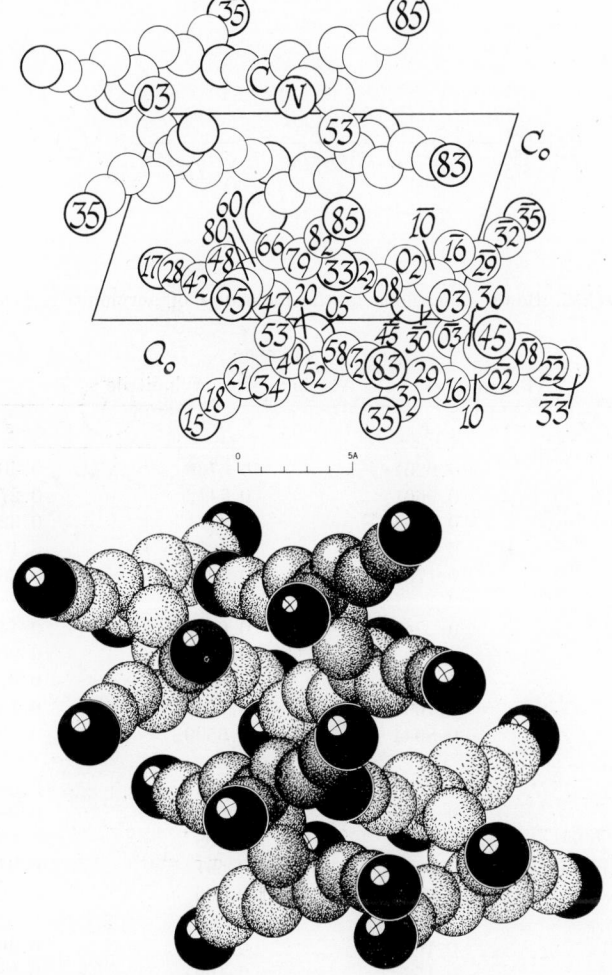

Fig. XIVC,56a (top). The monoclinic structure of acrylonitrile hexamer projected along its b_0 axis. Right-hand axes.

Fig. XIVC,56b (bottom). A packing drawing of the monoclinic structure of acrylonitrile hexamer seen along its b_0 axis. The nitrogen atoms are black, those of carbon line-and-dot shaded.

Fig. XIVC,57. Bond dimensions in the molecules of acrylonitrile hexamer.

TABLE XIVC,33

Parameters of the Atoms in Hexacrylonitrile

Atom	x	y	z
C(1)	0.9264	0.6766	0.4011
C(2)	0.9661	0.7112	0.2742
C(3)	0.0494	0.8451	0.3276
C(4)	0.0880	0.8970	0.1938
C(5)	0.1758	0.0256	0.2634
C(6)	0.2243	0.0797	0.1457
C(7)	0.2995	0.2258	0.2095
C(8)	0.0183	0.0293	0.0725
C(9)	0.1095	0.7043	0.1290
N(1)	0.8944	0.6500	0.4914
N(2)	0.3573	0.3337	0.2561
N(3)	0.1278	0.5526	0.0802
H(2)	0.911	0.808	0.196
H(2')	0.982	0.591	0.240
H(3)	0.024	0.006	0.363
H(3')	0.109	0.755	0.405
H(5)	0.152	0.169	0.281
H(5')	0.220	0.902	0.353
H(6)	0.176	0.191	0.047
H(6')	0.253	0.939	0.095
H(8)	0.999	0.194	0.117

The structure that results is shown in Figure XIVC,56. The molecules themselves have a center of symmetry midway along the double bond; bond dimensions of half the molecules are shown in Figure XIVC,57. The triple bonds, such as $N(1)$–$C(1)$–$C(2)$, are as usual almost straight (178°) and the atoms in the "vertical" direction from $N(3)$ to $N(3')$ through the double bond are coplanar. The double bond angle, $C(4)$–$C(8)$–$C(8') = 123.7°$. The C–H distances not shown in the figure range between 0.89 and 1.21 A. Between molecules the shortest atomic separations are N–C = 3.3 A., N–CH$_2$ = 3.36 A., and C–C = 3.31 A. The reliability index R = 0.110 if all the observed reflections are used.

XIV,c43. Crystals of *bis acetylacetone beryllium*, $Be[OC(CH_3)CHC-(CH_3)O]_2$, are monoclinic with a tetramolecular unit of the dimensions:

$$a_0 = 13.45 \text{ A.}; \quad b_0 = 11.30 \text{ A.}; \quad c_0 = 7.74 \text{ A.}; \quad \beta = 100°48'$$

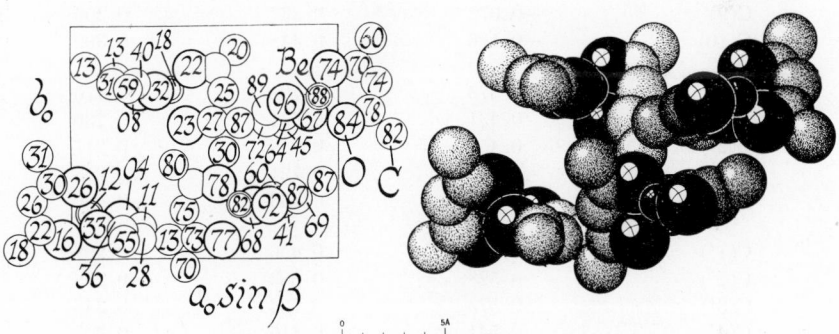

Fig. XIVC,58a (left). The monoclinic structure of $Be[OC(CH_3)CHC(CH_3)O]_2$ projected along its c_0 axis. Right-hand axes.

Fig. XIVC,58b (right). A packing drawing of the monoclinic structure of $Be[OC(CH_3)CHC(CH_3)O]_2$ seen along its c_0 axis. Atoms of oxygen are the large, those of beryllium the small black circles. The carbon atoms are short-line-and-dot shaded.

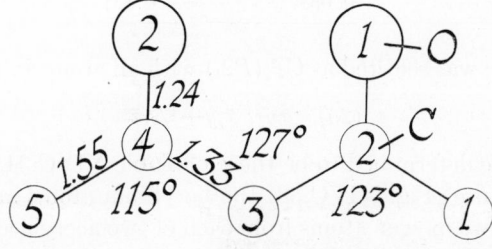

Fig. XIVC,59. Mean bond lengths in the crystallographically different acetylacetone molecules in their compound with beryllium.

TABLE XIVC, 34. Parameters of the Atoms in bis Acetylacetone Beryllium

Atom	x	y	z
Be(1)	0.068	0.195	0.122
O(1)	0.184	0.165	0.040
O(2)	0.101	0.133	0.329
O(3)	−0.033	0.090	0.160
O(4)	0.028	0.313	0.261
C(1)	0.192	0.110	0.553
C(2)	0.189	0.128	0.356
C(3)	0.262	0.093	0.278
C(4)	0.260	0.139	0.113
C(5)	0.372	0.091	0.126
C(6)	−0.201	0.038	0.184
C(7)	−0.119	0.135	0.216
C(8)	−0.150	0.253	0.258
C(9)	−0.072	0.333	0.296
C(10)	−0.126	0.448	0.309
Be(1′)	0.373	0.708	0.176
O(1′)	0.431	0.571	0.230
O(2′)	0.451	0.818	0.217
O(3′)	0.313	0.721	0.318
O(4′)	0.253	0.733	0.080
C(1′)	0.576	0.456	0.298
C(2′)	0.528	0.583	0.268
C(3′)	0.573	0.703	0.248
C(4′)	0.541	0.816	0.225
C(5′)	0.619	0.906	0.203
C(6′)	0.210	0.768	0.589
C(7′)	0.242	0.771	0.398
C(8′)	0.158	0.786	0.314
C(9′)	0.176	0.798	0.133
C(10′)	0.058	0.831	0.126

The space group was selected as C_2^2 ($P2_1$) with all atoms in the positions:

$$(2a) \quad xyz; \; \bar{x}, y+^1/_2, \bar{z}$$

Parameters were determined to be those of Table XIVC,34.

In this structure (Fig. XIVC,58) the beryllium atoms are tetrahedrally surrounded by two oxygen atoms from each of two acetylacetone molecules, with Be–O lying between 1.68 and 1.73 A. The mean bond lengths in the four crystallographically different planar molecules per cell are those of Figure XIVC,59.

XIV,c44. *Bis acetylacetone zinc monohydrate,* $Zn(C_5H_7O_2)_2 \cdot H_2O$, is monoclinic with a bimolecular cell of the dimensions:

$$a_0 = 10.480 \text{ A.}; \quad b_0 = 5.370 \text{ A.}; \quad c_0 = 10.935 \text{ A.}; \quad \beta = 93°48'$$

The space group was chosen as C_2^2 ($P2_1$) with all atoms in the positions:

$$(2a) \quad xyz; \; \bar{x}, y + {}^1\!/_2, \bar{z}$$

TABLE XIVC,35. Parameters of the Atoms in $Zn(C_5H_7O_2)_2 \cdot H_2O$

Atom	x	y	z
Zn	0.0892	0.0000	0.2156
$H_2O(1)$	0.0757	−0.2717	0.0904
O(2)	0.1961	0.2675	0.1410
O(3)	0.2507	−0.1235	0.3107
O(4)	−0.0144	−0.1198	0.3517
O(5)	−0.0610	0.2122	0.1577
C(2)	0.3746	0.4640	0.0583
C(3)	0.4678	−0.1889	0.3708
C(4)	−0.1584	−0.1246	0.5127
C(5)	−0.2346	0.5040	0.1733
C(22)	0.3203	0.2755	0.1400
C(23)	0.4021	0.1200	0.2125
C(33)	0.3667	−0.0596	0.2929
C(44)	−0.1139	−0.0261	0.3936
C(45)	−0.1822	0.1704	0.3354
C(55)	−0.1530	0.2806	0.2246
H(23)	0.503	0.181	0.214
H(2,2)	0.410	0.650	0.090
H(2,3)	0.330	0.500	−0.010
H(2,4)	0.500	0.430	0.040
H(3,5)	0.460	−0.375	0.350
H(3,6)	0.430	−0.160	0.440
H(3,7)	0.550	−0.150	0.350
H(45)	−0.260	0.256	0.384
H(5,9)	−0.280	0.380	0.090
H(5,10)	−0.180	0.670	0.200
H(5,11)	−0.370	0.524	0.230
H(4,12)	−0.180	−0.140	0.520
H(4,13)	−0.170	−0.325	0.560
H(4,14)	−0.120	−0.050	0.550
$H(H_2O,15)$	0.040	−0.225	0.010
$H(H_2O,16)$	0.160	−0.400	0.080

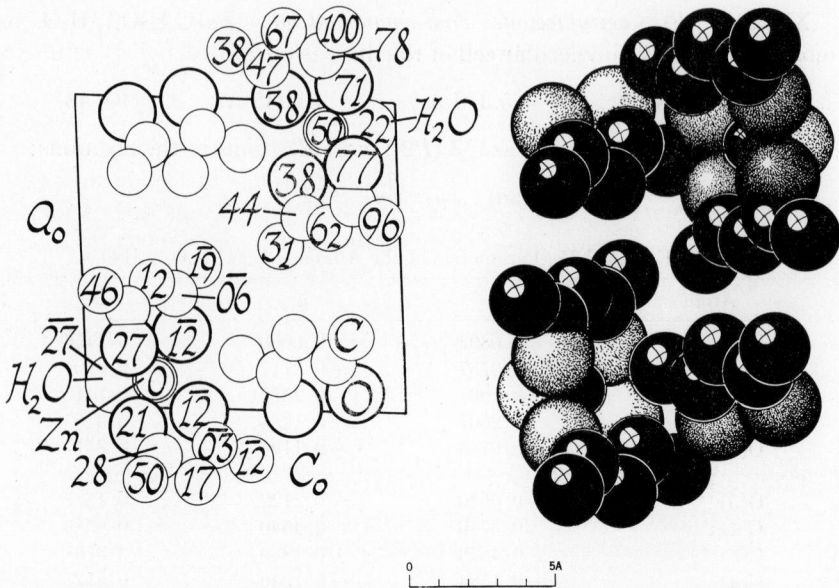

Fig. XIVC,60a (left). The monoclinic structure of $Zn(C_5H_7O_2)_2 \cdot H_2O$ projected along its b_0 axis. Right-hand axes.

Fig. XIVC,60b (right). A packing drawing of the monoclinic structure of $Zn(C_5H_7O_2)_2 \cdot H_2O$ seen along its b_0 axis. Both the carbon atoms with their circular highlights and the somewhat smaller zinc atoms are black. The oxygen atoms of the molecule are heavily outlined and short-line shaded; those of water are dotted and less strongly ringed.

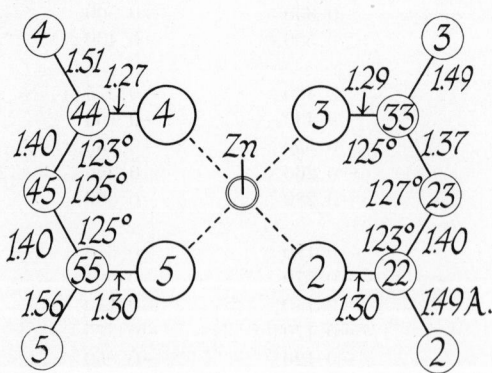

Fig. XIVC,61. Bond dimensions in crystals of $Zn(C_5H_7O_2)_2 \cdot H_2O$.

The parameters as given in the more recent study of this crystal are those of Table XIVC,35, the hydrogen positions being considered as approximate only.

The resulting structure is shown in Figure XIVC,60. Atoms of zinc are equally near to five oxygen atoms, one of which is from water; all Zn–O atomic separations are 2.02 A. The bond dimensions are given in Figure XIVC,61. The acetylacetone groups are almost planar, but each is turned ca. 12° from the normal to the Zn–H_2O line. The final $R = 0.072$.

In the earlier study the a_0 and c_0 axes are exchanged with respect to the foregoing. If the same axial sequence is used, there is approximate agreement with the parameters of Table XIVC,35, allowance being made for the choice of a different origin along b_0.

XIV,c45. *Bis acetylacetone cobalt(II) dihydrate,* $Co[OC(CH_3)CHC-(CH_3)O]_2 \cdot 2H_2O$, is monoclinic with a bimolecular unit of the dimensions:

$$a_0 = 10.91 \pm 0.04 \text{ A.}; \quad b_0 = 5.39 \pm 0.01 \text{ A.}; \quad c_0 = 11.19 \pm 0.03 \text{ A.}$$
$$\beta = 106°0' \pm 6'$$

The space group is C_{2h}^5 ($P2_1/c$) with the cobalt atoms in

$$(2a) \quad 000; 0\,{}^1/_2\,{}^1/_2$$

and all the other atoms in

$$(4e) \quad \pm(xyz; x,{}^1/_2-y,z+{}^1/_2)$$

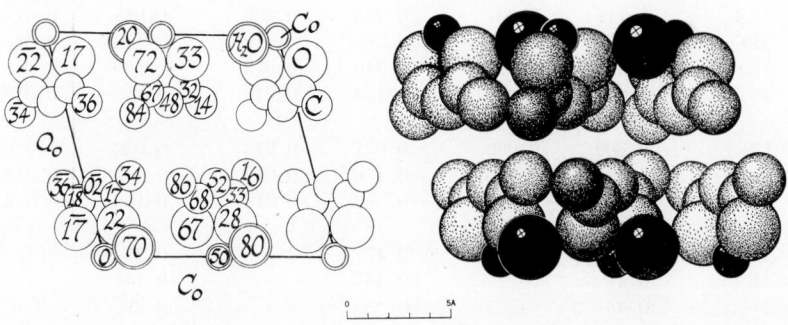

Fig. XIVC,62a (left). The monoclinic structure of $Co(C_5H_7O_2)_2 \cdot 2H_2O$ projected along its b_0 axis. Right-hand axes.

Fig. XIVC,62b (right). A packing drawing of the monoclinic structure of $Co(C_5H_7O_2)_2 \cdot 2H_2O$ seen along its b_0 axis. The water molecules are the large, the copper atoms the small black circles. Oxygen atoms are heavily outlined and hook shaded; the carbon atoms are fine-line shaded.

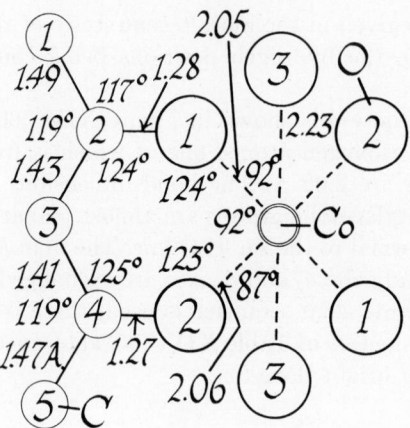

Fig. XIVC,63. Bond dimensions in crystals of Co(C₅H₇O₂)₂·2H₂O.

The parameters are stated in Table XIVC,36, the positions of hydrogen being chosen to give C–H = 1.09 A.

The structure is shown in Figure XIVC,62. The molecules of which it is built have the bond dimensions of Figure XIVC,63. In the acetylacetone group, C(2)–C(5), O(1), and O(2) are coplanar but C(1) lies 0.20 A. out-

TABLE XIVC,36. Parameters of the Atoms in Co(C₅H₇O₂)₂·2H₂O

Atom	x	σ_x, A.	y	σ_y, A.	z	σ_z, A.
O(1)	0.145	0.007	0.219	0.008	0.099	0.011
O(2)	0.117	0.007	−0.175	0.009	−0.088	0.011
H₂O(3)	0.050	0.008	−0.297	0.010	0.142	0.011
C(1)	0.356	0.010	0.340	0.013	0.199	0.016
C(2)	0.263	0.009	0.174	0.013	0.113	0.015
C(3)	0.309	0.009	−0.021	0.013	0.050	0.016
C(4)	0.238	0.010	−0.176	0.012	−0.045	0.015
C(5)	0.305	0.010	−0.362	0.013	−0.100	0.016
H(1)	0.31	—	0.47	—	0.25	—
H(2)	0.41	—	0.45	—	0.14	—
H(3)	0.43	—	0.23	—	0.27	—
H(4)	0.41	—	−0.03	—	0.07	—
H(5)	0.23	—	−0.47	—	−0.16	—
H(6)	0.37	—	−0.27	—	−0.15	—
H(7)	0.36	—	−0.48	—	−0.03	—
H(8)	0.13	—	−0.41	—	0.18	—
H(9)	0.00	—	−0.41	—	0.19	—

side this plane. The cobalt atom is 0.42 A. from it; its total coordination is a tetragonal distorted octahedron. There are hydrogen bonds tying the molecules into layers parallel to (100); their lengths are O(1)–H–O(3) = 2.90 A. and O(2)–H–O(3) = 2.91 A. The shortest intermolecular separations not involving a hydrogen bond are O–O(3) = 3.34 A., O(3)–CH₃ =

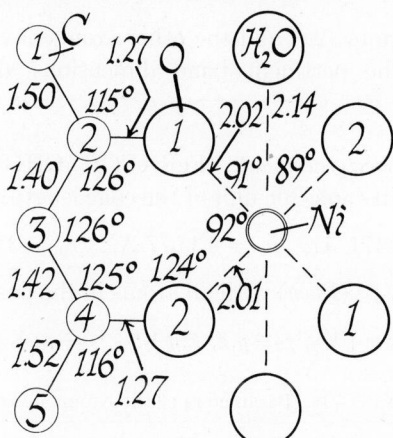

Fig. XIVC,64. Bond dimensions in crystals of Ni(C₅H₇O₂)₂·2H₂O.

TABLE XIVC,37. Parameters of the Atoms in Ni(C₅H₇O₂)₂·2H₂O

Atom	x	y	z
O(1)	0.1445	0.2147	0.1013
O(2)	0.1137	−0.1741	−0.0861
H₂O(3)	0.0453	−0.2849	0.1390
C(1)	0.3564	0.3371	0.2024
C(2)	0.2613	0.1696	0.1122
C(3)	0.3081	−0.0154	0.0485
C(4)	0.2346	−0.1755	−0.0458
C(5)	0.3024	−0.3645	−0.1019
H(1,C 1)	0.344	0.508	0.166
H(2,C 1)	0.340	0.337	0.285
H(3,C 1)	0.445	0.278	0.212
H(4,C 5)	0.283	−0.323	−0.184
H(5,C 5)	0.272	−0.536	−0.091
H(6,C 5)	0.396	−0.355	−0.062
H(7,C 3)	0.401	−0.022	0.079
H(8,H₂O)	0.088	−0.438	0.145
H(9,H₂O)	0.015	−0.246	0.213

3.52 A., C–CH$_3$ = 3.53 A., and CH$_3$–CH$_3$ = 3.56 A., all within these layers; between layers such separations exceed 4.0 A.

The *nickel* compound, Ni(C$_5$H$_7$O$_2$)$_2$·2H$_2$O, is isostructural with

$$a_0 = 10.954 \pm 0.01 \text{ A.}; \quad b_0 = 5.361 \pm 0.006 \text{ A.}; \quad c_0 = 11.245 \pm 0.01 \text{ A.}$$
$$\beta = 106°48' \pm 9'$$

The nickel atoms are in (2a) and the other atoms have the parameters of Table XIVC,37. The pertinent bond dimensions are given in Figure XIVC,64.

XIV,c46. *Ferric acetylacetonate*, Fe[OC(CH$_3$)CHC(CH$_3$)O]$_3$, is ortho-rhombic with an eight-molecule unit of the edge lengths:

$$a_0 = 15.471 \text{ A.}; \quad b_0 = 13.577 \text{ A.}; \quad c_0 = 16.565 \text{ A.}$$

The space group is V$_h^{15}$ (*Pbca*) with all atoms in the positions:

(8c) $\pm (xyz; \; x+\frac{1}{2},\frac{1}{2}-y,\bar{z}; \; \bar{x},y+\frac{1}{2},\frac{1}{2}-z; \; \frac{1}{2}-x,\bar{y},z+\frac{1}{2})$

TABLE XIVC,38. Parameters of the Atoms in Fe(C$_5$H$_7$O$_2$)$_3$

Atom	x	y	z
Fe	0.136	0.277	0.250
O(1)	0.039	0.182	0.250
O(2)	0.075	0.350	0.332
O(3)	0.075	0.350	0.166
O(4)	0.228	0.372	0.250
O(5)	0.195	0.200	0.332
O(6)	0.195	0.200	0.166
C(1)	0.244	0.132	0.051
C(2)	0.187	0.210	0.091
C(3)	0.136	0.277	0.051
C(4)	0.083	0.344	0.091
C(5)	0.028	0.420	0.051
C(6)	0.979	0.414	0.427
C(7)	0.003	0.336	0.370
C(8)	0.952	0.254	0.323
C(9)	0.972	0.180	0.273
C(10)	0.911	0.095	0.254
C(11)	0.300	0.139	0.427
C(12)	0.266	0.220	0.370
C(13)	0.322	0.299	0.323
C(14)	0.297	0.369	0.273
C(15)	0.354	0.468	0.254

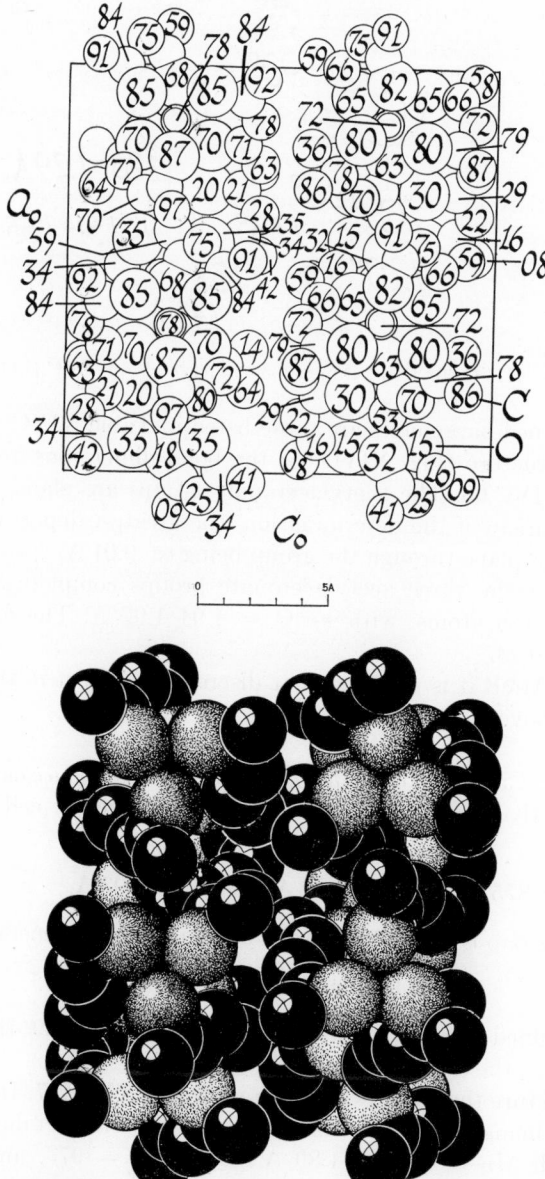

Fig. XIVC,65a (top). The orthorhombic structure of $Fe(C_5H_7O_2)_3$ projected along its b_0 axis. Iron atoms are the small, doubly ringed circles. Right-hand axes.

Fig. XIVC,65b (bottom). A packing drawing of the orthorhombic $Fe(C_5H_7O_2)_3$ arrangement seen along its b_0 axis. The carbon atoms are black; the oxygen atoms are line-and-dot shaded. Iron atoms scarcely show.

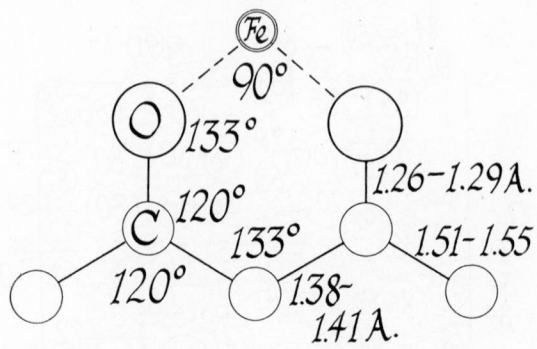

Fig. XIVC,66. Bond dimensions in crystals of $Fe(C_5H_7O_2)_3$.

The determined parameters are those listed in Table XIVC,38.

In this structure (Fig. XIVC,65) the bond dimensions are those shown in Figure XIVC,66. The acetylacetonate groups are planar to within the limit of accuracy of the determination, the greatest departure of an atom from the best plane through the group being ca. 0.01 A. Two oxygen atoms from each of the three acetylacetonate groups complete an octahedron around the iron atoms, with Fe–O = 1.94–1.97 A. The final R for this structure is 0.23.

In 1964: M&B it is reported that discrepancies exist in the structure as described above.

XIV,c47. Crystals of *manganese(III) tris acetylacetonate*, $Mn(CH_3-COCHCOCH_3)_3$, are monoclinic with a tetramolecular cell of the dimensions:

$$a_0 = 13.875 \text{ A.}; \quad b_0 = 7.467 \text{ A.}; \quad c_0 = 16.203 \text{ A.}; \quad \beta = 98°25.6'$$

The space group is C_{2h}^5 ($P2_1/c$) with all atoms in the general positions:

$$(4e) \quad \pm(xyz; \; x,^1/_2-y,z+^1/_2)$$

The determined parameters are listed in Table XIVC,39 those for hydrogen involving the assumption that C–H = 1.085 A.

The structure that results is shown in Figure XIVC,67. In each molecule the central manganese atom has around it a distorted octahedron of oxygen atoms, with Mn–O = 1.86–1.89 A., O–Mn–O = 97°, and Mn–O–C = 122–124°. In the acetylacetonate portion, atomic separations are C–O = 1.25–1.31 A., C–CH$_3$ = 1.49–1.55 A., and C–CH = 1.34–1.40 A. The final $R = 0.088$.

TABLE XIVC,39
Parameters of the Atoms in Mn(CH₃COCHCOCH₃)₃

Parameters of the Atoms in Mn($CH_3COCHCOCH_3$)$_3$

Atom	x	y	z
Mn	0.2418	0.2700	0.4687
O(1)	0.1235	0.3400	0.4081
O(2)	0.1946	0.1938	0.5645
O(3)	0.3660	0.1890	0.5190
O(4)	0.3023	0.3568	0.3804
O(5)	0.2189	0.0371	0.4271
O(6)	0.2516	0.5002	0.5141
C(1)	0.0424	0.3165	0.4322
C(1′)	−0.0434	0.3746	0.3683
C(12)	0.0263	0.2399	0.5075
C(2)	0.1020	0.1863	0.5690
C(2′)	0.0788	0.1212	0.6540
C(3)	0.4305	0.2919	0.5590
C(3′)	0.5254	0.2009	0.5887
C(36)	0.4150	0.4701	0.5797
C(6)	0.3291	0.5598	0.5580
C(6′)	0.3208	0.7555	0.5887
C(5′)	0.2085	−0.2057	0.3358
C(4)	0.3085	0.2633	0.3126
C(4′)	0.3536	0.3606	0.2477
C(45)	0.2760	0.0893	0.3007
C(5)	0.2353	−0.0130	0.3537
H(12)	−0.047	0.214	0.518
H(36)	0.473	0.543	0.615
H(45)	0.286	0.031	0.241
H(1′1)	−0.111	0.357	0.389
H(1′2)	−0.040	0.300	0.312
H(1′3)	−0.033	0.516	0.355
H(2′1)	0.001	0.118	0.657
H(2′2)	0.107	−0.008	0.665
H(2′3)	0.111	0.214	0.700
H(3′1)	0.579	0.288	0.622
H(3′2)	0.553	0.150	0.535
H(3′3)	0.511	0.094	0.629
H(4′1)	0.358	0.284	0.193

(continued)

TABLE XIVC,39 (continued)

Atom	x	y	z
H(4'2)	0.309	0.477	0.231
H(4'3)	0.425	0.401	0.275
H(5'1)	0.223	-0.245	0.275
H(5'2)	0.133	-0.221	0.339
H(5'3)	0.251	-0.286	0.382
H(6'1)	0.386	0.802	0.625
H(6'2)	0.303	0.837	0.535
H(6'3)	0.263	0.757	0.626

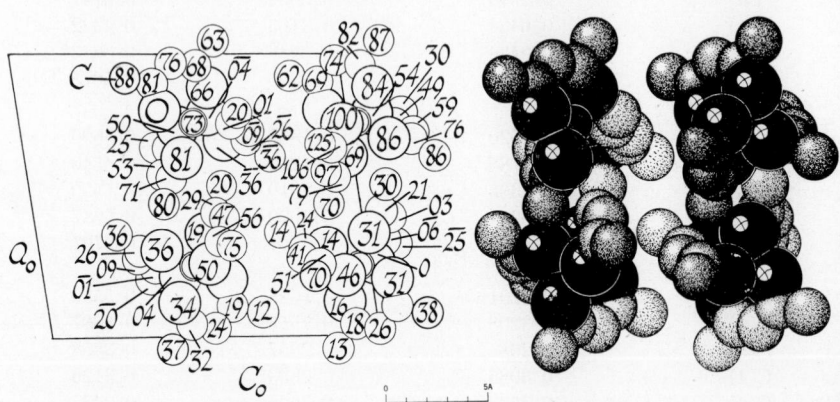

Fig. XIVC,67a (left). The monoclinic structure of $Mn(C_5H_7O_2)_3$ projected along its b_0 axis. Atoms of manganese are the small, doubly ringed circles. Right-hand axes.

Fig. XIVC,67b (right). A packing drawing of the monoclinic $Mn(C_5H_7O_2)_3$ arrangement viewed along its b_0 axis. The oxygen atoms are black; the carbons are short-line-and-dot shaded. Atoms of manganese do not show.

Structures have been published for the corresponding *chromium* and *cobalt* compounds which might be expected to be isostructural. The cell data are similar to those described above and the space group is the same $P2_1/c$, but the atomic parameters lead to significantly different atomic arrangements. For these two substances the cell dimensions are

$Cr(C_5H_7O_2)_3$:

$a_0 = 13.80$ A.; $b_0 = 7.58$ A.; $c_0 = 16.44$ A.; $\beta = 99°30'$

$Co(C_5H_7O_2)_3$:

$a_0 = 14.16$ A.; $b_0 = 7.48$ A.; $c_0 = 16.43$ A.; $\beta = 98°41'$

The parameters assigned the atoms in these two compounds are stated in Table XIVC,40.

TABLE XIVC,40
Parameters of the Atoms in $Cr(C_5H_7O_2)_3$ and for $Co(C_5H_7O_2)_3$
[in parentheses][a]

Atom	x	y	z
Cr	0.239 (0.272)	0.246 (0.168)	0.219 (0.234)
O(1)	0.290 (0.331)	0.026 (0.051)	0.195 (0.145)
O(2)	0.178 (0.202)	0.465 (0.276)	0.230 (0.140)
O(3)	0.192 (0.167)	0.155 (0.052)	0.314 (0.270)
O(4)	0.116 (0.204)	0.176 (0.281)	0.156 (0.332)
O(5)	0.288 (0.332)	0.334 (0.064)	0.127 (0.335)
O(6)	0.362 (0.374)	0.314 (0.323)	0.284 (0.221)
C(1)	0.114 (0.373)	0.070 (−0.012)	0.324 (0.051)
C(2)	0.104 (0.314)	0.024 (0.143)	0.415 (0.080)
C(3)	0.040 (0.262)	0.043 (0.272)	0.258 (0.052)
C(4)	0.044 (0.210)	0.099 (0.369)	0.176 (0.084)
C(5)	−0.048 (0.163)	0.059 (0.522)	0.116 (0.053)
C(6)	0.341 (0.116)	0.264 (0.402)	0.078 (0.452)
C(7)	0.383 (0.133)	0.325 (0.252)	0.007 (0.390)
C(8)	0.363 (0.082)	0.086 (0.163)	0.084 (0.364)
C(9)	0.354 (0.102)	−0.020 (0.030)	0.150 (0.312)
C(10)	0.382 (0.043)	−0.216 (−0.124)	0.140 (0.292)
C(11)	0.386 (0.432)	0.464 (−0.030)	0.318 (0.421)
C(12)	0.490 (0.392)	0.490 (0.146)	0.364 (0.381)
C(13)	0.308 (0.456)	0.594 (0.234)	0.312 (0.330)
C(14)	0.209 (0.426)	0.599 (0.380)	0.278 (0.286)
C(15)	0.155 (0.483)	0.771 (0.521)	0.278 (0.236)

[a] There is no assurance that the carbon and oxygen atoms in the two compounds have been correspondingly numbered.

From early work it would appear that the aluminum compound, Al-$(C_5H_7O_2)_3$, is isostructural. Its atomic arrangement has not been investigated, but its cell dimensions are the following:

$$a_0 = 14.25 \text{ A.}; \quad b_0 = 7.68 \text{ A.}; \quad c_0 = 16.17 \text{ A.}; \quad \beta = 99°22'$$

Evidently further work on these last three crystals is to be desired.

XIV,c48. *Zirconium bis acetylacetonate*, $Zr(C_5H_7O_2)_2$, forms monoclinic crystals. Its tetramolecular unit has the dimensions:

$$a_0 = 19.86 \pm 0.02 \text{ A.}; \quad b_0 = 8.38 \pm 0.01 \text{ A.}; \quad c_0 = 14.14 \pm 0.02 \text{ A.}$$
$$\beta = 102°50' \pm 10'$$

518   CRYSTAL STRUCTURES

The space group is C_{2h}^6 ($I2/c$) with zirconium atoms in the positions:

$$(4e) \quad \pm (0\ u\ ^1/_4;\ ^1/_2,^1/_2-u,^1/_4)$$
$$\text{with } u = -0.06556,\ \sigma = 0.00008$$

and all other atoms in the general positions:

$$(8f) \quad \pm (xyz;\ x,\bar{y},z+^1/_2;\ x+^1/_2,y+^1/_2,z+^1/_2;\ x+^1/_2,^1/_2-y,z)$$

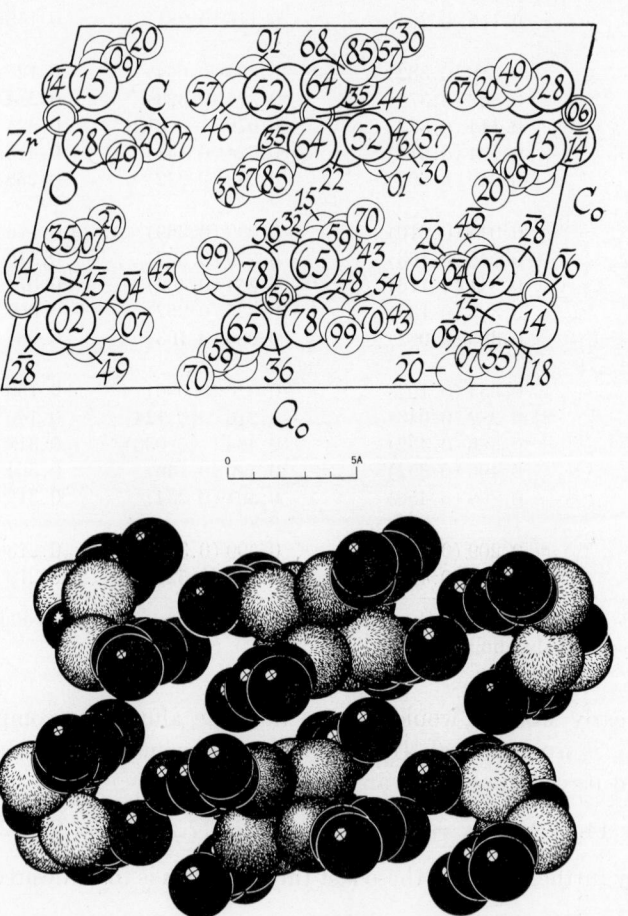

Fig. XIVC,68a (top). The monoclinic structure of $Zr(C_5H_7O_2)_2$ projected along its b_0 axis. Right-hand axes.

Fig. XIVC,68b (bottom). A packing drawing of the monoclinic structure of Zr-$(C_5H_7O_2)_2$ seen along its b_0 axis. Carbon atoms are the larger, zirconium atoms the somewhat smaller black circles. Atoms of oxygen are heavily outlined and line shaded.

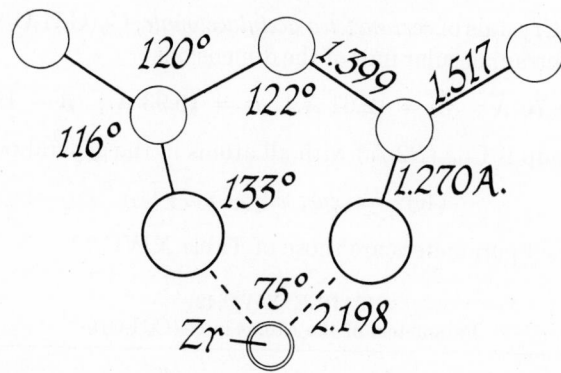

Fig. XIVC,69. Average bond dimensions in crystals of $Zr(C_5H_7O_2)_2$.

with the parameters of Table XIVC,41.

TABLE XIVC,41
Parameters of Atoms in $Zr(C_5H_7O_2)_2$

Atom	x	$\sigma(x)$	y	$\sigma(y)$	z	$\sigma(z)$
O(1)	0.0583	0.0003	−0.2761	0.0008	0.3197	0.0005
O(2)	0.0080	0.0003	0.1420	0.0007	0.1580	0.0002
O(3)	0.1051	0.0003	0.0220	0.0007	0.3082	0.0004
O(4)	0.0513	0.0003	−0.1550	0.0006	0.1393	0.0003
C(1)	0.1212	0.0005	−0.3161	0.0009	0.3314	0.0005
C(2)	0.0509	0.0004	0.1820	0.0009	0.1073	0.0005
C(3)	0.1743	0.0004	−0.2075	0.0014	0.3295	0.0006
C(4)	0.0949	0.0004	0.0729	0.0010	0.0752	0.0005
C(5)	0.1635	0.0003	−0.0435	0.0012	0.3213	0.0005
C(6)	0.0900	0.0003	−0.0919	0.0011	0.0883	0.0004
C(7)	0.2254	0.0005	0.0699	0.0019	0.3272	0.0015
C(8)	0.1290	0.0004	−0.2054	0.0012	0.0386	0.0005
C(9)	0.1353	0.0008	−0.4927	0.0015	0.3507	0.0010
C(10)	0.0519	0.0006	0.3542	0.0010	0.0769	0.0008

In this structure (Fig. XIVC,68) each zirconium atom is surrounded by eight oxygen atoms belonging to the four enveloping acetylacetone molecules, with Zr–O = 2.181–2.217 A.; their distribution is that of an approximately square antiprism. Within the ring formed by the zirconium atom and one of these molecules, the average bond dimensions are those of Figure XIVC,69; departures from these average values are scarcely significant.

XIV,c49. Crystals of *cerium tetra acetylacetonate*, $Ce(C_5H_7O_2)_4$, are monoclinic with a tetramolecular unit of the dimensions:

$$a_0 = 11.70 \text{ A.}; \quad b_0 = 12.64 \text{ A.}; \quad c_0 = 16.93 \text{ A.}; \quad \beta = 112°15'$$

The space group is C_{2h}^5 ($P2_1/c$) with all atoms in the general positions:

$$(4e) \quad \pm(xyz; \ x,{}^1\!/_2-y,z+{}^1\!/_2)$$

The determined parameters are those of Table XIVC,42.

TABLE XIVC,42
Parameters of the Atoms in $Ce(C_5H_7O_2)_4$

Atom	x	y	z
Ce	0.190	0.145	0.200
O(1)	0.332	0.145	0.128
O(2)	0.300	−0.019	0.235
O(3)	0.387	0.192	0.299
O(4)	0.188	0.121	0.338
O(5)	0.211	0.330	0.184
O(6)	0.005	0.232	0.194
O(7)	0.022	0.018	0.155
O(8)	0.054	0.182	0.054
C(1)	0.497	0.120	0.075
C(2)	0.425	0.085	0.130
C(3)	0.466	−0.001	0.187
C(4)	0.404	−0.052	0.233
C(5)	0.466	−0.154	0.282
C(6)	0.581	0.225	0.416
C(7)	0.442	0.200	0.380
C(8)	0.386	0.170	0.435
C(9)	0.268	0.129	0.416
C(10)	0.221	0.102	0.486
C(11)	0.194	0.521	0.163
C(12)	0.139	0.412	0.173
C(13)	0.019	0.412	0.177
C(14)	−0.045	0.323	0.186
C(15)	−0.180	0.321	0.180
C(16)	−0.158	−0.090	0.106
C(17)	−0.082	0.002	0.093
C(18)	−0.119	0.067	0.021
C(19)	−0.056	0.158	0.009
C(20)	−0.114	0.222	−0.073

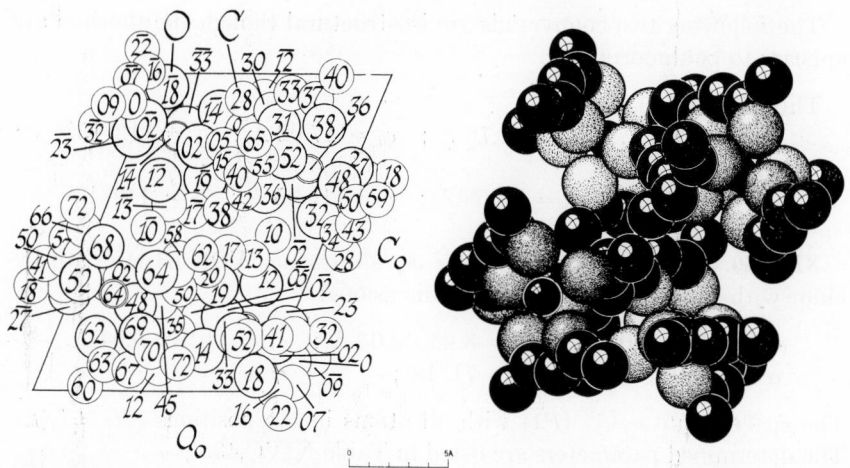

Fig. XIVC,70a (left). The monoclinic structure of Ce(C₅H₇O₂)₄ projected along its
b_0 axis. The cerium atoms are the small, doubly ringed circles. Right-hand axes.
Fig. XIVC,70b (right). A packing drawing of the monoclinic Ce(C₅H₇O₂)₄ arrangement viewed along its b_0 axis. The carbon atoms are the large, the cerium atoms the
somewhat smaller black circles. Atoms of oxygen are short-line-and-dot shaded.

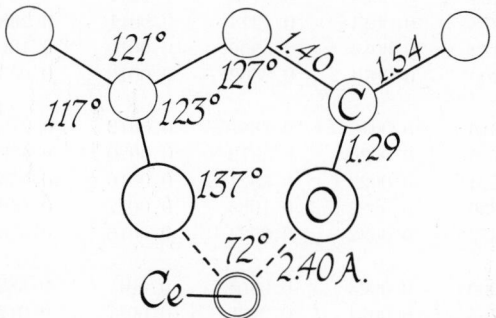

Fig. XIVC,71. Average bond dimensions in crystals of Ce(C₅H₇O₂)₄.

The atomic arrangement in the crystal is shown in Figure XIVC,70.
Each cerium atom is surrounded by an antiprism of eight oxygen atoms at
Ce–O distances lying between 2.36 and 2.43 A. The prism departs from
square by the fact that the O–O separation in a C₅H₇O₂ group is ca. 2.81 A.,
and between groups it is ca. 2.97 A. Bond dimensions in the essentially
planar C₅H₇O₂ molecules are shown in Figure XIVC,71. The final R =
0.17.

The following two compounds are isostructural though the thorium salt appears to be dimorphous:

$Th(C_5H_7O_2)_4$:

$a_0 = 11.72$ A.; $b_0 = 12.76$ A.; $c_0 = 17.02$ A.; $\beta = 112°15'$

$U(C_5H_7O_2)_4$:

$a_0 = 11.65$ A.; $b_0 = 12.68$ A.; $c_0 = 16.95$ A.; $\beta = 112°15'$

XIV,c50. Crystals of *vanadyl bis acetylacetonate*, $VO(C_5H_7O_2)_2$, are triclinic with a bimolecular cell of the dimensions:

$a_0 = 7.53 \pm 0.02$ A.; $b_0 = 8.23 \pm 0.03$ A.; $c_0 = 11.24 \pm 0.04$ A.

$\alpha = 73°0'$; $\beta = 71°18'$; $\gamma = 66°36'$

The space group is C_i^1 $(P\overline{1})$ with all atoms in the positions $(2i)$ $\pm(xyz)$. The determined parameters are listed in Table XIVC,43.

TABLE XIVC,43
Parameters of the Atoms in $VO(C_5H_7O_2)_2$

Atom	x	$\sigma(x)$	y	$\sigma(y)$	z	$\sigma(z)$
V(1)	0.1436	0.0004	0.2902	0.0003	0.2231	0.0002
O(2)	0.9404	0.0014	0.4142	0.0011	0.3592	0.0008
O(3)	0.9978	0.0014	0.1214	0.0011	0.2689	0.0008
O(4)	0.1472	0.0014	0.5359	0.0010	0.1367	0.0007
O(5)	0.2200	0.0014	0.2302	0.0010	0.0532	0.0007
O(6)	0.3316	0.0016	0.1887	0.0012	0.2793	0.0009
C(7)	0.2753	0.0026	0.5372	0.0020	0.4396	0.0014
C(8)	0.8224	0.0022	0.3521	0.0016	0.4561	0.0012
C(9)	0.7836	0.0022	0.1965	0.0017	0.4661	0.0012
C(10)	0.8672	0.0022	0.0924	0.0016	0.3700	0.0011
C(11)	0.7900	0.0023	0.9408	0.0017	0.3825	0.0012
C(12)	0.7765	0.0024	0.2166	0.0017	0.0125	0.0013
C(13)	0.2215	0.0022	0.5917	0.0016	0.0195	0.0011
C(14)	0.7039	0.0022	0.5090	0.0017	0.0751	0.0012
C(15)	0.7070	0.0021	0.6861	0.0015	0.0565	0.0011
C(16)	0.6307	0.0024	0.7855	0.0018	0.1659	0.0013

This is a molecular structure (Fig. XIVC,72) in which the individual molecules have the bond dimensions of Figure XIVC,73. The fivefold coordinated vanadium atom is at the approximate center of a nearly square pyramid.

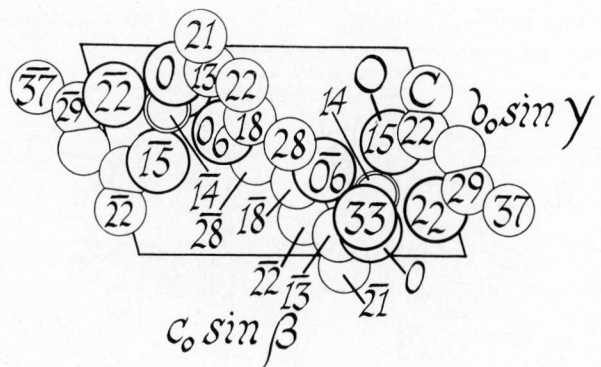

Fig. XIVC,72. The triclinic structure of $VO(C_5H_7O_2)_2$ projected along its a_0 axis. The vanadium atoms are the small, doubly ringed circles. Right-hand axes.

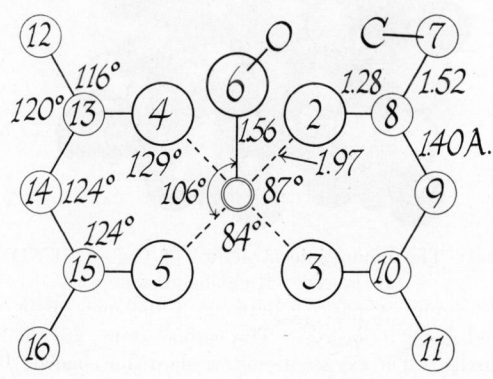

Fig. XIVC,73. Bond dimensions in crystals of $VO(C_5H_7O_2)_2$.

XIV,c51. Crystals of purple N,N'-*ethylene bis (acetylacetone iminato) copper(II)*, $Cu(C_6H_9NO)_2$, are monoclinic with a tetramolecular unit of the dimensions:

$$a_0 = 11.02 \pm 0.04 \text{ A.}; \quad b_0 = 8.97 \pm 0.03 \text{ A.}; \quad c_0 = 13.10 \pm 0.04 \text{ A.}$$
$$\beta = 94°0'$$

The space group is C_{2h}^5 $(P2_1/c)$ with all atoms in the general positions:

$$(4e) \quad \pm(xyz; \ x,{}^1/_2 - y, z + {}^1/_2)$$

Determined parameters are listed in Table XIVC,44.

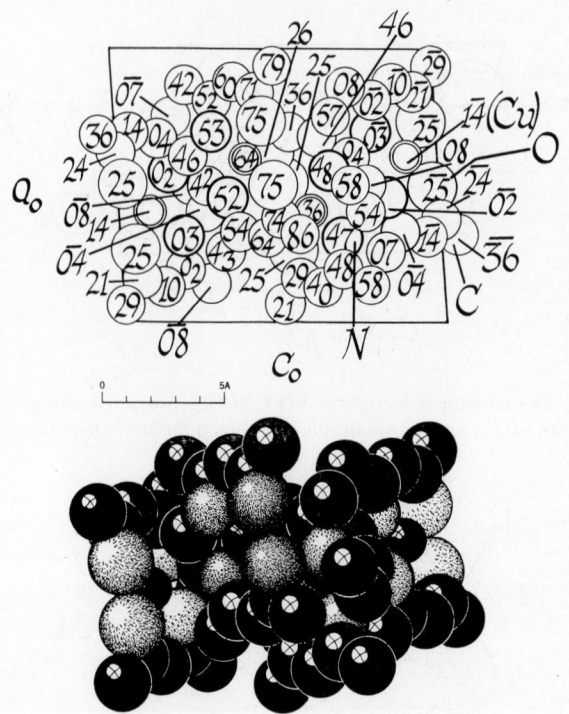

Fig. XIVC,74a (top). The monoclinic structure of Cu(—C₆H₉NO)₂ projected along its b_0 axis. Right-hand axes.

Fig. XIVC,74b (bottom). A packing drawing of the monoclinic structure of Cu-(—C₆H₉NO)₂ viewed along its b_0 axis. The carbon atoms are the larger, the coppers the smaller black circles. The oxygen atoms are short-line shaded; the nitrogen atoms are smaller, heavily ringed, and dotted.

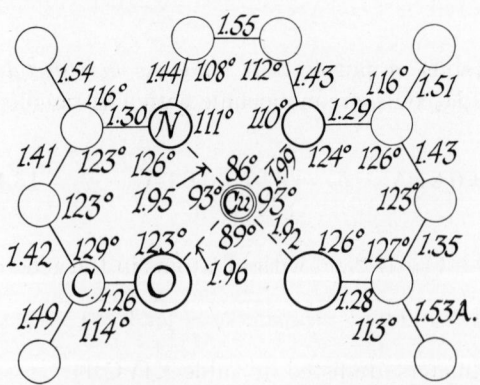

Fig. XIVC,75. Bond dimensions in crystals of Cu(—C₆H₉NO)₂.

TABLE XIVC,44
Parameters of the Atoms in $Cu(C_6H_9NO)_2$

Atom	x	y	z
Cu	0.4040	0.1376	0.1007
N(1)	0.3093	0.0270	0.1954
N(2)	0.5463	0.0233	0.1604
O(1)	0.2615	0.2465	0.0415
O(2)	0.5024	0.2530	0.0148
C(1)	0.3858	−0.0395	0.2760
C(2)	0.1912	0.0191	0.1929
C(3)	0.1375	−0.0840	0.2726
C(4)	0.1128	0.0970	0.1217
C(5)	0.1539	0.2099	0.0563
C(6)	0.0590	0.2910	−0.0083
C(7)	0.6166	0.2369	0.0055
C(8)	0.6727	0.3650	−0.0527
C(9)	0.6937	0.1421	0.0578
C(10)	0.6552	0.0400	0.1332
C(11)	0.7476	−0.0705	0.1780
C(12)	0.5070	−0.0845	0.2313

The structure is shown in Figure XIVC,74. The copper atom is coordinated with two oxygen and two nitrogen atoms of a molecule in an association that is essentially planar and has the bond dimensions of Figure XIVC,75. This molecule is flat but not strictly planar; each segment attached to the copper is planar, but these segments are tilted with respect to one another. Between molecules the shortest interatomic separations are of the order of 3.30 A. The final $R = 0.108$.

XIV,c52. Crystals of *nickel bis methyl ethyl glyoxime*, Ni[HONC-$(CH_3)C(C_2H_5)NO]_2$, are monoclinic with a bimolecular unit of the dimensions:

$$a_0 = 4.75 \pm 0.01 \text{ A.}; \quad b_0 = 11.75 \pm 0.03 \text{ A.}; \quad c_0 = 11.97 \pm 0.03 \text{ A.}$$
$$\beta = 92°0'$$

The space group is C_{2h}^5 ($P2_1/c$) with the nickel atoms in

$$(2a) \quad 000; \, 0\,^1/_2\,^1/_2$$

All other atoms are in the general positions:

$$(4e) \quad \pm(xyz; \, x,^1/_2-y,z+^1/_2)$$

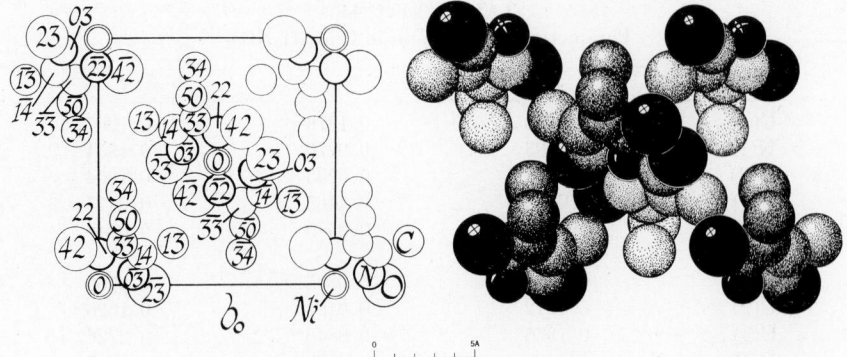

Fig. XIVC,76a (left). The monoclinic structure of nickel bis methyl ethyl glyoxime projected along its a_0 axis. Right-hand axes.

Fig. XIVC,76b (right). A packing drawing of the monoclinic structure of Ni[HONC-$(CH_3)C(C_2H_5)NO]_2$ viewed along its a_0 axis. Atoms of oxygen are the large, of nickel the smaller black circles. The carbon atoms are fine-line shaded; the nitrogen atoms are more heavily ringed and are dotted.

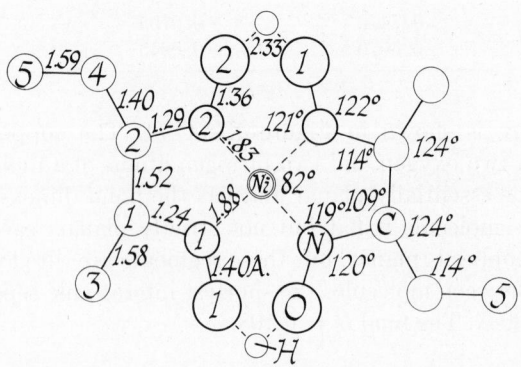

Fig. XIVC,77. Bond dimensions in crystals of nickel methyl ethyl glyoxime.

The assigned parameters are those of Table XIVC,45.

The atomic arrangement is shown in Figure XIVC,76. It is built up of planar molecules which have the bond dimensions of Figure XIVC,77. There is thus a nearly square coordination of four nitrogen atoms around the central metal atom. Between molecules the shortest atomic separations are given as C(2)–O(1) = 3.39 A., C(4)–N(1) = 3.41 A., and Ni–O(2) = 3.44 A. The short O(1)–O(2) between the two glyoxime components of a molecule (2.33 A.) presumably corresponds to a hydrogen bond.

TABLE XIVC,45. Parameters of Atoms in Nickel bis Methyl Ethyl Glyoxime

Atom	x	σ_x, A.	y	σ_y, A.	z	σ_z, A.
C(1)	0.136	0.020	0.185	0.005	0.120	0.034
C(2)	0.330	0.020	0.092	0.016	0.165	0.026
C(3)	0.136	0.020	0.311	0.022	0.164	0.032
C(4)	0.500	0.001	0.106	0.028	0.256	0.048
C(5)	0.342	0.003	0.090	0.036	0.371	0.021
N(1)	−0.030	0.005	0.152	0.005	0.045	0.004
N(2)	0.222	0.005	−0.001	0.018	0.112	0.008
O(1)	−0.231	0.002	0.228	0.010	−0.008	0.007
O(2)	0.422	0.002	−0.097	0.009	0.136	0.009

XIV,c53. Crystals of *cuprous nitrate bis glutaronitrile*, $[Cu(NC(CH_2)_3\text{-}CN)_2]NO_3$, are tetragonal with a bimolecular cell of the edge lengths:

$$a_0 = 8.25 \pm 0.01 \text{ A.}, \qquad c_0 = 9.71 \pm 0.01 \text{ A.}$$

The space group is V_d^4 ($P\bar{4}2_1c$) with atoms in the positions:

Cu: $(2a)$ $000; \frac{1}{2}\frac{1}{2}\frac{1}{2}$
N(2): $(2b)$ $0\ 0\ \frac{1}{2}; \frac{1}{2}\frac{1}{2}0$
C(3): $(4d)$ $0\ \frac{1}{2}\ u; \frac{1}{2}\ 0\ \bar{u}; 0,\frac{1}{2},u+\frac{1}{2}; \frac{1}{2},0,\frac{1}{2}-u$
 with $u = -0.362$

Other atoms are in the general positions:

$(8e)$ $xyz; \bar{x}\bar{y}z; \frac{1}{2}-x,y+\frac{1}{2},\frac{1}{2}-z; x+\frac{1}{2},\frac{1}{2}-y,\frac{1}{2}-z;$
 $\bar{y}x\bar{z}; y\bar{x}\bar{z}; y+\frac{1}{2},x+\frac{1}{2},z+\frac{1}{2}; \frac{1}{2}-y,\frac{1}{2}-x,z+\frac{1}{2}$

with the parameters:

Atom	x	y	z
N(1)	0.192	0.053	0.113
C(1)	0.297	0.088	0.183
C(2)	0.447	0.138	0.267

It is considered that the orientation of the nitrate ions must be disordered, with the oxygen atoms statistically distributed over a quarter of the positions of three sets of $(8e)$, the parameters being:

Atom	x	y	z
O(1)	0.150	0.000	0.500
O(2)	0.130	0.075	0.500
O(3)	0.075	0.130	0.500

As is usual, the cuprous atoms are tetrahedrally surrounded, in this case by nitrogen atoms from four different glutaronitrile groups, with Cu–N = 1.98 A. These groups have a twofold axis through the C(3) atom and the atoms C(2), C(1), N(1), and Cu are collinear. The distance C–N = 1.14 A. and C–C = 1.54 A.

XIV,c54. *Zinc bis(dipivaloylmethanido),* $Zn[(CH_3)_3CC(O)CC(O)C-(CH_3)_3]_2$, is tetragonal with a tetramolecular cell of the dimensions:

$$a_0 = 10.67 \text{ A.}, \qquad c_0 = 21.87 \text{ A.}$$

The space group has been chosen as C_{4h}^6 ($I4_1/a$) with atoms in the positions:

Zn: (4a) $0 \, ^1/_4 \, ^1/_8; 0 \, ^3/_4 \, ^7/_8$; B.C.
C(1): (8e) $\pm (0 \, ^1/_4 \, u; \, 0, ^1/_4, ^1/_4 - u)$; B.C.
          with $u = -0.0207$, $\sigma(u) = 0.0007$
H(1,C 1): (8e)   with $u = -0.068$

All other atoms are in the general positions:

(16f) $\pm (xyz; \qquad x,y+^1/_2,\bar{z};$
          $^3/_4 - y, x + ^1/_4, z + ^1/_4; \, y + ^1/_4, ^1/_4 - x, z + ^1/_4)$;   B.C.

with the parameters of Table XIVC,46.

TABLE XIVC,46
Parameters of Atoms in $Zn[(CH_3)_3CC(O)CC(O)C(CH_3)_3]_2$

Atom	x	$\sigma(x)$	y	$\sigma(y)$	z	$\sigma(z)$
O	0.1256	0.0006	−0.3029	0.0006	−0.0659	0.0003
C(2)	0.1087	0.0009	−0.2959	0.0008	−0.0081	0.0005
C(3)	0.2173	0.0011	−0.3417	0.0010	0.0309	0.0005
C(4)	0.3377	0.0016	−0.3578	0.0016	−0.0076	0.0007
C(5)	0.2428	0.0014	−0.2542	0.0014	0.0864	0.0007
C(6)	0.1792	0.0012	−0.4742	0.0012	0.0545	0.0006
H(2,C 4)	0.314	—	−0.408	—	−0.074	—
H(3,C 4)	0.357	—	−0.274	—	−0.026	—
H(4,C 4)	0.405	—	−0.395	—	0.022	—
H(5,C 5)	0.188	—	−0.157	—	0.079	—
H(6,C 5)	0.193	—	−0.288	—	0.127	—
H(7,C 5)	0.325	—	−0.213	—	0.092	—
H(8,C 6)	0.099	—	−0.466	—	0.089	—
H(9,C 6)	0.257	—	−0.523	—	0.079	—
H(10,C 6)	0.147	—	−0.536	—	0.023	—

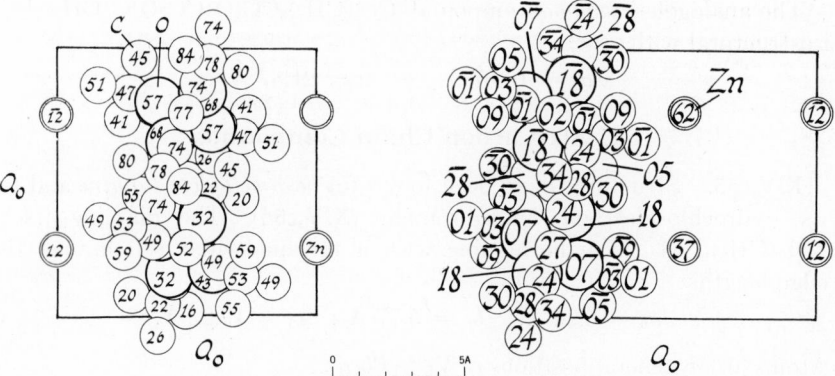

Fig. XIVC,78a (left). Half the molecules of zinc bis(dipivaloylmethanido) in its tetragonal unit projected along its c_0 axis. Right-hand axes.
Fig. XIVC,78b (right). The other half of the molecules of zinc bis(dipivaloylmethanido) in the tetragonal unit projected along c_0. Right-hand axes.

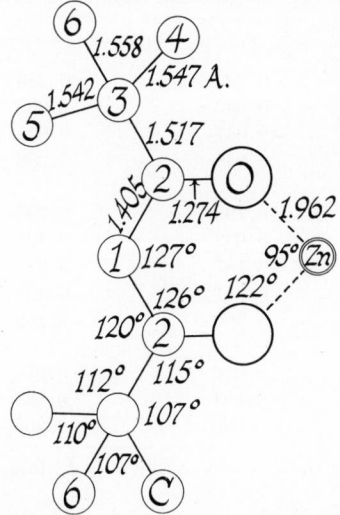

Fig. XIVC,79. Bond dimensions in crystals of $Zn[(CH_3)_3CC(O)CC(O)C(CH_3)_3]_2$.

The structure (Fig. XIVC,78) is built up of molecules having the bond dimensions of Figure XIVC,79. In them, the zinc atoms are tetrahedrally coordinated to four oxygen atoms, one of the dipivaloylmethanido parts being turned through 90° with respect to the other. The atoms O, C(1), C(2), and C(3) are coplanar.

The analogous *cobaltous* compound, $Co[(CH_3)_3CC(O)CC(O)C(CH_3)_3]_2$, is isostructural with

$$a_0 = 10.67 \text{ A.}, \qquad c_0 = 21.87 \text{ A.}$$

4. Six-Carbon Chain Compounds

XIV,c55. Structures have been found for *hexamethylenediamine* and for its hydrochloride and hydrobromide (**XIV,c56**). The diamine itself, $NH_2(CH_2)_6NH_2$, is orthorhombic with a tetramolecular cell having the edge lengths:

$$a_0 = 6.94 \text{ A.}; \quad b_0 = 5.77 \text{ A.}; \quad c_0 = 19.22 \text{ A.}$$

Atoms are in general positions of V_h^{15} (*Pbca*):

$$(8c) \quad \pm (xyz; \; x+^1/_2, ^1/_2-y, \bar{z}; \; \bar{x}, y+^1/_2, ^1/_2-z; \; ^1/_2-x, \bar{y}, z+^1/_2)$$

with the parameters of Table XIVC,47.

TABLE XIVC,47
Parameters of the Atoms in Hexamethylenediamine

Atom	x	y	z
C(1)	−0.030	0.286	0.145
C(2)	0.043	0.095	0.098
C(3)	−0.047	0.095	0.023
N	0.049	0.286	0.218
H(1)[N][a]	0.195	0.293	0.217
H(2)[N]	0.032	0.128	0.241
H(3)[C(1)]	−0.187	0.279	0.146
H(4)[C(1)]	0.011	0.451	0.121
H(5)[C(2)]	0.200	0.102	0.098
H(6)	0.016	−0.073	0.122
H(7)[C(3)]	−0.194	0.088	0.024
H(8)	0.000	0.260	−0.001

[a] Bracketed symbols refer to the atoms to which the hydrogen atoms are attached.

The resulting elongated planar molecule has the dimensions of Figure XIVC,80. The molecular packing in the crystal is indicated in Figure XIVC,81. The Fourier projections show definite indications of the hydrogen atoms. Using these and assuming C–H = 1.09 A. and N–H = 1.02 A., their parameters have been given values also listed in the table. The low melting point (42°C.) of this compound makes especially interesting the fact that even the hydrogen atoms are in positions prescribed by symmetry. The shortest intermolecular distance is N–N = 3.21 A.

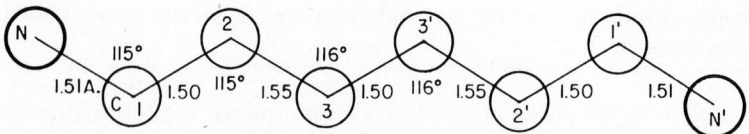

Fig. XIVC,80. Bond dimensions in the planar molecule of $NH_2(CH_2)_6NH_2$.

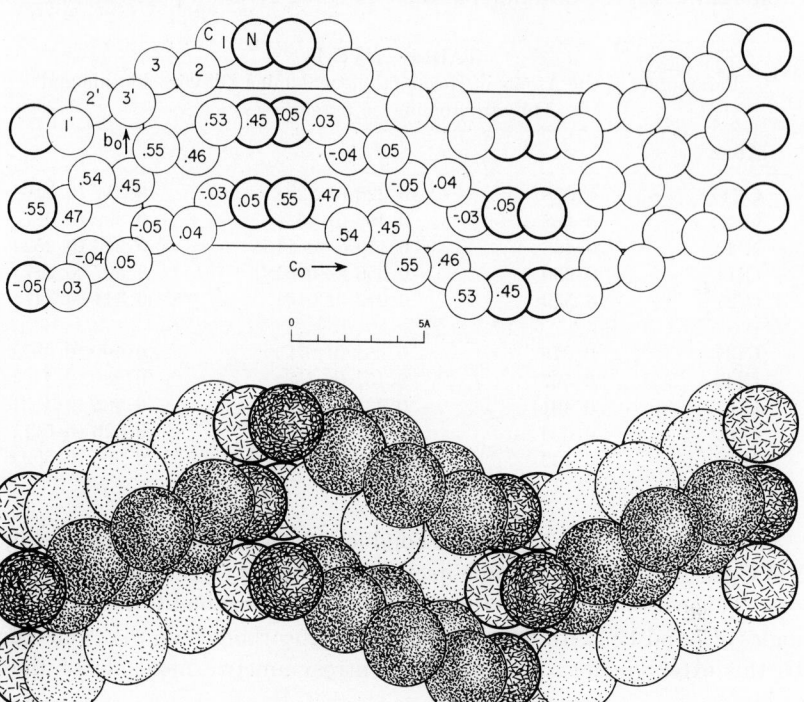

Fig. XIVC,81a (top). The orthorhombic structure of $NH_2(CH_2)_6NH_2$, projected along its a_0 axis.   Left-hand axes.

Fig. XIVC,81b (bottom). A packing drawing of the orthorhombic $NH_2(CH_2)_6NH_2$ arrangement seen along its a_0 axis.   The carbon atoms are dotted, the nitrogen atoms heavily outlined and line shaded.

XIV,c56.   Crystals of *hexamethylenediamine dihydrochloride*, $HCl \cdot NH_2$-$(CH_2)_6NH_2 \cdot HCl$, and of the corresponding *dihydrobromide*, are monoclinic. Their isomorphous tetramolecular units have the dimensions:

$(CH_2)_6(NH_3Br)_2$:

$a_0 = 4.68$ A.; $b_0 = 14.53$ A.; $c_0 = 16.21$ A.; $\beta = 91°$

$(CH_2)_6(NH_3Cl)_2$:

$a_0 = 4.60$ A.; $b_0 = 14.19$ A.; $c_0 = 15.68$ A.; $\beta = 90°48'$

The space group is C_{2h}^5 $(P2_1/c)$, and all atoms are in the general positions:

$$(4e) \quad \pm (xyz;\ x,{}^1\!/_2-y,z+{}^1\!/_2)$$

A complete set of parameters has been determined for the hydrochloride, and y and z but not x parameters for the hydrobromide. They are listed in Table XIVC,48, the bromide parameters being given in parentheses.

TABLE XIVC,48
Parameters of the Atoms in Hexamethylenediamine Dihydrochloride and
Dihydrobromide (in parentheses)

Atom	x	y	z
Cl(1)	0.078	0.226 (0.229)	0.217 (0.216)
Cl(2)	0.128	0.460 (0.462)	0.392 (0.387)
N(1)	0.468	−0.129 (−0.115)	0.243 (0.252)
C(1)	0.330	−0.056 (−0.048)	0.307 (0.317)
C(2)	0.539	0.012 (0.016)	0.344 (0.347)
C(3)	0.375	0.081 (0.079)	0.400 (0.397)
C(4)	0.567	0.157 (0.153)	0.436 (0.443)
C(5)	0.401	0.233 (0.223)	0.486 (0.483)
C(6)	0.603	0.303 (0.287)	0.526 (0.511)
N(2)	0.422	0.380 (0.390)	0.565 (0.552)

The cation that results has the bond lengths and angles shown in Figure XIVC,82. As with the diamine itself (**XIV,c55**), its C–N bond is unusually long. This cation is not strictly planar since the terminal C–N bonds make angles of 7 and 10° with the plane of the hydrocarbon chain that joins them. To this extent it deviates from the centrosymmetric molecule of the diamine.

The way the molecules pack in the crystal is illustrated in Figure XIVC,83. The shortest intermolecular distances are N–Cl separations that range from 3.10 A. upwards. The N(1) atom has three chlorine neighbors at distances of 3.10, 3.12, and 3.27 A., with a fourth chloride ion at 3.58 A. Around the N(2) atoms are three chloride ions (3.21, 3.25, and 3.26 A.) with a fourth nearly as close at 3.37 A. The chloride ions themselves are, as the figure indicates, grouped in octahedral clusters, the shortest Cl–Cl distance being 4.29 A.

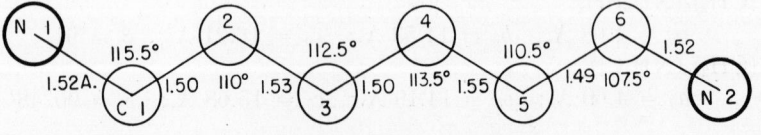

Fig. XIVC,82. Bond dimensions of the $[NH_3(CH_2)_6NH_3]^{2+}$ cation in its chloride.

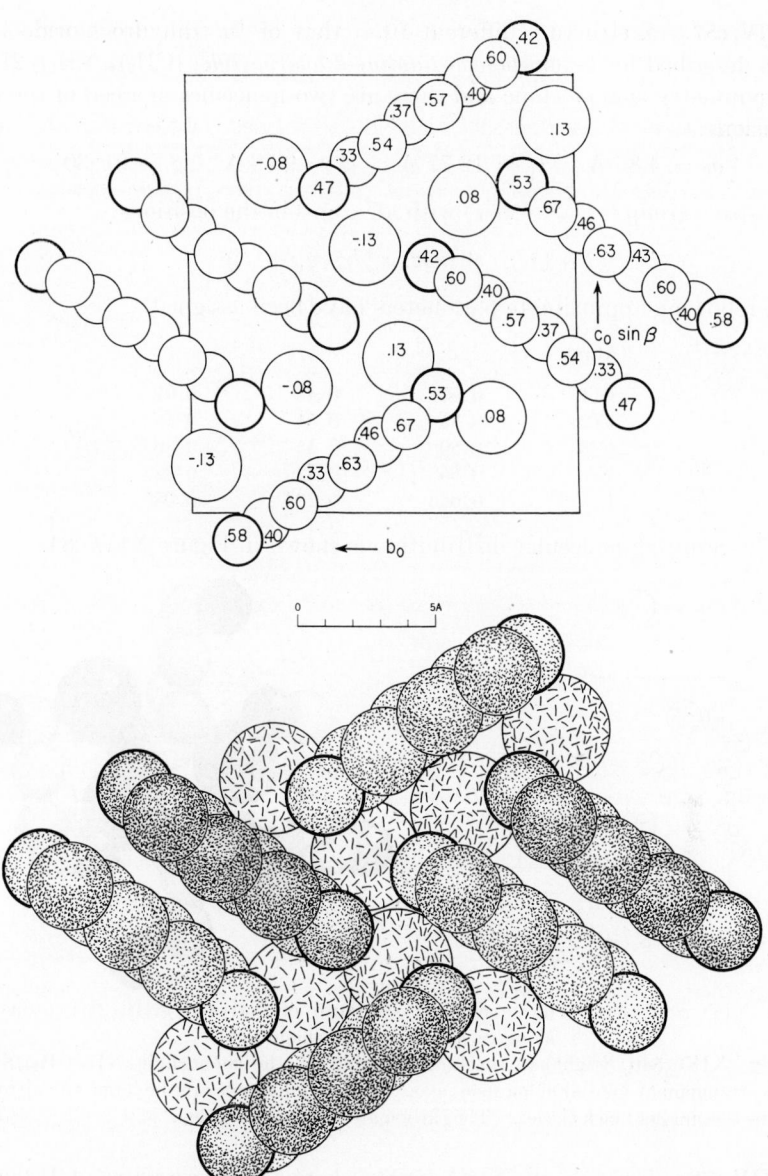

Fig. XIVC,83a (top). The monoclinic structure of $NH_2(CH_2)_6NH_2 \cdot 2HCl$ projected along its a_0 axis. The largest circles are chlorine. Of the smaller ones the nitrogen atoms are heavily ringed. Left-hand axes.

Fig. XIVC,83b (bottom). A packing drawing of the monoclinic structure of $NH_2(CH_2)_6NH_2 \cdot 2\,HCl$ viewed along its a_0 axis. The chlorine atoms are large and line shaded. Of the dotted circles the terminal nitrogen atoms are heavily ringed.

XIV,c57. A structure different from that of the dihydrochloride has been described for *hexamethylenediamine dihydroiodide*, $(CH_2)_6(NH_2)_2 \cdot 2HI$. Its symmetry is monoclinic and there are two molecules in a cell of the dimensions:

$$a_0 = 4.85 \text{ A.}; \quad b_0 = 12.77 \text{ A.}; \quad c_0 = 9.73 \text{ A.}; \quad \beta = 91°30'$$

The space group is C_{2h}^5 $(P2_1/c)$ with all atoms in the positions:

$$(4e) \quad \pm(xyz; \ x,{}^1/_2-y,z+{}^1/_2)$$

The following approximate parameters have been assigned:

Atom	x	y	z
C(1)	0.57	0.48	0.07
C(2)	0.35	0.51	0.19
C(3)	0.50	0.45	0.30
N	0.55	0.33	0.32
I	0.046	0.204	0.085

The resulting molecular distribution is shown in Figure XIVC,84.

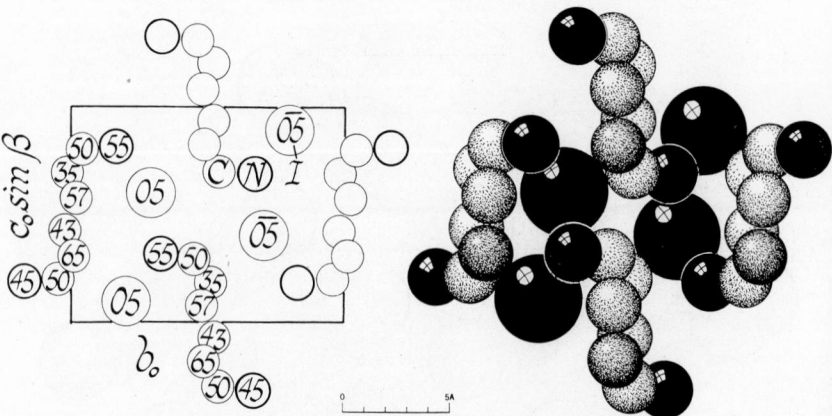

Fig. XIVC,84a (left). The monoclinic structure of $NH_2(CH_2)_6NH_2 \cdot 2HI$ projected along its a_0 axis. Right-hand axes.

Fig. XIVC,84b. (right). A packing drawing of the monoclinic $NH_2(CH_2)_6NH_2 \cdot$ 2HI arrangement viewed along its a_0 axis. The iodine atoms are the large, the nitrogen atoms the smaller black circles. The carbon atoms are line shaded.

XIV,c58. Crystals of *N,N'-diacetyl hexamethylenediamine*, $CH_3C(O)$-$NH(CH_2)_6NHC(O)CH_3$, are triclinic with a unimolecular cell of the dimensions:

$$a_0 = 12.35 \text{ A.}; \quad b_0 = 5.44 \text{ A.}; \quad c_0 = 4.93 \text{ A.}$$
$$\alpha = 116°37'; \quad \beta = 99°4'; \quad \gamma = 94°18'$$

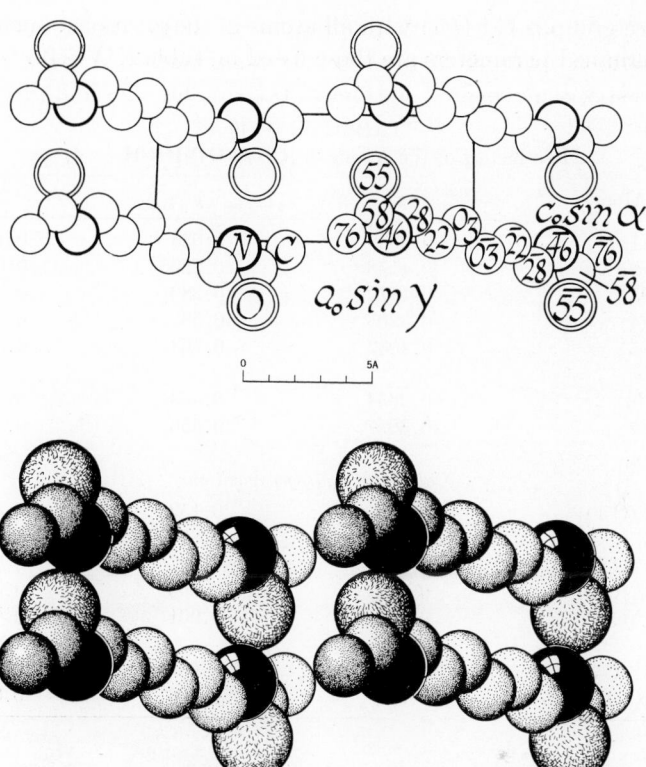

Fig. XIVC,85a (top). The triclinic structure of diacetyl hexamethylene diamine projected along its b_0 axis. Right-hand axes.

Fig. XIVC,85b (bottom). A packing drawing of the triclinic structure of $CH_3C(O)$-$NH(CH_2)_6NHC(O)CH_3$ seen along its b_0 axis. The nitrogen atoms are black, the oxygen atoms line shaded. The smaller carbon atoms are dotted.

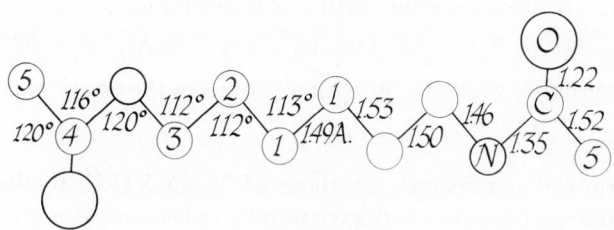

Fig. XIVC,86. Bond dimensions in the molecule of diacetyl hexamethylene diamine.

The space group is C_i^1 ($P\overline{1}$) with all atoms in the positions: $(2i)$ $\pm(xyz)$. The determined parameters are those listed in Table XIVC,49.

TABLE XIVC,49
Parameters of the Atoms in $[CH_3C(O)NH(CH_2)_3\text{—}]_2$

Atom	x	y	z
C(1)	0.0338	0.033	0.114
C(2)	0.1178	0.220	0.024
C(3)	0.1872	0.283	0.210
C(4)	0.3208	0.583	0.230
C(5)	0.4032	0.763	0.073
N(6)	0.2654	0.463	0.075
O(7)	0.3097	0.550	0.485
	Calculated hydrogen positions		
H'(C 1)	0.079	0.152	0.303
H''(C 1)	0.980	0.855	0.190
H'(C 2)	0.171	0.897	0.822
H''(C 2)	0.072	0.600	0.709
H'(C 3)	0.234	0.901	0.401
H''(C 3)	0.134	0.604	0.288
H(N)	0.282	0.506	0.866

The structure is shown in Figure XIVC,85. Its molecules, which have the bond dimensions of Figure XIVC,86, have a central coplanar hexamethylene group and two terminal groups which are planar and parallel, but inclined $11°28'$ to the central plane. Each of these contains the atoms —$CH_2NHC(O)CH_3$. The shortest intermolecular distances are C–C = 3.57 A., and C–O = 3.68 A.

XIV,c59. Crystals of *N,N'-hexamethylene bis propionamide*, $[C_2H_5C(O)\text{-}$ $NH(CH_2)_3\text{—}]_2$, are monoclinic with a bimolecular cell of the dimensions:

$$a_0 = 18.60 \text{ A.}; \quad b_0 = 4.96 \text{ A.}; \quad c_0 = 7.49 \text{ A.}; \quad \beta = 97°15'$$

The space group is C_{2h}^5 ($P2_1/a$) with all atoms in the positions:

$$(4e) \quad \pm(xyz;\ x+{}^1/_2,{}^1/_2-y,z)$$

The determined parameters are those of Table XIVC,50, the hydrogen positions having been chosen to give tetrahedral carbon atoms with C–H = 1.1 A. and trigonal nitrogen atoms with N–H = 1.0 A.

The resulting structure (Fig. XIVC,87) is made up of molecules that have the bond dimensions shown in Figure XIVC,88. The hexamethylene

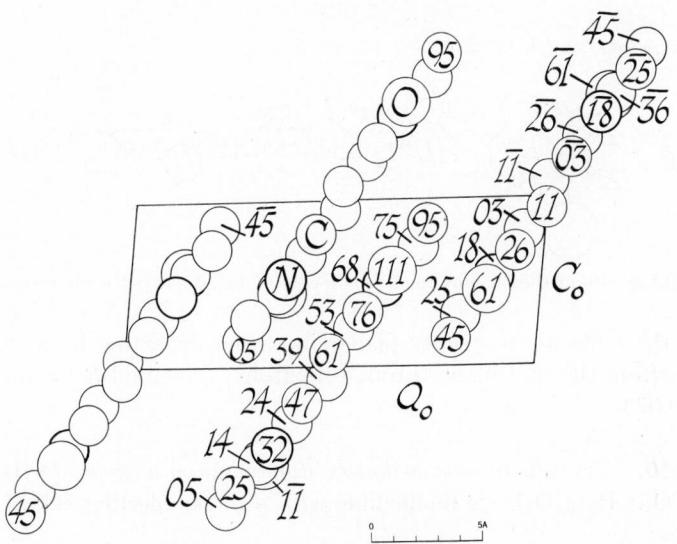

Fig. XIVC,87. The monoclinic structure of $[C_2H_5C(O)NH(CH_2)_3—]_2$ projected along its b_0 axis. Right-hand axes.

TABLE XIVC,50. Parameters of the Atoms in $[C_2H_5C(O)NH(CH_2)_3—]_2$

Atom	x	y	z	σ, A.
O	0.1329	0.6070	0.4490	0.0021
N	0.1125	0.1805	0.5323	0.0025
C(1)	0.0080	0.1151	0.9378	0.0031
C(2)	0.0551	0.0296	0.7939	0.0031
C(3)	0.0695	0.2652	0.6724	0.0031
C(4)	0.1413	0.3628	0.4314	0.0031
C(5)	0.1885	0.2476	0.2960	0.0031
C(6)	0.2087	0.4580	0.1626	0.0031
H(1)	0.044	0.807	0.130	0.051
H(2)	0.036	0.276	0.020	0.051
H(3)	0.027	0.869	0.711	0.051
H(4)	0.107	0.952	0.861	0.051
H(5)	0.018	0.346	0.608	0.051
H(6)	0.099	0.423	0.754	0.051
H(7)	0.121	0.984	0.512	0.051
H(8)	0.159	0.083	0.222	0.051
H(9)	0.239	0.168	0.371	0.051
H(10)	0.159	0.532	0.087	0.051
H(11)	0.239	0.617	0.237	0.051
H(12)	0.243	0.370	0.065	0.051

Fig. XIVC,88. Bond dimensions in the molecule of hexamethylene bis propionamide.

part of the molecule is strictly planar, no atom departing by as much as 0.002 A. from the best plane through its atoms. The final R for this structure is 0.073.

XIV,c60, Crystals of *tetramethylene diammonium adipate*, $[NH_3(CH_2)_4-NH_3][CO_2(CH_2)_4CO_2]$, are monoclinic with a tetramolecular cell of the dimensions:

$$a_0 = 15.36 \text{ A.}; \quad b_0 = 5.97 \text{ A.}; \quad c_0 = 14.93 \text{ A.}; \quad \beta = 110°30'$$

TABLE XIVC,51. Parameters of the Atoms in $[NH_3(CH_2)_4NH_3][CO_2(CH_2)_4CO_2]$

Atom	x	σ_x, A.	y	σ_y, A.	z	σ_z, A.
C(1)	0.1869	0.0029	0.8117	0.0033	0.3879	0.0027
C(2)	0.2477	0.0028	0.9992	0.0029	0.4474	0.0025
C(3)	0.0659	0.0025	0.3488	0.0032	0.1735	0.0024
C(4)	0.9979	0.0028	0.1802	0.0039	0.1089	0.0031
C(5)	0.0346	0.0028	0.0796	0.0034	0.0348	0.0025
N	0.1790	0.0026	0.8302	0.0029	0.2856	0.0021
O(1)	0.1437	0.0018	0.2732	0.0021	0.2298	0.0021
O(2)	0.0438	0.0026	0.5507	0.0026	0.1692	0.0025
H(1)	0.160	—	0.978	—	0.260	—
H(2)	0.129	—	0.725	—	0.240	—
H(3)	0.230	—	0.814	—	0.279	—
H(4)	0.218	—	0.677	—	0.417	—
H(5)	0.124	—	0.822	—	0.393	—
H(6)	0.219	—	0.151	—	0.416	—
H(7)	0.317	—	0.982	—	0.449	—
H(8)	0.937	—	0.251	—	0.073	—
H(9)	0.987	—	0.050	—	0.149	—
H(10)	0.089	—	0.977	—	0.074	—
H(11)	0.054	—	0.203	—	0.000	—

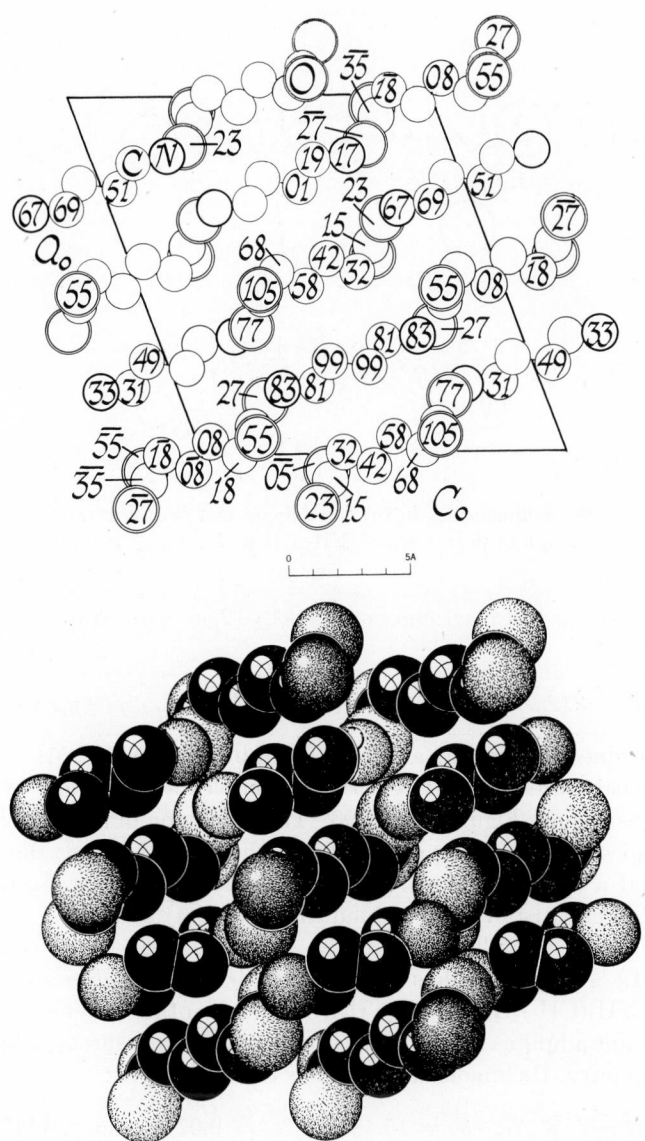

Fig. XIVC,89a (top). The monoclinic structure of tetramethylene diammonium adipate projected along its b_0 axis. Right-hand axes.

Fig. XIVC,89b (bottom). A packing drawing of the monoclinic structure of [NH$_3$-CH$_2$)$_4$NH$_3$][OOC(CH$_2$)$_4$COO] projected along its b_0 axis. The carbon atoms are black. Nitrogen atoms are heavily ringed and dotted; the larger oxygen atoms are short-line-and-dot shaded.

Fig. XIVC,90. Bond dimensions in (a) the adipate and (b) the tetramethylene dia𝗆 monium ions in crystals of [NH₃(CH₂)₄NH₃][CO₂(CH₂)₄CO₂].

The space group has been chosen as C_{2h}^6 $(A2/a)$ with atoms in the pos tions:

(8f) $\pm(xyz;\ x+^1/_2,\bar{y},z;\ x,y+^1/_2,z+^1/_2;\ x+^1/_2,^1/_2-y,z+^1/_2)$

The determined parameters are those listed in Table XIVC,51.

In this structure (Fig. XIVC,89), the complex anions and cations ha𝗏 the dimensions of Figure XIVC,90. The diammonium cation is near planar and so is the zigzag chain part of the adipate anion; the carbox𝗒 groups and α carbon atoms of the latter are strictly planar. In this ani𝖼 there is a large twist of 67.5° around the C(3)–C(4) bond.

XIV,c61. A structure has been found for *hexamethylene diammoniu* *adipate,* [NH₃(CH₂)₆NH₃][CO₂(CH₂)₄CO₂]. Like other polymethylene a𝗇 diammonium adipates whose unit cells have been measured, it has mon𝖾 clinic symmetry. Its bimolecular unit has the dimensions:

$$a_0 = 5.57\ \text{A.};\quad b_0 = 15.48\ \text{A.};\quad c_0 = 9.07\ \text{A.};\quad \beta = 114°$$

The space group is C_{2h}^5 $(P2_1/n)$ with all atoms in the general positions:

(4e) $\pm(xyz;\ x+^1/_2,^1/_2-y,z+^1/_2)$

The chosen parameters are listed in Table XIVC,52.

The hexamethylene diammonium ion is planar except for hydrogen. T𝖍 central part of the adipate ion is also planar, but the planes formed by t𝖍

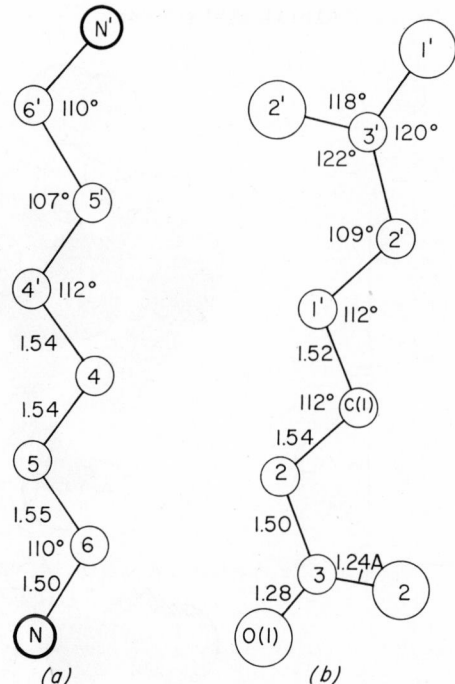

Fig. XIVC,91. Bond dimensions in (a) the hexamethylene diammonium and (b) the adipate ions in crystals of $[NH_3(CH_2)_6NH_3][CO_2(CH_2)_4CO_2]$.

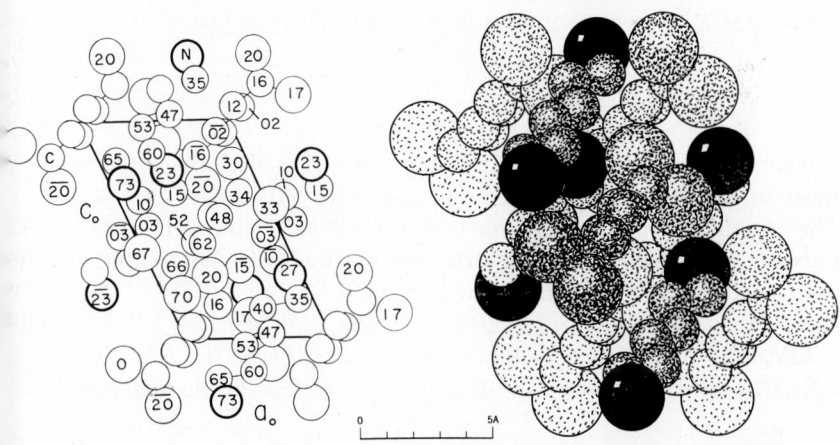

Fig. XIVC,92a (left). The monoclinic structure of hexamethylene diammonium adipate projected along its b_0 axis. Oxygen atoms are the larger of the lightly ringed circles. Left-hand axes.

Fig. XIVC,92b (right). A packing drawing of the monoclinic structure of hexamethylene diammonium adipate viewed along its b_0 axis. The nitrogen atoms are black; the oxygen atoms, of equal size, are dotted. The smaller dotted circles are carbon.

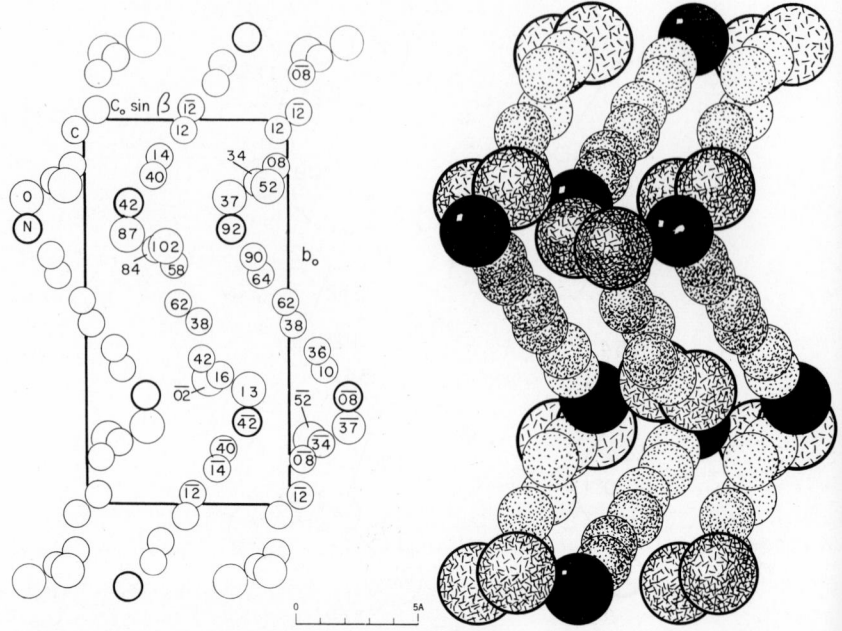

Fig. XIVC,93a (left). The monoclinic structure of hexamethylene diammonium adipate projected along its a_0 axis. This brings out better than does the the b_0 projection of the preceding drawing the zigzag distribution of the elongated ions. Left-hand axes.

Fig. XIVC,93b (right). A packing drawing of the a_0 projection of the monoclinic hexamethylene diammonium adipate structure. The nitrogen atoms are black; the oxygen atoms here are heavily outlined and line shaded. The smaller carbon atoms are dotted.

oxygen atoms and the terminal carbons are twisted through 19° with respect to the rest. Bond lengths and angles are given in Figure XIVC,91. The packing within the crystal is indicated in Figures XIVC,92 and 93. Between ions throughout the structure are hydrogen bonds that involve N–O separations of 2.72, 2.75, and 2.88 A.

XIV,c62. Crystals of *cuprous nitrate di adiponitrile*, [Cu(NC(CH$_2$)$_4$ CN)$_2$]NO$_3$, are orthorhombic with a bimolecular unit of the edge lengths:

$$a_0 = 9.41 \pm 0.02 \text{ A.}; \quad b_0 = 13.73 \pm 0.02 \text{ A.}; \quad c_0 = 5.85 \pm 0.01 \text{ A.}$$

The space group has been chosen as V_h^2 (*Pnnn*) with atoms in the positions:

$$\text{Cu: } (2a) \quad 000; \; {}^1/_2 \, {}^1/_2 \, {}^1/_2$$
$$\text{N(2): } (2b) \quad 0 \, {}^1/_2 \, {}^1/_2; \; {}^1/_2 \, 0 \, 0$$

TABLE XIVC,52
Parameters of the Atoms in Hexamethylene Diammonium Adipate

Atom	x	y	z
C(1)	0.125	0.023	0.054
C(2)	0.083	0.121	0.058
C(3)	0.342	0.162	0.158
C(4)	0.617	0.471	0.017
C(5)	0.642	0.400	0.142
C(6)	0.900	0.350	0.175
N	0.925	0.275	0.286
O(1)	0.375	0.200	0.292
O(2)	0.517	0.171	0.108
Calculated hydrogen positions			
H(1,C 1)	0.260	−0.010	0.017
H(2,C 1)	0.196	−0.002	0.177
H(3,C 2)	0.013	0.147	−0.063
H(4,C 2)	−0.057	0.134	0.111
H(5,C 4)	0.575	0.439	−0.103
H(6,C 4)	0.808	0.495	0.050
H(7,C 5)	0.473	0.357	0.092
H(8,C 5)	0.633	0.433	0.238
H(9,C 6)	0.900	0.325	0.067
H(10,C 6)	0.067	0.392	0.200
H(11,N)	0.792	0.228	0.225
H(12,N)	0.042	0.305	0.400
H(13,N)	0.092	0.258	0.267

Other atoms are in the general positions:

$(8m)$ $xyz; \bar{x}\bar{y}z; {}^1/_2-x, {}^1/_2-y, {}^1/_2-z; x+{}^1/_2, y+{}^1/_2, {}^1/_2-z;$
$x\bar{y}\bar{z}; \bar{x}y\bar{z}; {}^1/_2-x, y+{}^1/_2, z+{}^1/_2; x+{}^1/_2, {}^1/_2-y, z+{}^1/_2$

with the parameters:

Atom	x	y	z
N(1)	0.129	0.076	0.200
C(1)	0.194	0.110	0.341
C(2)	0.292	0.164	0.509
C(3)	0.201	0.223	0.667

The nitrate ions are considered to be statistically oriented with their oxygen atoms distributed over a quarter of the following positions:

$O(3): (8m)$ with $x = 0.000, y = 0.455, z = 0.317$

O(4): (8m)   with $x = 0.000$, $y = 0.422$, $z = 0.394$
O(1): (4i) $\pm (^1/_2\,u\,0; 0,u+^1/_2,^1/_2)$ with $u = 0.090$
O(2): (4l) $\pm (0\,^1/_2\,u; \,^1/_2,0,u+^1/_2)$ with $u = 0.288$

In this structure the cuprous atoms are tetrahedrally surrounded by nitrogen atoms from four different adiponitrile groups, with Cu–N = 1.98 A. Each of these groups is bound to two copper atoms in such a way that the whole is tied together into a three-dimensional net, in pores of which the nitrate ions are to be found.

XIV,c63. *Pentamethonium iodide*, $(CH_2)_5N_2(CH_3)_6I_2$, is orthorhombic with a large unit containing eight molecules and having the edges:

$$a_0 = 12.37 \text{ A.}; \quad b_0 = 24.36 \text{ A.}; \quad c_0 = 11.94 \text{ A.}$$

The space group is V_h^{16} (*Pnam*) with all atoms in the general positions:

(4c) $\pm (u\,v\,^1/_4; u+^1/_2,^1/_2-v,^1/_4)$
(8d) $\pm (xyz; \,^1/_2-x,y+^1/_2,z+^1/_2; x,y,^1/_2-z; x+^1/_2,^1/_2-y,z)$

The determined parameters are those of Table XIVC,53, the standard de-

TABLE XIVC,53
Parameters of the Atoms in $(CH_2)_5N_2(CH_3)_6I_2$

Atom	Position	x	y	z
I(1)	(4c)	0.1059	0.0454	$^1/_4$
I(2)	(4c)	0.0933	0.2770	$^1/_4$
I(3)	(4c)	0.0604	0.5192	$^1/_4$
I(4)	(4c)	0.3069	0.7896	$^1/_4$
C(1)	(8d)	0.064	0.681	0.095
C(2)	(8d)	0.178	0.711	0.924
C(3)	(8d)	0.121	0.618	0.933
C(4)	(8d)	0.234	0.647	0.080
C(5)	(8d)	0.355	0.654	0.013
C(6)	(8d)	0.437	0.628	0.095
C(7)	(8d)	0.543	0.616	0.002
C(8)	(8d)	0.650	0.598	0.089
C(9)	(8d)	0.681	0.521	0.941
C(10)	(8d)	0.809	0.558	0.086
C(11)	(8d)	0.767	0.614	0.937
N(2)	(8d)	0.150	0.670	0.006
N(4)	(8d)	0.724	0.577	0.004

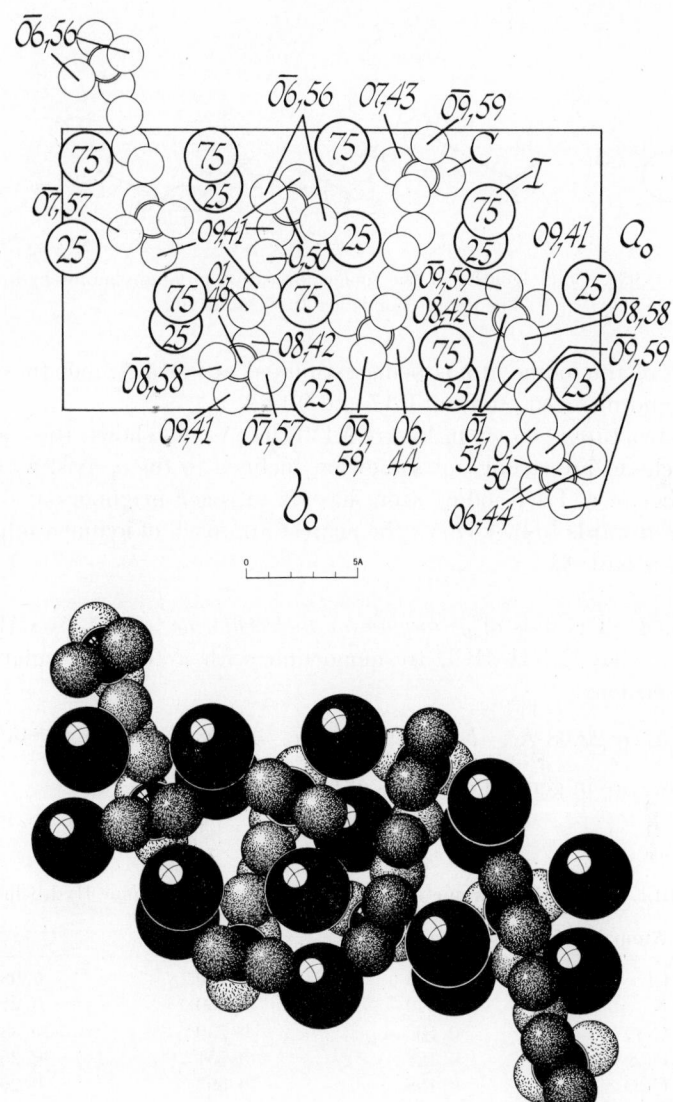

Fig. XIVC,94a (top). The orthorhombic structure of pentamethonium iodide projected along its c_0 axis. The nitrogen atoms are doubly ringed. Right-hand axes.

Fig. XIVC,94b (bottom). A packing drawing of the $(CH_2)_5N_2(CH_3)_6I_2$ arrangement seen along its c_0 axis. The iodine atoms are the large, the nitrogen atoms the smaller black circles. Atoms of carbon are short-line-and-dot shaded.

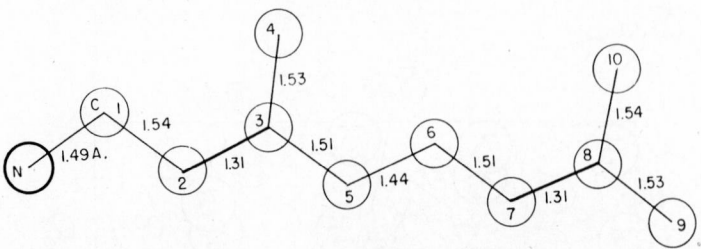

Fig. XIVC,95a. Bond lengths in the molecular cation of geranylamine hydrochloride
The double bonds are the heavier lines.

viations of the iodine atoms being estimated at 0.015 A. and those of the
carbon and nitrogen atoms at 0.17 and 0.14 A.

The structure is shown in Figure XIVC,94. As this shows, the essentially
planar chains are strung out somewhat inclined to the a_0 axis, with iodine
atoms between. Each iodine atom has six nitrogen neighbors at distances
ranging upwards from 4.47 A.; the nearest approach of iodine atoms to one
another is 5.61 A.

XIV,c64. Crystals of *geranylamine hydrochloride*, $(CH_3)_2C{=}CH(CH_2)_2$
$C(CH_3){=}CHCH_2NH_2{\cdot}HCl$, are monoclinic with a tetramolecular unit of
the dimensions:

$$a_0 = 22.68 \text{ A.}; \quad b_0 = 5.94 \text{ A.}; \quad c_0 = 8.98 \text{ A.}; \quad \beta = 98°48'$$

All atoms are in general positions of C_{2h}^5 $(P2_1/c)$:

$$(4e) \quad \pm(xyz; \ x,{}^1/_2{-}y,z{+}{}^1/_2)$$

TABLE XIVC,54.   Parameters of the Atoms in Geranylamine Hydrochloride

Atom	x	y	z
Cl	0.053	0.250	0.080
N	0.049	0.750	0.216
C(1)	0.103	0.750	0.333
C(2)	0.158	0.751	0.255
C(3)	0.198	0.593	0.264
C(4)	0.198	0.379	0.358
C(5)	0.253	0.625	0.182
C(6)	0.304	0.700	0.287
C(7)	0.357	0.746	0.207
C(8)	0.394	0.916	0.212
C(9)	0.447	0.918	0.118
C(10)	0.392	0.120	0.315

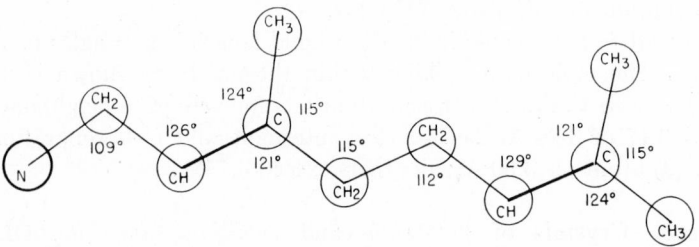

Fig. XIVC,95b. Bond angles in the molecular cation of geranylamine hydrochloride.

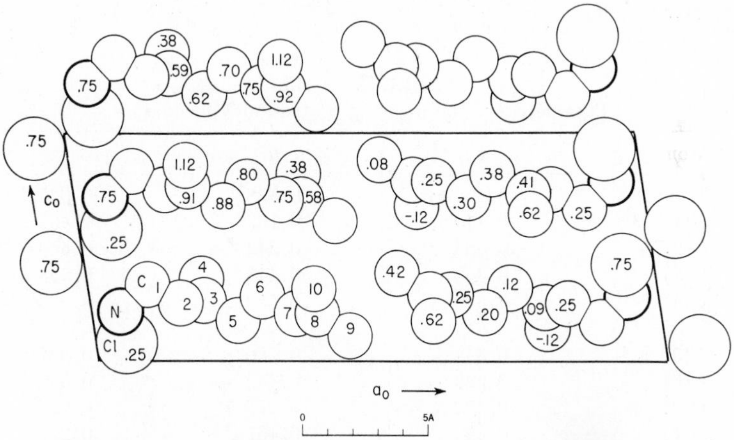

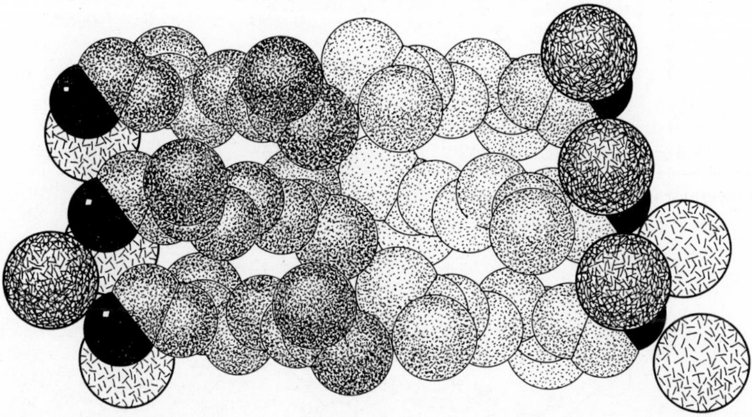

Fig. XIVC,96a (top). The monoclinic structure of geranylamine hydrochloride projected along its b_0 axis. Left-hand axes.

Fig. XIVC,96b (bottom). A packing drawing of the monoclinic structure of geranylamine hydrochloride viewed along its b_0 axis. The nitrogen atoms are black; carbon atoms, of the same size, are dotted. Atoms of chlorine are larger and line shaded.

with the parameters of Table XIVC,54.

These lead to a molecule having the bond lengths and angles of Figure XIVC,95. The molecular packing within the crystal is shown in Figure XIVC,96. Each terminal nitrogen atom has four chlorine neighbors, with N–Cl = 3.17 or 3.24 A. The shortest intermolecular C–C separation, between C(9) atoms, is 3.60 A.; all others exceed 3.75 A.

XIV,c65. Crystals of *potassium* and *rubidium gluconate*, $OHCH_2$-$(CHOH)_4CO_2R$, are orthorhombic with tetramolecular units having the edges:

$KC_6H_{11}O_7$: $a_0 = 23.63$ A.; $b_0 = 7.46$ A.; $c_0 = 5.05$ A.

TABLE XIVC,55
Parameters of the Atoms in Potassium Gluconate

Atom	x	y	z
K	0.465	0.828	0 035
C(1)	0.440	0.162[a]	0.467
C(2)	0.383	0.242	0.560
C(3)	0.388	0.448	0.558
C(4)	0.330[a]	0.550[a]	0.583[a]
C(5)	0.344	0.750	0.537[a]
C(6)	0.286	0.840	0.558
O(0)	0.472	0.077	0.627
O(1)	0.449	0.167[a]	0.227
O(2)	0.372	0.180	0.825
O(3)	0.435	0.509	0.702
O(4)	0.321	0.545[a]	0.873
O(5)	0.361	0.772	0.268
O(6)	0.242	0.758	0.407

[a] These parameters are of lower accuracy than the rest.

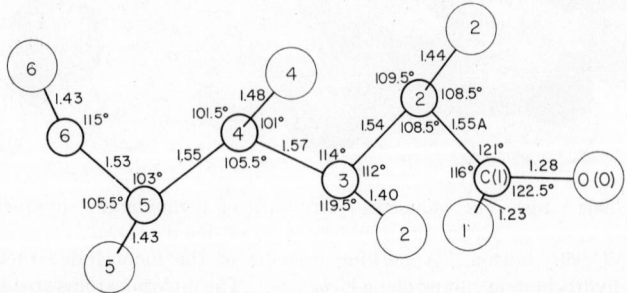

Fig. XIVC,97. Bond dimensions of the gluconate anion in its potassium salt.

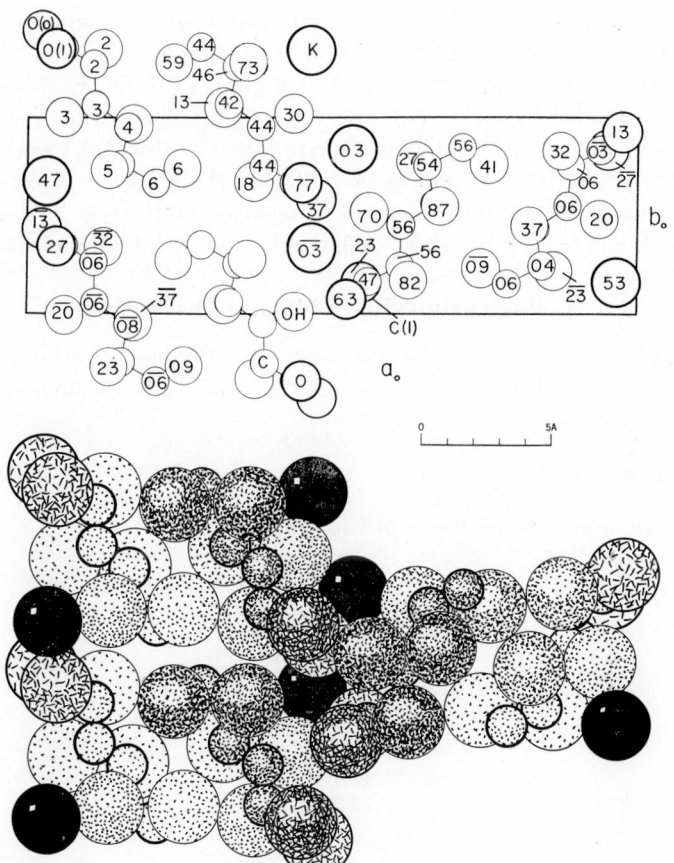

Fig. XIVC,98a (top). The orthorhombic structure of potassium gluconate projected along its c_0 axis. In the anions the carboxyl oxygens are more heavily ringed than the hydroxyls. Left-hand axes.

Fig. XIVC,98b (bottom). A packing drawing of the orthorhombic structure of $OHCH_2(CHOH)_4COOK$ seen along its c_0 axis. The potassium atoms are black. Atoms of carbon are small, heavily ringed, and dotted. Of the large oxygen circles those for hydroxyls are dotted while the carboxyls are more heavily outlined and line shaded.

$RbC_6H_{11}O_7$: $a_0 = 23.94$ A.; $b_0 = 7.52$ A.; $c_0 = 5.63$ A.

All atoms are in general positions of V^4 $(P2_12_12_1)$:

(4a) $xyz;$ $^1/_2-x,\bar{y},z+^1/_2;$ $x+^1/_2,^1/_2-y,^1/_2-z;$ $\bar{x},y+^1/_2,\bar{z}$

with the parameters, for the potassium salt, listed in Table XIVC,55.

These lead to a gluconate ion (Fig. XIVC,97) having as backbone a zigzag and almost planar chain of carbon atoms. It can be converted into an α-glucose molecular ring (with appropriate atomic suppressions) by rotations about the C(2)–C(3) and C(3)–C(4) bonds.

The crystal structure is illustrated in Figure XIVC,98. Each potassium ion is surrounded by an irregular octahedron of oxygen atoms, with K–O between 2.71 and 3.03 A. The gluconate ions are also tied together in all directions by hydrogen bonds between carboxyl and hydroxyl oxygens and between hydroxyl and other hydroxyl oxygens (O–H–O = 2.71–2.80 A.). Each carboxyl oxygen has three such bonds, and each hydroxyl two.

5. Polycarboxylic Acids and Derivatives

XIV,c66. *Maleic acid*, the *cis* form of ethylene dicarboxylic acid, $HCO_2CH{=}CHCO_2H$, forms monoclinic crystals. Its tetramolecular unit has the dimensions:

$$a_0 = 7.47 \text{ A.}; \quad b_0 = 10.15 \text{ A.}; \quad c_0 = 7.65 \text{ A.}; \quad \beta = 123°30'$$

Atoms are in general positions of C_{2h}^5 ($P2_1/c$):

$$(4e) \quad \pm(xyz; \ x,{}^1/_2{-}y,z{+}{}^1/_2)$$

with the parameters of Table XIVC,56.

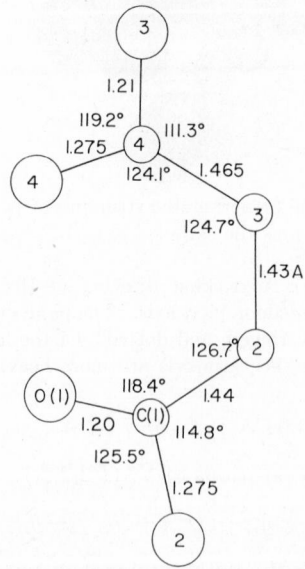

Fig. XIVC,99. Bond dimensions in the molecule of maleic acid.

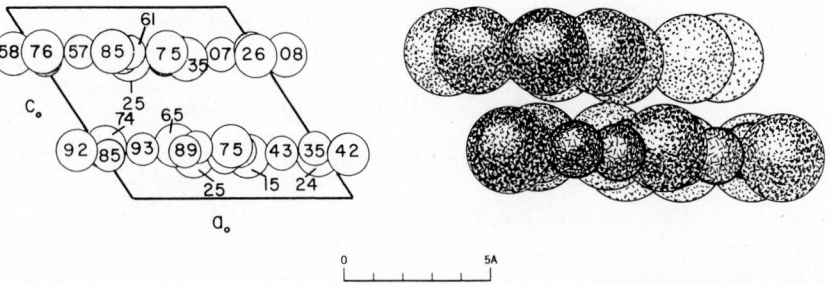

Fig. XIVC,100a (left). The monoclinic structure of maleic acid projected along its
b_0 axis. The larger circles are the oxygen atoms. Left-hand axes.
Fig. XIVC,100b (right). A packing drawing of the monoclinic structure of maleic
acid seen along its b_0 axis. The oxygen atoms are dotted; the carbon atoms are smaller,
heavily outlined, and line shaded.

TABLE XIVC,56
Parameters of the Atoms in Maleic Acid

Atom	x	y	z
C(1)	0.966	0.352	0.254
C(2)	0.807	0.431	0.257
C(3)	0.624	0.390	0.262
C(4)	0.565	0.253	0.245
O(1)	0.962	0.238	0.248
O(2)	0.112	0.418	0.243
O(3)	0.404	0.253	0.244
O(4)	0.654	0.153	0.242

The resulting molecule is nearly planar, with the bond dimensions of
Figure XIVC,99. The close approach of O(1) and O(4), 2.46 A., is thought
to indicate an intramolecular hydrogen bond (a chelation) creating a kind
of six-membered ring within the molecule. The molecular arrangement in
the crystal is indicated by Figure XIVC,100. The molecules are connected
into strings along a_0 by hydrogen bonds [O(2)–O(3) = 2.75 A.], and these
strings are tied together to form sheets parallel to the c-face by hydrogen
bonds [O(2)–O(4) = 2.98 A.].

XIV,c67. *Potassium acid maleate*, K[OOCCH=CHCOOH], is orthorhombic with a tetramolecular cell of the edge lengths:

$$a_0 = 4.578 \pm 0.002 \text{ A.}; \quad b_0 = 7.791 \pm 0.004 \text{ A.}; \quad c_0 = 15.953 \pm 0.005 \text{ A.}$$

The space group is V_h^{11} (*Pbcm*) with atoms in the positions:

$$K: (4c) \quad \pm (u \, ^1/_4 \, 0; \, u \, ^1/_4 \, ^1/_2)$$
$$\text{with } u = -0.2615, \ \sigma(u) = 0.002 \text{ A.}$$

The other atoms are in the general positions:

$$(8e) \quad \pm (xyz; \ x,y,^1/_2-z; \ \bar{x},y+^1/_2,z; \ x,^1/_2-y,z+^1/_2)$$

Their parameters are:

Atom	x	y	z	σ, A.
O(1)	0.4851	0.2958	0.1737	0.0017
O(2)	0.2360	0.4200	0.0719	0.0017
C(1)	0.2845	0.3969	0.1471	0.0019
C(2)	0.0904	0.4885	0.2078	0.0019

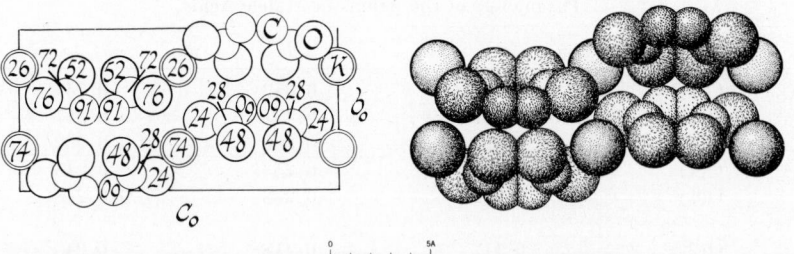

Fig. XIVC,101a (left). The orthorhombic structure of potassium acid maleate projected along its a_0 axis. Right-hand axes.

Fig. XIVC,101b (right). A packing drawing of the orthorhombic potassium acid maleate structure seen along its a_0 axis. The potassium atoms are heavily outlined and dotted. Oxygen atoms are fine-line shaded; the carbon atoms are smaller and hook shaded.

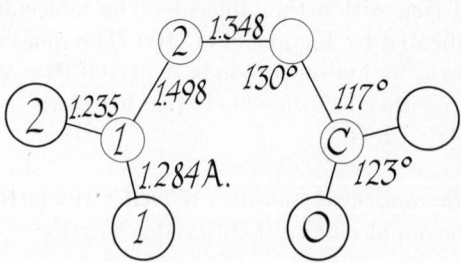

Fig. XIVC,102. Bond dimensions of the acid maleate anion in its potassium salt.

In this structure (Fig. XIVC,101) each potassium atom has around it eight close oxygen atoms: six $O(2)$, with K–O = 2.817–2.891 A. and two $O(1)$ at a distance of 3.025 A. The acid maleate anion is planar to within 0.01 A.; its dimensions are those of Figure XIVC,102. As Figure XIVC,101 indicates, the K^+ and acid maleate anions are arranged in rows that alternate along the c_0 axis. Positions are discussed for the hydrogen atoms, but parameters are not stated.

XIV,c68. Crystals of *acetylene dicarboxylic acid dihydrate*, HOOC—C≡ C—COOH·$2H_2O$, are monoclinic with a bimolecular cell of the dimensions:

$$a_0 = 11.0516 \text{ A.}; \quad b_0 = 3.8609 \text{ A.}; \quad c_0 = 7.9728 \text{ A.}; \quad \beta = 97°42.4'$$

The space group is C_{2h}^5 ($P2_1/a$) with all atoms in the general positions:

$$(4e) \quad \pm (xyz; \ x+\tfrac{1}{2}, \tfrac{1}{2}-y, z)$$

The determined parameters are as follows:

Atom	x	y	z
C(1)	−0.032	−0.006	0.054
C(2)	−0.110	−0.018	0.182
O(1)	−0.213	0.130	0.159
O(2)	−0.071	−0.181	0.315
H_2O	−0.341	0.000	0.396

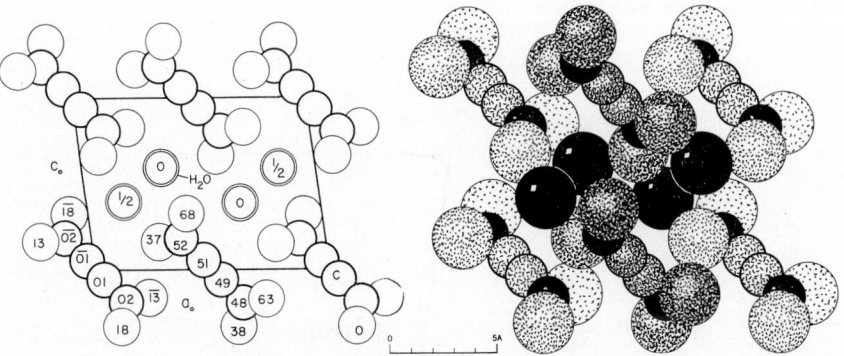

Fig. XIVC,103a (left). The monoclinic structure of HOOC—C≡C—COOH·$2H_2O$ projected along its b_0 axis. Left-hand axes.

Fig. XIVC,103b (right). A packing drawing of the monoclinic structure of acetylene dicarboxylic acid dihydrate viewed along its b_0 axis. The carbon atoms are the small circles, the carboxyl carbons being black. Carboxyl oxygens are large and dotted; the water molecules, of the same large size, are black.

The structure that results is shown in Figure XIVC,103. Its molecules lie in planes that are inclined 133° to the a_0c_0 plane of the projection. Within a molecule the significant atomic separations are C(1)–C(1) = 1.19 A., C(1)–C(2) = 1.43 A., C(2)–O(1) = 1.27 A., C(2)–O(2) = 1.26 A., and O(1)–O(2) = 2.22 A. A water molecule is separated from an O(1) atom by the distance 2.56 A., while H_2O–O(2) = 2.82 or 2.89 A.

Accurate measurements have been made of the cell dimensions of the corresponding *deuterium* acid DOOC—C≡C—COOD·$2D_2O$. They are

$$a_0 = 11.0484 \text{ A.}; \quad b_0 = 3.8740 \text{ A.}; \quad c_0 = 7.9894 \text{ A.}; \quad \beta = 98°0.2'$$

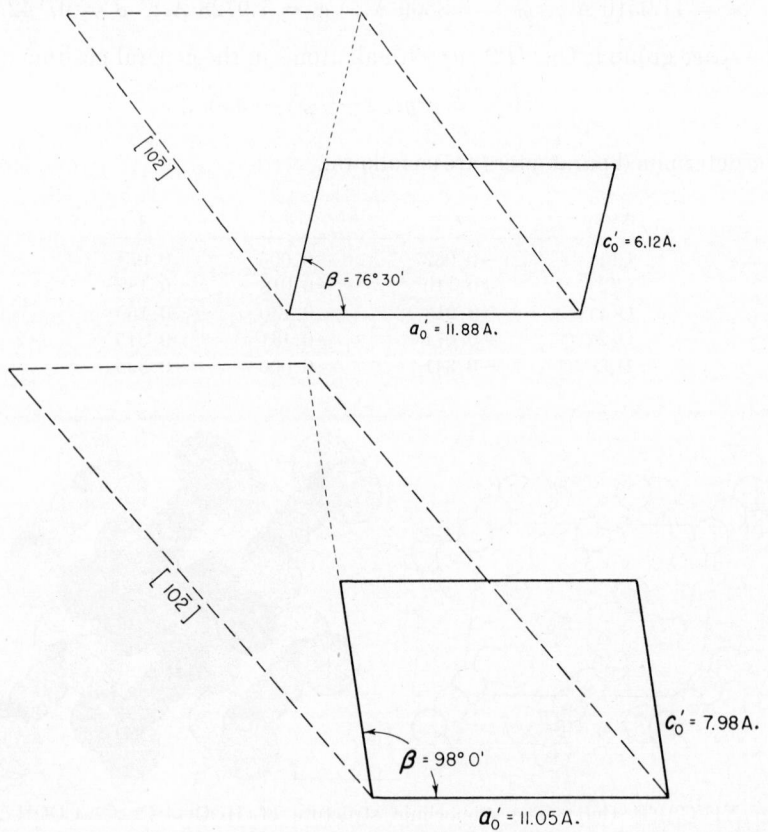

Fig. XIVC,104. The unit cells of oxalic acid dihydrate (above) and of acetylene dicarboxylic acid dihydrate, projected along their b_0 axes. The units are shown as full lines with the a_0 and c_0 axes of oxalic acid exchanged. Their similar pseudocells appear as broken lines.

This structure is closely related to that of oxalic acid dihydrate (**XIV,-b119**). This is best brought out by describing the oxalic acid in terms of a cell having the dimensions:

$$a_0' = 11.88 \text{ A.}; \quad b_0' = 3.60 \text{ A.}; \quad c_0' = 6.12 \text{ A.}; \quad \beta' = 76°30'$$

In this cell the axial orientation corresponds to $P2_1/a$ (rather than the $P2_1/n$ used in **XIV,b119**). All atoms will then be in the positions:

$$(4e) \quad \pm(xyz; x+{}^1/_2,{}^1/_2-y,z)$$

with the following parameters:

Atom	x	y	z
C	−0.051	−0.033	0.090
O(1)	−0.147	−0.057	0.065
O(2)	−0.036	0.213	0.255
H$_2$O	−0.321	0.108	0.367

The relation of these axes to those of the cells that are similar for the two structures is shown in Figure XIVC,104.

XIV,c69. Crystals of *diacetylene dicarboxylic acid dihydrate*, HOOC—C≡C—C≡C—COOH·2H$_2$O, are monoclinic with a tetramolecular unit having the dimensions:

$$a_0 = 11.15 \text{ A.}; \quad b_0 = 3.75 \text{ A.}; \quad c_0 = 20.18 \text{ A.}; \quad \beta = 107°0'$$

All atoms are in the general positions of C_{2h}^6 ($I2/c$):

$$(8f) \quad \pm(xyz; x,\bar{y},z+{}^1/_2); \text{ B.C.}$$

with the parameters of Table XIVC,57.

TABLE XIVC,57
Parameters of the Atoms in HO$_2$C—C≡C—C≡C—CO$_2$H·2H$_2$O

Atom	x	y	z
C(1)	0.030	0.029	0.226
C(2)	0.082	0.029	0.183
C(3)	0.147	0.029	0.130
O(1)	0.256	−0.096	0.144
O(2)	0.094	0.182	0.074
H$_2$O	0.354	0.000	0.046

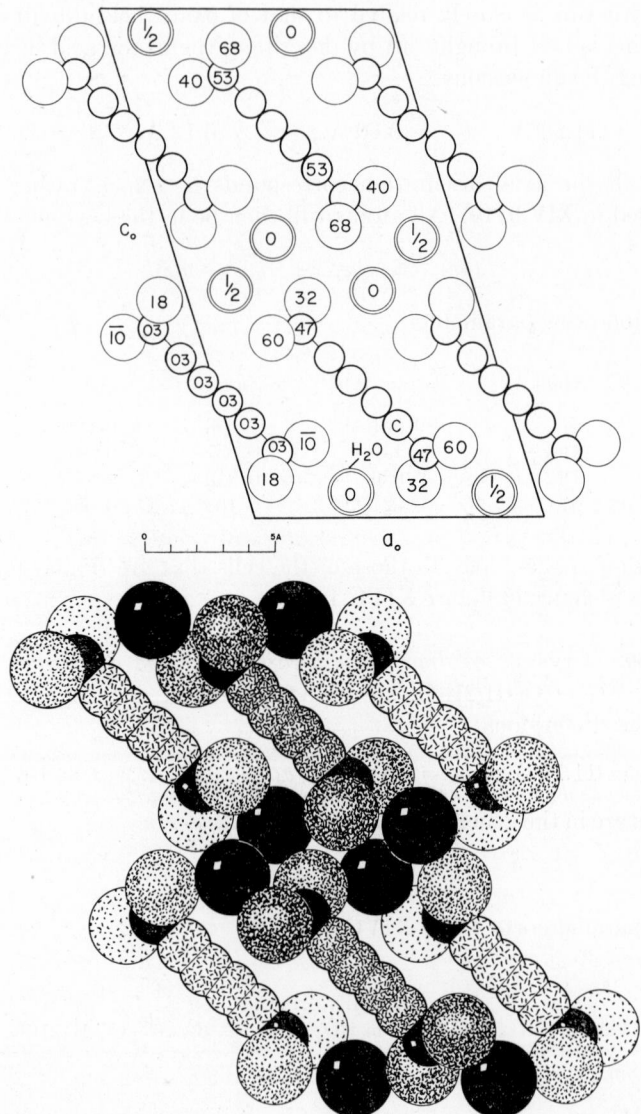

Fig. XIVC,105a (top). The monoclinic structure of diacetylene dicarboxylic acid dihydrate projected along its b_0 axis. Carboxyl oxygen atoms are the larger, lightly outlined circles. Left-hand axes.

Fig. XIVC,105b (bottom). A packing drawing of the monoclinic structure of $HOOCC{\equiv}C{-}C{\equiv}CCOOH \cdot 2H_2O$ viewed along its b_0 axis. The water molecules are the large black, the carboxyl carbon atoms the small black circles. The other carbon atoms are small and line shaded; the carboxyl oxygens are larger and dotted.

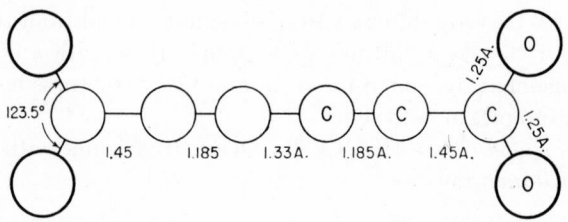

Fig. XIVC,106. Bond dimensions in the molecule of HOOCC≡C—C≡CCOOH.

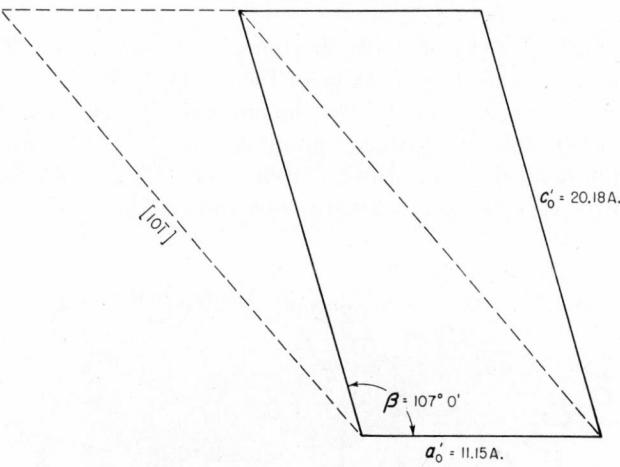

Fig. XIVC,107. A projection along b_0 of the monoclinic axial system of diacetylene dicarboxylic acid dihydrate (full lines) for comparison with the cells of Figure XIVC,104. The value of the a_0 axis is practically the same for the three compounds but for the present crystal the c_0 axis is doubled. The pseudocell to be compared with those of Figure XIVC,104 is indicated by the broken lines.

The structure is given in Figure XIVC,105. The planes of the two carboxyl groups of a molecule make an angle of 57° with one another; other bond dimensions are those of Figure XIVC,106.

This structure is closely related to those of oxalic acid dihydrate (**XIV,-b119**) and of acetylene dicarboxylic acid dihydrate (**XIV,c68**). This is shown by comparing Figures XIVC,104 and 107. In both, the similar pseudocells are dashed.

XIV,c70. The long-chain dicarboxylic acids, HOOC·(CH₂)ₙ·COOH, are, like the fatty acids, polymorphic. The first two members of the series, malonic acid, HOOCCH₂COOH, and succinic acid, HOOC(CH₂)₂COOH, have structures that are different from one another and from the rest. The higher acids, with $n > 2$, fall into two groups: those with n odd have one atomic arrangement, those with n even, another. Structures have been determined of examples of both series.

Crystals of *malonic acid*, $HOOCCH_2COOH$, are triclinic with a bimolecular unit of the dimensions:

$$a_0 = 5.33 \text{ A.}; \quad b_0 = 5.14 \text{ A.}; \quad c_0 = 11.25 \text{ A.}$$
$$\alpha = 102°42'; \quad \beta = 135°10'; \quad \gamma = 85°10'$$

The space group is C_i^1 $(P\bar{1})$ with all atoms in the positions: $(2i)$ $\pm(xyz)$. The determined parameters are those of Table XIVC,58.

In this structure (Fig. XIVC,108) the molecule has the bond dimensions of Figure XIVC,109; one carboxyl group is turned 13°, the other 90° out of the plane defined by the three carbon atoms. The molecules are tied together by O–H $\cdots$ O bonds of lengths 2.68 and 2.71 A.

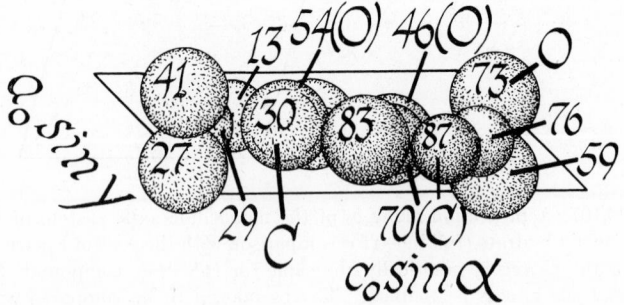

Fig. XIVC,108. The triclinic structure of malonic acid projected along its b_0 axis. Right-hand axes.

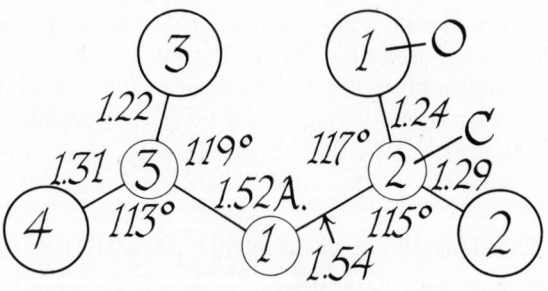

Fig. XIVC,109. Bond dimensions in the molecule of $HOOCCH_2COOH$.

TABLE XIVC,58.   Parameters of the Atoms in Malonic Acid

Atom	x	y	z
C(1)	0.620	0.129	0.250
C(2)	0.565	0.297	0.361
O(1)	0.565	0.542	0.371
O(2)	0.570	0.170	0.449
C(3)	0.557	0.290	0.134
O(3)	0.815	0.415	0.185
O(4)	0.220	0.274	−0.013

XIV,c71. One, β, of the three modifications of *succinic acid*, HOOC-(CH$_2$)$_2$COOH, has been investigated in detail. Its monoclinic unit, containing two molecules, has the dimensions:

$$a_0 = 5.126 \pm 0.01 \text{ A.}; \quad b_0 = 8.880 \pm 0.007 \text{ A.}; \quad c_0 = 7.619 \pm 0.006 \text{ A.}$$
$$\beta = 133°36' \pm 6'$$

TABLE XIVC,59.   Parameters of the Atoms in β-Succinic Acid

Atom	x	y	z
C(1)	0.05352	0.06651	0.08258
C(2)	0.02642	0.03450	0.26070
O(1)	−0.12482	−0.07780	0.25386
O(2)	0.16813	0.14041	0.42508
H(1)	−0.1231	0.1610	−0.0301
H(2)	0.3312	0.0968	0.1852
H(3)	0.1510	0.1158	0.5516

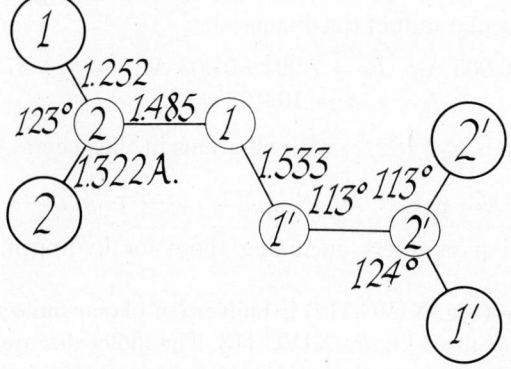

Fig. XIVC,110. The bond dimensions in the molecule of the β form of succinic acid.

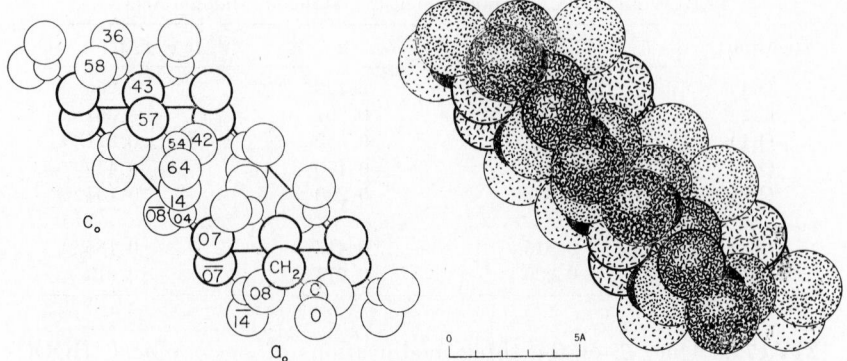

Fig. XIVC,111a (left). The monoclinic structure of the β form of succinic acid projected along its b_0 axis. Left-hand axes.

Fig. XIVC,111b (right). A packing drawing of the monoclinic structure of β-succinic acid viewed along its b_0 axis. The carboxyl carbon atoms are black; the other carbon atoms are line shaded and heavily ringed. The carboxyl oxygens are large and dotted.

All atoms are in general positions of C_{2h}^5 $(P2_1/a)$:

$$(4e) \quad \pm (xyz;\ x+{}^1/_2,{}^1/_2-y,z)$$

with the redetermined parameters of Table XIVC,59.

The molecules have the shape indicated in Figure XIVC,110, but the oxygen atoms are rotated ca. 11° about the line CH_2–C out of the approximate plane of the carbon atoms. The structure as a whole is shown in Figure XIVC,111. Between molecules there is one short O(1)–O(2) = 2.66 A.; other intermolecular separations are from 3.14 A. upwards.

XIV,c72. Crystals of *succinamide*, $[NH_2C(O)CH_2—]_2$, are monoclinic with a tetramolecular unit of the dimensions:

$$a_0 = 6.932 \pm 0.003 \text{ A.}; \quad b_0 = 7.994 \pm 0.003 \text{ A.}; \quad c_0 = 9.878 \pm 0.004 \text{ A.}$$
$$\beta = 102°28' \pm 5'$$

The space group is C_{2h}^6 $(C2/c)$ with all atoms in the general positions:

$$(8f) \quad \pm (xyz;\ x,\bar{y},z+{}^1/_2;\ x+{}^1/_2,y+{}^1/_2,z;\ x+{}^1/_2,{}^1/_2-y,z+{}^1/_2)$$

The determined parameters, including those for hydrogen, are listed in Table XIVC,60.

This structure (Fig. XIVC,112) is built up of planar molecules that have the bond dimensions of Figure XIVC,113. The molecules are held together by N–H–O bonds of length 2.94 A., there being a tendency to form sheets parallel to the (001) plane.

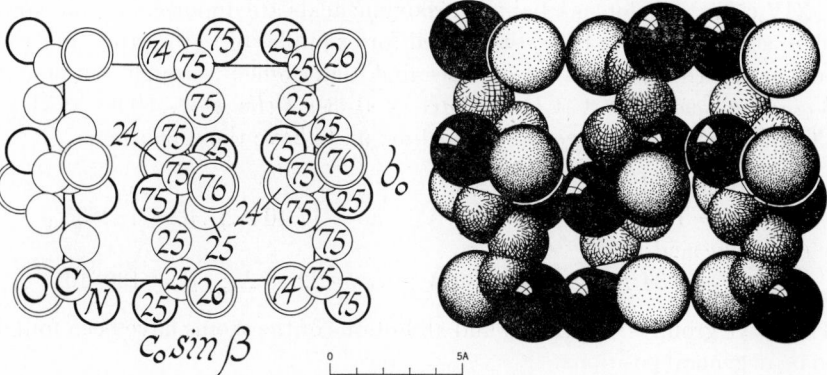

Fig. XIVC,112a (left). The monoclinic structure of succinamide projected along its a_0 axis. Right-hand axes.

Fig. XIVC,112b (right). A packing drawing of the monoclinic succinamide arrangement viewed along its a_0 axis. The nitrogen atoms are black, the oxygens heavily outlined and dotted. The smaller carbon atoms are line shaded.

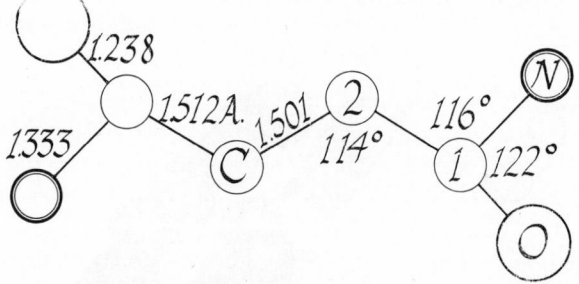

Fig. XIVC,113. Bond dimensions in the molecule of succinamide.

TABLE XIVC,60
Parameters of the Atoms in Succinamide[a]

Atom	x	y	z
C(1)	0.2496	−0.1943	0.0611
C(2)	0.2537	−0.0097	0.0285
N	0.2532	0.0933	0.1346
O	0.2559	0.0429	−0.0890
H(1)	0.125	−0.225	0.110
H(2)	0.375	−0.225	0.110
H(3)	0.250	0.203	0.117
H(4)	0.250	0.055	0.220

[a] Average $\sigma(x) = 0.022$ A., $\sigma(y) = \sigma(z) = 0.0014$ A.

XIV,c73.   The longer-chain dicarboxylic acids are dimorphous and complete structures have been established for examples of both forms. For the odd series the β modification was first determined through studies of *β-glutaric acid*, $HOOC(CH_2)_3COOH$, and *β-p.melic acid*, $HOOC(CH_2)_5$-COOH. Their tetramolecular monoclinic units have the dimensions:

β-Glutaric acid:
   $a_0 = 10.06$ A.; $b_0 = 4.87$ A.; $c_0 = 17.40$ A.; $\beta = 132°35'$
β-Pimelic acid:
   $a_0 = 9.84$ A.; $b_0 = 4.89$ A.; $c_0 = 22.43$ A.; $\beta = 130°40'$

The space group is C_{2h}^6 ($I2/a$) and all but one of the atoms have been found to be in general positions:

$$(8f)   \pm(xyz;\ x+{}^1/_2,\bar{y},z)\ ;\ \text{B.C.}$$

The central carbon atom in each compound is in the special positions:

$$(4e)   \pm({}^1/_4\ u\ 0)\ ;\ \text{B.C.}$$

Parameters are listed in Table XIVC,61.

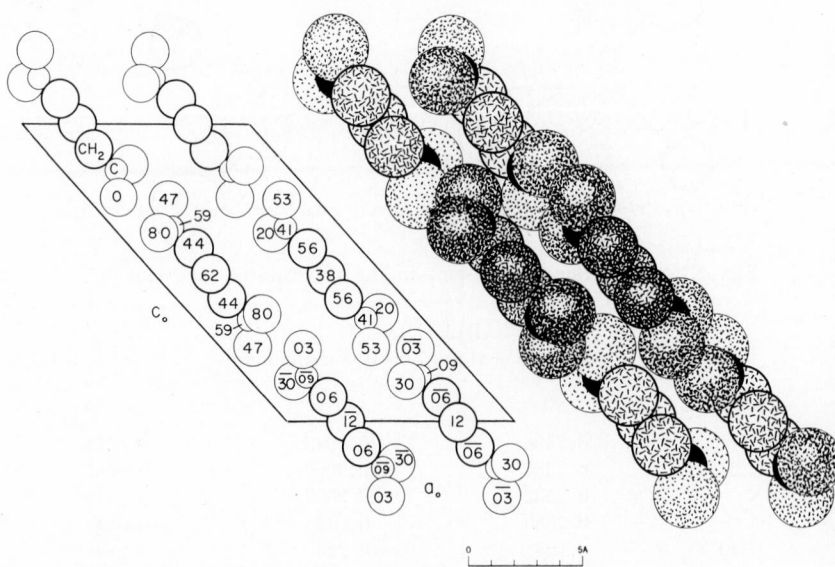

Fig. XIVC,114a (left). The monoclinic structure of β-glutaric acid projected along its b_0 axis. Left-hand axes.

Fig. XIVC,114b (right). A packing drawing of the monoclinic structure of β-glutaric acid seen along its b_0 axis. The carboxyl carbon atoms are black; the others are heavily outlined and line shaded. Carboxyl oxygens are dotted.

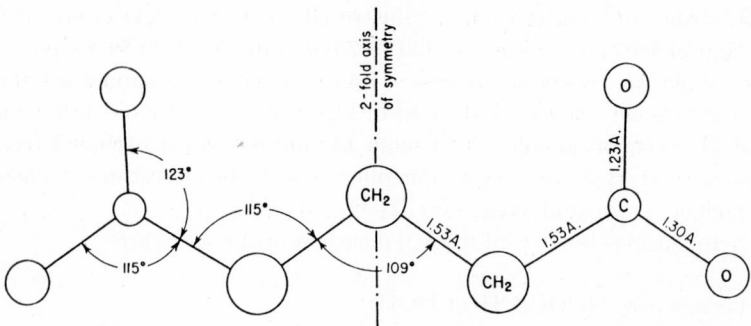

Fig. XIVC,115. Bond dimensions in the molecule of β-glutaric acid.

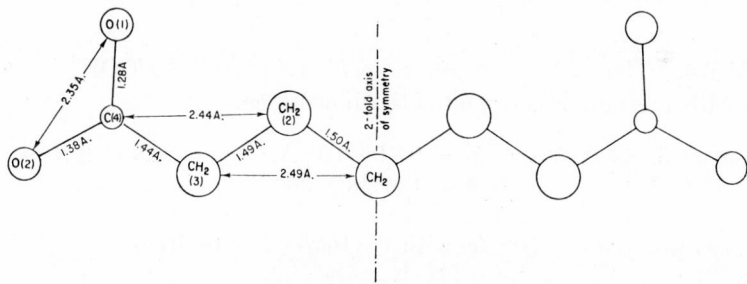

Fig. XIVC,116. Bond dimensions in the molecule of β-pimelic acid.

TABLE XIVC,61

Positions and Parameters of the Atoms in β-Glutaric and β-Pimelic Acids

Atom	Position	x	y	z
		β-Glutaric Acid		
$CH_2(1)$	(4e)	$^1/_4$	−0.122	0
$CH_2(2)$	(8f)	0.220	0.059	−0.082
C	(8f)	0.233	−0.086	−0.154
O(1)	(8f)	0.315	−0.303	−0.132
O(2)	(8f)	0.144	0.030	−0.244
		β-Pimelic Acid		
$CH_2(1)$	(4e)	$^1/_4$	0.008	0
$CH_2(2)$	(8f)	0.267	−0.150	0.063
$CH_2(3)$	(8f)	0.292	0.033	0.122
C	(8f)	0.278	−0.106	0.178
O(1)	(8f)	0.198	−0.330	0.161
O(2)	(8f)	0.366	0.036	0.244

The structure that results, as illustrated by Figure XIVC,114, differs from that of the even series acids (**XIV,c76**) mainly in the alternate positions of its nonplanar, twisted molecules. The molecular shapes and sizes of glutaric and pimelic acids are indicated in Figures XIVC,115 and 116. Closest contact between molecules is through hydrogen-bonded carboxyl oxygen atoms, with O–H–O = 2.69 A. For pimelic acid, the closest intermolecular approach of carbon and oxygen is CH_2–O = 3.59 A.

Cell dimensions for two additional members of the series are

β-*Azelaic acid*, $HOOC(CH_2)_7COOH$:
$$a_0 = 9.72 \text{ A.}; \quad b_0 = 4.83 \text{ A.}; \quad c_0 = 27.14 \text{ A.}; \quad \beta = 129°30'$$
Brassylic acid, $HOOC(CH_2)_{11}COOH$:
$$a_0 = 9.63 \text{ A.}; \quad b_0 = 4.82 \text{ A.}; \quad c_0 = 37.95 \text{ A.}; \quad \beta = 128°20'$$

XIV,c74. The α form of *pimelic acid*, $HOOC(CH_2)_5COOH$, is monoclinic with a tetramolecular unit of the dimensions:

$$a_0 = 5.68 \pm 0.05 \text{ A.}; \quad b_0 = 9.71 \pm 0.02 \text{ A.}; \quad c_0 = 22.45 \pm 0.10 \text{ A.}$$
$$\beta = 136°48' \pm 30'$$

The space group is C_{2h}^5 ($P2_1/c$) with all atoms in the positions:

$$(4e) \quad \pm (xyz; \; x, \tfrac{1}{2} - y, z + \tfrac{1}{2})$$

The determined parameters are those listed in Table XIVC,62.

TABLE XIVC,62
Parameters of the Atoms in α-Pimelic Acid

Atom	x	y	z
C(1)	0.466	0.074	0.0742
C(2)	0.439	0.147	0.1284
C(3)	0.534	0.062	0.2002
C(4)	0.531	0.151	0.2571
C(5)	0.564	0.064	0.3209
C(6)	0.570	0.157	0.3764
C(7)	0.547	0.085	0.4299
O(1)	0.243	0.102	−0.0060
O(2)	0.700	−0.009	0.1071
O(3)	0.514	−0.037	0.4309
O(4)	0.586	0.168	0.4844

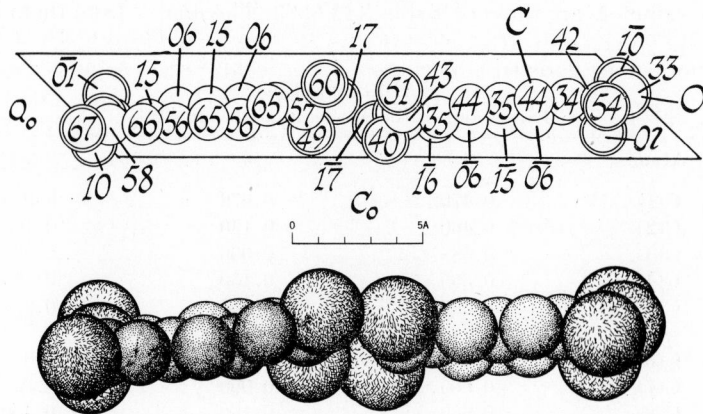

Fig. XIVC,117a (top). The monoclinic structure of the α modification of pimelic acid projected along its b_0 axis. Right-hand axes.

Fig. XIVC,117b (bottom). A packing drawing of the monoclinic structure of α-pimelic acid seen along its b_0 axis. The carbon atoms are dotted; the oxygen atoms are larger, fine-line shaded, and more heavily ringed.

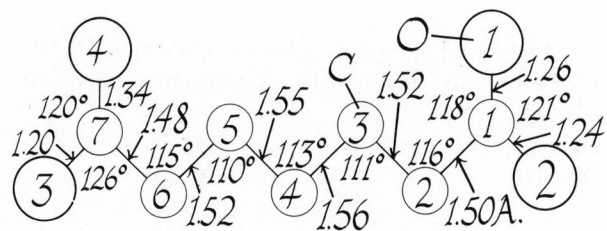

Fig. XIVC,118. Bond dimensions in the molecule of α-pimelic acid.

The resulting structure is shown in Figure XIVC,117; its molecules have the bond dimensions of Figure XIVC,118. They are tied together into chains extending along the c_0 axis by hydrogen bonds between the carboxyl oxygens, with O–H–O = 2.68 A. Their C–O–O angles lie between 113 and 125°. The shortest separation between chains is an O–O = 3.07 A. The final $R = 0.20$.

A study has been made of the corresponding, α, form of *azelaic acid*, $HOOC(CH_2)_7COOH$. Its unit has the dimensions:

$$a_0 = 5.67 \pm 0.01 \text{ A.}; \quad b_0 = 9.60 \pm 0.02 \text{ A.}; \quad c_0 = 27.35 \pm 0.04 \text{ A.}$$
$$\beta = 137°$$

The parameters are those of Table XIVC,63. The final R from them is 0.15.

TABLE XIVC,63
Parameters of the Atoms in α-Azelaic Acid

Atom	x	y	z
C(1)	0.470	0.070	0.065
C(2)	0.460	0.150	0.105
C(3)	0.585	0.060	0.165
C(4)	0.555	0.150	0.210
C(5)	0.590	0.055	0.265
C(6)	0.605	0.155	0.310
C(7)	0.605	0.060	0.355
C(8)	0.630	0.160	0.400
C(9)	0.595	0.085	0.445
O(1)	0.240	0.100	−0.005
O(2)	0.705	−0.015	0.090
O(3)	0.495	−0.040	0.440
O(4)	0.565	0.165	0.485

XIV,c75. Crystals of *meso α,α'-dimethyl glutaric acid*, $HO_2CCH(CH_3)$-$CH_2CH(CH_3)CO_2H$, are triclinic with a bimolecular cell of the dimensions:

$a_0 = 9.90$ A.; $b_0 = 8.35$ A.; $c_0 = 7.32$ A.,   all ±0.05 A.
$\alpha = 119°\pm30'$; $\beta = 72°\pm1'$; $\gamma = 126°30'$

The space group is C_i^1 $(P\bar{1})$ with all atoms in the positions $(2i)$ $\pm(xyz)$. The chosen parameters are listed in Table XIVC,64, those for hydrogen being calculated.

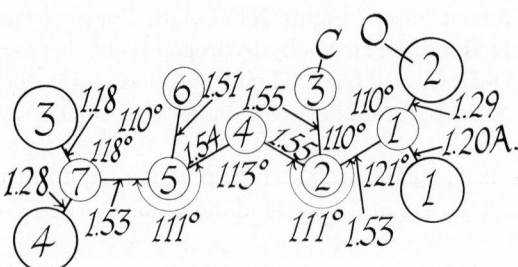

Fig. XIVC,119. Bond dimensions in the molecule of dimethyl glutaric acid.

TABLE XIVC,64
Parameters of the Atoms in Dimethyl Glutaric Acid

Atom	x	y	z
C(1)	0.166	0.285	0.120
C(2)	0.310	0.517	0.220
C(3)	0.404	0.537	0.369
C(4)	0.247	0.677	0.345
C(5)	0.177	0.700	0.202
C(6)	0.298	0.786	0.050
C(7)	0.101	0.838	0.330
O(1)	0.029	0.239	0.195
O(2)	0.222	0.169	0.974
O(3)	0.184	0.012	0.459
O(4)	0.951	0.761	0.307
H(1)	0.118	0.000	0.908
H(2)	0.389	0.548	0.083
H(3)	0.319	0.500	0.497
H(3')	0.442	0.418	0.280
H(3'')	0.510	0.695	0.423
H(4)	0.148	0.616	0.451
H(4')	0.346	0.830	0.433
H(5)	0.080	0.530	0.115
H(6)	0.229	0.774	0.936
H(6')	0.351	0.681	0.945
H(6'')	0.395	0.950	0.127
H(7)	0.891	0.856	0.418

The molecule that results has the bond dimensions of Figure XIVC,119. In the a_0b_0 plane, the closest approach of different molecules is through C–C = 3.54 and 3.58 A. The molecular layers along c_0 are in contact through O–O = 3.38 and 3.40 A., and through O–C = 3.42, 3.52, and 3.56 A.

XIV,c76. A determination has been made of the structure of *sebacic acid*, $HOOC(CH_2)_8COOH$, as an example of the even series of dicarboxylic acids. The bimolecular monoclinic unit of the α form has the dimensions:

$$a_0 = 10.05 \text{ A.}; \quad b_0 = 4.96 \text{ A.}; \quad c_0 = 15.02 \text{ A.}; \quad \beta = 133°50'$$

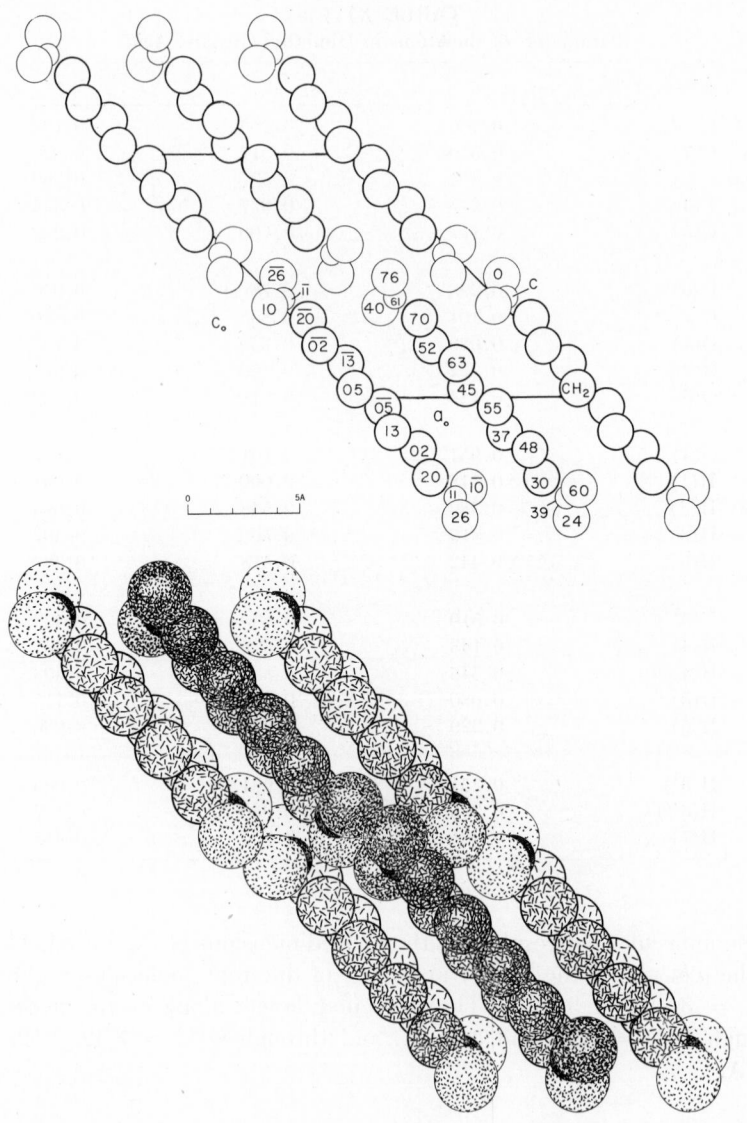

Fig. XIVC,120a (top). The monoclinic structure of sebacic acid projected along its b_0 axis. Left-hand axes.

Fig. XIVC,120b (bottom). A packing drawing of the monoclinic structure of sebacic acid viewed along its b_0 axis. The carboxyl carbon atoms are black; the others are line shaded. Atoms of oxygen are dotted.

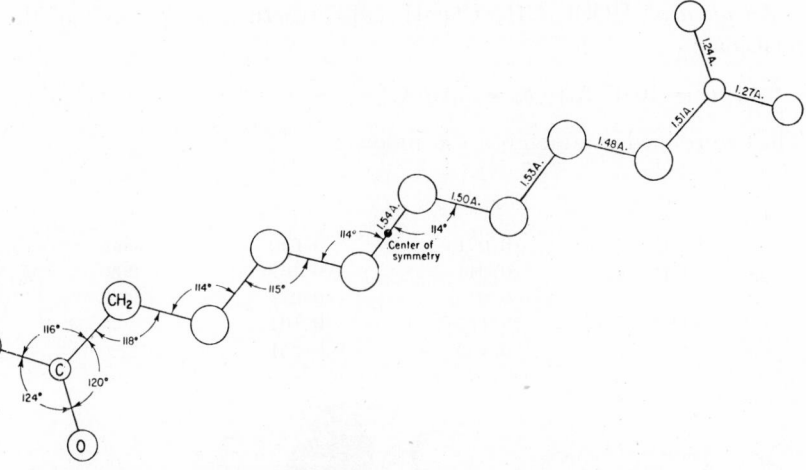

Fig. XIVC,121. Bond dimensions in the molecule of sebacic acid.

TABLE XIVC,65
Parameters of the Atoms in Sebacic Acid

Atom	x	y	z
CH$_2$(1)	−0.022	0.054	0.037
CH$_2$(2)	0.038	−0.126	0.138
CH$_2$(3)	0.003	−0.019	0.215
CH$_2$(4)	0.064	−0.200	0.315
C	0.035	−0.113	0.396
O(1)	−0.050	0.096	0.372
O(2)	0.103	−0.263	0.487
H(1)	−0.138	0.008	0.152
H(2)	0.061	0.169	0.247
H(3)	0.002	−0.389	0.280
H(4)	0.208	−0.228	0.378

All atoms are in the general positions of C_{2h}^5 ($P2_1/a$):

$$(4e) \quad \pm (xyz; \ x+^1/_2, ^1/_2-y, z)$$

with the parameters of Table XIVC,65.

The structure is shown in Figure XIVC,120. It is built up of molecules which are nearly planar and have a center of symmetry (Fig. XIVC,121). The closest approaches of these molecules are through hydrogen-bonded carboxyl oxygen atoms, with O–H–O = 2.68 A. and through a short CH$_2$–O = 3.32 A.

Adipic acid, $HOOC(CH_2)_4COOH$, is isostructural with a unit of the dimensions:

$$a_0 = 10.07 \text{ A.}; \quad b_0 = 5.16 \text{ A.}; \quad c_0 = 10.00 \text{ A.}; \quad \beta = 137°5'$$

The determined parameters are as follows:

Atom	x	y	z
$CH_2(1)$	−0.021	0.040	0.060
$CH_2(2)$	0.044	−0.162	0.202
C	0.027	−0.087	0.333
O(1)	−0.171	0.104	0.293
O(2)	0.102	−0.251	0.472

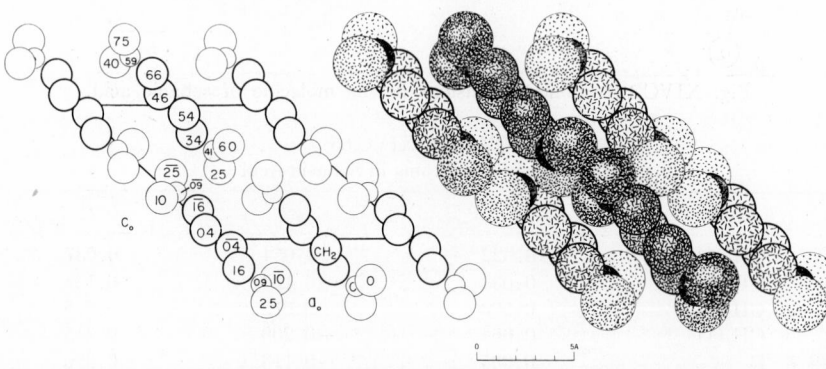

Fig. XIVC,122a (left). The monoclinic structure of adipic acid projected along its b_0 axis. Left-hand axes.

Fig. XIVC,122b (right). A packing drawing of the monoclinic adipic acid arrangement seen along its b_0 axis. The carboxyl carbon atoms are black; the others are heavily outlined and line shaded. Carboxyl oxygens are dotted.

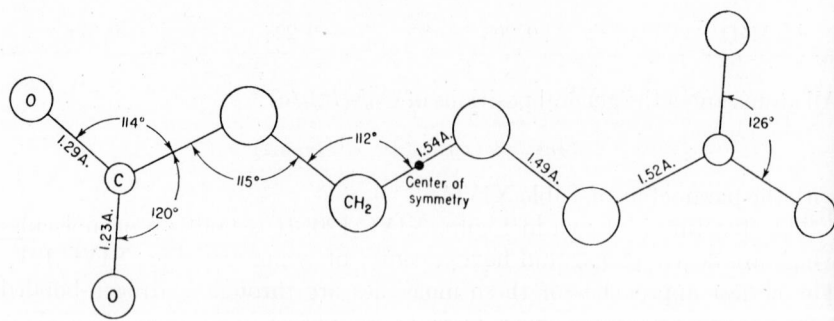

Fig. XIVC,123. Bond dimensions in the molecule of adipic acid.

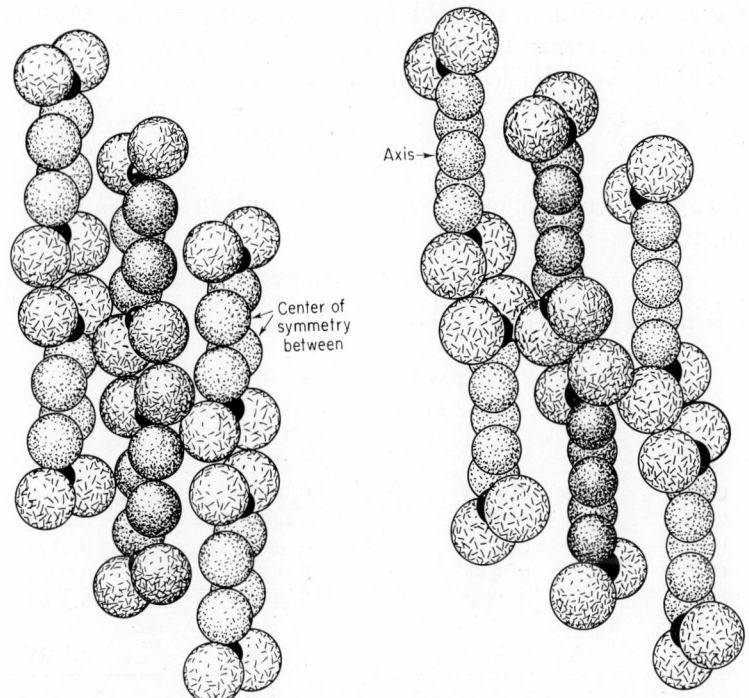

Fig. XIVC,124a (left). A general view of the molecular arrangement in crystals of adipic acid for comparison with the arrangement in β-pimelic acid shown in Figure XIVC,124b. In the even-numbered acids such as adipic there is a center of symmetry on the line joining the two central carbon atoms. In such odd-numbered acid structures as that of β-pimelic there is a twofold axis through the central carbon atom.
Fig. XIVC,124b (right). A general view of the molecular arrangement in crystals of β-pimelic acid for comparison with the adipic acid structure of Figure XIVC,124a.

The resulting atomic arrangement (Fig. XIVC,122) leads to molecules of the dimensions of Figure XIVC,123. In the drawings of Figure XIVC,124 the molecular arrangements are shown for crystals of adipic and pimelic (**XIV,c73**) acids. They indicate the way structures differ for the even and odd series of acids.

Cell dimensions have been measured for the following additional members of this series:

α-*Suberic acid*, $HOOC(CH_2)_6COOH$:
$a_0 = 10.12$ A.; $b_0 = 5.06$ A.; $c_0 = 12.58$ A.; $β = 135°$
Hexadecane dicarboxylic acid, $HOOC(CH_2)_{16}COOH$:
$a_0 = 9.76$ A.; $b_0 = 4.92$ A.; $c_0 = 25.10$ A.; $β = 131°10'$

XIV,c77. *Suberic acid,* $HOOC(CH_2)_6COOH$, is the member of the even-series dicarboxylic acids for which a β form has been analyzed. Its monoclinic crystals have a bimolecular cell of the dimensions:

$$a_0 = 8.98 \text{ A.}; \quad b_0 = 5.06 \text{ A.}; \quad c_0 = 10.12 \text{ A.}, \qquad \text{all } \pm 0.01 \text{ A.}$$
$$\beta = 97°50'$$

The space group has been found as C_{2h}^5 ($P2_1/c$) with all atoms in the general positions:

$$(4e) \quad \pm (xyz; \ x,{}^1/_2 - y,z + {}^1/_2)$$

The determined parameters are those of Table XIVC,66.

TABLE XIVC,66
Parameters of the Atoms in β-Suberic Acid

Atom	x	y	z
C(1)	0.0625	0.0860	0.0300
C(2)	0.1490	0.0070	0.1590
C(3)	0.2745	0.1865	0.2130
C(4)	0.3725	0.0980	0.3385
O(1)	0.3400	0.9040	0.4010
O(2)	0.4855	0.2695	0.3810

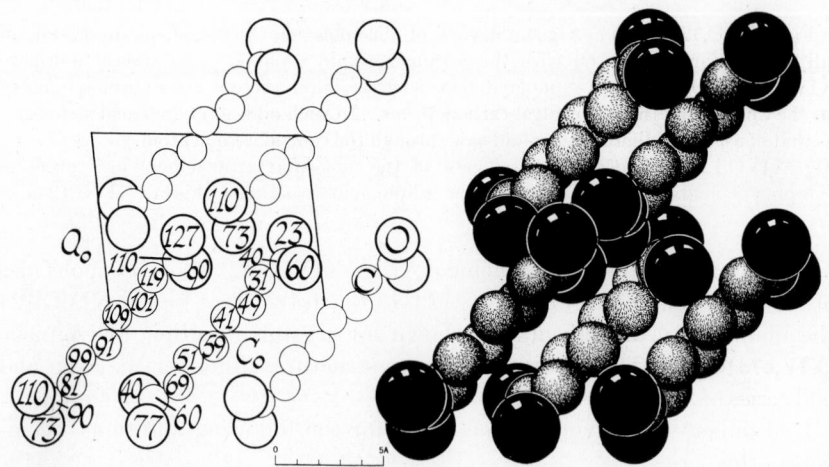

Fig. XIVC,125a (left). The monoclinic structure of the second, β, form of suberic acid projected along its b_0 axis. Right-hand axes.

Fig. XIVC,125b (right). A packing drawing of the monoclinic structure of the β form of suberic acid viewed along its b_0 axis. The oxygen atoms are black, the atoms of carbon smaller and fine-line shaded.

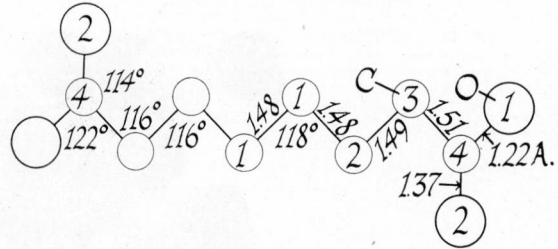

Fig. XIVC,126. Bond dimensions in the molecule of β-suberic acid.

The atomic arrangement is shown in Figure XIVC,125. The molecules that make it up have the dimensions of Figure XIVC,126. They are held together by hydrogen bonds between oxygen atoms, with O–H–O = 2.67 A. The final R = 0.10.

XIV,c78. The optically active form of *tartaric acid*, COOH(CHOH)₂-COOH, is monoclinic with a bimolecular unit of the dimensions:

$$a_0 = 7.72 \text{ A.}; \quad b_0 = 6.00 \text{ A.}; \quad c_0 = 6.20 \text{ A.}; \quad \beta = 100°10'$$

The space group is C_2^2 ($P2_1$) with all atoms in the general positions:

$$(2a) \quad xyz; \ \bar{x},y+{}^1/_2,\bar{z}$$

Chosen parameters are those of Table XIVC,67.

The structure that results is shown in Figure XIVC,127. As this suggests, the molecule is not planar. Atoms of each half are, however, within less than 0.1 A. of being coplanar, the plane of one half making 63° with that of the other (COOH·CHOH—). The bond lengths and angles, given

TABLE XIVC,67
Parameters of the Atoms in D-Tartaric Acid

Atom	x	y	z
C(1)	0.290	0.003	0.269
C(2)	0.208	0.210	0.170
C(3)	0.458	0.003	0.183
C(4)	0.034	0.240	0.250
O(1)	0.332	0.013	0.510
O(2)	0.312	0.413	0.252
O(3)	0.430	−0.010	−0.015
O(3′)	0.606	−0.014	0.300
O(4)	−0.060	0.086	0.182
O(4′)	−0.017	0.397	0.334

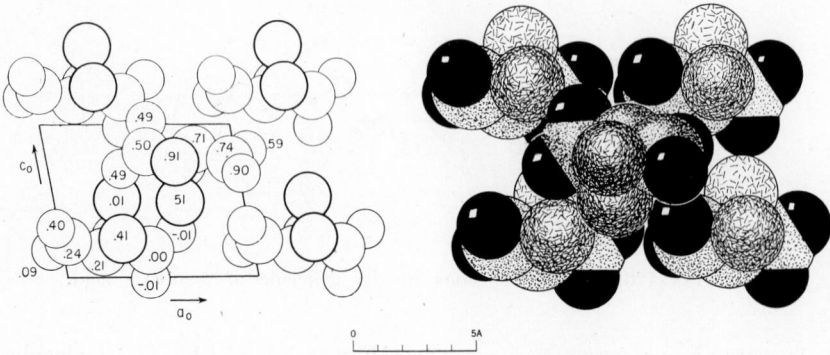

Fig. XIVC,127a (left). The monoclinic structure of optically active tartaric acid projected along its b_0 axis. In this drawing the carboxyl oxygen atoms are smaller than the atoms of carbon; the hydroxyls are larger and heavily outlined. Left-hand axes.

Fig. XIVC,127b (right). A packing drawing of the monoclinic structure of active tartaric acid seen along its b_0 axis. The carboxyl oxygens are black; the hydroxyls are line shaded. Carbon atoms are large and dotted.

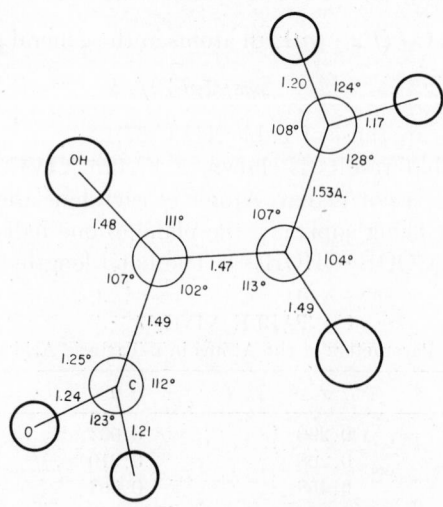

Fig. XIVC,128. Bond dimensions in the molecule of active tartaric acid.

in Figure XIVC,128, are substantially the same as those in the Rochelle salt ion and in the ion of its DL-isomer (Fig. XIVC,134). Between molecules the shortest distances are O–O = 2.75–2.92 A. Some of these are between carboxyl oxygen atoms, others between a carboxyl and a hydroxyl oxygen.

XIV,c79. *Lithium ammonium tartrate monohydrate,* $LiNH_4C_4H_4O_6 \cdot H_2O$, is orthorhombic with a tetramolecular unit of the dimensions:

$$a_0 = 7.878 \text{ Å.}; \quad b_0 = 14.642 \text{ Å.}; \quad c_0 = 6.426 \text{ Å.}$$

The space group is V^3 ($P2_12_12$) with ammonium ions in the special positions:

$$NH_4(1): \quad (2a) \quad 00u; \ ^1/_2 \ ^1/_2 \ \bar{u} \quad \text{with } u = 0.939$$
$$NH_4(2): \quad (2b) \quad 0 \ ^1/_2 \ u; \ ^1/_2 \ 0 \ \bar{u} \quad \text{with } u = 0.924$$

All other atoms have been placed in the general positions:

$$(4c) \quad xyz; \ \bar{x}\bar{y}z; \ x+^1/_2, ^1/_2-y, \bar{z}; \ ^1/_2-x, y+^1/_2, \bar{z}$$

with the parameters of Table XIVC,68.

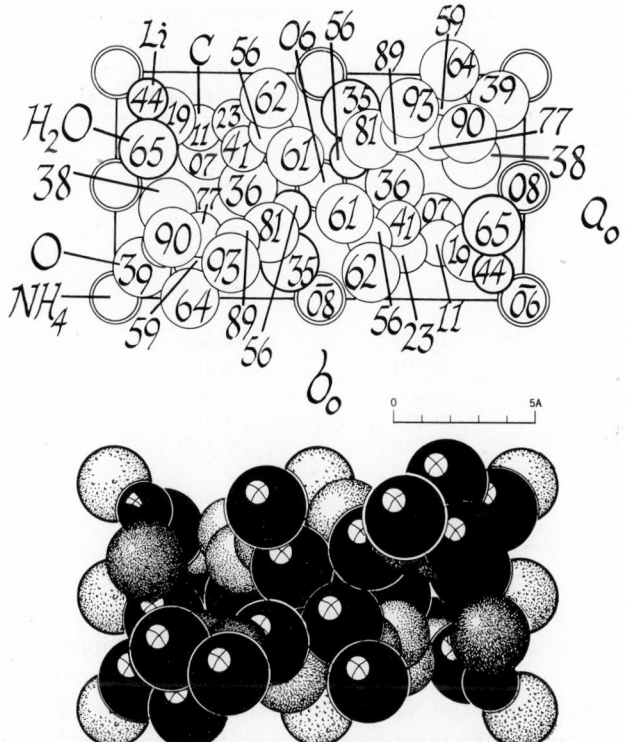

Fig. XIVC,129a (top). The orthorhombic structure of $LiNH_4C_4H_4O_6 \cdot H_2O$ projected along its c_0 axis. Right-hand axes.

Fig. XIVC,129b (bottom). A packing drawing of the orthorhombic structure of $LiNH_4 \cdot C_4H_4O_6 \cdot H_2O$ viewed along its c_0 axis. The oxygen atoms are the large, the lithium atoms the smaller black circles. The ammonium ions are heavily outlined and hook shaded. Water molecules are fine-line-and-dot shaded; the carbon atoms are dotted.

576   CRYSTAL STRUCTURES

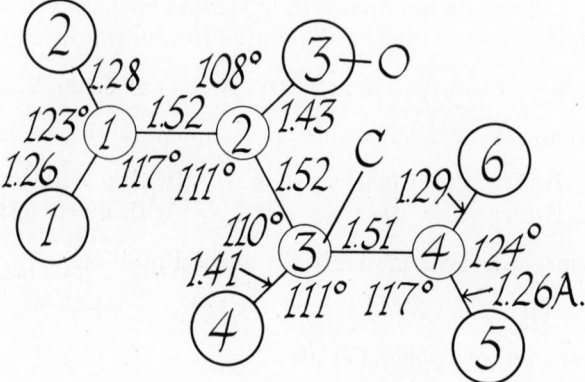

Fig. XIVC,130. Bond dimensions of the tartrate anion in LiNH$_4$C$_4$H$_4$O$_6$·H$_2$O.

TABLE XIVC,68
Parameters of the Atoms in LiNH$_4$C$_4$H$_4$O$_6$·H$_2$O

Atom	x	y	z
C(1)	0.272	0.375	0.556
C(2)	0.333	0.299	0.414
C(3)	0.217	0.289	0.226
C(4)	0.254	0.203	0.106
O(1)	0.121	0.370	0.621
O(2)	0.375	0.439	0.608
O(3)	0.504	0.318	0.356
O(4)	0.233	0.367	0.099
O(5)	0.314	0.211	0.074
O(6)	0.218	0.126	0.192
H$_2$O	0.325	0.074	0.650
Li	0.129	0.078	0.444

The resulting structure is shown in Figure XIVC,129. Its tartrate anions have the dimensions of Figure XIVC,130. The NH$_4$(1) ions are surrounded by tartrate oxygen atoms, the NH$_4$(2) ions by both tartrate and water oxygens; in both cases the NH$_4$–O separations lie between 2.89 and 3.27 A. The lithium atoms have five nearest oxygen neighbors, with Li–O = 1.90–2.22 A. The shortest H$_2$O–O = 2.68 A.

Other isostructural compounds have the following cell dimensions:

	a_0, A.	b_0, A.	c_0, A.
LiKC$_4$H$_4$O$_6$·H$_2$O	7.854	14.347	6.339
LiRbC$_4$H$_4$O$_6$·H$_2$O	7.89	14.65	6.36
LiTlC$_4$H$_4$O$_6$·H$_2$O	7.88	14.63	6.40

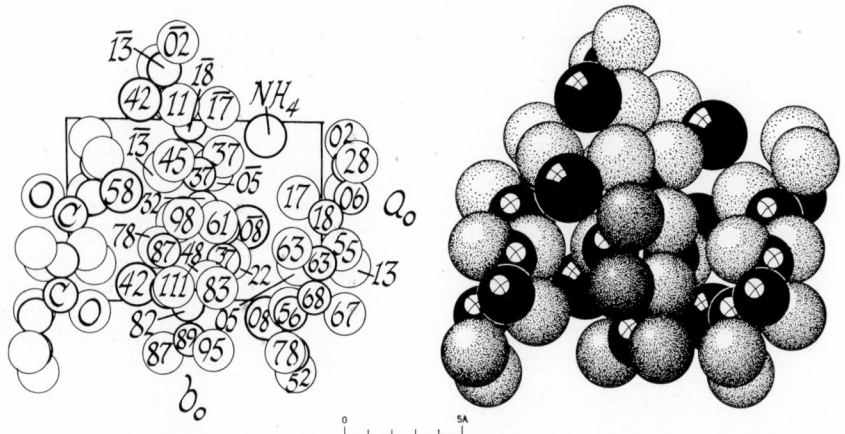

Fig. XIVC,131a (left). The orthorhombic structure of $NH_4HC_4H_4O_6$ projected along its c_0 axis. Right-hand axes.

Fig. XIVC,131b (right). A packing drawing of the orthorhombic $NH_4HC_4H_4O_6$ arrangement seen along its c_0 axis. The large black circles are ammonium, the smaller ones carbon. The larger oxygen atoms are dot-and-short-line shaded.

XIV,c80. *Ammonium acid* D-*tartrate*, $NH_4HC_4H_4O_6$, is orthorhombic with a tetramolecular cell of the edge lengths:

$$a_0 = 7.648 \text{ A.}; \quad b_0 = 11.066 \text{ A.}; \quad c_0 = 7.843 \text{ A.}, \qquad \text{all} \pm 0.003 \text{ A.}$$

The space group is V^4 ($P2_12_12_1$) with all atoms in the positions:

$$(4a) \quad xyz; \; {}^1\!/_2-x,\bar{y},z+{}^1\!/_2; \; x+{}^1\!/_2,{}^1\!/_2-y,\bar{z}; \; \bar{x},y+{}^1\!/_2,{}^1\!/_2-z$$

Accurately determined parameters are those of Table XIVC,69. Positions were found for the tartrate hydrogen atoms, but there was no indication

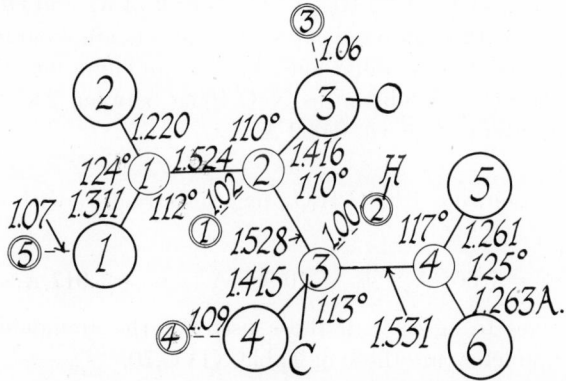

Fig. XIVC,132. Bond dimensions in the tartrate anion in $NH_4HC_4H_4O_6$.

TABLE XIVC,69
Parameters of the Atoms in $NH_4HC_4H_4O_6$

Atom	x	y	z
O(1)	0.7949	0.3835	0.3715
O(2)	0.7679	0.5758	0.4526
O(3)	0.4388	0.5898	0.3266
O(4)	0.4319	0.4172	0.6122
O(5)	0.2062	0.3604	0.2184
O(6)	0.1085	0.3991	0.4806
N	0.9160	0.2161	0.0716
C(1)	0.7078	0.4844	0.3896
C(2)	0.5202	0.4749	0.3255
C(3)	0.4186	0.3864	0.4376
C(4)	0.2293	0.3803	0.3747
H(1)	0.528	0.445	0.203
H(2)	0.475	0.306	0.431
H(3)	0.417	0.600	0.195
H(4)	0.378	0.506	0.645
H(5)	0.931	0.395	0.400

in the data for positions of the hydrogen atoms of the ammonium ions; accordingly, it was concluded that these cations are spherically symmetrical (i.e., "rotating").

The structure is shown in Figure XIVC,131; its tartrate anions have the bond dimensions of Figure XIVC,132. The equality of the C(4)–O(6) and C(4)–O(5) distances in contrast to the different distances of the oxygen atoms from C(1) suggests that the former is the ionized part of the anion. These ions are bound together into a sheet by hydrogen bonds O(1)–H(5)···O(6′) = 2.55 A., O(3)–H(3)···O(6′) = 2.74 A., and O(4)–H(4)··· O(5′) = 2.80 A. As the figure indicates, these sheets are separated in the b_0 direction by layers of ammonium ions. Each ammonium ion is surrounded by eight oxygen atoms at distances N–O lying between 2.85 and 3.15 A. For the heavier atoms, σ = ca. 0.002 A.

Rubidium acid tartrate, $RbHC_4H_4O_6$, has this structure with a unit of the cell edges:

$$a_0 = 7.665 \text{ A.}; \quad b_0 = 10.980 \text{ A.}; \quad c_0 = 7.917 \text{ A.}$$

Choosing its axes to agree with those used for the ammonium salt, the determined parameters are those of Table XIVC,70.

TABLE XIVC,70. Parameters of the Atoms in Rubidium Acid Tartrate

Atom	x	y	z
C(1)	0.713	0.485	0.389
C(2)	0.526	0.476	0.325
C(3)	0.421	0.383	0.435
C(4)	0.228	0.378	0.372
O(1)	0.783	0.386	0.367
O(2)	0.771	0.576	0.450
O(3)	0.440	0.593	0.326
O(4)	0.435	0.419	0.614
O(5)	0.219	0.363	0.218
O(6)	0.119	0.393	0.478
Rb	0.917	0.217	0.080

The *potassium* salt also has this structure with a cell of the dimensions:

$$a_0 = 7.64 \text{ A.}; \quad b_0 = 10.62 \text{ A.}; \quad c_0 = 7.75 \text{ A.}$$

No determination of parameters has been made.

For the isostructural *cesium* compound:

$$a_0 = 7.66 \text{ A.}; \quad b_0 = 11.58 \text{ A.}; \quad c_0 = 8.03 \text{ A.}$$

XIV,c81. *Rochelle salt*, $KNaC_4H_4O_6 \cdot 4H_2O$, and the isomorphous *rubidium* and *ammonium sodium tartrates* form orthorhombic crystals with tetramolecular units having the edges:

$KNaC_4H_4O_6 \cdot 4H_2O$:
$$a_0 = 11.93 \text{ A.}; \quad b_0 = 14.30 \text{ A.}; \quad c_0 = 6.17 \text{ A.}$$
$NH_4NaC_4H_4O_6 \cdot 4H_2O$:
$$a_0 = 12.15 \text{ A.}; \quad b_0 = 14.40 \text{ A.}; \quad c_0 = 6.18 \text{ A.}$$
$RbNaC_4H_4O_6 \cdot 4H_2O$:
$$a_0 = 12.05 \text{ A.}; \quad b_0 = 14.40 \text{ A.}; \quad c_0 = 6.21 \text{ A.}$$

Their space group is V^3 ($P2_12_12$) with atoms in the positions:

$(2a) \quad 0\ 0\ u;\ ^1/_2\ ^1/_2\ \bar{u}$

$(2b) \quad ^1/_2\ 0\ v;\ 0\ ^1/_2\ \bar{v}$

$(4c) \quad xyz;\ \bar{x}\bar{y}z;\ x+^1/_2, ^1/_2-y, \bar{z};\ ^1/_2-x, y+^1/_2, \bar{z}$

Approximate positions were determined years ago for Rochelle salt. Since then, the x and y, but not the z, parameters have been more accurately established at room temperature and at $-64\,^\circ$C. With the low-temperature values in parentheses they are stated in Table XIVC,71.

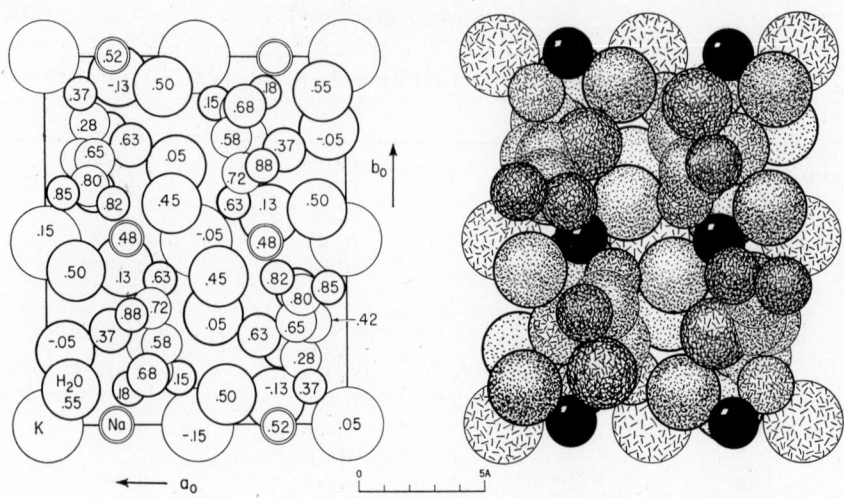

Fig. XIVC,133a (left). The orthorhombic structure of Rochelle salt projected along its c_0 axis. Carboxyl oxygens are the small, hydroxyls the middle sized, and water the largest, heavily outlined circles. Left-hand axes.

Fig. XIVC,133b (right). A packing drawing of the orthorhombic structure of KNa-$C_4H_4O_6 \cdot 4H_2O$ seen along its c_0 axis. The sodium atoms are black; the potassium are large, lightly ringed, and line shaded. Water molecules are heavily outlined and dotted. Carboxyl and hydroxyl oxygens also are heavily outlined but line shaded, the hydroxyls being the larger. Atoms of carbon are large, lightly outlined, and dotted.

TABLE XIVC,71

Parameters of the Atoms in Rochelle Salt (low-temperature values in parentheses)

Atom	x	y	z
K(1)	0	0	0.05
K(2)	0	$^1/_2$	0.15
Na	0.232 (0.232)	0.992 (0.992)	0.52
C(1)	0.153 (0.153)	0.188 (0.188)	0.28
C(2)	0.127 (0.127)	0.271 (0.271)	0.42
C(3)	0.175 (0.175)	0.267 (0.267)	0.65
C(4)	0.170 (0.170)	0.356 (0.356)	0.80
O(1)	0.121 (0.121)	0.109 (0.109)	0.37
O(2)	0.209 (0.211)	0.202 (0.202)	0.12
O(3)	0.235 (0.232)	0.406 (0.407)	0.82
O(4)	0.054 (0.050)	0.361 (0.359)	0.85
O(5)	0.154 (0.154)	0.356 (0.356)	0.32
O(6)	0.297 (0.298)	0.249 (0.249)	0.63
O(7)	0.397 (0.397)	0.084 (0.084)	0.50
O(8)	0.248 (0.244)	0.040 (0.040)	0.87
O(9)	0.440 (0.440)	0.301 (0.305)	0.05
O(10)	0.427 (0.426)	0.396 (0.395)	0.45

The structure that results is shown in Figure XIVC,133. The tartrate ion has substantially the same shape as in the more accurately determined tartrates just described. The coordination of the alkali atoms is indefinite. One potassium atom has two oxygen and two water molecules as closest neighbors; the other has oxygen and hydroxyl from the tartrate ions as well as water molecules.

XIV,c82. The *racemic potassium sodium tartrate tetrahydrate*, DL-$KNaC_4H_4O_6 \cdot 4H_2O$, has lower symmetry than its optically active isomer, Rochelle salt (**XIV,c81**). Its triclinic unit, containing two molecules, has the dimensions:

$$a_0 = 9.80 \text{ A.}; \quad b_0 = 9.66 \text{ A.}; \quad c_0 = 8.21 \text{ A.}$$
$$\alpha = 110°52'; \quad \beta = 101°28'; \quad \gamma = 119°44'$$

Atoms are in general positions of C_i^1 ($P\bar{1}$): $(2i)$ $\pm(xyz)$ with the parameters of Table XIVC,72.

TABLE XIVC,72
Parameters of the Atoms in Potassium Sodium DL-Tartrate Tetrahydrate

Atom	x	y	z
C(1)	0.125	0.848	0.904
C(2)	0.960	0.740	0.938
C(3)	0.995	0.842	0.151
C(4)	0.833	0.737	0.190
O(1)	0.270	0.940	0.033
O(2)	0.105	0.910	0.793
O(3)	0.745	0.797	0.226
O(4)	0.820	0.606	0.208
OH(1)	0.795	0.640	0.775
OH(2)	0.050	0.022	0.200
K	0.680	0.173	0.471
Na	0.590	0.723	0.900
$H_2O(1)$	0.580	0.480	0.892
$H_2O(2)$	0.643	0.948	0.596
$H_2O(3)$	0.362	0.575	0.604
$H_2O(4)$	0.406	0.770	0.001

The tartrate ion, as shown in Figure XIVC,134, has substantially the same bond lengths and angles as in other tartrates. The structure, too, re-

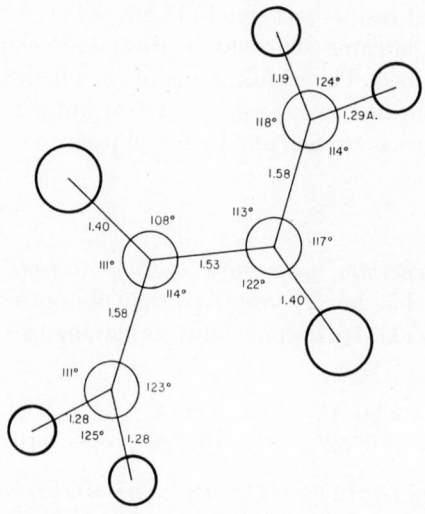

Fig. XIVC,134. Bond dimensions in the tartrate ion in racemic $KNaC_4H_4O_6 \cdot 4H_2O$.

sembles that of Rochelle salt but with the difference that in this racemic form, D- and L-tartrate ions occur in pairs. It is layered, like $D-NH_4HC_4H_4O_6$ (**XIV,c80**) but in this case D and L layers alternate.

XIV,c83. Crystals of DL-*ammonium antimony tartrate dihydrate*, DL-$NH_4SbC_4H_4O_6 \cdot 2H_2O$, are monoclinic with an eight-molecule cell of the dimensions:

$$a_0 = 13.81 \pm 0.02 \text{ A.;} \quad b_0 = 8.37 \pm 0.02 \text{ A.;} \quad c_0 = 16.01 \pm 0.03 \text{ A.}$$
$$\beta = 94°0' \pm 30'$$

The space group is C_{2h}^6 ($C2/c$) with all atoms in the positions:

$$(8f) \quad \pm (xyz; \; x,\bar{y},z+{}^1/_2; \; x+{}^1/_2,y+{}^1/_2,z; \; x+{}^1/_2,{}^1/_2-y,z+{}^1/_2)$$

The determined parameters are listed in Table XIVC,73.

The structure that results is shown in Figure XIVC,135. The tartrate radicals have the dimensions of Figure XIVC,136. The antimony atoms are distant from their oxygen atoms by 2.04–2.16 A. and it would appear that they combine to form ring-like anions of the composition $[Sb_2(C_4H_4O_6)_2]^{2-}$.

The *rubidium* salt, DL-$RbSbC_4H_4O_6 \cdot 2H_2O$, is isostructural with a unit of the dimensions:

$$a_0 = 13.87 \text{ A.;} \quad b_0 = 8.54 \text{ A.;} \quad c_0 = 15.73 \text{ A.;} \quad \beta = 94°0'$$

TABLE XIVC,73
Parameters of the Atoms in $NH_4SbC_4H_4O_6 \cdot 2H_2O$

Atom	x	y	z
Sb	0.125	0.332	0.136
O(1)	0.155	0.120	0.211
O(2)	0.314	0.537	0.158
O(3)	0.382	0.923	0.247
O(4)	0.492	0.725	0.124
O(5)	0.375	0.103	0.075
O(6)	0.033	0.532	0.104
$H_2O(1)$	0.179	0.002	0.019
$H_2O(2)$	0.123	0.763	0.244
C(1)	0.335	0.647	0.201
C(2)	0.347	0.812	0.180
C(3)	0.407	0.828	0.109
C(4)	0.442	−0.003	0.091
NH_4	0.023	0.892	0.098

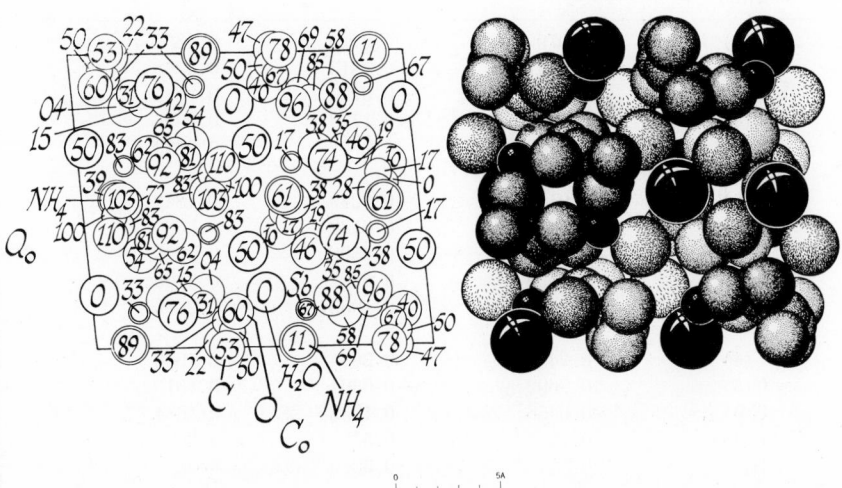

Fig. XIVC,135a (left). The monoclinic structure of DL-$NH_4SbC_4H_4O_6 \cdot 2H_2O$ projected along its b_0 axis. Right-hand axes.

Fig. XIVC,135b (right). The monoclinic structure of DL-$NH_4SbC_4H_4O_6 \cdot 2H_2O$ viewed along its b_0 axis. Ammonium ions are the large, antimony atoms the small black circles. Water molecules are heavily outlined and fine-line shaded. Atoms of oxygen are dotted; the carbon atoms, which show but poorly, are heavily outlined and hook shaded.

Fig. XIVC,136. Bond dimensions of the tartrate anion in $NH_4SbC_4H_4O_6 \cdot 2H_2O$.

XIV,c84. Anhydrous *citric acid*, $HC_6H_7O_7$, forms monoclinic crystals which have a tetramolecular cell ot the dimensions:

$$a_0 = 12.821 \pm 0.026 \text{ A.}; \quad b_0 = 5.622 \pm 0.012 \text{ A.}; \quad c_0 = 11.545 \pm 0.023 \text{ A.}$$
$$\beta = 111°10' \pm 6'$$

TABLE XIVC,74. Parameters of the Atoms in Citric Acid

Atom	x	y	z
O(1)	0.0790	0.6724	0.4358
O(2)	0.0447	0.2813	0.4268
O(3)	0.3777	−0.0773	0.1885
O(4)	0.1989	−0.1773	0.0905
O(5)	0.0142	0.2722	0.0682
O(6)	−0.0113	−0.0249	0.1849
O(7)	0.1994	0.0132	0.3520
C(1)	0.0898	0.4568	0.3998
C(2)	0.1598	0.4382	0.3195
C(3)	0.1627	0.1955	0.2616
C(4)	0.2489	0.2084	0.1960
C(5)	0.2692	−0.0304	0.1515
C(6)	0.0458	0.1332	0.1684
H(1)	0.15	0.55	0.25
H(2)	0.25	0.50	0.38
H(3)	0.21	0.30	0.12
H(4)	0.32	0.27	0.24
H(5)	0.06	0.68	0.51
H(6)	0.38	−0.23	0.16
H(7)	−0.06	0.25	0.02
H(8)	(0.14)	(−0.10)	(0.37) Assumed

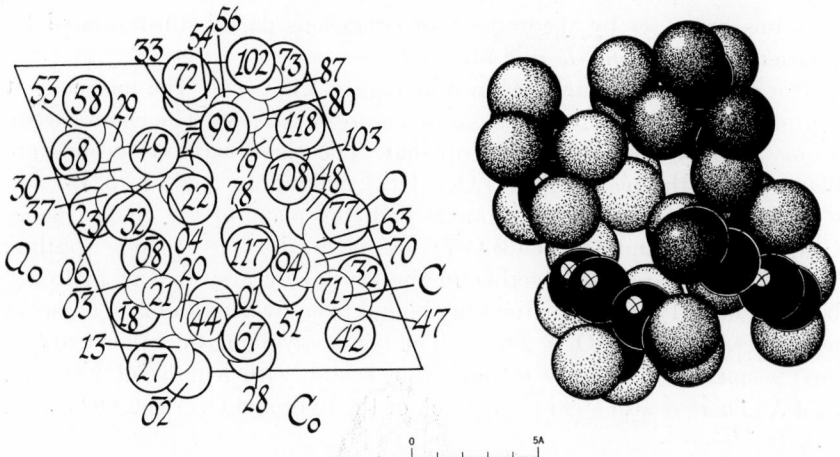

Fig. XIVC,137a (left). The monoclinic structure of citric acid projected along its b_0 axis. Right-hand axes.

Fig. XIVC,137b (right). A packing drawing of the monoclinic citric acid arrangement seen along its b_0 axis. The carbon atoms are black, the atoms of oxygen dot-and-short-line shaded.

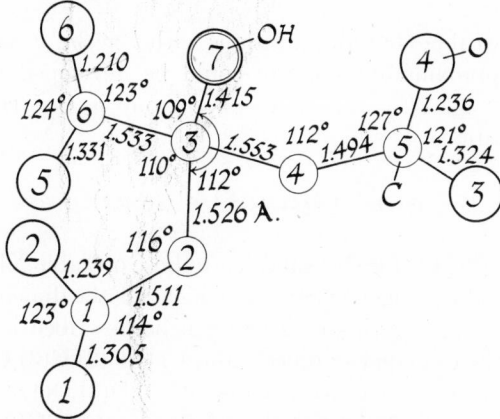

Fig. XIVC,138. Bond dimensions of the molecule in citric acid.

The space group is C_{2h}^5 ($P2_1/a$) with atoms in the positions:

$$(4e) \quad \pm(xyz; \; x+^1/_2, {}^1/_2-y, z)$$

The determined parameters are those of Table XIVC,74, the hydrogen positions being set by theoretical considerations partly substantiated by Fourier data. The final $R = 0.14$.

The resulting structure is shown in Figure XIVC,137. The molecules it contains have the bond dimensions of Figure XIVC,138. The chain carbon atoms [C(1)–C(5)] are nearly coplanar. The three carboxyl groups are planar, with the plane of C(1),O(1),O(2) making an angle of $3°18'$ and C(5),O(3),O(4) nearly a right angle with the plane of the chain. Planes C(6),O(5),O(6) and C(6),C(3),O(7) are inclined $11°18'$ to one another. The molecules are tied together in the crystal by a system of hydrogen bonds. The O(1) and O(2) atoms are bound to equivalent atoms in adjacent molecules, with O–H–O $= 2.63$ A. The bonds involving the other two carboxyls spiral continuously through the crystal, with lengths of 2.66 and 2.73 A. There is also a close approach of the hydroxyl O(7) and O(1) [2.84 A.].

XIV,c85. *Rubidium dihydrogen citrate*, $RbH_2[OOCCH_2C(OH)(COO)-CH_2COO]$, forms monoclinic crystals whose large 12-molecule cell has the dimensions:

$$a_0 = 14.924 \pm 0.03 \text{ A.}; \quad b_0 = 9.710 \pm 0.02 \text{ A.}; \quad c_0 = 19.145 \pm 0.04 \text{ A.}$$
$$\beta = 108°38'$$

Because of the weakness of all reflections with $l \neq 3n$, it has been considered that an approximate structure could be developed from a subcell having $c_0/3 = 6.382$ A. $= c_0'$. The space group is C_{2h}^5 ($P2_1/a$) with all atoms in the positions:

$$(4e) \quad \pm(xyz; x+\tfrac{1}{2},\tfrac{1}{2}-y,z)$$

The parameters selected for this subcell are those of Table XIVC,75.

They lead to a dihydrogen citrate ion which has the dimensions of Figure XIVC,139. The shape of this molecule can be described in terms of four planes: that of the principal chain contains C(1),C(2),C(3),C(4),O(1),O(2). Both the other terminal carboxyl planes [C(5),O(3),O(4) and the central C(6),O(5),O(6)] are turned out of this plane as is C(6),C(3),O(7) containing the hydroxyl radical. The rubidium atom has nine oxygen neighbors at distances between 2.90 and 3.25 A. There are several close approaches of oxygen atoms to one another corresponding to hydrogen bonds: between O(3) and O(5), with O–H–O $= 2.48$ A., between O(1) and O(6) [2.53 A.], between O(6) and O(2) [2.84 A.], and between carboxyl O(4) and hydroxyl O(7) [2.74 A.]. In the same molecule, O(6)–O(7) $= 2.70$ A.

TABLE XIVC,75. Parameters of the Atoms in $RbH_2C_6H_5O_7$

Atom	x	y	z
Rb	0.122	0.123	0.061
O(1)	0.356	−0.071	0.953
O(2)	0.316	0.131	0.018
O(3)	0.530	0.249	0.568
O(4)	0.473	0.401	0.289
O(5)	0.191	0.208	0.546
O(6)	0.238	0.396	0.784
O(7)	0.421	0.331	0.850
C(1)	0.352	0.056	0.894
C(2)	0.373	0.109	0.695
C(3)	0.355	0.261	0.679
C(4)	0.365	0.323	0.469
C(5)	0.465	0.320	0.431
C(6)	0.250	0.292	0.672

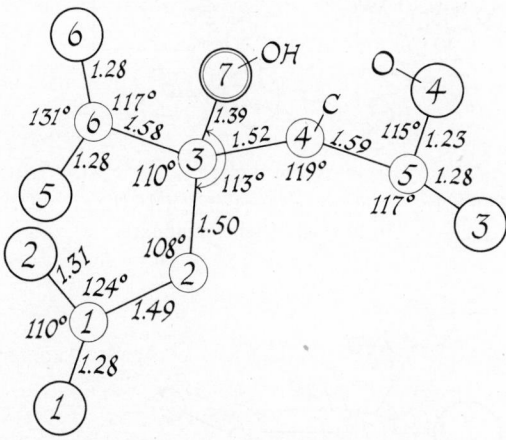

Fig. XIVC,139. Bond dimensions of the dihydrogen citrate anion in $RbH_2C_6H_5O_7$.

XIV,c86. A structure has been described for the mixed silico-hydrocarbon chain compound *2,4-di hydroxy-2,4-di methyl-2,4-di silapentane*, [OH-$(CH_3)_2Si$—]$_2CH_2$. It is orthorhombic with a tetramolecular cell of the edge lengths:

$$a_0 = 14.50 \text{ A.}; \quad b_0 = 11.31 \text{ A.}; \quad c_0 = 6.14 \text{ A.}$$

The space group is V^4 ($P2_12_12_1$) with all atoms in the positions:

(4a) $xyz;$ $^1/_2-x,\bar{y},z+^1/_2;$ $x+^1/_2,^1/_2-y,\bar{z};$ $\bar{x},y+^1/_2,^1/_2-z$

Atoms have been given the parameters of Table XIVC,76.

In this structure (Fig. XIVC,140) the elongated molecules have the approximate bond dimensions of Figure XIVC,141. These molecules are tied together into chains along the c_0 axis by hydrogen bonds between hydroxyl groups, with O–H–O = 2.64 A. Between molecules the shortest CH_3–CH_3 = 3.75 A., and the shortest C–O = 3.50 A.

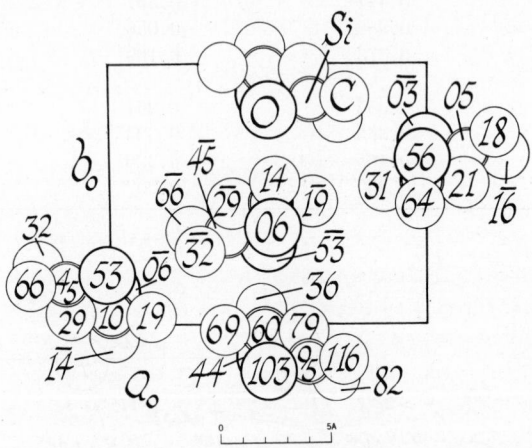

Fig. XIVC,140. The orthorhombic structure of $[OH(CH_3)_2Si—]_2CH_2$ projected along its c_0 axis. Right-hand axes.

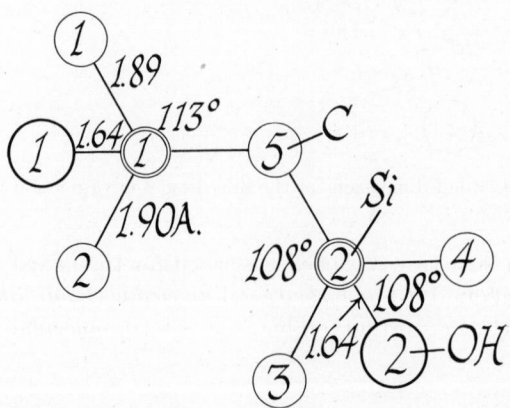

Fig. XIVC,141. Bond dimensions in the molecules of $[OH(CH_3)_2Si—]_2CH_2$.

TABLE XIVC,76
Parameters of the Atoms in $[OH(CH_3)_2Si—]_2CH_2$

Atom	x	y	z
Si(1)	0.018	0.0416	0.105
Si(2)	-0.125	0.157	0.453
O(1)	0.018	0.131	-0.0583
O(2)	0.008	0.189	0.535
C(1)	0.140	0.0283	0.188
C(2)	0.020	-0.0584	-0.141
C(3)	-0.233	0.146	0.658
C(4)	-0.210	0.225	0.325
C(5)	-0.115	0.051	0.294

6. Fatty Acids and Derivatives

The extreme pleomorphism of many of the long-chain compounds as well as the fragility of their crystals has made it very difficult to gain a thorough understanding of their structures. Work carried out in recent years has finally led to complete determinations of atomic positions for a number of these compounds as well as to partial structures for several others. There is, however, much more to be done to provide knowledge of all the various modifications of representative groups of these substances. In the following paragraphs the structures that have been fully analyzed and those for which a projection gives two coordinates will be described in detail, but it has not been considered appropriate to do more than refer in passing to less complete work. Hydrocarbons will be considered first and these will be followed by the fatty acids and their simple derivatives and finally by their salts.

XIV,c87. The low-temperature form of the even-numbered saturated hydrocarbons is triclinic with a unimolecular cell. For normal *octane*, C_8H_{18}, which has been studied in detail, it has the dimensions:

$a_0 = 4.16$ A.; $b_0 = 4.75$ A.; $c_0 = 11.00$ A., all ±0.02 A.
$\alpha = 94°48'$; $\beta = 85°30'$; $\gamma = 105°6'$, all $\pm18'$

The space group is C_i^1 ($P\bar{1}$) with all atoms in the general positions (2i) $\pm(xyz)$. The determined parameters, including those for the hydrogen atoms, are listed in Table XIVC,77.

The resulting structure is shown in Figure XIVC,142. In the molecule the important bond distances are C(1)–C(2) = C(3)–C(4) = 1.542 A., C(2)–C(3) = 1.519 A., and C(4)–C(5) = 1.514 A. The bond angles lie between 111.8 and 112.3°. The C–H separations are 0.93–1.32 A.

TABLE XIVC,77
Parameters of the Atoms in $N\text{-}C_8H_{18}$

Atom	x	y	z
C(1)	0.3037	0.2447	0.3811
C(2)	0.1715	0.0161	0.2777
C(3)	0.1417	0.1585	0.1620
C(4)	0.0176	-0.0699	0.0576
H(1)	0.325	0.150	0.449
H(2)	0.550	0.380	0.360
H(3)	0.117	0.379	0.390
H(4)	0.390	-0.126	0.267
H(5)	-0.083	-0.120	0.297
H(6)	0.387	0.293	0.140
H(7)	-0.033	0.295	0.166
H(8)	0.242	-0.223	0.050
H(9)	-0.235	-0.191	0.075

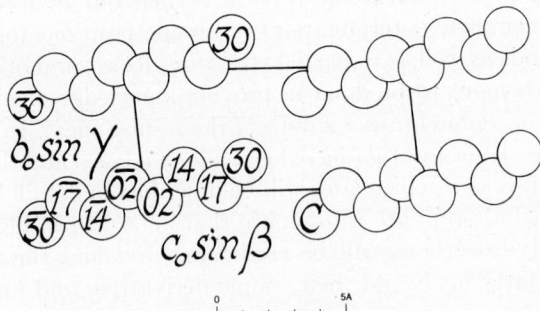

Fig. XIVC,142. The triclinic structure of C_8H_{18} projected along its a_0 axis. Left-hand axes.

A structure has been established for $N\text{-}hexane$, C_6H_{14}, which is isostructural. The unit has the dimensions:

$$a_0 = 4.19 \text{ A.;} \quad b_0 = 4.75 \text{ A.;} \quad c_0 = 8.62 \text{ A.,} \quad \text{all } +0.02 \text{ A.}$$
$$\alpha = 97°0'; \quad \beta = 85°36'; \quad \gamma = 105°0', \quad \text{all } \pm18'$$

The parameters determined for its carbon atoms are as follows:

Atom	x	y	z
C(1)	0.2795	0.2322	0.3482
C(2)	0.1366	-0.0065	0.2153
C(3)	0.0723	0.1180	0.0664

Some, but not all, of the parameters have been found for the hydrogen atoms. In the molecule, $C(1)-C(2) = 1.551$ A., $C(2)-C(3) = 1.547$ A., and $C(3)-C(4) = 1.545$ A.; bond angles are $114°$.

XIV,c88. Normal *octadecane*, $N\text{-}C_{18}H_{38}$, is triclinic with a unit cell of the dimensions:

$$a_0 = 4.361 \text{ A.}; \quad b_0 = 4.893 \text{ A.}; \quad c_0 = 25.04 \text{ A.}$$
$$\alpha = 83°6'; \quad \beta = 67°4'; \quad \gamma = 111°24'$$

The space group is C_i^1 ($P\bar{1}$) with all atoms in the general positions ($2i$) $\pm(xyz)$. In accordance with an accepted bond distance of $C-C = 1.54$ A. and a bond angle of $114°$, the carbon parameters have been taken to be

For C_{2n-1}:
$$x = -0.0281n+0.621, \, y = 0.0427n+0.178, \, z = 0.1042n-0.046$$
For C_{2n}:
$$x = -0.0281n+0.660, \, y = 0.0427n+0.389, \, z = 0.1042n+0.003$$

Parameters were also proposed for the atoms of hydrogen.

XIV,c89. The monoclinic form of *N-hexatriacontane*, $N\text{-}C_{36}H_{74}$, has a bimolecular unit of the dimensions:

$$a_0 = 5.57 \pm 0.01 \text{ A.}; \quad b_0 = 7.42 \pm 0.01 \text{ A.}; \quad c_0 = 48.35 \pm 0.08 \text{ A.}$$
$$\beta = 119°6' \pm 4'$$

The space group is C_{2h}^5 ($P2_1/a$) with all atoms in the positions:

$$(4e) \quad \pm(xyz; \, x+^1/_2, ^1/_2-y, z)$$

The determined parameters are those of Table XIVC,78.

The resulting structure is shown in Figure XIVC,143. In the molecule, $C-C$ ranges from 1.50 to 1.57 A. and the $C-C-C$ angle lies between 110.6 and $113.3°$; there is, however, no indication of a regular alternation of bond dimensions along the chain.

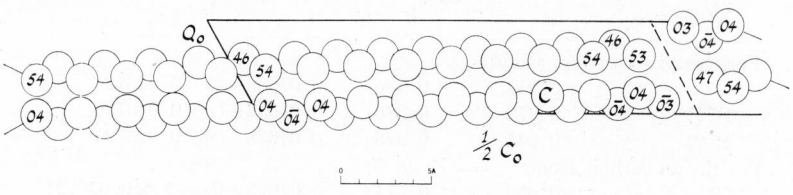

Fig. XIVC,143. The monoclinic structure of $C_{36}H_{74}$ projected along its b_0 axis. Right-hand axes.

TABLE XIVC,78

Parameters of the Atoms in N-$C_{36}H_{74}$

Atom	x	y	z
C(1)	0.070	0.038	0.0173
C(2)	−0.056	−0.038	0.0364
C(3)	0.084	0.037	0.0705
C(4)	−0.041	−0.040	0.0898
C(5)	0.102	0.038	0.1243
C(6)	−0.025	−0.040	0.1432
C(7)	0.115	0.036	0.1775
C(8)	−0.009	−0.040	0.1970
C(9)	0.131	0.038	0.2312
C(10)	0.005	−0.040	0.2506
C(11)	0.145	0.036	0.2843
C(12)	0.020	−0.038	0.3040
C(13)	0.160	0.038	0.3377
C(14)	0.040	−0.039	0.3580
C(15)	0.178	0.039	0.3911
C(16)	0.057	−0.036	0.4114
C(17)	0.196	0.038	0.4444
C(18)	0.074	−0.034	0.4647

XIV,c90. The orthorhombic form of *hexatriacontane*, $C_{36}H_{74}$, has a tetra-molecular cell of the edge lengths:

$$a_0 = 7.42 \pm 0.01 \text{ A.}; \quad b_0 = 4.96 \pm 0.01 \text{ A.}; \quad c_0 = 95.14 \pm 0.20 \text{ A.}$$

All atoms are in the general positions of C_{2v}^5 ($Pca2_1$):

$$(4a) \quad xyz; \ \bar{x},\bar{y},z+\tfrac{1}{2}; \ \tfrac{1}{2}-x,y,z+\tfrac{1}{2}; \ x+\tfrac{1}{2},\bar{y},z$$

with parameters for both the carbon and hydrogen atoms that can be given the abbreviated description:

Atom	x	y	z
For the $2n+1$ carbon atom:			
C	0.014	0.186	$0.01638+(n \times 0.02672)$
H(1)	0.872	0.286	$0.01638+(n \times 0.02672)$
H(2)	0.045	0.976	$0.01638+(n \times 0.02672)$
For the $2n$ carbon atom:			
C′	0.092	0.314	$0.02973+[(n-1) \times 0.02672]$
H(1′)	0.061	0.524	$0.02973+[(n-1) \times 0.02672]$
H(2′)	0.234	0.294	$0.02973+[(n-1) \times 0.02672]$

These parameters correspond to a C–C $= 1.533$ A. and a C–C–C $= 111°54'$. The side-by-side packing of the planar chains of this structure is very similar to that of the monoclinic modification (**XIV,c89**).

It would appear that this form is the one most common among the higher normal hydrocarbons. It was pointed out many years ago that compounds C_nH_{2n+2} are isostructural and that the dimensions of their tetramolecular units are described by the relation:

$$a_0 = 7.45 \text{ A.}; \quad b_0 = 4.86 \text{ A.}; \quad c_0 = 2.54n+4.0 \text{ A.}$$

XIV,c91. A partial structure for the so-called A form of *lauric acid*, $CH_3(CH_2)_{10}COOH$, has been described in terms of a large triclinic cell having 12 molecules in the unit. This has the dimensions:

$$a_0 = 5.41\pm0.01 \text{ A.}; \quad b_0 = 26.27\pm0.07 \text{ A.}; \quad c_0 = 35.42\pm0.13 \text{ A.}$$
$$\alpha = 69°36'\pm30'; \quad \beta = 113°9'\pm26'; \quad \gamma = 121°21'\pm34'$$

The space group is $C_i{}^1$ $(P\bar{1})$ with all atoms in the general positions $(2i)$ $\pm(xyz)$. Parameters proposed for y and z, but not for x, are given in the original publication. Further work, however, is desirable to be certain that a triclinic cell as big as this actually is primitive.

XIV,c92. *Lauric acid*, $CH_3(CH_2)_{10}COOH$, has a so-called A_1 form which, like the other A forms, is triclinic. It has a bimolecular cell of the dimensions:

$$a_0 = 7.45 \text{ A.}; \quad b_0 = 5.40 \text{ A.}; \quad c_0 = 17.47 \text{ A.}$$
$$\alpha = 96°53'; \quad \beta = 113°8'; \quad \gamma = 81°7'$$

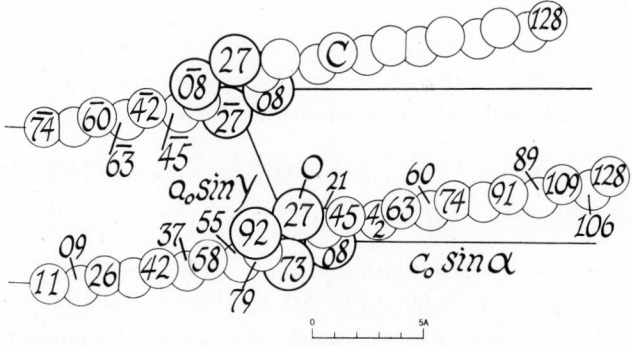

Fig. XIVC,144. The triclinic structure of the A_1 form of lauric acid projected along its b_0 axis. Right-hand axes.

All atoms have been placed in the general positions of C_i^1 $(P\bar{1})$: $(2i)$ $\pm(xyz)$, with the parameters of Table XIVC,79.

TABLE XIVC,79
Parameters of the Atoms in A_1-Lauric Acid

Atom	x	y	z
C(1)	0.0727	0.2067	0.0928
C(2)	0.1597	0.4509	0.1598
C(3)	0.1256	0.4174	0.2429
C(4)	0.2138	0.6293	0.3138
C(5)	0.1802	0.5974	0.3862
C(6)	0.2664	0.7433	0.4625
C(7)	0.2364	0.7443	0.5373
C(8)	0.3199	0.9135	0.6093
C(9)	0.3029	0.8857	0.6905
C(10)	0.3834	0.0958	0.7617
C(11)	0.3619	0.0592	0.8385
C(12)	0.4320	0.2804	0.8981
O(1)	0.1661	0.2665	0.0381
O(2)	0.9866	0.0841	0.0927

The resulting structure is shown in Figure XIVC,144. In the molecules, which are tied together in pairs by hydrogen bonds between facing oxygen atoms, the average C–C distance is 1.505 A. and the average distance between alternate carbon atoms of the chain is 2.538 A. The C–O separations are given as 1.45 and 0.99 A. The hydrogen-bond length between molecules is O–H–O = 2.77 A.

XIV,c93. Crystals of the A′ form of *N-pentadecanoic acid*, $CH_3(CH_2)_{13}$-COOH, are triclinic with a bimolecular unit of the dimensions:

$$a_0 = 4.25\pm0.02 \text{ A.}; \quad b_0 = 5.01\pm0.02 \text{ A.}; \quad c_0 = 42.76\pm0.17 \text{ A.}$$
$$\alpha = 89°50'\pm25'; \quad \beta = 111°5'\pm20'; \quad \gamma = 112°10'\pm20'$$

The space group is C_i^1 $(P\bar{1})$ with all atoms in the positions: $(2i)$ $\pm(xyz)$. Parameters, as listed in Table XIVC,80, were determined for the y and z but not for the x coordinates. In the original, suggested parameters were also given for the hydrogen atoms. According to this determination, the carbon atoms of the molecule are not coplanar.

TABLE XIVC,80
Parameters of the Atoms in A'-$C_{14}H_{29}COOH$

Atom	y	z
O(1)	0.244	0.0041
O(2)	0.621	0.0434
C(1)	0.401	0.0350
C(2)	0.306	0.0540
C(3)	0.459	0.0926
C(4)	0.341	0.1160
C(5)	0.516	0.1532
C(6)	0.373	0.1798
C(7)	0.575	0.2131
C(8)	0.410	0.2433
C(9)	0.630	0.2735
C(10)	0.454	0.3062
C(11)	0.681	0.3337
C(12)	0.497	0.3681
C(13)	0.725	0.3941
C(14)	0.547	0.4308
C(15)	0.767	0.4546

XIV,c94.   The B modification of *stearic acid*, $C_{17}H_{35}COOH$, is monoclinic with a tetramolecular cell of the dimensions:

$$a_0 = 5.591 \pm 0.011 \text{ A.}; \quad b_0 = 7.404 \pm 0.008 \text{ A.}; \quad c_0 = 49.38 \pm 0.010 \text{ A.}$$
$$\beta = 117°22' \pm 7'$$

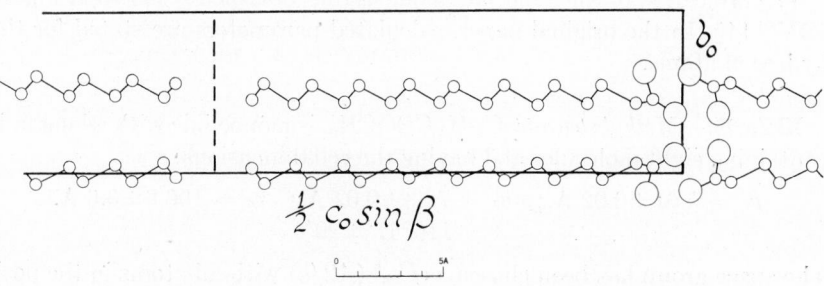

Fig. XIVC,145. The monoclinic structure of the B form of stearic acid projected along its a_0 axis. The oxygens are the larger circles. Right-hand axes.

TABLE XIVC,81
Parameters of the Atoms in B-Stearic Acid

Atom	y	z
O(1)	0.113	0.0087
O(2)	0.898	0.0379
C(1)	0.020	0.0320
C(2)	0.040	0.0507
C(3)	0.960	0.0866
C(4)	0.040	0.1026
C(5)	0.960	0.1385
C(6)	0.040	0.1545
C(7)	0.960	0.1904
C(8)	0.040	0.2064
C(9)	0.960	0.2423
C(10)	0.040	0.2583
C(11)	0.960	0.2942
C(12)	0.040	0.3102
C(13)	0.960	0.3461
C(14)	0.040	0.3621
C(15)	0.960	0.3980
C(16)	0.040	0.4140
C(17)	0.960	0.4499
C(18)	0.040	0.4659

The space group is C_{2h}^5 ($P2_1/a$) with all atoms in the positions:

$$(4e) \quad \pm (xyz; x+{}^1/_2, {}^1/_2-y, z)$$

The shape of the crystals was such that the x parameters could not be established, but the values for y and z are listed in Table XIVC,81.

The projection of the molecules along a_0 thus obtained is shown in Figure XIVC,145. In the original paper, calculated parameters are stated for the hydrogen atoms.

XIV,c95. *Methyl stearate*, $C_{17}H_{35}COOCH_3$, is monoclinic with a long unit containing eight molecules and having the cell dimensions:

$$a_0 = 5.61 \pm 0.02 \text{ A.}; \quad b_0 = 7.33 \pm 0.02 \text{ A.}; \quad c_0 = 106.6 \pm 0.6 \text{ A.}$$
$$\beta = 116°47' \pm 20'$$

The space group has been chosen as C_{2h}^6 ($A2/a$) with all atoms in the positions:

$$(8f) \quad \pm (xyz; x+{}^1/_2, \bar{y}, z; x, y+{}^1/_2, z+{}^1/_2; x+{}^1/_2, {}^1/_2-y, z+{}^1/_2)$$

The parameters as selected are listed in Table XIVC,82.

TABLE XIVC,82
Parameters of the Atoms in Methyl Stearate

Atom	x	y	z
C(1)	0.133	0.247	0.0342
C(2)	0.019	0.207	0.0433
C(3)	0.156	0.290	0.0584
C(4)	0.031	0.209	0.0674
C(5)	0.170	0.291	0.0821
C(6)	0.042	0.211	0.0913
C(7)	0.172	0.290	0.1062
C(8)	0.057	0.211	0.1150
C(9)	0.188	0.288	0.1302
C(10)	0.066	0.209	0.1391
C(11)	0.188	0.291	0.1538
C(12)	0.082	0.207	0.1629
C(13)	0.197	0.295	0.1778
C(14)	0.096	0.201	0.1869
C(15)	0.209	0.297	0.2019
C(16)	0.106	0.201	0.2110
C(17)	0.229	0.298	0.2258
C(18)	0.121	0.207	0.2355
CH_3	0.141	0.258	0.0119
O(1)	0.001	0.221	0.0213
O(2)	0.357	0.338	0.0383

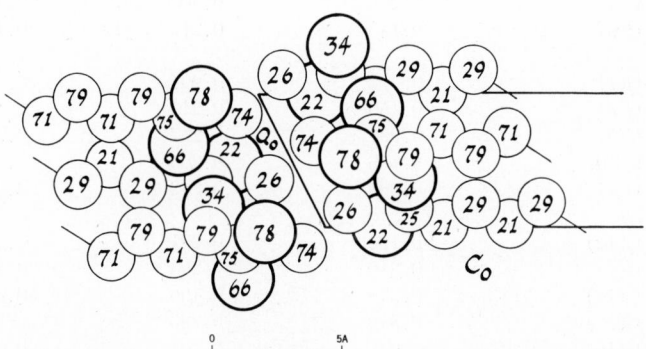

Fig. XIVC,146. The monoclinic structure of methyl stearate projected along its b_0 axis. The cell is so big that only the ends of molecules can be shown. The larger, more heavily ringed circles are oxygen. Right-hand axes.

This structure (Fig. XIVC,146) resembles that of the B form of stearic acid (**XIV,c94**) but with a cell twice as long in the c_0 direction. The ester ends of the molecules face one another, with $CH_3-CH_3 = 3.3$ A. The average C–C separation in the planar chains is 1.54 A. The final value of R is 0.18.

XIV,c96. *Ethyl stearate*, $C_{17}H_{35}COOC_2H_5$, forms monoclinic crystals which have a tetramolecular unit of the dimensions:

$$a_0 = 5.59 \pm 0.02 \text{ A.}; \quad b_0 = 7.40 \pm 0.02 \text{ A.}; \quad c_0 = 57.1 \pm 0.4 \text{ A.}$$
$$\beta = 118°30'$$

The space group has been selected as C_s^4 (Aa) with all atoms in the positions:

$$(4a) \quad xyz; \ x+{}^1/_2,\bar{y},z; \ x,y+{}^1/_2,z+{}^1/_2; \ x+{}^1/_2,{}^1/_2-y,z+{}^1/_2$$

The chosen parameters are those of Table XIVC,83.

TABLE XIVC,83
Parameters of the Atoms in Ethyl Stearate

Atom	x	y	z
O(1)	−0.015	0.247	0.0555
O(2)	0.375	0.320	0.0905
C(e1)	−0.293	0.253	0.0420
C(e2)	−0.525	0.245	0.0138
C(1)	0.182	0.233	0.0810
C(2)	0.074	0.217	0.1010
C(3)	0.248	0.271	0.1320
C(4)	0.141	0.207	0.1500
C(5)	0.289	0.300	0.1774
C(6)	0.143	0.212	0.1922
C(7)	0.272	0.297	0.2200
C(8)	0.159	0.213	0.2387
C(9)	0.291	0.297	0.2660
C(10)	0.176	0.205	0.2840
C(11)	0.298	0.282	0.3100
C(12)	0.198	0.201	0.3295
C(13)	0.316	0.293	0.3560
C(14)	0.218	0.196	0.3730
C(15)	0.319	0.295	0.3992
C(16)	0.217	0.205	0.4187
C(17)	0.353	0.312	0.4444
C(18)	0.266	0.210	0.4633

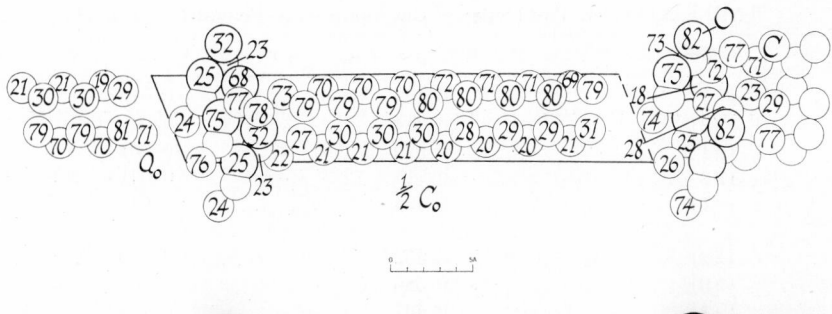

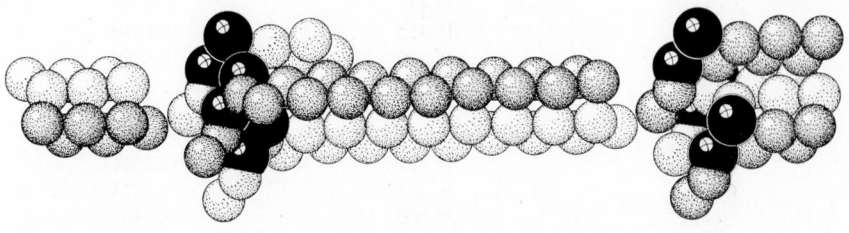

Fig. XIVC,147a (top). The monoclinic structure of ethyl stearate projected along its b_0 axis. Right-hand axes.

Fig. XIVC,147b (bottom). A packing drawing of the monoclinic structure of ethyl stearate viewed along its b_0 axis. The oxygen atoms are black, the carbons dot-and-short-line shaded.

The structure is shown in Figure XIVC,147. In this crystal, unlike the acids, the carboxyl end of one molecule faces the inert end of the next. The atoms of a chain, except for C(2) and C(3), are coplanar to within 0.08 A. The C(2), C(1), O(1), and O(2), and C(1), O(1), C(e1), and C(e2) atoms are also coplanar, but these three planes are twisted with respect to one another. Along the c_0 direction, the separation between the ethyl carbon of one molecule and the methyl end group of the next is 3.66 A. In the molecule, the mean C–C separation in a chain is 1.54 A.; C(1)–O(1) = 1.36 A. and C(1)–O(2) = 1.15 A. The distance between O(1) and C(e1) is 1.37 A. and C(e1)–C(e2) = 1.52 A.

XIV,c97. A partial structure has been described for the B′ form of *N-pentadecanoic acid*, $CH_3(CH_2)_{13}COOH$. Its triclinic unit containing four molecules has the dimensions:

$$a_0 = 5.543 \pm 0.010 \text{ A.}; \quad b_0 = 8.061 \pm 0.030 \text{ A.}; \quad c_0 = 42.58 \pm 0.14 \text{ A.}$$
$$\alpha = 114°18' \pm 11'; \quad \beta = 114°13' \pm 13'; \quad \gamma = 80°37' \pm 6'$$

Parameters were determined for y and z, but not for x (Table XIVC,84). In the original, positions were also proposed for the atoms of hydrogen.

TABLE XIVC,84. Parameters of the Atoms in B′-Pentadecanoic Acid

Atom	y	z
O(1)	0.428	0.013
O(2)	0.314	0.040
O(1′)	0.920	0.004
O(2′)	0.814	0.031
C(1)	0.424	0.041
C(2)	0.452	0.066
C(3)	0.380	0.1008
C(4)	0.467	0.1242
C(5)	0.412	0.1619
C(6)	0.500	0.1853
C(7)	0.444	0.2230
C(8)	0.532	0.2464
C(9)	0.477	0.2841
C(10)	0.564	0.3075
C(11)	0.509	0.3452
C(12)	0.597	0.3686
C(13)	0.541	0.4063
C(14)	0.629	0.4297
C(15)	0.574	0.4674
C(1′)	0.915	0.032
C(2′)	0.948	0.076
C(3′)	0.856	0.0890
C(4′)	0.961	0.1345
C(5′)	0.890	0.1504
C(6′)	0.995	0.1960
C(7′)	0.924	0.2119
C(8′)	0.029	0.2574
C(9′)	0.959	0.2733
C(10′)	0.063	0.3189
C(11′)	0.993	0.3348
C(12′)	0.098	0.3803
C(13′)	0.027	0.3962
C(14′)	0.132	0.4418
C(15′)	0.061	0.4577

XIV,c98. Crystals of the C (or α) form of *lauric acid*, $CH_3(CH_2)_{10}COOH$, are monoclinic. In the original study, a tetramolecular unit was chosen which has the dimensions:

$$a_0 = 9.524 \pm 0.020 \text{ A.}; \quad b_0 = 4.965 \pm 0.010 \text{ A.}; \quad c_0 = 35.39 \pm 0.07 \text{ A.}$$
$$\beta = 129°13'$$

The space group is C_{2h}^5 $(P2_1/a)$ with all atoms in the positions:

$$(4e) \quad \pm (xyz;\ x+{}^1/_2,{}^1/_2-y,z)$$

In terms of this unit, atoms have the parameters of Table XIVC,85.

TABLE XIVC,85
Parameters of the Atoms in C-Lauric Acid
(for cell having $\beta = 129°13'$)

Atom	x	y	z
O(1)	0.3933	0.3000	0.0067
O(2)	0.5457	0.6533	0.0538
C(1)	0.4513	0.4667	0.0429
C(2)	0.4000	0.3300	0.0750
C(3)	0.4900	0.5267	0.1208
C(4)	0.4217	0.3917	0.1483
C(5)	0.5167	0.5600	0.1958
C(6)	0.4467	0.4100	0.2242
C(7)	0.5433	0.5667	0.2739
C(8)	0.4767	0.4333	0.2992
C(9)	0.5690	0.5733	0.3483
C(10)	0.4983	0.4567	0.3742
C(11)	0.5867	0.5633	0.4237
C(12)	0.5167	0.4533	0.4492

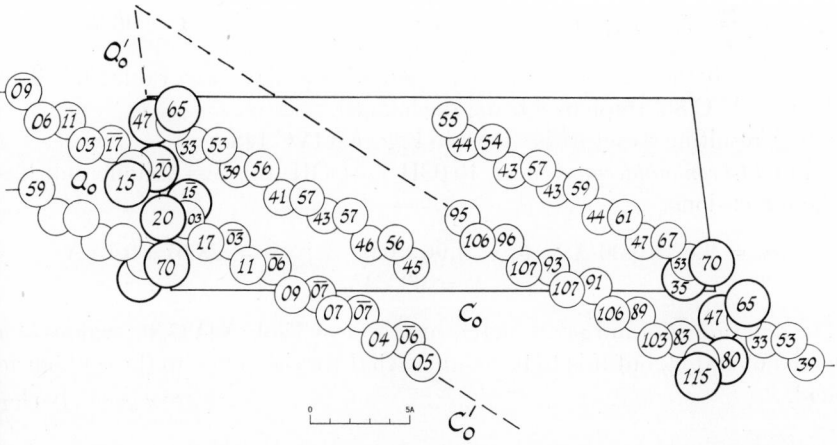

Fig. XIVC,148. The monoclinic structure of the C modification of lauric acid projected along its b_0 axis. The nearly orthogonal axes are the full lines; the dashed lines are the more oblique axes of the original description. The larger, heavily ringed circles are oxygen. Right-hand axes.

TABLE XIVC,86
Parameters of the Atoms in 11-Bromoundecanoic and C-Lauric Acids
(referred to approximately orthogonal axes)

Atom	11-Br(CH$_2$)$_{10}$COOH		CH$_3$(CH$_2$)$_{10}$COOH	
	x	z	x	z
O(1)	0.8725	0.0060	0.8799	0.0067
O(2)	0.9495	0.0520	0.9381	0.0538
C(1)	0.8703	0.0385	0.8655	0.0429
C(2)	0.7680	0.0741	0.7500	0.0750
C(3)	0.7570	0.1160	0.7483	0.1208
C(4)	0.6384	0.1517	0.6251	0.1483
C(5)	0.6296	0.1931	0.6251	0.1958
C(6)	0.5314	0.2258	0.4983	0.2242
C(7)	0.5150	0.2701	0.4955	0.2739
C(8)	0.4055	0.3020	0.3783	0.2992
C(9)	0.3948	0.3436	0.3724	0.3483
C(10)	0.2691	0.3791	0.2499	0.3742
C(11)	0.2552	0.4215	0.2393	0.4237
C(12) or Br	0.1116	0.4531	0.1183	0.4492

In a later study of the isostructural C form of 11-bromoundecanoic acid, Br(CH$_2$)$_{10}$COOH, a different c_0 axis has been selected in order to have a value of β closer to 90°. According to this tetramolecular cell the axial lengths are

$$a_0 = 9.524 \text{ A.}; \quad b_0 = 4.965 \text{ A.}; \quad c_0' = 27.37 \text{ A.}; \quad \beta = 96°55'$$

Referred to these axes, the x and z parameters have the values listed in Table XIVC,86, those of y being unchanged.

The resulting structure is shown in Figure XIVC,148.

For *11-bromoundecanoic acid*, Br(CH$_2$)$_{10}$COOH, this less oblique unit has the dimensions:

$$a_0 = 9.66 \pm 0.06 \text{ A.}; \quad b_0 = 4.96 \pm 0.03 \text{ A.}; \quad c_0 = 28.07 \pm 0.15 \text{ A.}$$
$$\beta = 99°54'$$

The determined x and z parameters are those of Table XIVC,86; values of y were not stated, but it is to be assumed that they are close to those of lauric acid.

Cell dimensions have been determined for the following longer-chain acids that are isostructural.

Compound	a_0, A.	b_0, A.	c_0, A.	β
$CH_3(CH_2)_{12}COOH$ myristic acid	9.509	4.968	40.71	129°7'
$CH_3(CH_2)_{14}COOH$ palmitic acid	9.41	5.00	45.9	129°10'
$CH_3(CH_2)_{16}COOH$ C-stearic acid	9.357	4.956	50.76	128°14'
$CH_3(CH_2)_{20}COOH$ behenic acid	9.292	4.953	60.87	127°37'
$CH_3(CH_2)_{22}COOH$ lignocerinic acid	9.0	4.97	66.5	127°30'
$CH_3(CH_2)_{24}COOH$ cerotic acid	9.249	4.954	71.18	127°19'

XIV,c99. The C' form of *N-hendecanoic acid*, $C_{10}H_{21}COOH$, is monoclinic with a tetramolecular cell of the dimensions:

$$a_0 = 9.822 \pm 0.030 \text{ A.}; \quad b_0 = 4.915 \pm 0.008 \text{ A.}; \quad c_0 = 34.18 \pm 0.10 \text{ A.}$$
$$\beta = 131°17' \pm 15'$$

The space group is C_{2h}^5 ($P2_1/a$) with atoms in the positions:

$$(4e) \quad \pm (xyz; \ x+{}^1/_2, {}^1/_2-y, z)$$

Parameters for the x and z but not for the y coordinates were established. They are given in Table XIVC,87.

TABLE XIVC,87
Parameters of the Atoms in C'-$C_{10}H_{21}COOH$

Atom	x	z
O(1)	0.914	0.0061
O(2)	0.067	0.0591
C(1)	0.979	0.0443
C(2)	0.918	0.0749
C(3)	0.016	0.1249
C(4)	0.983	0.1592
C(5)	0.067	0.2060
C(6)	0.031	0.2411
C(7)	0.108	0.2860
C(8)	0.074	0.3204
C(9)	0.140	0.3655
C(10)	0.120	0.3995
C(11)	0.175	0.4432

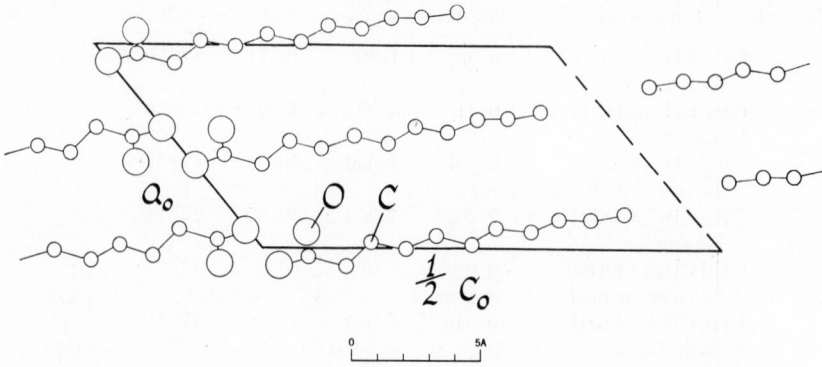

Fig. XIVC,149. The monoclinic structure of $C_{10}H_{21}COOH$ projected along its b_0 axis. Right-hand axes.

The type of arrangement to which these point is indicated by the projection of Figure XIVC,149.

XIV,c100. Crystals of *decanamide*, $CH_3(CH_2)_8CONH_2$, are monoclinic with a tetramolecular unit of the dimensions:

$$a_0 = 9.830 \pm 0.013 \text{ A.}; \quad b_0 = 5.555 \pm 0.003 \text{ A.}; \quad c_0 = 21.224 \pm 0.033 \text{ A.}$$
$$\beta = 103°27' \pm 6'$$

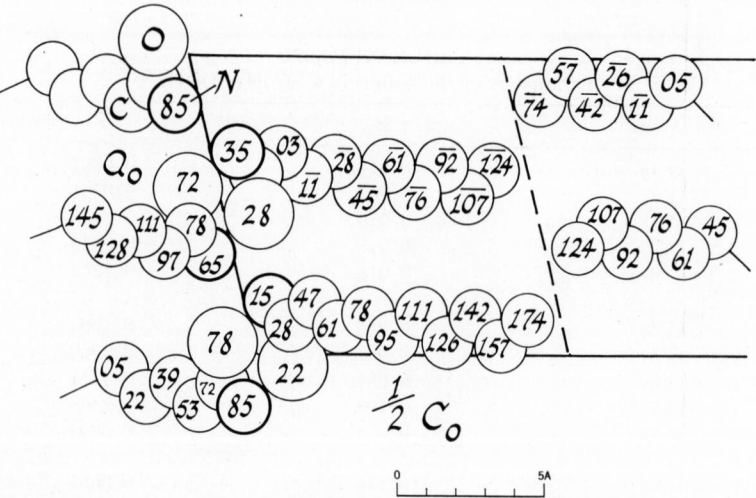

Fig. XIVC,150. The monoclinic structure of decanamide projected along its b_0 axis. Right-hand axes.

The space group is $C_{2h}{}^5$ $(P2_1/a)$ with all atoms in the positions:

$$(4e) \quad \pm (xyz; \ x+{}^1/_2, {}^1/_2-y, z)$$

The determined parameters are those of Table XIVC,88.

TABLE XIVC,88
Parameters of the Atoms in Decanamide

Atom	x	σ_x, A.	y	σ_y, A.	z
N	0.1711	0.0045	0.1460	0.008	0.0375
O	−0.0380	0.0035	0.2243	0.008	0.0507
C(1)	0.0917	0.0047	0.2827	0.010	0.0645
C(2)	0.1639	0.0057	0.4682	0.010	0.1096
C(3)	0.0663	0.0060	0.6095	0.010	0.1384
C(4)	0.1481	0.0067	0.7801	0.010	0.1913
C(5)	0.0530	0.0066	0.9471	0.010	0.2223
C(6)	0.1346	0.0067	0.1110	0.012	0.2760
C(7)	0.0369	0.0087	0.2619	0.017	0.3075
C(8)	0.1203	0.0131	0.4208	0.017	0.3624
C(9)	0.0237	0.0143	0.5737	0.017	0.3925
C(10)	0.1100	0.0218	0.7397	0.025	0.4446
H(1)	0.125	—	0.017	—	0.007
H(2)	0.272	—	0.191	—	0.042
H(3)	0.217	—	0.588	—	0.084
H(4)	0.239	—	0.382	—	0.148
H(5)	0.000	—	0.714	—	0.101
H(6)	0.004	—	0.489	—	0.160
H(7)	0.215	—	0.892	—	0.170
H(8)	0.211	—	0.672	—	0.229
H(9)	−0.008	—	0.058	—	0.185
H(10)	−0.015	—	0.835	—	0.243
H(11)	0.199	—	0.231	—	0.256
H(12)	0.200	—	0.001	—	0.313
H(13)	−0.027	—	0.375	—	0.271
H(14)	−0.029	—	0.143	—	0.327
H(15)	0.188	—	0.537	—	0.343
H(16)	0.182	—	0.307	—	0.399
H(17)	−0.042	—	0.681	—	0.356
H(18)	−0.040	—	0.458	—	0.414
H(19)	0.174	—	0.855	—	0.423
H(20)	0.176	—	0.632	—	0.482
H(21)	0.042	—	0.848	—	0.466

The atomic arrangement is shown in Figure XIVC,150. Hydrogen positions involve the assumption that C–H = 1.075 A. and N–H = 1.005 A.

In these molecules the average C–C bond length [except for C(1)–C(2) = 1.472 A. and C(2)–C(3) = 1.478 A.] is 1.540 A. (σ = 0.014 A.). Other separations within the molecule are C(1)–N = 1.312 A. and C(1)–O = 1.282 A. The hydrogen bonds N–H–O between molecules = 2.88 and 2.90 A.

The *tetradecanamide*, $CH_3(CH_2)_{12}CONH_2$, is isostructural with a cell of the dimensions:

$$a_0 = 9.83 \text{ A.}; \quad b_0 = 5.61 \text{ A.}; \quad c_0 = 27.98 \text{ A.}, \quad \text{all } \pm 0.03 \text{ A.}$$
$$\beta = 95°6' \pm 30'$$

The parameters are those of Table XIVC,89.

TABLE XIVC,89
Parameters of the Atoms in $CH_3(CH_2)_{12}CONH_2$

Atom	x	y	z
N	0.161	0.165	0.029
O	−0.048	0.212	0.039
C(1)	0.069	0.286	0.047
C(2)	0.128	0.466	0.082
C(3)	0.027	0.596	0.103
C(4)	0.086	0.764	0.143
C(5)	−0.014	0.916	0.163
C(6)	0.043	0.085	0.205
C(7)	−0.058	0.233	0.227
C(8)	0.004	0.404	0.269
C(9)	−0.100	0.557	0.291
C(10)	−0.035	0.724	0.333
C(11)	−0.139	0.877	0.356
C(12)	−0.083	0.012	0.396
C(13)	−0.178	0.156	0.421
C(14)	−0.112	0.318	0.462

In this structure the molecules of the carbon chains are planar within the limit of accuracy of the determination, with C–C = 1.59–1.43 A., C–O = 1.23 A., and C–N = 1.26 A. The average C–C–C angle is 115°. The molecules can be thought of as united into pairs by two N–H⋯O bonds of length 2.99 A. between their amide groups, and there is another such hydrogen bond of length 2.93 A. between adjacent pairs. The N, O, C(1), and C(2) atoms of each pair are coplanar, but these polar groups are twisted from the plane of the carbon chains. The final R of this determination is ca. 0.245 though for $(00l)$ it is half as large.

Other amides that appear to have this structure have the following cell dimensions:

	a_0, A.	b_0, A.	c_0, A.	β
Propanamide	9.76	5.78	8.97	113.5°
Butanamide	9.94	5.79	10.02	100.9°
Pentanamide	9.84	6.13	11.08	100.7°
Hexanamide	9.78	5.65	14.11	101.6°
Heptanamide	9.90	5.65	15.29	105.0°
Octanamide	9.78	5.69	17.39	91.9°
Nonamide	9.86	5.59	39.15	103.8°
Hendecanamide	9.85	5.59	21.96	92.8°
Dodecanamide	9.83	5.57	25.10	103.4°
Tridecanamide	9.79	5.59	25.90	93.8°
Hexadecanamide	9.84	5.54	31.44	93.4°

XIV,c101. Crystals of 2D-*methyl octadecanoic acid*, $C_{16}H_{33}CH(CH_3)$-COOH, are monoclinic with a bimolecular cell of the dimensions:

$$a_0 = 9.08 \pm 0.03 \text{ A.}; \quad b_0 = 5.01 \pm 0.02 \text{ A.}; \quad c_0 = 24.0 \pm 0.2 \text{ A.}$$
$$\beta = 116°37' \pm 30'$$

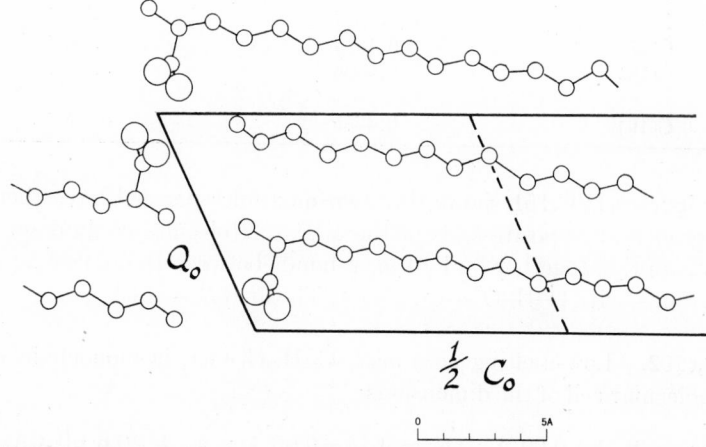

Fig. XIVC,151. The monoclinic structure of 2D-methyl octadecanoic acid projected along its b_0 axis. As usual, the larger circles are oxygen. Right-hand axes.

Atoms are in the general positions:

$$(2a) \quad xyz; \; \bar{x}, y + \tfrac{1}{2}, \bar{z}$$

of C_2^2 $(P2_1)$ with the parameters for x and z as listed in Table XIVC,90.

TABLE XIVC,90
Parameters of the Atoms in 2D-Methyl Octadecanoic Acid

Atom	x	z
O(1)	0.1920	0.0320
O(2)	0.1126	0.0500
C(1)	0.2240	0.0541
C(2)	0.3925	0.0970
C(3)	0.4195	0.1550
C(4)	0.3500	0.1985
C(5)	0.3832	0.2591
C(6)	0.3115	0.2968
C(7)	0.3439	0.3585
C(8)	0.2702	0.3950
C(9)	0.3040	0.4560
C(10)	0.2270	0.4924
C(11)	0.2597	0.5534
C(12)	0.1836	0.5899
C(13)	0.2154	0.6519
C(14)	0.1402	0.6873
C(15)	0.1772	0.7494
C(16)	0.0977	0.7863
C(17)	0.1313	0.8468
C(18)	0.0556	0.8849
C(19)	0.4785	0.0595

As Figure XIVC,151 shows, the two molecules per cell have their carboxyl groups at opposite ends of the c_0 axis. Molecules of different levels along c_0 are tied together by hydrogen bonds between the oxygen atoms of these groups, with O–H–O = ca. 2.80 A. The final R = 0.16.

XIV,c102. Low-melting *oleic acid*, $C_{17}H_{33}COOH$, is monoclinic with a tetramolecular cell of the dimensions:

$$a_0 = 9.51 \pm 0.09 \text{ A.}; \quad b_0 = 4.74 \pm 0.05 \text{ A.}; \quad c_0 = 40.6 \pm 0.3 \text{ A.}$$
$$\beta = 90°$$

TABLE XIVC,91
Parameters of the Atoms in Oleic Acid

Atom	x	y	z
C(1)	0.8670	0.6757	0.0305
C(2)	0.7687	0.7954	0.0557
C(3)	0.7086	0.5866	0.0798
C(4)	0.6238	0.7578	0.1054
C(5)	0.5629	0.5643	0.1324
C(6)	0.4872	0.7356	0.1577
C(7)	0.4122	0.5505	0.1833
C(8)	0.3347	0.7351	0.2093
C(9)	0.2569	0.5317	0.2311
C(10)	0.2562	0.5343	0.2653
C(11)	0.3370	0.7378	0.2869
C(12)	0.4225	0.5602	0.3118
C(13)	0.4909	0.7536	0.3386
C(14)	0.5695	0.5900	0.3642
C(15)	0.6468	0.7849	0.3891
C(16)	0.7178	0.6160	0.4174
C(17)	0.7970	0.7955	0.4416
C(18)	0.8736	0.6376	0.4684
O(1)	0.8789	0.7881	0.0029
O(2)	0.9247	0.4232	0.0358

For C, average $\sigma_x = \sigma_z = 0.024$ A., $\sigma_y = 0.036$ A.
For O, average $\sigma_x = \sigma_z = 0.014$ A., $\sigma_y = 0.022$ A.

H(1)	0.692	0.925	0.041
H(2)	0.846	0.925	0.068
H(3)	0.629	0.443	0.067
H(4)	0.783	0.443	0.094
H(5)	0.548	0.881	0.092
H(6)	0.702	0.881	0.119
H(7)	0.485	0.411	0.119
H(8)	0.639	0.411	0.145
H(9)	0.411	0.899	0.144
H(10)	0.565	0.899	0.171
H(11)	0.337	0.396	0.170
H(12)	0.490	0.396	0.197
H(13)	0.259	0.912	0.196
H(14)	0.413	0.912	0.222
H(15)	0.258	0.910	0.300

(continued)

TABLE XIVC,91 (*continued*)

Atom	x	y	z
H(16)	0.412	0.910	0.273
H(17)	0.346	0.421	0.327
H(18)	0.500	0.421	0.300
H(19)	0.415	0.902	0.351
H(20)	0.569	0.902	0.325
H(21)	0.489	0.440	0.379
H(22)	0.644	0.440	0.352
H(23)	0.574	0.909	0.403
H(24)	0.728	0.909	0.376
H(25)	0.641	0.476	0.430
H(26)	0.795	0.476	0.403
H(27)	0.727	0.928	0.455
H(28)	0.871	0.928	0.428
H(29)	0.797	0.497	0.481
H(30)	0.951	0.497	0.454
H(31)	0.930	0.770	0.486
H(32)	0.178	0.392	0.218
H(33)	0.175	0.392	0.279
H(34)	0.953	0.699	0.012

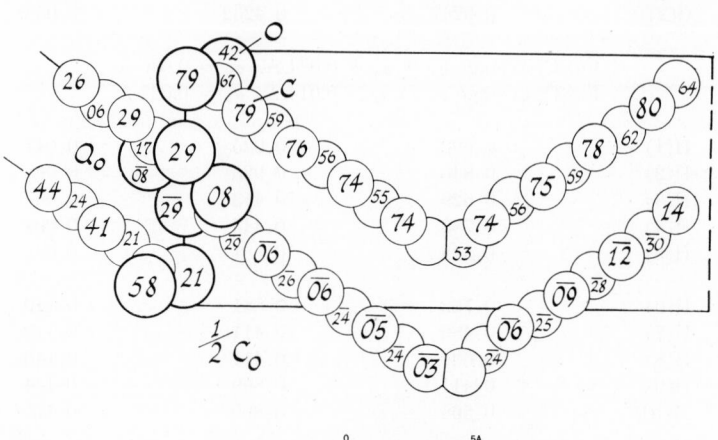

Fig. XIVC,152. The monoclinic structure of oleic acid projected along its b_0 axis.
Right-hand axes.

The selected space group is C_{2h}^5 ($P2_1/a$) and all atoms have been placed in its general positions:

$$(4e) \quad \pm (xyz; \ x+^1/_2, ^1/_2-y, z)$$

The parameters as determined are given in Table XIVC,91.

The structure that results is shown in Figure XIVC,152. The length of the double bond between C(9) and C(10) is 1.39 A.; the other C–C distances lie between 1.50 and 1.56 A. Atoms C(12)–C(18) are coplanar to within 0.05 A., but C(11) is 0.08 A. and C(10) 0.21 A. out of this plane. The chain from C(2)–C(8) is also planar, with C(1) 0.18 A. and C(9) 0.20 A. away. The two parts of the molecule make 113° with one another. Atoms C(8)–C(11) across the double bond are planar, with the bonds C(8)–C(7) and C(11)–C(12) rotated ca. 47° out of the plane. As is usual, the carboxyl groups of adjacent molecules are tied together by hydrogen bonds, with O–O = 2.64 A.

XIV,c103. The monoclinic form of *N-hexadecanol*, $C_{16}H_{33}OH$, has a long unit that contains eight molecules and has the dimensions:

$$a_0 = 8.95 \pm 0.03 \text{ A.}; \quad b_0 = 4.93 \pm 0.02 \text{ A.}; \quad c_0 = 88.1 \pm 0.3 \text{ A.}$$
$$\beta = 122°23'$$

The space group is C_{2h}^6 ($A2/a$) with all atoms in the general positions:

(8f) $\pm (xyz; x+\frac{1}{2}, \bar{y}, z; x, y+\frac{1}{2}, z+\frac{1}{2}; x+\frac{1}{2}, \frac{1}{2}-y, z+\frac{1}{2})$

The determined parameters are those listed in Table XIVC,92. Their stand-

TABLE XIVC,92. Parameters of the Atoms in *N*-Hexadecanol

Atom	x	y	z
O	0.1400	0.1750	0.0060
C(1)	0.2010	0.3130	0.0227
C(2)	0.1145	0.1890	0.0319
C(3)	0.1993	0.3100	0.0513
C(4)	0.1042	0.1900	0.0604
C(5)	0.1863	0.3090	0.0794
C(6)	0.0962	0.1912	0.0885
C(7)	0.1770	0.3080	0.1077
C(8)	0.0866	0.1922	0.1170
C(9)	0.1668	0.3078	0.1365
C(10)	0.0755	0.1925	0.1455
C(11)	0.1604	0.3070	0.1651
C(12)	0.0641	0.1933	0.1739
C(13)	0.1519	0.3058	0.1938
C(14)	0.0544	0.1945	0.2027
C(15)	0.1425	0.3044	0.2221
C(16)	0.0429	0.1958	0.2305

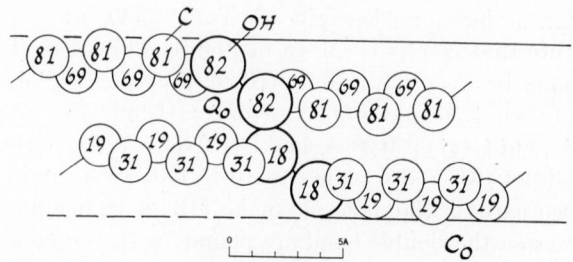

Fig. XIVC,153. Part of the monoclinic structure of hexadecanol projected along its b_0 axis. Right-hand axes.

ard deviations are $\sigma_x = \sigma_y = 0.01$ A., $\sigma_z = 0.015$ A. In the molecules, O–C(1) = 1.44 A. and C–C = 1.52–1.59 A.; all angles lie between 109 and 113°.

As Figure XIVC,153 indicates, the molecules are arranged in pairs so that the chain ends are in contact; the hydroxyl groups also make contact between adjacent chains and bind the crystal together. These hydrogen bonds, extending indefinitely along the a_0 axis, are O–H–O = 2.74 or 2.69 A. At the other ends of the molecules, methyl–methyl separations are 3.93 or 4.24 A.

XIV,c104. The *1,3-diglyceride* of *3-thiododecanoic acid*, $CH_3(CH_2)_8SCH_2C(O)OCH_2C(O)CH_2OC(O)CH_2S(CH_2)_8CH_3$, forms orthorhombic crystals which have a tetramolecular cell of the edge lengths:

$$a_0 = 9.15 \text{ A.}; \quad b_0 = 4.99 \text{ A.}; \quad c_0 = 63.1 \text{ A.}$$

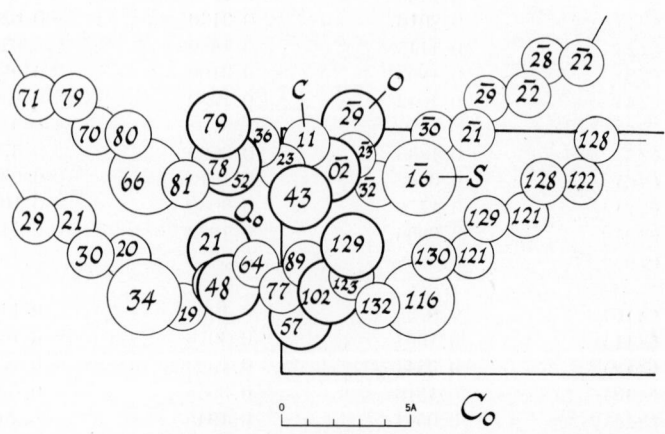

Fig. XIVC,154. Part of the orthorhombic structure of the 1,3-diglyceride of 3-thiododecanoic acid projected along its b_0 axis. Right-hand axes.

The space group has been chosen as C_{2v}^5 ($Pca2_1$) with all atoms in the positions:

(4a) $xyz;\ \bar{x},\bar{y},z+{}^1/_2;\ {}^1/_2-x,y,z+{}^1/_2;\ x+{}^1/_2,\bar{y},z$

The parameters have been determined to be those of Table XIVC,93,

TABLE XIVC,93
Parameters of the Atoms in $C_{25}H_{46}O_5S_2$

Atom	x	y	z
S(1)	0.3070	0.160	0.0844
S(2)	0.3013	0.343	−0.0844
O(1)	0.3303	0.021	0.0305
O(2)	0.5229	0.288	0.0444
O(3)	0.3500	0.477	−0.0316
O(4)	0.5100	0.212	−0.0379
O(5)	0.2400	0.566	0.0120
C(1)	0.3995	0.227	0.0438
C(2)	0.2776	0.325	0.0590
C(3)	0.4708	0.302	0.0945
C(4)	0.4940	0.210	0.1168
C(5)	0.6331	0.294	0.1272
C(6)	0.6460	0.223	0.1505
C(7)	0.7840	0.285	0.1602
C(8)	0.8065	0.221	0.1840
C(9)	0.9528	0.281	0.1938
C(10)	0.9616	0.216	0.2165
C(11)	0.1076	0.290	0.2264
C(12)	0.3770	0.223	−0.0400
C(13)	0.2754	0.194	−0.0596
C(14)	0.4714	0.197	−0.0944
C(15)	0.4900	0.298	−0.1170
C(16)	0.6367	0.214	−0.1275
C(17)	0.6463	0.286	−0.1502
C(18)	0.7669	0.222	−0.1611
C(19)	0.8082	0.297	−0.1836
C(20)	0.9368	0.189	−0.1944
C(21)	0.9604	0.283	−0.2172
C(22)	0.0917	0.211	−0.2293
C(23)	0.4390	0.886	0.0161
C(24)	0.3290	0.770	0.0013
C(25)	0.4340	0.637	−0.0156

with an accuracy estimated as 0.01 A. for sulfur, 0.03 A. for oxygen, and 0.04 A. for carbon in the x and z directions, and twice these values along y.

The atomic arrangement is shown in Figure XIVC,154. The C–C separations range from 1.44 to 1.58 A., and C–S = 1.82 A.; C(1)–O(2) = 1.17 A. and C(12)–O(4) = 1.23 A. All other C–O distances are 1.40–1.50 A. Between molecules the shortest CH_3–CH_3 = 3.93 A.

XIV, c105. Crystals of *2DL-methyl octadecanoic acid*, $C_{16}H_{33}CH(CH_3)$-COOH, are triclinic with a bimolecular cell of the dimensions:

$$a_0 = 5.07 \pm 0.02 \text{ A.}; \quad b_0 = 5.67 \pm 0.02 \text{ A.}; \quad c_0 = 52.5 \pm 0.3 \text{ A.}$$
$$\alpha = 133°45' \pm 30'; \quad \beta = 87°26' \pm 30'; \quad \gamma = 109°14' \pm 30'$$

All atoms are in the general positions $(2i) \pm (xyz)$ of C_i^1 $(P\bar{1})$ with the y and z parameters of Table XIVC,94.

The planar molecules, repeating themselves every cell length in the a_0 direction, are distributed in pairs along the c_0 axis, with carboxyl groups facing one another and united by hydrogen bonds. The final $R = 0.16$.

TABLE XIVC,94

Parameters of the Atoms in 2DL-Methyl Octadecanoic Acid

Atom	y	z
O(1)	0.053	0.4733
O(2)	0.340	0.4705
C(1)	0.085	0.4630
C(2)	0.674	0.4318
C(3)	0.528	0.3984
C(4)	0.609	0.3775
C(5)	0.516	0.3489
C(6)	0.606	0.3275
C(7)	0.497	0.2985
C(8)	0.592	0.2799
C(9)	0.481	0.2500
C(10)	0.576	0.2311
C(11)	0.470	0.2005
C(12)	0.557	0.1812
C(13)	0.447	0.1519
C(14)	0.548	0.1332
C(15)	0.421	0.1017
C(16)	0.529	0.0846
C(17)	0.400	0.0525
C(18)	0.512	0.0367
C(19)	0.471	0.4370

XIV,c106. Crystals of *14DL-methyl octadecanoic acid*, $CH_3(CH_2)_3$-$CH(CH_3)(CH_2)_{12}COOH$, are triclinic with a bimolecular cell of the dimensions:

$$a_0 = 5.06 \pm 0.03 \text{ A.}; \quad b_0 = 6.01 \pm 0.04 \text{ A.}; \quad c_0 = 54.5 \pm 0.5 \text{ A.}$$
$$\alpha = 135°6' \pm 45'; \quad \beta = 91°42' \pm 45'; \quad \gamma = 107°18' \pm 45'$$

The space group is C_i^1 ($P\bar{1}$) with all atoms in the positions $(2i)$ $\pm (xyz)$. The determined y and z parameters are those of Table XIVC,95.

The structure, like that of the 2DL compound (**XIV,c105**), consists of pairs of molecules along the c_0 direction held together by hydrogen bonds between the carboxyl oxygens. These molecules are considered to be tilted by 41° to the (001) plane to accommodate the methyl-substituted end of the chains.

TABLE XIVC,95.   Parameters of the Atoms in 14DL-Methyl Octadecanoic Acid

Atom	y	z
O(1)	0.375	0.0205
O(2)	0.666	0.0272
C(1)	0.537	0.0405
C(2)	0.535	0.0635
C(3)	0.430	0.0800
C(4)	0.493	0.1085
C(5)	0.405	0.1235
C(6)	0.469	0.1560
C(7)	0.377	0.1715
C(8)	0.474	0.2045
C(9)	0.354	0.2155
C(10)	0.472	0.2515
C(11)	0.361	0.2655
C(12)	0.475	0.3020
C(13)	0.353	0.3125
C(14)	0.511	0.3525
C(15)	0.835	0.3850
C(16)	0.055	0.4215
C(17)	0.400	0.4365
C(18)	0.489	0.4655
C(19)	0.305	0.3535

XIV,c107. Crystals of *9DL-methyl octadecanoic acid*, $CH_3(CH_2)_8CH$-$(CH_3)(CH_2)_7COOH$, are triclinic with a bimolecular cell of the dimensions:

$$a_0 = 4.97 \text{ A.}; \quad b_0 = 14.77 \text{ A.}; \quad c_0 = 15.07 \text{ A.}$$
$$\alpha = 115°30'; \quad \beta = 90°; \quad \gamma = 90°$$

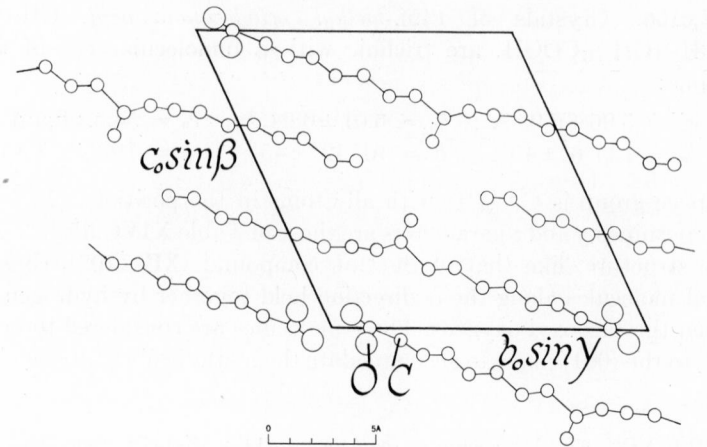

Fig. XIVC,155. The triclinic structure of 9DL-methyl octadecanoic acid projected along its a_0 axis. Right-hand axes.

TABLE XIVC,96. Parameters of the Atoms in 9DL-Methyl Octadecanoic Acid

Atom	y	z
O(1)	0.085	0.037
O(2)	0.922	0.055
C(1)	0.893	0.015
C(2)	0.805	0.031
C(3)	0.756	0.085
C(4)	0.669	0.092
C(5)	0.619	0.145
C(6)	0.533	0.153
C(7)	0.483	0.206
C(8)	0.397	0.213
C(9)	0.339	0.273
C(10)	0.248	0.248
C(11)	0.176	0.283
C(12)	0.089	0.268
C(13)	0.024	0.317
C(14)	0.949	0.321
C(15)	0.886	0.373
C(16)	0.811	0.375
C(17)	0.748	0.428
C(18)	0.673	0.430
C(19)	0.408	0.365

The space group is C_i^1 ($P\bar{1}$) with all atoms in the positions: $(2i)$ $\pm (xyz)$. Parameters were determined for y and z, but not for $x;$ they are listed in Table XIVC,96.

A projection of the resulting structure is shown in Figure XIVC,155.

XIV,c108. Crystals of *16DL-methyl octadecanoic acid*, $CH_3CH_2CH(CH_3)$-$(CH_2)_{14}COOH$, are triclinic with a bimolecular cell of the dimensions:

$$a_0 = 5.40 \pm 0.02 \text{ A.}; \quad b_0 = 7.54 \pm 0.03 \text{ A.}; \quad c_0 = 51.8 \pm 0.3 \text{ A.}$$
$$\alpha = 145°38' \pm 30'; \quad \beta = 105°42' \pm 30'; \quad \gamma = 60°18' \pm 30'$$

The space group is C_i^1 ($P\bar{1}$) with all atoms in $(2i)$ $\pm (xyz)$. Because of the shapes of the crystals, data could only be obtained from $(0kl)$ planes; from them the determined y and z parameters have the values of Table XIVC,97. As with other acids of similar type, the molecules are arranged in pairs along the c_0 direction, the pairs being tied together by hydrogen bonds between the carboxyl oxygens. The plane of the chain is tilted through 33° with respect to the carboxyl planes to permit packing of the side chains.

TABLE XIVC,97. Parameters of Atoms in 16DL-Methyl Octadecanoic Acid

Atom	y	z
O(1)	0.727	0.0390
O(2)	0.370	0.0165
C(1)	0.580	0.0470
C(2)	0.611	0.0700
C(3)	0.511	0.0775
C(4)	0.600	0.1115
C(5)	0.469	0.1270
C(6)	0.583	0.1590
C(7)	0.443	0.1710
C(8)	0.558	0.2050
C(9)	0.438	0.2215
C(10)	0.567	0.2580
C(11)	0.425	0.2720
C(12)	0.550	0.3055
C(13)	0.417	0.3200
C(14)	0.579	0.3580
C(15)	0.428	0.3680
C(16)	0.562	0.4050
C(17)	0.371	0.4105
C(18)	0.466	0.4435
C(19)	0.956	0.4475

TABLE XIVC,98
Parameters of Atoms in 17-Methyl Octadecanoic Acid

Atom	y	z
O(1)	0.192	0.458
O(2)	0.404	0.479
O(3)	0.724	0.482
O(4)	0.884	0.500
C(1)	0.312	0.465
C(2)	0.390	0.444
C(3)	0.288	0.415
C(4)	0.347	0.392
C(5)	0.267	0.365
C(6)	0.334	0.344
C(7)	0.240	0.316
C(8)	0.326	0.296
C(9)	0.230	0.268
C(10)	0.312	0.245
C(11)	0.220	0.219
C(12)	0.305	0.200
C(13)	0.213	0.170
C(14)	0.295	0.150
C(15)	0.213	0.120
C(16)	0.283	0.102
C(17)	0.197	0.072
C(18)	0.274	0.054
C(19)	0.099	0.040
C(20)	0.800	0.464
C(21)	0.890	0.449
C(22)	0.783	0.416
C(23)	0.848	0.391
C(24)	0.762	0.365
C(25)	0.834	0.342
C(26)	0.740	0.316
C(27)	0.826	0.295
C(28)	0.730	0.268
C(29)	0.812	0.245
C(30)	0.720	0.219
C(31)	0.805	0.200

(continued)

TABLE XIVC,98 (*continued*)

Atom	y	z
C(32)	0.713	0.170
C(33)	0.795	0.150
C(34)	0.713	0.120
C(35)	0.783	0.102
C(36)	0.697	0.072
C(37)	0.774	0.054
C(38)	0.599	0.040

XIV,c109.   The *17-methyl octadecanoic acid*, $(CH_3)_2CH(CH_2)_{15}COOH$, is triclinic with a tetramolecular cell of the dimensions:

$a_0 = 5.63 \pm 0.02$ A.; $b_0 = 9.66 \pm 0.04$ A.; $c_0 = 53.3 \pm 0.3$ A.
$\alpha = 93°30'$; $\beta = 134°0'$; $\gamma = 101°12'$,   all $\pm 30'$

The space group is C_i^1 ($P\bar{1}$) with all atoms in the positions: $(2i) \pm (xyz)$. Parameters assigned the y and z coordinates are those of Table XIVC,98.

The long cell found for these crystals (Fig. XIVC,156) is perhaps to be attributed to the fact that the two kinds of molecule present differ from one another in the shape of their carboxyl groups.

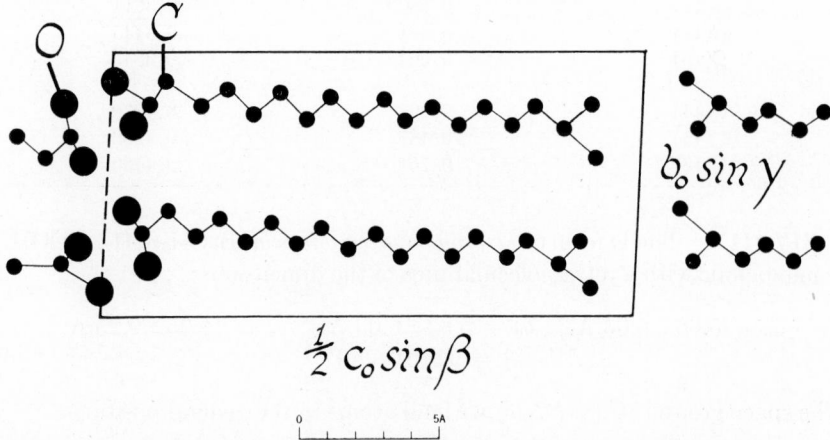

Fig. XIVC,156. The triclinic structure of 17-methyl octadecanoic acid projected along its a_0 axis.

XIV,c110.   A partial structure has been described for *isopalmitic acid*, $C_{15}H_{31}COOH$. Its crystals are triclinic with a bimolecular unit of the dimensions:

$$a_0 = 5.09 \text{ A.}; \quad b_0 = 5.68 \text{ A.}; \quad c_0 = 48.1 \text{ A.}$$
$$\alpha = 140°0'; \quad \beta = 111°6'; \quad \gamma = 72°42'$$

Parameters have been assigned to y and z, but not to x; they are given in Table XIVC,99.

TABLE XIVC,99
Parameters of the Atoms in Isopalmitic Acid

Atom	y	z
OH	0.421	0.0058
O	0.681	0.0583
C(1)	0.514	0.0450
C(2)	0.421	0.0608
C(3)	0.540	0.1017
C(4)	0.453	0.1182
C(5)	0.533	0.1575
C(6)	0.461	0.1750
C(7)	0.540	0.2143
C(8)	0.453	0.2284
C(9)	0.553	0.2725
C(10)	0.461	0.2858
C(11)	0.580	0.3314
C(12)	0.481	0.3445
C(13)	0.594	0.3845
C(14)	0.493	0.4030
C(15)	0.487	0.4260
C(16)	0.701	0.4430

XIV,c111.   The D form of *11-bromoundecanoic acid*, $BrCH_2(CH_2)_9COOH$, is monoclinic with a tetramolecular unit of the dimensions:

$$a_0 = 5.63 \pm 0.03 \text{ A.}; \quad b_0 = 5.33 \pm 0.03 \text{ A.}; \quad c_0 = 44.05 \pm 0.25 \text{ A.}$$
$$\beta = 92°10' \pm 40'$$

The space group is C_{2h}^5 $(P2_1/c)$ with all atoms in the general positions:

$$(4e) \quad \pm (xyz; \; x, 1/2 - y, z + 1/2)$$

The parameters are those of Table XIVC,100.

TABLE XIVC,100
Parameters of the Atoms in D-11 Bromoundecanoic Acid[a]

Atom	x	y	z
C(1)	0.664	0.208	0.0230
C(2)	0.801	0.013	0.0384
C(3)	0.927	0.135	0.0644
C(4)	0.047	0.870	0.0807
C(5)	0.221	0.960	0.1043
C(6)	0.321	0.679	0.1165
C(7)	0.487	0.746	0.1452
C(8)	0.638	0.497	0.1505
C(9)	0.815	0.566	0.1798
C(10)	0.940	0.292	0.1884
C(11)	0.119	0.373	0.2147
Br	0.345	0.093	0.2272
O(1)	0.273	0.734	0.0025
O(2)	0.478	0.348	0.0346

[a] $\sigma(\mathrm{Br}) = 0.01$ A., $\sigma(\mathrm{O}) = 0.05$ A., $\sigma(\mathrm{C}) = 0.08$ A.

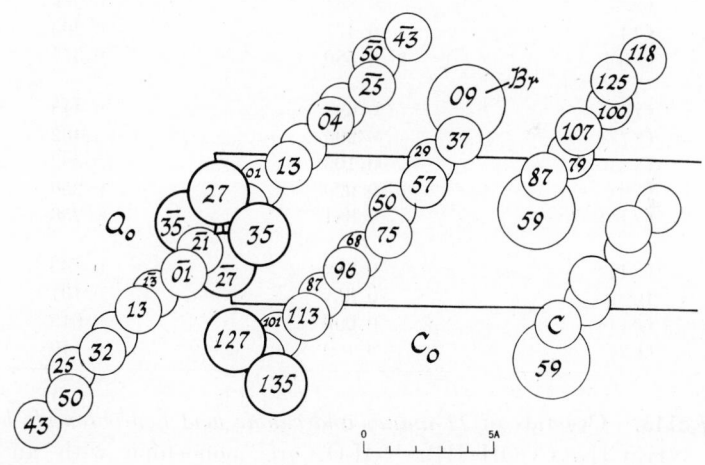

Fig. XIVC,157. Part of the monoclinic structure of D 11-bromo undecanoic acid projected along its b_0 axis. The heavily ringed circles are oxygen. Right-hand axes.

As Figure XIVC,157 indicates, the molecules are arranged obliquely with respect to the long axis. They are tied together into pairs by hydrogen bonds of O–H–O = 2.73 A. In the chains, the C–C distances range between 1.71 and 1.45 A., and C(11)–Br = 2.02 A. The terminal C–COO group is planar, as usual, with $C(1)-O(1) = 1.23$ A. and $C(1)-O(2) = 1.40$ A. The shortest separations between nonhydrogen-bonded molecules are $Br-Br = 3.77$ A. and $Br-C(11) = 3.95$ A.

XIV,c112. The E form of *11-bromoundecanoic acid*, $BrCH_2(CH_2)_9COOH$, is triclinic with a bimolecular unit of the dimensions:

$$a_0 = 4.80 \pm 0.04 \text{ A.}; \quad b_0 = 11.72 \pm 0.10 \text{ A.}; \quad c_0 = 12.41 \pm 0.10 \text{ A.}$$
$$\alpha = 107°11' \pm 60'; \quad \beta = 92°34' \pm 60'; \quad \gamma = 81°26' \pm 60'$$

The space group is C_i^1 ($P\bar{1}$) with all atoms in the positions: $(2i) \pm (xyz)$. The y and z parameters are those of Table XIVC,101; no x parameters could be determined.

The projection along the a_0 axis of Figure XIVC,158 gives an idea of the structure.

TABLE XIVC,101. Parameters of the Atoms in E-11 Bromoundecanoic Acid

Atom	y	z
C(1)	0.165	0.088
C(2)	0.277	0.155
C(3)	0.361	0.288
C(4)	0.477	0.293
C(5)	0.580	0.377
C(6)	0.688	0.445
C(7)	0.206	0.462
C(8)	0.105	0.392
C(9)	0.990	0.256
C(10)	0.881	0.256
C(11)	0.782	0.165
Br	0.651	0.101
O(1)	0.099	0.015
O(2)	0.109	0 119

XIV,c113. Crystals of *11-amino undecanoic acid hydrobromide hemihydrate*, $NH_2(CH_2)_{10}COOH \cdot HBr \cdot 1/2H_2O$, are monoclinic with an eight-molecule unit of the dimensions:

$$a_0 = 11.08 \pm 0.03 \text{ A.}; \quad b_0 = 5.27 \pm 0.02 \text{ A.}; \quad c_0 = 50.60 \pm 0.20 \text{ A.}$$
$$\beta = 90°42' \pm 10'$$

The space group is C_{2h}^6 ($A2/a$) with all atoms except the water oxygens in the positions:

$$(8f) \quad \pm (xyz; \ x+1/2, \bar{y}, z; \ x, y+1/2, z+1/2; \ x+1/2, 1/2-y, z+1/2)$$

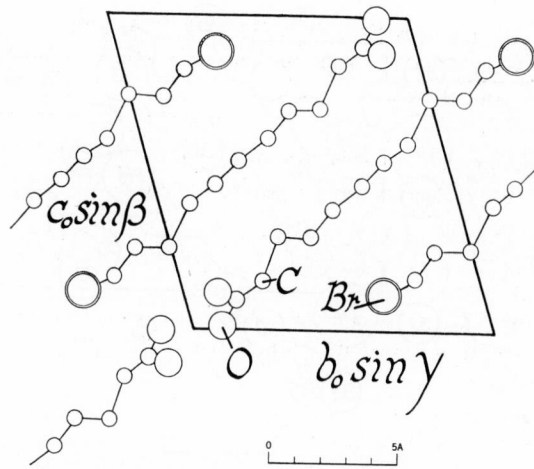

Fig. XIVC,158. The triclinic structure of the E form of 11-bromoundecanoic acid projected along its a_0 axis. Right-hand axes.

The water oxygens are in the special positions:

$$(4e) \quad \pm(^1/_4\, u\; 0;\; ^1/_4, u+^1/_2, ^1/_2) \qquad \text{with } u = 0.300$$

Determined parameters are listed in Table XIVC,102.

TABLE XIVC,102
Parameters of the Atoms in $NH_2(CH_2)_{10}COOH \cdot HBr \cdot ^1/_2H_2O$

Atom	x	y	z
Br	−0.087	0.140	0.0328
O(1)	0.942	0.520	0.2333
O(2)	0.860	0.870	0.2400
C(1)	0.860	0.667	0.2290
C(2)	0.780	0.625	0.2050
C(3)	0.705	0.864	0.1970
C(4)	0.626	0.813	0.1730
C(5)	0.556	0.055	0.1650
C(6)	0.486	0.008	0.1400
C(7)	0.417	0.238	0.1300
C(8)	0.340	0.200	0.1050
C(9)	0.270	0.450	0.0950
C(10)	0.197	0.410	0.0700
C(11)	0.130	0.660	0.0630
N	0.103	0.623	0.0342

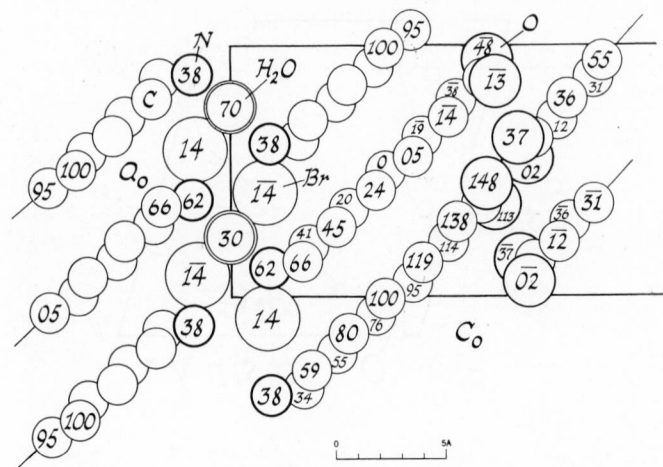

Fig. XIVC,159. Part of the monoclinic structure of 11-amino undecanoic acid hydrobromide hemihydrate projected along its b_0 axis. Right-hand axes.

In this structure (Fig. XIVC,159) the carbon atoms are coplanar to within
the limit of accuracy of the determination, but the nitrogen atoms are about
0.59 A. outside this plane.   The carboxyl oxygens, too, are outside the
plane to give a carboxyl plane turned with respect to the principal plane of
the molecule. As usual, the molecules are paired through O–H–O bonds of
length 2.64 A. The amino ends of the molecules are joined to one another
by the bromide ions and the water molecules. The nitrogen atoms are considered to form three hydrogen bonds: two to bromine, with N–H–Br =
3.30 and 3.44 A., and one to H_2O, with N–H–O = 2.92 A. The water molecules would have four hydrogen bonds, two to nitrogen and two to bromine,
with O–H–Br = 3.38 A.; their distribution is approximately tetrahedral.
The C–C distances in the chain, with an average length of 1.539 A., range
between 1.51 and 1.61 A., and the mean bond angle is 112°44′.

XIV,c114.   Crystals of *3-thiododecanoic acid*, $CH_3(CH_2)_8SCH_2COOH$, are
triclinic with a bimolecular cell of the dimensions:

$$a_0 = 4.69 \text{ A.}; \quad b_0 = 5.10 \text{ A.}; \quad c_0 = 29.3 \text{ A.}$$
$$\alpha = 95°8'; \quad \beta = 92°36'; \quad \gamma = 114°48'$$

The space group is C_i^1 ($P\bar{1}$) with all atoms in the positions: $(2i) \pm (xyz)$
and the parameters of Table XIVC,103. The hydrogen atoms were intro-

TABLE XIVC,103
Parameters of the Atoms in $CH_3(CH_2)_8SCH_2COOH$

Atom	x	y	z
S(3)	0.0602	0.5290	0.1352
O(1)	0.7169	0.0425	0.0422
O(2)	0.1962	0.3348	0.0227
C(1)	0.9322	0.2762	0.0454
C(2)	0.8952	0.5134	0.0802
C(4)	0.7652	0.2091	0.1577
C(5)	0.8429	0.2230	0.2082
C(6)	0.6012	0.9578	0.2299
C(7)	0.6859	0.9797	0.2798
C(8)	0.4570	0.7180	0.3038
C(9)	0.5314	0.7363	0.3538
C(10)	0.3025	0.4880	0.3780
C(11)	0.3743	0.5025	0.4279
C(12)	0.1509	0.2530	0.4539
H(1)	0.957	0.723	0.072
H(2)	0.650	0.472	0.086
H(3)	0.762	0.013	0.140
H(4)	0.534	0.203	0.151
H(5)	0.077	0.236	0.215
H(6)	0.855	0.423	0.226
H(7)	0.586	0.757	0.212
H(8)	0.368	0.949	0.225
H(9)	0.922	0.986	0.285
H(10)	0.712	0.175	0.298
H(11)	0.438	0.514	0.287
H(12)	0.219	0.695	0.297
H(13)	0.768	0.757	0.361
H(14)	0.554	0.942	0.371
H(15)	0.277	0.282	0.361
H(16)	0.065	0.467	0.371
H(17)	0.614	0.533	0.435
H(18)	0.403	0.710	0.445
H(19)	0.122	0.046	0.437
H(20)	−0.089	0.223	0.447
H(21)	0.204	0.264	0.491

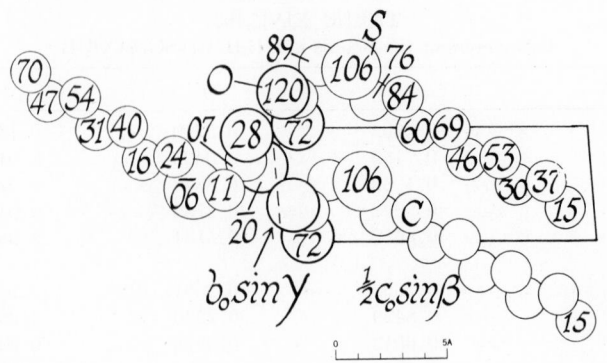

Fig. XIVC,160. The triclinic structure of 3-thiododecanoic acid projected along its a_0 axis. Right-hand axes.

duced into the structure on the assumptions that the distribution around carbon was tetrahedral and that C–H $= 1.10$ A. From this determination, σ for sulfur was considered to be 0.006 A., for oxygen 0.01 A., and for carbon 0.03 A. The final $R = 0.11$.

In this structure (Fig. XIVC,160), C(1)–O(1) $= 1.20$ A., C(1)–O(2) $=$ 1.37 A., C(2)–S $= 1.74$ A., C(4)–S $= 1.85$ A., and the mean C–C $= 1.53$ A. The angle C(2)–S–C(4) $= 103°$ and the S–C(2) bond is rotated by 12° out of the plane of the chain C(12) through sulfur. Between carboxyl groups of adjacent molecules, O–H–O $= 2.74$ A.

XIV,c115. Crystals of *N-nonanoic acid hydrazide*, $CH_3(CH_2)_7C(O)$-$NHNH_2$, are monoclinic with a unit containing eight molecules and having the dimensions:

$$a_0 = 7.44 \text{ A.}; \quad b_0 = 4.87 \text{ A.}; \quad c_0 = 58.73 \text{ A.}; \quad \beta = 95°4'$$

The space group has been chosen as C_{2h}^6 ($A2/a$) with atoms in the positions:

$$(8f) \quad \pm (xyz; \ x+\tfrac{1}{2},\bar{y},z; \ x,y+\tfrac{1}{2},z+\tfrac{1}{2}; \ x+\tfrac{1}{2},\tfrac{1}{2}-y,z+\tfrac{1}{2})$$

The parameters, as determined by two- and three-dimensional analyses, are those of Table XIVC,104.

This structure, drawn for the isotypic dodecanoic compound, is shown in Figure XIVC,161. The molecules that are present have the bond lengths of Figure XIVC,162. Angles between the chain carbon atoms range from 111 to 114°. The chain atoms from C(2) through C(8) are strictly planar and C(1) is displaced by only 0.022 A. from this plane. The terminal atoms

TABLE XIVC,104
Parameters of the Atoms in $C_8H_{17}C(O)NHNH_2$

Atom	x	y	z	σ, A.
O	0.6284	0.5727	0.2031	0.0026
N(1)	0.7000	0.1435	0.2144	0.0029
N(2)	0.8053	0.2269	0.2345	0.0029
C(1)	0.4801	0.2946	0.0266	0.0037
C(2)	0.4132	0.1856	0.0486	0.0037
C(3)	0.5150	0.3053	0.0701	0.0037
C(4)	0.4510	0.1913	0.0920	0.0037
C(5)	0.5499	0.3094	0.1136	0.0037
C(6)	0.4856	0.1941	0.1356	0.0037
C(7)	0.5865	0.3153	0.1570	0.0037
C(8)	0.5180	0.1999	0.1788	0.0037
C(9)	0.6187	0.3199	0.1996	0.0037
H(1)	0.407	0.208	0.011	0.039
H(2)	0.464	0.519	0.026	0.039
H(3)	0.624	0.243	0.026	0.039
H(4)	0.250	0.232	0.049	0.039
H(5)	0.424	−0.019	0.049	0.039
H(6)	0.496	0.500	0.070	0.039
H(7)	0.664	0.275	0.070	0.039
H(8)	0.304	0.228	0.092	0.039
H(9)	0.465	−0.025	0.092	0.039
H(10)	0.526	0.525	0.114	0.039
H(11)	0.703	0.272	0.114	0.039
H(12)	0.339	0.242	0.136	0.039
H(13)	0.503	−0.029	0.136	0.039
H(14)	0.578	0.541	0.157	0.039
H(15)	0.733	0.274	0.157	0.039
H(16)	0.381	0.252	0.179	0.039
H(17)	0.538	−0.026	0.179	0.039
H(18)	0.685	−0.075	0.211	0.039
H(19)	0.922	0.275	0.230	0.039
H(20)	0.728	0.350	0.242	0.039

C(8), C(9), O, N(1), and N(2) also are in one plane which makes an angle of 56°2′ to the other. The shortest intermolecular distances involving hydrogen bondings are O–H–N = 2.896 A. and N(2)–H(20)–N(2) = 3.188 A.; others are 3.44 A. and more.

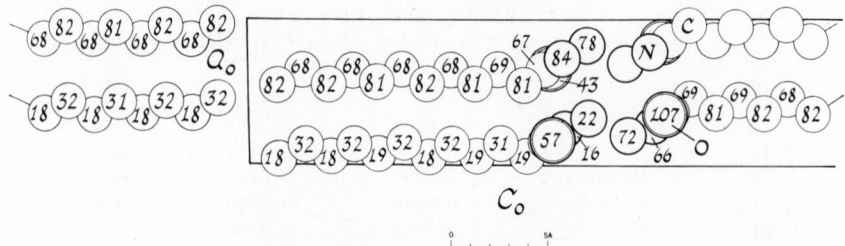

Fig. XIVC,161. Part of the monoclinic structure of dodecanoic acid hydrazide projected along its b_0 axis. Right-hand axes.

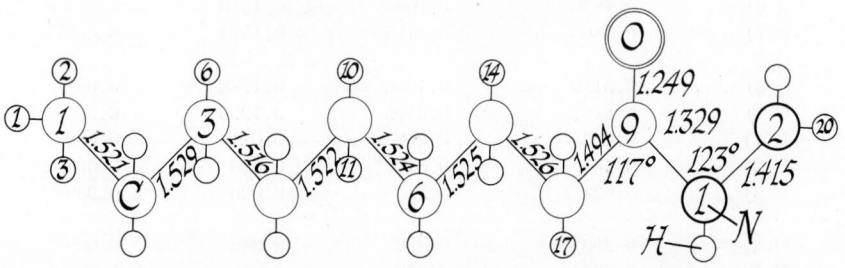

Fig. XIVC,162. Bond dimensions in the molecule of nonanoic acid hydrazide.

The longer-chain N-*dodecanoic acid hydrazide*, $CH_3(CH_2)_{10}C(O)NHNH_2$, is isostructural with the unit:

$$a_0 = 7.46 \text{ A.}; \quad b_0 = 4.88 \text{ A.}; \quad c_0 = 74.04 \text{ A.}; \quad \beta = 91°2'$$

The chosen parameters are those of Table XIVC,105.

In this structure, (Fig. XIVC,161), as in many other long-chain compounds, the ribbon-like molecules are essentially planar, with C–C = 1.55 ± 0.01 A. At the end of the chains, C–O = 1.21 A., C–N = 1.33 A., and N–N = 1.39 A. Between molecules, N–H–O = 3.00 A. and N–H–N = 3.17 A. The final R = ca. 0.09.

The following shorter-chain compounds are substantially isostructural:

Hexanoic acid hydrazide, $C_5H_{11}C(O)NHNH_2$:
$$a_0 = 7.49 \text{ A.}; \quad b_0 = 4.88 \text{ A.}; \quad c_0 = 43.47 \text{ A.}; \quad \beta = 91°16'$$
Heptanoic acid hydrazide, $C_6H_{13}C(O)NHNH_2$:
$$a_0 = 7.44 \text{ A.}; \quad b_0 = 4.88 \text{ A.}; \quad c_0 = 48.60 \text{ A.}; \quad \beta = 96°14'$$
Octanoic acid hydrazide, $C_7H_{15}C(O)NHNH_2$:
$$a_0 = 7.46 \text{ A.}; \quad b_0 = 4.87 \text{ A.}; \quad c_0 = 53.47 \text{ A.}; \quad \beta = 91°20'$$

TABLE XIVC,105
Parameters of the Atoms in $CH_3(CH_2)_{10}C(O)NHNH_2$

Atom	x	y	z
C(1)	0.0349	0.1823	0.0220
C(2)	0.1192	0.3181	0.0395
C(3)	0.0381	0.1819	0.0565
C(4)	0.1229	0.3156	0.0736
C(5)	0.0425	0.1856	0.0907
C(6)	0.1264	0.3162	0.1080
C(7)	0.0475	0.1848	0.1251
C(8)	0.1340	0.3170	0.1422
C(9)	0.0534	0.1860	0.1595
C(10)	0.1410	0.3146	0.1766
C(11)	0.0568	0.1922	0.1938
C(12)	0.1472	0.3251	0.2096
O	0.1546	0.5671	0.2133
N(1)	0.2221	0.1593	0.2219
N(2)	0.3123	0.2218	0.2380

XIV,c116. Crystals of *N-dodecyl ammonium bromide*, $CH_3(CH_2)_{11}$-NH_3Br, are monoclinic with a tetramolecular cell of the dimensions:

$$a_0 = 6.06 \text{ A.}; \quad b_0 = 7.02 \text{ A.}; \quad c_0 = 35.8 \text{ A.}; \quad \beta = 91°36'$$

TABLE XIVC,106. Parameters of the Atoms in $C_{12}H_{25}NH_3Br$

Atom	x	z
Br[a]	0.256	0.0156
N	0.161	0.0278
C(1)	0.1968	0.0650
C(2)	0.4003	0.0860
C(3)	0.3737	0.1258
C(4)	0.5563	0.1520
C(5)	0.5270	0.1902
C(6)	0.7205	0.2165
C(7)	0.2911	0.2438
C(8)	0.1129	0.2178
C(9)	0.1179	0.1800
C(10)	0.9315	0.1550
C(11)	0.9320	0.1163
C(12)	0.7580	0.0900

[a] For Br, $y = 0.225$.

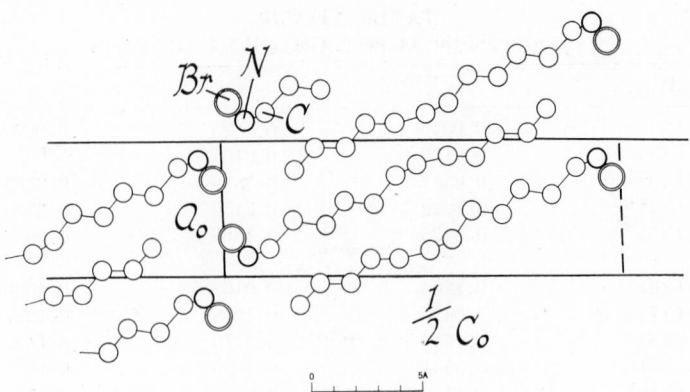

Fig. XIVC,163. Part of the monoclinic structure of dodecyl ammonium bromide projected along its b_0 axis. Right-hand axes.

The space group is C_{2h}^5 ($P2_1/c$) with all atoms in the general positions:

$$(4e) \quad \pm (xyz;\ x,{}^1/_2-y,z+{}^1/_2)$$

Parameters as stated for the bromine atom and x and z for carbon are those of Table XIVC,106. A projection of this partial structure is shown in Figure XIVC,163.

The *chloride* has a cell which, except for a c_0 half that of the bromide, is like the foregoing. Further work is needed to be sure it does not in fact have the structure described above. Its cell dimensions are

$$a_0 = 5.68 \text{ A.}; \quad b_0 = 7.16 \text{ A.}; \quad c_0 = 17.86 \text{ A.}; \quad \beta = 91°12'$$

The following two longer-chain chlorides are isomorphous with the foregoing. They, too, may well have the arrangement assigned the bromide.

For *tetradecyl ammonium chloride*, $CH_3(CH_2)_{13}NH_3Cl$:
$$a_0 = 5.67 \text{ A.}; \quad b_0 = 7.20 \text{ A.}; \quad c_0 = 20.13 \text{ A.}; \quad \beta = 95°52'$$
For *hexadecyl ammonium chloride*, $CH_3(CH_2)_{15}NH_3Cl$:
$$a_0 = 5.71 \text{ A.}; \quad b_0 = 7.24 \text{ A.}; \quad c_0 = 22.56 \text{ A.}; \quad \beta = 98°21'$$

XIV,c117. The triclinic β form of the triglyceride *trilaurin*, $C_{39}H_{74}O_6$, has two molecules in a cell of the dimensions:

$$a_0 = 12.31 \pm 0.10 \text{ A.}; \quad b_0 = 5.40 \pm 0.06 \text{ A.}; \quad c_0 = 31.77 \pm 0.10 \text{ A.}$$
$$\alpha = 94°16' \pm 33'; \quad \beta = 96°52' \pm 18'; \quad \gamma = 99°12' \pm 106'$$

A partial determination established the x and z but not the y parameters; their values are stated in Table XIVC,107.

TABLE XIVC,107
Parameters of the Atoms in β-Trilaurin

Atom	x	z
C(1)	0.1792	0.0260
C(2)	0.1824	0.0554
C(3)	0.1251	0.0950
C(4)	0.1283	0.1244
C(5)	0.0710	0.1640
C(6)	0.0742	0.1934
C(7)	0.0169	0.2330
C(8)	0.0200	0.2624
C(9)	0.9628	0.3020
C(10)	0.9660	0.3314
C(11)	0.9087	0.3710
C(12)	0.9119	0.4004
C(13)	0.5429	0.4464
C(14)	0.5393	0.4170
C(15)	0.5966	0.3774
C(16)	0.5934	0.3480
C(17)	0.6507	0.3084
C(18)	0.6475	0.2790
C(19)	0.7048	0.2394
C(20)	0.7016	0.2100
C(21)	0.7589	0.1704
C(22)	0.7557	0.1410
C(23)	0.8130	0.1014
C(24)	0.8098	0.0720
C(25)	0.7728	0.5766
C(26)	0.7760	0.6060
C(27)	0.7187	0.6456
C(28)	0.7219	0.6750
C(29)	0.6646	0.7146
C(30)	0.6678	0.7440
C(31)	0.6105	0.7836
C(32)	0.6137	0.8130
C(33)	0.5564	0.8526
C(34)	0.5596	0.8820
C(35)	0.5023	0.9216
C(36)	0.5055	0.9410

XIV,c118.   Salts of the fatty acids like the acids themselves are polymorphic, and in spite of many x-ray observations made over the years, it is certain that much additional experimentation will be needed to establish their structures with any approach to completeness.

A thorough determination of atomic positions has, however, been made on one form of *potassium caprate*, $K(C_9H_{19}COO)$. This, the so-called A form, has been observed in potassium salts of the fatty acids $K(C_nH_{2n+1}COO)$, when n is odd and lies between 3 and 11. These crystals are monoclinic with a unit containing four molecules and a structure based on the space group C_{2h}^5 $(P2_1/a)$.

For $K(C_9H_{19}COO)$, the cell dimensions are

$$a_0 = 8.119 \text{ A.}; \quad b_0 = 5.650 \text{ A.}; \quad c_0 = 28.907 \text{ A.}; \quad \beta = 108°2'$$

All atoms are in general positions:

$$(4e) \quad \pm (xyz; x+{}^1/_2, {}^1/_2-y, z)$$

with the assigned parameters listed in Table XIVC,108.

TABLE XIVC,108
Parameters of the Atoms in the A Form of $K(C_9H_{19}COO)$

Atom	x	y	z
K	0.148	0.250	0.039
O(1)	0.425	0.545	0.059
O(2)	0.412	0.960	0.051
C(1)	0.445	0.722	0.073
C(2)	0.442	0.743	0.130
C(3)	0.480	0.490	0.150
C(4)	0.476	0.509	0.211
C(5)	0.510	0.256	0.232
C(6)	0.500	0.277	0.290
C(7)	0.530	0.022	0.312
C(8)	0.540	0.045	0.369
C(9)	0.562	0.790	0.389
C(10)	0.566	0.811	0.450

The structure that results (Fig. XIVC,164) is sheet-like with chains along the c_0 axis oriented alternately so that their carboxyl ends approach one another and are adjacent to the potassium atoms.

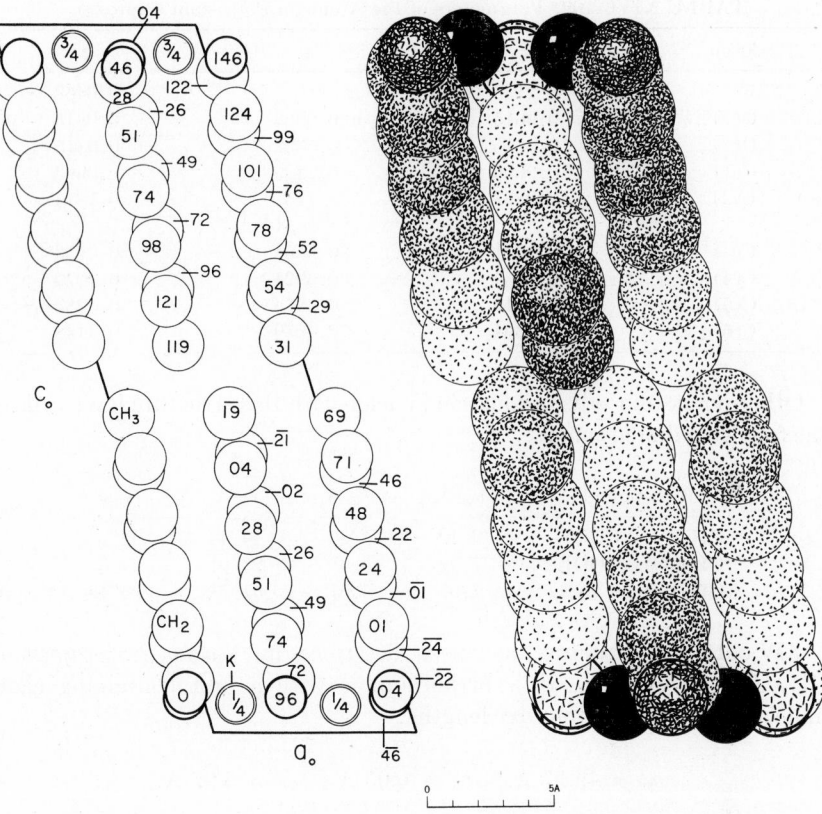

Fig. XIVC,164a (left). The monoclinic structure of the A form of potassium caprate projected along its b_0 axis. Left-hand axes.

Fig. XIVC,164b (right). A packing drawing of the monoclinic $KC_9H_{19}COO$ arrangement seen along its b_0 axis. The potassium atoms are black; the atoms of oxygen are line shaded. Carbon atoms are large and dotted.

The shorter-chained *potassium caproate*, $CH_3(CH_2)_4COOK$, is isostructural with the tetramolecular monoclinic cell:

$$a_0 = 8.00 \text{ A.}; \quad b_0 = 5.74 \text{ A.}; \quad c_0 = 18.94 \text{ A.}, \qquad \text{all } \pm 0.02 \text{ A.}$$
$$\beta = 91°54'$$

Parameters determined for the atoms in this compound are listed in Table XIVC,109.

TABLE XIVC,109. Parameters of the Atoms in Potassium Caproate

Atom	x	y	z
K	0.2126	$1/4$	0.0463
O(1)	0.9615	0.9668	0.0815
O(2)	0.9615	0.5332	0.0815
C(1)	0.9656	0.7429	0.0963
C(2)	0.9827	0.7757	0.1814
C(3)	0.9911	0.5554	0.2153
C(4)	0.0064	0.5924	0.2979
C(5)	0.0144	0.3673	0.3283
C(6)	0.0304	0.4039	0.4148

Other potassium salts of long-chain acids with this structure have cells of the following dimensions:

Compound	a_0, A.	b_0, A.	c_0, A.	β
$CH_3(CH_2)_2COOK$	8.12	5.69	14.42	92°
$CH_3(CH_2)_6COOK$	7.92	5.68	23.01	92°6'
$CH_3(CH_2)_{10}COOK$	7.97	5.67	31.78	95°42'

XIV,c119. A structure has been described for *sodium-2-oxocaprylate*, $CH_3(CH_2)_5COCOONa$. It is orthorhombic with a unit containing eight molecules and having the edge lengths:

$$a_0 = 49.57 \text{ A.}; \quad b_0 = 6.05 \text{ A.}; \quad c_0 = 5.97 \text{ A.}$$

TABLE XIVC,110. Parameters of the Atoms in Sodium-2-Oxocaprylate

Atom	x	y	z
C(1)	0.2118	0.2557	0.527
C(2)	0.1828	0.2334	0.409
C(3)	0.1577	0.2992	0.545
C(4)	0.1319	0.2132	0.434
C(5)	0.1077	0.2992	0.546
C(6)	0.0808	0.2109	0.436
C(7)	0.0559	0.2849	0.563
C(8)	0.0287	0.2343	0.434
O(1)	0.2285	0.1579	0.427
O(2)	0.1817	0.1589	0.202
O(3)	0.2109	0.3773	0.682
Na	0.2274	0.2315	0.029

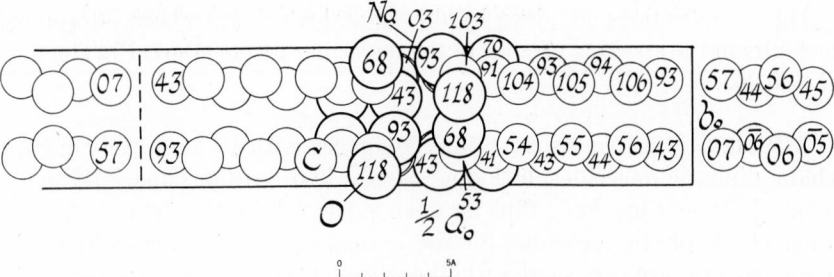

Fig. XIVC,165. Part of the orthorhombic structure of $CH_3(CH_2)_5C(O)COONa$ projected along its c_0 axis.   Left-hand axes.

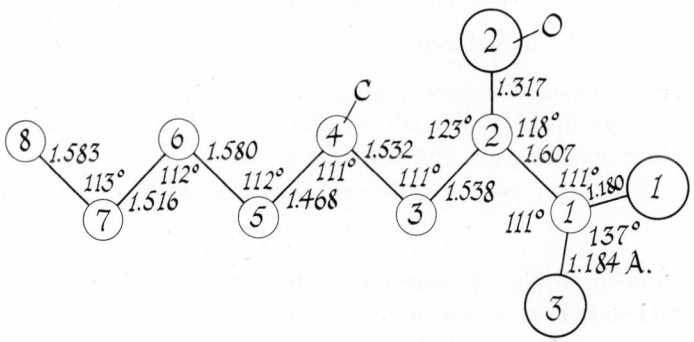

Fig. XIVC,166. Bond dimensions of the anion in $CH_3(CH_2)_5C(O)COONa$.

The space group has been chosen as V_h^{14} (*Pbcn*) with all atoms in the general positions:

$$(8d) \quad \pm (xyz; \; x+1/2, y+1/2, 1/2-z; \; 1/2-x, y+1/2, z; \; x, \bar{y}, z+1/2)$$

The chosen parameters are those of Table XIVC,110.

This structure (Fig. XIVC,165) is of the same type as that of sodium-2-oxobutyrate (**XIV,c19**). The dimensions of its anion are those of Figure XIVC,166. Atoms C(2), C(1), O(1), and O(2) are coplanar, as are C(2), C(1), O(2), and O(3), the angle between the two planes being 17°18′. The atoms C(2)–C(7) are coplanar to within a few hundredths of an angstrom unit, while the terminal C(8) departs by ca. 0.2 A. from this plane. The angle between the plane through the chain and that through C(1),C(2), O(2),O(3) is 14°42′.

Diffuse reflections are given by this crystal which have been interpreted as faults in the stacking of some of the molecular layers (1964: P). The final $R = 0.262$.

XIV,c120. Urea forms crystalline complexes with a wide variety of long-chain aliphatic molecules, including normal hydrocarbons and fatty acids. The cell dimensions have thus far been substantially the same, no matter what the aliphatic molecule; for the normal hydrocarbon complexes, the x-ray patterns are very similar whether there are eight or 50 carbon atoms in the chain.

A detailed study of single crystals was made for the *1,10-dibromodecane-urea* complex. It has hexagonal symmetry with

$$a_0 = 8.230 \text{ A.}, \qquad c_0 = 11.005 \text{ A.}$$

This cell contains six molecules of urea.

The space group is considered to be D_6^2 ($P6_122$), or its enantiomorph, with the carbon and oxygen atoms of urea in

$(6b)$ $u \ \bar{u} \ ^1/_2$; $2u \ u \ ^1/_4$; $u \ 2u \ ^5/_{12}$;
$\bar{u} \ u \ ^7/_{12}$; $2\bar{u} \ \bar{u} \ ^3/_4$; $\bar{u} \ 2\bar{u} \ ^{11}/_{12}$

the origin being displaced along the c_0 axis from its position in the International Tables. For carbon, $u = 0.4094$; for oxygen, $u = 0.3193$. The nitrogen atoms are in general positions:

$(12c)$ $x,y,z+^1/_{12}$; $\bar{y},\bar{x},^1/_{12}-z$; $x-y,x,z+^1/_4$; $x-y,\bar{y},^1/_4-z$;
$\bar{y},x-y,z+^5/_{12}$; $x,x-y,^5/_{12}-z$; $\bar{x},\bar{y},z+^7/_{12}$; $y,x,^7/_{12}-z$;
$y-x,\bar{x},z+^3/_4$; $y-x,y,^3/_4-z$; $y,y-x,z+^{11}/_{12}$; $\bar{x},y-x,^{11}/_{12}-z$
with $x = 0.4415$, $y = 0.5225$, $z = 0.1035$

In the resulting structure there are vertical channels large enough to contain the aliphatic chain molecules. This is, then, a skeletal structure somewhat resembling that formed by *p*-dinitrodiphenyl in its combination with other diphenyl derivatives (loose-leaf **XIV,a10**).

XIV,c121. It has been hard to decide whether or not the several long-chain polymers that have been studied in considerable detail with x-rays should be included in this compilation. The data they provide are necessarily limited compared with those required to establish exact atomic positions in single crystals of corresponding chemical complexity. It would obviously be beyond the scope of the present survey to discuss in detail the probable correctness of the structures that have been proposed for each polymer for which x-ray measurements have been made. For this reason and because information about them has already been included in the Structure Reports, they will be omitted. Nevertheless the structure assigned two of them, the isostructural *polyethylene adipate* and *suberate*, will be described as examples of the general type of atomic arrangement to be expected from polymeric fibers.

The cold-drawn fibers of these polyethylene derivatives have crystalline regions whose x-ray effects are explicable in terms of monoclinic symmetry. On this basis there are, for the adipate, two repeating units of composition $[-CO(CH_2)_4COO(CH_2)_2O-]_n$ in a cell of the dimensions:

$$a_0 = 5.47 \pm 0.03 \text{ A.}; \quad b_0 = 7.23 \pm 0.02 \text{ A.}; \quad c_0 = 11.72 \pm 0.04 \text{ A.}$$
$$\beta = 113°30'$$

For the suberate, of composition $[-CO(CH_2)_6COO(CH_2)_2O-]_n$, the cell dimensions are

$$a_0 = 5.51 \pm 0.03 \text{ A.}; \quad b_0 = 7.25 \pm 0.02 \text{ A.}; \quad c_0 = 14.28 \pm 0.04 \text{ A.}$$
$$\beta = 114°30'$$

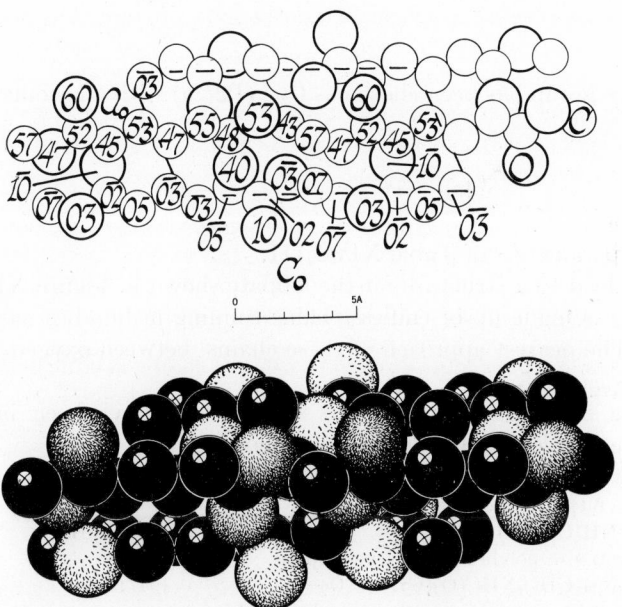

Fig. XIVC,167a (top). The monoclinic structure assigned polyethylene adipate projected along its b_0 axis. Right-hand axes.

Fig. XIVC,167b (bottom). A packing drawing of the monoclinic structure of polyethylene adipate seen along its b_0 axis. The carbon atoms are black, the oxygen atoms larger, heavily outlined, and line shaded.

638 CRYSTAL STRUCTURES

TABLE XIVC,111
Parameters of Atoms in Polyethylene Adipate and in Polyethylene Suberate
(in parentheses when different)

Atom	x	y	z
C(1)	0.059	0.069	0.468 (0.474)
O(1)	0.102 (0.101)	−0.035	0.374 (0.398)
C(2)	−0.060	0.025	0.258 (0.301)
C(3)	0.068	−0.050	0.173 (0.232)
C(4)	−0.063	0.035	0.044 (0.125)
H(1)	0.248 (0.246)	−0.125	0.535 (0.530)
H(2)	−0.081	0.182	0.426 (0.438)
O(2)	−0.280 (−0.278)	0.105	0.230 (0.276)
H(3)	0.281 (0.279)	−0.014	0.214 (0.268)
H(4)	0.044	−0.202	0.167 (0.227)
H(5)	−0.043	0.186	0.051 (0.131)
H(6)	−0.276 (−0.275)	0.000	0.004 (0.091)
C(5)	— (0.064)	— (−0.039)	— (0.055)
H(7)	— (0.277)	— (−0.002)	— (0.089)
H(8)	— (0.041)	— (−0.189)	— (0.051)

The space group has been chosen as C_{2h}^5 ($P2_1/a$) with all atoms in the positions:

$$(4e) \quad \pm (xyz; x+^1/_2, {}^1/_2-y,z)$$

and the parameters of Table XIVC,111.

These lead to a structure for the adipate shown in Figure XIVC,167. It evidently is made up of endless chains running in bundles parallel to the c_0 axis. The nearest approach of these chains, between oxygen and carbon atoms, is ca. 3.15 A.

Among other polymeric fibrous solids which have been investigated, special reference may be made to the following:

Nylon 6 (polycaproamide)
  [—NH(CH₂)₅CO—]ₙ   1955: H,B&S
Nylon 6.6 = polyhexamethylene adipate
  [—NH(CH₂)₆NHC(O)(CH₂)₄C(O)—]ₙ   1947: B&G
Nylon 6.10 = polyhexamethylene sebacamide
  [—NH(CH₂)₆·NH·C(O)(CH₂)₈·C(O)—]ₙ   1947: B&G
Nylon 77 (polyheptamethylene pimelamide) 1959: K
Polyvinylchloride
  (—CH₂CHCl—)ₙ   1956: N&C

Polyvinyl alcohol	
$[-CHOH \cdot CH_2-]_n$	1948: B
Polyketone (1:1 ethylene–carbon monoxide copolymer)	
$[C_2H_4 \cdot CO]_n$	1961: C,T,M,S&N
Polyoxymethylene	1963: C&M
Polypropylene	
$[-CH_2 \cdot CH(CH_3)-]_n$	1956: N,C&C
Poly-α-butene	
$[-CH_2 \cdot CH(CH_2 \cdot CH_3)-]_n$	1956: N,C&B
1,2-Polybutadiene	
$[-CH_2 \cdot CH(CH{:}CH_2)-]_n$	1956: N&C; 1957: N,C&B
1,4-*cis* Polybutadiene	
$[-CH_2 \cdot CH{:}CH \cdot CH_2-]_n$	1956: N&C
1,4-*trans* Poly(1-ethyl butadiene)	1963: P&B
1,4-*trans* Poly(1-propyl butadiene)	1964: N,B&P
Polyisoprene	
$[-CH_2C(CH_3){=}CHCH_2-]_n$	1942: B; 1954: N

BIBLIOGRAPHY TABLE, CHAPTER XIVC

Compound	Paragraph	Literature
Acetoxime		
$(CH_3)_2C{=}NOH$	c9	1951: B&L
Acetylene dicarboxylic acid dihydrate		
$HOOC-C{\equiv}C-COOH \cdot 2H_2O$	c68	1947: D&R; 1954: G,U&W
di-Acetylene dicarboxylic acid dihydrate		
$HOOC-C{\equiv}C-C{\equiv}C-COOH \cdot 2H_2O$	c69	1947: D&R
N,N'-di-Acetylhexamethylene diamine		
$CH_3CONH(CH_2)_6NHCOCH_3$	c58	1955: B
Acrylic acid		
$CH_2{=}CH-COOH$	c11	1963: C,S&N; H&S
Acrylonitrile hexamer		
$[NC(CH_2)_2]_2C(CN)CH{=}CHC(CN)[NC(CH_2)_2]_2$	c42	1964: K&H
Adipic acid		
$COOH \cdot (CH_2)_4 \cdot COOH$	c76	1925: T; 1928: C; 1941: MG; 1949: M&R
di-Allylsilanediol		
$(CH_2{=}CHCH_2SiOH)_2$	c15	1954: K&K
Aluminum trisacetylacetonate		
$Al(C_5H_7O_2)_3$	c47	1926: A; S; 1956: S&S

(continued)

640   CRYSTAL STRUCTURES

BIBLIOGRAPHY TABLE, CHAPTER XIVC (*continued*)

Compound	Paragraph	Literature
11-Amino undecanoic acid hydrobromide hemihydrate $NH_2(CH_2)_{10}COOH \cdot HBr \cdot {}^1/_2H_2O$	c113	1955: S
Ammonium acid D-tartrate $NH_4HC_4H_4O_6$	c80	1953: vB; 1958: vB&B
DL-Ammonium antimony tartrate dihydrate $DL\text{-}NH_4SbC_4H_4O_6 \cdot 2H_2O$	c83	1964: K,G&B
Ammonium sodium tartrate tetrahydrate $NH_4NaC_4H_4O_6 \cdot 4H_2O$	c81	1941: B&H
tetra iso-Amyl ammonium fluoride hydrate $(C_5H_{11})_4NF \cdot 38H_2O$	c38	1961: F&J
Angelic acid $cis\text{-}CH_3CH:C(CH_3)COOH$	c41	1959: P&R
Azelaic acid $HOOC(CH_2)_7COOH$	c73, c74	1925: T; 1928: C; 1929: C; 1936: DL; 1964: H&H
Barium dicalcium propionate $BaCa_2(C_2H_5COO)_6$	c2	1935: B&H; N&W
Behenic acid $CH_3(CH_2)_{20}COOH$	c98	1954: A&vS
Beryllium bis acetylacetone $Be(C_5H_7O_2)_2$	c43	1960: A,P&S
Brassylic acid $HOOC(CH_2)_{11}COOH$	c73	1928: C
1,10-di-Bromodecane-urea	c120	1950: S; 1952: S
11-Bromoundecanoic acid $Br(CH_2)_{10}COOH$	c98, c111, c112	1962: L; 1963: L
Butadiene iron tricarbonyl $Fe(CO)_3 \cdot C_4H_6$	c35	1960: M&R; 1963: M&R
Butanamide $CH_3(CH_2)_2CONH_2$	c100	1955: T&L
tri-N-Butyl sulfonium fluoride hydrate $(C_4H_9)_3SF \cdot 20H_2O$	c21	1962: J&MM
tetra-N-Butyl ammonium copper bis (maleonitrile dithiolate) $(C_4H_9)_4NCu(S_2C_4N_2)_2$	c33	1964: F,Z&T

(*continued*)

BIBLIOGRAPHY TABLE, CHAPTER XIVC (*continued*)

Compound	Paragraph	Literature
tetra-*N*-Butyl ammonium fluoride hydrate		
$(C_4H_9)_4NF \cdot 32.8H_2O$	c20	1963: MM,B&J
di-(tetra-*N*-Butyl ammonium) cobalt(II) bis		
(maleonitrile dithiolate)		
$[(C_4H_9)_4N]_2Co(S_2C_4N_2)_2$	c34	1964: F,Z&T
N-tert-Butyl propylamine hydrochloride		
$C_3H_7NHC(CH_3)_3 \cdot HCl$	c30	1963: T&C
N-Butyric acid		
$CH_3(CH_2)_2COOH$	c18	1962: S&T
Calcium arabonate pentahydrate		
$Ca(C_5O_6H_9)_2 \cdot 5H_2O$	c39	1962: F&H
tetra-Carbonyl (acrylonitrile) iron		
$Fe(CO)_4(CH_2{=}CHCN)$	c12	1961: L&T; 1962: L&T
Cerium tetra-acetylacetonate		
$Ce(C_5H_7O_2)_4$	c49	1959: G&M; 1963: M&G
Cerotic acid		
$CH_3(CH_2)_{24}COOH$	c98	1954: A&vS
Cesium acid D-tartrate		
$CsHC_4H_4O_6$	c80	1931: H
trans-di-Chloro bis-L-propylenediamine cobaltic		
chloride hydrochloride dihydrate		
$[Co(NH_2CH(CH_3)CH_2NH_2)_2Cl_2]Cl \cdot HCl \cdot 2H_2O$	c17	1962: S&I
1,1,1-tri-Chloro-2-methylpropyl alcohol		
hemihydrate		
$CCl_3C(CH_3)_2OH \cdot {}^1/_2H_2O$	c22	1963: P&vS
Chromic acetylacetonate		
$Cr(C_5H_7O_2)_3$	c47	1926: A; 1956: S&S; 1960: S&S; 1964: M&B
Citric acid		
$HC_6H_7O_7$	c84	1960: N,W&P
Cobalt trisacetylacetonate		
$Co(C_5H_7O_2)_3$	c47	1926: A; 1958: P; 1961: S&S
Cobalti bis dimethyl glyoximino diammine		
nitrate		
$Co[(-C(CH_3)NOH)_2]_2NO_3 \cdot 2NH_3$	c26	1961: V&K
Cobaltous bis acetylacetone dihydrate		
$Co(C_5H_7O_2)_2 \cdot 2H_2O$	c45	1959: B

(*continued*)

BIBLIOGRAPHY TABLE, ChAPTER XIVC (*continued*)

Compound	Paragraph	Literature
Cobaltous bis(dipivaloylmethanido)		
$Co[(CH_3)_3CC(O)CC(O)C(CH_3)_3]_2$	c54	1964: C&W
Copper dimethyl glyoxime		
$Cu[(CH_3)_2C_2N_2O_2H]_2$	c24	1936: C,S,W&W; 1951: B,B&S; B&S; 1958: F,B,Z&M; F,Z,B,B&G; 1959: F,B&B
Copper-N,N-di propyl dithiocarbamate		
$Cu[(C_3H_7)_2NCS_2]_2$	c13	1943: P; 1959: P&P; 1962: P&P
penta-Cupric chloride di propylalcohol		
$5CuCl_2 \cdot 2C_3H_7OH$	c5	1964: W&R
Cuprous nitrate di adiponitrile		
$[Cu(NC(CH_2)_4CN)_2]NO_3$	c62	1959: K,M,H&S
Cuprous nitrate bis glutaronitrile		
$[Cu(NC(CH_2)_3CN)_2]NO_3$	c53	1959: K,M&S
Cuprous nitrate bis succinonitrile		
$[Cu(NC(CH_2)_2CN)_2]NO_3$	c32	1959: K,M&S
Decanamide		
$CH_3(CH_2)_8CONH_2$	c100	1958: B&L
Deuterium acetylene dicarboxylic acid dideuterate		
$DOOC—C{\equiv}C—COOD \cdot 2D_2O$	c68	1939: R&U
Dodecanamide		
$CH_3(CH_2)_{10}CONH_2$	c100	1955: T&L
N-Dodecanoic acid hydrazide		
$CH_3(CH_2)_{10}CONHNH_2$	c115	1956: J
N-Dodecyl ammonium bromide		
$CH_3(CH_2)_{11}NH_3Br$	c116	1953: G,S&V
N-Dodecyl ammonium chloride		
$CH_3(CH_2)_{11}NH_3Cl$	c116	1950: C&H; 1953: G,S&V
Erythritol		
$CH_2OH(CHOH)_2CH_2OH$	c29	1959: B&P; S
Ethyl stearate		
$C_{17}H_{35}COOC_2H_5$	c96	1962: A
N,N'-Ethylene bis(acetylacetone iminato) copper(II)		
$Cu(C_6H_9NO)_2$	c51	1962: H,R&W; 1963 H,R&W
Ferric acetylacetonate		
$Fe(C_5H_7O_2)_3$	c46	1926: A; 1956: R; 1964: M&B

(*continued*)

BIBLIOGRAPHY TABLE, CHAPTER XIVC *(continued)*

Compound	Paragraph	Literature
Geranylamine hydrochloride $(CH_3)_2C\!=\!CH(CH_2)_2C(CH_3)\!=\!CHCH_2NH_2 \cdot HCl$	c64	1943: B&J; 1945: J; 1946: J
β-Glutaric acid $HOOC(CH_2)_3COOH$	c73	1932: D; 1948: MG,H&S; 1949: M&R
Hendecanamide $CH_3(CH_2)_9CONH_2$	c100	1955: T&L
N-Hendecanoic acid $C_{10}H_{21}COOH$	c99	1955: vS; 1956: vS
Heptanamide $CH_3(CH_2)_5CONH_2$	c100	1955: T&L
Heptanoic acid hydrazide $C_6H_{13}C(O)NHNH_2$	c115	1953: J&L
Hexadecanamide $CH_3(CH_2)_{14}CONH_2$	c100	1955: T&L
Hexadecane dicarboxylic acid $COOH \cdot (CH_2)_{16} \cdot COOH$	c76	1928: C; 1938: S
N-Hexadecanol $C_{16}H_{33}OH$	c103	1932: B; 1938: S 1944: O; 1960: A,L&vS
Hexadecyl ammonium chloride $CH_3(CH_2)_{15}NH_3Cl$	c116	1950: C&H
N,N'-Hexamethylene bis propionamide $[C_2H_5C(O)NH(CH_2)_3\!-\!]_2$	c59	1957: J,K,P&W; 1962: J
Hexamethylene diammonium adipate $[NH_3(CH_2)_6NH_3][CO_2(CH_2)_4CO_2]$	c61	1954: H,O&N
Hexamethylenediamine $NH_2(CH_2)_6NH_2$	c55	1950: B&R
Hexamethylenediamine dihydrobromide $HBr \cdot NH_2(CH_2)_6NH_2 \cdot HBr$	c56	1949: B&R
Hexamethylenediamine dihydrochloride $HCl \cdot NH_2(CH_2)_6NH_2 \cdot HCl$	c56	1949: B&R
Hexamethylenediamine dihydroiodide $(CH_2)_6(NH_2)_2 \cdot 2HI$	c57	1963: H
Hexanamide $CH_3(CH_2)_4CONH_2$	c100	1955: T&L
N-Hexane C_6H_{14}	c87	1927: ML&P; 1960: N&M; 1961: N&M

(continued)

BIBLIOGRAPHY TABLE, CHAPTER XIVC *(continued)*

Compound	Paragraph	Literature
Hexanoic acid hydrazide		
$C_5H_{11}C(O)NHNH_2$	c115	1953: J&L
2,4-di-Hydroxy-2,4-di methyl-2,4-di silapentane		
$[OH(CH_3)_2Si—]_2CH_2$	c86	1955: K
Isopalmitic acid		
$C_{15}H_{31}COOH$	c110	1952: S,V&S
Hexatriacontane		
$C_{36}H_{74}$	c89, c90	1928: H; 1953: V; 1956:S&V;1959:T
Lauric acid		
$CH_3(CH_2)_{10}COOH$	c91, c92, c98	1921: B&J; 1923: M; 1925: T; 1926: P; P&C; 1928: B&M; 1951: V,M&L; 1954: A&vS; 1956: vS; 1963: L
Lignocerinic acid		
$CH_3(CH_2)_{22}COOH$	c98	1936: C
Lithium ammonium tartrate monohydrate		
$LiNH_4C_4H_4O_6 \cdot H_2O$	c79	1954: S; 1956: S; Z,U,B,E&Z
Lithium potassium tartrate monohydrate		
$LiKC_4H_4O_6 \cdot H_2O$	c79	1956: Z,U,B,E&Z
Lithium rubidium tartrate monohydrate		
$LiRbC_4H_4O_6 \cdot H_2O$	c79	1956: Z,U,B,E&Z
Lithium thallium tartrate monohydrate		
$LiTlC_4H_4O_6 \cdot H_2O$	c79	1956: Z,U,B,E&Z
Maleic acid		
$HCO_2CH{=}CHCO_2H$	c66	1921: B&J; 1925: Y; 1952: S
Malonic acid		
$HOOCCH_2COOH$	c70	1928: G,M&R; 1957: G&MG
Manganese trisacetylacetonate		
$Mn(CH_3COCHCOCH_3)_3$	c47	1926: A; 1964: M&B
penta-Methonium iodide		
$(CH_2)_5N_2(CH_3)_6I_2$	c63	1963: C
Methyl stearate		
$C_{17}H_{35}COOCH_3$	c95	1960: A&vS
meso-α,α'-di-Methyl glutaric acid		
$HO_2CCH(CH_3)CH_2CH(CH_3)CO_2H$	c75	1963: C,D,G&P; 1964: G,P&T

(continued)

BIBLIOGRAPHY TABLE, CHAPTER XIVC *(continued)*

Compound	Paragraph	Literature
Dimethyl glyoxime		
$HON=C(CH_3)\cdot C(CH_3)=NOH$	c23	1952: M&L; 1961: H
penta-Methyl ethyl alcohol (2,3,3-trimethyl butanol) hemihydrate		
$C(CH_3)_3C(CH_3)_2OH\cdot{}^1/_2H_2O$	c22	1963: P&vS
2-Methyl octadecanoic acid		
$C_{16}H_{33}CH(CH_3)COOH$	c101, c105	1959: A
9DL-Methyl octadecanoic acid		
$CH_3(CH_2)_8CH(CH_3)(CH_2)_7COOH$	c107	1956: A
14DL-Methyl octadecanoic acid		
$CH_3(CH_2)_3CH(CH_3)(CH_2)_{12}COOH$	c106	1959: A
16DL-Methyl octadecanoic acid		
$CH_3CH_2CH(CH_3)(CH_2)_{14}COOH$	c108	1958: A
17-Methyl octadecanoic acid		
$(CH_3)_2CH(CH_2)_{15}COOH$	c109	1951: C&C; 1959: A
Myristic acid		
$CH_3(CH_2)_{12}COOH$	c98	1921: B&J; 1923: M; 1925: T; 1954: A&vS
Nickel bis acetylacetone dihydrate		
$Ni(C_5H_7O_2)_2\cdot 2H_2O$	c45	1964: M&L
Nickel dimethyl glyoxime		
$Ni[(CH_3)_2C_2N_2O_2H]_2$	c25	1938: M; 1953: G&R; W,W&R
Nickel bis methyl ethyl glyoxime		
$Ni[HONC(CH_3)C(C_2H_5)NO]_2$	c52	1960: F&P
Nickel-N,N-di propyl dithiocarbamate		
$Ni[(C_3H_7)_2NCS_2]_2$	c14	1940: P; 1941: P
β-Nitropropionic acid		
$NO_2\cdot CH_2\cdot CH_2\cdot COOH$	c1	1954: S,C&L
Nitrosoisobutane, *trans* dimer		
$[(CH_3)_2CHCH_2NO{-}]_2$	c27	1961: D&H
Nonamide		
$CH_3(CH_2)_7CONH_2$	c100	1955: T&L
N-Nonanoic acid hydrazide		
$C_8H_{17}C(O)NHNH_2$	c115	1953: J&L; 1961: J&L
N-Octadecane		
N-$C_{18}H_{38}$	c88	1962: H
Octanamide		
$CH_3(CH_2)_6CONH_2$	c100	1955: T&L

(continued)

BIBLIOGRAPHY TABLE, CHAPTER XIVC *(continued)*

Compound	Paragraph	Literature
N-Octane N-C_8H_{18}	c87	1927: ML&P; 1960: N&M; 1961: N&M
Octanoic acid hydrazide $C_7H_{15}C(O)NHNH_2$	c115	1953: J&L
Oleic acid $C_{17}H_{33}COOH$	c102	1962: A&RN
Palladium dimethyl glyoxime $Pd[(CH_3)_2C_2N_2O_2H]_2$	c25	1953: G&R; 1959: P,F&Z; W,W&R
Palmitic acid $CH_3(CH_2)_{14}COOH$	c98	1921: B&J; 1923: M; 1925: T; W,H&M; 1926: P; P&C; 1927: dB; 1932: D; T&D; 1938: S
$PdClC_3H_5$	c16	1962: L&PK
N-Pentadecanoic acid $CH_3(CH_2)_{13}COOH$	c93, c97	1954: vS; 1955: vS
Pentanamide $CH_3(CH_2)_3CONH_2$	c100	1955: T&L
Pentane C_5H_{12}	c36	1927: ML&P; 1960: N&M; 1964: N&M
Pimelic acid $HOOC(CH_2)_5COOH$	c73, c74	1925: T; 1928: C; 1935: DL; 1936: DL; 1948: MG,H&S; 1958: K&K
Platinum dimethyl glyoxime $Pt[HONC(CH_3)C(CH_3)NOH]_2$	c25	1959: F,P&Z
α-Platinum (tripropylphosphine) chloride thiocyanate α-$Pt[P(C_3H_7)_3]Cl(SCN)$	c8	1960: O&R
Polyethylene adipate $[—CO(CH_2)_4COO(CH_2)_2O—]_n$	c121	1962: TJ&B
Polyethylene suberate $[—CO(CH_2)_6COO(CH_2)_2O—]_n$	c121	1962: TJ&B

(continued)

BIBLIOGRAPHY TABLE, CHAPTER XIVC (*continued*)

Compound	Paragraph	Literature
Potassium acid maleate		
K[OOCCH=CHCOOH]	c67	1958: P&L; 1961: D&C
Potassium acid D-tartrate		
$KHC_4H_4O_6$	c80	1932: F&C
Potassium butyrate		
$CH_3(CH_2)_2COOK$	c118	1952: L
Potassium caprate		
$K(C_9H_{19}COO)$	c118	1947: V,L&L; 1949: V,L&L; 1952: L
Potassium caproate		
$CH_3(CH_2)_4COOK$	c118	1952: L
Potassium caprylate		
$CH_3(CH_2)_6COOK$	c118	1952: L
Potassium gluconate		
$KC_6H_{11}O_7$	c65	1953: L
Potassium laurate		
$CH_3(CH_2)_{10}COOK$	c118	1952: L
Potassium sodium tartrate tetrahydrate (Rochelle Salt)		
$KNaC_4H_4O_6 \cdot 4H_2O$	c81	1933: S; 1940: B&H; 1941: B&H; M: 1946: U&W; 1957; M,J&P; 1964: C
Potassium sodium DL-tartrate tetrahydrate		
$DL\text{-}KNaC_4H_4O_6 \cdot 4H_2O$	c82	1950: S
Propanamide		
$CH_3CH_2CONH_2$	c100	1955: T&L
mono-Propyl ammonium chloride		
$NH_3C_3H_7Cl$	a102, c3	1950: K&L
di-*N*-Propyl gold cyanide		
$(C_3H_7)_2AuCN$	c6	1939: P&P
tetra-*N*-Propyl ammonium bromide		
$(C_3H_7)_4NBr$	c4	1957: Z
Rubidium acid D-tartrate		
$RbHC_4H_4O_6$	c80	1953: vB
DL-Rubidium antimony tartrate dihydrate		
$DL\text{-}RbSbC_4H_4O_6 \cdot 2H_2O$	c83	1964: K,G&B
Rubidium dihydrogen citrate		
$RbH_2[OOCCH_2C(OH)(COO)CH_2COO]$	c85	1960: N,W&P

(*continued*)

BIBLIOGRAPHY TABLE, CHAPTER XIVC (*continued*)

Compound	Paragraph	Literature
Rubidium gluconate $RbC_6H_{11}O_7$	c65	1953: L
Rubidium sodium tartrate tetrahydrate $RbNaC_4H_4O_6 \cdot 4H_2O$	c81	1941: B&H; 1949; B; 1951: P,vB&B
Sebacic acid $HOOC(CH_2)_8COOH$	c76	1925: T; 1928: C; 1938: S; 1949: M&R
Silver perfluorobutyrate $CF_3(CF_2)_2COOAg$	c28	1956: B&H
Silver thiocyanate tripropylphosphine $AgSCN \cdot P(C_3H_7)_3$	c7	1960: T,P&F; 1963: P&F
Sodium-α-ketobutyrate $C_2H_5C(O)COONa$	c19	1963: T,P&B
Sodium-2-oxocaprylate $CH_3(CH_2)_5COCOONa$	c119	1964: P; T,P&B
Sodium pyruvate $CH_3C(O)COONa$	c10	1961: T,P&B
Stearic acid $CH_3(CH_2)_{16}COOH$	c94, c98	1921: B&J; 1923: M; 1925: M; T 1926: P&C 1927: dB; M; 1932: D; T&D 1950: V&P; 1954: A&vS; 1955: vS
Strontium arabonate pentahydrate $Sr(C_5O_6H_9)_2 \cdot 5H_2O$	c39	1962: F&H
Suberic acid $HOOC(CH_2)_6COOH$	c76, c77	1925: T; 1928: C; 1964: H&H
Succinamide $[NH_2C(O)CH_2—]_2$	c72	1953: H&MC; P; 1956: D&P
β-Succinic acid $HOOC(CH_2)_2COOH$	c71	1921: B&J; 1924: Y; 1925; T; 1938: V&MG; 1939: V&MG; 1949: M&R; 1959: B,C,M,R&S

(*continued*)

BIBLIOGRAPHY TABLE, CHAPTER XIVC *(continued)*

Compound	Paragraph	Literature
D-Tartaric acid		
$COOH(CHOH)_2COOH$	c78	1921: B&J; 1923: A; 1928: R&S; 1948: B&S; 1950: S&B; 1951: P,vB&B
Tetradecanamide		
$CH_3(CH_2)_{12}CONH_2$	c100	1955: T&L
Tetradecyl ammonium chloride		
$CH_3(CH_2)_{13}NH_3Cl$	c116	1950: C&H
Tetramethylene diammonium adipate		
$[NH_3(CH_2)_4NH_3][CO_2(CH_2)_4CO_2]$	c60	1961: H&A; 1962: H&A
Tetramethylene diammonium chloride		
$NH_2(CH_2)_4NH_2 \cdot 2HCl$	c31	1963: A&H
3-Thiododecanoic acid		
$CH_3(CH_2)_8SCH_2COOH$	c114	1963: A&W
3-Thiododecanoic acid-1,3-diglyceride		
$C_{25}H_{46}O_5S_2$	c104	1963: L
Thorium tetra acetylacetonate		
$Th(C_5H_7O_2)_4$	c49	1958: G&M; 1959: G&M
Tiglic acid		
$trans\text{-}CH_3CH:C(CH_3)COOH$	c40	1959: P&R
Tridecanamide		
$CH_3(CH_2)_{11}CONH_2$	c100	1955: T&L
Trilaurin		
$C_{39}H_{74}O_6$	c117	1951: V&B
Uranium tetra acetylacetonate		
$U(C_5H_7O_2)_4$	c49	1959: G&M
Valeric acid		
C_4H_9COOH	c37	1962: S&S
Vanadyl bis acetylacetonate		
$VO(C_5H_7O_2)_2$	c50	1958: D; 1961: D, T&Z
Zinc bis acetylacetone monohydrate		
$Zn(C_5H_7O_2)_2 \cdot H_2O$	c44	1960: L&T; 1963: M&L
Zinc bis(dipivaloylmethanido)		
$Zn[(CH_3)_3CC(O)CC(O)C(CH_3)_3]_2$	c54	1964: C&W
Zirconium bis acetylacetonate		
$Zr(C_5H_7O_2)_2$	c48	1963: S&H

D. AMINO ACIDS AND RELATED COMPOUNDS

XIV,d1. A redetermination has confirmed the structure originally given *α-glycine*, NH_2CH_2COOH, and refined its parameters. The symmetry is monoclinic with a tetramolecular cell of the newly determined dimensions:

$$a_0 = 5.1020 \pm 0.0008 \text{ A.}; \quad b_0 = 11.9709 \pm 0.0017 \text{ A.}; \quad c_0 = 5.4575 \pm 0.0015 \text{ A.}$$
$$\beta = 111°42.3' \pm 1'$$

TABLE XIVD,1.   Parameters of the Atoms in α-Glycine

Atom	x	y	z
C(1)	0.07542	0.12478	0.06605
C(2)	0.06536	0.14499	0.78711
N	0.30135	0.08980	0.74113
O(1)	0.30583	0.09427	0.23553
O(2)	0.85224	0.14154	0.10711
H(1)	0.286	0.100	0.570
H(2)	0.457	0.119	0.837
H(3)	0.298	0.020	0.763
H(4)	0.080	0.220	0.771
H(5)	0.898	0.118	0.671

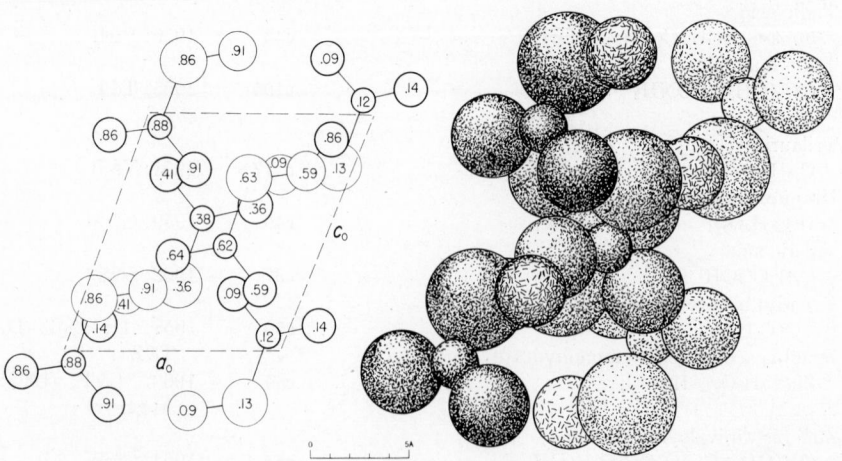

Fig. XIVD,1a (left). The monoclinic structure of α-glycine projected along its b_0 axis. The b_0 parameters shown here are those of the original determination. The larger, heavily ringed circles are oxygen; the smaller are the carboxyl carbon atoms. The largest, lightly outlined circles are CH_2; those that are somewhat smaller are NH_2. Left-hand axes.

Fig. XIVD,1b (right). A packing drawing of the monoclinic structure of α-glycine seen along its b_0 axis. The NH_2 radicals are line shaded; other atoms have the relative sizes of Fig. XIVD,1a.

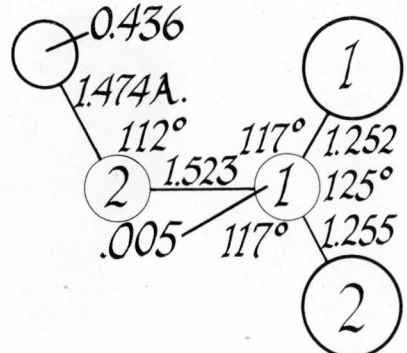

Fig. XIVD,2. Bond dimensions of the molecule in crystals of α-glycine. Fractional numbers attached to the C(1) and N atoms give their distance outside the best plane through the others.

The space group is C_{2h}^5 ($P2_1/n$) with all atoms in the positions:

$$(4e) \quad \pm(xyz; \; x+\tfrac{1}{2}, \tfrac{1}{2}-y, z+\tfrac{1}{2})$$

The final values of the atomic parameters are those of Table XIVD,1. For them the standard deviations of the x and z parameters are ca. 0.0007 A. and those of the y parameters, 0.0010 A. For the hydrogen atoms, the estimated error in atomic positions is 0.06 A.; their mean distances from the nitrogen and carbon atoms are the rather short 0.87 and 0.91 A.

The resulting structure is shown in Figure XIVD,1. The molecule according to the new determination has the bond dimensions of Figure XIVD,2. Its carbon and oxygen atoms are strictly planar but the nitrogen atom is 0.436 A. outside this plane. In the a_0c_0 plane, the molecules are held together by N–H–O bonds (2.768 and 2.850 A.) involving the H(1) and H(2) atoms. Other N–H–O bonds equal 2.949 and 3.074 A. The final $R = 0.063$.

XIV,d2. The unstable, β, form of *glycine*, NH_2CH_2COOH, is monoclinic with a bimolecular unit of the dimensions:

$$a_0 = 5.077 \text{ A.}; \quad b_0 = 6.268 \text{ A.}; \quad c_0 = 5.380 \text{ A.}; \quad \beta = 113°12'$$

The space group has been chosen as $C_2^2(P2_1)$ with all atoms in the positions:

$$(2a) \quad xyz; \; \bar{x}, y+\tfrac{1}{2}, \bar{z}$$

The determined parameters, including those assigned the hydrogen atoms, are listed in Table XIVD,2.

The resulting structure is shown in Figure XIVD,3. It is made up of molecules having the bond dimensions of Figure XIVD,4, very like those of

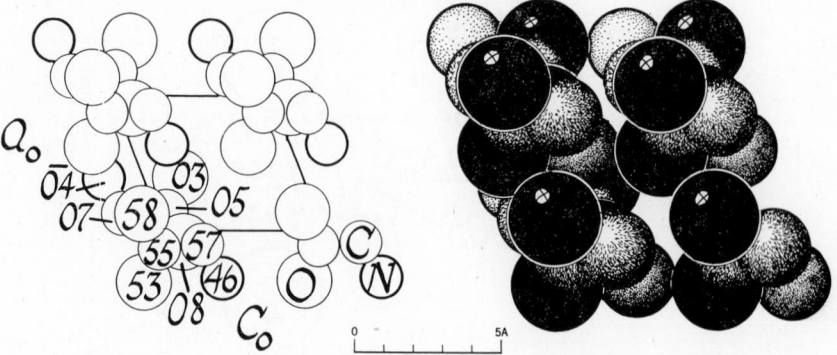

Fig. XIVD,3a (left). The monoclinic structure of the β modification of glycine projected along its b_0 axis. Right-hand axes.

Fig. XIVD,3b (right). A packing drawing of the monoclinic β-glycine structure viewed along its b_0 axis. The oxygen atoms are black, the carbon atoms fine-line shaded, and the nitrogens are more heavily ringed and dotted.

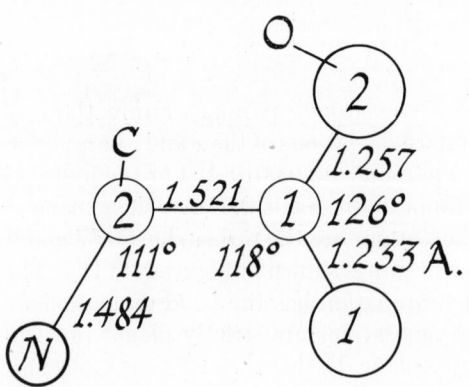

Fig. XIVD,4. Bond dimensions in the glycine molecule in crystals of its β modification.

the α modification (**XIV,d1**). These molecules are tied together by hydrogen bonds N–H–O = 2.758, 2.833, 3.002, and 3.022 A.

XIV,d3. The γ modification of *glycine*, NH_2CH_2COOH, is hexagonal with a trimolecular cell of the edge lengths:

$$a_0 = 7.037 \text{ A.,} \qquad c_0 = 5.483 \text{ A.}$$

The space group is C_3^3 ($P3_2$) [or the enantiomorphous C_3^2] with atoms in the positions:

$$(3a) \quad xyz; \; \bar{y}, x-y, z+{}^2/_3; \; y-x, \bar{x}, z+{}^1/_3$$

TABLE XIVD,2
Parameters of the Atoms in β-Glycine

Atom	x	σ_x, A.	y	σ_y, A.	z	σ_z, A.
N	0.3522	0.008	−0.0440	0.017	−0.2619	0.008
O(1)	0.3772	0.008	0.0270	0.017	0.2420	0.008
O(2)	−0.0896	0.008	0.0773	0.017	0.0970	0.008
C(1)	0.1378	0.008	0.0532	0.017	0.0633	0.008
C(2)	0.1145	0.008	0.0719	0.017	−0.2265	0.008
H(N,1)	0.540	—	0.012	—	−0.135	—
H(N,2)	0.340	—	−0.031	—	−0.449	—
H(N,3)	0.337	—	−0.203	—	−0.216	—
H(C,1)	0.125	—	0.241	—	−0.274	—
H(C,2)	−0.091	—	0.008	—	−0.362	—

The parameters are those of Table XIVD,3, the hydrogen positions being selected on the assumptions that H–C = 1.05 A. and H–N = 1.00 A., and that the distribution about carbon is tetrahedral.

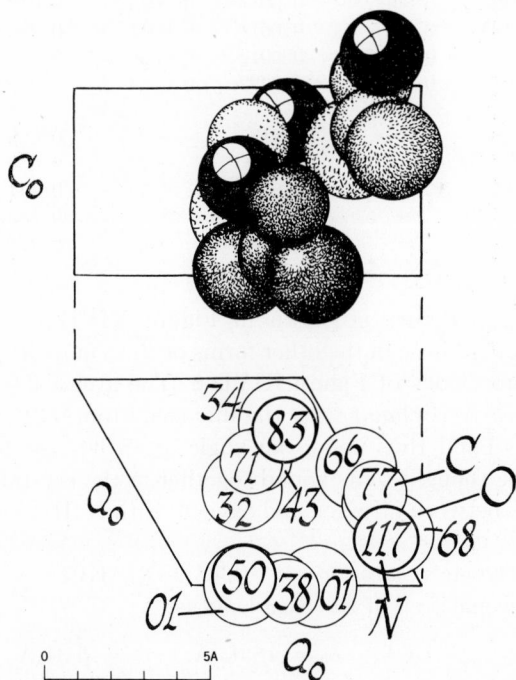

Fig. XIVD,5. Two projections of the hexagonal γ modification of glycine. In the upper projection the nitrogen atoms are black and the largest short-line shaded circles are oxygen. The dotted carbon atoms are of the same size as nitrogen.

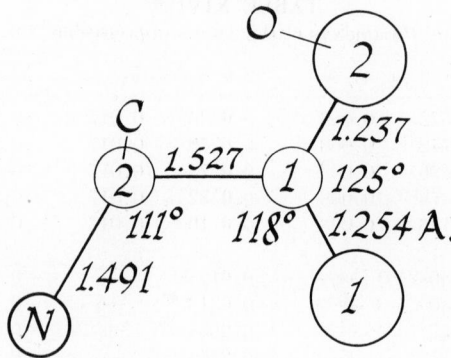

Fig. XIVD,6. Bond dimensions of the glycine molecule in crystals of its γ form.

TABLE XIVD,3.   Parameters of the Atoms in γ-Glycine

Atom	x	$\sigma(x)$	y	$\sigma(y)$	z	$\sigma(z)$
N	0.2414	0.0074	0.0263	0.0071	0.5035	0.0091
O(1)	0.2325	0.0066	0.0083	0.0063	0.0139	0.0073
O(2)	0.5425	0.0075	0.0011	0.0071	−0.0150	0.0069
C(1)	0.3929	0.0086	0.0012	0.0088	0.1033	0.0081
C(2)	0.4010	0.0084	−0.0222	0.0084	0.3794	0.0076
H(1)	0.248	—	0.013	—	0.686	—
H(2)	0.274	—	0.183	—	0.458	—
H(3)	0.084	—	−0.079	—	0.441	—
H(4)	0.567	—	0.089	—	0.446	—
H(5)	0.369	—	−0.186	—	0.427	—

The resulting structure is shown in Figure XIVD,5. The molecules, which are similar to those in the other forms of this compound (**XIV,d1 and 2**), have the dimensions of Figure XIVD,6. The atoms C(1), C(2), O(1), and O(2) are nearly coplanar with the nitrogen atom 0.31 A. outside this plane; the N–C bond thus makes an angle of 18°36′ to this plane. The molecules can be thought of as bound together in the crystal by a series of hydrogen bonds between nitrogen and oxygen, with N–H–O = 2.80–3.06 A.

XIV,d4. *Diglycine hydrochloride*, $(C_2H_5NO_2)_2 \cdot HCl$, is orthorhombic with a tetramolecular cell of the edge lengths:

$$a_0 = 8.15 \text{ A.}; \quad b_0 = 18.03 \text{ A.}; \quad c_0 = 5.34 \text{ A.}$$

The space group is V^4 $(P2_12_12_1)$ and all atoms are therefore in the positions:

$$(4a) \quad xyz; \ ^1/_2-x,\bar{y},z+^1/_2; \ x+^1/_2,^1/_2-y,\bar{z}; \ \bar{x},y+^1/_2,^1/_2-z$$

TABLE XIVD,4

Parameters of the Atoms in Diglycine HCl (and, in parentheses, for the Hydrobromide)

Atom	x	y	z
Cl	0.4252 (0.427)	0.0337 (0.035)	0.8368 (0.832)
C(1)	0.3557 (0.340)	0.3812 (0.380)	0.8617 (0.849)
C(2)	0.3640 (0.355)	0.4340 (0.427)	0.0804 (0.080)
C(3)	0.7263 (0.719)	0.2731 (0.270)	0.5121 (0.506)
C(4)	0.3220 (0.317)	0.1669 (0.170)	0.3595 (0.381)
N(1)	0.5324 (0.531)	0.4349 (0.427)	0.1874 (0.168)
N(2)	0.4007 (0.394)	0.1909 (0.193)	0.1184 (0.129)
O(1)	0.7214 (0.719)	0.1117 (0.112)	0.2711 (0.301)
O(2)	0.4619 (0.454)	0.3346 (0.335)	0.8236 (0.833)
O(3)	0.7204 (0.717)	0.2103 (0.210)	0.6189 (0.610)
O(4)	0.6632 (0.662)	0.2887 (0.286)	0.3025 (0.304)
H(1)	0.277	0.414	0.221
H(2)	0.332	0.491	0.009
H(3)	0.541	0.468	0.334
H(4)	0.570	0.384	0.224
H(5)	0.612	0.456	0.051
H(6)	0.236	0.122	0.320
H(7)	0.419	0.150	0.492
H(8)	0.319	0.220	0.013
H(9)	0.437	0.147	0.020
H(10)	0.500	0.224	0.155
H(11)	0.720	0.149	0.403

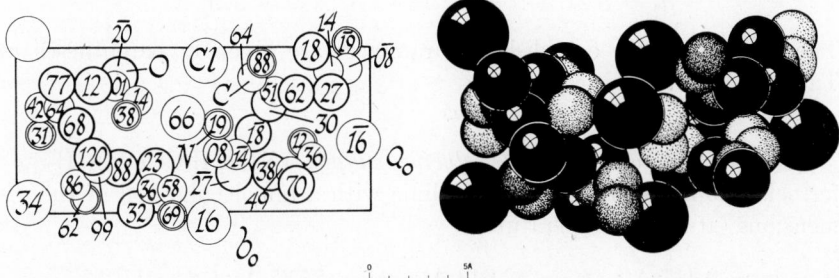

Fig. XIVD,7a (left). The orthorhombic structure of diglycine hydrochloride projected along its c_0 axis. Right-hand axes.

Fig. XIVD,7b (right). A packing drawing of the orthorhombic diglycine hydrochloride arrangement seen along its c_0 axis. The large black circles are the Cl⁻ ions; the somewhat smaller black circles with circular highlights are oxygen. Carbon atoms are hook-and-dot shaded; nitrogens are heavily outlined and short-line-and-dot shaded.

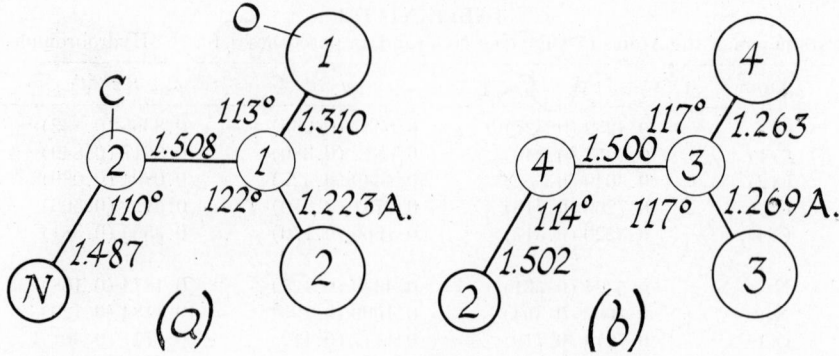

Fig. XIVD,8a (left). Bond dimensions in half the molecular cations of diglycine hydrochloride.
Fig. XIVD,8b (right). Bond dimensions in the other half the molecules of diglycine
hydrochloride.

According to a more recent refinement, the parameters have the values of
Table XIVD,4; the hydrogen parameters are, however, those proposed on
the basis of the original determination.

The structure as a whole is shown in Figure XIVD,7. It contains two
crystallographically different but practically identical glycine groups for
which the bond dimensions are those of Figure XIVD,8. The shortest
intermolecular separations, which undoubtedly involve hydrogen bonds,
are O–O = 2.57 A. and N–O = 2.91 A.

The *hydrobromide* has the same atomic arrangement with a unit that has
the dimensions:

$$a_0 = 8.21 \text{ A.}; \quad b_0 = 18.42 \text{ A.}; \quad c_0 = 5.40 \text{ A.}$$

The less accurately established parameters are given in parentheses in
Table XIVD,4.

XIV,d5. The compound *tri-glycine sulfate*, $(NH_2CH_2COOH)_3 \cdot H_2SO_4$,
ferroelectric below 47°C., is monoclinic with a bimolecular cell of the dimensions (at room temperature):

$$a_0 = 9.417 \text{ A.}; \quad b_0 = 12.643 \text{ A.}; \quad c_0 = 5.735 \text{ A.}; \quad \beta = 110°23'$$

Because of the ferroelectricity, the space group has been chosen as C_2^2
$(P2_1)$ [rather than the higher symmetry $P2_1/m$] and all atoms are in the
general positions:

$$(2a) \quad xyz; \ \bar{x}, y + \tfrac{1}{2}, \bar{z}$$

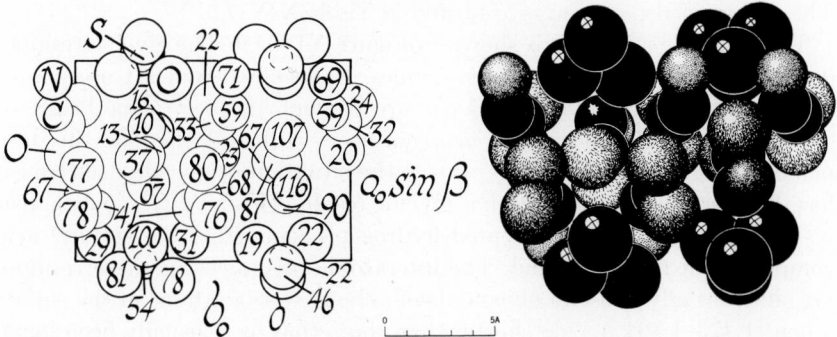

Fig. XIVD,9a (left). The monoclinic structure of triglycine sulfate projected along its c_0 axis. Right-hand axes.

Fig. XIVD,9b (right). A packing drawing of the monoclinic structure of triglycine sulfate viewed along its c_0 axis. The sulfate oxygens are the large black circles; atoms of sulfur do not show. The smaller black atoms are carbon; nitrogens, of the same size, are heavily outlined and dotted. Oxygen atoms are fine-line shaded.

TABLE XIVD,5. Parameters of the Atoms in tri-Glycine Sulfate

Atom	x	$\sigma(x)$	y	$\sigma(y)$	z	$\sigma(z)$
S	0.9995	0.0002	0.2500	0.0001	0.2250	0.0002
O(1)	0.8583	0.0005	0.2447	0.0006	0.0051	0.0009
O(2)	0.9669	0.0006	0.2437	0.0006	0.4572	0.0009
O(3)	0.0920	0.0005	0.1565	0.0006	0.2234	0.0009
O(4)	0.0769	0.0006	0.3469	0.0006	0.1941	0.0009
Glycine I						
O(1)	0.6064	0.0007	0.2393	0.0005	0.0746	0.0010
O(2)	0.4935	0.0009	0.2718	0.0008	0.6668	0.0011
C(1)	0.4905	0.0010	0.2472	0.0010	0.8727	0.0017
C(2)	0.3348	0.0010	0.2361	0.0010	0.9049	0.0017
N	0.3595	0.0013	0.2110	0.0012	0.1639	0.0018
Glycine II						
O(1')	0.2218	0.0006	0.4975	0.0006	0.7646	0.0009
O(2')	0.4596	0.0008	0.5397	0.0008	0.7988	0.0011
C(1')	0.3153	0.0010	0.5331	0.0010	0.6797	0.0016
C(2')	0.2675	0.0012	0.5734	0.0011	0.4070	0.0017
N'	0.0939	0.0008	0.5800	0.0008	0.3063	0.0012
Glycine III						
O(1'')	0.7824	0.0007	0.4931	0.0007	0.2229	0.0011
O(2'')	0.5454	0.0007	0.4825	0.0007	0.2317	0.0011
C(1'')	0.6937	0.0007	0.4749	0.0007	0.3281	0.0011
C(2'')	0.7440	0.0007	0.4320	0.0007	0.5906	0.0011
N''	0.9068	0.0006	0.4331	0.0006	0.7059	0.0009

The determined parameters are listed in Table XIVD,5.

The resulting structure is shown in Figure XIVD,9. The glycine residues differ in the planarity of their molecules. Molecules I and III are almost completely planar, the largest departure through the best plane being ca. 0.04 A. In the II residue, the nitrogen atom is 0.30 A. outside this best plane through the other atoms; this is the situation in glycine itself and it has therefore been concluded that glycine residue II is a zwitter ion whereas residues I and III have accepted hydrogen atoms from the sulfuric acid component of the compound. The interatomic distances in these residues are substantially those in glycine itself. The S–O separations in the sulfate anion (1.473–1.494 A.) are shorter than those that have usually been found for inorganic sulfates.

XIV,d6. In the structure described for *silver glycine*, $AgOOC \cdot CH_2NH_2$, its monoclinic crystals have a unit said to contain eight molecules. The cell dimensions are

$$a_0 = 9.0 \text{ A.}; \quad b_0 = 6.4 \text{ A.}; \quad c_0 = 18.0 \text{ A.}; \quad \beta = 126°20'$$

The arrangement within this cell, which may be larger than necessary, has been described as having all atoms in general positions of C_{2h}^1 ($P2/m$):

$$(4o) \quad \pm(xyz; \; x\bar{y}z)$$

with the parameters of Table XIVD,6.

TABLE XIVD,6
Parameters of the Atoms in Silver Glycine

Atom	x	y	z
Ag(1)	0.58	0.17	0.22
Ag(2)	0.45	0.325	0.31
C(1)	0.20	0.205	0.14
C(2)	0.80	0.30	0.36
$CH_2(1)$	0.19	0.125	0.06
$CH_2(2)$	0.80	0.38	0.44
$NH_2(1)$	0.27	0.155	0.01
$NH_2(2)$	0.71	0.35	0.48
O(1a)	0.38	0.125	0.22
O(1b)	0.09	0.125	0.16
O(2a)	0.64	0.38	0.29
O(2b)	0.94	0.38	0.37

XIV,d7. *Copper glycinate monohydrate,* $Cu(NH_2CH_2CO_2)_2 \cdot H_2O$, is orthorhombic with a tetramolecular unit of the edge lengths:

$$a_0 = 10.866 \pm 0.017 \text{ A.}; \quad b_0 = 5.220 \pm 0.007 \text{ A.}; \quad c_0 = 13.502 \pm 0.021 \text{ A.}$$

TABLE XIVD,7
Parameters of the Atoms in $Cu(NH_2CH_2CO_2)_2 \cdot H_2O$

Atom	x	$\sigma(x)$	y	$\sigma(y)$	z	$\sigma(z)$
Cu	0.1031	0.0002	0.3403	0.0004	0.4013	0.0001
O(1)	0.0813	0.0008	0.5772	0.0017	0.2903	0.0006
O(2)	0.1242	0.0009	0.6286	0.0019	0.1314	0.0006
O(3)	0.0104	0.0010	0.5568	0.0018	0.4919	0.0006
O(4)	−0.0664	0.0010	0.5797	0.0022	0.6444	0.0007
H_2O	0.2878	0.0010	0.5558	0.0028	0.4530	0.0008
N(1)	0.1842	0.0009	0.1152	0.0021	0.3025	0.0007
N(2)	0.1196	0.0011	0.0989	0.0024	0.5176	0.0008
C(1)	0.1938	0.0013	0.2519	0.0025	0.2074	0.0010
C(2)	0.1285	0.0009	0.5047	0.0024	0.2085	0.0009
C(3)	0.0623	0.0012	0.2162	0.0034	0.6065	0.0009
C(4)	−0.0045	0.0011	0.4676	0.0026	0.5803	0.0008

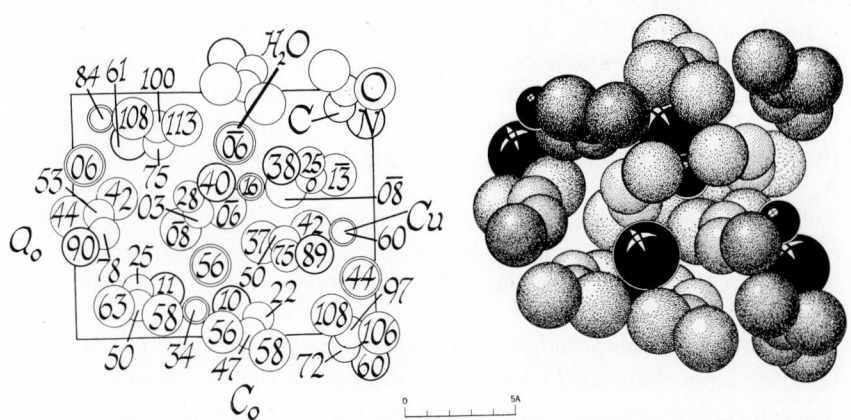

Fig. XIVD,10a (left). The orthorhombic structure of copper glycinate monohydrate projected along its b_0 axis. Right-hand axes.

Fig. XIVD,10b (right). A packing drawing of the orthorhombic structure of $Cu(NH_2CH_2COO)_2 \cdot H_2O$ viewed along its b_0 axis. Water molecules are the large, copper atoms the small black circles. The carbon atoms are dotted; the nitrogens are more heavily outlined and hook shaded. The larger carboxyl oxygen atoms are fine-line shaded.

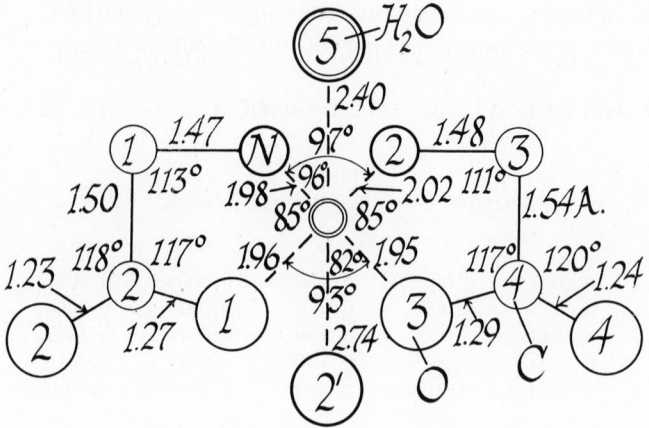

Fig. XIVD,11. Bond dimensions within crystals of copper glycinate monohydrate.

The space group is V^4 ($P2_12_12_1$) with all atoms in the positions:

$$(4a) \quad xyz; \; {}^1/_2-x,\bar{y},z+{}^1/_2; \; x+{}^1/_2,{}^1/_2-y,\bar{z}; \; \bar{x},y+{}^1/_2,{}^1/_2-z$$

Two determinations have led to concordant results with the more exact parameters of Table XIVD,7.

In the structure (Fig. XIVD,10), the two crystallographically different glycine groups have the bond dimensions of Figure XIVD,11. Copper has its usual approximately square coordination made distortedly octahedral by two (more distant) oxygen atoms. There are five hydrogen bonds; two are between water and carboxyl oxygens, with O–H–O = 2.76 or 2.80 A., and three between oxygen and nitrogen, with lengths of 2.98, 3.03, and 3.09 A. The carboxyl groups and their attached α carbon atoms are each coplanar and there is an angle of 5.4° between the two planes; N(1) and N(2) are 0.103 and 0.162 A. out of their appropriate carbon–oxygen planes.

XIV,d8. Crystals of *cadmium glycinate monohydrate*, $Cd(NH_2CH_2CO_2)_2 \cdot H_2O$, are monoclinic with a tetramolecular unit of the dimensions:

$$a_0 = 14.862 \pm 0.005 \text{ A.}; \quad b_0 = 5.297 \pm 0.005 \text{ A.}; \quad c_0 = 10.006 \pm 0.003 \text{ A.}$$
$$\beta = 90°24'$$

The space group is C_{2h}^6 ($I2/a$) with cadmium atoms in the special positions:

$$(4a) \quad 000; \; {}^1/_2\,0\,0; \; {}^1/_2\,{}^1/_2\,{}^1/_2; \; 0\,{}^1/_2\,{}^1/_2$$

The oxygen atoms of the water molecules are in the positions:

$$(4e) \quad {}^1/_4\, u\, 0;\; {}^3/_4\, \bar{u}\, 0;\; \text{B.C.} \qquad \text{with}\quad u = 0.491$$

The other atoms are in the general positions:

$$(8f) \quad \pm(xyz;\; x + {}^1/_2, \bar{y}, z);\; \text{B.C.}$$

with the approximate parameters listed below:

Atom	x	y	z
C(1)	0.092	0.321	0.252
C(2)	0.149	0.566	0.285
N	0.108	0.736	0.383
O(1)	0.039	0.226	0 331
O(2)	0.114	0.212	0.143

As Figure XIVD,12 indicates, the cadmium atoms in this structure are octahedrally surrounded by one nitrogen and one oxygen atom from each of two glycinate groups and by oxygen atoms from two other such groups. The chelated nitrogen and oxygen bonds to cadmium are about 2.3 A. long while the other two Cd–O = ca. 2.5 A. These together produce sheets that repeat themselves along the a_0 axis and are held together by water molecules through the hydrogen bonds they form.

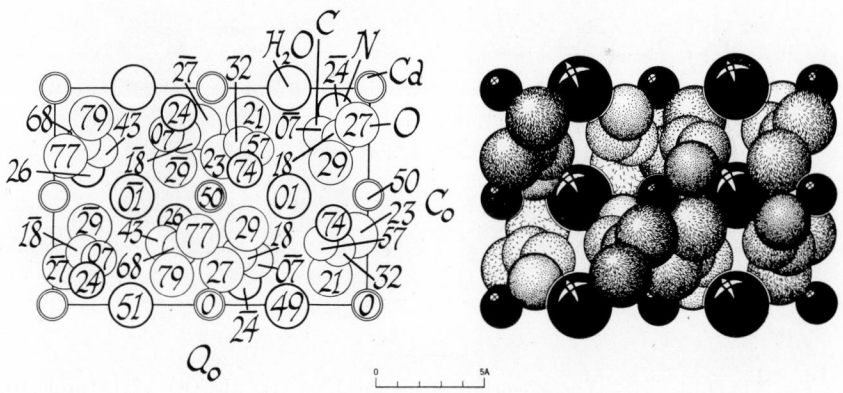

Fig. XIVD,12a (left). The monoclinic structure of cadmium glycinate dihydrate projected along its b_0 axis. Right-hand axes.

Fig. XIVD,12b (right). A packing drawing of the monoclinic $Cd(NH_2CH_2COO)_2 \cdot 2H_2O$ arrangement seen along its b_0 axis. Water molecules are the large, cadmium atoms the small black circles. Carbon atoms are hook shaded; the atoms of nitrogen are heavily outlined and dotted. Carboxyl oxygens are fine-line shaded.

CRYSTAL STRUCTURES

The corresponding zinc compound is triclinic, but with a cell having dimensions that suggest that its structure may be a superstructure on the foregoing arrangement.

XIV,d9. A structure has been described for *nickel glycine dihydrate*, $Ni(NH_2CH_2CO_2)_2 \cdot 2H_2O$. The monoclinic unit, containing two molecules, has the dimensions:

$$a_0 = 7.60 \text{ A.}; \quad b_0 = 6.60 \text{ A.}; \quad c_0 = 9.63 \text{ A.}; \quad \beta = 63°25'$$

TABLE XIVD,8
Parameters of Atoms in Nickel Glycine Dihydrate

Atom	x	y	z
C	0.232	0.345	0.871
O(1)	0.074	0.272	0.883
O(2)	0.308	0.505	0.790
CH_2	0.322	0.233	0.959
NH_2	0.276	0.025	0.996
H_2O	0.132	0.859	0.779

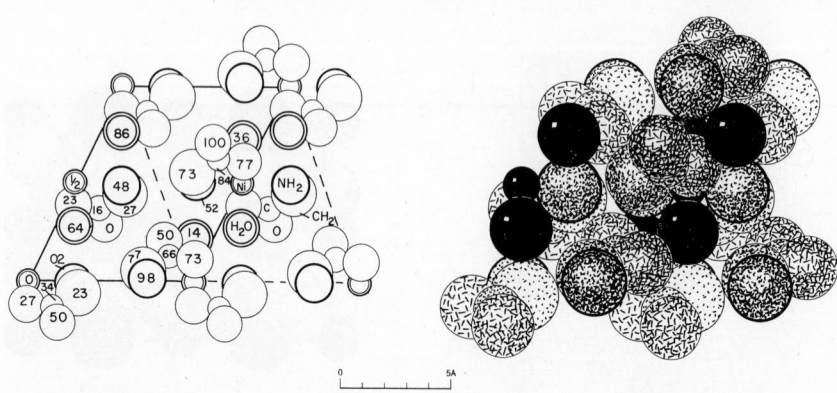

Fig. XIVD,13a (left). The monoclinic structure of $Ni(NH_2CH_2COO)_2 \cdot 2H_2O$ projected along its b_0 axis. The β angle used in the description of the text, and in the original, is an unusually acute one. The projection of a unit based on the conventional obtuse angle is indicated by the dashed lines. Left-hand axes.

Fig. XIVD,13b (right). A packing drawing of the monoclinic structure of $Ni(NH_2CH_2COO)_2 \cdot 2H_2O$ viewed along its b_0 axis. The smaller of the black circles are the nickel atoms; the larger are water molecules. Carboxyl oxygens are line shaded. The heavily ringed, dotted circles are nitrogen; other dotted circles are carbon, the larger being the CH_2 radicals.

The two nickel atoms have been placed in special positions:

$$(2a) \quad 000; 0 \, ^1/_2 \, ^1/_2$$

of C_{2h}^5 $(P2_1/c)$; all other atoms are in general positions:

$$(4e) \quad \pm (xyz; \ x, ^1/_2 - y, z + ^1/_2)$$

with the parameters of Table XIVD,8.

This leads to a structure (Fig. XIVD,13) composed of glycine residues having substantially the same dimensions as those in glycine itself: C–O = 1.25 and 1.29 A., C–CH$_2$ = 1.50 A., CH$_2$–NH$_2$ = 1.42 A., O–C–O = 122°, O–C–CH$_2$ = 123 or 115°, and C–CH$_2$–NH$_2$ = 103°. Each nickel atom has about it two carboxyl oxygen atoms at 2.08 A., two amino nitrogens at 2.09 A., and two water oxygens at 2.12 A. arranged at the corners of a distorted octahedron.

XIV,d10. A preliminary description has been given of the structure of crystals of the addition compound of *ferrous sulfate pentahydrate* with *glycine*, FeSO$_4$·HO$_2$CCH$_2$NH$_2$·5H$_2$O. The symmetry is triclinic with a bimolecular cell of the dimensions:

$$a_0 = 6.86 \text{ A.}; \quad b_0 = 13.6 \text{ A.}; \quad c_0 = 6.07 \text{ A.}$$
$$\alpha = 96°6'; \quad \beta = 96°48'; \quad \gamma = 92°30'$$

TABLE XIVD,9
Parameters of the Atoms in Ferrous Sulfate Pentahydrate–Glycine

Atom	x	y	z
O(1)	0.178	0.055	0.708
O(2)	0.104	0.133	0.205
O(3)	0.759	0.085	0.958
O(4)	0.488	0.123	0.287
O(5)	0.770	0.140	0.560
O(6)	0.655	0.285	0.403
O(7)	0.460	0.205	0.670
O(8)	0.169	0.383	0.535
O(9)	0.633	0.463	0.681
O(10)	0.169	0.343	0.067
O(11)	0.663	0.472	0.869
C(1)	0.023	0.322	0.858
C(2)	0.149	0.398	0.737
N	0.003	0.345	0.100
S	0.588	0.1885	0.487

All atoms except iron have been placed in the general positions $(2i)$ $\pm (xyz)$ of the space group C_i^1 $(P\bar{1})$ with the approximate parameters listed in Table XIVD,9. The atoms of iron are in the positions:

$$Fe(1): (1a) \quad 000$$
$$Fe(2): (1e) \quad {}^1/_2\,{}^1/_2\,0$$

According to this arrangement, the glycine molecules are bound to one of the two iron atoms to form the complex cation $[Fe(H_2O)_4(O_2CCH_2NH_3)_2]^{2+}$, the other iron is surrounded by six water molecules to yield $[Fe(H_2O)_6]^{2+}$.

XIV,d11. Crystals of *N-acetylglycine*, $CH_3C(O)NHCH_2COOH$, are monoclinic with a tetramolecular cell of the dimensions:

$$a_0 = 4.86 \text{ A.}; \quad b_0 = 11.54 \text{ A.}; \quad c_0 = 14.63 \text{ A.}; \quad \beta = 138°12'$$

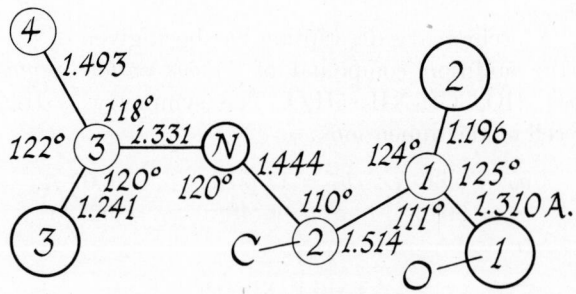

Fig. XIVD,14. Bond dimensions in molecules of acetylglycine.

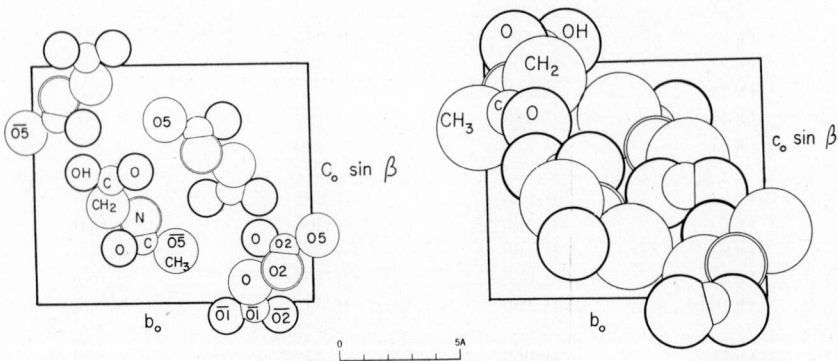

Fig. XIVD,15a (left). The monoclinic structure of *N*-acetylglycine projected along its a_0 axis. Left-hand axes.

Fig. XIVD,15b (right). An unshaded packing drawing of the monoclinic acetylglycine structure viewed along its a_0 axis.

TABLE XIVD,10
Parameters of the Atoms in N-Acetylglycine

Atom	x	$\sigma(x)$	y	$\sigma(y)$	z	$\sigma(z)$
C(1)	0.0113	0.0012	0.7169	0.0003	0.5170	0.0004
C(2)	0.0039	0.0012	0.7295	0.0004	0.4122	0.0004
C(3)	−0.0178	0.0013	0.6084	0.0004	0.2730	0.0004
C(4)	−0.0546	0.0019	0.4900	0.0004	0.2234	0.0006
N	−0.0178	0.0010	0.6171	0.0003	0.3637	0.0003
O(1)	0.0089	0.0009	0.8184	0.0003	0.5565	0.0003
O(2)	0.0199	0.0011	0.6255	0.0003	0.5576	0.0003
O(3)	0.0052	0.0011	0.6968	0.0003	0.2306	0.0003
H(1,C 2)	−0.240	0.013	0.776	0.005	0.340	0.004
H(2,C 2)	0.262	0.014	0.776	0.005	0.458	0.005
H(3,N)	−0.001	0.012	0.549	0.005	0.398	0.004
H(4,O 1)	0.007	0.012	0.810	0.004	0.620	0.004
$^1/_4$H(5,C 4)	−0.053	—	0.496	—	0.155	—
$^1/_4$H(6,C 4)	0.091	—	0.489	—	0.198	—
$^1/_4$H(7,C 4)	0.195	—	0.474	—	0.245	—
$^1/_4$H(8,C 4)	0.230	—	0.457	—	0.284	—
$^1/_4$H(9,C 4)	0.187	—	0.442	—	0.303	—
$^1/_4$H(10,C 4)	0.078	—	0.432	—	0.297	—
$^1/_4$H(11,C 4)	−0.068	—	0.431	—	0.269	—
$^1/_4$H(12,C 4)	−0.212	—	0.438	—	0.226	—
$^1/_4$H(13,C 4)	−0.315	—	0.453	—	0.179	—
$^1/_4$H(14,C 4)	−0.350	—	0.470	—	0.141	—
$^1/_4$H(15,C 4)	−0.308	—	0.486	—	0.122	—
$^1/_4$H(16,C 4)	−0.199	—	0.495	—	0.127	—

The space group is C_{2h}^5 $(P2_1/c)$ with all atoms in the general positions:

$$(4e) \quad \pm (xyz; x,^1/_2-y,z+^1/_2)$$

A recent reexamination has led to parameters which are in close agreement with those found earlier. They are listed in Table XIVD,10.

The molecule has the bond lengths and angles shown in Figure XIVD,14. It is nearly but not exactly planar, heavier atoms being as much as 0.08 A. away from the common plane. The arrangement of the molecules in the unit cell of the crystal is shown in Figure XIVD,15. The shortest intermolecular distance, indicative of hydrogen bonding, is O(1)–O(3) = 2.558 A. The next shortest separation, between N and O(2), is 3.030 A.

The new study has assigned parameters to the hydrogen atoms, as indicated in the table. These have definite positions except for those about the C(4) atom. From the analysis it is evident that there is disorder in the orientation of these methyl groups which is best expressed by saying that there is $1/4$ H in each of the positions listed in the table.

XIV,d12. A structure has been found for DL-*alanine*, $NH_2CH(CH_3)$-COOH. Its tetramolecular orthorhombic cell has the edges:

$$a_0 = 12.06 \text{ A.}; \quad b_0 = 6.05 \text{ A.}; \quad c_0 = 5.82 \text{ A.}$$

TABLE XIVD,11. Parameters of the Atoms in DL-Alanine

Atom	x	y	z
$C(CH_3)$	0.0908	0.0192	0.4457
$C(CH)$	0.1632	0.2192	0.4027
N	0.1397	0.3955	0.5762
C	0.1428	0.3154	0.1622
O(1)	0.0897	0.4832	0.1355
O(2)	0.1841	0.1985	0.0019
Assumed hydrogen positions			
$H(1,NH_2)$	0.060	0.447	0.566
$H(2,NH_2)$	0.192	0.521	0.548
$H(3,NH_2)$	0.157	0.332	0.733
$H(4,CH)$	0.249	0.185	0.428
$H(5,CH_3)$	0.004	0.063	0.448
$H(6,CH_3)$	0.114	0.940	0.607
$H(7,CH_3)$	0.105	0.900	0.305

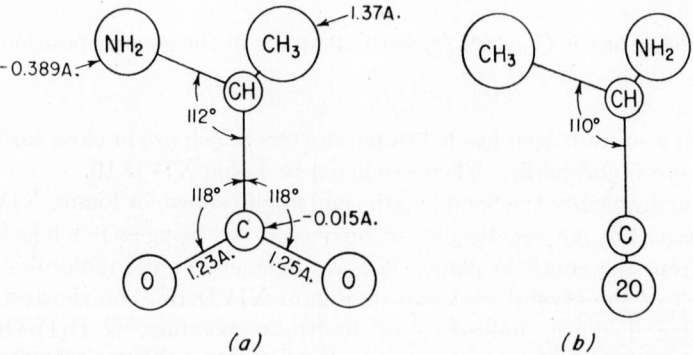

Fig. XIVD,16. Approximate dimensions of the alanine molecule in DL-alanine, according to the original determination. Distances indicated by arrows show the departures of atoms from the plane determined by the CH carbon and the oxygen atoms. According to the more recent study, the NH_2–CH separation is 1.496 A.

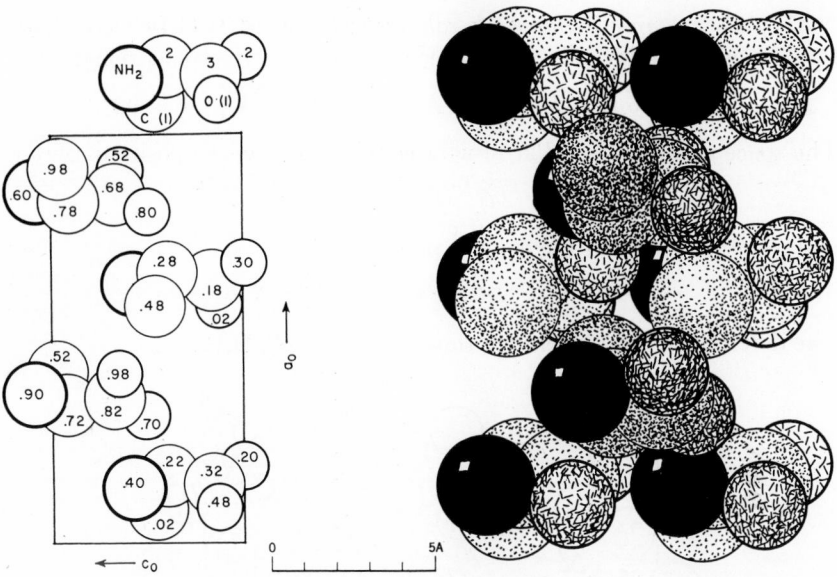

Fig. XIVD,17a (left). A projection along its b_0 axis of the orthorhombic structure of DL-alanine. Left-hand axes.

Fig. XIVD,17b (right). A packing drawing of the orthorhombic structure of DL-alanine seen along its b_0 axis. The nitrogen atoms are black. In this drawing the carboxyl oxygens are small and line shaded; carbon atoms are large and dotted.

All atoms are in the general positions of C_{2v}^9 $(Pna2_1)$:

(4a) $xyz;\ \bar{x},\bar{y},z+{}^1/_2;\ {}^1/_2-x,y+{}^1/_2,z+{}^1/_2;\ x+{}^1/_2,{}^1/_2-y,z$

with the refined parameters of Table XIVD,11.

The shape of the resulting molecule and the distances between its atoms are shown in Figure XIVD,16. The rather complicated way these molecules pack together to form the crystal is suggested by the drawing of Figure XIVD,17. They most closely approach one another through NH_2–O separations of 2.78–2.88 A.; these undoubtedly involve hydrogen bonds and tie all the molecules together into a three-dimensional network. The other significant close approaches of molecules are through CH_3–O and CH–O distances of 3.30 and 3.44 A., and through CH_3–CH_3 = 3.64 A. In the original study it was found that agreement between observed and calculated intensities could be materially improved by assigning definite positions to the hydrogen atoms in accordance with the requirements that the bond angles be approximately tetrahedral and that the distances C–H = 1.09 A. and N–H = 1.00 A. Such hydrogen positions are also listed in Table XIVD,11.

XIV,d13.   *Copper β-alanine hexahydrate,* $Cu(NH_2CH_2CH_2COO)_2 \cdot 6H_2O$, forms monoclinic crystals. Its bimolecular unit has the dimensions:

$$a_0 = 5.46 \text{ A.}; \quad b_0 = 7.71 \text{ A.}; \quad c_0 = 18.11 \text{ A.}; \quad \beta = 92°$$

The space group is C_{2h}^5 ($P2_1/c$) with copper atoms in the positions:

$$(2a) \quad 000; \, 0 \, {}^1/_2 \, {}^1/_2$$

and all other atoms in the general positions:

$$(4e) \quad \pm (xyz; \, x, {}^1/_2 - y, z + {}^1/_2)$$

The determined parameters are those of Table XIVD,12.

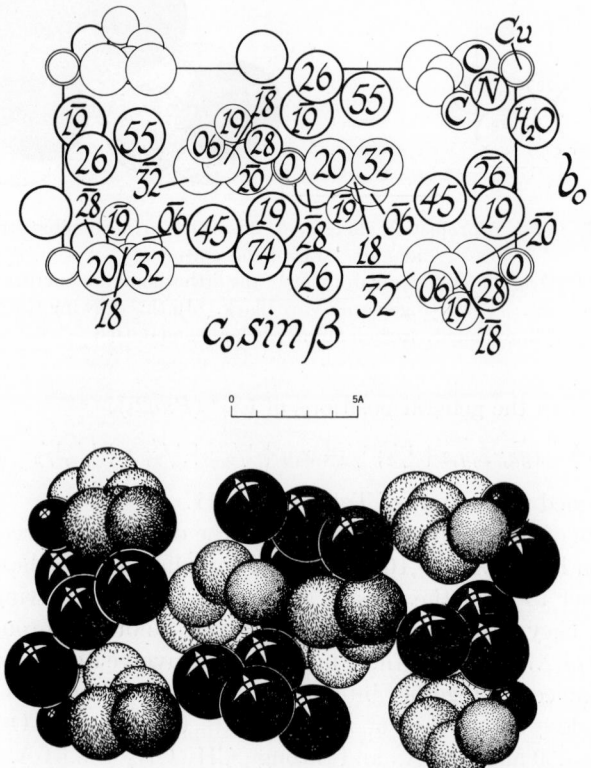

Fig. XIVD,18a (left). The monoclinic structure of $Cu[NH_2(CH_2)_2COO]_2 \cdot 6H_2O$ projected along its a_0 axis.   Right-hand axes.

Fig. XIVD,18b (right). A packing drawing of the monoclinic $Cu[NH_2(CH_2)_2COO]_2 \cdot 6H_2O$ arrangement seen along its a_0 axis.   Water molecules are the large, copper atoms the small black circles.   The carbon atoms are hook shaded; the nitrogens are heavily outlined and dotted.   Atoms of oxygen are somewhat larger and short-line shaded.

Fig. XIVD,19. Bond dimensions in crystals of $Cu[NH_2(CH_2)_2COO]_2 \cdot 6H_2O$.

TABLE XIVD,12
Parameters of Atoms in $Cu(C_3H_6NO_2)_2 \cdot 6H_2O$

Atom	x	y	z
C(1)	−0.185	−0.025	0.158
C(2)	0.065	−0.118	0.185
C(3)	0.187	−0.228	0.130
O(1)	−0.205	0.005	0.095
O(2)	−0.325	0.005	0.206
H₂O(3)	0.187	0.280	0.040
H₂O(4)	0.450	0.322	0.172
H₂O(5)	−0.258	0.445	0.060
N	0.278	−0.135	0.058

The structure as a whole is shown in Figure XIVD,18. The bond dimensions of the β-alanine molecular anions and their association with the copper atom are shown in Figure XIVD,19; in addition to the square coordination of two oxygen and two nitrogen atoms shown in this figure, there are two water oxygens above and below this plane, with $Cu-OH_2 =$ 2.53 A. The other water molecules may be thought of as providing hydrogen bonds which tie the various alanine molecules of the crystal together.

XIV,d14. *Nickel β-alanine dihydrate,* $Ni[NH_2(CH_2)_2CO_2]_2 \cdot 2H_2O$, is triclinic with a unimolecular cell of the dimensions:

$$a_0 = 8.48 \text{ A.}; \quad b_0 = 6.77 \text{ A.}; \quad c_0 = 4.93 \text{ A.}$$
$$\alpha = 103°; \quad \beta = 95°12'; \quad \gamma = 102°18'$$

The space group is C_i^1 ($P\bar{1}$) with the nickel atom in ($1a$) 000 and all other atoms in ($2i$) $\pm(xyz)$. The parameters are given in Table XIVD,13.

The resulting arrangement is shown in Figure XIVD,20. The alanine ions it contains have the bond dimensions of Figure XIVD,21. The nickel atoms are octahedrally surrounded by two water and two carboxyl oxygens

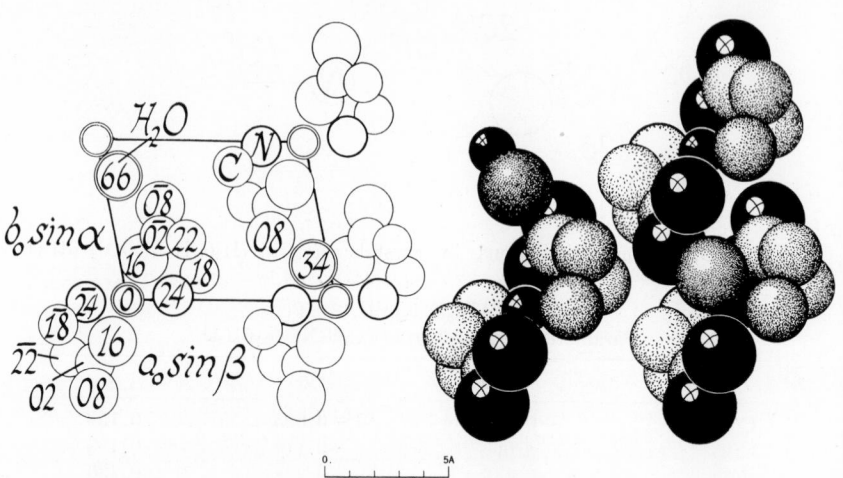

Fig. XIVD,20a (left). The triclinic structure of $Ni[NH_2(CH_2)_2COO]_2 \cdot 2H_2O$ projected along its c_0 axis. The small, doubly ringed circles are nickel. Right-hand axes.

Fig. XIVD,20b (right). A packing drawing of the triclinic $Ni[NH_2(CH_2)_2COO]_2 \cdot 2H_2O$ structure seen along its c_0 axis. Carboxyl oxygens are the large, nickel atoms the small black circles. Water molecules are large and dotted. The smaller dot-and-hook shaded atoms are nitrogen; the carbon atoms are lightly outlined and short-line-and-dot shaded.

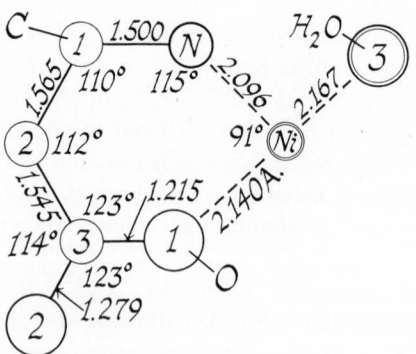

Fig. XIVD,21. Bond dimensions in crystals of $Ni[NH_2(CH_2)_2COO]_2 \cdot 2H_2O$.

TABLE XIVD,13
Parameters of Atoms in $Ni[NH_2(CH_2)_2CO_2]_2 \cdot 2H_2O$

Atom	x	y	z
N	0.2219	0.0206	0.2428
C(1)	0.3675	0.1638	0.1813
C(2)	0.3373	0.3878	0.2173
C(3)	0.2155	0.4038	-0.0247
O(1)	0.1095	0.2555	-0.1631
O(2)	0.2456	0.5811	-0.0847
H_2O	-0.0690	0.2176	0.3355

and by two nitrogen atoms, with $Ni-H_2O$ = 2.167 A., $Ni-O(1)$ = 2.40 A., and $Ni-N$ = 2.096 A. There are $N-O(2)$ and $N-H_2O$ distances of 3.11 and 3.12 A. which are considered to involve hydrogen bonds as well as two $H_2O-O(1)$ = 2.96 and 3.03 A. The structure has a definite layered character with a stacking of the layers normal to the (100) plane.

XIV,d15. Crystals of DL-*serine*, $OHCH_2CH(NH_2)COOH$, are monoclinic with a tetramolecular unit of the dimensions:

$$a_0 = 10.72 \text{ A.}; \quad b_0 = 9.14 \text{ A.}; \quad c_0 = 4.825 \text{ A.}; \quad \beta = 106°27'$$

TABLE XIVD,14
Parameters of the Atoms in DL-Serine

Atom	x	y	z
C(1)	0.2514	0.4047	0.1704
C(2)	0.2547	0.2789	0.3808
C(3)	0.3885	0.2100	0.4715
N	0.1532	0.1686	0.2460
O(1)	0.1631	0.4021	-0.0684
O(2)	0.3340	0.5054	0.2533
OH	0.4316	0.1683	0.2294
Calculated hydrogen positions			
H(1,C 2)	0.2336	0.3222	0.5728
H(2,C 3)	0.3858	0.1147	0.6056
H(3,C 3)	0.4572	0.2889	0.5994
H(4,N)	0.1562	0.1182	0.0699
H(5,N)	0.1560	0.0988	0.4076
H(6,N)	0.0692	0.2232	0.2122
H(7,OH)	0.3990	0.0708	0.1704

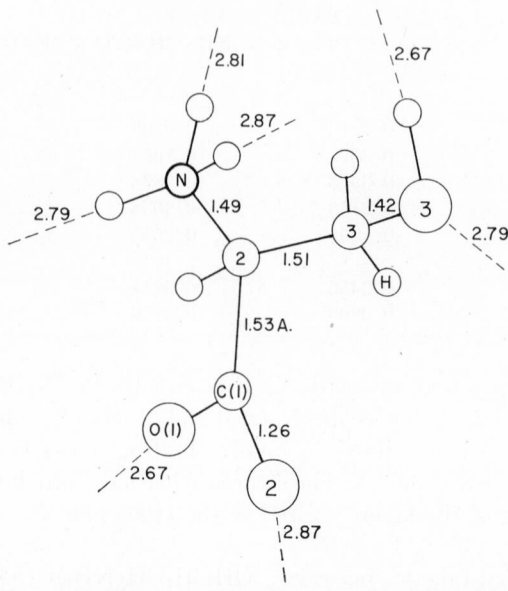

Fig. XIVD,22. Bond dimensions in crystals of DL-serine. The intermolecular hydrogen bonds are indicated by the dashed lines.

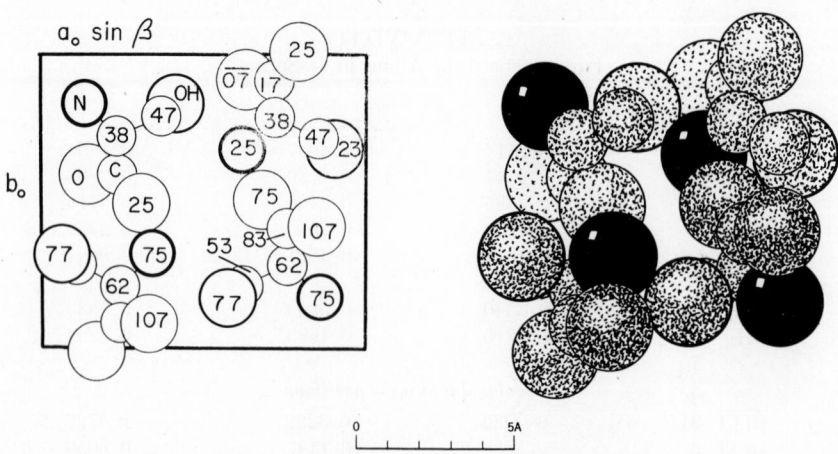

Fig. XIVD,23a (left). The monoclinic structure of DL-serine projected along its c_0 axis. Left-hand axes.

Fig. XIVD,23b (right). A packing drawing of the monoclinic DL-serine structure seen along its c_0 axis. The nitrogen atoms are black. Oxygen atoms are the large dotted circles, the hydroxyls being the more heavily outlined. Atoms of carbon are smaller and dotted.

Atoms are in general positions of C_{2h}^5 $(P2_1/a)$:

$$(4e) \quad \pm(xyz; x+{}^1/_2,{}^1/_2-y,z)$$

with the parameters of Table XIVD,14.

The general configuration of the resulting molecule and its bond dimensions are indicated in Figure XIVD,22. The molecular distribution within the crystal is illustrated in Figure XIVD,23. As is usual in amino acid crystals, the molecules are tied together through a system of hydrogen bonds. Each amino nitrogen has three of these, two to carboxyl (N–O = 2.81 and 2.87 A.) and one to a hydroxyl oxygen (N–O = 2.79 A.). The N–O (carboxyl) hydrogen bonds link the molecules together into sheets parallel to (010); the N–O (hydroxyl) bond can be thought of as tying these sheets to one another. The positions for the hydrogen atoms, including those involved in these bonds, have been computed on the assumptions that C–H = 1.09 A., N–H = 1.01 A., O–H = 0.97 A., and that the angles of their bonds are close to tetrahedral.

XIV,d16. Crystals of L-*threonine*, $CH_3CH(OH)CH(NH_2)COOH$, are orthorhombic with a tetramolecular cell having the edge lengths:

$$a_0 = 13.611 \text{ A.}; \quad b_0 = 7.738 \text{ A.}; \quad c_0 = 5.142 \text{ A.}$$

The space group is V^4 $(P2_12_12_1)$ with all atoms in the positions:

$$(4a) \quad xyz; \; {}^1/_2-x,\bar{y},z+{}^1/_2; \; x+{}^1/_2,{}^1/_2-y,\bar{z}; \; \bar{x},y+{}^1/_2,{}^1/_2-z$$

The parameters are those of Table XIVD,15.

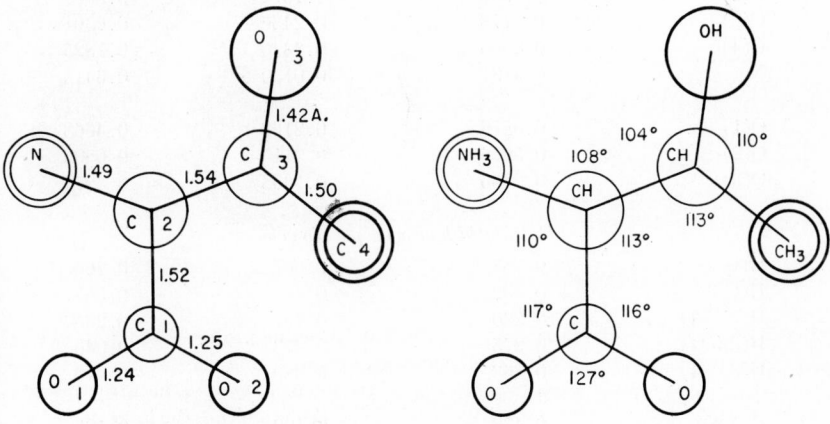

Fig. XIVD,24a (left). Bond lengths in the nonplanar molecule of threonine.
Fig. XIVD,24b (right). Bond angles in the molecule of threonine.

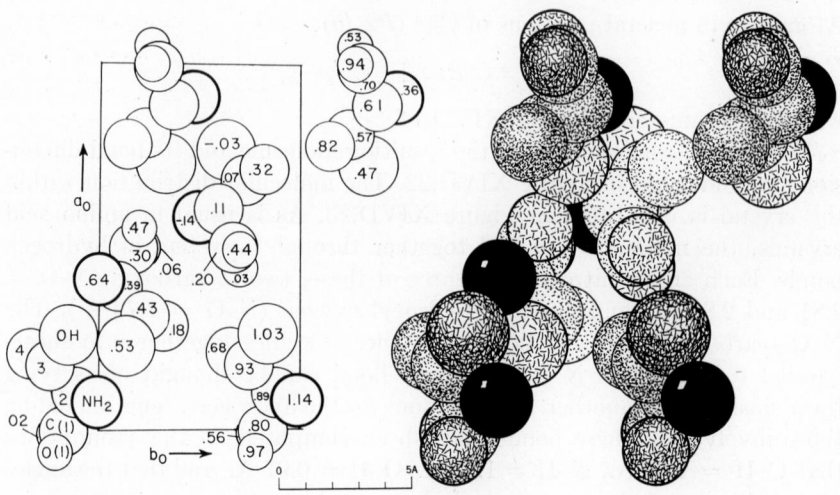

Fig. XIVD,25a (left). The orthorhombic structure of threonine projected along its c_0 axis. Left-hand axes.

Fig. XIVD,25b (right). A packing drawing of the orthorhombic threonine structure seen along its c_0 axis. The nitrogen atoms are black; those of oxygen are line shaded, the carboxyl oxygens being the smaller. Atoms of carbon are large and dotted.

TABLE XIVD,15
Parameters of the Atoms in L-Threonine

Atom	x	y	z
C(1)	0.4956	0.1836	0.2984
C(2)	0.3990	0.1074	0.3908
C(3)	0.3178	0.2436	0.4301
C(4)	0.2905	0.3346	0.1825
N	0.4131	0.0139	0.6415
O(1)	0.5576	0.2168	0.4668
O(2)	0.5026	0.2113	0.0590
O(3)	0.2366	0.1468	0.5274
Calculated hydrogen positions			
H(C 2)	0.387	0.007	0.260
H(C 3)	0.332	0.353	0.555
H(1,C 4)	0.230	0.427	0.235
H(2,C 4)	0.275	0.253	0.040
H(3,C 4)	0.363	0.380	0.090
H(1,N)	0.432	−0.100	0.585
H(2,N)	0.488	0.050	0.760
H(3,N)	0.340	−0.057	0.710
H(O 3)	0.155	0.226	0.535

They lead to molecules which have the dimensions of Figure XIVD,24, and are arranged as indicated in Figure XIVD,25. Between molecules the shortest atomic separations are N–O(2) = 2.80 and 2.90 A. and O(1)–O(3) = 2.66 A., indicative of hydrogen bonding. Other near approaches between molecules give N–O(1) = 3.08 A., N–O(3) = 3.10 A., O(1)–O(2) = 3.14 A., and C(2)–O(3) = 3.28 A.

Positions for the hydrogen atoms were chosen making use of the expected bond dimensions and of certain minor peaks in Fourier projections. Their parameters are appended to Table XIVD,15.

XIV,d17. An approximate structure has been described for DL-*aspartic acid*, $HO_2CCH_2CH(NH_2)CO_2H$. It is monoclinic with a unit containing eight molecules and having the dimensions:

$$a_0 = 9.18 \text{ A.}; \quad b_0 = 7.49 \text{ A.}; \quad c_0 = 15.79 \text{ A.}; \quad \beta = 96°$$

The space group is C_{2h}^6 ($I2/a$) with atoms in the positions:

$$(8f) \quad \pm(xyz; \ x+\tfrac{1}{2},\bar{y},z); \text{ B.C.}$$

The assigned parameters are those listed in Table XIVD,16.

TABLE XIVD,16
Parameters of the Atoms in DL-Aspartic Acid

Atom	x	y	z
C(1)	0.038	0.161	0.049
C(2)	0.079	0.094	0.134
C(3)	0.000	−0.017	0.199
C(4)	0.027	0.047	0.290
O(1)	0.106	0.114	−0.011
O(2)	−0.092	0.236	0.094
O(3)	0.134	−0.060	0.308
O(4)	−0.058	0.154	0.312
N	−0.133	0.081	0.208

Interatomic distances resulting from these parameters, such as C–O = 1.56 A., are such as to suggest that further study should be carried out.

XIV,d18. A preliminary announcement, which does not seem to have been followed up with a complete paper, has been made of a structure for *asparagine monohydrate*, $NH_2C(O)CH_2CH(NH_2)COOH \cdot H_2O$. The symmetry is orthorhombic with a tetramolecular cell of the edge lengths:

$$a_0 = 5.582 \text{ A.}; \quad b_0 = 9.812 \text{ A.}; \quad c_0 = 11.796 \text{ A.}$$

The space group is V^4 ($P2_12_12_1$) with atoms in its general positions. There appears to be error in the parameters as stated, and a detailed description of this work seems premature.

XIV,d19. Crystals of $(+)$-*S-methyl-L-cysteine sulfoxide*, $CH_3S(O)$-$CH_2CH(NH_2)COOH$, are orthorhombic with a tetramolecular cell of the edge lengths:

$a_0 = 5.214 \pm 0.002$ A.; $b_0 = 7.410 \pm 0.002$ A.; $c_0 = 16.548 \pm 0.008$ A.

The space group is V^4 ($P2_12_12_1$) with atoms in the positions:

$$(4a) \quad xyz; \; {}^1/_2-x,\bar{y},z+{}^1/_2; \; x+{}^1/_2,{}^1/_2-y,\bar{z}; \; \bar{x},y+{}^1/_2,{}^1/_2-z$$

The determined parameters are those of Table XIVD,17.

TABLE XIVD,17
Parameters of the Atoms in $(+)$-S-Methyl-L-Cysteine Sulfoxide

Atom	x	$\sigma(x)$	y	$\sigma(y)$	z	$\sigma(z)$
C(1)	−0.0388	0.0038	0.0896	0.0023	−0.1197	0.0010
C(2)	0.2388	0.0042	0.0064	0.0024	−0.1188	0.0010
C(3)	0.2250	0.0042	−0.1898	0.0030	−0.1389	0.0010
C(4)	0.4163	0.0042	−0.5303	0.0021	−0.1533	0.0010
O(1)	−0.1122	0.0027	0.1504	0.0015	−0.0538	0.007
O(2)	−0.1355	0.0027	0.0982	0.0017	−0.1865	0.0007
O(3)	0.6703	0.0027	−0.2727	0.0017	−0.0739	0.0007
N	0.3633	0.0029	0.0295	0.0013	−0.0367	0.0007
S	0.5363	0.0010	−0.3006	0.0005	−0.1523	0.0002
H(1)[a]	0.317	—	−0.073	—	−0.002	—
H(2)	0.294	—	0.138	—	−0.010	—
H(3)	0.550	—	0.003	—	−0.041	—
H(4)	0.361	—	0.071	—	−0.164	—
H(5)	0.129	—	−0.254	—	−0.089	—
H(6)	0.121	—	−0.201	—	−0.194	—

[a] No parameters were given for H(7)–(9).

The molecular arrangement in the crystals is shown in Figure XIVD,26. These molecules have the bond dimensions of Figure XIVD,27. The molecules form paired sheets that are repeated along the c_0 axis. Within the sheets, the molecules are held together by three types of hydrogen bond: N–O(1) = 2.808 and 2.892 A. and N–O(3) = 2.826 A. The shortest non-bridged contacts between separate molecules are C(4)–O(2) = 3.18 A. and C(3)–O(2) = 3.32 A.; all others exceed 3.4 A.

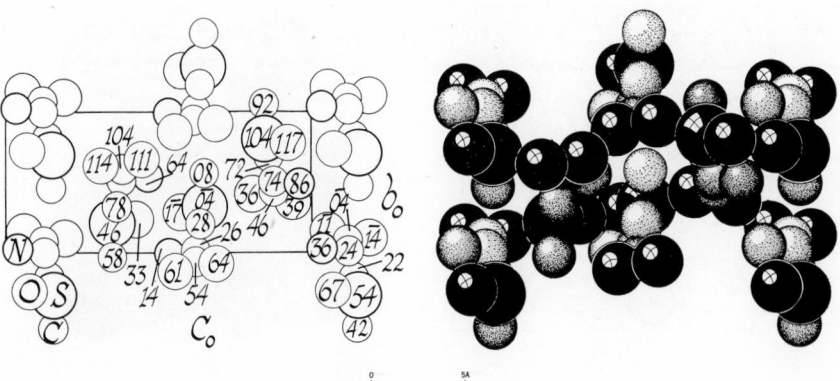

Fig. XIVD,26a (left). The orthorhombic structure of $CH_3S(O)CH_2CH(NH_2)COOH$ projected along its a_0 axis. Right-hand axes.

Fig. XIVD,26b (right). A packing drawing of the orthorhombic $CH_3S(O)CH_2CH-(NH_2)COOH$ arrangement viewed along its a_0 axis. Both the sulfur and the oxygen atoms are black, the sulfur being the larger. Of the smaller atoms, the nitrogens are the more heavily outlined and the carbons are short-line-and-dot shaded.

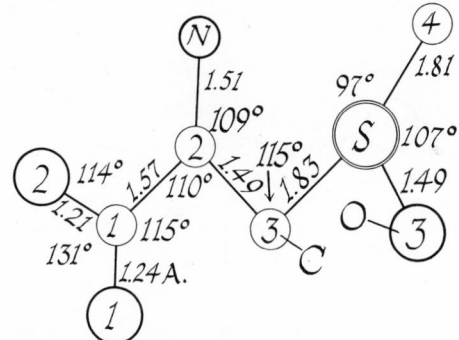

Fig. XIVD,27. Bond dimensions in the molecule of $CH_3S(O)CH_2CH(NH_2)COOH$.

XIV,d20. Crystals of *α-amino isobutyric acid*, $(CH_3)_2C(NH_2)COOH$, are monoclinic with a unit containing eight molecules and having the dimensions:

$$a_0 = 10.61 \text{ A}; \quad b_0 = 8.99 \text{ A.}; \quad c_0 = 11.36 \text{ A.}; \quad \beta = 94°$$

All atoms have been placed in general positions of C_{2h}^6 ($C2/c$):

(8f) $\quad \pm(xyz; x+1/2, y+1/2, z; x, \bar{y}, z+1/2; x+1/2, 1/2-y, z+1/2)$

with the parameters of Table XIVD,18.

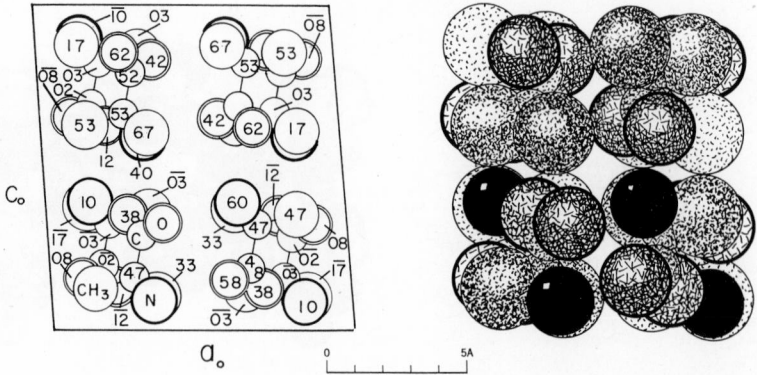

Fig. XIVD,28a (left). The monoclinic structure of $(CH_3)_2C(NH_2)COOH$ projected along its b_0 axis. Left-hand axes.

Fig. XIVD,28b (right). A packing drawing of the monoclinic structure of $(CH_3)_2C-(NH_2)COOH$ viewed along its b_0 axis. The nitrogen atoms are black; the oxygens are heavily outlined and line shaded. The dotted terminal methyl carbons are larger than the others.

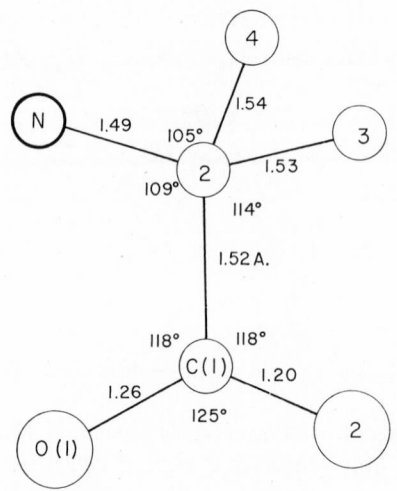

Fig. XIVD,29. Bond dimensions in the molecule of $(CH_3)_2C(NH_2)COOH$.

In the resulting structure (Fig. XIVD,28), the molecules have the bond dimensions of Figure XIVD,29. There are two hydrogen bonds of length 2.82 A. and one of length 2.88 A. between nitrogen and oxygen atoms of neighboring molecules. Methyl carbons of adjacent molecules come as close as 3.64 A.

TABLE XIVD,18. Parameters of the Atoms in α-Amino Isobutyric Acid

Atom	x	y	z
N	0.154	0.100	0.388
C(1)	0.178	−0.025	0.199
C(2)	0.212	−0.031	0.331
C(3)	0.355	−0.027	0.365
C(4)	0.148	−0.167	0.382
O(1)	0.112	0.083	0.159
O(2)	0.224	−0.121	0.138

XIV,d21. *Nickel di α-amino isobutyrate tetrahydrate,* $Ni[(CH_3)_2C(NH_2)$-$COO]_2\cdot 4H_2O$, forms monoclinic crystals. Their tetramolecular cells have the dimensions:

$$a_0 = 9.92\pm0.02 \text{ A.};\quad b_0 = 14.24\pm0.05 \text{ A.};\quad c_0 = 11.42\pm0.02 \text{ A.}$$
$$\beta = 107°30'\pm30'$$

The space group is C_{2h}^5 $(P2_1/a)$ with all atoms in the general positions:

$$(4e) \quad \pm(xyz;\ x+{}^1/_2, {}^1/_2-y, z)$$

TABLE XIVD,19. Parameters of the Atoms in $Ni[(CH_3)_2NH_2CCO_2]_2\cdot 4H_2O$

Atom	x	y	z
N(1)	0.285	0.250	0.326
O(2)	0.395	0.152	0.252
$H_2O(3)$	0.387	0.363	0.263
O(4)	0.175	0.140	0.392
$H_2O(5)$	0.177	0.350	0.383
O(6)	0.397	0.092	0.075
O(7)	0.233	0.032	0.532
$H_2O(8)$	0.017	0.077	0.630
$H_2O(9)$	0.170	0.475	0.135
N(10)	0.133	0.238	0.152
N(11)	0.439	0.237	0.498
C(12)	0.338	0.149	0.132
C(13)	0.192	0.182	0.071
C(14)	0.218	0.229	0.963
C(15)	0.095	0.100	0.005
C(16)	0.263	0.105	0.490
C(17)	0.392	0.165	0.572
C(18)	0.496	0.091	0.652
C(19)	0.328	0.218	0.667

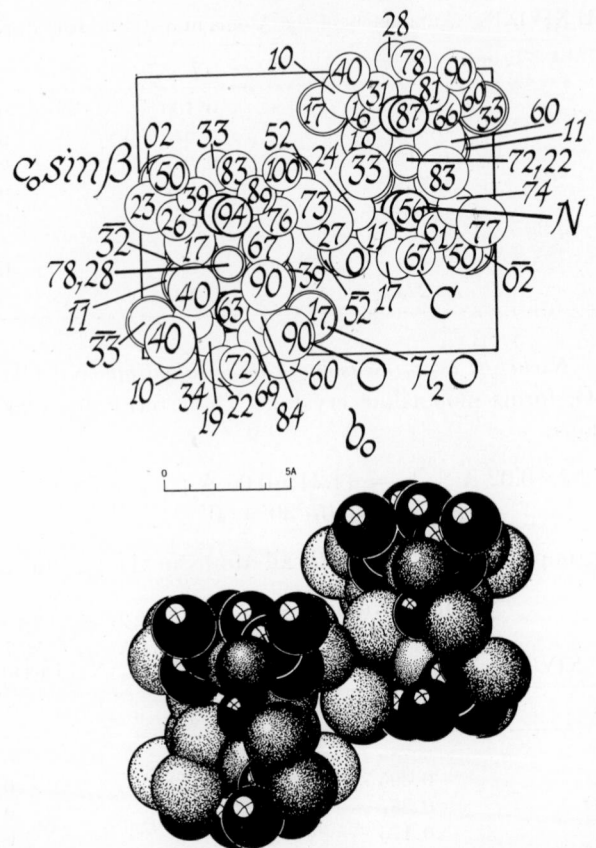

Fig. XIVD,30a (top). The monoclinic structure of Ni[(CH₃)₂C(NH₂)COO]₂·4H₂O projected along its a_0 axis. The small, doubly ringed circles are the nickel atoms. Right-hand axes.

Fig. XIVD,30b (bottom). A packing drawing of the monoclinic structure of Ni[(CH₃)₂C-(NH₂)COO]₂·4H₂O seen along its a_0 axis. The carbon atoms are the larger, the nickels the smaller of the black circles. Water molecules are heavily outlined and dotted. Atoms of nitrogen are dotted. The oxygen atoms are slightly larger and short-line-and-dot shaded.

The parameters are listed in Table XIVD,19.

The structure is shown in Figure XIVD,30. Each nickel atom is octa-hedrally surrounded by four oxygen and two nitrogen atoms, at distances of Ni–O = 2.01–2.24 A. and Ni–N = 2.10 and 2.11 A. Two of the oxygen neighbors are from carboxyl groups, the other two from water molecules; one nitrogen and one oxygen atom are furnished by each of the two crystal-lographically different anions. Each of the two water molecules not associated with the nickel atoms has three carboxylic oxygen and one nitrogen neighbor at distances of N–O = 3.04 or 3.40 A. and of O–O = 2.72–3.31 A.

XIV,d22. *Copper(II) di β-aminobutyrate dihydrate,* $Cu[NH_2CH(CH_3)CH_2COO]_2\cdot 2H_2O$, is triclinic with a unimolecular cell of the dimensions:

$$a_0 = 6.69 \pm 0.02 \text{ A.;} \quad b_0 = 5.06 \pm 0.02 \text{ A.;} \quad c_0 = 9.87 \pm 0.04 \text{ A.}$$
$$\alpha = 87°30'; \quad \beta = 104°55'; \quad \gamma = 105°45'$$

The space group was taken as C_i^1 ($P\bar{1}$) with the copper atom in ($1a$) 000 and all other atoms in ($2i$) $\pm(xyz)$; their parameters are those of Table XIVD,20.

TABLE XIVD,20
Parameters of the Atoms in $Cu[NH_2CH(CH_3)CH_2COO]_2\cdot 2H_2O$

Atom	x	y	z	σ, A.
N	0.0252	0.1842	0.1803	0.018
O(1)	0.2427	−0.1636	0.0882	0.014
O(2)	0.5765	−0.1175	0.2156	0.014
H$_2$O	0.2365	0.4025	−0.0791	0.014
C(1)	0.4073	−0.0527	0.1910	0.023
C(2)	0.3940	0.1556	0.2882	0.023
C(3)	0.1735	0.1283	0.3046	0.022
C(4)	0.1795	0.3090	0.4221	0.023

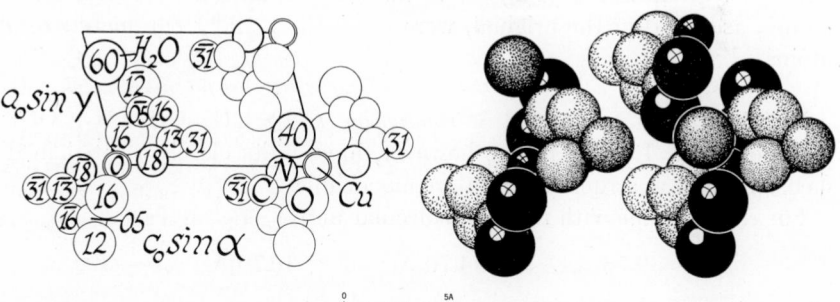

Fig. XIVD,31a (left).   The triclinic structure of $Cu[NH_2CH(CH_3)CH_2COO]_2\cdot 2H_2O$ projected along its b_0 axis.   Right-hand axes.

Fig. XIVD,31b (right).   A packing drawing of the triclinic structure of cupric β-amino butyrate dihydrate seen along its b_0 axis.   Oxygen atoms are the large, copper atoms the small black circles.   Water molecules are heavily outlined and hook shaded.   Atoms of nitrogen are slightly smaller, heavily outlined, and dotted.   The carbon atoms, of the same size, are lightly outlined and short-line-and-dot shaded.

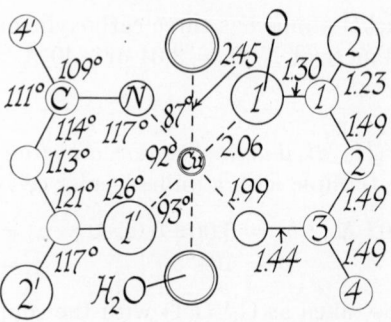

Fig. XIVD,32. Bond dimensions in crystals of $Cu[NH_2CH(CH_3)CH_2COO]_2 \cdot 2H_2O$.

The copper atom in this structure (Fig. XIVD,31) is octahedrally surrounded by two nitrogen atoms, two carboxyl oxygen atoms, and two water molecules; their distances and the various bonds of the amino butyrate anions are shown in Figure XIVD,32. The molecules are to be thought of as held together by hydrogen bonds of lengths 2.73 and 2.79 A. between water and carboxyl oxygens, and of a length 3.06 A. between nitrogen and both kinds of oxygen.

A partial structure has also been described for the *copper* DL-*α-amino-butyrate*, $Cu(OOC \cdot C_3H_6NH_2)_2$. Its bimolecular monoclinic cell has the dimensions:

$$a_0 = 11.09 \text{ A.;} \quad b_0 = 5.06 \text{ A.;} \quad c_0 = 9.45 \text{ A.;} \quad \beta = 87°51'$$

The space group is C_{2h}^5 ($P2_1/c$) and though y parameters were not found, values, as stated in the original, were given the x and z parameters of all atoms.

XIV,d23. The amino acid DL-*methionine*, $CH_3S(CH_2)_2CH(NH_2)COOH$, is dimorphous. Its two forms, however, have units closely related in dimensions, and in structure; both are monoclinic.

For *α-methionine* with its tetramolecular unit:

$$a_0 = 9.76 \text{ A.;} \quad b_0 = 4.70 \text{ A.;} \quad c_0 = 16.70 \text{ A.;} \quad \beta = 102°$$

The space group is C_{2h}^5 ($P2_1/a$) with all atoms in the general positions:

$$(4e) \quad \pm (xyz; \ x+\tfrac{1}{2}, \tfrac{1}{2}-y, z)$$

The determined parameters are those of Table XIVD,21.

These lead to a molecule that has the dimensions of Figure XIVD,33. Here the five atoms C–C(4) and S are practically planar; the N and C(5) atoms lie at considerable distances from this plane.

TABLE XIVD,21
Parameters of the Atoms in α-Methionine and in the β form (in parentheses)

Atom	x	y	z
S	0.328 (0.017)	−0.056 (0.172)	0.128 (0.059)
O(1)	0.007 (−0.270)	−0.094 (0.192)	0.378 (0.187)
O(2)	0.174 (−0.091)	−0.342 (−0.078)	0.435 (0.216)
N	0.364 (0.099)	0.053 (0.306)	0.406 (0.203)
C(1)	0.125 (−0.142)	−0.144 (0.122)	0.392 (0.194)
C(2)	0.218 (−0.049)	0.028 (0.306)	0.353 (0.175)
C(3)	0.225 (−0.051)	0.096 (0.169)	0.268 (0.131)
C(4)	0.311 (0.038)	0.093 (0.336)	0.224 (0.109)
C(5)	0.164 (0.151)	0.047 (0.350)	0.068 (0.043)

The arrangement of these molecules in the crystal is indicated in Figure XIVD,34. The shortest intermolecular distances involve a series of hydrogen bonds between nitrogen and oxygen atoms; their lengths are 2.59, 2.80, and 2.92 A. Between other parts of the molecules the shortest distances are C(5)–C(5) = 3.55 and 3.87 A., and S–C(5) = 3.78 A.

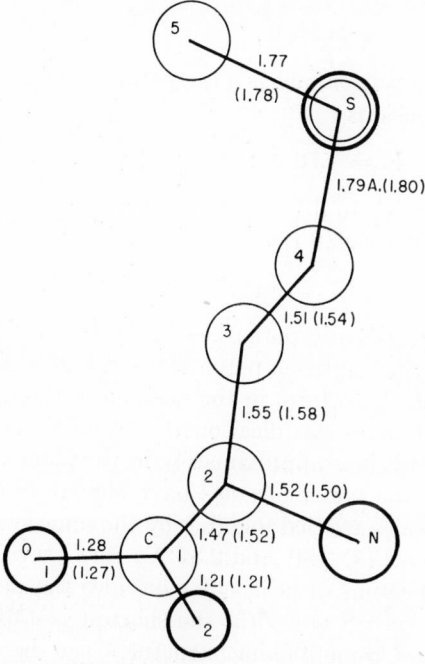

Fig. XIVD,33. Bond lengths of the molecules in the two forms of methionine, those for the β modification being given in parentheses.

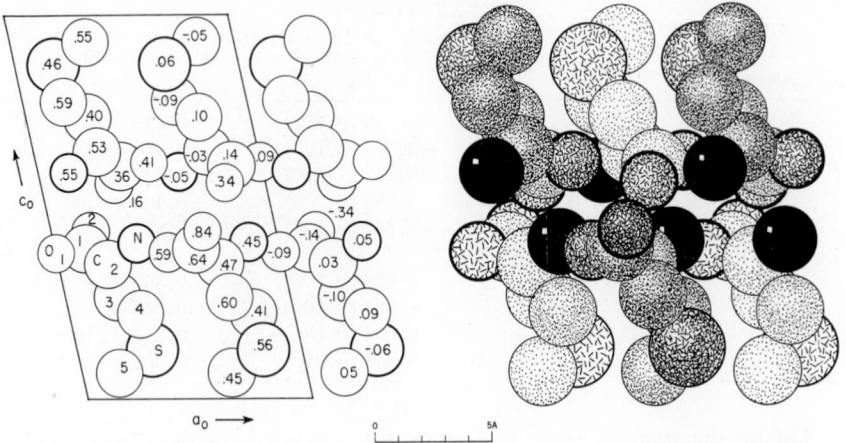

Fig. XIVD,34a (left). The monoclinic structure of the α form of methionine projected along its b_0 axis. Left-hand axes.

Fig. XIVD,34b (right). A packing drawing of the monoclinic structure of α-methionine seen along its b_0 axis. The nitrogen atoms are black, the carbon atoms large and dotted. The sulfurs are the large line shaded circles; the oxygen atoms are smaller, line shaded, and heavily ringed.

The β form of DL-*methionine* has a larger monoclinic cell containing eight molecules. Its dimensions

$$a_0 = 9.94 \text{ A.}; \quad b_0 = 4.70 \text{ A.}; \quad c_0 = 33.40 \text{ A.}; \quad \beta = 106°36'$$

bear an obvious relation to those of the α form. The space group is, however, C_{2h}^6 ($I2/a$) with all atoms in the general positions:

$$(8f) \quad \pm(xyz; \ x+\tfrac{1}{2},\bar{y},z); \text{ B.C.}$$

The parameters are given in parentheses in Table XIVD,21.

As indicated by the lengths in parentheses in Figure XIVD,33, the bonds differ from those in the α form in the position of the terminal C(5) atom. In the molecule of the β modification this atom is turned about the line C(4)–S as an axis till it is approximately in the plane of the other carbon atoms; otherwise, the two molecules have similar configurations. In the β crystal the molecules are tied together by the same system of N–O hydrogen bonds (N–O = 2.78, 2.80, and 2.82 A.) that prevails in α-methionine. Between the C(5) atoms of adjacent molecules the distances are 3.78 A. and upwards; between S and C(5) the shortest is 4.01 A. The difference between the α- and β-methionine structures can be seen by comparing Figures XIVD,34 and 35 with the preceding.

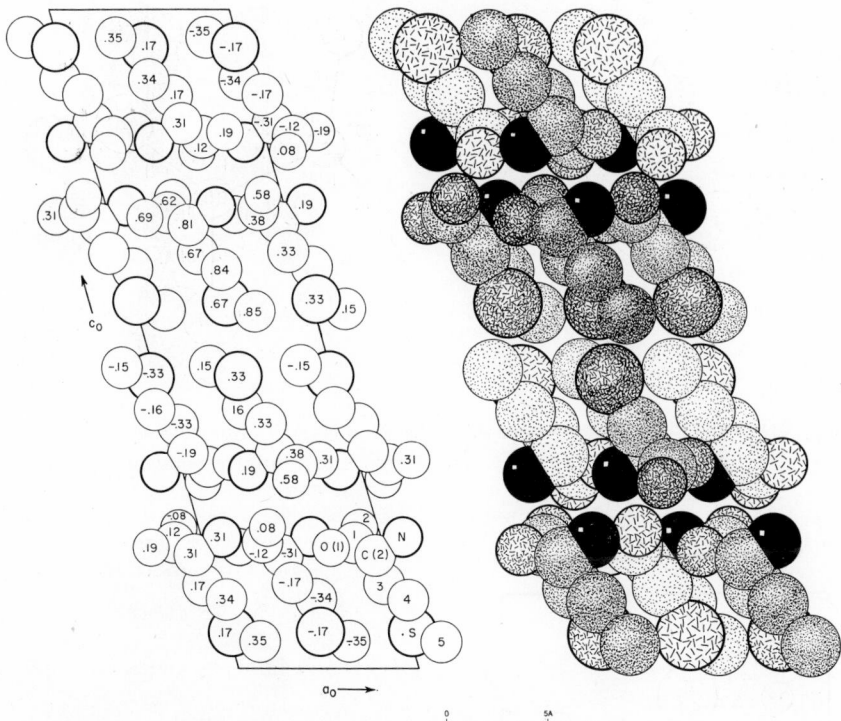

Fig. XIVD,35a (left). The monoclinic structure of the β modification of methionine projected along its b_0 axis. Left-hand axes.

Fig. XIVD,35b (right). A packing drawing of the monoclinic structure of β-methionine
viewed along its b_0 axis.   As in the previous figure, the nitrogens are black, the carbons
large and dotted.   Sulfur atoms are the large, oxygens the small line shaded circles.

XIV,d24.   Crystals of DL-*glutamic acid hydrochloride*, HOOCCH(NH₂·
HCl)(CH₂)₂COOH, are orthorhombic with the tetramolecular cell:

$$a_0 = 5.16 \text{ A.}; \quad b_0 = 11.80 \text{ A.}; \quad c_0 = 13.80 \text{ A.}$$

Atoms are in general positions of V^4 $(P2_12_12_1)$:

(4a) $xyz; \ ^1/_2-x,\bar{y},z+^1/_2; \ x+^1/_2,^1/_2-y,\bar{z}; \ \bar{x},y+^1/_2,^1/_2-z$

with the parameters of Table XIVD,22.

The bond dimensions of the resulting molecule are shown in Figure
XIVD,36. The molecular packing is illustrated in Figure XIVD,37. As in
other amino acids the molecules are tied together by a system of hydrogen
bonds. Each nitrogen has two such bonds joining it to chlorine (N–H–Cl =

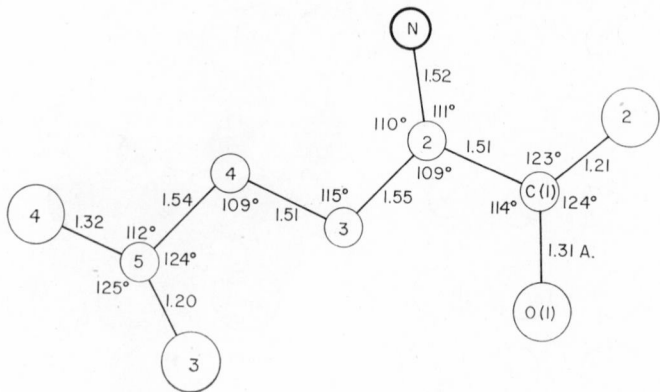

Fig. XIVD,36. Bond dimensions of the molecule in crystals of DL-glutamic acid hydrochloride.

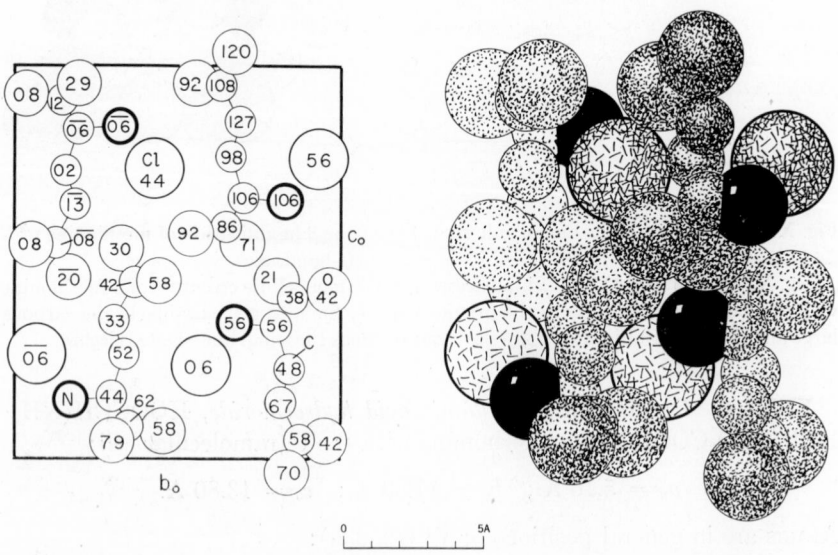

Fig. XIVD,37a (left). The orthorhombic structure of DL-glutamic acid hydrochloride projected along its a_0 axis. Left-hand axes.

Fig. XIVD,37b (right). A packing drawing of the orthorhombic DL-glutamic acid hydrochloride arrangement seen along its a_0 axis. The oxygens are the large, the carbons the small dotted circles. Atoms of nitrogen are black. Chloride ions are line shaded and heavily ringed.

TABLE XIVD,22
Parameters of the Atoms in DL-Glutamic Acid Hydrochloride

Atom	x	y	z
C(1)	0.379	0.144	0.410
C(2)	0.562	0.195	0.334
C(3)	0.477	0.157	0.227
C(4)	0.669	0.183	0.145
C(5)	0.580	0.126	0.047
O(1)	0.422	0.036	0.428
O(2)	0.209	0.198	0.451
O(3)	0.425	0.051	0.044
O(4)	0.701	0.168	−0.032
N	0.560	0.323	0.341
Cl	0.060	0.427	0.238

3.18 A.) and one to oxygen (N–H–O = 2.89 A.). The carboxyl oxygens of adjacent molecules are connected by hydrogen bonds of length 2.57 A. There are also close nonhydrogen-bonded contacts between molecules [O(2)–O(2) = 3.13 A., O(1)–O(4) = 3.16 A., and other O–O = 3.26 and 3.35 A.]. There is likewise a short N–O(3) = 3.10 A.

XIV,d25. The β form of *glutamic acid*, $HOOC(CH_2)_2CH(NH_2)COOH$, is orthorhombic with a tetramolecular cell of the edge lengths:

$$a_0 = 5.17 \text{ A.}; \quad b_0 = 17.34 \text{ A.}; \quad c_0 = 6.95 \text{ A.}, \qquad \text{all} \pm 0.2\%$$

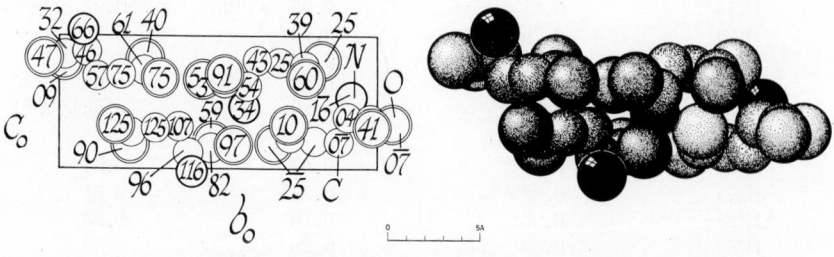

Fig. XIVD,38a (left). The orthorhombic structure of the β form of glutamic acid projected along its a_0 axis. Right-hand axes.

Fig. XIVD,38b (right). A packing drawing of the orthorhombic structure of β-glutamic acid viewed along its a_0 axis. The nitrogen atoms are black. Oxygen atoms are large, heavily outlined, and dotted; carbons are line shaded.

Fig. XIVD,39. Bond dimensions of the glutamic acid molecule in crystals of its β form.

TABLE XIVD,23
Parameters of the Atoms in β-Glutamic Acid

Atom	x	y	z
O(1)	0.092	0.016	0.820
O(2)	0.475	−0.045	0.842
C(3)	0.325	0.013	0.863
C(4)	0.463	0.091	0.899
C(5)	0.571	0.123	0.706
C(6)	0.750	0.194	0.719
C(7)	0.607	0.273	0.754
O(8)	0.746	0.325	0.692
O(9)	0.400	0.276	0.830
N(10)	0.663	0.079	0.043
H(1)	0.81	0.06	0.97
H(2)	0.76	0.13	0.08
H(3)	0.57	0.05	0.14
H(4)	0.67	0.37	0.68
H(5)	0.32	0.13	0.96
H(6)	0.67	0.07	0.63
H(7)	0.41	0.13	0.61
H(8)	0.89	0.18	0.83
H(9)	0.86	0.20	0.59

The space group is V^4 ($P2_12_12_1$) with all atoms in the positions:

$$(4a) \quad xyz; \ {}^1\!/_2-x,\bar{y},z+{}^1\!/_2; \ x+{}^1\!/_2,{}^1\!/_2-y,\bar{z}; \ \bar{x},y+{}^1\!/_2,{}^1\!/_2-z$$

The parameters, considered to be accurate to ±0.003, are those of Table XIVD,23. The hydrogen positions were assigned on the assumptions that C–H = 1.09 A., N–H = 1.01 A., and O–H = 0.97 A.

In this structure (Fig. XIVD,38) the nonplanar molecules have the bond dimensions shown in Figure XIVD,39. They are associated together in the b_0 direction by O–H$\cdots$O bonds of length 2.54 A. and in other directions by N–H$\cdots$O = 2.86–2.94 A.

XIV,d26. Crystals of L-*glutamine*, $C_5H_{10}O_3N_2$, are orthorhombic with a tetramolecular cell of the dimensions:

$$a_0 = 16.01 \text{ A.}; \quad b_0 = 7.76 \text{ A.}; \quad c_0 = 5.10 \text{ A.}$$

Atoms are in the general positions of V^4 ($P2_12_12_1$):

$$(4a) \quad xyz; \ \bar{x},\bar{y},z+{}^1\!/_2; \ x+{}^1\!/_2,{}^1\!/_2-y,{}^1\!/_2-z; \ {}^1\!/_2-x,y+{}^1\!/_2,\bar{z}$$

with the parameters of Table XIVD,24. The hydrogen positions are based on the assumption of a distribution that is tetrahedral about carbon (with C–H = 1.05 A.) and threefold about nitrogen.

The resulting molecule, which is of course nonplanar, has the bond dimensions of Figure XIVD,40. The molecular packing is indicated in Figure XIVD,41. Each N(2) atom has three close oxygen neighbors at

TABLE XIVD,24
Parameters of the Atoms in L-Glutamine

Atom	x	y	z
C(1)	0.048	0.234	0.696
C(2)	0.114	0.377	0.706
C(3)	0.091	0.530	0.861
C(4)	0.160	0.661	0.880
C(5)	0.242	0.588	0.982
N(1)	0.055	0.117	0.521
N(2)	0.172	0.747	0.618
O(1)	−0.009	0.234	0.869
O(2)	0.242	0.540	0.209
O(3)	0.298	0.564	0.812
H(1)	0.130	0.419	0.521
H(2)	0.169	0.331	0.778
H(3)	0.073	0.487	0.047
H(4)	0.036	0.581	0.776
H(5)	0.138	0.756	0.000
H(6)	0.036	−0.007	0.459
H(7)	0.114	0.093	0.402
H(8)	0.112	0.747	0.523
H(9)	0.203	0.849	0.682
H(10)	0.200	0.665	0.467

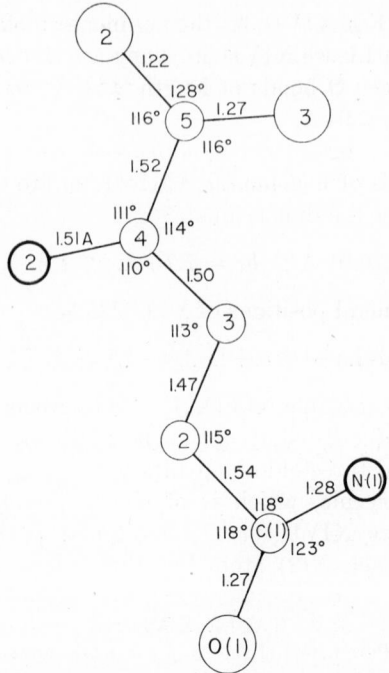

Fig. XIVD,40. Bond dimensions in the molecule of L-glutamine.

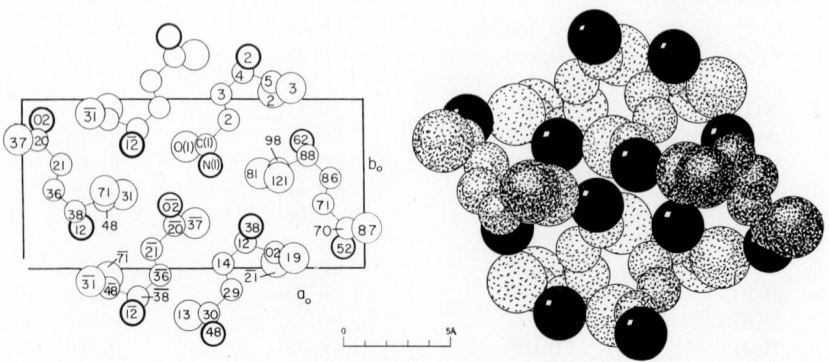

Fig. XIVD,41a (left). The orthorhombic structure of L-glutamine projected along its c_0 axis.   Left-hand axes.

Fig. XIVD,41b (right). A packing drawing of the orthorhombic L-glutamine arrangement viewed along its c_0 axis.   The nitrogen atoms are black.   Oxygen atoms are the larger, carbon atoms the smaller dotted circles.

distances (2.79, 2.85, and 2.91 A.) which indicate hydrogen bonding and support the assignment of three hydrogens to it. There are also two hydrogen bonds linking each N(1) with O(3) and O(1) [2.91 and 2.94 A.].

XIV,d27. The α form of DL-*norleucine* (DL-α-*aminocaproic acid*), $CH_3(CH_2)_3CHNH_2COOH$, is monoclinic with a tetramolecular cell having the dimensions:

$$a_0 = 9.84 \text{ A.}; \quad b_0 = 4.74 \text{ A.}; \quad c_0 = 16.56 \text{ A.}; \quad \beta = 104°30'$$

All atoms are in general positions of C_{2h}^5 ($P2_1/a$):

$$(4e) \quad \pm (xyz; x+\tfrac{1}{2}, \tfrac{1}{2}-y, z)$$

with the parameters of Table XIVD,25.

This is substantially the same structure as that found for DL-methionine (**XIV,d23**), $CH_3S(CH_2)_2CH(NH_2)COOH$. It leads to a molecule having the bond dimensions of Figure XIVD,42 and to a molecular distribution practically identical with that shown in Figure XIVD,34. As was the case

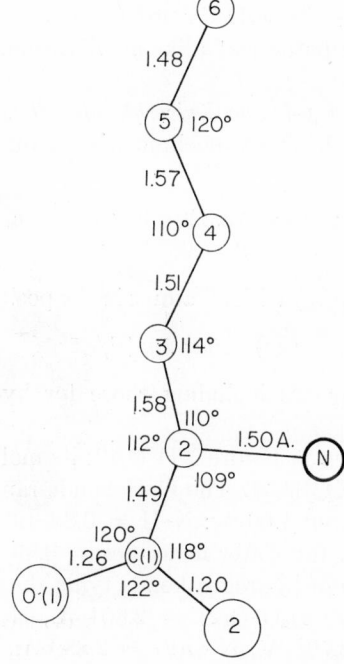

Fig. XIVD,42. Bond dimensions in the molecule of DL-norleucine.

TABLE XIVD,25. Parameters of the Atoms in DL-Norleucine

Atom	x	y	z
O(1)	−0.012	−0.067	0.375
O(2)	0.169	−0.317	0.432
N	0.356	0.050	0.404
C(1)	0.118	−0.114	0.392
C(2)	0.209	0.050	0.350
C(3)	0.207	−0.069	0.261
C(4)	0.301	0.089	0.217
C(5)	0.288	−0.036	0.128
C(6)	0.394	0.036	0.083

with α-methionine, the molecules are tied together to form double layers by hydrogen bonds between amino nitrogen and carboxyl oxygen atoms, with N–O = 2.73 and 2.87 A.

No crystals having a form of norleucine analogous to the β form of methionine were found but a so-called "super lattice form" was observed. It had the same values of a_0, b_0, and β as the α form and a c_0 four times as great (66.24 A.). Its diffraction pattern contains diffuse regions which are considered to point to an imperfectly ordered stacking of molecular layers involving the α and a hypothetical β-like modification.

XIV,d28. Crystals of L-*lysine hydrochloride dihydrate*, $NH_2(CH_2)_4CH$-$(NH_2)COOH \cdot HCl \cdot 2H_2O$, are monoclinic with a bimolecular unit of the dimensions:

$$a_0 = 7.492 \pm 0.001 \text{ A.}; \quad b_0 = 13.320 \pm 0.004 \text{ A.}; \quad c_0 = 5.879 \pm 0.001 \text{ A.}$$
$$\beta = 97°47'$$

The space group is C_2^2 ($P2_1$) with all atoms in the positions:

$$(2a) \quad xyz; \ \bar{x}, y + 1/2, \bar{z}$$

The determined parameters including those for hydrogen are given in Table XIVD,26.

The structure is shown in Figure XIVD,43; its molecules have the bond dimensions of Figure XIVD,44. The C–H bonds range between 0.91 and 1.18 A. with an average of 1.06 A.; N–H = 0.82–1.01 A. with an average of 0.94 A., and O–H in the water molecules = 0.89–1.02 A. with 0.98 A. as average. The system of hydrogen bonds tying the molecules together is a complicated one, with H_2O–O(2) = 2.804 A., H_2O–OH_2 = 2.712 A.; N–O(1) = 2.789 and 2.791 A., N–O(2) = 2.889 A., N–OH_2 = 2.813 A.; H_2O–Cl = 3.218 and 3.245 A.; N–Cl = 3.224, 3.166, and 3.346 A.

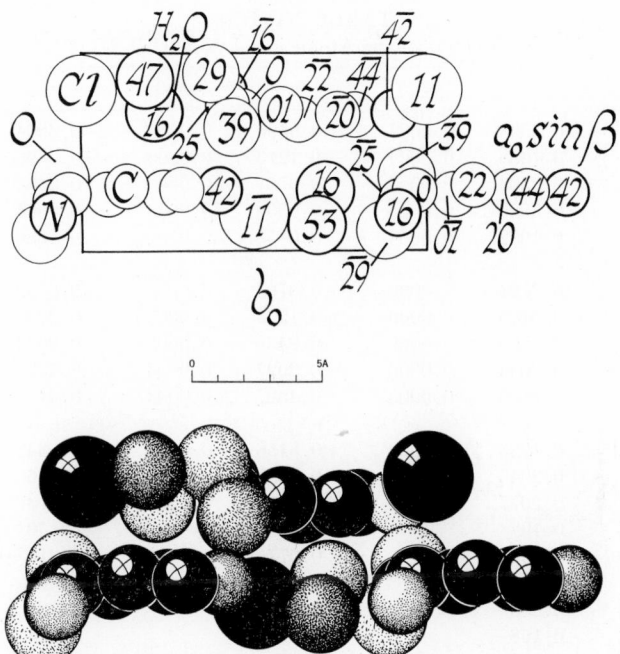

Fig. XIVD,43a (top). The monoclinic structure of L-lysine hydrochloride dihydrate projected along its c_0 axis. Right-hand axes.

Fig. XIVD,43b (bottom). A packing drawing of the monoclinic structure of $NH_2(CH_2)_4CH(NH_2)COOH \cdot HCl \cdot 2H_2O$ viewed along its c_0 axis. Chlorine atoms are the large, carbon atoms the smaller black circles. Atoms of nitrogen are heavily ringed and dotted. Water molecules are the larger, heavily outlined, and hook shaded circles. Carboxyl oxygens are short-line-and-dot shaded.

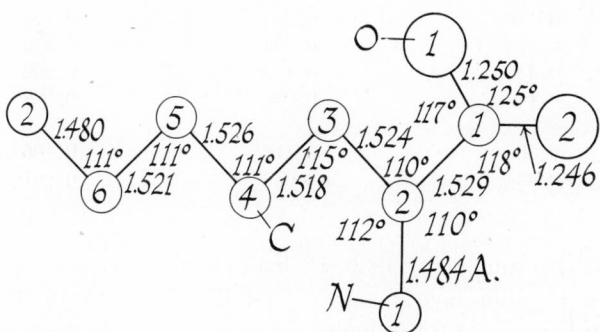

Fig. XIVD,44. Bond dimensions in the lysine molecule in crystals of L-lysine $HCl \cdot 2H_2O$.

TABLE XIVD,26

Parameters of the Atoms in L-Lysine HCl·2H$_2$O

Atom	x	$\sigma(x)$	y	$\sigma(y)$	z	$\sigma(z)$
O(1)	0.3568	0.0004	0.0596	0.0003	−0.3904	0.0005
O(2)	0.1038	0.0004	0.1213	0.0003	−0.2893	0.0005
C(1)	0.2503	0.0005	0.0761	0.0003	−0.2483	0.0006
C(2)	0.3026	0.0005	0.0339	0.0003	−0.0066	0.0006
N(1)	0.1982	0.0004	0.0855	0.0002	0.1560	0.0005
C(3)	0.2721	0.0006	−0.0793	0.0003	−0.0082	0.0007
C(4)	0.3126	0.0006	−0.1293	0.0003	0.2250	0.0007
C(5)	0.2920	0.0006	−0.2430	0.0003	0.2044	0.0008
C(6)	0.3034	0.0006	−0.2927	0.0003	0.4389	0.0007
N(2)	0.2800	0.0005	−0.4027	0.0003	0.4158	0.0006
H$_2$O(1)	0.1283	0.0008	0.3156	0.0003	0.5342	0.0008
H$_2$O(2)	0.2945	0.0007	0.2897	0.0003	0.1563	0.0007
Cl	0.2077	0.0001	0.5003	0.0001	−0.1137	0.0002
H(1,N 1)	0.198	—	0.155	—	0.140	—
H(2,N 1)	0.266	—	0.077	—	0.318	—
H(3,N 1)	0.083	—	0.070	—	0.106	—
H(4,N 2)	0.168	—	−0.420	—	0.276	—
H(5,N 2)	0.394	—	−0.431	—	0.397	—
H(6,N 2)	0.267	—	−0.429	—	0.566	—
H(7,H$_2$O)	0.110	—	0.356	—	0.664	—
H(8,H$_2$O)	0.108	—	0.254	—	0.602	—
H(9,H$_2$O)	0.301	—	0.353	—	0.068	—
H(10,H$_2$O)	0.252	—	0.315	—	0.312	—
H(11,C 6)	0.436	—	−0.282	—	0.537	—
H(12,C 6)	0.213	—	−0.252	—	0.532	—
H(13,C 5)	0.418	—	−0.274	—	0.123	—
H(14,C 5)	0.170	—	−0.264	—	0.125	—
H(15,C 4)	0.228	—	−0.104	—	0.352	—
H(16,C 4)	0.438	—	−0.112	—	0.292	—
H(17,C 3)	0.140	—	−0.090	—	−0.088	—
H(18,C 3)	0.366	—	−0.116	—	−0.106	—
H(19,C 2)	0.424	—	0.039	—	0.061	—

An earlier determination (1959: R) led to a structure similar to the foregoing (when due allowance is made for a difference in origin). The positions
of the chloride ions and water molecules and the general shape of the molecule is the same in the two arrangements though there are significant differ-

ences in the coordinates of some of the atoms; unless there is an error in the tabulated parameters, the C(1) and N(1) atoms are impossibly far apart in the earlier structure.

XIV,d29. The structures of D(−)-*isoleucine hydrochloride monohydrate* and of the corresponding *hydrobromide* have been established in such a way as to confirm that their absolute configurations correspond to the usual chemical convention.

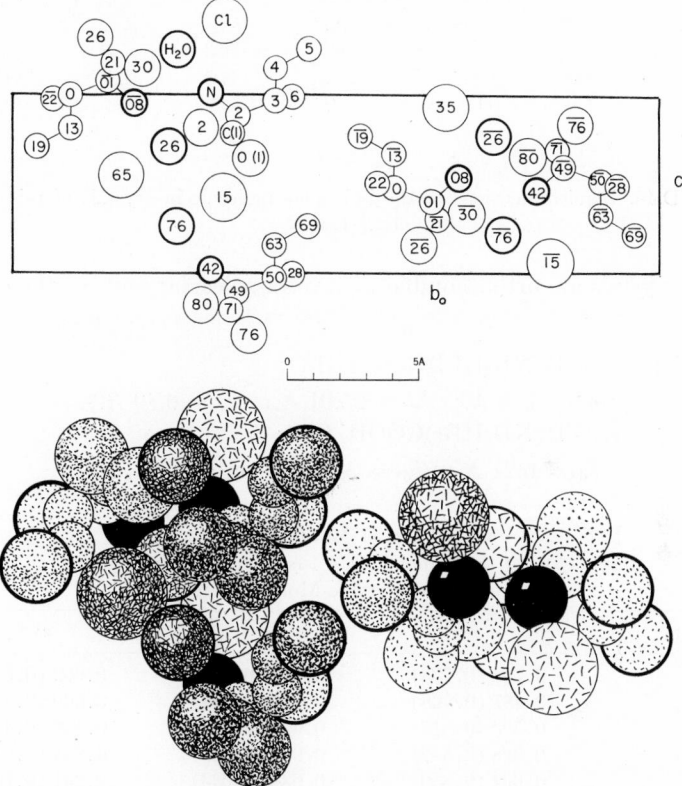

Fig. XIVD,45a (top). The orthorhombic structure of D-isoleucine HCl·H₂O projected along its a_0 axis.   Left-hand axes.

Fig. XIVD,45b (bottom). A packing drawing of the orthorhombic structure of D-isoleucine HCl·H₂O viewed along its a_0 axis.   The nitrogen atoms are black and the chloride ions large and line shaded.   The carboxyl oxygens, of intermediate size, are lightly outlined and dotted; the water molecules are heavily ringed and line shaded.   The terminal methyl radicals are dotted and heavily outlined; the other carbons are smaller, lightly outlined, and dotted.

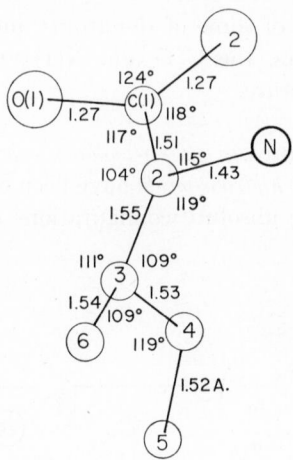

Fig. XIVD,46. Bond dimensions of the isoleucine molecule in crystals of D-isoleucine HCl·H$_2$O.

Their crystals are orthorhombic with tetramolecular units of the dimensions:

$C_2H_5CH(CH_3)CH(NH_2HCl)COOH \cdot H_2O$:

$a_0 = 6.13$ A.; $b_0 = 25.01$ A.; $c_0 = 6.79$ A.

$C_2H_5CH(CH_3)CH(NH_2HBr)COOH \cdot H_2O$:

$a_0 = 6.21$ A.; $b_0 = 24.4$ A.; $c_0 = 7.00$ A.

TABLE XIVD,27

Parameters of the Atoms in D($-$)-Isoleucine Hydrochloride
and Hydrobromide Monohydrate

Atom	$-x$	y	z
C(1)	0.708 (0.733)	0.158 (0.156)	0.692 (0.678)
C(2)	0.487 (0.528)	0.148 (0.144)	0.600 (0.575)
C(3)	0.503 (0.517)	0.090 (0.087)	0.522 (0.478)
C(4)	0.628 (0.542)	0.090 (0.039)	0.328 (0.622)
C(5)	0.694 (0.350)	0.038 (0.033)	0.231 (0.761)
C(6)	0.276 (0.692)	0.065 (0.092)	0.489 (0.325)
O(1)	0.758 (0.761)	0.129 (0.122)	0.839 (0.825)
O(2)	0.799 (0.856)	0.203 (0.194)	0.660 (0.650)
N	0.417 (0.467)	0.190 (0.192)	0.467 (0.467)
Cl (or Br)	0.153 (0.186)	0.168 (0.161)	0.072 (0.072)
H$_2$O	0.756 (0.794)	0.243 (0.242)	0.226 (0.242)

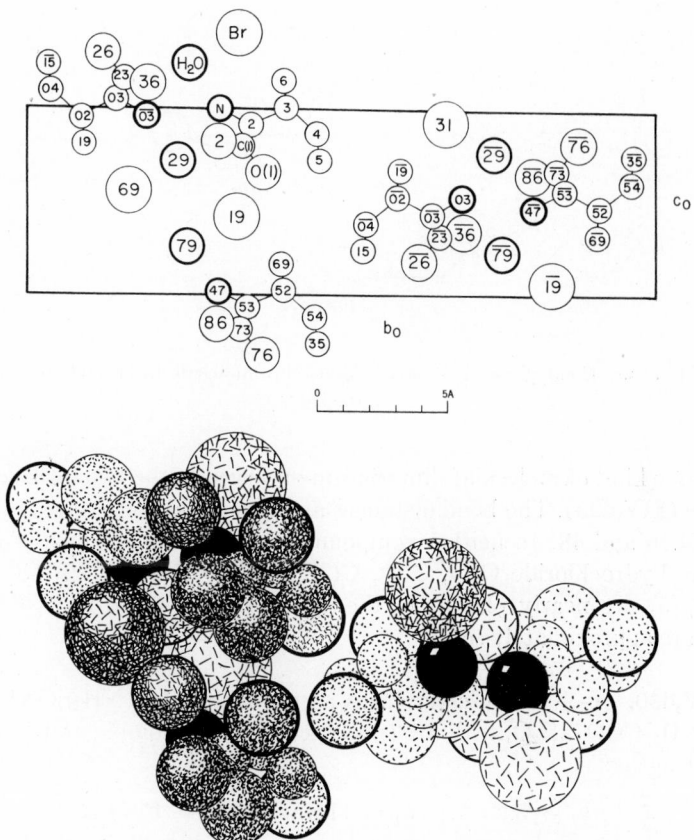

Fig. XIVD,47a (top). The orthorhombic structure of D-isoleucine HBr·H₂O projected along its a_0 axis. Left-hand axes.

Fig. XIVD,47b (bottom). A packing drawing of the orthorhombic structure of D-isoleucine HBr·H₂O seen along its a_0 axis. The atoms are shaded as in Fig. XIVD,45b for the hydrochloride.

The space group is V^4 ($P2_12_12_1$) with all atoms in the general positions:

$$(4a) \quad xyz; \; {}^1\!/_2-x,\bar{y},z+{}^1\!/_2; \; x+{}^1\!/_2,{}^1\!/_2-y,\bar{z}; \; \bar{x},y+{}^1\!/_2,{}^1\!/_2-z$$

The established parameters are listed in Table XIVD,27, the values for the bromide being in parentheses.

The two compounds evidently are not strictly isomorphous though they are similar in structure. The differences that exist, evident in Figures XIVD,45 and 47, can be related to the different ways in which the C_2H_5CH-(CH_3) part of the molecule is turned about the C(2)–C(3) bond. This is

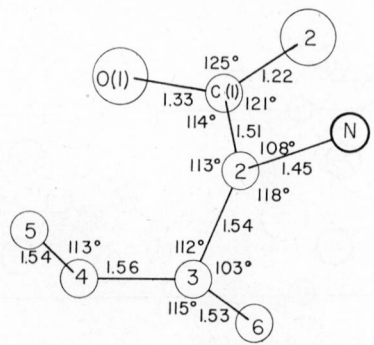

Fig. XIVD,48. Bond dimensions of the isoleucine molecule in crystals of D-isoleucine HBr·H₂O.

the same kind of molecular dimorphism observed in the two forms of methionine (**XIV,d23**). The bond distances and bond angles are stated in Figures XIVD,46 and 48. In neither compound are the carbon atoms in a plane. In the hydrochloride C(1), C(2), C(3), and C(6), and C(2), C(3), C(4), and C(5) are nearly planar; in the hydrobromide C(3), C(4), C(5), and C(6) are approximately so.

XIV,d30. L-*Arginine* crystallizes as the *dihydrate*, $(NH_2)_2CNH(CH_2)_3$-$CH(NH_2)COO·2H_2O$, with a tetramolecular orthorhombic unit having the edge lengths:

$$a_0 = 5.68 \text{ A.}; \quad b_0 = 11.87 \text{ A.}; \quad c_0 = 15.74 \text{ A.}$$

Atoms are in the general positions of V^4 ($P2_12_12_1$): ·

$$(4a) \quad xyz; \; {}^1\!/_2 - x, \bar{y}, z + {}^1\!/_2; \; x + {}^1\!/_2, {}^1\!/_2 - y, \bar{z}; \; \bar{x}, y + {}^1\!/_2, {}^1\!/_2 - z$$

with the parameters of Table XIVD,28.

The structure as a whole is shown in Figure XIVD,49. The individual molecules have the bond dimensions of Figure XIVD,50. Atoms O(1), O(2), C(3), and C(4) are coplanar and N(5) is 0.280 A. outside this plane. Atoms N(9), C(10), N(11), and N(12) are also coplanar; the five hydrogen atoms H(10)–(14) belonging to these atoms are likewise in this plane, thus making the entire guanidyl part of the molecule planar. The carbon atoms C(6)–(8) also lie nearly in this plane, which makes a dihedral angle of ca. 74° with the acid O_2C_2 plane.

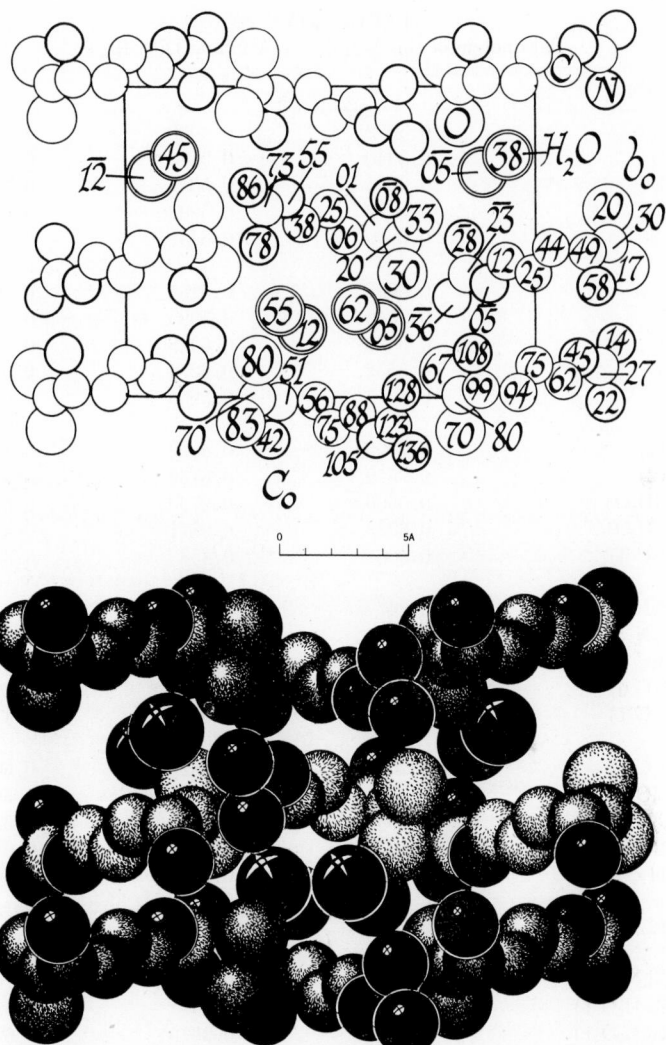

Fig. XIVD,49a (top). The orthorhombic structure of L-arginine dihydrate projected along its a_0 axis. Right-hand axes.

Fig. XIVD,49b (bottom). A packing drawing of the orthorhombic L-arginine dihydrate arrangement viewed along its a_0 axis. The larger black circles are the water molecules; the smaller, with the circular highlights, are nitrogens. Atoms of carbon are line shaded; those of carboxyl oxygens are larger and hook shaded.

TABLE XIVD,28
Parameters of the Atoms in L-Arginine Dihydrate

Atom	x	y	z
O(1)	0.3022	0.4036	0.3231
O(2)	0.3340	0.5832	0.2830
C(3)	0.2353	0.5047	0.3220
C(4)	0.0072	0.5293	0.3724
N(5)	−0.0839	0.6432	0.3524
C(6)	0.0622	0.5186	0.4680
C(7)	0.2546	0.5995	0.5002
C(8)	0.3822	0.5478	0.5753
N(9)	0.5532	0.6274	0.6118
C(10)	0.7315	0.5955	0.6633
N(11)	0.7796	0.4862	0.6760
N(12)	0.8579	0.6739	0.7016
O(1,H$_2$O)	0.5506	0.2790	0.6214
O(2,H$_2$O)	0.6222	0.2871	0.4356
H(1,C 4)	0.908	0.458	0.350
H(2,N 5)	0.847	0.650	0.328
H(3,N 5)	0.938	0.645	0.362
H(4,C 6)	0.076	0.442	0.475
H(5,C 6)	0.933	0.525	0.495
H(6,C 7)	0.366	0.645	0.450
H(7,C 7)	0.208	0.682	0.502
H(8,C 8)	0.388	0.478	0.589
H(9,C 8)	0.253	0.512	0.623
H(10,N 9)	0.525	0.703	0.617
H(11,N 11)	0.722	0.433	0.660
H(12,N 11)	0.862	0.472	0.705
H(13,N 12)	0.933	0.663	0.719
H(14,N 12)	0.822	0.741	0.700
H(15,H$_2$O 1)	0.533	0.207	0.629
H(16,H$_2$O 1)	0.459	0.267	0.628
H(17,H$_2$O 2)	0.558	0.335	0.412
H(18,H$_2$O 2)	0.558	0.275	0.463

The molecules are bound together in the crystal with a system of hydrogen bonds, some of which involve the water molecules. There are N–H–N, N–H–O, and O–H–O bonds between 2.709 and 2.954 A. in length. The shortest approach of nonhydrogen-bonded atoms of different molecules is ca. 3.4 A.

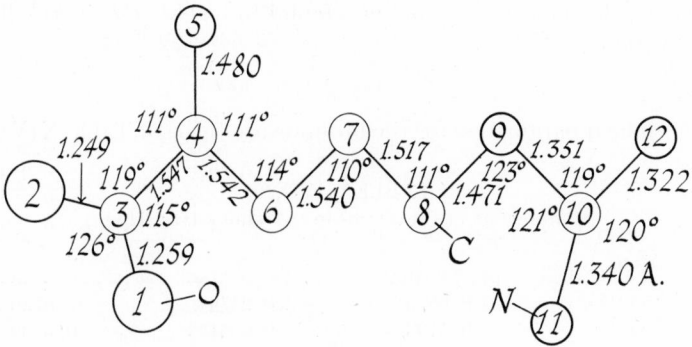

Fig. XIVD,50. Bond dimensions of the arginine molecule in its dihydrate.

XIV,d31. Crystals of L-*arginine hydrobromide monohydrate*, NH_2C-$(NH)NH(CH_2)_3CH(NH_2)COOH \cdot HBr \cdot H_2O$, and the isomorphous *hydrochloride* are monoclinic with a tetramolecular unit of the dimensions:

$$a_0 = 11.26 \text{ A.}; \quad b_0 = 8.65 \text{ A.}; \quad c_0 = 11.25 \text{ A.}; \quad \beta = 91°30'$$

for the bromide and

$$a_0 = 11.22 \text{ A.}; \quad b_0 = 8.50 \text{ A.}; \quad c_0 = 11.07 \text{ A.}; \quad \beta = 91°0'$$

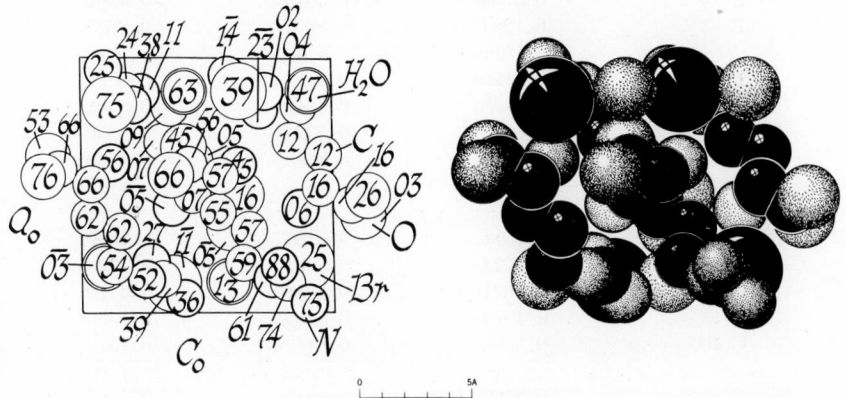

Fig. XIVD,51a (left). The monoclinic structure of L-arginine $HBr \cdot H_2O$ projected along its b_0 axis. Right-hand axes.

Fig. XIVD,51b (right). A packing drawing of the monoclinic structure of L-arginine $HBr \cdot H_2O$ seen along its b_0 axis. The bromide ions are the large, the carbon atoms the small black circles. All oxygens are lightly outlined, the water oxygens being hook and the carboxyl oxygens short-line shaded. Nitrogen atoms are heavily outlined and dotted.

for the chloride. The space group was found to be C_2^2 ($P2_1$) with all atoms
in the positions:

$$(2a) \quad xyz; \ \bar{x},y+{}^1/_2,\bar{z}$$

The determined parameters for the bromide are those of Table XIVD,29.

TABLE XIVD,29
Parameters of the Atoms in L-Arginine HBr·H₂O

Atom	x	y	z
Br(1)	0.1679	0.2474	0.8726
H₂O(3)	0.1198	0.1341	0.5890
O(2)	0.5415	0.6654	0.3578
O(1)	0.6491	0.4493	0.4095
N(4)	0.0422	0.7509	0.9054
N(3)	0.1826	0.8771	0.7878
N(2)	0.1320	0.6117	0.7560
N(1)	0.5758	0.4511	0.6357
C(6)	0.1221	0.7405	0.8141
C(5)	0.1943	0.5955	0.6441
C(4)	0.3288	0.5681	0.6702
C(3)	0.3876	0.5510	0.5452
C(2)	0.5225	0.5719	0.5561
C(1)	0.5789	0.5610	0.4302
Br(2)	0.1284	0.8902	0.3623
H₂O(6)	0.1199	0.9724	0.0827
O(5)	0.5458	0.7647	0.8506
O(4)	0.6093	0.5324	0.8850
N(8)	0.0607	0.3648	0.4089
N(7)	0.2098	0.2665	0.2875
N(6)	0.1371	0.5186	0.2580
N(5)	0.5724	0.5601	0.1226
C(12)	0.1321	0.3888	0.3170
C(11)	0.1922	0.5435	0.1406
C(10)	0.3138	0.6172	0.1576
C(9)	0.3722	0.6213	0.0285
C(8)	0.5072	0.6632	0.0424
C(7)	0.5601	0.6569	0.9163

The resulting structure is shown in Figure XIVD,51. The two crystallographically unlike molecules it contains have the bond dimensions of
Figure XIVD,52. These molecules are tied to one another, to the halogen

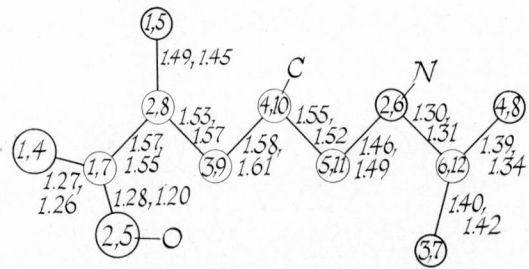

Fig. XIVD,52. Bond lengths of the two crystallographically different arginine molecules in its hydrobromide monohydrate.

atoms, and to the water molecules by the following system of hydrogen bonds:

$$\text{O–H}\cdots\text{Br} = 3.22 \text{ and } 3.38 \text{ A.}$$
$$\text{N–H}\cdots\text{O} = 2.79\text{–}3.06 \text{ A.}$$
$$\text{N–H}\cdots\text{Br} = 3.32\text{–}3.41 \text{ A.}$$
$$\text{O(water)–H}\cdots\text{O} = 3.02 \text{ and } 3.09 \text{ A.}$$

XIV,d32. Crystals of L-*cystine dihydrochloride*, [HOOCCH(NH$_2$HCl)-CH$_2$S—]$_2$, are monoclinic with a bimolecular unit of the dimensions:

$$a_0 = 18.61 \text{ A.}; \quad b_0 = 5.25 \text{ A.}; \quad c_0 = 7.23 \text{ A.}; \quad \beta = 103°36'$$

TABLE XIVD,30. Parameters of the Atoms in L-Cystine Dihydrochloride

Atom	x	y	z
C(1)	0.0642	0.7660	0.2406
C(2)	0.1465	0.6790	0.3048
C(3)	0.1552	0.4670	0.4418
N	0.1765	0.6100	0.1388
O(1)	0.1924	0.2750	0.4292
O(2)	0.1195	0.4790	0.5772
S	0.0003	0.5000	0.1415
Cl	0.1532	0.0890	0.8848
H(1)	0.067	0.910	0.100
H(2)	0.057	0.900	0.367
H(3)	0.188	0.850	0.377
H(4)	0.150	0.460	0.033
H(5)	0.172	0.825	0.033
H(6)	0.233	0.460	0.167
H(7)	0.133	0.360	0.667

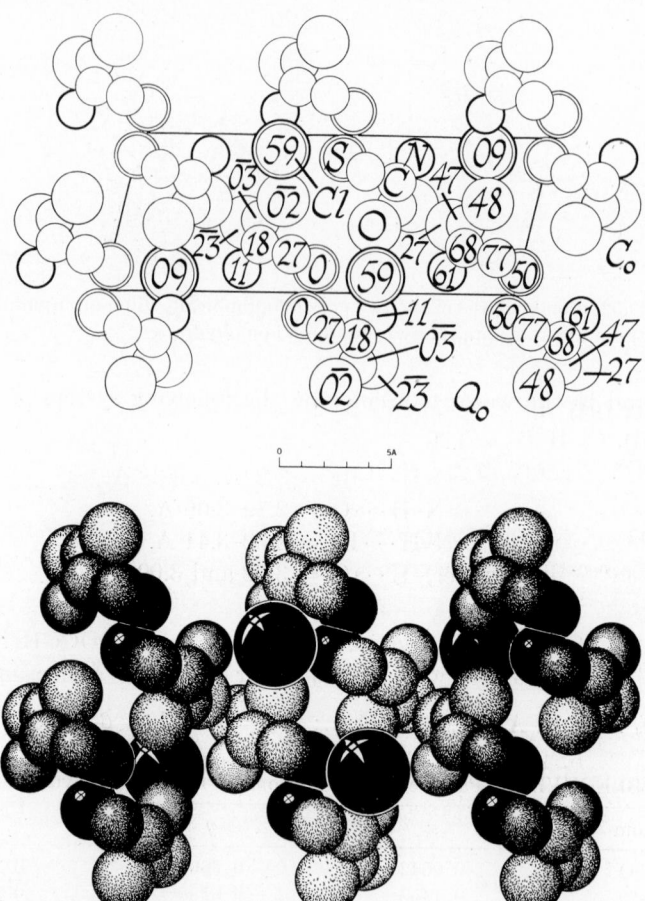

Fig. XIVD,53a (top). The monoclinic structure of L-cystine dihydrochloride projected along its b_0 axis. Right-hand axes.

Fig. XIVD,53b (bottom). A packing drawing of the monoclinic L-cystine dihydrochloride arrangement viewed along its b_0 axis. Chlorine atoms are the large, sulfur atoms the smaller black circles. Nitrogen atoms are heavily ringed and dotted; carbons, of similar size, are line shaded. The somewhat larger oxygen atoms are hook shaded.

The space group is C_2^3 ($C2$) with all atoms in the general positions:

$$(4c) \quad xyz;\ \bar{x}y\bar{z};\ x+{}^1\!/_2,y+{}^1\!/_2,z;\ {}^1\!/_2-x,y+{}^1\!/_2,\bar{z}$$

The atomic parameters are listed in Table XIVD,30, those for hydrogen not having been established experimentally.

The resulting structure is shown in Figure XIVD,53. The molecules from which it is built have the bond dimensions of Figure XIVD,54. It is

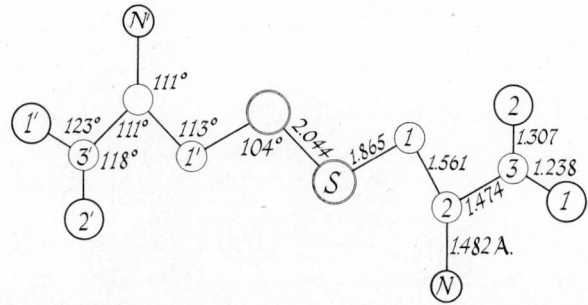

Fig. XIVD,54. Bond dimensions of the cystine molecule in its dihydrochloride.

considered that the Cl^- ions are hydrogen bonded to a carboxyl oxygen atom, with Cl–H–O = 2.98 A., and to three nitrogen atoms, with Cl–H–N = 3.08, 3.25, and 3.27 A.

XIV,d33. Crystals of L-*cystine dihydrobromide*, [—$SCH_2CH(NH_2HBr)$-COOH]$_2$, are orthorhombic with a bimolecular unit of the edge lengths:

$$a_0 = 17.85 \text{ A.}; \quad b_0 = 5.35 \text{ A.}; \quad c_0 = 7.48 \text{ A.}$$

The space group is V^3 ($P2_12_12_1$) with all atoms in the positions:

$$(4c) \quad xyz; \; \bar{x}y\bar{z}; \; {}^1\!/_2-x,\bar{y},z+{}^1\!/_2; \; x+{}^1\!/_2,\bar{y},{}^1\!/_2-z$$

TABLE XIVD,31
Parameters of the Atoms in L-Cystine 2HBr

Atom	x	y	z	σ, A.
C(1)	0.0606	0.8820	0.2050	0.021
C(2)	0.1394	0.7905	0.2320	0.021
C(3)	0.1457	0.5745	0.3608	0.021
N	0.1795	0.7210	0.0641	0.018
O(1)	0.1795	0.3845	0.3233	0.017
O(2)	0.1128	0.6097	0.5091	0.017
S	0.9963	0.6254	0.1350	0.0048
Br	0.1579	0.2046	0.7984	0.0018
H(1)	0.059	0.010	0.080	—
H(2)	0.042	0.040	0.307	—
H(3)	0.165	0.950	0.277	—
H(4)	0.162	0.560	0.973	—
H(5)	0.165	0.900	0.980	—
H(6)	0.232	0.680	0.145	—
H(7)	0.125	0.450	0.625	—

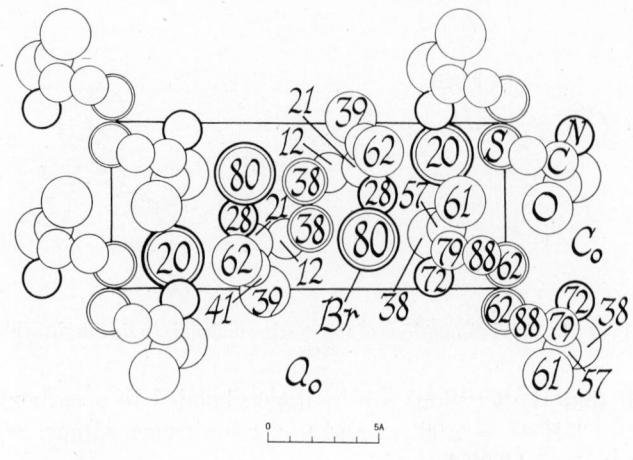

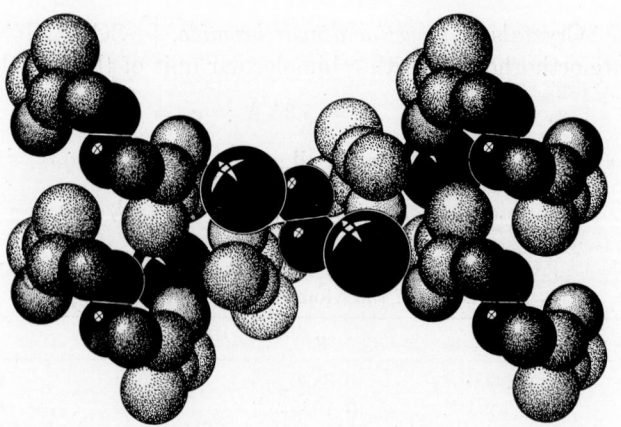

Fig. XIVD,55a (top). The orthorhombic structure of L-cystine·2HBr projected along its b_0 axis. Right-hand axes.

Fig. XIVD,55b (bottom). A packing drawing of the orthorhombic L-cystine dihydrobromide arrangement seen along its b_0 axis. Bromide ions are the large, sulfur atoms the smaller black circles. Atoms of nitrogen are heavily outlined and dotted. The carbons are line shaded; the oxygens are slightly larger and hook shaded.

The determined parameters are given in Table XIVD,31. Positions for hydrogen were found from the projections.

The structure is shown in Figure XIVD,55. It is composed of molecules having the bond dimensions of Figure XIVD,56. There are hydrogen bridges between the bromide ion and both nitrogen and oxygen: Br–

Fig. XIVD,56. Bond dimensions of the cystine molecule in its dihydrobromide.

$H(7)-O(2) = 3.17$ A., $Br-H(5)-N = 3.28$ A., $Br-H(4)-N = 3.42$ A., and $Br-H(6)-N = 3.41$ A. The shortest intermolecular separations are $N-O(1) = 3.15$ A., $S-O(2) = 3.30$ A., $C(2)-O(1) = 3.33$ A., and $C(1)-O(1) = 3.54$ A.

This arrangement is closely related to that for the hydrochloride (**XIV,d32**). Though the cell is of nearly the same size for the two salts, the molecular arrangements are evidently different.

XIV,d34. Hexagonal L-*cystine*, $[SCH_2CH(NH_2)COOH]_2$, has a six-molecule cell of the edge lengths:

$$a_0 = 5.422 \text{ A.}, \qquad c_0 = 56.275 \text{ A.}$$

TABLE XIVD,32
Parameters of the Atoms in L-Cystine

Atom	x	$\sigma(x)$	y	$\sigma(y)$	z	$\sigma(z)$
S	0.19866	0.00051	0.03138	0.00055	0.41278	0.00004
C(1)	0.0886	0.0024	0.7310	0.0024	0.3927	0.0002
C(2)	0.0515	0.0021	0.7836	0.0021	0.3670	0.0002
C(3)	0.7967	0.0019	0.8301	0.0019	0.3623	0.0001
N	0.3165	0.0017	0.0337	0.0017	0.3569	0.0002
O(1)	0.8493	0.0015	0.0732	0.0015	0.3561	0.0001
O(2)	0.5586	0.0017	0.6200	0.0017	0.3658	0.0002
H(1)	0.223	—	0.675	—	0.392	—
H(2)	0.926	—	0.562	—	0.396	—
H(3)	0.010	—	0.620	—	0.356	—
H(4)	0.493	—	0.015	—	0.354	—
H(5)	0.383	—	0.215	—	0.367	—
H(6)	0.283	—	0.985	—	0.339	—

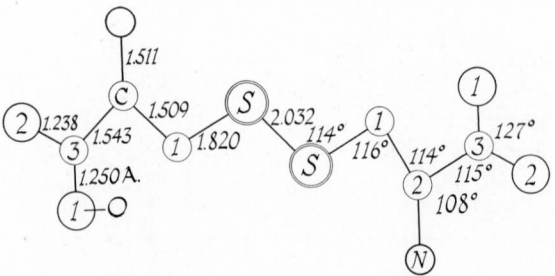

Fig. XIVD,57. Bond dimensions of the cystine molecule in its hexagonal crystals.

The space group is D_6^2 ($P6_122$) with all atoms in the positions:

$$(12c) \quad xyz; \qquad \bar{y},x-y,z+{}^1/_3; \quad y-x,\bar{x},z+{}^2/_3;$$
$$\bar{x},\bar{y},z+{}^1/_2; \quad y,y-x,z+{}^5/_6; \quad x-y,x,z+{}^1/_6;$$
$$y,x,{}^1/_3-z; \quad \bar{x},y-x,{}^2/_3-z; \quad x-y,\bar{y},\bar{z};$$
$$\bar{y},\bar{x},{}^5/_6-z; \quad x,x-y,{}^1/_6-z; \quad y-x,y,{}^1/_2-z$$

The determined parameters are those listed in Table XIVD,32.

In this rather complicated structure, the molecules are stacked in six layers around the c_0 axis and are tied together both in and between layers by hydrogen bonds joining nitrogen and oxygen (2.789, 2.809, and 2.865 A.). The bond dimensions of a molecule are shown in Figure XIVD,57.

XIV,d35. A structure has been determined for the dipeptide *β-gly-cylglycine*, $NH_2CH_2C(O)NHCH_2COOH$. Its monoclinic unit containing eight molecules has the dimensions:

$$a_0 = 17.93 \text{ A.}; \quad b_0 = 4.62 \text{ A.}; \quad c_0 = 17.09 \text{ A.}; \quad \beta = 125°10'$$

TABLE XIVD,33
Parameters of the Atoms in β-Glycylglycine

Atom	x	y	z
$NH_2(1)$	0.155	0.390	0.091
$CH_2(1)$	0.093	0.440	0.123
C(2)	0.132	0.290	0.219
O(1)	0.179	0.070	0.241
NH(2)	0.105	0.400	0.268
$CH_2(3)$	0.134	0.280	0.362
C(4)	0.095	0.460	0.406
O(2)	0.047	0.670	0.367
O(3)	0.124	0.375	0.490

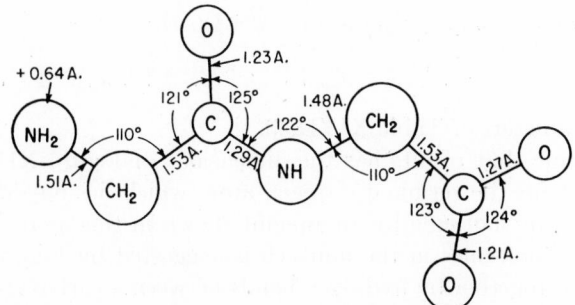

Fig. XIVD,58. Bond dimensions of the molecule in β-glycylglycine.

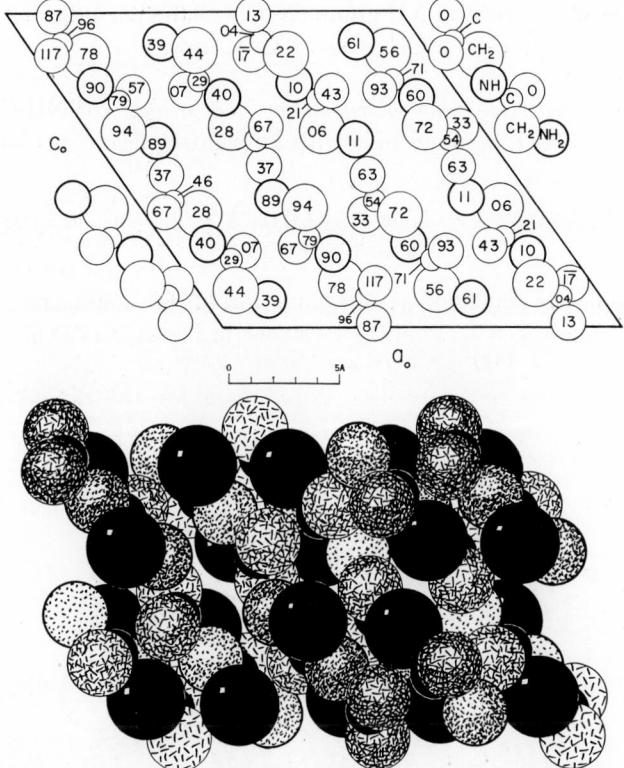

Fig. XIVD,59a (top). The monoclinic structure of β-glycylglycine projected along its b_0 axis. Left-hand axes.

Fig. XIVD,59b (bottom). A packing drawing of the monoclinic structure of β-glycyl-glycine viewed along its b_0 axis. The carbon atoms are black, the larger circles representing the CH_2 groups. Atoms of nitrogen are dotted and heavily outlined; the oxygens are line shaded.

All atoms are in general positions of C_{2h}^6 $(A2/a)$:

$$(8f) \quad \pm (xyz; \; x+\tfrac{1}{2},\bar{y},z; \; x,y+\tfrac{1}{2},z+\tfrac{1}{2}; \; x+\tfrac{1}{2},\tfrac{1}{2}-y,z+\tfrac{1}{2})$$

with the parameters of Table XIVD,33.

The molecule that results has the dimensions of Figure XIVD,58; it is planar except for the terminal nitrogen atom which lies outside the plane of the rest of the molecule by an amount shown in this figure. The distribution of the molecules in the unit cell is suggested by Figure XIVD,59. They are held together by hydrogen bonds between a carboxyl oxygen and terminal amino groups (N–O = 2.68 and 2.80 A.), between these amino groups and a chain oxygen (N–O = 2.81 A.), and by others joining carboxyl oxygen with a central NH group (N–O = 3.07 A.). Other intermolecular distances range from a CH_2–O = 3.32 A. upward.

XIV,d36. *Copper(II) monoglycylglycine trihydrate*, $Cu[NH_2CH_2C(O)\text{-}NCH_2COOH]\cdot 3H_2O$, is monoclinic with an eight-molecule cell having the dimensions:

$$a_0 = 14.95 \pm 0.02 \text{ A.}; \quad b_0 = 7.54 \pm 0.03 \text{ A.}; \quad c_0 = 15.80 \pm 0.02 \text{ A.}$$
$$\beta = 102°13' \pm 7'$$

The space group is C_{2h}^5 $(P2_1/n)$ with all atoms in the positions:

$$(4e) \quad \pm (xyz; \; x+\tfrac{1}{2},\tfrac{1}{2}-y,z+\tfrac{1}{2})$$

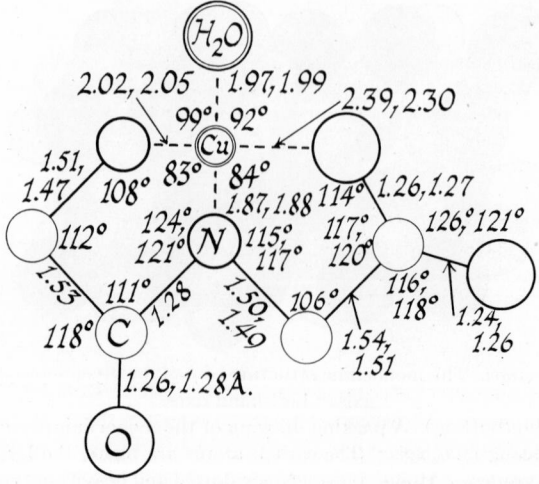

Fig. XIVD,60. Bond dimensions in crystals of cupric monoglycylglycine trihydrate.

The determined parameters are listed in Table XIVD,34.

TABLE XIVD,34

Parameters of the Atoms in Cupric Monoglycylglycine Trihydrate

Atom	x	σ_x, A.	y	σ_y, A.	z	σ_z, A.
			Molecule A			
Cu	0.0215	0.002	0.1855	0.002	0.3712	0.002
N(1)	0.9114	0.011	0.2537	0.017	0.2792	0.011
C(1)	0.8345	0.010	0.2991	0.015	0.3237	0.013
C(2)	0.8663	0.010	0.3148	0.013	0.4218	0.011
O(1)	0.8092	0.010	0.3686	0.012	0.4647	0.010
N(2)	0.9500	0.010	0.2687	0.012	0.4464	0.004
C(3)	0.9966	0.013	0.2723	0.016	0.5399	0.013
C(4)	0.0921	0.011	0.1892	0.015	0.5469	0.009
O(2)	0.1407	0.008	0.1794	0.011	0.6205	0.008
O(3)	0.1135	0.008	0.1398	0.011	0.4781	0.007
O(4)	0.0834	0.008	0.0256	0.010	0.3038	0.007
O(5)	0.1129	0.010	0.4133	0.012	0.3264	0.010
O(6)	0.8963	0.009	0.6530	0.012	0.2337	0.009
			Molecule B			
Cu(2)	0.8783	0.002	0.3156	0.002	0.0385	0.002
N(3)	0.7887	0.010	0.2887	0.015	0.9219	0.009
C(5)	0.8357	0.015	0.1982	0.021	0.8612	0.013
C(6)	0.9391	0.010	0.1889	0.011	0.8966	0.009
O(7)	0.9886	0.008	0.1276	0.010	0.8467	0.009
N(4)	0.9643	0.009	0.2389	0.011	0.9751	0.009
C(7)	0.0622	0.015	0.2458	0.015	0.0196	0.014
C(8)	0.0635	0.008	0.3198	0.015	0.1089	0.010
O(8)	0.1398	0.009	0.3481	0.012	0.1587	0.008
O(9)	0.9893	0.007	0.3665	0.009	0.1293	0.008
O(10)	0.8003	0.007	0.4650	0.010	0.0973	0.008
O(11)	0.8438	0.008	0.0632	0.010	0.1065	0.009
O(12)	0.3043	0.008	0.2642	0.012	0.1224	0.008

With eight molecules per cell there are two crystallographically different molecules which, however, as Figure XIVD,60 indicates, do not greatly differ in dimensions. A drawing showing the distribution of some of the molecules in the structure is given in Figure XIVD,61; the other molecules are superimposed on these in such a way that they would not show if included. The copper atom has its usual approximately square coordination

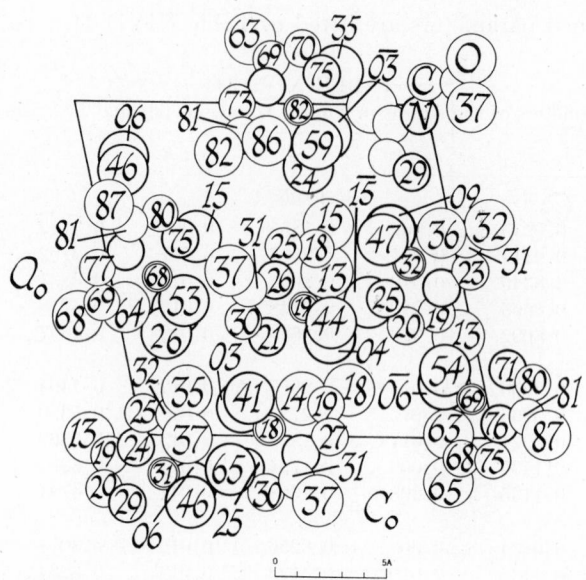

Fig. XIVD,61. Part of the monoclinic structure of $Cu[NH_2CH_2C(O)NCH_2COOH] \cdot 3H_2O$ projected along its b_0 axis. Copper atoms are the doubly ringed circles; the water molecules are large and heavily outlined. Right-hand axes.

with two corners occupied by nitrogen and two by water oxygens; there is also an additional water oxygen normal to the plane of the square, with $Cu-O(2) = 2.30$ or 2.39 A. There are probably several hydrogen bonds corresponding to O–O distances of 2.66–2.85 A. and N–O of 3.06–3.14 A.

XIV,d37. Crystals of *glycyl-L-asparagine*, $NH_2CH_2C(O)NHCH(COOH)CH_2CONH_2$, are orthorhombic with a tetramolecular cell of the dimensions:

$$a_0 = 4.81 \text{ A.}; \quad b_0 = 12.85 \text{ A.}; \quad c_0 = 13.52 \text{ A.}$$

All atoms are in general positions of V^4 $(P2_12_12_1)$:

$$(4a) \quad xyz; \; 1/2-x,\bar{y},z+1/2; \; x+1/2,1/2-y,\bar{z}; \; \bar{x},y+1/2,1/2-z$$

with the parameters of Table XIVD,35.

The structure is illustrated in Figure XIVD,62. Bond angles and lengths in the molecule are given in Figure XIVD,63. The atoms C(3), N(2), C(5), and O(4) are planar within 0.01 A.; the N(3)–C(6) bond makes only 7° with this plane. The other half of the molecule is also nearly planar, the angle between O(1), C(1), N(1), C(2) on the one hand and the carboxyl

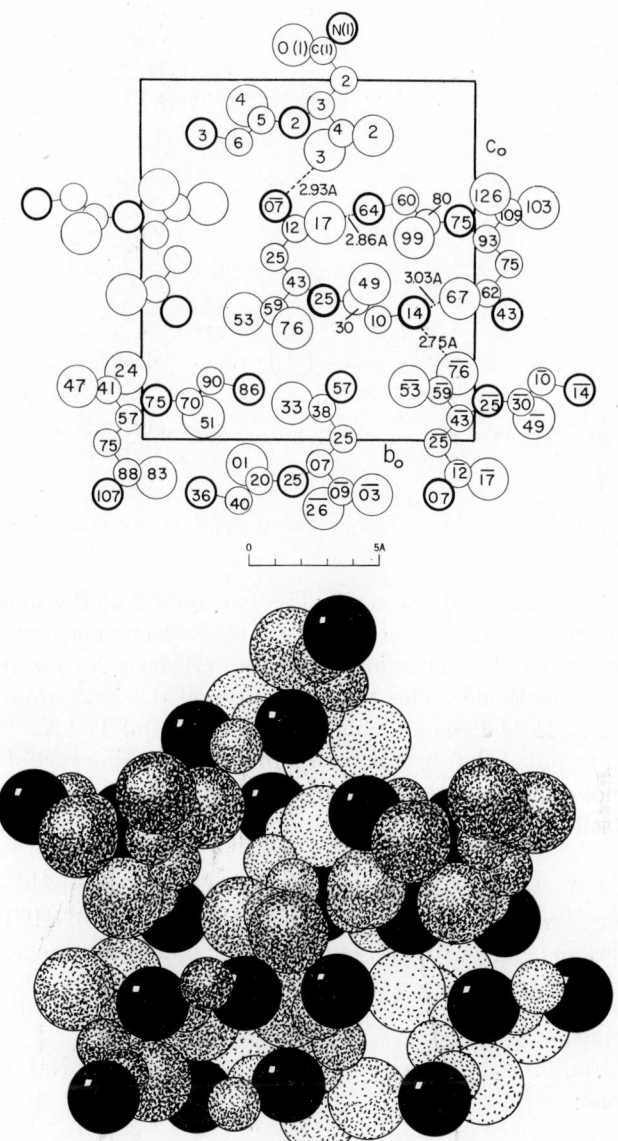

Fig. XIVD,62a (top). The orthorhombic structure of glycyl-L-asparagine projected along its a_0 axis. Left-hand axes.

Fig. XIVD,62b (bottom). A packing drawing of the orthorhombic glycyl-L-asparagine arrangement seen along its a_0 axis. The nitrogen atoms are black; carbons are the small dotted circles. Oxygen atoms are large and dotted.

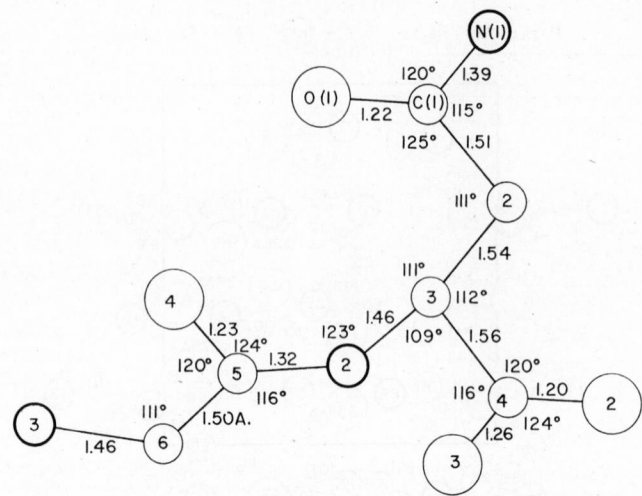

Fig. XIVD,63. Bond dimensions in the molecule of glycyl-L-asparagine.

group on the other being about 3°. The two halves of the molecule are, as indicated in the figures, about at right angles to one another. There are three hydrogen bonds connecting the amino nitrogen N(3) with oxygens of neighboring molecules. The N(1) has two and the N(2) atom one such bond. These N–H–O distances range between 2.75 and 3.03 A.

It should be noted that in this structure the asparagine part of the molecule does not exhibit the cyclic character that has been proposed for asparagine itself.

XIV,d38. A structure has been described for the dipeptide derivative *cysteylglycine hemi sodium iodide*, $SHCH_2CH(NH_2)C(O)NHCH_2COOH \cdot {}^1/_2NaI$. It is monoclinic with four molecules in a unit of the dimensions:

$$a_0 = 11.11 \text{ A.}; \quad b_0 = 5.12 \text{ A.}; \quad c_0 = 16.04 \text{ A.}; \quad \beta = 90°57'$$

The space group is C_2^3 ($A2$) with all atoms, except those of NaI, in the general positions:

$$(4c) \quad xyz; \ \bar{x}y\bar{z}; \ x,y+{}^1/_2,z+{}^1/_2; \ \bar{x},y+{}^1/_2,{}^1/_2-z$$

The sodium and iodine atoms are in the special positions:

$$\text{I: } (2a) \quad 000; \ 0\,{}^1/_2\,{}^1/_2$$
$$\text{Na: } (2b) \quad {}^1/_2\,{}^1/_2\,{}^1/_2; \ {}^1/_2\,0\,0$$

TABLE XIVD,35
Parameters of the Atoms in Glycyl-L-Asparagine

Atom	x	y	z
C(1)	0.1215	0.5377	0.5899
C(2)	0.251	0.6013	0.5084
C(3)	0.426	0.5327	0.4383
C(4)	0.588	0.5989	0.3604
C(5)	0.2975	0.3560	0.3891
C(6)	0.096	0.2891	0.3330
N(1)	−0.072	0.5896	0.6481
N(2)	0.251	0.4573	0.3876
N(3)	0.1445	0.1787	0.3528
O(1)	0.166	0.4456	0.6044
O(2)	0.533	0.6895	0.3491
O(3)	0.7565	0.5499	0.3063
O(4)	0.4895	0.3151	0.4347

Assumed hydrogen positions (peaks found in difference Fouriers are in parentheses)

H(1,C 2)	0.387 (0.39)	0.660 (0.66)	0.543 (0.54)
H(2,C 2)	0.087 (0.08)	0.640 (0.64)	0.467 (0.46)
H(3,C 3)	0.590 (0.58)	0.495 (0.48)	0.481 (0.48)
H(4,C 6)	0.122 (0.13)	0.303 (0.31)	0.254 (0.27)
H(5,C 6)	−0.115 (−0.13)	0.309 (0.30)	0.356 (0.36)
H(6,N 1)	−0.170 —	0.552 —	0.703 —
H(7,N 1)	−0.105 —	0.666 —	0.636 —
H(8,N 2)	0.080 —	0.486 —	0.353 —
H(9,N 3)	0.181 (0.10)	0.139 (0.13)	0.289 (0.30)
H(10,N 3)	−0.024 (0.08)	0.147 (0.14)	0.386 (0.41)
H(11,N 3)	0.311 (0.30)	0.169 (0.17)	0.398 (0.38)

The other atomic parameters are listed in Table XIVD,36.

The structure that results is shown in Figure XIVD,64. As it indicates, the molecule is far from planar. Its bond lengths and angles are given on the projected molecule of Figure XIVD,65. There are several unusually large bond angles.

Between molecules the shortest interatomic distance, $N(2)-O(2) = 2.55$ A., is indicative of an exceptionally strong hydrogen bond. It joins the ends of adjacent molecules that follow one another in the direction of the long axis of the molecular chain. Between the same nitrogen and oxygen

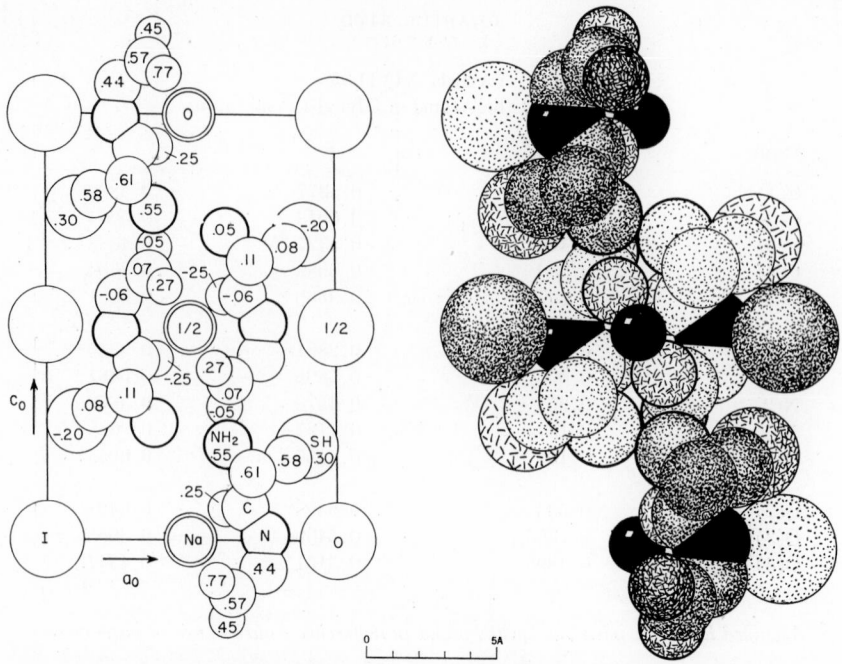

Fig. XIVD,64a (left). The monoclinic structure of cysteylglycine·$^1/_2$NaI projected along its b_0 axis. Left-hand axes.

Fig. XIVD,64b (right). A packing drawing of the monoclinic structure of cysteylglycine·$^1/_2$NaI seen along its b_0 axis. The iodide ions are the largest dotted circles; the SH radicals are next smaller and line shaded. The amino nitrogens are dotted and heavily outlined; the central nitrogens appear as black segments. Small black circles are sodium. Carboxyl oxygen atoms are small and line shaded.

TABLE XIVD,36

Parameters of Atoms in Cysteylglycine–Sodium Iodide

Atom	x	y	z
C(1)	0.255	0.943	0.569
C(2)	0.362	0.070	0.623
C(3)	0.293	0.113	0.343
C(4)	0.167	0.077	0.308
C(5)	0.315	0.943	0.420
O(1)	0.415	0.269	0.596
O(2)	0.382	0.950	0.688
O(3)	0.379	0.755	0.424
N(1)	0.247	0.940	0.487
N(2)	0.375	0.053	0.279
S	0.109	0.797	0.277

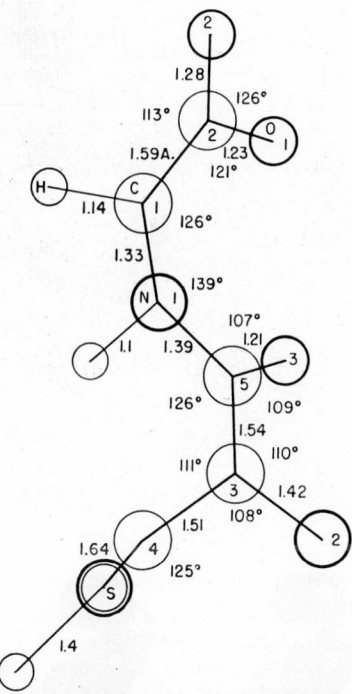

Fig. XIVD,65. A representation of the molecule of cysteylglycine giving bond dimensions.

atoms of parallel molecules in the structure N–O = 3.03 A. Each sodium ion has around it two O(1) atoms, with Na–O(1) = 2.16 A., and two O(3) atoms, with Na–O(3) = 2.23 A.

XIV,d39. The *copper chloride* complex of the tripeptide *diglycylglycine* which crystallizes with $^3/_2$ molecule of water, $CuCl[NH_2CH_2C(O)NHCH_2C(O)NHCH_2COO]\cdot{}^3/_2H_2O$, is monoclinic with an eight-molecule cell of the dimensions:

$$a_0 = 21.36 \text{ A.;} \quad b_0 = 6.72 \text{ A.;} \quad c_0 = 15.64 \text{ A.;} \quad \beta = 98°15'$$

The space group is C_{2h}^6 ($C2/c$). One water, $H_2O(2)$, is in

(4e) $\pm(0 \ u \ ^1/_4; \ ^1/_2, u+^1/_2, ^1/_4)$ with $u = 0.2953$

All other atoms are in the general positions:

(8f) $\pm(xyz; \ x,\bar{y},z+^1/_2; \ x+^1/_2,y+^1/_2,z; \ x+^1/_2,^1/_2-y,z+^1/_2)$

The chosen parameters are listed in Table XIVD.37.

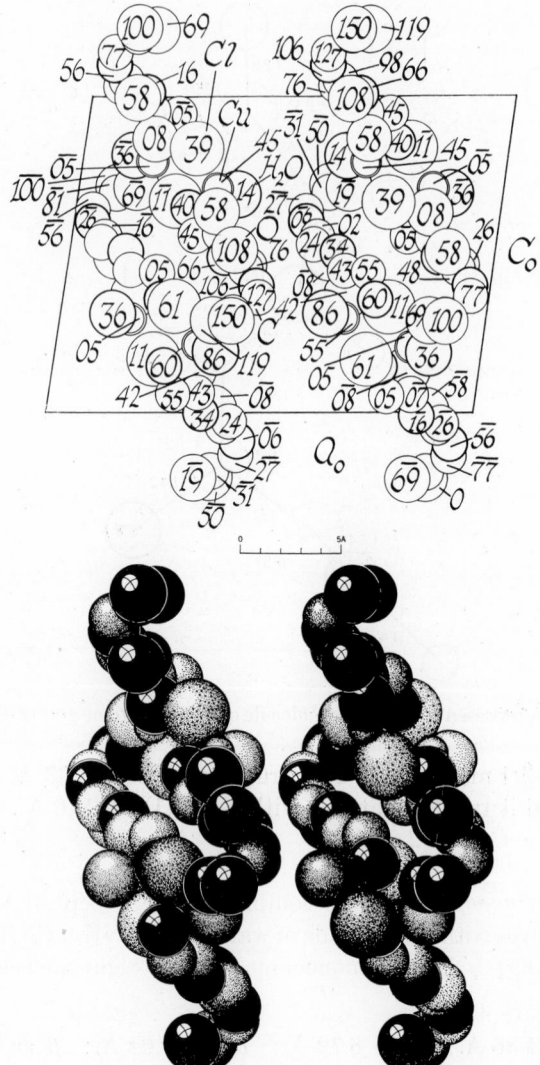

Fig. XIVD,66a (top). The monoclinic structure of copper (diglycylglycine) chloride hydrate projected along its b_0 axis.   The smaller, heavily ringed circles are nitrogen. Right-hand axes.

Fig. XIVD,66b (bottom). A packing drawing of the monoclinic copper (diglycylglycine) chloride hydrate arrangement seen along its b_0 axis.   Of the black circles those with the round "windows" are oxygen; the others are nitrogen.   The heavily outlined, smallest dotted circles are copper; the larger are water.   The largest dot-and-cross shaded circles are chlorine; the carbons, of intermediate size, are short-line shaded.

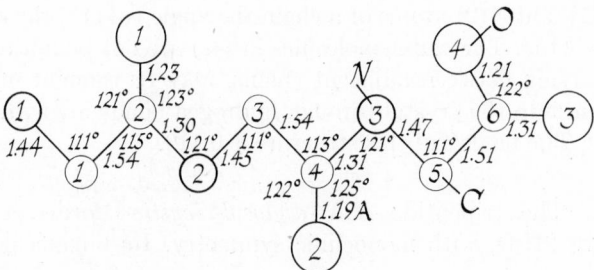

Fig. XIVD,67. Bond dimensions of the diglycylglycine molecule in its CuCl complex.

TABLE XIVD,37
Parameters of the Atoms in CuCl($C_6H_{10}N_3O_4$)·$^3/_2H_2O$

Atom	x	y	z
Cu	0.1642	−0.0541	0.7222
Cl	0.2310	−0.1109	0.8440
C(1)	0.2108	−0.0507	0.5602
C(2)	0.1497	0.0743	0.5563
C(3)	0.0700	0.2604	0.4657
C(4)	0.0783	0.4860	0.4546
C(5)	0.0428	0.7708	0.3662
C(6)	0.0940	0.8097	0.3110
N(1)	0.2324	−0.0967	0.6496
N(2)	0.1308	0.1610	0.4825
N(3)	0.0407	0.5583	0.3885
O(1)	0.1195	0.0796	0.6175
O(2)	0.1130	0.5822	0.5048
O(3)	0.0946	0.9946	0.2842
O(4)	0.1323	0.6896	0.2939
$H_2O(1)$	0.1215	−0.3617	0.6834

The structure is shown in Figure XIVD,66. Bond lengths in the tripeptide chain that results are stated in Figure XIVD,67. Each of the peptide segments of the chain is essentially planar, with adjacent segments roughly but not exactly normal to one another. The copper atoms have five closest neighbors; they are a chlorine, with Cu–Cl = 2.24 A., a water, with Cu–OH₂ = 2.30 A., the nitrogen and oxygen of a peptide chain, with Cu–N(1) = 1.99 A. = Cu–O(1), and a carboxyl oxygen of an adjacent chain with Cu–O(3′) = 1.93 A. The next nearest atom is the other carboxyl oxygen (4′) of this chain at a distance of 2.82 A. Between the copper atom and the N(1) and N(2) atoms of a chain the angle is 111°; the angle N(1)–Cu–O(1) = 113°. The water molecules in (4e) are not bound to the metal atoms but, lying between adjacent chains, may be thought of as helping to bind them into the crystal through hydrogen bonds involving N(3) and O(3) atoms. The final R of this structure is 0.119.

XIV,d40. The tripeptide N,N'-*diglycyl*-L-*cystine* forms a *dihydrate*, $C_{10}N_4O_6H_{18}S_2 \cdot 2H_2O$, with monoclinic symmetry. Its bimolecular unit has the dimensions:

$$a_0 = 12.26 \text{ A.}; \quad b_0 = 4.84 \text{ A.}; \quad c_0 = 17.17 \text{ A.}; \quad \beta = 124°24'$$

Atoms, in general positions of C_2^3 (*A*2):

$$(4c) \quad xyz; \ \bar{x}y\bar{z}; \ x,y+\frac{1}{2},z+\frac{1}{2}; \ \bar{x},y+\frac{1}{2},\frac{1}{2}-z$$

have been given the parameters of Table XIVD,38.

The resulting molecules have the configuration and bond dimensions of Figure XIVD,68. Their packing in the crystal is indicated in Figure

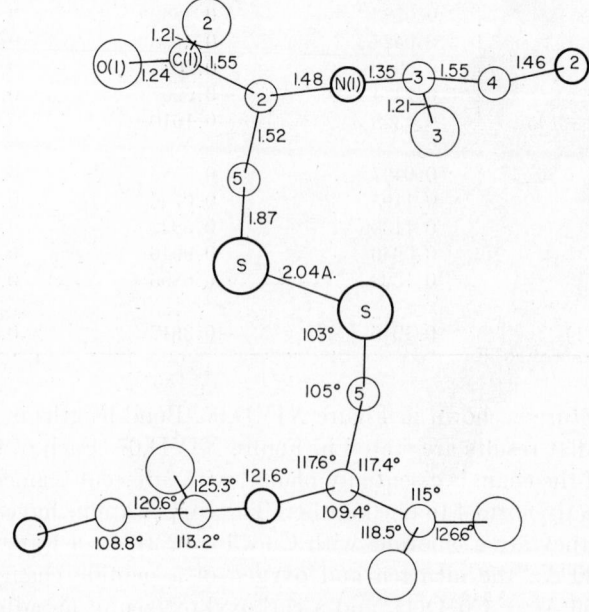

Fig. XIVD,68. Bond dimensions of the molecule of diglycyl-L-cystine.

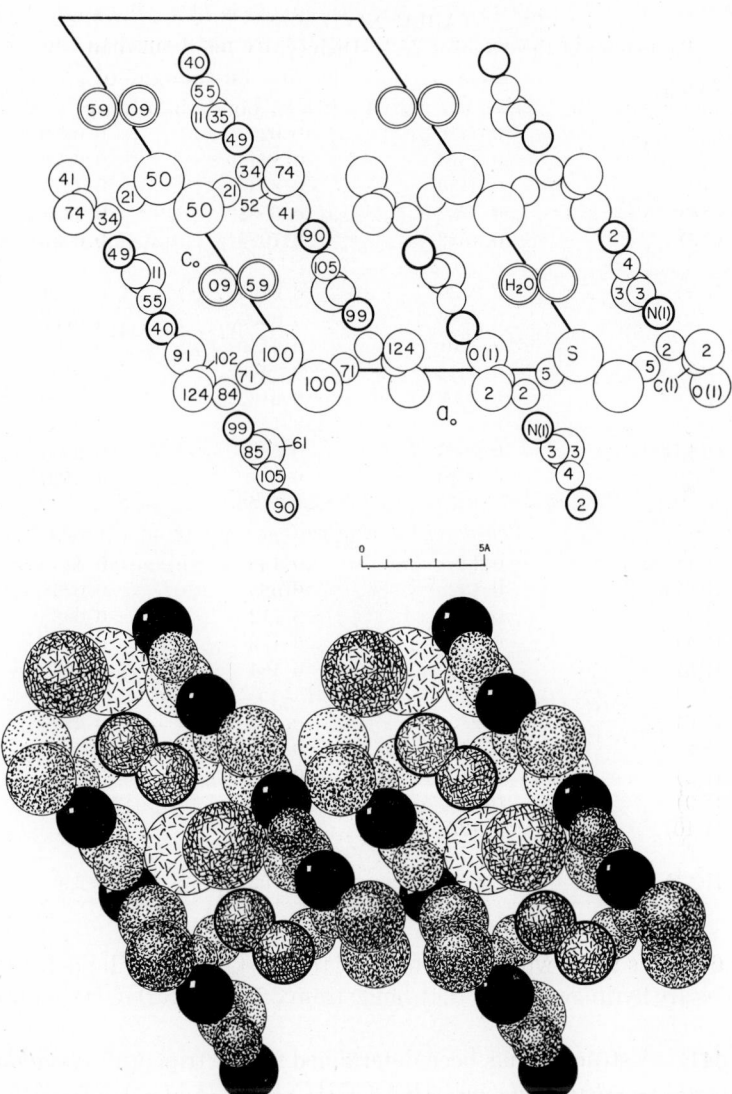

Fig. XIVD,69a (top). The monoclinic structure of diglycyl-L-cystine projected along its b_0 axis. Left-hand axes.

Fig. XIVD,69b (bottom). A packing drawing of the monoclinic structure of diglycyl-L-cystine viewed along its b_0 axis. Sulfur atoms and water molecules are line shaded, the sulfurs being smaller and heavily outlined. Oxygens are the large, carbons the small dotted circles. Atoms of nitrogen are black.

TABLE XIVD,38
Parameters of the Atoms in N,N'-Diglycyl-L-Cystine Dihydrate

Atom	x	y	z
C(1)	0.3406	0.0209	0.0200
C(2)	0.2873	0.8449	0.0673
C(3)	0.3156	0.8539	0.2219
C(4)	0.3232	0.0538	0.2954
C(5)	0.1532	0.7090	0.0032
N(1)	0.3139	0.9895	0.1524
N(2)	0.3546	0.8962	0.3777
O(1)	0.3301	0.4130	0.4508
O(2)	0.3912	0.2403	0.0553
O(3)	0.2908	0.6132	0.2210
O(4,H$_2$O)	0.0663	0.5891	0.2560
S	0.0316	0.4989	0.4563
	Calculated hydrogen positions		
H(1)	0.145	0.130	0.453
H(2)	0.132	0.585	0.033
H(3)	0.349	0.712	0.098
H(4)	0.324	0.172	0.158
H(5)	0.382	0.184	0.307
H(6)	0.245	0.127	0.280
H(7)	0.443	0.867	0.405
H(8)	0.310	0.750	0.378
H(9)	0.348	0.036	0.410
H(10)	0.127	0.594	0.247
H(11)	0.040	0.752	0.256

XIVD,69. As is usual with compounds of this sort, the chief links between molecules are hydrogen bonds that range from 2.75 A. [N(2)–O(1)] upward.

XIV,d41. A structure has been determined for the tripeptide *glutathione*, γ-L-*glutamyl*-L-*cysteinyl glycine*, HOOCCH(NH$_2$)(CH$_2$)$_2$C(O)NHCH(CH$_2$-SH)C(O)NHCH$_2$COOH. Its symmetry is orthorhombic with a tetramolecular cell of the edge lengths:

$$a_0 = 28.05 \pm 0.02 \text{ A.}; \quad b_0 = 8.802 \pm 0.002 \text{ A.}; \quad c_0 = 5.630 \pm 0.002 \text{ A.}$$

The space group is V^4 ($P2_12_12_1$) with atoms in the positions:

$$(4a) \quad xyz; \; {}^1\!/_2-x,\bar{y},z+{}^1\!/_2; \; x+{}^1\!/_2,{}^1\!/_2-y,\bar{z}; \; \bar{x},y+{}^1\!/_2,{}^1\!/_2-z$$

The chosen parameters are listed in Table XIVD,39.

The resulting structure is shown in Figure XIVD,70. Its molecules have the bond dimensions of Figure XIVD,71. In them the two peptide linkages

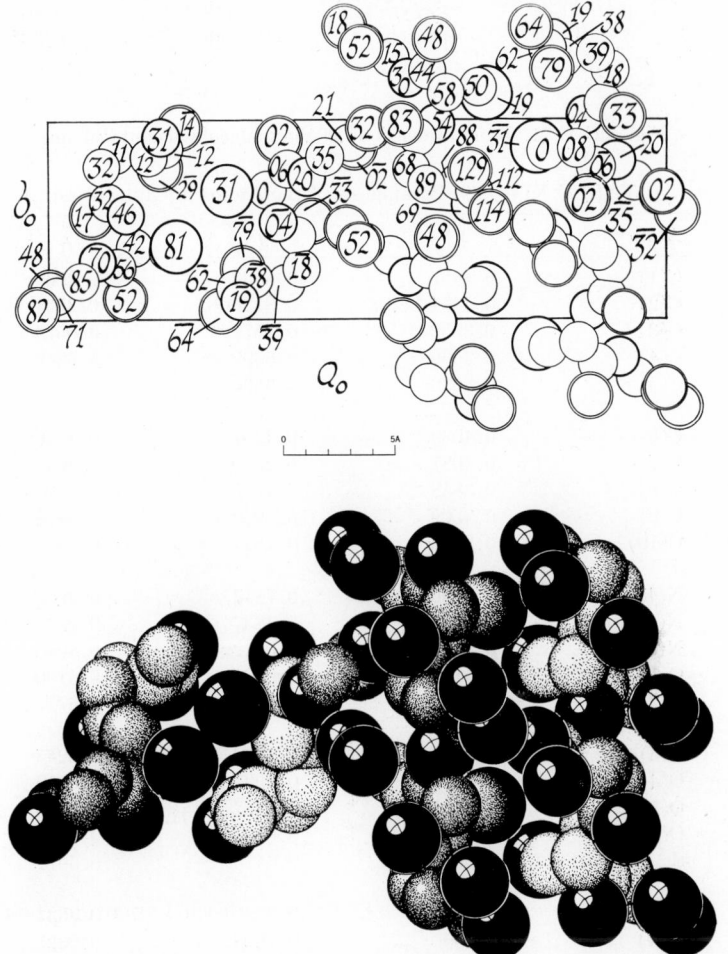

Fig. XIVD,70a (top). The orthorhombic structure of glutathione projected along its c_0 axis. The oxygen atoms are doubly ringed. Of the heavily outlined circles the sulfur are the larger, the nitrogen the smaller. Atoms of carbon are of the same size as nitrogen.

Fig. XIVD,70b (bottom). A packing drawing of the orthorhombic glutathione arrangement seen along its c_0 axis. The black circles are oxygen. Atoms of nitrogen are heavily outlined and hook-and-dot shaded; the carbons are short-line-and-dot shaded.

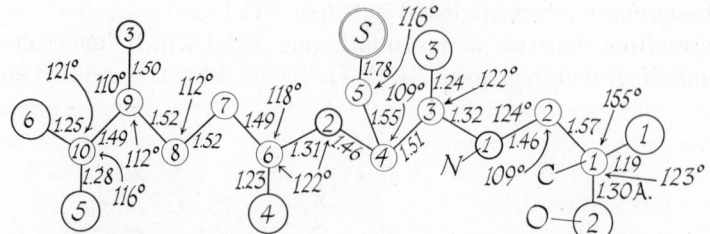

Fig. XIVD,71. Bond dimensions in the molecule of glutathione.

TABLE XIVD,39. Parameters of the Atoms in Glutathione

Atom	x	y	z
C(1)	0.4897	0.8833	0.2066
C(2)	0.4474	0.8126	0.3531
C(3)	0.3852	0.7577	0.0574
C(4)	0.3566	0.6399	0.9240
C(5)	0.3036	0.6526	0.9977
C(6)	0.4039	0.4207	0.8166
C(7)	0.4087	0.2520	0.8231
C(8)	0.3848	0.1775	0.6100
C(9)	0.3309	0.1909	0.6178
C(10)	0.3088	0.1541	0.3832
N(1)	0.4196	0.7117	0.2000
N(2)	0.3724	0.4837	0.9574
N(3)	0.3109	0.0912	0.8086
O(1)	0.5186	0.9506	0.3190
O(2)	0.4920	0.8699	0.9777
O(3)	0.3761	0.8938	0.0245
O(4)	0.4270	0.4937	0.6712
O(5)	0.3161	0.2487	0.2140
O(6)	0.2829	0.0404	0.3560
S	0.2910	0.6415	0.3072
H(3)[a]	0.360	0.659	0.733
H(4)	0.289	0.760	0.932
H(5)	0.284	0.561	0.911
H(6)	0.394	0.209	0.989
H(7)	0.446	0.223	0.823
H(8)	0.400	0.223	0.447
H(9)	0.394	0.057	0.619
H(10)	0.322	0.309	0.657

[a] Positions were not found for the remaining hydrogen atoms.

are planar, with C(2), N(1), C(3), O(3), and C(4) not departing by as much as 0.05 A. from the best plane through them, and with C(4), C(6), C(7), O(4), and N(2) defining the second plane, N(2) departing 0.12 A. from it. The molecules are held together in the crystal by a complex system of hydrogen bonds such as O(2)–H–O(4) = 2.66 A., N(1)–H–O(1) = 2.88 A., N(2)–H–O(5) = 2.98 A., and N(3)–H–O = 2.81, 2.68, and 2.89 A. There is a short nonbonded C(6)–O(1) = 3.00 A.; S–O separations range upwards from 3.53 A. The final $R = 0.253$; the σ of the bond lengths is considered to be ca. 0.028 A.

BIBLIOGRAPHY TABLE, CHAPTER XIVD

Compound	Paragraph	Literature
N-Acetylglycine		
$CH_3C(O)NHCH_2COOH$	d11	1950: C&D; 1962: D&M
DL-Alanine		
$NH_2CH(CH_3)COOH$	d12	1931: B; 1941: L&C; 1950: D
α-Amino isobutyric acid		
$(CH_3)_2C(NH_2)COOH$	d20	1952: H,K&N; 1958: H&A
L-Arginine dihydrate		
$(NH_2)_2CNH(CH_2)_3CH(NH_2)COO \cdot$	d30	1964: K&K
$2H_2O$		
L-Arginine hydrobromide monohydrate		
$NH_2C(NH)NH(CH_2)_3CH(NH_2)-$	d31	1964: M&S
$COOH \cdot HBr \cdot H_2O$		
L-Arginine hydrochloride monohydrate		
$NH_2C(NH)NH(CH_2)_3CH(NH_2)-$	d31	1964: M&S
$COOH \cdot HCl \cdot H_2O$		
Asparagine monohydrate		
$NH_2C(O)CH_2CH(NH_2)COOH \cdot H_2O$	d18	1955: S,CC&P; 1961: K&dV
DL-Aspartic acid		
$HO_2CCH_2CH(NH_2)CO_2H$	d17	1955: A&R
Cadmium glycinate monohydrate		
$Cd(NH_2CH_2CO_2)_2 \cdot H_2O$	d8	1959: L,H&R
Copper β-alanine hexahydrate		
$Cu(NH_2CH_2CH_2COO)_2 \cdot 6H_2O$	d13	1961: T
Copper DL-α-aminobutyrate		
$Cu(OOC \cdot C_3H_6NH_2)_2$	d22	1945: S
Copper chloride diglycylglycine		
$CuCl[NH_2CH_2C(O)NHCH_2C(O)-$	d39	1964: F,R&S
$NHCH_2COO] \cdot {}^3/_2H_2O$		
Copper(II) di β-amino butyrate dihydrate		
$Cu[NH_2CH(CH_3)CH_2COO]_2 \cdot 2H_2O$	d22	1961: B,P&T

(continued)

BIBLIOGRAPHY TABLE, CHAPTER XIVD (*continued*)

Compound	Paragraph	Literature
Copper glycinate monohydrate		
$Cu(NH_2CH_2CO_2)_2 \cdot H_2O$	d7	1961: T&N; 1964: F,S,N&T
Copper(II) monoglycylglycine trihydrate		
$Cu[NH_2CH_2C(O)NCH_2COOH] \cdot 3H_2O$	d36	1961: S,L&R
Cysteylglycine hemi sodium iodide		
$SHCH_2CH(NH_2)C(O)NHCH_2COOH \cdot$	d38	1951: D
$^1/_2NaI$		
L-Cystine		
$[SCH_2CH(NH_2)COOH]_2$	d34	1957: O&H; 1959: O&H
L-Cystine dihydrobromide		
$[-SCH_2CH(NH_2HBr)COOH]_2$	d33	1960: P,S&J; 1964: A&S
L-Cystine dihydrochloride		
$[HOOCCH(NH_2HCl)CH_2S-]_2$	d32	1956: C,S&F; 1958: S,P&J
Ferrous sulfate pentahydrate–glycine		
$FeSO_4 \cdot HO_2CCH_2NH_2 \cdot 5H_2O$	d10	1960: L&R
Glutamic acid		
$HOOC(CH_2)_2CH(NH_2)COOH$	d25	1955: H
DL-Glutamic acid hydrochloride		
$HOOCCH(NH_2 \cdot HCl)(CH_2)_2COOH$	d24	1953: D
L-Glutamine		
$C_5H_{10}O_3N_2$	d26	1952: C&P
Glutathione		
$HOOCCH(NH_2)(CH_2)_2C(O)NHCH-$	d41	1958: W
$(CH_2SH)C(O)NHCH_2COOH$		
Glycine		
NH_2CH_2COOH	d1, d2, d3	1931: B; H&L; 1936: K; 1939: A&C; 1958: I; M; 1960: I; 1961: I
di-Glycine hydrobromide		
$(C_2H_5NO_2)_2 \cdot HBr$	d4	1956: B,B&H; 1959: H
di-Glycine hydrochloride		
$(C_2H_5NO_2)_2 \cdot HCl$	d4	1957: H&B; 1960: H
tri-Glycine sulfate		
$(NH_2CH_2COOH)_3 \cdot H_2SO_4$	d5	1957: W&H; 1959: H,O&P
Glycyl-L-asparagine		
$NH_2CH_2C(O)NHCH(COOH)CH_2CONH_2$	d37	1954: P,K&C
β-Glycylglycine		
$NH_2CH_2C(O)NHCH_2COOH$	d35	1931: B; 1932: L; 1942: H&M; 1949: H&M
N,N'-di-Glycyl-L-cystine dihydrate		
$C_{10}N_4O_6H_{18}S_2 \cdot 2H_2O$	d40	1952: Y&H; 1954: Y&H
D(−)-Isoleucine hydrobromide monohydrate		
$C_2H_5CH(CH_3)CH(NH_2HBr)COOH \cdot H_2O$	d29	1954: T&B

(*continued*)

BIBLIOGRAPHY TABLE, CHAPTER XIVD (*continued*)

Compound	Paragraph	Literature
D(−)-Isoleucine hydrochloride monohydrate		
$C_2H_5CH(CH_3)CH(NH_2HCl)COOH \cdot H_2O$	d29	1953: T; 1954: T; T&B
L-Lysine hydrochloride dihydrate		
$NH_2(CH_2)_4CH(NH_2)COOH \cdot HCl \cdot 2H_2O$	d28	1959: R; 1962: W&M
DL-Methionine		
$CH_3S(CH_2)_2CH(NH_2)COOH$	d23	1951: D&M; 1952: M
(+)-S-Methyl-L-cysteine sulfoxide		
$CH_3S(O)CH_2 \cdot CH(NH_2)COOH$	d19	1962: H
Nickel β-alanine dihydrate		
$Ni[NH_2(CH_2)_2CO_2]_2 \cdot 2H_2O$	d14	1964: J,P&B
Nickel di α-amino isobutyrate tetra-hydrate		
$Ni[(CH_3)_2C(NH_2)COO]_2 \cdot 4H_2O$	d21	1962: N
Nickel glycine dihydrate		
$Ni(NH_2CH_2COO)_2 \cdot 2H_2O$	d9	1945: S
DL-Norleucine		
$CH_3(CH_2)_3CHNH_2COOH$	d27	1951: D&M; 1953: M
L-Threonine		
$CH_3CH(OH)CH(NH_2)COOH$	d16	1950: S,D,S&C
DL-Serine		
$OHCH_2CH(NH_2)COOH$	d15	1953: S,B,D&L
Silver glycine		
$AgOOC \cdot CH_2NH_2$	d6	1939: K&K
Zinc glycinate monohydrate		
$Zn(NH_2CH_2CO_2)_2 \cdot H_2O$	d8	1959: L,H&R

BIBLIOGRAPHY, CHAPTER XIV

1917

Vegard, L., "Results of Crystal Analysis IV," *Phil. Mag.*, **33**, 395.

1921

Becker, K., and Jancke, W., "X-Ray Spectroscopic Investigations with Organic Compounds I and II," *Z. Physik. Chem.*, **99**, 242, 267.

1922

Niggli, P., "On the Crystal Structure of Tetramethylammonium Iodide," *Z. Krist.*, **56**, 213.
Wyckoff, R. W. G., "On the Symmetry and Crystal Structure of Sodium Hydrogen Acetate," *Am. J. Sci.*, **4**, 193.

1923

Astbury, W. T., "The Crystalline Structure and Properties of Tartaric Acid," *Proc. Roy. Soc. (London)*, **102A**, 506.
Astbury, W. T., "The Crystalline Structure of Anhydrous Racemic Acid," *Proc. Roy. Soc. (London)*, **104A**, 219.
Bragg, W. H., "The Crystal Structure of Basic Beryllium Acetate," *Nature*, **111**, 532.
Bragg, W. H., and Morgan, G. T., "Crystal Structure and Chemical Constitution of Basic Beryllium Acetate and Propionate," *Proc. Roy. Soc. (London)*, **104A**, 437.
Mark, H., and Weissenberg, K., "X-Ray Determination of the Structure of Urea and of Tin Tetraiodide," *Z. Physik*, **16**, 1.
Mark, H., and Weissenberg, K., "The Crystal Structure of Pentaerythritol and a Graphical Interpretation of Interference Diagrams," *Z. Physik*, **17**, 301.
Müller, A., "The X-Ray Investigation of Fatty Acids," *J. Chem. Soc.*, **123**, 2043.

1924

Hoffman, H., and Mark, H., "Crystal Structure of Oxalic Acid," *Z. Physik. Chem.*, **111**, 321.
Mark, H., "The Application of X-Ray Crystal Analysis to the Problem of the Structure of Organic Compounds," *Chem. Ber.*, **57B**, 1820.
Yardley, K., "The Crystalline Structure of Succinic Acid, Succinic Anhydride and Succinimide," *Proc. Roy. Soc. (London)*, **105A**, 451.

1925

Knaggs, I. E., "The Crystalline Structure of Pentaerythritol Tetranitrate," *Mineral. Mag.*, **20**, 346.
Mark, H., Basche, W., and Pohland, E., "Determination of the Structure of Some Simple Inorganic Substances," *Z. Elektrochem.*, **31**, 523.
Mark, H., and Pohland, E., "The Structures of Ethane and Diborane," *Z. Krist.*, **62**, 103.
Meisel, K., Dissertation, Hannover.
Müller, A., "Structure of Stearic and Stearolic Acids," *Nature*, **116**, 45.
Trillat, J.-J., "Study of Fatty Acids and of Dibasic Acids by Means of X-Rays," *Compt. Rend.*, **180**, 1329.

Wyckoff, R. W. G., Hunt, F. L., and Merwin, H. E., "On the X-Ray Diffraction Effects from Solid Fatty Acids," *Science*, **61**, 613; *Z. Krist.*, **62**, 553.
Yardley, K., "An X-Ray Examination of Maleic and Fumaric Acids," *J. Chem. Soc.*, **127**, 2207.
Yardley, K., "An X-Ray Examination of Calcium Formate," *Mineral. Mag.*, **20**, 296.

1926

Astbury, W. T., "The Structure and Isotrimorphism of the Tervalent Metallic Acetylacetones," *Proc. Roy. Soc. (London)*, **112A**, 448.
Huggins, M. L., "The Crystal Structure of $[N(CH_3)_4]_2PtCl_6$," *Phys. Rev.*, **27**, 638.
Huggins, M. L., and Hendricks, S. B., "Confirmation of the Presence of a Non-tetrahedral Carbon Atom in Crystals of Pentaerythritol," *J. Am. Chem. Soc.*, **48**, 164.
Morgan, G. T., and Astbury, W. T., "Crystal Structure and Chemical Constitution of Basic Beryllium Acetate and Its Homologues," *Proc. Roy. Soc. (London)*, **112A**, 441.
Nitta, I., "The Crystal Structure of Iodoform," *Sci. Papers Inst. Phys. Chem. Res. (Tokyo)*, **4**, 49.
Nitta, I., "Crystal Structure of Pentaerythritol," *Bull. Chem. Soc. Japan*, **1**, 62.
Prins, J. A., "Examination of Fatty Acid Crystals with X-Rays," *Physica*, **6**, 305.
Prins, J. A., and Coster, D., "Higher Order X-Ray Reflections from Fatty Acids," *Nature*, **118**, 83.
Sarkar, A. N., "X-Ray Examination of the Crystal Structure of Certain Compounds," *Phil. Mag.*, **2**, 1153.
Vegard, L., and Berge, T., "Continued Investigations on the Structure of Tetramethyl Ammonium Iodide," *Skrifter Norske Videnskaps-Akad. Oslo I Mat. Naturv. Kl.*, **1926**, No. 10.
Westenbrink, H. G. K., and van Melle, F. A., "The Crystal Structure of Pentaerythritol," *Z. Krist.*, **64**, 548.
Wood, J. F., "The Crystal Structure of Some Oxalates," *Proc. Univ. Durham Phil. Soc.*, **7**, 111.

1927

Boer, G. M. de, "X-Ray Evidence for the Existence of Different Modifications of Fatty Acids," *Nature*, **119**, 50, 634.
Burgers, W. G., "Investigations of the Molecular Arrangement of Uniaxial Optically Active Crystals," *Proc. Roy. Soc. (London)*, **116A**, 553.
Greenwood, G., "Rotating-Crystal X-Ray Photographs," *Mineral. Mag.*, **21**, 258.
Hendricks, S. B., "The Molecular Symmetry of Pentaerythritol," *Z. Krist.*, **66**, 131.
Mark, H., and Weissenberg, K., "The Structure of Pentaerythritol," *Z. Krist.*, **65**, 499.
McLennan, J. C., and Plummer, W. G., "The Crystal Structures of the *n*-Paraffins, Octane, Hexane and Pentane," *Trans. Roy. Soc. Can. Sect. III*, **21**, 99.
Müller, A., "An X-Ray Investigation of Certain Long-Chain Compounds," *Proc. Roy. Soc. (London)*, **114A**, 542.
Schleede, A., and Schneider, E., "The Tetrahedral Carbon Atom and the Crystal Structure of Pentaerythritol," *Naturwissenschaften*, **15**, 970; *Z. Anorg. Allgem. Chem.*, **168**, 313 (1928).
Seifert, H., "The Symmetry of Crystals of Pentaerythritol," *Sitzber. Preuss. Akad. Wiss.*, **1927**, 289; *Z. Anorg. Allgem. Chem.*, **174**, 318 (1928).

Vegard, L., and Sollesnes, K., "The Structure of the Isomorphic Substances $N(CH_3)_4I$, $N(CH_3)_4Br$, $N(CH_3)_4Cl$," *Phil. Mag.*, **4**, 985; *Skrifter Norske Videnskaps-Akad Oslo I Mat. Naturv. Kl.*, **1927**, No. 10.

Weissenberg, K., "The Tetrahedral Carbon Atom and the Crystal Structure of Pentaerythritol," *Naturwissenschaften*, **15**, 995.

Wood, J. F., "The Crystal Structure of Iodoform," *Proc. Univ. Durham Phil. Soc.*, **7**, 168.

1928

Brill, R., and Meyer, K. H., "X-Ray Investigation of Lauric Acid," *Z. Krist.*, **67**, 570.

Caspari, W. A., "Crystallography of Aliphatic Dicarboxylic Acids," *J. Chem. Soc.*, **1928**, 3235.

Demény, L., and Nitta, I., "The Crystal Structure of Thiourea," *Bull. Chem. Soc. Japan*, **3**, 128.

Ernst, E., "The Crystal Class of Pentaerythritol," *Z. Krist.*, **68**, 139.

Gerstäcker, A., Möller, H., and Reis, A., "The Crystal Structure of the Tetraacetate and Tetranitrate of Pentaerythritol," *Z. Krist.*, **66**, 355.

Gerstäcker, A., Möller, H., and Reis, A., "X-Ray Investigation of Several Triclinic-Pinacoidal Crystals," *Z. Krist.*, **66**, 421.

Hendricks, S. B., "The Crystal Structures of the Monomethyl Ammonium Halides," *Z. Krist.*, **67**, 106.

Hendricks, S. B., "The Crystal Structure of Monoethyl Ammonium Bromide and Iodide," *Z. Krist.*, **67**, 119.

Hendricks, S. B., "The Crystal Structure of the *n*-Monopropyl Ammonium Halides," *Z. Krist.*, **67**, 465.

Hendricks, S. B., "The Crystal Structure of the Triethyl Ammonium Halides," *Z. Krist.*, **67**, 472.

Hendricks, S. B., "The Crystal Structure of the *n*-Mono Butyl, Amyl, Hexyl and Heptyl Ammonium Halides," *Z. Krist.*, **68**, 189.

Hendricks, S. B., "Crystal Structure of Urea and the Molecular Symmetry of Thiourea," *J. Am. Chem. Soc.*, **50**, 2455.

Hengstenberg, J., "X-Ray Investigation of the Structure of the Carbon Chains in Hydrocarbons (C_nH_{2n+2})," *Z. Krist.*, **67**, 583.

Hettich, A., Schleede, A., and Schneider, E., "The Tetrahedral Carbon Atom and the Crystal Structure of Pentaerythritol," *Naturwissenschaften*, **16**, 547.

Knaggs, I. E., "The Form of the Carbon Atom in Crystal Structure," *Nature*, **121**, 616.

Mark, H., and Susich, G. v., "The X-Ray Investigation of Pentaerythritol," *Z. Krist.*, **69**, 105.

Mark, H., and Weissenberg, K., "Pyro- and Piezoelectric Phenomena with Pentaerythritol," *Z. Physik*, **47**, 301.

McLennan, J. C., and Plummer, W. G., "The Crystal Structure of Solid Methane," *Nature*, **122**, 571; *Phil. Mag.*, **7**, 761 (1929).

Melle, F. A. van, and Schurink, H. B. J., "The Crystal Structure of Pentaerythritol, Pentaerythritol Tetraacetate and Dibenzal-Pentaerythritol," *Z. Krist.*, **69**, 1.

Möller, H., and Reis, A., "The Crystal Structure of Pentaerythritol Tetraacetate," *Z. Krist.*, **68**, 385.

Müller, A., "A Further X-Ray Investigation of Long-Chain Compounds (*n*-Hydrocarbon)," *Proc. Roy. Soc. (London)*, **120A**, 437.

Nehmitz, A., "The Crystal Structure of Pentaerythritol," *Z. Krist.*, **66**, 408.

Nitta, I., "On the Crystal Structure of Tetraethyl Ammonium Iodide," *Proc. Imp. Acad. Japan*, **4**, 292.

Nitta, I., "The Crystal Structure of Some Rhombic Formates," *Sci. Papers Inst. Phys. Chem. Res. (Tokyo)*, **9**, 151.

Reis, A., and Schneider, W., "The Crystal Structure of Tartaric Acid, Isohydrobenzoin and Rubidium Tartrate," *Z. Krist.*, **69**, 62.

Schleede, A., and Hettich, A., "The Crystal Class of Pentaerythritol and the Tetrahedral Carbon Atom," *Z. Anorg. Allgem. Chem.*, **172**, 121.

Wyckoff, R. W. G., "The Crystal Structures of Monomethyl Ammonium Chlorostannate and Chloroplatinate," *Am. J. Sci.*, **16**, 349.

Wyckoff, R. W. G., "The Crystal Structure of the Tetramethyl Ammonium Halides," *Z. Krist.*, **67**, 91.

Wyckoff, R. W. G., "On the Crystal Structure of Tetraethyl Ammonium Iodide," *Z. Krist.*, **67**, 550.

Wyckoff, R. W. G., "The Crystal Structure of Monoethyl Ammonium Chlorostannate $(NH_3C_2H_5)_2SnCl_6$," *Z. Krist.*, **68**, 231.

Yardley, K., "An X-Ray Study of Some Simple Derivatives of Ethane I and II," *Proc. Roy. Soc. (London)*, **118A**, 449, 485.

Zachariasen, W. H., "The Crystal Structure of Tetramethyl Ammonium Iodide," *Norsk Geol. Tidsskr.*, **10**, 14.

1929

Caspari, W. A., "Dimorphism in the Aliphatic Dicarboxylic Acid Series (Azelaic Acid)," *J. Chem. Soc.*, **1929**, 2709.

Hassel, O., and Luzanski, N., "The Space Lattice of Trigonal Acetamide," *Z. Physik. Chem.*, **3B**, 282.

Herrmann, K., and Ilge, W., "The Structure of Tetramethyl Ammonium Perchlorate and Permanganate," *Z. Krist.*, **71**, 47

Knaggs, I. E., "The Space Group of Pentaerythritol Tetraacetate," *Z. Krist.*, **70**, 185.

Knaggs, I. E., "The Form of the Central Carbon Atom in Pentaerythritol Tetraacetate as Shown by X-Ray Crystal Analysis," *Proc. Roy. Soc. (London)*, **122A**, 69.

Müller, A., "A Hydrocarbon Model," *Trans. Faraday Soc.*, **25**, 347.

Wyckoff, R. W. G., and Corey, R. B., "The Crystal Structure of Trimethyl Ethyl Ammonium Chlorostannate," *Am. J. Sci.*, **17**, 239.

Wyckoff, R. W. G., and Corey, R. B., "The Crystal Structure of Dimethyl Diethyl Ammonium Chlorostannate," *Am. J. Sci.*, **18**, 138.

Wyckoff, R. W. G., and Corey, R. B., "The Crystal Structure of Tetramethyl, Trimethyl and Triethyl Methyl Ammonium Clorostannates," *Am. J. Sci.*, **18**, 437.

1930

Corey, R. B., and Wyckoff, R. W. G., "The Crystal Structures of Trimethyl and Dimethyl Ethyl Sulfonium Chlorostannates and of Methyl Triethyl Phosphonium Chlorostannate," *Radiology*, **15**, 241.

Dupré la Tour, F., "The Polymorphism of the Saturated Diacids of the Aliphatic Series as a Function of Temperature," *Compt. Rend.*, **191**, 1348.

Hendricks, S. B., "The Crystal Structure of Primary Amyl Ammonium Chloride," *Z. Krist.*, **74**, 29.

Hendricks, S. B., "Molecular Rotation in the Solid State," *Nature*, **126**, 167.

Jong, W. F. de, "The Crystal Structure of Sodium Uranylacetate," *Physica*, **10**, 101.

Wyckoff, R. W. G., "X-Ray Diffraction Data from Several Mono-Alkyl Substituted Ammonium Iodides," *Z. Krist.*, **74**, 25.
Wyckoff, R. W. G., "A Powder Spectrometric Study of the Structure of Urea," *Z. Krist.*, **75**, 529.

1931

Bernal, J. D., "The Crystal Structure of the Natural Amino Acids and Related Compounds," *Z. Krist.*, **78**, 363.
Dupré la Tour, F., "The Polymorphism of Malonic, Succinic and Glutaric Acids as a Function of the Temperature," *Compt. Rend.*, **193**, 180.
Hémon, Y., "Crystallographic Study of Cesium Bitartrate," *Bull. Soc. Franc. Mineral.*, **54**, 47.
Hengstenberg, J., and Lenel, F. V., "The Structure of Glycine," *Z. Krist.*, **77**, 424.
Hertel, E., "Structure of Compounds of Sulfur with Iodides I. Compounds of Sulfur with Triiodides," *Z. Physik. Chem.*, **15B**, 51.
Huggins, M. L., and Noble, B. A., "The Crystal Structure of Iodoform,' *Am. Mineralogist*, **16**, 519.
Kabraji, K. J., "Crystal Structure of the Hydrates of Copper Formate I. Copper Formate Tetrahydrate," *Indian J. Phys.*, **6**, 81.
Mooy, H. H., "Crystal Structure of Methane," *Nature*, **127**, 707; *Proc. Acad. Sci. Amsterdam*, **34**, 550, 660.

1932

Bernal, J. D., "Rotation of Carbon Chains in Crystals," *Z. Krist.*, **83**, 153.
Bernal, J. D., "Rotation of Molecules in Crystals," *Nature*, **129**, 870.
Dupré la Tour, F., "X-Ray Study of the Polymorphism of Normal Saturated Fatty Acids," Thesis, Univ. Paris.
Dupré la Tour, F., "Study of the Dimorphism of Normal Saturated Aliphatic Dicarboxylic Acids as a Function of Temperature," *Compt. Rend.*, **194**, 622.
Dupré la Tour, F., "X-Ray Study of the Polymorphism of the Normal Saturated Acids of the Aliphatic Series," *Ann. Phys. (Paris)*, **18**, 199.
Ferrari, A., and Curti, R., "The Habit and Crystal Structure of Potassium Bitartrate," *Z. Krist.*, **84**, 8.
Halla, F., and Zimmermann, L., "The Structure of Lead Formate," *Z. Krist.*, **83**, 497.
Lenel, F. V., "The Structures of Some Simple Glycine Polypeptides," *Z. Krist.*, **81**, 224.
Thibaud, J., and Dupré la Tour, F., "The Polymorphism of Saturated Long Chain Monobasic Acids. Influence of the Temperature," *J. Chim. Phys.*, **29**, 153.
Thibaud, J., and Dupré la Tour, F., "Study of the Polymorphism of Fatty Acids as a Function of the Temperature," *J. Phys. Radium*, **3**, 378.
Wyckoff, R. W. G., "Some Single Crystal Spectrometric Data on Urea," *Z. Krist.*, **81**, 102.
Wyckoff, R. W. G., and Corey, R. B., "The Crystal Structure of Thiourea," *Z. Krist.*, **81**, 386.

1933

Corey, R. B., and Wyckoff, R. W. G., "On the Structure of Methyl Urea," *Z. Krist.*, **85**, 132.
Greenwood, G., "On the 'Correct' Setting of Crystals," *Z. Krist.*, **85**, 420.

Southard, J. C., Milner, R. T., and Hendricks, S. B., "Low Temperature Specific Heats III. Molecular Rotation in Crystalline Primary Normal Amyl Ammonium Chloride," *J. Chem. Phys.*, **1**, 95.

Staub, H., "Investigation of the Dielectric Properties of Seignette Salt by Means of X-Rays," *Physikal. Z.*, **34**, 292.

1934

Corey, R. B., "The Crystal Structure of Tetramethyl Ammonium Fluosilicate," *Z. Krist.*, **89**, 10.

Corey, R. B., and Wyckoff, R. W. G., "The Crystal Structure of Dimethyl-Ammonium Chlorostannate," *Z. Krist.*, **89**, 469.

Hull, R. B., "Crystal Structure of Nickelous Acetate Tetrahydrate," *Phys. Rev.*, **46**, 329.

Pauling, L., and Sherman, J., "Structure of the Carboxyl Group II. Crystal Structure of Basic Beryllium Acetate," *Proc. Natl. Acad. Sci. U.S.*, **20**, 340.

Powell, H. M., and Crowfoot, D., "The Crystal Structures of Dimethyl Thallium Halides," *Z. Krist.*, **87**, 370.

West, C. D., "The Crystal Structure of Hexamethylethane and of Cubic Hexachloroethane," *Z. Krist.*, **88**, 195.

Wyckoff, R. W. G., and Corey, R. B., "Spectrometric Measurements on Hexamethylene Tetramine and Urea," *Z. Krist.*, **89**, 462.

Zachariasen, W. H., "The Crystal Lattice of Oxalic Acid Dihydrate $H_2C_2O_4 \cdot 2H_2O$ and the Structure of the Oxalate Radical," *Z. Krist.*, **89A**, 442.

1935

Biefeld, L. P., and Harris, P. M., "The Crystal Structure of Dicalcium Barium Propionate," *J. Am. Chem. Soc.*, **57**, 396.

Cox, E. G., Wardlaw, W., and Webster, K. C., "The Planar Configuration for Quadricovalent Nickel, Palladium and Platinum. Dithiooxalate Derivatives," *J. Chem. Soc.*, **1935**, 1475.

Cox, E. G., and Webster, K. C., "The Structure of Trimethyl Platinic Chloride, $(CH_3)_3$-PtCl," *Z. Krist.*, **90A**, 561.

Cox, E. G., and Webster, K. C., "An X-Ray Investigation of Some Non-Planar Coordination Compounds of Bivalent Nickel," *Z. Krist.*, **92A**, 478.

Cox, E. G., and Webster, K. C., "The Planar Structure of Quadricovalent Cupric Compounds," *J. Chem. Soc.*, **1935**, 731.

Dupré la Tour, F., "Polymorphism in the Normal Dibasic Acid Series," *Compt. Rend.*, **201**, 479.

Fankuchen, I., "Crystal Structure of Sodium Uranyl Acetate," *Z. Krist.*, **91A**, 473.

Hendricks, S. B., "The Orientation of the Oxalic Group in Oxalic Acid and Some of Its Salts," *Z. Krist.*, **91A**, 48.

Jaeger, F. M., and Beintema, J., "On the Symmetry and the Structure of the Crystals of the Hydrochlorides of Triamino-Triethylamine," *Proc. Acad. Sci. Amsterdam*, **38**, 243.

Keesom, W. H., and Taconis, K. W., "An X-Ray Goniometer for the Investigation of the Crystal Structure of Solidified Gases," *Physica*, **2**, 463.

Klug, H. P., "A Study of the Molecular Structure of Diiodoethane," *J. Chem. Phys.*, **3**, 747.

Klug, H. P., "A Study of the Crystal Structures of Sym-Diiodoethane and Sym-Di-iodoethylene," Z. Krist., **90A**, 495.

Nitta, I., and Watanabé, T., "The Crystal Structure of Barium Dicalcium Propionate," Sci. Papers Inst. Phys. Chem. Res. (Tokyo), **26**, 164.

Theilacker, W., "Calculation of the Refractive Index of Guanidonium Iodide," Z. Krist., **90A**, 77.

Theilacker, W., "The Crystal Structure of Guanidonium Halides II. The Structure of Guanidonium Iodide," Z. Krist., **90A**, 51.

Theilacker, W., "The Crystal Structure of Guanidonium Halides III. The Structure of Guanidonium Bromide," Z. Krist., **90A**, 256.

1936

Cox, E. G., Wardlaw, W., and Webster, K. C., "The Stereochemistry of Quadrico-valent Atoms. Copper and Silver," J. Chem. Soc., **1936**, 775.

Cox, E. G., Sharratt, E., Wardlaw, W., and Webster, K. C., "The Planar Configuration of Quadricovalent Compounds of Bivalent Copper and Nickel," J. Chem. Soc., **1936**, 129.

Dupré la Tour, F., "Inverted Polymorphism in the Series of Normal Saturated Dibasic Acids," Compt. Rend., **202**, 1935.

Fankuchen, I., "Structure of Silver Uranyl Acetate," Z. Krist., **94A**, 212.

Hendricks, S. B., and Jefferson, M. E., "Electron Distribution in $(NH_4)_2C_2O_4 \cdot H_2O$ and the Structure of the Oxalate Group," J. Chem. Phys., **4**, 102; Phys. Rev., **49**, 200.

Kitaiigorodskii, A., "On the Structure of Glycine," Acta Physicochim. URSS, **5**, 749.

Mann, F. G., Purdie, D., and Wells, A. F., "The Constitution of Complex Metallic Salts IV. The Constitution of the Phosphine and Arsine Derivatives of Cuprous Iodide. The Configuration of the Coordinated Cuprous Complex," J. Chem. Soc., **1936**, 1503.

Minder, W., and Stocker, E., "On Trithiourea Cuprous Chloride," Z. Krist., **94A**, 137.

Robertson, J. M., and Woodward, J., "The Structure of the Carboxyl Group. A Quantitative Investigation of Oxalic Acid Dihydrate by Fourier Synthesis from X-Ray Crystal Data," J. Chem. Soc., **1936**, 1817.

Wells, A. F., "The Crystal Structures of Alkyl Metallic Complexes," Z. Krist., **94A**, 447.

1937

Elliot, N., "The Unit of Structure and Space Group of Chloral Hydrate," Z. Krist., **98A**, 180.

Ketelaar, J. A. A., "The Crystal Structure of the Ethyl Sulphates of the Rare Earths and Yttrium," Physica, **4**, 619.

Llewellyn, F. J., Cox, E. G., and Goodwin, T. H., "The Crystalline Structure of the Sugars IV. Pentaerythritol and the Hydroxyl Bond," J. Chem. Soc., **1937**, 883.

Thiessen, P. A., and Schoon, T., "Electron Diffraction by Natural Surfaces of Single Organic Crystals," Z. Physik. Chem., **36B**, 216.

Wells, A. F., "The Crystal Structures of the Mercury-n-Alkyl-Mercaptides," Z. Krist., **96A**, 435.

West, C. D., "Sulfur-Iodide Crystals $RI_3 \cdot 3S_8$: Structure Unit and Optical Properties," Z. Krist., **96A**, 459.

1938

Bonati, S., "Crystal Structure of Nickel Xanthate," *Atti Soc. Toscana Sci. Nat. Pisa, Mem.*, **47**, 71.

Bonati, S., and Derenzini, T., "Determination of the Crystal Structure of the Salts of Xanthic Acid," *Atti Soc. Toscana Sci. Nat. Pisa, Proc. Verbali Mem.*, **47**, 7.

Derenzini, T., and Rossoni, P., "Crystal Structure of Antimony Xanthate," *Atti Soc. Toscana Sci. Nat. Pisa, Proc. Verbali Mem.*, **47**, 67.

Goodwin, T. H., and Hardy, R., "The Crystal Structure of Pentaerythritol Tetra-acetate," *Proc. Roy. Soc. (London)*, **164A**, 369.

Mann, F. G., and Wells, A. F., "The Constitution of Complex Metallic Salts VII. The Structure and Configuration of the Bridged Derivatives of Trimethyl Arsine with Palladous Halides," *J. Chem. Soc.*, **1938**, 702.

Milone, M., "X-Ray Analyses of Complex Nickel Glyoximes," *Atti Congr. Intern. Chim. 10 Rome*, **II**, 346.

Misch, L. C., and Wyk, A. J. A. van der, "Structure of Oxamide," *Compt. Rend. Soc. Phys. Nat. Genève*, **55**, 96 .

Nitta, I., and Watanabé, T., "X-Ray Investigation of the Cubic Modification of Pent-aerythritol, $C(CH_2OH)_4$," *Bull. Chem. Soc. Japan*, **13**, 28.

Schoon, T., "Polymorphic Forms of Crystalline Long Chain Carbon Compounds," *Z. Physik. Chem.*, **39B**, 385.

Taconis, K. W., "Investigations of the Structures of Solidified Gases at Very Low Temperatures," Dissertation, Leiden.

Verweel, H. J., and MacGillavry, C. H., "The Crystal Structure of Succinic Acid COOH—CH_2—CH_2—COOH," *Nature*, **142**, 161; *Z. Krist.*, **102A**, 60 (1939).

Wells, A. F., "The Crystal Structure of Certain Bridged Palladium Compounds," *Proc. Roy. Soc. (London)*, **167A**, 169.

Wells, A. F., "The Crystal Structure of the Trimethyl-Stibine Dihalides $(CH_3)_3SbX_2$," *Z. Krist.*, **99A**, 367.

Zachariasen, W. H., "Crystal Structure of Sodium Formate, $NaHCO_2$," *Phys. Rev.*, **53**, 917.

1939

Albrecht, G., and Corey, R. B., "The Structure of Glycine," *J. Am. Chem. Soc.*, **61**, 1087.

Brill, R., Hermann, C., and Peters, C., "Studies of Chemical Bonding by Fourier Analysis IV. The Oxygen Bond in Oxalic Acid Dihydrate," *Naturwissenschaften*, **27**, 677.

Johansson, A., "X-Ray Studies on Chloro-Mercury Alkyl Mercaptides," *Arkiv Kemi Mineral. Geol.*, **13A**, 1.

Kitaigorodskii, A. I., and Kozhin, V. M., "Structure of Silver Aminoacetate," *Zh. Eksperim. i Teor. Fiz.*, **9**, 1127.

Mellor, D. P., "The Unit Cell and Space Group of Cs_2CuCl_4," *Z. Krist.*, **101A**, 160.

Mooney, R. C. L., "An X-Ray Determination of the Structure of Tetramethyl Ammonium Dichloriodide Crystals $N(CH_3)_4ICl_2$," *Z. Krist.*, **100A**, 519.

Müller, A., and Schallamach, A., "Crystal Structure of Methane at the Transition Point, 20.4°K," *Nature*, **143**, 375.

Phillips, R. F., and Powell, H. M., "The Crystal Structure of Di-*n*-Propyl-Monocyano-Gold," *Proc. Roy. Soc. (London)*, **173A**, 147.

Robertson, J. M., and Ubbelohde, A. R., "Structure and Thermal Properties Associated with Some Hydrogen Bonds in Crystals," *Proc. Roy. Soc. (London)*, **170A**, 222, 241.

Schallamach, A., "X-Ray Investigation of the Structure Transition of Methane at the λ Point," *Proc. Roy. Soc. (London)*, **171A**, 569.

Verweel, H. J., and MacGillavry, C. H., "The Crystal Structure of Succinic Acid," *Z. Krist.*, **102A**, 60.

1940

Beevers, C. A., and Hughes, W., "Crystal Structure of Rochelle Salt," *Nature*, **146**, 96.

Hughes, E. W., "The Crystal Structure of Dicyandiamide," *J. Am. Chem. Soc.*, **62**, 1258.

Mussgnug, F., "Relation between Triethylsulfonium Iodide and Triethylammonium Iodide," *Naturwissenschaften*, **28**, 366.

Peyronel, G., "Crystal Structure of Nickel *N,N*-Dipropyl-Dithiocarbamate I," *Z. Krist.*, **103A**, 139.

Senti, F., and Harker, D., "The Crystal Structure of Rhombohedral Acetamide," *J. Am. Chem. Soc.*, **62**, 2008.

Zachariasen, W. H., "The Crystal Structure of Sodium Formate," *J. Am. Chem. Soc.*, **62**, 1011.

1941

Beevers, C. A., and Hughes, W., "The Crystal Structure of Rochelle Salt," *Proc. Roy. Soc. (London)*, **177A**, 251.

Levy, H. A., and Corey, R. B., "The Crystal Structure of DL-Alanine," *J. Am. Chem. Soc.*, **63**, 2095.

Lonsdale, K., "Molecular Anisotropy of Urea $CO(NH_2)_2$ and of Related Compounds," *Proc. Roy. Soc. (London)*, **177A**, 272.

Lu, C. S., Hughes, E. W., and Giguere, P. A., "The Crystal Structure of the Urea–Hydrogen Peroxide Addition Compound $CO(NH_2)_2 \cdot H_2O_2$," *J. Am. Chem. Soc.*, **63**, 1507.

MacGillavry, C. H., "Crystal Structure of Adipic Acid," *Rec. Trav. Chim.*, **60**, 605.

Miyake, S., "Effect of Temperature Variation and Electric Field on X-Ray Intensity Reflected from Rochelle Salt Crystals I. Temperature Effect," *Proc. Phys.-Math. Soc. Japan*, **23**, 377.

Mussgnug, F., "Trimethylammonium Iodide and Trimethylsulfonium Iodide," *Naturwissenschaften*, **29**, 256.

Peyronel, G., "Crystal Structure of Ni-*N,N*-Dipropyl-Dithiocarbamate II," *Z. Krist.*, **103A**, 157.

Yamada, T., and Mizuno, K., "Crystal Structure of Accelerators of the Thiuram Series," *Nippon Gomu Kyokaishi*, **15**, 436; *J. Soc. Chem. Ind. Japan*, **44**, 708.

1942

Brill, R., Hermann, C., and Peters, C., "Röntgen Fourier Synthesis of Oxalic Acid Dihydrate," *Ann. Physik*, **42**, 357.

Bunn, C. W., "Molecular Structure and Rubber-Like Elasticity I. The Crystal Structures of β-Gutta Percha, Rubber, and Polychloroprene," *Proc. Roy. Soc. (London)*, **180A**, 40.

Hughes, E. W., and Moore, W. J., "Crystal Structure of β-Glycylglycine," *J. Am. Chem. Soc.*, **64**, 2236.

Snaauw, G. J., and Wiebenga, E. H., "Crystal Structure of Hexabromoethane," *Rec. Trav. Chim.*, **61**, 253.

1943

Bateman, L., and Jeffrey, G. A., "Structure of Geranylamine-HCl," *Nature*, **152**, 446.

Beevers, C. A., "Change of Symmetry of Basic Beryllium Acetate," *Nature*, **152**, 447.

Griffith, R. L., "The Crystal Structure of $Ag_2C_2O_4$," *J. Chem. Phys.*, **11**, 499.

Pabst, A., "Crystal Structure of Gadolinium Formate, $Gd(OOCH)_3$," *J. Chem. Phys.*, **11**, 145.

Perdok, W. G., and Terpstra, P., "Order-Disorder Transformations in the Lattice of Organic Molecules II. The Crystal Structure of $C(SCH_3)_4$ Below 23.2°," *Rec. Trav. Chim.*, **62**, 687.

Peyronel, G., "Crystal Structure of Cu-N,N-Dipropyldithiocarbamate I," *Gazz. Chim Ital.*, **73**, 89.

1944

Bunn, C. W., "Crystal Structure of Ethylene," *Trans. Faraday Soc.*, **40**, 23.

Ott, H., "Röntgen-Ray Investigation of Normal Primary Alcohols with Long Carbon Chains," *Z. Physik. Chem.*, **193A**, 218.

Rieck, G. D., "The Crystal Structure of α-Succinic Acid," *Rec. Trav. Chim.*, **63**, 170.

1945

Jeffrey, G. A., "The Structure of Polyisoprenes I. The Crystal Structure of Geranylamine Hydrochloride," *Proc. Roy. Soc. (London)*, **183A**, 388.

Powell, H. M., and Bartindale, G. W. R., "Structure of Compounds of Ferrocyanide Type I. Crystal Structure of Hexamethylisocyanidoferrous Chloride," *J. Chem. Soc.*, **1945**, 799.

Stosick, A. J., "The X-Ray Investigation of Cu DL-α-Aminobutyrate," *J. Am. Chem. Soc.*, **67**, 362.

Stosick, A. J., "The Crystal Structure of Ni Glycine Dihydrate," *J. Am. Chem. Soc.*, **67**, 365.

1946

Hughes, E. W., and Lipscomb, W. N., "The Crystal Structure of Methylammonium Chloride," *J. Am. Chem. Soc.*, **68**, 1970.

Jeffrey, G. A., "The Structure of Polyisoprenes I. The Crystal Structure of Geranylamine Hydrochloride," *Rubber. Chem. Technol.*, **19**, 351.

Perdok, W. G., and Terpstra, P. "Order-Disorder Transformations in the Lattice of Organic Molecules III. The Crystal Structure of Tetramethyl-Orthothiocarbonate above 23.2°," *Rec. Trav. Chim.*, **65**, 493.

Perutz, M. F., and Weisz, O., "Crystal Structure of Tribromo (Trimethylphosphine) Gold," *J. Chem. Soc.*, **1946**, 438.

Ubbelohde, A. R., and Woodward, I., "Structure and Thermal Properties of Crystals VI. The Role of Hydrogen Bonds in Rochelle Salt," *Proc. Roy. Soc. (London)*, **185A**, 448.

1947

Booth, A. D., "The Accuracy of Atomic Coordinates Derived from Fourier Series in X-Ray Structure Analysis IV. The Two Dimensional Projection of Oxalic Acid," *Proc. Roy. Soc. (London)*, **190A**, 490.

Booth, A. D., and Llewellyn, F. J., "The Crystal Structure of Pentaerythritol Tetranitrate," *J. Chem. Soc.*, **1947**, 837.

Bunn, C. W., and Garner, E. V., "The Crystal Structure of Two Polyamides ('Nylons')," *Proc. Roy. Soc. (London)*, **189A**, 39.

Dunitz, J. D., and Robertson, J. M., "The Crystal and Molecular Structure of Certain Dicarboxylic Acids I. Oxalic Acid Dihydrate," *J. Chem. Soc.*, **1947**, 142.

Dunitz, J. D., and Robertson, J. M., "The Crystal and Molecular Structure of Certain Dicarboxylic Acids II. Acetylenedicarboxylic Acid Dihydrate," *J. Chem. Soc.*, **1947**, 148.

Dunitz, J. D., and Robertson, J. M., "The Crystal and Molecular Structure of Certain Dicarboxylic Acids III. Diacetylenedicarboxylic Acid Dihydrate," *J. Chem. Soc.*, **1947**, 1145.

Rundle, R. E., and Sturdivant, J. H., "The Crystal Structure of Trimethylplatinum Chloride and Tetramethylplatinum," *J. Am. Chem. Soc.*, **69**, 1561.

Vand, V., Lomer, T. R., and Lang, A., "Crystal Structure of a Crossed-Chain Potassium Soap," *Nature*, **159**, 507.

1948

Beevers, C. A., and Stern, F., "Crystal Structure of D-Tartaric Acid," *Nature*, **162**, 854.

Bunn, C. W., "Crystal Structure of Polyvinyl Alcohol," *Nature*, **161**, 929.

Dawson, I. M., Mathieson, A. McL., and Robertson, J. M., "The Structure of Certain Polysulfides and Sulfonyl Sulfides I. A Preliminary X-Ray Survey," *J. Chem. Soc.*, **1948**, 322.

Dawson, I. M., and Robertson, J. M., "The Structure of Certain Polysulfides and Sulfonyl Sulfides II. The Crystal Structure of 2,2'-Diiododiethyl Trisulfide," *J. Chem. Soc.*, **1948**, 1256.

Doll, J., and Grison, E., "Crystallographic Study of Nitroguanidine," *Compt. Rend.*, **226**, 679.

Lewellyn, F. J., and Whitmore, F. E., "Crystal Structure of *S*-Bisnitraminoethane," *J. Chem. Soc.*, **1948**, 1316.

MacGillavry, C. H., Hoogschagen, G., and Sixma, F. L. J., "Crystal Structure of Glutaric Acid and Pimelic Acid. Alternation of Properties in the Series of Dicarboxylic Acids," *Rec. Trav. Chim.*, **67**, 869.

Muller, A., and Lonsdale, K., "The Low-Temperature Form of Octadecane," *Acta Cryst.*, **1**, 129.

Nitta, I., and Osaki, K., "The Crystal Structure of Calcium Formate," *X-Sen (X-Rays)*, **5**, 37.

Roth, W. L., and Harker, D., "The Crystal Structure of Octamethyl Spiro [5·5] Pentasiloxane," *Acta Cryst.*, **1**, 34.

1949

Bijvoet, J. M., "Phase Determination in Direct Fourier Synthesis of Crystal Structure," *Koninkl. Ned. Akad. Wetenschap. Proc.*, **52**, 513.

Binnie, W. P., and Robertson, J. M., "The Crystal Structure of Hexamethylenediamine and Its Dihalides. Hexamethylenediamine Dihydrobromide," *Acta Cryst.*, **2**, 116.

Binnie, W. P., and Robertson, J. M., "The Crystal Structure of Hexamethylenediamine Dihydrochloride," *Acta Cryst.*, **2**, 180.

Hughes, E. W., and Moore, W. J., "The Crystal Structure of β-Glycylglycine," *J. Am. Chem. Soc.*, **71**, 2618.

Morrison, J. D., and Robertson, J. M., "Crystal and Molecular Structure of Certain Dicarboxylic Acids IV. β-Succinic Acid," *J. Chem. Soc.*, **1949**, 980.

Morrison, J. D., and Robertson, J. M., "Crystal and Molecular Structure of Certain Dicarboxylic Acids V. Adipic Acid," *J. Chem. Soc.*, **1949**, 987.

Morrison, J. D., and Robertson, J. M., "Crystal and Molecular Structure of Certain Dicarboxylic Acids VI. Sebacic Acid," *J. Chem. Soc.*, **1949**, 993.

Morrison, J. D., and Robertson, J. M., "Crystal and Molecular Structure of Certain Dicarboxylic Acids VII. β-Glutaric Acid," *J. Chem. Soc.*, **1949**, 1001.

Nitta, I., and Saito, Y., "The Crystal Structure of Anhydrous Strontium Formate," *X-Sen (X-Rays)*, **5**, 89.

Sørum, H., and Foss, O., "Structure of bis(Methylsulfonyl) Disulfide," *Acta Chem. Scand.*, **3**, 987.

Vand, V., Lomer, T. R., and Lang, A., "The Crystal Structure of Form A of Potassium Caprate," *Acta Cryst.*, **2**, 214.

Watanabé, T., and Saito, T., "Polymorphism of Beryllium Oxyacetate," *Nature*, **163**, 225.

Watanabé, T., Saito, Y., and Koyama, H., "Order-Disorder Transition in Beryllium Oxyacetate Crystals," *Nature*, **164**, 1046.

Zachariasen, W. H., "Crystal Chemical Studies of the 5f-Series of Elements XII. New Compounds Representing Known Structure Types," *Acta Cryst.*, **2**, 388.

1950

Binnie, W. P., and Robertson, J. M., "The Crystal Structure of Hexamethylenediamine," *Acta Cryst.*, **3**, 424.

Carpenter, G. B., and Donohue, J., "The Crystal Structure of N-Acetylglycine," *J. Am. Chem. Soc.*, **72**, 2315.

Clark, G. L., and Hudgens, C. R., "Preliminary Results on the Crystal Structure of Some Ammonium Salts with Substituted Aliphatic Chains," *Science*, **112**, 309.

Donohue, J., "The Crystal Structure of DL-Alanine II. Revision of Parameters by Three-dimensional Fourier Analysis," *J. Am. Chem. Soc.*, **72**, 949.

Donohue, J., "The Crystal Structure of 2,2'-Diiododiethyl Trisulfide," *J. Am. Chem. Soc.*, **72**, 2701.

Geller, S., and Hoard, J. L., "Structures of Molecular Addition Compounds I. Mono-methylamine–Boron Trifluoride, $CH_3NH_2 \cdot BF_3$," *Acta Cryst.*, **3**, 121.

Hoard, J. L., Owen, T. B., Buzzell, A., and Salmon, O. N., "Structures of Molecular Addition Compounds II. Methyl Cyanide–Boron Trifluoride, $CH_3CN \cdot BF_3$," *Acta Cryst.*, **3**, 130.

King, M. V., and Lipscomb, W. N., "The Structure of the n-Propylammonium Halides at Room Temperature," *Acta Cryst.*, **3**, 222.

King, M. V., and Lipscomb, W. N., "The Low-Temperature Modification of n-Propylammonium Chloride," *Acta Cryst.*, **3**, 227.

Kondo, S., and Nitta, I., "Crystal Structure of Chloral Hydrate," *X-Sen (X-Rays)*, **6**, 53.

Niekirk, J. N. van, and Schoening, F. R. L., "Structure of Transpotassium Dioxalato Diaquo Chromiate," *Nature*, **166**, 108.

Rundle, R. E., and Snow, A. I., "Electron-Deficient Compounds V. The Structure of Dimethyl Beryllium," *J. Chem. Phys.*, **18**, 1125.

Sadanaga, R., "The Crystal Structure of Potassium Sodium *dl*-Tartrate Tetrahydrate, $KNaC_4H_4O_6 \cdot 4H_2O$," *Acta Cryst.*, **3**, 416.

Scouloudi, H., and Carlisle, C. H., "Crystal Structure of $[Cu(en)_2][Hg(SCN)_4]$," *Nature*, **166**, 357.

Shoemaker, D. P., Donohue, J., Schomaker, V., and Corey, R. B., "The Crystal Structure of L_s-Threonine," *J. Am. Chem. Soc.*, **72**, 2328.

Smith, A. E., "The Crystal Structure of Urea–Hydrocarbon and Thiourea–Hydrocarbon Complexes," *J. Chem. Phys.*, **18**, 150.

Stern, F., and Beevers, C. A., "The Crystal Structure of Tartaric Acid," *Acta Cryst.*, **3**. 341.

Vainshtein, B. K., and Pinsker, Z. G., "Determination of the Position of Hydrogen in the Crystal Lattice of Paraffin," *Dokl. Akad. Nauk SSSR*, **72**, 53.

1951

Bezzi, S., Bua, E., and Schiavinato, G., "Structural Analysis of Copper Dimethyl Glyoxime III. Preliminary Fourier-Bragg Analysis," *Gazz. Chim. Ital.*, **81**, 856.

Bierlein, T. K., and Lingafelter, E. C., "The Crystal Structure of Acetoxime," *Acta Cryst.*, **4**, 450.

Bua, E., and Schiavinato, G., "Structural Analysis of Copper Dimethyl Glyoxime," *Gazz. Chim. Ital.*, **81**, 212, 847.

Christ., C. L., "X-Ray Crystallography of Cyanamide, H_2NCN," *Acta Cryst.*, **4**, 77.

Clark, G. L., and Chu, C.-C., "X-Ray Diffraction Study of Several Branched-Chain Fatty Acids," *Acta Cryst.*, **4**, 470.

Dawson, B., and Mathieson, A. McL., "The Crystal Structure of Some Alpha-Amino Acids. A Preliminary X-Ray Examination," *Acta Cryst.*, **4**, 475.

Dyer, H. B., "The Crystal Structure of Cisteylglycine–Sodium Iodide," *Acta Cryst.*, **4**, 42.

Foss, O., Furberg, S., and Hadler, E., "The Crystal Structure of the Methane Thiosulfonates of Bivalent S, Se and Te," *Acta Chem. Scand.*, **5**, 1417.

Geller, S., and Hoard, J. L., "Structures of Molecular Addition Compounds IV. Trimethylamine–Boron Trifluoride," *Acta Cryst.*, **4**, 399.

Geller, S., Hughes, R. E., and Hoard, J. L., "Note on the Crystalline Structure of Trimethylamine–Borine, $(CH_3)_3N \cdot BH_3$," *Acta Cryst.*, **4**, 380.

Geller, S., and Salmon, O. N., "Note on the Crystalline Structures of Methyl Cyanide–Boron Trihalides," *Acta Cryst.*, **4**, 379.

Hach, R. J., and Rundle, R. E., "The Structure of Tetramethylammonium Pentaiodide," *J. Am. Chem. Soc.*, **73**, 4321.

Hoard, J. L., Geller, S., and Owen, T. B., "Structures of Molecular Addition Compounds V. Comparison of Four Related Structures," *Acta Cryst.*, **4**, 405.

Kitaigorodskii, A. I., Khotsyanova, T. L., and Struchkov, Y. T., "Crystal Structure of Iodoform," *Dokl. Akad. Nauk SSSR*, **78**, 1161.

Milberg, M. E., and Lipscomb, W. N., "Crystal Structure of 1,2-Dichloroethane at $-50°$," *Acta Cryst.*, **4**, 369.

Niekirk, J. N. van, and Schoening, F. R. L., "The Crystal Structure of *trans*-Potassium Dioxalato Diaquo Chromiate $K[Cr(C_2O_4)_2(H_2O)_2] \cdot 3H_2O$," *Acta Cryst.*, **4**, 35.

Niekirk, J. N. van, and Schoening, F. R. L., "Preliminary X-Ray Investigation of Potassium and Ammonium Trioxalato Chromate III," *Acta Cryst.*, **4**, 381.

Peerdeman, A. F., Bommel, A. J. van, and Bijvoet, J. M., "Determination of Absolute Configuration of Optically Active Compounds by Means of X-Rays," *Koninkl. Ned. Akad. Wetenschap. Proc.*, **54B**, 16.

Saito, Y., "The Transition which Occurs in Crystals of Beryllium Oxyacetate at about 40°," *Sci. Papers Osaka Univ.*, No. 36, 14 pp.

Snow, A. I., and Rundle, R. E., "Structure of Dimethyl Beryllium," *Acta Cryst.*, **4**, 348.

Sugawara, T., Kakudo, M., Saito, Y., and Nitta, I., "Crystal Structures of Rhombic Formates," Appendix *X-Sen*, **6**, 85.

Vand, V., and Bell, I. P., "A Direct Determination of the Crystal Structure of the β-Form of Trilaurin," *Acta Cryst.*, **4**, 465.

Vand, V., Morley, W. M., and Lomer, T. R., "Crystal Structure of Lauric Acid," *Acta Cryst.*, **4**, 324.

Watanabé, T., and Atoji, M., "Crystal Structure of Ethylene Diamine Complex," *Kagaku (Tokyo)*, **21**, 301.

1952

Cochran, W., and Penfold, B. R., "The Crystal Structure of L-Glutamine," *Acta Cryst.*, **5**, 644.

Cox, E. G., Dougill, M. W., and Jeffrey, G. A., "Structure of α-Oxalic Acid and of the Carboxyl Group," *J. Chem. Soc.*, **1952**, 4854.

Dreyfus-Alain, B., and Viallard, R., "Lattice Arrangement of Methanol Crystallized between 159° and 175°," *Compt. Rend.*, **234**, 536.

Hardt, H. D., and Hendus, H., "The Monoclinic Modification of Basic Beryllium Acetate," *Z. Anorg. Allgem. Chem.*, **270**, 298.

Hirokawa, S., Kuribayashi, S., and Nitta, I., "Crystal Structure of α-Aminoiso-Butyric Acid," *Bull. Chem. Soc. Japan*, **25**, 192.

Holtzberg, F., Post, B., and Fankuchen, I., "Crystal Structure of Formic Acid," *J. Chem. Phys.*, **20**, 198.

Jeffrey, G. A., and Parry, G. S., "Structure of the Oxalate Ion," *J. Chem. Soc.*, **1952**, 4864.

Jeffrey, G. A., and Rollett, J. S., "The Structure of Dimethyl Triacetylene," *Proc. Roy. Soc. (London)*, **213A**, 86.

Lomer, T. R., "The Unit Cell Dimensions of Potassium Soaps," *Acta Cryst.*, **5**, 11.

Lomer, T. R., "The Application of Booth's Method of Steepest Descents to the Determination of the Structure of Potassium Caproate," *Acta Cryst.*, **5**, 14.

Mathieson, A. McL., "The Crystal Structure of the Dimorphs of DL-Methionine," *Acta Cryst.*, **5**, 332.

Merritt, L. L., Jr., and Lanterman, E., "The Crystal Structure of Dimethyl Glyoxime," *Acta Cryst.*, **5**, 811.

Nakahara, A., Saito, Y., and Kuroya, H., "The Crystal Structure of *trans*-Dichloro-Diethylene Diamine Cobalt(III) Chloride Hydrochloride Dihydrate [Co(en)$_2$Cl$_2$]·Cl·HCl·2H$_2$O," *Bull. Chem. Soc. Japan*, **25**, 331.

Niekirk, J. N. van, and Schoening, F. R. L., "The Crystal Structure of Potassium Trioxalatochromate(III), K$_3$[Cr(C$_2$O$_4$)$_3$]·3H$_2$O," *Acta Cryst.*, **5**, 196.

Niekirk, J. N. van, and Schoening, F. R. L., "The Structure of Rubidium Trioxalato Chromate(III), Rb$_3$[Cr(C$_2$O$_4$)$_3$]·xH$_2$O, and its Relation to the Corresponding K and NH$_4$ Structures," *Acta Cryst.*, **5**, 475, 499.

Padmanabhan, V. M., "Space Group of Magnesium and Sodium Acetates," *Current Sci. (India)*, **21**, 97.

Saito, Y., "X-Ray Investigation of Thermal Transition in Some Molecular Crystals," X-Sen, **7**, 9.

Shahat, M., "The Crystal and Molecular Structure of Maleic Acid," Acta Cryst., **5**, 763.

Smith, A. E., "The Crystal Structure of the Urea–Hydrocarbon Complexes," Acta Cryst., **5**, 224.

Staritzky, E., and Singer, J., "Optical and X-Ray Data on a Group of Isostructural Uranium and Plutonium Compounds," Acta Cryst., **5**, 536.

Stenhagen, E., Vand, V., and Sim, A., "The Crystal Structure of Isopalmitic Acid," Acta Cryst., **5**, 695.

Struchkov, Y. T., Kitaigorodskii, A. I., and Khotsyanova, T. L., "The Crystal Structure of Tris (2-Chlorovinyl) Dichlorostibines," Zh. Fiz. Khim., **26**, 530.

Sugawara, T., and Kanda, E., "The Crystal Structure of Acetylene I," Sci. Rept. Res. Inst. Tohoku Univ., **4A**, 607.

Tauer, K. J., and Lipscomb, W. N., "The Crystal Structures, Residual Entropy and Dielectric Anomaly of Methanol," Acta Cryst., **5**, 606.

Vaughan, P., and Donohue, J., "The Structure of Urea. Interatomic Distances and Resonance in Urea and Related Compounds," Acta Cryst., **5**, 530.

Yakel, H. L., Jr., and Hughes, E. W., "The Structure of N,N'-Diglycyl-L-Cystine Dihydrate," J. Am. Chem. Soc., **74**, 6302.

Zaslow, B., Atoji, M., and Lipscomb, W. N., "The Crystal Structure of Phosgene," Acta Cryst., **5**, 833.

1953

Abe, H., and Shimada, J., "Anomalous Magnetic Resonance Absorption of Copper Acetate at 40 K Mc/sec," Phys. Rev., **90**, 316.

Ahmed, F. R., and Cruickshank, D. W. J., "A Refinement of the Crystal Structure Analyses of Oxalic Acid Dihydrate," Acta Cryst., **6**, 385.

Allentoff, N., and Wright, G. F., "Disodium Ethane-bis-Nitraminate," Acta Cryst., **6**, 1.

Atoji, M., and Lipscomb, W. N., "The Crystal Structures of Methyl Amine," Acta Cryst., **6**, 770.

Atoji, M., Oda, T., and Watanabé, T., "Crystal Structure of Cubic Hexachlorethane," Acta Cryst., **6**, 868.

Bommel, A. J. van, "Crystal Structure of D-Rubidium Acid Tartrate and Its Absolute Configuration," Koninkl. Ned. Akad. Wetenschap. Proc., **56B**, 268.

Burbank, R. D., "The Crystal Structure of Methyl Chloride at $-125°C$.," J. Am. Chem. Soc., **75**, 1211.

Dawson, B., "The Crystal Structure of dl-Glutamic Acid Hydrochloride," Acta Cryst., **6**, 81.

Dougill, M. W., and Jeffrey, G. A., "The Structure of Dimethyl Oxalate," Acta Cryst., **6**, 831.

Foss, O., "Rotational Isomers of Pentathionic Compounds in Crystals," Acta Chem. Scand., **7**, 1221.

Godycki, L. E., and Rundle, R. E., "The Symmetry of Nickel Dimethylglyoxime," Acta Cryst., **6**, 487.

Gordon, M., Stenhagen, E., and Vand, V., "The Crystal Structure of n-Dodecyl-Ammonium Chloride and Bromide," Acta Cryst., **6**, 739.

Hannan, R. B., and Collin, R. L., "The Crystal Structure of Dicyanoacetylene," Acta Cryst., **6**, 350.

Hinch, R. J., Jr., and McCrone, W. C., "Succinamide," *Anal. Chem.*, **25**, 675.

Holtzberg, F., Post, B., and Fankuchen, I., "Crystal Structure of Formic Acid," *Acta Cryst.*, **6**, 127.

Jarvis, J. A. J., "Crystal Structure of Potassium Ethyl Sulfate," *Acta Cryst.*, **6**, 327.

Jensen, L. H., and Lingafelter, E. C., "Crystal Structures of *n*-Aliphatic Acid Hydrazides," *Acta Cryst.*, **6**, 300.

Khotsyanova, T. L., Kitaigorodskii, A. I., and Struchkov, Y. T., "Crystal Structure of Iodoform," *Zh. Fiz. Khim.*, **27**, 647.

Lewis, P. H., and Rundle, R. E., "Electron-Deficient Compounds VII. The Structure of the Trimethyl Aluminum Dimer," *J. Chem. Phys.*, **21**, 986.

Littleton, C. D., "A Structure Determination of the Gluconate Ion," *Acta Cryst.*, **6**, 775.

Mathieson, A. McL., "Polymorphism of *dl*-Norleucine," *Acta Cryst.*, **6**, 399.

Niekirk, J. N. van, and Schoening, F. R. L., "A New Type of Copper Complex as Found in the Crystal Structure of Cupric Acetate, $Cu_2(CH_3COO)_4 \cdot 2H_2O$," *Acta Cryst.*, **6**, 227.

Niekirk, J. N. van, and Schoening, F. R. L., "The Crystal Structures of Nickel Acetate and Cobalt Acetate," *Acta Cryst.*, **6**, 609.

Niekirk, J. N. van, and Schoening, F. R. L., "X-Ray Evidence for Metal-to-Metal Bonds in Cupric and Chromous Acetate," *Nature*, **171**, 36.

Niekirk, J. N. van, Schoening, F. R. L., and Talbot, J. H., "The Crystal Structure of Zinc Acetate Dihydrate, $Zn(CH_3COO)_2 \cdot 2H_2O$," *Acta Cryst.*, **6**, 720.

Niekirk, J. N. van, Schoening, F. R. L., and Wet, J. F. de, "The Structure of Crystalline Chromous Acetate Revealing Paired Chromium Atoms," *Acta Cryst.*, **6**, 501.

Pasternak, R. A., "The Crystal Structure of Succinamide, $(CH_2CONH_2)_2$," *Acta Cryst.*, **6**, 808.

Reed, T. B., and Lipscomb, W. N., "The Crystal and Molecular Structure of 1,2-Dichloroethane at $-140°C$.," *Acta Cryst.*, **6**, 45.

Romers, C., "The Structure of Oxamide," *Acta Cryst.*, **6**, 429.

Sasada, Y., and Atoji, M., "Crystal Structure and Lattice Energy of Orthorhombic Hexachloroethane," *J. Chem. Phys.*, **21**, 145.

Scouloudi, H., "The Crystal Structure of Mercury Tetrathiocyanate-Copper Diethylene Diamine $[Hg(SCN)_4][Cu(en)_2]$," *Acta Cryst.*, **6**, 651.

Shoemaker, D. P., Barieau, R. E., Donohue, J., and Lu, C., "The Crystal Structure of *dl*-Serine," *Acta Cryst.*, **6**, 241.

Simonsen, S. H., and Ho, J. W., "The Unit Cell Dimensions and Space Group of Zinc Diethyl Dithiocarbamate," *Acta Cryst.*, **6**, 430.

Sørum, H., "The Crystal Structure of Dimethane Sulfonyl Disulfide," *Acta Chem. Scand.*, **7**, 1.

Struchkov, Y. T., and Khotsyanova, T. L., "X-Ray Investigation of *trans, trans, trans,* tris (2-Chlorovinyl) Dichlorostibine," *Dokl. Akad. Nauk SSSR*, **91**, 565.

Trommel, J., "Crystal Structure of $d(-)$ Isoleucine Hydrochloride Monohydrate I. The [100] Projection," *Koninkl. Ned. Akad. Wetenschap., Proc.*, **56B**, 272.

Vand, V., "Density and Unit Cell of Hexatriacontane," *Acta Cryst.*, **6**, 797.

1954

Abrahamson, S., and Sydow, E. v., "Variation of Unit Cell Dimensions of a Crystal Form of Long Normal Chain Carboxylic Acids," *Acta Cryst.*, **7**, 591.

Ayerst, E. M., and Duke, J. R. C., "Refinement of the Crystal Structure of Oxamide," *Acta Cryst.*, **7**, 588.

Burns, D. M., and Iball, J., "Unit Cells and Space Groups of Citric Acid and Some Potassium and Sodium Citrates," *Acta Cryst.*, **7**, 137.

Cromer, D. T., and Kline, R. J., "The Crytal Structure of $(Me_4N)_2CeCl_6$," *J. Am. Chem. Soc.*, **76**, 5282.

Foss, O., and Vihovde, E. H., "Crystal Structure of Tellurium Dimethane-Sulfonate," *Acta Chem. Scand.*, **8**, 1032.

Gallagher, K. J., Ubbelohde, A. R., and Woodward, I., "The Hydrogen Bond in Crystals IX. The Isotope Effect in Acetylene Dicarboxylic Acid Dihydrate," *Proc. Roy. Soc. (London)*, **222A**, 195.

Garrett, B. S., "Crystal Structures of Oxalic Acid Dihydrate and α-Iodic Acid by Neutron Diffraction," *U.S. At. Energy Comm.*, ORNL-1745, 149 pp.

Hirokawa, S., Ohashi, T., and Nitta, I., "The Crystal Structure of Some Polymethylene Diammonium Adipates I. Hexamethylene Diammonium Adipate," *Acta Cryst.*, **7**, 87.

Jeffrey, G. A., and Parry, G. S., "The Crystal Structure of Sodium Oxalate," *J. Am. Chem. Soc.*, **76**, 5283.

Kasai, N., and Kakudo, M., "The Crystal Structure of Diallylsilanediol," *Bull. Chem. Soc. Japan*, **27**, 605.

Kiriyama, R., Ibamoto, H., and Matsuo, K., "The Crystal Structure of Cupric Formate Tetrahydrate, $Cu(HCOO)_2 \cdot 4H_2O$," *Acta Cryst.*, **7**, 482.

Koyama, H., and Saito, Y., "Crystal Structure of Zinc Oxyacetate $Zn_4O(CH_3COO)_6$," *Bull. Chem. Soc. Japan*, **27**, 112.

Ladell, J., and Post, B., "The Crystal Structure of Formamide," *Acta Cryst.*, **7**, 559.

Lazzarini, E., and Mazzi, F., "Crystallographic and Structural Investigations on the Cupric Bromide–Methylamine Complex Compound," *Periodico Mineral. (Rome)*, **23**, 195.

Mendel, H., and Hodgkin, D. C., "The Crystal Structure of Creatine Monohydrate," *Acta Cryst.*, **7**, 443.

Nyburg, S. C., "A Statistical Structure for Crystalline Rubber," *Acta Cryst.*, **7**, 385.

Pasternak, R. A., Katz, L., and Corey, R. B., "The Crystal Structure of Glycyl-L-Asparagine," *Acta Cryst.*, **7**, 225.

Pringle, G. E., "X-Ray Analysis of Hydrogen Bonding in the Structure of Oxalic Acid Dihydrate," *Acta Cryst.*, **7**, 716.

Semenenko, K. N., Simanov, Y. P., and Novoselova, A. V., "Basic Beryllium Acetate I. Monoclinic High Temperature Modification of Basic Beryllium Acetate," *Vestn. Mosk. Univ.*, **9**, No. 2; *Ser. Fiz. Mat. Estestv. Nauk*, No. 1, 61.

Sprenkels, A. J. J., "The Crystal Structure of Some Tartrates and Corresponding Racemates," *Koninkl. Ned. Akad. Westenschap. Proc.*, **57B**, 524.

Sutor, D. J., Calvert, L. D., and Llewellyn, F. J., "The Crystal Structure of β-Nitropropionic Acid," *Acta Cryst.*, **7**, 767.

Sutor, D. J., Llewellyn, F. J., and Maslen, H. S., "The Crystal Structure of Dipotassium Nitroacetate," *Acta Cryst.*, **7**, 145.

Sydow, E. v., "The Structure of the Crystal Form A′ of Pentadecanoic Acid," *Acta Cryst.*, **7**, 529.

Trommel, J., "Crystal Structure of $D(-)$ Isoleucine Hydrochloride Monohydrate II. The (001) Projection and Final Structure," *Koninkl. Ned. Akad. Wetenschap. Proc.*, **57B**, 364.

Trommel, J., and Bijvoet, J. M., "Crystal Structure and Absolute Configuration of the Hydrochloride and Hydrobromide of $D(-)$ Isoleucine," *Acta Cryst.*, **7**, 703.

Wheatley, P. J., "The Stereochemistry of Molecules Containing the C:C:N Group: the Crystal Structure of N-Methyl-2,2-bis (Methyl Sulfonyl) Vinylidineamine," *Acta Cryst.*, **7**, 68.

Wunderlich, J. A., and Mellor, D. P., "The Crystal Structure of Zeise's Salt," *Acta Cryst.*, **7**, 130.

Yakel, H. L., Jr., and Hughes, E. W., "The Crystal Structure of N,N'-Diglycyl-L-Cystine Dihydrate," *Acta Cryst.*, **7**, 291.

1955

Amirthalingam, V., and Ramachandran, G. N., "Structure of DL-Aspartic Acid," *Current Sci.* (*India*), **24**, 294.

Bailey, M., "The Crystal Structure of N,N'-Diacetylhexamethylenediamine," *Acta Cryst.*, **8**, 575.

Becker, K. A., Grosse, G., and Plieth, K., "X-Ray Structure Study of $\frac{1}{6}$(CoCl$_2$(en)$_2$)-Cl," *Naturwissenschaften*, **42**, 254.

Curtis, R. M., and Pasternak, R. A., "The Crystal Structure of Methylguanidinium Nitrate," *Acta Cryst.*, **8**, 675.

Dejace, J., "Crystal Structure of Chloroacetamide," *Acta Cryst.*, **8**, 851.

Dempsey, J. N., and Baenziger, N. C., "The Crystal Structure of an Ethylene–Palladium Chloride Complex," *J. Am. Chem. Soc.*, **77**, 4984.

Dohlen, W. C. v., and Carpenter, G. B., "The Crystal Structure of Isocyanic Acid," *Acta Cryst.*, **8**, 646.

Hamilton, W. C., "The Crystal Structure of Dimethylphosphinoborine Trimer," *Acta Cryst.*, **8**, 199.

Havinga, E. E., and Wiebenga, E. H., "The Crystal Structure of N(C$_2$H$_5$)$_4$I$_7$," *Koninkl. Ned. Akad. Westenschap. Proc.*, **58B**, 412.

Hirokawa, S., "A New Modification of L-Glutamic Acid and Its Crystal Structure," *Acta Cryst.*, **8**, 637.

Holmes, D. R., Bunn, C. W., and Smith, D. J., "The Crystal Structure of Polycapro-amide: Nylon 6," *J. Polymer Sci.*, **17**, 159.

James, W. H., Hach, R. J., French, D., and Rundle, R. E., "The Structure of Tetra-methylammonium Enneaiodide," *Acta Cryst.*, **8**, 814.

Jensen, L. H., "The Crystal Structure of Creatine Monohydrate," *Acta Cryst.*, **8**, 237.

Kakudo, M., "The Crystal Structure of 2,4-Dihydroxy-2,4-Dimethyl-2,4-Disilapentane," *Technol. Rept. Osaka Univ.*, **5**, 211.

Pignataro, E., and Post, B., "The Crystal Structure of Dimethyl Acetylene at −50°C.," *Acta Cryst.*, **8**, 672.

Saito, Y., Cano-Corona, O., and Pepinsky, R., "X-Ray Examination of Molecular Configuration of Asparagine in Crystalline L-Asparagine Monohydrate," *Science*, **121**, 435.

Saito, Y., Nakatsu, K., Shiro, M., and Kuroya, H., "Determination of the Absolute Configuration of the Optically Active Complex Ion [Co(en)$_3$]$^{3+}$ by Means of X-Rays," *Acta Cryst.*, **8**, 729.

Sim, G. A., "The Crystal Structure of 11-Amino-Undecanoic Acid Hydrobromide Hemihydrate," *Acta Cryst.*, **8**, 833.

Steinfink, H., Post, B., and Fankuchen, I., "The Crystal Structure of Octamethyl Cyclotetrasiloxane," *Acta Cryst.*, **8**, 420.

Sydow, E. v., "On the Structure of the Crystal Form B of Stearic Acid," *Acta Cryst.*, **8**, 557.

Sydow, E. v., "The Structure of the Crystal Form C′ of n-Hendecanoic Acid," *Acta Cryst.*, **8**, 810.

Sydow, E. v., "On the Structure of the Crystal Form A′ of n-Pentadecanoic Acid," *Acta Cryst.*, **8**, 845.

Truter, M. R., "An Accurate Determination of the Structure of Sodium Hydroxy-methanesulphinate (Rongalite)," *J. Chem. Soc.*, **1955**, 3064.

Turner, J. D., and Lingafelter, E. C., "The X-Ray Crystallography of the n-Aliphatic Amides," *Acta Cryst.*, **8**, 549.

Turner, J. D., and Lingafelter, E. C., "The Crystal Structure of Tetradecanamide," *Acta Cryst.*, **8**, 551.

Wunderlich, J. A., and Mellor, D. P., "A Correction and a Supplement to a Note on the Crystal Structure of Zeise's Salt," *Acta Cryst.*, **8**, 57.

1956

Abrahamsson, S., "On the Crystal Structure of 9DL-Methyloctadecanoic Acid," *Acta Cryst.*, **9**, 663.

Blakeslee, A. E., and Hoard, J. L., "The Structure of Silver Perfluorobutyrate," *J. Am. Chem. Soc.*, **78**, 3029.

Bryden, J. H., Burkhardt, L. A., Hughes, E. W., and Donohue, J., "The Crystal Structure of Nitroguanidine," *Acta Cryst.*, **9**, 573.

Buerger, M. J., Barney, E., and Hahn, T., "The Crystal Structure of Diglycine Hydrobromide," *Z. Krist.*, **108**, 130.

Cavalca, L., Nardelli, M., and Braibanti, A., "The Crystal Structure of Tetrathiourea-nickel(II) Chloride," *Gazz. Chim. Ital.*, **86**, 942.

Corsmit, A. F., Schuyff, A., and Feil, D., "The Crystal Structure of L-Cystine Hydrochloride," *Koninkl. Ned. Akad. Wetenschap. Proc.*, **59B**, 470.

Davies, D. R., and Pasternak, R. A., "A Refinement of the Crystal Structure of Succinamide," *Acta Cryst.*, **9**, 334.

Foss, O., and Tjomsland, O., "Solvates of Barium Pentathionates with Acetone and Tetrahydrofuran," *Acta Chem. Scand.*, **10**, 424.

Grenville-Wells, H. J., "Anisotropic Temperature Vibrations in Crystals II. The Effect of Changes in Atomic Scattering Factors and Temperature Parameters on the Accuracy of the Determination of the Structure of Urea," *Acta Cryst.*, **9**, 709.

Hall, D., and Llewellyn, F. J., "The Crystal Structure of Formamidoxime," *Acta Cryst.*, **9**, 108.

Jensen, L. H., "The Crystal Structure of n-Dodecanoic Acid Hydrazide," *J. Am. Chem. Soc.*, **78**, 3993.

Jones, R. E., "The Crystal Structure of Acetic Acid," *U.S. At. Energy Comm.*, UCRL-3641, 18 pp.

Katayama, M., "The Crystal Structure of an Unstable Form of Chloroacetamide," *Acta Cryst.*, **9**, 986.

Moore, E. B., Jr., and Lipscomb, W. N., "The Crystal and Molecular Structure of $Cl_2BC_2H_4BCl_2$," *Acta Cryst.*, **9**, 668.

Nakatsu, K., Saito, Y., and Kuroya, H., "Crystals of Metallic tris(Ethylenediamine) Complexes I. The Crystal Structure of DL-tris(Ethylenediamine)Cobalt(III) Chloride Trihydrate, $[Co(en)_3]Cl_3 \cdot 3H_2O$," *Bull. Chem. Soc. Japan*, **29**, 428.

Nardelli, M., Cavalca, L., and Braibanti, A., "Fourfold Coordination Complexes of Bivalent Metals with Thiourea." *Gazz. Chim. Ital.*, **86**, 867.

Natta, G., and Corradini, P., "The Structure of 1,2-Polybutadiene and of Other 'Syndiotactic Polymers', " *J. Polymer Sci.*, **20**, 251.

Natta, G., and Corradini, P., "The Crystal Structure of 1,4-*cis*-Polybutadiene and of 1,4-*cis*-Polyisoprene," *Angew. Chem.*, **68**, 615.

Natta, G., Corradini, P., and Bassi, I. W., "The Crystal Structure of the Isotactic Poly-α-Butene," *Makromol. Chem.*, **21**, 240.

Natta, G., Corradini, P., and Cesari, M., "The Crystal Structure of Isotactic Polypropylene," *Atti Accad. Nazl. Lincei Rend. Classe Sci. Fis. Mat. Nat.*, **21**, 365.

Penfold, B. R., and Simpson, W. S., "The Crystal Structure of Chloroacetamide," *Acta Cryst.*, **9**, 831.

Roof, R. B., Jr., "The Crystal Structure of Ferric Acetylacetonate," *Acta Cryst.*, **9**, 781.

Senko, M. E., "Crystal Structure of a Triazole and Choline Chloride," *U.S. At. Energy Comm.*, UCRL-3521, 47 pp.

Shearer, H. M. M., and Vand, V., "The Crystal Structure of the Monoclinic Form of *n*-Hexatriacontane," *Acta Cryst.*, **9**, 379.

Shugam, E. A., and Shkolnikova, L. M., "The Crystal Structure of Acetonylacetonates of Aluminum and Chromium," *Kristallografiya*, **1**, 478.

Sprenkels, A. J. J., "Crystal Structure of Lithium Ammonium Tartrate Monohydrate," *Koninkl. Ned. Akad. Wetenschap. Proc.*, **59B**, 221.

Sydow, E. v., "The Normal Fatty Acids in Solid State, a Crystal Structure Investigation," *Arkiv Kemi*, **9**, 231.

Sydow, E. v., "Structure of Crystal Form A of Lauric Acid," *Acta Chem. Scand.*, **10**, 1.

Zhdanov, G. S., Umanskii, M. M., Barfolomeeva, L. A., Ezhkova, Z. I., and Zolina, Z. K., "X-Ray Determination of the Elementary Cells and Space Groups of Piezoelectric Crystals," *Kristallografiya*, **1**, 271.

1957

Broekema, J., Havinga, E. E., and Wiebenga, E. H., "Refinement of the Crystal Structure of N(CH₃)₄I₅," *Acta Cryst.*, **10**, 596.

Bryden, J. H., "The Crystal Structure of Aminoguanidine Hydrochloride," *Acta Cryst.*, **10**, 677.

Bryden, J. H., "The Crystal Structure of N-Methylurea Nitrate," *Acta Cryst.*, **10**, 714.

Burns, J. H., and Waser, J., "The Crystal Structure of Arsenomethane," *J. Am. Chem. Soc.*, **79**, 859.

Dejace, J., "Crystalline Structure of Chloracetamide," *Acta Cryst.*, **10**, 240.

Ferrari, A., Nardelli, M., and Tani, E., "Structural Relations between Alkali Metal Uranyl Acetates and Propionates," *Gazz. Chim. Ital.*, **87**, 1203.

Foss, O., and Hordvik, A., "Nature of the Sulphur–Sulphur Bond in Thiosulphate and Thiosulphonate Ions," *Acta Chem. Scand.*, **11**, 1443.

Foss, O., and Johnsen, J., "Structure of Formamidine Disulphide," *Acta Chem. Scand.*, **11**, 189.

Goedkoop, J. A., and MacGillavry, C. H., "The Crystal Structure of Malonic Acid," *Acta Cryst.*, **10**, 125.

Hahn, T., and Buerger, M. J., "The Crystal Structure of Diglycine Hydrochloride, 2(C₂H₅O₂N)·HCl," *Z. Krist.*, **108**, 419.

Hulme, R., and Powell, H. M., "Structure of Compounds of Ferrocyanide Type II. Crystal Structure of β-Tetramethyl Ferrocyanide," *J. Chem. Soc.*, **1957**, 719.

Jensen, L. H., Krimm, S., Parrish, R. G., and Wood, D. L., "The Crystal Structure of *N,N′*-Hexamethylenebispropionamide," *Acta Cryst.*, **10**, 528.

Katayama, M., "Polymorphism of Chloroacetamide," *Acta Cryst.*, **10**, 468.

Mazzi, F., and Garavelli, C., "Structure of Oxalite, $FeC_2O_4 \cdot 2H_2O$," *Periodico Mineral.* (*Rome*), **26**, 269.

Mazzi, F., Jona, F., and Pepinsky, R., "Preliminary X-Ray Study of the Non-Ferro-electric Phases of Rochelle Salt," *Z. Krist.*, **108**, 359.

Nakatsu, K., Shiro, M., Saito, Y., and Kuroya, H., "Studies on Crystals of Metallic tris-Ethylenediamine-Complexes II. The Crystal Structure of Sodium D-tris-Ethylenediamine-Cobalt(III) Chloride Hexahydrate, $2D\text{-}[Co(en)_3] \cdot NaCl \cdot 6H_2O$," *Bull. Chem. Soc. Japan*, **30**, 158.

Nardelli, M., Braibanti, A., and Fava, G., "Structure of Nickel Thiocyanate-bis-Thiourea," *Gazz. Chim. Ital.*, **87**, 1209.

Nardelli, M., Cavalca, L., and Braibanti, A., "Structure of bis(Thiourea) Cadmium Chloride," *Gazz. Chim. Ital.*, **87**, 137.

Nardelli, M., Cavalca, L., and Braibanti, A., "Thiocyanate Complexes of Bivalent Metals with Thiourea," *Gazz. Chim. Ital.*, **87**, 917.

Nardelli, M., Cavalca, L., and Fava, G., "Structure of Cadmium Chloride Diurea," *Gazz. Chim. Ital.*, **87**, 1232.

Natta, G., Corradini, P., and Bassi, I. W., "The Crystalline Structure of Isotactic Poly(1,2-Butadiene)," *Atti Accad. Nazl. Lincei Rend. Classe Sci. Fis. Mat. Nat.*, **23**, 363.

Okaya, Y., Ahmed, M. S., Pepinsky, R., and Vand, V., "X-Ray Crystallographic Study of Room Temperature Modification of Monomethylammonium Aluminum Sulfate Alum, $(CH_3NH_3)[Al(H_2O)_6](SO_4)_2 \cdot 6H_2O$," *Z. Krist.*, **109**, 367.

Okaya, Y., and Pepinsky, R., "Crystal Structure of Triaminoguanidinium Chloride, $(NH_2 \cdot NH)_3C \cdot Cl$," *Acta Cryst.*, **10**, 681.

Oughton, B. M., and Harrison, P. M., "The Crystal Structure of Hexagonal L-Cystine," *Acta Cryst.*, **10**, 479.

Rérat, C., "Crystallographic Study of Trimethylamine *N*-Oxide Hydrochloride," *Compt. Rend.*, **245**, 704.

Saito, Y., Nakatsu, K., Shiro, M., and Kuroya, H., "Studies on Crystals of Metallic tris-Ethylenediamine Complexes III. The Determination of the Absolute Configuration of Optically Active Complex Ion $[Co(en)_3]^{3+}$ by Means of X-Rays," *Bull. Chem. Soc. Japan*, **30**, 795.

Shankar, K., Khubchandani, P. G., and Padmanabhan, V. M., "The Crystal Structure of Magnesium Acetate-Tetrahydrate $Mg(CH_3COO)_2 \cdot 4H_2O$," *Proc. Indian Acad. Sci.*, **45A**, 117.

Sundera Rao, R. V. G., Turley, J. W., and Pepinsky, R., "The Crystal Structure of Urea Phospate," *Acta Cryst.*, **10**, 435.

Wood, E. A., and Holden, A. N., "Monoclinic Glycine Sulfate: Crystallographic Data," *Acta Cryst.*, **10**, 145.

Worsham, J. E., Jr., Levy, H. A., and Peterson, S. W., "The Positions of Hydrogen Atoms in Urea by Neutron Diffraction," *Acta Cryst.*, **10**, 319.

Zalkin, A., "The Crystal Structure of Tetra-*n*-Propyl Ammonium Bromide," *Acta Cryst.*, **10**, 557.

1958

Abrahamsson, S., "On the Crystal Structure of 16DL-Methyloctadecanoic Acid," *Acta Cryst.*, **11**, 270.

Amirthalingam, V., and Padmanabhan, V. M., "The Crystal Structure of Lithium Acetate Dihydrate, $CH_3COOLi \cdot 2H_2O$," *Acta Cryst.*, **11**, 896.

Amma, E. L., and Rundle, R. E., "Electron Deficient Compounds VIII. The Crystal and Molecular Structure of Trimethylindium," *J. Am. Chem. Soc.*, **80**, 4141.

Bommel, A. J. van, and Bijvoet, J. M., "The Crystal Structure of Ammonium Hydrogen D-Tartrate," *Acta Cryst.*, **11**, 61.

Brathovde, J. R., and Lingafelter, E. C., "The Crystal Structure of Decanamide," *Acta Cryst.*, **11**, 729.

Bringeland, R., and Foss, O., "The Crystal and Molecular Structure of Ethylene Thiocyanate," *Acta Chem. Scand.*, **12**, 79.

Bryden, J. H., "The Crystal Structure of Azo-bis-N-Chloroformamidine," *Acta Cryst.*, **11**, 158.

Christofferson, G. D., Sparks, R. A., and McCullough, J. D., "The Crystal Structure of α-Dimethyltellurium Dichloride," *Acta Cryst.*, **11**, 782.

Dodge, R. P., "The Crystal Structure of Vanadyl Bisacetylacetonate," *U.S. At. Energy Comm.*, UCRL-8225, 27 pp.

Foss, O., Johnsen, J., and Tvedten, O., "The Constitution of the Formamidinium Disulphide Ion from the Crystal Structures of the Diiodide and Dibromide," *Acta Chem. Scand.*, **12**, 1782.

Foss, O., and Tjomsland, O., "Structure of a Solvate of Barium Pentathionate with Acetone," *Acta Chem. Scand.*, **12**, 44.

Frasson, E., Bardi, R., Zannetti, R., and Mammi, M., "Use of Low Temperature and a Particular Fourier Series for Refinement of Complex Organic Structures. Structure of Copper Dimethyl Glyoxime," *Ann. Chim. (Rome)*, **48**, 1007.

Frasson, E., Zannetti, R., Bardi, R., Bezzi, S., and Giacometti, G., "Nonplanar Coordination in a Cu^{2+} Complex," *J. Inorg. Nucl. Chem.*, **8**, 452.

Grdenić, D., and Matković, B., "Coordination in Thorium(IV) Acetylacetonate," *Nature*, **182**, 465.

Hall, D., and Woulfe, M. D., "The Structure of the Nickel-$\beta\beta'\beta''$-Triaminotriethylamine Complex," *Proc. Chem. Soc. (London)*, **1958**, 346.

Havinga, E. E., and Wiebenga, E. H., "The Crystal Structure of $N(C_2H_5)_4I_7$ at $-175°C.$," *Acta Cryst.*, **11**, 733.

Herpin, P., "Crystalline Structure of the Complex Trioxalates of Potassium II. Structure of Racemic Potassium Trisoxalatoferrate(III) and Comparison with the Complex Active Trioxalates," *Bull. Soc. Franc. Minéral. Crist.*, **81**, 245.

Hirokawa, S., and Ashida, T., "A Remark on the Crystal Structure of α-Amino-Isobutyric Acid," *Bull. Chem. Soc. Japan*, **31**, 142.

Hvoslef, J., "A Neutron Diffraction Study of Pentaerythritol," *Acta Cryst.*, **11**, 383.

Iitaka, Y., "The Crystal Structure of γ-Glycine," *Acta Cryst.*, **11**, 225.

Jellinek, F., "Crystal Structure of the Low Temperature Form of Monoethylamine Hydrobromide," *Acta Cryst.*, **11**, 626.

Jones, R. E., and Templeton, D. H., "The Crystal Structure of Acetic Acid," *Acta Cryst.*, **11**, 484.

Kay, M. I., and Katz, L., "The Crystal Structure of α-Pimelic Acid," *Acta Cryst.*, **11**, 289.

Kunchur, N. R., and Truter, M. R., "A Detailed Refinement of the Crystal and Molecular Structure of Thiourea," *J. Chem. Soc.*, **1958**, 2551.

Kunchur, N. R., and Truter, M. R., "The Crystal Structure of Dichlorobisthioureazinc," *J. Chem. Soc.*, **1958**, 3478.

Marsh, R. E., "A Refinement of the Crystal Structure of Glycine," *Acta Cryst.*, **11**, 654.

Nardelli, M., and Fava, G., "Structure of bis Thiourealead(II) Chloride," *Gazz. Chim. Ital.*, **88**, 536.

Osaki, K., "The Crystal Structure of Strontium Formate Dihydrate, $Sr(HCOO)_2 \cdot 2H_2O$," *Ann. Rept. Sci. Works, Fac. Sci. Osaka Univ.*, **6**, 13.

Padmanabhan, V. M., "The Crystal Structure of Cobaltic Acetylacetonate, $Co(C_5H_7O_2)_3$," *Proc. Indian Acad. Sci.*, **47A**, 329.

Peterson S. W., and Levy, H. A., "Structure of Potassium Hydrogen Maleate by Neutron Diffraction," *J. Chem. Phys.*, **29**, 948.

Rasmussen, S. E., "On the Crystal and Molecular Structure of Ni-"tren" $(SCN)_2$," *J. Inorg. Nucl. Chem.*, **8**, 441.

Scatturin, V., and Turco, A., "Crystal Structure of $NiBr_2 \cdot 2P(C_2H_5)_3$ and Preliminary X-Ray Data on $Ni(NO_3)_2 \cdot 2P(C_2H_5)_3$ and $NiBr_3 \cdot 2P(C_2H_5)_3$," *J. Inorg. Nucl. Chem.*, **8**, 447.

Shallcross, F. V., and Carpenter, G. B., "The Crystal Structure of Cyanoacetylene," *Acta Cryst.*, **11**, 490.

Shiono, R., Cruickshank, D. W. J., and Cox, E. G., "A Refinement of the Crystal Structure of Pentaerythritol," *Acta Cryst.*, **11**, 389.

Shugam, E. A., and Shkolnikova, L. M., "Internal Complex Compounds Containing the Me—S Bond," *Kristallografiya*, **3**, 749.

Steinrauf, L. K., Peterson, J., and Jensen, L. H., "The Crystal Structure of L-Cystine Hydrochloride," *J. Am. Chem. Soc.*, **80**, 3835.

Sundara Rao, R. V. G., Sundaramma, K., and Sivasankara Rao, G., "The Structure of Dibarium Cupric Formate Tetrahydrate, $CuBa_2(COOH)_6 \cdot 4H_2O$," *Z. Krist.*, **110**, 231.

Thomas, J. T., Robertson, J. H., and Cox, E. G., "The Crystal Structure of Roussin's Red Ethyl Ester," *Acta Cryst.*, **11**, 599.

Tomiie, Y., Koo, C.-H., and Nitta, I., "The Crystal and Molecular Structure of Diformylhydrazine, OHC—HN—NH—CHO I. The X-Ray Analysis," *Acta Cryst.*, **11**, 774.

Truter, M. R., "A Detailed Refinement of the Crystal Structure of Potassium Ethyl Sulphate," *Acta Cryst.*, **11**, 680.

Varfolomeeva, L. A., Zhdanov, G. S., and Umanskii, M. M., "Principle of Determination of the Structure of Isomorphous Compounds of the Type $[C(NH_2)_3][M(H_2O)_6]-[EO_4]_2$, M = Al, Cr, E = S, Se," *Kristallografiya*, **3**, 368.

Wait, E., and Powell, H. M., "The Crystal and Molecular Structure of Tetraethylammonium Iodide," *J. Chem. Soc.*, **1958**, 1872.

Wright, W. B., "The Crystal Structure of Glutathione," *Acta Cryst.*, **11**, 632.

Zvonkova, Z. V., and Tashpulatov, Y., "New Determination of Crystal Structure of Thiourea," *Kristallografiya*, **3**, 553.

1959

Abrahamsson, S., "On the Crystal Structure of 17-Methyloctadecanoic Acid," *Arkiv Kemi*, **14**, 49.

Abrahamsson, S., "The Crystal Structure of 14DL-Methyloctadecanoic Acid," *Acta Cryst.*, **12**, 206.

Abrahamsson, S., "On the Crystal Structure of 2DL-Methyloctadecanoic Acid," *Acta Cryst.*, **12**, 301.

Abrahamsson, S., "Crystal Structure of 2D-Methyloctadecanoic Acid," *Acta Cryst.*, 12, 304.

Becker, K. A., Grosse, G., and Plieth, K., "X-Ray Structure Analysis of *trans*-Dichlorodiethylenediaminecobalt(III) Chloride," *Z. Krist.*, 112, 375.

Bekoe, A., and Powell, H. M., "Crystal Structure of *i*-Erythritol and Its Relation to Some Derived D and L Racemic Substances," *Proc. Roy. Soc. (London)*, 250A, 301.

Broadley, J. S., Cruickshank, D. W. J., Morrison, J. D., Robertson, J. M., and Shearer, H. M. M., "Three-Dimensional Refinement of the Structure of β-Succinic Acid," *Proc. Roy. Soc. (London)*, 251A, 441.

Bryden, J. H., "The Crystal Structure of the Dipotassium Salt of Methylene-bis-Nitrosohydroxylamine, $CH_2(N_2O_2K)_2$," *Acta Cryst.*, 12, 581.

Bullen, G. J., "The Crystal Structure of Cobalt(II) Bisacetylacetone Dihydrate," *Acta Cryst.*, 12, 703.

Carić, S., "Refinement of the Structure of Humboldtine ($FeC_2O_4 \cdot 2H_2O$)," *Bull. Soc. Franc. Minéral. Crist.*, 82, 50.

Dejace, J., "Crystal Structure of Bromacetamide," *Bull. Soc. Franc. Mineral. Crist.*, 82, 12.

Dietrich, H., "The Crystal Structure of Ethyllithium," *Z. Naturforsch.*, 146, 739.

Fitzwater, D. R., and Rundle, R. E., "Crystal Structure of Hydrated Erbium, Yttrium, and Praseodymium Ethyl Sulfates," *Z. Krist.*, 112, 362.

Fleming, J. E., and Lynton, H., "Crystal Structure of the Uranyl Nitrate-Triethyl Phosphate Complex," *Chem. Ind. (London)*, 1959, 1409.

Frasson, E., Bardi, R., and Bezzi, S., "Structure of Copper Dimethylglyoxime at Low Temperature," *Acta Cryst.*, 12, 201.

Frasson, E., Panattoni, C., and Zannetti, R., "X-Ray Studies on the Metal Complexes with the Glyoximes II. Structure of the Pt-Dimethyl-Glyoxime," *Acta Cryst.*, 12, 1027.

Geller, S., and Booth, D. P., "The Crystal Structure of Guanidinium Gallium Sulfate Hexahydrate, $[C(NH_2)_3]Ga(SO_4)_2 \cdot 6H_2O$," *Z. Krist.*, 111, 117.

Goldsmith, G. J., and White, J. G., "Ferroelectric Behavior of Thiourea," *J. Chem. Phys.*, 31, 1175.

Grdenić, D., and Matković, B., "Dimorphism and Isomorphism of Zirconium(IV), Cerium(IV), Thorium(IV), and Uranium(IV) Acetylacetonates," *Acta Cryst.*, 12, 817.

Hahn, T., "Refinement of Crystal Structure with Nonresolved Projections," *Z. Krist.*, 111, 161.

Hoshino, S., Okaya, Y., and Pepinsky, R., "Crystal Structure of the Ferroelectric Phase of $(Glycine)_3 \cdot H_2SO_4$," *Phys. Rev.*, 115, 323.

Kinoshita, Y., "The Crystal Structure of Polyheptamethylene Pimelamide (Nylon 77)," *Makromol. Chem.*, 33, 21.

Kinoshita, Y., Matsubara, I., Higuchi, T., and Saito, Y., "The Crystal Structure of bis(Adiponitrilo)copper(I) Nitrate," *Bull. Chem. Soc. Japan*, 32, 1221.

Kinoshita, Y., Matsubara, I., and Saito, Y., "The Crystal Structure of bis(Succininitrilo)copper(I) Nitrate," *Bull. Chem. Soc. Japan*, 32, 741.

Kinoshita, Y., Matsubara, I., and Saito, Y., "The Crystal Structure of bis(Glutaronitrilo)copper(I) Nitrate," *Bull. Chem. Soc. Japan*, 32, 1216.

Knobler, C. B., Okaya, Y., and Pepinsky, R., "The Crystal Structure of tris(Thiourea) Copper(I) Chloride, $Cu(SCN_2H_4)_3Cl$," *Z. Krist.*, 111, 385.

Krishna Murti, G. S. R., "On the Crystal Structure of Methanol at −180°C.," *Indian J. Phys.*, **33**, 458.

Lindqvist, I., and Brändén, C. I., "The Crystal Structure of $SbCl_5 \cdot POCl_3$," *Acta Cryst.*, **12**, 642.

Low, B. W., Hirshfeld, F. L., and Richards, F. M., "Glycinate Complexes of Zinc and Cadmium," *J. Am. Chem. Soc.*, **81**, 4412.

Meerssche, M. van, and Germain, G., "Structure of *trans*-$(CH_3NO)_2$," *Acta Cryst.*, **12**, 818; *Bull. Soc. Chim. Belges*, **68**, 244.

Meerssche, M. van, and Léonard, A., "Structure of $SP(C_2H_5)_3$ and $SeP(C_2H_5)_3$," *Acta Cryst.*, **12**, 1053.

Meerssche, M. van, and Léonard, A., "Refinement of the Structure of $SP(C_2H_5)_3$," *Bull. Soc. Chim. Belges*, **68**, 683.

Miksic, M. G., Segerman, E., and Post, B., "The Solid Phase Transformation in Dimethylacetylene at −110°C.," *Acta Cryst.*, **12**, 390.

Morosin, B., and Lingafelter, E. C., "The Crystal Structure of Tetramethylammonium Tetrachlorozincate and Tetrachlorocobaltate II," *Acta Cryst.*, **12**, 611.

Nardelli, M., and Fava, G., "The Crystal Structure of Mono(Thiourea)Lead(II) Acetate," *Proc. Chem. Soc.*, **1959**, 194.

Nardelli, M., and Fava, G., "The Crystal Structure of bis-Thiourea-Lead(II) Chloride," *Acta Cryst.*, **12**, 727.

Ooi, S., Komiyama, Y., Saito, Y., and Kuroya, H., "The Crystal Structure of *trans*-Dibromo-bisethylenediaminecobalt(III) Bromide Hydrobromide Dihydrate," *Bull. Chem. Soc. Japan*, **32**, 263.

Oughton, B. M., and Harrison, P. M., "The Crystal Structure of Hexagonal L-Cystine," *Acta Cryst.*, **12**, 396.

Panattoni, C., Frasson, E., and Zannetti, R., "X-Ray Studies of the Metal Complexes with the Glyoximes III. Structure of Palladium Dimethylglyoxime," *Gazz. Chim. Ital.*, **89**, 2132.

Peyronel, G., and Pignedoli, A., "Crystal Structure of Cu-N,N-di-n-Propyldithiocarbamate III. Determination of the Signs by the Sayre Method and ($hk0$) Fourier Projection," *Ric. Sci.*, **29**, 1505.

Peyronel, G., and Pignedoli, A., "Crystal Structure of Cu-N,N-di-n-Propyldithiocarbamate II. Patterson Synthesis ($hk0$) and its Interpretation by the Buerger Method of Minimum Function," *Ric. Sci.*, **29**, 1218.

Porte, A. L., and Robertson, J. M., "The Crystal and Molecular Structures of Tiglic and Angelic Acids I. Tiglic Acid," *J. Chem. Soc.*, **1959**, 817.

Porte, A. L., and Robertson, J. M., "The Crystal and Molecular Structures of Tiglic and Angelic Acids II. Angelic Acid," *J. Chem. Soc.*, **1959**, 825.

Raman, S., "Determination of the Structure and Absolute Configuration of L(+)-Lysine Hydrochloride Dihydrate by the Anomalous-Dispersion Method," *Z. Krist.*, **111**, 301.

Rasmussen, S. E., "The Crystal and Molecular Structure of 2,2′,2″-Triaminotriethylamine-Ni(II)-dithiocyanate," *Acta Chem. Scand.*, **13**, 2009.

Reddy, J. van der M., and Lipscomb, W. N., "Molecular Structure of $B_{10}H_{12}(CH_3CN)_2$," *J. Am. Chem. Soc.*, **81**, 754; *J. Chem. Phys.*, **31**, 610.

Sasada, Y., and Atoji, M., "Crystal Structure and Lattice Energy of Orthorhombic Hexachloroethane," *J. Chem. Phys.*, **30**, 1103.

Shimada, A., "Crystal Structure and Lattice Energy of i-Erythritol," *Bull. Chem. Soc. Japan*, **32**, 325.

Smith, G. S., and Hoard, J. L., "The Structure of Dihydrogen Ethylenediaminetetra-acetatoaquonickel(II)," *J. Am. Chem. Soc.*, **81**, 556.

Smith, G. S., and Hoard, J. L., "The Structure of the Cyclic Tetramer of Dimethyl-gallium Hydroxide," *J. Am. Chem. Soc.*, **81**, 3907.

Sørum, H., "The Crystal and Molecular Structure of Acetyl Choline Bromide," *Acta Chem. Scand.*, **13**, 345.

Speakman, J. C., "The Crystal Structure of, and the Hydrogen Bond in, Sodium Hydrogen Diacetate," *Proc. Chem. Soc.*, **1959**, 316.

Strømme, K. O., "An X-Ray Analysis of the 1:1 Compound Trimethyl Amine-Iodine," *Acta Chem. Scand.*, **13**, 268.

Strømme, K. O., "Crystal Structure of the Compounds Dimethylammonium Bromide–Bromine (2:1) and Dimethylammonium Chloride-Iodine (2:1)," *Acta Chem. Scand.*, **13**, 2089.

Teare, P. W., "The Crystal Structure of Orthorhombic Hexatriacontane, $C_{36}H_{74}$," *Acta Cryst.*, **12**, 294.

Trefonas, L. M., and Lipscomb, W. N., "Molecular Structure of $(Me_2N)_3(BH_2)_3$," *J. Am. Chem. Soc.*, **81**, 4435.

Tulinsky, A., "Basic Beryllium Acetate: III. Evidence for Chemical Bonding; Assessment of Accuracy," *Acta Cryst.*, **12**, 634.

Tulinsky, A., and Worthington, C. R., "Basic Beryllium Acetate: II. The Structure Analysis," *Acta Cryst.*, **12**, 626.

Tulinsky, A., Worthington, C. R., and Pignataro, E., "Basic Beryllium Acetate: I. The Collection of Intensity Data," *Acta Cryst.*, **12**, 623.

Weakliem, H. A., and Hoard, J. L., "The Structures of Ammonium and Rubidium Ethylenediaminetetraacetatocobaltate(III)," *J. Am. Chem. Soc.*, **81**, 549.

Williams, D. E., Wohlauer, G., and Rundle, R. E., "Crystal Structures of Nickel and Palladium Dimethylglyoximes," *J. Am. Chem. Soc.*, **81**, 755.

Zachariasen, W. H., and Plettinger, H. A., "Crystal Chemical Studies of the 5f-Series of Elements XXV. The Crystal Structure of Sodium Uranyl Acetate," *Acta Cryst.*, **12**, 526.

Zuccaro, D. E., and McCullough, J. D., "The Crystal Structure of Trimethylsulfonium Iodide," *Z. Krist.*, **112**, 401.

1960

Abrahamsson, S., Larsson, G., and Sydow. E. v., "The Crystal Structure of the Monoclinic Form of *n*-Hexadecanol," *Acta Cryst.*, **13**, 770.

Alderman, P. R. H., Owston, P. G., and Rowe, J. M., "The Crystal Structure of the Ethylene Complex *trans*-[Pt(C_2H_4)($NH(CH_3)_2$)Cl_2]," *Acta Cryst.*, **13**, 149.

Aleby, S., and Sydow, E. v., "The Crystal Structure of Methyl Stearate," *Acta Cryst.*, **13**, 487.

Amirthalingam, V., Padmanabhan, V. M., and Shankar, J., "The Crystal Structure of bis-Acetylacetone Beryllium," *Acta Cryst.*, **13**, 201.

Bekoe, D. A., and Trueblood, K. N., "The Crystal Structure of Tetracyanoethylene," *Z. Krist.*, **113**, 1.

Brown, I. D., and Dunitz, J. D., "The Crystal Structure of the Cuprous Chloride Azomethane Complex," *Acta Cryst.*, **13**, 28.

Bullen, G. J., "Structure of Tetrameric Phosphonitrilic Dimethylamide $P_4N_4(NMe_2)_8$," *Proc. Chem. Soc.*, **1960**, 425.

Carrai, G., and Gottardi, G., "The Crystal Structure of Xanthates I. Arsenious Zanthate," Z. Krist., 113, 373.

Cavalca, L., Nardelli, M., and Branchi, G., "The Crystal Structure of mono-Thiosemicarbazide-Zinc Chloride," Acta Cryst., 13, 688.

Cavalca, L., Nardelli, M., and Fava, G., "The Crystal Structure of bis-Biuret-Cadmium Chloride," Acta Cryst., 13, 594.

Dvoryankin, V. F., and Vainshtein, B. K., "Electron Diffraction Study of Thiourea," Kristallografiya, 5, 589.

Fisher, P. J., Taylor, N. E., and Harding, M. M., "The Crystal Structure of Cuprous Iodide-Methyl Isocyanide," J. Chem. Soc., 1960, 2303.

Fleming, J. E., and Lynton, H., "Crystal Structure of the Complex Uranyl Nitrate-Triethyl Phosphate," Chem. Ind. (London), 1960, 1415.

Frasson, E., and Panattoni, C., "X-Ray Studies on the Metal Complexes with the Glyoximes IV. Structure of Ni-Methyl-Ethyl-Glyoxime," Acta Cryst., 13, 893.

Geller, S., "On the Structure of Guanidinium Gallium Sulfate Hexahydrate," Z. Krist., 114, 148.

Grønbäk, R., and Rasmussen, S. E., "Crystal Data of Nickel(II) Dithiosemicarbazide-Sulphate," Acta Chem. Scand., 14, 782.

Hahn, T., "Systematic Influences on the Refinement of the Crystal Structure of Diglycine Hydrochloride," Z. Krist., 113, 403.

Hassel, O., and Hope, H., "Crystal Structure of the 1:1 Addition Compound Trimethylamine-Iodo-Monochloride," Acta Chem. Scand., 14, 391.

Iitaka, Y., "The Crystal Structure of β-Glycine," Acta Cryst., 13, 35.

Katz, J. L., and Post, B., "The Crystal Structure and Polymorphism of N-Methyl Acetamide," Acta Cryst., 13, 624.

Larsson, K., "The Crystal Structure of Octa-(Methylsilsesquioxane), $(CH_3SiO_{1.5})_8$," Arkiv Kemi, 16, 203.

Lindqvist, I., and Rosenstein, R., "The Crystal Structure of Iron(II) Sulphate Pentahydrate Glycine," Acta Chem. Scand., 14, 1228.

Lippert, E. L., and Truter, M. R., "The Stereochemistry of Acetylacetone Complexes of Zinc I. The Crystal Structure of Monoaquobisacetylacetonezinc," J. Chem Soc., 1960, 4996.

Meerssche, M. van and Léonard, A., "The Structure of $SeP(C_2H_5)_3$," Bull. Soc. Chim. Belges, 69, 45.

Mills, O. S., and Robinson, G., "The Structure of Butadiene Iron Tricarbonyl," Proc. Chem. Soc., 1960, 421.

Nardelli, M., and Chierici, I., "Biuret Complexes of Bivalent Metal Chlorides," J. Chem. Soc., 1960, 1952.

Nardelli, M., Fava, G., and Branchi, G., "The Crystal Structure of mono-Thiourea-Lead(II) Acetate," Acta Cryst., 13, 898.

Nordman, C. E., Weldon, A. S., and Patterson, A. L., "X-Ray Analysis of the Substrates of Aconitase I. Rubidium Dihydrogen Citrate," Acta Cryst., 13, 414.

Nordman, C. E., Weldon, A. S., and Patterson, A. L., "X-Ray Analysis of the Substrates of Aconitase II. Anhydrous Citric Acid," Acta Cryst., 13, 418.

Norman, N., and Mathisen, H., "Structure of Linear Polymers—Lower n-Hydrocarbons," U.S. Dept. Comm., Office Tech. Serv., P. B. Rept. 171, 181, 93 pp.

Ooi, S., Kawase, T., Nakatsu, K., and Kuroya, H., "The Crystal Structure of Tetra-Thioureapalladium(II) Chloride, $[Pd(SCN_2H_4)_4]Cl_2$," Bull. Chem. Soc. Japan, 33, 861.

Ooi, S., Komiyama, Y., and Kuroya, H., "The Crystal Structure of *trans*-[Cr(en)₂-Cl₂] Cl·HCl·2H₂O," *Bull. Chem. Soc. Japan*, **33**, 354.

Owston, P. G., Partridge, J. M., and Rowe, J. M., "The Crystal Structure of a Complex Hydride of Platinum, [Pt((C₂H₅)₃P)₂HBr]," *Acta Cryst.*, **13**, 246.

Owston, P. G., and Rowe, J. M., "The Crystal Structure of the Binuclear Thiocyanate Complex α-[Pt₂(SCN)₂Cl₂(P(C₃H₇)₃)₂]," *Acta Cryst.*, **13**, 253.

Peterson, J., Steinrauf, L. K., and Jensen, L. H., "Direct Determination of the Structure of ʟ-Cystine Dihydrobromide," *Acta Cryst.*, **13**, 104.

Rérat, C., "Crystal Structure of Trimethylamine Oxide Hydrochloride," *Acta Cryst.*, **13**, 63.

Rimsky, A., "Atomic Structure of Mixed Crystals NH₄Cl-Urea, Adsorption, and Reciprocal Epitaxy, Syntaxy," *Bull. Soc. Franc. Mineral. Crist.*, **83**, 187.

Semenenko, K. N., "The Study of the High Temperature Form of Beryllium Oxyacetate, *Zh. Strukt. Khim.*, **1**, 442.

Senko, M. E., and Templeton, D. H., "Unit Cells of Choline Halides and Structure of Choline Chloride," *Acta Cryst.*, **13**, 281.

Shintani, R., "The Crystal and Molecular Structure of Anhydrous Diacetylhydrazine (—NHCOCH₃)₂," *Acta Cryst.*, **13**, 609.

Shkolnikova, L. M., and Shugam, E. A., "Crystalline and Molecular Structure of Chromium Acetylacetonate," *Kristallografiya*, **5**, 32.

Shugam, E. A., and Levina, V. M., "π-Bonding in the Nickel Diethyldithiocarbamate Molecule," *Kristallografiya*, **5**, 257.

Swink, L. N., and Atoji, M., "The Crystal Structure of Triethylenediaminenickel(II) Nitrate, Ni(NH₂CH₂CH₂NH₂)₃(NO₃)₂," *Acta Cryst.*, **13**, 639.

Tashpulatov, Y., "Crystal Structure of Tetramethylthiuram Monosulfide," *Usbek. Khim. Zh.*, **1960**, 35.

Truter, M. R., "An Accurate Determination of the Crystal Structure of Thioacetamide," *J. Chem. Soc.*, **1960**, 997.

Turco, A., Panattoni, C., and Frasson, E., "Structural Configuration of the Group SCN in AgSCN·P(*n*-C₃H₇)₃," *Nature*, **187**, 772; *Ric. Sci.*, **30**, 1071.

Wheatley, P. J., "Structure of Lithium Methoxide," *Nature*, **185**, 681.

Zvonkova, Z. V., Astakhova, L. I., and Glushkova, V. P., "The Atomic Structure of Tetramethylthiourea," *Kristallografiya*, **5**, 547.

1961

Barclay, G. A., and Kennard, C. H. L., "The Crystal Structure of Anhydrous Copper(II) Formate," *J. Chem. Soc.*, **1961**, 3289.

Berthold, H. J., "The Crystal Structure of Perchlorodimethyltrisulfide, Cl₃C—S₃—CCl₃," *Z. Krist.*, **116**, 290.

Brändén, C. I., and Lindqvist, I., "The Crystal Structure of SbCl₅·PO(CH₃)₃," *Acta Chem. Scand.*, **15**, 167.

Bryan, R. F., Poljak, R. J., and Tomita, K., "The Crystal and Molecular Structures of bis-(β-Aminobutyrato)copper(II) Dihydrate, Cu(C₄H₈NO₂)₂·2H₂O," *Acta Cryst.*, **14**, 1125.

Bryden, J. H., "The Crystal Structure of Azodicarbonamide," *Acta Cryst.*, **14**, 61.

Chatani, Y., Takizawa, T., Murahashi, S., Sakata, Y., and Nishimura, Y., "Crystal Structure of Polyketone (1:1 Ethylene Carbon Monoxide Copolymer)," *J. Polymer Sci.*, **55**, 811.

Cocco, G., "The Structure of Whewellite," *Atti Accad. Nazl. Lincei Rend. Classe Sci. Fis. Mat. Nat.*, **31**, 292.

Craven, B. M., and Hall, D., "The Crystal Structure of Wolffram's Red Salt," *Acta Cryst.*, **14**, 475.

Darlow, S. F., and Cochran, W., "The Crystal Structure of Potassium Hydrogen Maleate," *Acta Cryst.*, **14**, 1250.

Dietrich, H., and Hodgkin, D. C., "Crystal Structure of the *trans*-Dimer of Nitrosoisobutane," *J. Chem. Soc.*, **1961**, 3686.

Dodge, R. P., Templeton, D. H., and Zalkin, A., "Crystal Structure of Vanadyl bis-Acetylacetonate. Geometry of Vanadium in Fivefold Coordination," *J. Chem. Phys.*, **35**, 55.

Dougill, M. W., "Phosphonitrilic Derivatives IX. The Crystal Structure of Octamethylcyclotetraphosphonitrile," *J. Chem. Soc.*, **1961**, 5471.

Dunken, H., and Krausse, J., "Structure Study of Lithium Methylate," *Z. Chem.*, **1**, 27.

Dutta, S. N., and Woolfson, M. M., "The Crystal and Moleculer Structure of Tetraethyl Diphosphine Disulphide," *Acta Cryst.*, **14**, 178.

Dvoryankin, V. F., and Vainshtein, B. K., "An Electron Diffraction Study of the Low Temperature Ferroelectric Form of Thiourea," *Kristallografiya*, **6**, 949.

Feil, D., and Jeffrey, G. A., "The Polyhedral Clathrate Hydrates II. Structure of the Hydrate of Tetra IsoAmyl Ammonium Fluoride," *J. Chem. Phys.*, **35**, 1863.

Fletcher, R. O. W., and Steeple, H., "Low Temperature Transitions in Methyl Ammonium Alum," *Acta Cryst.*, **14**, 891.

Freeman, H. C., Smith, J. E. W. L., and Taylor, J. C., "Crystallographic Studies of the Biuret Reaction I. Potassium bis-Biureto Cuprate(II) Tetrahydrate, $K_2[Cu(NHCONHCONH)_2] \cdot 4H_2O$," *Acta Cryst.*, **14**, 407.

Gabe, E. J., "The Crystal Structure of Methylammonium Bromide," *Acta Cryst.*, **14**, 1296.

Galigné, J. L., and Falgueirettes, J., "Atomic Structure of Strontium Formate Dihyrate," *Compt. Rend.*, **253**, 994.

Gottardi, G., "The Crystal Structure of Xanthates II. Antimonious Xanthate," *Z. Krist.*, **115**, 451.

Groth, P., and Hassel, O., "Charge-transfer Bonds in Solids: Crystal Structure of Oxalyl Bromide," *Proc. Chem. Soc.*, **1961**, 343.

Hamilton, W. C., "A Neutron Diffraction Refinement of the Crystal Structure of Dimethylglyoxime," *Acta Cryst.*, **14**, 95.

Haussühl, S., "Physical Crystallography of the Orthorhombic Formates of Calcium and Cadmium," *Fortschr. Mineral.*, **39**, 345.

Hirokawa, S., and Ashida, T., "The Crystal Structures of Some Polymethylenediammonium Adipates II. Tetramethylenediammonium Adipate," *Acta Cryst.*, **14**, 1004.

Hoard, J. L., Glen, G. L., and Silverton, J. V., "The Configuration of $Zr(C_2O_4)^{-4}$ and the Stereochemistry of Discrete Eight-Coordination," *J. Am. Chem. Soc.*, **83**, 4293.

Hock, A. A., and Mills, O. S., "Studies of Some Carbon Compounds of the Transition Metals II. The Structure of $MeC \equiv CMe \cdot H_2Fe_2(CO)_8$," *Acta Cryst.*, **14**, 139.

Hoffmann, W., "The Crystal Structure of Whewellite, $Ca(C_2O_4) \cdot H_2O$," *Fortschr. Mineral.*, **39**, 346.

Hughes, E. W., Yakel, H. L., and Freeman, H. C., "The Crystal Structure of Biuret Hydrate," *Acta Cryst.*, **14**, 345.

Iitaka, Y., "The Crystal Structure of γ-Glycine," *Acta Cryst.*, **14**, 1.

Jensen, L. H., and Lingafelter, E. C., "Refinement of the Structure of *n*-Nonanoic Acid Hydrazide," *Acta Cryst.*, **14**, 507.

Kartha, G., and Vries, A. de, "Structure of Asparagine Monohydrate," *Nature*, **192**, 862.

Kraut, J., "The Crystal Structure of 2-Amino-Ethanol Phosphate," *Acta Cryst.*, **14**, 1146.

Lipscomb, W. N., Wang, F. E., May, W. R., and Lippert, E. L., Jr., "Comments on the Structure of 1,2-Dichloroethane and of N_2O_2," *Acta Cryst.*, **14**, 1100.

Lobachev, A. N., and Vainshtein, B. K., "An Electron Diffraction Study of Urea," *Kristallografiya*, **6**, 395.

Luxmoore, A. R., and Truter, M. R., "The Crystal Structure of Tetracarbonylacrylonitrileiron," *Proc. Chem. Soc.*, **1961**, 466.

Morosin, B., and Lingafelter, E. C., "The Configuration of the Tetrachlorocuprate(II) Ion," *J. Phys. Chem.*, **65**, 50.

Natta, G., Allegra, G., Perego, G., and Zambelli, A., "A New Coordination Type Around Fluorine Atoms," *J. Am. Chem. Soc.*, **83**, 5033.

Norman, N., and Mathisen, H., "The Crystal Structure of Lower *n*-Paraffins I. *n*-Octane," *Acta Chem. Scand.*, **15**, 1747.

Penfold, B. R., and Lipscomb, W. N., "The Molecular and Crystal Structure of Hydrogen Cyanide Tetramer (Diaminomaleonitrile)," *Acta Cryst.*, **14**, 589.

Ryan, T. D., and Rundle, R. E., "The Crystal Structure of Ethylenediaminetribromoplatinum $(C_2N_2H_8)PtBr_3$," *J. Am. Chem. Soc.*, **83**, 2814.

Sakurai, K., "A Direct Determination of the Crystal Structure of Ethylenediammonium Sulfate," *J. Phys. Soc. Japan*, **16**, 1205.

Shanley, P., and Collin, R. L., "The Crystal Structure of the High Temperature Form of Choline Chloride," *Acta Cryst.*, **14**, 79.

Shkolnikova, L. M., and Shugam, E. A., "Crystal Structure of Cobalt(III) Acetylacetonate," *Zh. Strukt. Khim.*, **2**, 72.

Sklar, N., Senko, M. E., and Post B., "Thermal Effects in Urea: the Crystal Structure at $-140°C.$ and at Room Temperature," *Acta Cryst.*, **14**, 716.

Speakman, J. C., and Mills, H. H., "The Crystal Structures of the Acid Salts of Some Monobasic Acids VI. Sodium Hydrogen Diacetate," *J. Chem. Soc.*, **1961**, 1164.

Spencer, C. J., and Lipscomb, W. N., "The Molecular and Crystal Structure of $(PCF_3)_5$," *Acta Cryst.*, **14**, 250; **15**, 509 (1962).

Strandberg, B., Lindqvist, I., and Rosenstein, R., "The Crystal Structure of Copper(II)-monoglycylglycine Trihydrate, $Cu(NH_2CH_2CONCH_2COO) \cdot 3H_2O$," *Z. Krist.*, **116**, 266.

Tavale, S. S., Pant, L. M., and Biswas, A. B., "The Crystal Structure of Sodium Pyruvate," *Acta Cryst.*, **14**, 1281.

Tomita, K., "The Copper Salts of the ω-Amino Acids III. The Crystal Structure of the Copper Salt of β-Alanine," *Bull. Chem. Soc. Japan*, **34**, 297.

Tomita, K., and Nitta, I., "On the Copper Salts of ω-Amino Acids II. The Crystal Structure of Copper Glycine Monohydrate," *Bull. Chem. Soc. Japan*, **34**, 286.

Trefonas, L. M., Mathews, F. S., and Lipscomb, W. N., "The Molecular and Crystal Structure of $(BH_2)_3[N(CH_3)_2]_3$," *Acta Cryst.*, **14**, 273.

Viswanathan, K. S., and Kunchur, N. R., "Rotational Disorder in the Crystal Lattice of Cobalt(III) bis Dimethyl Glyoximino Diammine Nitrate. An X-Ray Investigation," *Acta Cryst.*, **14**, 675.

Wang, F. E., Simpson, P. G., and Lipscomb, W. N., "Molecular Structure of B_9H_{13}-(CH_3CN)," *J. Chem. Phys.*, **35**, 1335.

Zvonkova, Z. V., and Khvatkina, A. N., "The Atomic Structure of Cyanamide," *Kristallografiya*, **6**, 184.

1962

Abrahamsson, S., and Ryderstedt-Nahringbauer, I., The Crystal Structure of the Low Melting Form of Oleic Acid," *Acta Cryst.*, **15**, 1261.

Alderman, P. R. H., Owston, P. G., and Rowe, J. M., The Crystal Structure of Nitroso-(dimethyldithiocarbonato)cobalt, [$Co(NO)(S_2C \cdot N(CH_3)_2)_2$]," *J. Chem. Soc.*, **1962**, 668.

Aleby, S., The Crystal Structure of Ethyl Stearate," *Acta Cryst.*, **15**, 1248.

Allegra, G., and Perego, G., "X-Ray Diffraction Structure of Crystalline ($AlCl_2CH_3)_2$," *Atti Accad. Nazl. Lincei Rend. Classe Sci. Fis. Mat. Nat.*, **33**, 450.

Atovmyan, L. O., and Bokii, G. B., "Structure of the Compound $NH_4Na[MoO_3C_2O_4] \cdot 2H_2O$ and Its Place in the Classification of Mo Compounds," *Dokl. Akad. Nauk SSSR*, **143**, 342.

Bjorvatten, T., "Crystal Structure of the 1:3 Addition Compound Iodoform–Sulphur ($CHI_3 \cdot 3S_8$)," *Acta Chem. Scand.*, **16**, 749.

Bradley, D. C., and Kunchur, N. R., "The Polymeric Structure of Mercury Methyl Mercaptide," *Chem. Ind. (London)*, **1962**, 1240.

Bullen, G. J., "The Crystal and Molecular Structure of Tetrameric Phosphonitrilic Dimethylamide," *J. Chem. Soc.*, **1962**, 3193.

Caron, A., and Donohue, J., "Refinement of the Crystal Structure of Trimethylamine Oxide Hydrochloride, ($CH_3)_3NO \cdot HCl$," *Acta Cryst.*, **15**, 1052.

Cavalca, L., Nardelli, M., and Fava, G., "The Crystal and Molecular Structure of bis-Thiosemicarbazidato-Nickel(II), (Red Crystals)," *Acta Cryst.*, **15**, 1139.

Cocco, G., and Sabelli, C., "Refinement of the Structure of Whewellite with Electronic Computer," *Atti Soc. Toscana Sci. Nat. Pisa Proc. Verbali Mem.*, **69A**, 289.

Coppens, P., MacGillavry, C. H., Hovenkamp, S. G., and Douwes, H., "The Crystal Structure of Potassium-O,O-Dimethyl-Phosphordithioate," *Acta Cryst.*, **15**, 765.

Donohue, J., and Marsh, R. E., "A Refinement of the Structure of N-Acetylglycine," *Acta Cryst.*, **15**, 941.

Ferrier, W. G., Lindsay, A. R., and Young, D. W., "Confirmation of the Crystal Structure of 2-Amino-Ethanol Phosphate," *Acta Cryst.*, **15**, 616.

Furberg, S., and Helland, S., "The Crystal Structure of the Calcium and Strontium Salts of Arabonic Acid," *Acta Chem. Scand.*, **16**, 2373.

Geller, S., and Katz, H., "An X-Ray Diffraction Study of the Structure of Guanidinium Aluminum Sulfate Hexahydrate," *Bell System Tech. J.*, **41**, 425.

Gobillon, Y., Piret, P., and Meerssche, M. van, "Structure of the Complex $NaI \cdot 3$-(Dimethylformamide)," *Bull. Soc. Chim. France*, **1962**, 551.

Grønbäk, R., and Rasmussen, S. E., "The Structure of Nickel(II) Dithiosemicarbazide Sulphate Trihydrate," *Acta Chem. Scand.*, **16**, 2325.

Groth, P., and Hassel, O., "Crystal Structure of Oxalyl Bromide and Oxalyl Chloride," *Acta Chem. Scand.*, **16**, 2311.

Hall, D., Rae, A. D., and Waters, T. N., "The Structure of N,N'-Ethylenebis(acetyl-acetoneiminato)copper(II)," *Proc. Chem. Soc.*, **1962**, 143.

Hayashida, T., "Crystal Structure of Triclinic Form of *n*-Octadecane," *J. Phys. Soc. Japan*, **17**, 306.

Hine, R., "The Crystal Structure and Molecular Configuration of (+)-*S*-Methyl-L-Cysteine Sulfoxide," *Acta Cryst.*, **15**, 635.

Hirokawa, S., and Ashida, T., "Crystal Structure of Tetramethylenediammonium Adipate," *Mem. Defense Acad. Math. Phys. Chem. Eng. (Yokosuka, Japan)*, **2**, 95.

Hughes, D. O., and Small, R. W. H., "The Crystal and Molecular Structure of Monofluoroacetamide," *Acta Cryst.*, **15**, 933.

Isakov, I. V., and Zvonkova, Z. V., "Crystal Structure of the Complex of Zinc Chloride and Acetonitrile," *Dokl. Akad. Nauk SSSR*, **145**, 801.

Jeffrey, G. A., and McMullan, R. K., "Polyhedral Clathrate Hydrates IV. The Structure of the Tributylsulfonium Fluoride Hydrate," *J. Chem. Phys.*, **37**, 2231.

Jensen, L. H., "Refinement of the Structure of N,N'-Hexamethylenebispropionamide," *Acta Cryst.*, **15**, 433.

Jordan, T., Smith, W., and Lipscomb, W. N., "$(CH_3)_2NSO_2N(CH_3)_2$ as a Model for α-Sulfonyl Carbanions," *Tetrahedron Letters*, **1962**, 37.

Larsson, K., "On the Structure of the Crystal Form C of 11-Bromoundecanoic Acid," *Acta Chem. Scand.*, **16**, 1751.

Levdik, V. F., and Porai-Koshits, M. A., "Crystal and Molecular Structure of $PdClC_3H_5$," *Zh. Strukt. Khim.*, **3**, 472.

Luxmoore, A. R., and Truter, M. R., "The Crystal Structure of Tetracarbonyl (Acrylonitrile) Iron," *Acta Cryst.*, **15**, 1117.

Margulis, T. N., and Templeton, D. H., "Crystal and Molecular Structure of Triethylscarphane," *J. Chem. Phys.*, **36**, 2311.

Mayer, I., Steinberg, M., Feigenblatt, F., and Glasner, A., "The Preparation of Some Rare Earth Formates and Their Crystal Structures," *J. Phys. Chem.*, **66**, 1737.

Nakatsu, K., "The Crystal Structure of D-tris-Ethylenediamine-cobalt(III) Bromide Monohydrate, D-$[Co(en)_3]Br_3 \cdot H_2O$, and the Absolute Configuration of the D-$[Co(en)_3]^{3+}$ Ion," *Bull. Chem. Soc. Japan*, **35**, 832.

Nardelli, M., Fava, G., and Boldrini, P., "Complexes with Sulfur Bridges: Structure of bis(Thiourea)Cadmium Formate," *Gazz. Chim. Ital.*, **92**, 1392.

Noguchi, T., "The Crystal Structure of Nickel DialphaAmino Isobutyrate Tetrahydrate," *Bull. Chem. Soc. Japan*, **35**, 99.

Palenik, G. J., and Donohue, J., "The Molecular and Crystal Structure of $(PCF_3)_4$," *Acta Cryst.*, **15**, 564.

Pignedoli, A., and Peyronel, G., "Crystalline and Molecular Structure of the Cu(II)-bis(N,N-Dipropyldithiocarbamate)," *Gazz. Chim. Ital.*, **92**, 745.

Saito, Y., and Iwasaki, H., "The Crystal Structure of *trans*-Dichloro-bis-L-Propylenediamine Cobalt(III) Chloride Hydrochloride Dihydrate and the Absolute Configuration of the Complex Ion, $[Co\text{-}L\text{-}pn_3Cl_2]^+$," *Bull. Chem. Soc. Japan*, **35**, 1131.

Sands, D. E., and Zalkin, A., "The Crystal Structure of $B_{10}H_{12}[S(CH_3)_2]_2$," *Acta Cryst.*, **15**, 410.

Scheuerman, R. F., and Sass, R. L., "The Crystal Structure of Valeric Acid," *Acta Cryst.*, **15**, 1244.

Stam, C. H., "The Crystal Structure of *S*-Methylisothiourea Sulphate $[CH_3SC(NH_2)_2]_2{}^+\text{-}SO_4{}^{2-}$," *Acta Cryst.*, **15**, 317.

Stephenson, N. C., "The Crystal Structure of Diammine bis(Acetamidine)platinum(II) Chloride Monohydrate," *J. Inorg. Nucl. Chem.*, **24**, 801.

Strieter, F. J., and Templeton, D. H., "Crystal Structure of Butyric Acid," *Acta Cryst.*, **15**, 1240.

Sullivan, R. A. L., and Hargreaves, A., "The Crystal and Molecular Structure of Thiourea Dioxide," *Acta Cryst.*, **15**, 675.

Truter, M. R., "The Crystal Structure of Potassium Methylenedisulfonate," *J. Chem. Soc.*, **1962**, 3393.

Truter, M. R., "A Detailed Refinement of the Crystal Structure of Sodium Hydroxy-methanesulphinate Dihydrate (Rongalite)," *J. Chem. Soc.*, **1962**, 3400.

Truter, M. R., and Rutherford, K. W., "The Crystal Structure of Tetrakisthioacet-amidecopper(I) Chloride," *J. Chem. Soc.*, **1962**, 1748.

Turner-Jones, A., and Bunn, C. W., "The Crystal Structure of Polyethylene Adipate and Polyethylene Suberate," *Acta Cryst.*, **15**, 105.

Viswamitra, M. A., "The Crystal Structure of Copper Potassium Oxalate Dihydrate, $CuK_2(C_2O_4)_2 \cdot 2H_2O$," *Z. Krist.*, **117**, 437.

Viswamitra, M. A., "Crystal Structure of Copper Ammonium Oxalate Dihydrate, $Cu(NH_4)_2(C_2O_4)_2 \cdot 2H_2O$," *J. Chem. Phys.*, **37**, 1408.

Wheatley, P. J., "An X-Ray Diffraction Determination of the Crystal and Molecular Structure of 'Methyl Metadithiophosphonate,' $(CH_3 \cdot PS_2)_2$," *J. Chem. Soc.*, **1962**, 300.

Wheatley, P. J., "An X-Ray Diffraction Determination of the Crystal and Molecular Structure of Tetramethyl-N,N'-bistrimethylsilylcyclodisilazane," *J. Chem. Soc.*, **1962**, 1721.

Wright, D. A., and Marsh, R. E., "The Crystal Structure of L-Lysine Monohydro-chloride Dihydrate," *Acta Cryst.*, **15**, 54; **16**, 431 (1963).

1963

Abrahamsson, S., and Westerdahl, A., "The Crystal Structure of 3-Thiododecanoic Acid," *Acta Cryst.*, **16**, 404.

Allegra, G., and Perego, G., "The Crystal Structure of the $KF \cdot 2Al(C_2H_5)_3$ Complex," *Acta Cryst.*, **16**, 185.

Allegra, G., Perego, G., and Immirzi, A., "The Crystal Structure of $AlCl_2CH_3$ Dimer," *Makromol. Chem.*, **61**, 69.

Ashida, T., and Hirokawa, S., "The Crystal Structure of Ethylenediammonium Chloride," *Bull. Chem. Soc. Japan*, **36**, 704.

Ashida, T., and Hirokawa, S., "The Crystal Structure of Tetramethylenediammonium Chloride," *Bull. Chem. Soc. Japan*, **36**, 1086.

Atovmyan, L. O., and Bokii, G. B., "Investigation of the Structure of $NaNH_4$-$[MoO_3C_2O_4] \cdot 2H_2O$," *Zh. Strukt. Khim.*, **4**, 576.

Bally, R., "The Structure of Copper Diethyldithiocarbamate," *Compt. Rend.*, **257**, 425.

Beagley, B., and Small, R. W. H., "An Accurate Determination of the Structure of Ammonium Oxamate," *Proc. Roy. Soc. (London)*, **275A**, 469.

Bonamico, M., Dessy, G., Mazzone, G., Mugnoli, A., Vaciago, A., and Zambonelli, L., "Pentacoordination of Bivalent Copper and Zinc in Their Diethyldithio-carbamates," *Atti Accad. Nazl. Lincei Rend. Classe Sci. Fis. Mat. Nat.*, **35**, 338.

Brown, B. W., and Lingafelter, E. C., "The Crystal Structure of *trans*-bis(Ethylene-diamine)-bis(Isothiocyanato)nickel(II)," *Acta Cryst.*, **16**, 753.

Bukowska-Strzyzewska, M., "Crystal Structure of Diamminecopper(II) Actate," *Roczniki Chem.*, **37**, 1335.

Camerman, N., and Trotter, J., "Stereochemistry of Arsenic VIII. Cyanodimethyl-arsine (Cacodyl Cyanide)," *Can. J. Chem.*, **41**, 460.

Camerman, N., and Trotter, J., "Stereochemistry of Arsenic IX. Diiodomethylarsine," *Acta Cryst.*, **16**, 922.

Canepa, F. G., "The Crystal Structure of Pentamethonium Iodide," *Acta Cryst.*, **16**, 145.

Carazzolo, G., and Mammi, M., "Crystal Structure of a New Form of Polyoxymethylene," *J. Polymer Sci.*, **1A**, 965.

Catesby, C. G. C., "The Crystal Structure of Urea Ammonium Bromide," *Acta Cryst.*, **16**, 392.

Chatani, Y., Sakata, Y., and Nitta, I., "Crystal Structures of Monomers Polymerizable in Their Solid States I. Acrylic Acid," *J. Polymer Sci.*, **1B**, 419.

Clark, H. C., O'Brien, R. J., and Trotter, J., "The Crystal Structure of Trimethyltin Fluoride," *Proc. Chem. Soc.*, **1963**, 85.

Collins, E., Sutor, D. J., and Mann, F. G., "The Crystal Structure of Tetramethylarsonium Bromide," *J. Chem. Soc.*, **1963**, 4051.

Corradini, P., Diana, G., Ganis, P., and Pedone, C., "Crystal Structure of Low Molecular Weight Model Compounds having Structural Features Similar to those of High Molecular Weight Stereoregular Polymers: the Crystalline Structure of meso-α,α'-Dimethyl Glutaric Acid," *Makromol. Chem.*, **61**, 242.

Cotton, F. A., and Soderberg, R. H., "The Crystal and Molecular Structure of bis-(Trimethylphosphine Oxide)Cobalt(II) Dinitrate," *J. Am. Chem. Soc.*, **85**, 2402.

Coulter, C. L., Gantzel, P. K., and McCullough, J. D., "The Crystal Structure of Trimethyloxosulfonium Perchlorate [$(CH_3)_3SO$]$^+ClO_4^-$," *Acta Cryst.*, **16**, 676.

Dahl, L. F., and Chin-Hsuan Wei, "Structure and Nature of Bonding of [C_2H_5S-$Fe(CO)_3$]$_2$," *Inorg. Chem.*, **2**, 328.

Danielsen, J., and Rasmussen, S. E., "Crystallographic Proof for a Dichlorophosphate Compound. The Structure of $Mn(PO_2Cl_2)_2(CH_3COOC_2H_5)_2$," *Acta Chem. Scand.*, **17**, 1971.

Dietrich, H., "The Crystal Structure of Ethyllithium," *Acta Cryst.*, **16**, 681.

Dunitz, J. D., "The Supposed Existence of Two Molecular Forms in Crystals of Acetyl Choline Bromide," *Acta Chem. Scand.*, **17**, 1471.

Franzini, M., "The Crystal Structure of Nickelious Xanthate," *Z. Krist.*, **118**, 393.

Franzini, M., and Schiaffino, L., "Pseudosymmetry through Twinning in Crystals of bis(N,N-Diethyldithiocarbamato)nickel(II)," *Atti Accad. Nazl. Lincei Rend. Classe Sci. Fis. Mat. Nat.*, **34**, 670.

Germain, G., Piret, P., and Meerssche, M. van, "Structure of *cis*-Di-Nitrosomethane," *Acta Cryst.*, **16**, 109.

Glen, G. L., Silverton, J. V., and Hoard, J. L., "Stereochemistry of Discrete Eight-Coordination III. Tetrasodium Tetrakisoxalatozirconate(IV) Trihydrate," *Inorg. Chem.*, **2**, 250.

Groth, P., and Hassel, O., "A Crystal Structure Exhibiting both Hydrogen and Halogen Molecule Bridges between Oxygen Atoms (2Methanol·Br_2)," *Mol. Phys.*, **6**, 543.

Hall, D., Rae, A. D., and Waters, T. N., "The Colour, Isomerism and Structure of Some Copper, Coordination Compounds V. The Crystal Structure of N,N'-Ethylene-bis(Acetylacetoneiminato)copper(II)," *J. Chem. Soc.*, **1963**, 5897.

Han, K.-S., "The Crystal Structure of Hexamethylenediamine Dihydroiodide," *J. Korean Chem. Soc.*, **7**, 74.

Haussühl, S., "Physical Crystallography of Calcium and Cadmium Formates," *Z. Krist.*, **118**, 33.

Heitsch, C. W., Nordman, C. E., and Parry, R. W., "The Crystal Structure and Dipole Moment in Solution of the Compound $AlH_3 \cdot 2N(CH_3)_3$," *Inorg. Chem.*, **2**, 508.

Hess, H., "The Crystal and Molecular Structure of the Dimer of Dimethylamino-Borodichloride $(BCl_2N(CH_3)_2]_2$," *Z. Krist.*, **118**, 361.

Hesse, R., "The Crystal Structure of Cu(I) Diethyldithiocarbamate and its Interpretation; Application of Chemical Topology," *Arkiv Kemi*, **20**, 481.

Higgs, M. A., and Sass, R. L., "The Crystal Structure of Acrylic Acid," *Acta Cryst.*, **16**, 657.

Jordan, T., Smith, H. W., Lohr, L. L., Jr., and Lipscomb, W. N., "X-Ray Structure Determination of $(CH_3)_2NSO_2N(CH_3)_2$ and LCAO-MO Study of Multiple Bonding in Sulfones," *J. Am. Chem. Soc.*, **85**, 846.

Koo, C. H., Kim, M. I., and Yoo, C. S., "The Crystal Structure of Ethylenediamine Dihydrochloride," *J. Korean Chem. Soc.*, **7**, 293.

Krogmann, K., and Mattes, R., "The Crystal Structure of Nickel Formate, $Ni(HCOO)_2 \cdot 2H_2O$," *Z. Krist.*, **118**, 291.

Larsson, K., "The Crystal Structure of the 1,3-Diglyceride of 3-Thiododecanoic Acid," *Acta Cryst.*, **16**, 741.

Larsson, K., "The Crystal Structure of the D-Form of 11-Bromoundecanoic Acid," *Acta Chem. Scand.*, **17**, 199.

Larsson, K., "On the Structure of the Crystal Form E of 11-Bromoundecanoic Acid," *Acta Chem. Scand.*, **17**, 215.

Lewin, R., Simpson, P. G., and Lipscomb, W. N., "Molecular and Crystal Structure of $C_2H_5NH_2B_8H_{11}NHC_2H_5$," *J. Chem. Phys.*, **39**, 1532; *J. Am. Chem. Soc.*, **85**, 478.

Lomer, T. R., "The Crystal and Molecular Structure of Lauric Acid (Form A_1)," *Acta Cryst.*, **16**, 984.

Lopez-Castro, A., and Truter, M. R., "The Crystal and Molecular Structure of Dichlorotetrakisthioureanickel, $[(NH_2)_2CS]_4NiCl_2$," *J. Chem. Soc.*, **1963**, 1309.

McMullan, R. K., Bonamico, M., and Jeffrey, G. A., "Polyhedral Clathrate Hydrates V. Structure of the Tetra-n-Butyl Ammonium Fluoride Hydrate," *J. Chem. Phys.*, **39**, 3295.

Matković, B., and Grdenić, D., "The Crystal Structure of Cerium(IV) Acetylacetonate," *Acta Cryst.*, **16**, 456.

Mazzi, F., and Tadini, C., "The Crystal Structure of Potassium Xanthate $KS_2COC_2H_5$ (with an Appendix on Rubidium Xanthate $RbS_2COC_2H_5$)," *Z. Krist.*, **118**, 378.

Meyer, H.-J., "The Crystal Structure of Tetramethylammonium-Diargentic Iodide," *Acta Cryst.*, **16**, 788.

Mills, O. S., and Robinson, G., "Studies of Some Carbon Compounds of the Transition Metals IV. The Structure of Butadiene Irontricarbonyl," *Acta Cryst.*, **16**, 758.

Montgomery, H., and Lingafelter, E. C., "The Crystal Structure of Monoaquobis-acetylacetonatozinc," *Acta Cryst.*, **16**, 748.

Ogawa, K., "Redetermination of the Crystal and Molecular Structure of Chloral Hydrate," *Bull. Chem. Soc. Japan*, **36**, 610.

Ooi, S., and Kuroya, H., "The Crystal Structure of *trans*-Dichloro-bisethylenediamine-cobalt(III) Nitrate, $[Co(en)_2Cl_2]NO_3$," *Bull. Chem. Soc. Japan*, **36**, 1083.

Osaki, K., Nakai, Y., and Watanabé, T., "The Crystal Structure of Monoclinic Formate Dihydrates," *J. Phys. Soc. Japan*, **18**, 919.

Pachler, K., and Stackelberg, M. v., "The Crystal Structure of Some Alcohol Hemihydrates," *Z. Krist.*, **119**, 15.

Panattoni, C., and Frasson, E., "Crystal Structure of $AgSCN \cdot P(C_3H_7)_3$," *Gazz. Chim. Ital.*, **93**, 601; *Acta Cryst.*, **16**, 1258.

Parkes, A. S., and Hughes, R. E., "The Crystal Structure of Cyanogen," *Acta Cryst.*, **16**, 734.

Perego, G., and Bassi, I. W., "The Crystal Structure of Isotactic *trans*-1,4-Poly-1-Ethyl-Butadiene," *Makromol. Chem.*, **61**, 198.

Piret, P., Gobillon, Y., and Meerssche, M. van, "Structure of Sodium Iodide Triacetonate," *Bull. Soc. Chim. France*, **1963**, 205.

Rasmussen, S. E., and Grønbäk, R., "The Structure of 2,2′,2″-Triamino-triethylamine-trihydrochloride," *Acta Chem. Scand.*, **17**, 832.

Sands, D. E., "The Crystal Structure of Dimethyl Sulfone," *Z. Krist.*, **119**, 245.

Shriver, D. F., and Nordman, C. E., "The Crystal Structure of Trimethylamine Gallane, $(CH_3)_3NGaH_3$," *Inorg. Chem.*, **2**, 1298.

Silverton, J. V., and Hoard, J. L., "Stereochemistry of Discrete Eight-Coordination II. The Crystal and Molecular Structure of Zirconium (IV) Acetylacetonate," *Inorg. Chem.*, **2**, 243.

Simonov, Y. A., Ablov, A. V., and Malinovskii, T. I., "Crystal Structure of Diacetato-diamminocopper," *Kristallografiya*, **8**, 270.

Streib, W. E., Boer, F. P., and Lipscomb, W. N., "Molecular Structure of $B_4H_6C_2$-$(CH_3)_2$," *J. Am. Chem. Soc.*, **85**, 2331.

Sutherland, H. H., and Young, D. W., "The Crystal and Molecular Structure of Taurine," *Acta Cryst.*, **16**, 897.

Takano, T., Kasai, N., and Kakudo, M., "Crystal Structure of Bistetramethyldisilanilenedioxide, $[(CH_3)_4Si_2O]_2$," *Bull. Chem. Soc. Japan*, **36**, 585.

Tavale, S. S., Pant, L. M., and Biswas, A. B., "The Crystal and Molecular Structure of Sodium α-Ketobutyrate," *Acta Cryst.*, **16**, 566.

Trefonas, L. M., and Couvillion, J., "The Crystal Structure of *N*-Tertiary-Butyl Propylamine," *Acta Cryst.*, **16**, 576.

Trotter, J., "Bond Lengths and Angles in Pentaerythritol Tetranitrate," *Acta Cryst.*, **16**, 698.

Weiss, E., "Crystal Structure of Potassium Methylate," *Helv. Chim. Acta*, **46**, 2051.

Weiss, E., and Büchner, W., "The So-Called Alkali-Carbonyls I. Crystal Structure of Potassium Acetylenediolate, KOC≡COK," *Helv. Chim. Acta*, **46**, 1121.

Wheatley, P. J., "An X-Ray Diffraction Determination of the Crystal and Molecular Structure of Tetramethylstibonium Tetrakistrimethylsiloxyaluminate," *J. Chem. Soc.*, **1963**, 3200.

White, J. G., "The Crystal Structure of Tetramethylammonium Mercury Tribromide, $N(CH_3)_4HgBr_3$," *Acta Cryst.*, **16**, 397.

Zimmermann, I. C., Barlow, M., and McCullough, J. D., "The Crystal Structure of Trimethyloxosulfonium Fluoborate, $[(CH_3)_3SO]^+BF_4^-$," *Acta Cryst.*, **16**, 883.

1964

Ananthakrishnan, N., and Srinivasan, R., "Structure of L-Cystine Hydrobromide," *Indian J. Pure Appl. Phys.*, **2**, 62.

Anderson, P., and Thurmann-Moe, T., "The Crystal Structure of the Addition Compound between Diethylether and Monobromodichloromethane at $-130°C$.," *Acta Chem. Scand.*, **18**, 433.

Beagley, B., and Small, R. W. H., "The Structure of Lithium Oxalate," *Acta Cryst.*, **17**, 783.

Blake, A. B., Cotton, F. A., and Wood, J. S., "The Crystal, Molecular and Electronic Structures of a Binuclear Oxomolybdenum(V) Xanthate Complex," *J. Am. Chem. Soc.*, **86**, 3024.

Bradley, D. C., and Kunchur, N. R., "Structures of Mercury Mercaptides I. X-Ray Structure Analysis of Mercury Methylmercaptide," *J. Chem. Phys.*, **40**, 2258.

Bränden, C. I., "Crystal Structure of $2HgCl_2 \cdot (C_2H_5)_2S$," *Arkiv Kemi*, **22**, 83.

Brown, B. W., and Lingafelter, E. C., "The Crystal Structure of Bis(ethylenediamine)-copper(II) Thiocyanate," *Acta Cryst.*, **17**, 254.

Camerman, N., and Trotter, J., "Stereochemistry of Arsenic XI. 'Cacodyl Disulfide,' Dimethylarsino Dimethyldithioarsinate," *J. Chem. Soc.*, **1964**, 219.

Caron, A., and Donohue, J., "Three-dimensional Refinement of Urea," *Acta Cryst.*, **17**, 544.

Caron, A., Palenik, G. J., Goldish, E., and Donohue, J., "The Molecular and Crystal Structure of Trimethylamine Oxide, $(CH_3)_3NO$," *Acta Cryst.*, **17**, 102.

Clark, H. C., O'Brien, R. J., and Trotter, J., "The Structure of Trimethyltin Fluoride," *J. Chem. Soc.*, **1964**, 2332.

Clark, J. R., "Comment on Three Proposed Hydrate Structures," *Acta Cryst.*, **17**, 459.

Cotton, F. A., Dunne, T. G., and Wood, J. S., "The Structure of the Deca(methyliso-nitrile)dicobalt(II) Cation; an Isostere of Dimanganese Decacarbonyl," *Inorg. Chem.*, **3**, 1495.

Cotton, F. A., and Elder, R. C., "Crystal and Molecular Structure of Trioxo(diethyl-enetriamine)molybdenum(VI)," *Inorg. Chem.*, **3**, 397.

Cotton, F. A., Morehouse, S. M., and Wood, J. S., "The Identification and Characterization by X-Ray Diffraction of a New Binuclear Molybdenum(VI) Oxalate Complex," *Inorg. Chem.*, **3**, 1603.

Cotton, F. A., and Wood, J. S., "The Crystal and Molecular Structure of Bis(dipivaloyl-methanido)zinc(II)," *Inorg. Chem.*, **3**, 245.

Cruickshank, D. W. J., Jones, D. W., and Walker, G., "The Crystal Structure of Ammonium Trifluoroacetate," *J. Chem. Soc.*, **1964**, 1303.

Donohue, J., "Hydrogen Bonding and the Zwitterion Structure of Taurine," *Acta Cryst.*, **17**, 761.

Fehér, F., and Linke, K. H., "Chemistry of Sulfur LXVII. The Crystal Structure of Dicyanogentrisulfane, $NC-S_3-CN$," *Z. Anorg. Allgem. Chem.*, **327**, 151.

Fletcher, R. O. W., and Steeple, H., "The Crystal Structure of the Low Temperature Phase of Methylammonium Alum," *Acta Cryst.*, **17**, 290.

Forrester, J. D., Zalkin, A., and Templeton, D. H., "The Crystal and Molecular Structure of Di(tetra-*n*-butylammonium)cobalt(II) bis(maleonitrile dithiolate) and the Geometry of the Divalent Cobalt(II) bis(maleonitrile dithiolate) Ion," *Inorg. Chem.*, **3**, 1500.

Forrester, J. D., Zalkin, A., and Templeton, D. H., "The Crystal and Molecular Structure of Tetra-*n*-butylammoniumcopper(III) bis(maleonitrile dithiolate) and the Geometry of the Monovalent Copper(III) bis(maleonitrile dithiolate) Ion," *Inorg. Chem.*, **3**, 1507.

Foss, O., and Hordvik, A., "The Crystal Structure of Sodium Methanethiosulphonate Monohydrate," *Acta Chem. Scand.*, **18**, 619.

Freeman, H. C., Robinson, G., and Schoone, J. C., "Crystallographic Studies of Metal-Peptide Complexes I. Glycylglycylglycinocopper(II) Chloride Sesquihydrate," *Acta Cryst.*, **17**, 719.

Freeman, H. C., Snow, M. R., Nitta, I., and Tomita, K., "A Refinement of the Structure of Bisglycino-Copper(II) Monohydrate, $Cu(NH_2CH_2COO)_2 \cdot H_2O$," *Acta Cryst.*, **17**, 1463.

Fukushima, F., Iwasaki, H., and Saito, Y., "A New Modification of Deuterated Oxalic Acid Dihydrate, $(COOD)_2 \cdot 2D_2O$," *Acta Cryst.*, **17**, 1472.

Ganis, P., Pedone, C., and Temussi, P. A., "Crystal Structure of meso-α,α'-Dimethylglutaric Acid." *Atti Accad. Nazl. Lincei Rend. Classe Sci. Fis. Mat. Nat.*, **36**, 510.

Gerteis, R. L., Dickerson, R. E., and Brown, T. L., "The Crystal Structure of Lithium Aluminum Tetraethyl," *Inorg. Chem.*, **3**, 872.

Groth, P., and Hassel, O., "The 2:1 Addition Compound Methanol-Bromine, a Crystalline Substance Exhibiting both Hydrogen and Halogen Molecule Bridges between Oxygen Atoms," *Acta Chem. Scand.*, **18**, 402.

Haas, D. J., "The Crystal Structure of Potassium Tetroxalate, $K(HC_2O_4)(H_2C_2O_4) \cdot 2H_2O$," *Acta Cryst.*, **17**, 1511.

Housty, J., and Hospital, M., "Crystalline Structure of Azelaic Acid," *Compt. Rend.*, **258**, 1551.

Housty, J., and Hospital, M., "Crystalline Structure of Suberic Acid, $COOH(CH_2)_6COOH$," *Acta Cryst.*, **17**, 1387.

Jelenić, I., Grdenić, D., and Bezjak, A., "The Crystal Structure of Uranium(IV) Acetate," *Acta Cryst.*, **17**, 758.

Jose, P., Pant, L. M., and Biswas, A. B., "The Crystal Structure of Nickel β-Alanine Dihydrate," *Acta Cryst.*, **17**, 24.

Karle, I. L., and Karle, J., "An Application of the Symbolic Addition Method to the Structure of L-Arginine Dihydrate," *Acta Cryst.*, **17**, 835.

Kiosse, G. A., Golovastikov, N. I., and Belov, N. V., "The Crystal Structure of the Mixed DL-Ammonium Antimony Tartrate, DL-$(NH_4)_2[Sb_2(C_4H_4O_6)_2] \cdot 4H_2O$," *Dokl. Akad. Nauk SSSR*, **155**, 545.

Komiyama, Y., and Lingafelter, E. C., "The Crystal Structure of Bisethylenediaminecopper(II) Nitrate," *Acta Cryst.*, **17**, 1145.

Kornblau, M. J., and Hughes, R. E., "The Crystal and Molecular Structure of Hexacrylonitrile," *Acta Cryst.*, **17**, 1033.

Laurent, A., and Rérat, C., "Crystal Structure of O-Methylhydroxylamine Hydrochloride," *Acta Cryst.*, **17**, 277.

McCullough, J. D., "The Crystal Structure of Tetramethylammonium Perchlorate," *Acta Cryst.*, **17**, 1067.

Mazumdar, S. K., and Srinivasan, R., "X-Ray Analysis of L-Arginine Hydrohalides," *Current Sci. (India)*, **33**, 573.

Montgomery, H., and Lingafelter, E. C., "The Crystal Structure of Diaquobisacetylacetonatonickel(II)," *Acta Cryst.*, **17**, 1481.

Morosin, B., and Brathovde, J. R., "The Crystal Structure and Molecular Configuration of Trisacetylacetonatomanganese(III)," *Acta Cryst.*, **17**, 705.

Natta, G., Bassi, I. W., and Perego, G., "Crystal Structure of Isotactic 1,4-*trans*-poly(1-propylbutadiene)," *Atti Accad. Nazl. Lincei Rend. Classe Sci. Fis. Mat. Nat.*, **36**, 291.

Norman, N., and Mathisen, H., "Crystal Structure of Lower Paraffins III. Pentane," *Acta Chem. Scand.*, **18**, 353.

Okaya, Y., and Knobler, C. B., "Refinement of the Crystal Structure of tris(Thiourea)-copper(I) Chloride," *Acta Cryst.*, **17**, 928.

Osaki, K., Nakai, Y., and Watanabé, T., "The Crystal Structures of Magnesium Formate Dihydrate and Manganous Formate Dihydrate," *J. Phys. Soc. Japan*, **19**, 717.

Palenik, G. J., "The Crystal Structure of the Aluminum Hydride-N,N,N',N'-Tetramethylenediamine Adduct," *Acta Cryst.*, **17**, 1573.

Pant, L. M., "One-Dimensional Disorder in the Structure of Sodium 2-Oxocaprylate," *Acta Cryst.*, **17**, 219.

Perloff, A., "The Crystal Structure of 1-Ethyldecaborane," *Acta Cryst.*, **17**, 332.

Rannev, N. V., and Ozerov, R. P., "Neutron Diffraction Determination of the Position of H Atoms in the Structure of Dicyandiamide," *Dokl. Akad. Nauk SSSR*, **155**, 1415.

Tavale, S. S., Pant, L. M., and Biswas, A. B., "Crystal and Molecular Structure of Sodium 2-Oxocaprylate," *Acta Cryst.*, **17**, 215.

Visser, G. J., and Vos, A., "The Length of the I–Cl Bond in Tetramethylammonium Dichloroiodide," *Acta Cryst.*, **17**, 1336.

Weiss, E., "The Crystal Structure of Dimethylmagnesium," *J. Organometal. Chem. (Amsterdam)*, **2**, 314.

Weiss, E., and Büchner, W., "Crystal Structure of Rubidium and Cesium Acetylenediolate," *Z. Anorg. Allgem. Chem.*, **330**, 251.

Weiss, E., and Lucken, E. A. C., "The Crystal and Electronic Structure of Methyllithium," *J. Organometal. Chem. (Amsterdam)*, **2**, 197.

Willett, R. D., and Rundle, R. E., "Crystal Structure of $Cu_2Cl_4(CH_3CN)_2$, $Cu_3Cl_6(CH_3CN)_2$ and $Cu_5Cl_{10}(C_3H_7OH)_2$," *J. Chem. Phys.*. **40**, 838.

NAME INDEX

A

Acetamide, 312–313, 432
Acetic acid, 311–312, 432
di-Acetonitrile dodecahydrodecaborane, 38–39, 211
Acetoxime, 453–455, 639
Acetyl choline bromide, 340–342, 432
N-Acetylglycine, 664–666, 725
N,N'-di-Acetylhexamethylene diamine, 534–536, 639
di-Acetyl hydrazine, 321–323, 432
Acetylene, 265, 432
Acetylene dicarboxylic acid dihydrate, 553–555, 639
di-Acetylene dicarboxylic acid dihydrate, 555–557, 639
Acrylic acid, 456–458, 639
Acrylonitrile hexamer, 502–505, 639
Adipic acid, 570–571, 639
DL-Alanine, 666–667, 725
di-Allylsilanediol, 462–463, 639
Aluminum hydride di trimethylamine, 95, 211
Aluminum methyl dichloride, 70, 72, 211
Aluminum triethyl potassium fluoride, 274–275, 432
Aluminum trisacetylacetonate, 517, 639
DL-α-Aminocaproic acid, 691–692, 727
2-Amino ethanol phosphate, 256–258, 432
α-Amino isobutyric acid, 677–679, 725
2,2',2'' tri-Amino triethylamine–nickel thiocyanate, 347–349, 432
2,2',2'' tri-Amino-triethylamine tri-hydrochloride, 251, 432
11-Amino undecanoic acid hydrobromide hemihydrate, 622–624, 640
Aminoguanidine hydrochloride, 206–207, 211
tri-Aminoguanidinium chloride, 201–202, 211
di-Aminomaleonitrile, 260–262, 432
Ammonium acid D-tartrate, 577–578, 640
DL-Ammonium antimony tartrate di-hydrate, 582–584, 640
Ammonium cobaltic ethylene diamine tetraacetate dihydrate, 364–365, 432
Ammonium oxalate monohydrate, 412, 432
Ammonium oxamate, 413–414, 432
Ammonium sodium tartrate tetra-hydrate, 579, 640
Ammonium trifluoroacetate, 323–324, 432
Ammonium trioxalatochromiate di-hydrate, 421–422, 432
Ammonium uranyl propionate, 334, 432
mono-Amyl ammonium bromide, 137, 212
mono-Amyl ammonium chloride, 137, 142, 212
mono-Amyl ammonium iodide, 137, 212
di-Amyl mercury mercaptan, 272, 432
tetra iso-Amyl ammonium fluoride hydrate, 497–498, 640
Angelic acid, 501–502, 640
Antimonious xanthate, 304–305, 432
Antimony pentachloride phosphorus oxytrichloride, 90, 212
Antimony pentachloride phosphorus oxytrimethyl, 89–90, 212
L-Arginine dihydrate, 698–701, 725
L-Arginine hydrobromide monohydrate, 701–703, 725
L-Arginine hydrochloride monohydrate, 701–702, 725
Arsenious xanthate, 306, 433
Arsenomethane, 103–104, 212
Asparagine monohydrate, 675–676, 725
DL-Aspartic acid, 675, 725
Auric bromide trimethyl phosphine, 96–98, 212
Azelaic acid, 564–566, 640
Azochloramide, 28–29, 212
Azodicarbonamide, 28, 30–31, 212

B

Barium dicalcium propionate, 446, 640
Barium formate, 128–129, 212
Barium pentathionate–acetone mono-hydrate, 310–311, 433

FORMULA INDEX*

* In this bulk formula index carbon atoms come first and the formulas are arranged in order of the increasing number of these atoms. In each formula carbon is followed by hydrogen, oxygen, nitrogen, etc., each of which is ordered according to increasing number. Names are added only to distinguish between compounds having the same total formula.

779